FUNDAMENTALS OF GEOBIOLOGY

FUNDAMENTALS OF GEOBIOLOGY

EDITED BY

ANDREW H. KNOLL, DONALD E. CANFIELD

AND

KURT O. KONHAUSER

WILEY-BLACKWELL

A John Wiley & Sons, Ltd., Publication

This edition first published 2012 © 2012 by Blackwell Publishing Ltd

Blackwell Publishing was acquired by John Wiley & Sons in February 2007. Blackwell's publishing program has been merged with Wiley's global Scientific, Technical and Medical business to form Wiley-Blackwell.

Registered Office
John Wiley & Sons, Ltd, The Atrium, Southern Gate, Chichester, West Sussex, PO19 8SQ, UK

Editorial Offices
9600 Garsington Road, Oxford, OX4 2DQ, UK
The Atrium, Southern Gate, Chichester, West Sussex, PO19 8SQ, UK
111 River Street, Hoboken, NJ 07030-5774, USA

For details of our global editorial offices, for customer services and for information about how to apply for permission to reuse the copyright material in this book please see our website at www.wiley.com/wiley-blackwell.

The right of the author to be identified as the author of this work has been asserted in accordance with the UK Copyright, Designs and Patents Act 1988.

Library of Congress Cataloging-in-Publication Data

Fundamentals of geobiology / edited by Andrew H. Knoll, Donald E. Canfield & Kurt O. Konhauser.
    p. cm.
  Includes index.
  ISBN 978-1-4051-8752-7 (pbk.) – ISBN 978-1-1182-8081-2 (hardcover)
1. Geobiology. I. Knoll, Andrew H. II. Canfield, Donald E. III. Konhauser, Kurt.
  QH343.4.F86 2012
  577–dc23

2011046005

A catalogue record for this book is available from the British Library.

Wiley also publishes its books in a variety of electronic formats. Some content that appears in print may not be available in electronic books.

Set in 9.5/12pt Palatino by SPi Publisher Services, Pondicherry, India
Printed and bound by CPI Group (UK) Ltd, Croydon, CR0 4YY
C9781405187527_221123

Contents

Colour plate pages fall between pp. 228 and 229

COMPANION WEBSITE

This book has a companion website:

www.wiley.com/go/knoll/geobiology

with Figures and Tables from the book for downloading.

Contributors

GIOVANNI ALOISI *UMR CNRS 7159LOCEAN, Université Pierre et Marie Curie, Paris, France*

ARIEL D. ANBAR *School of Earth and Space Exploration and Department of Chemistry and Biochemistry, Arizona State University, Tempe AZ 85287, USA*

DAVID J. BEERLING *Department of Animal and Plant Sciences, University of Sheffield, Sheffield S10 2TN, UK*

ROGER BUICK *Department of Earth & Space Sciences and Astrobiology Program, University of Washington, Seattle WA 98195, USA*

NICHOLAS J. BUTTERFIELD *Department of Earth Sciences, University of Cambridge, Cambridge, CB2 2EQ, UK*

SUSAN L. BRANTLEY *Center for Environmental Kinetics Analysis, Earth and Environmental Systems Institute, Pennsylvania State University, University Park PA 16802, USA*

DONALD E. CANFIELD *Institute of Biology Nordic Center for Earth Evolution, University of Southern Denmark, Campusvej 55, DK-5230 Odense M, Denmark*

PATRICIA M. DOVE *Department of Geosciences, Virginia Polytechnic Institute and State University, Blacksburg VA 24061, USA*

PAUL G. FALKOWSKI *Department of Earth and Planetary Sciences and Institute of Marine and Coastal Sciences, Rutgers University, New Brunswick NJ 08901, USA*

JAMES FARQUHAR *Department of Geology and ESSIC, University of Maryland, College Park MD 20742, USA*

W.W. FISCHER *Division of Geological and Planetary Sciences, California Institute of Technology, Pasadena, CA 91125, USA*

LAURA M. HAMM *Department of Geosciences, Virginia Polytechnic Institute and State University, Blacksburg VA 24061, USA*

ELISABETH M. HAUSRATH *Department of Geosciences, University of Nevada, Las Vegas, 4505 S. Maryland Parkway, Las Vegas, NV 89154, USA*

ROBERT M. HAZEN *Geophysical Laboratory, Carnegie Institution of Washington, 5251 Broad Branch Road NW, Washington, DC 20015, USA*

D.T. JOHNSTON *Department of Earth and Planetary Sciences, Harvard University, Cambridge MA 02138, USA*

ANDREAS KAPPLER *Geomicrobiology, Center for Applied Geosciences, University of Tübingen, Sigwartstrasse 10, 72076, Tübingen, Germany*

JAMES F. KASTING *Department of Geosciences, Pennsylvania State University, University Park, PA 16802, USA*

BRIAN KENDALL *School of Earth and Space Exploration, Arizona State University, Tempe AZ 85287, USA*

ANDREW H. KNOLL *Department of Organismic and Evolutionary Biology, Harvard University, Cambridge MA 02138, USA*

KURT O. KONHAUSER *Department of Earth and Atmospheric Sciences, University of Alberta, Edmonton, AB T6G 2E3, Canada*

MARINA LEBEDEVA *Center for Environmental Kinetics Analysis, Earth and Environmental Systems Institute, Pennsylvania State University, University Park PA 16802, USA*

JONATHAN R. LLOYD *Williamson Research Centre for Molecular Environmental Science and School of Earth, Atmospheric and Environmental Science, University of Manchester, Manchester M13 9PL, UK*

SARA A. LINCOLN *Massachusetts Institute of Technology, Department of Earth and Planetary Sciences, 77 Massachusetts Ave., Cambridge MA 02139, USA*

GORDON LOVE *Department of Earth Sciences, University of California, Riverside CA 92521, USA*

TIMOTHY W. LYONS *Department of Earth Sciences, University of California, Riverside CA 92521, USA*

DIANNE K. NEWMAN *Howard Hughes Medical Institute, California Institute of Technology, 1200 E. California Blvd., Pasadena, CA 91125, USA*

VICTORIA J. ORPHAN *Division of Geological and Planetary Sciences, California Institute of Technology, Pasadena, CA 91125, USA*

DOMINIC PAPINEAU *Department of Earth and Environmental Sciences, Boston College, 140 Commonwealth Avenue, Chestnut Hill, MA 02467, USA*

CHRISTOPHER T. REINHARD *Department of Earth Sciences, University of California, Riverside CA 92521, USA*

ANNA-LOUISE REYSENBACH *Portland State University, Portland, OR 97207, USA*

ROBERT RIDING *Department of Earth & Planetary Sciences, University of Tennessee, Knoxville, TN 37996, USA*

DANIEL P. SCHRAG *Department of Earth and Planetary Sciences, Harvard University, Cambridge, MA 02138, USA*

STEVEN M. STANLEY *Department of Geology and Geophysics, University of Hawaii, 1680 East-West Road, Honolulu HI 96822, USA*

ROGER E. SUMMONS *Massachusetts Institute of Technology, Department of Earth and Planetary Sciences, 77 Massachusetts Ave., Cambridge MA 02139, USA*

DAVID J. VUAGHAN, *Williamson Research Centre for Molecular Environmental Science and School of Earth, Atmospheric and Environmental Science, University of Manchester, Manchester M13 9PL, UK*

ADAM F. WALLACE *Earth Sciences Division, Lawrence Berkeley National Laboratory, Berkeley CA 94720, USA*

KLAUS WALLMANN *Leibniz Institute for Marine Sciences (IFM-GEOMAR), Wischhofstrasse, 1-3; 24148, Kiel, Germany*

DONGBO WANG *Department of Geosciences, Virginia Polytechnic Institute and State University, Blacksburg VA 24061, USA*

BESS WARD *Department of Geosciences, Princeton University, Princeton NJ 08540 USA*

SHUHAI XIAO *Department of Geosciences, Virginia Polytechnic Institute and State University, Blacksburg VA 24061, USA*

1
WHAT IS GEOBIOLOGY?

Andrew H. Knoll[1], Donald E. Canfield[2], and Kurt O. Konhauser[3]

[1] Department of Organismic and Evolutionary Biology, Harvard University, Cambridge MA 02138, USA
[2] Nordic Center for Earth Evolution, University of Southern Denmark, Campusvej 55, DK-5230 Odense M, Denmark
[3] Department of Earth and Atmospheric Sciences, University of Alberta, Edmonton, AB T6G 2E3 Canada

1.1 Introduction

Geobiology is a scientific discipline in which the principles and tools of biology are applied to studies of the Earth. In concept, geobiology parallels geophysics and geochemistry, two longer established disciplines within the Earth sciences. Beginning in the 1940s, and accelerating through the remainder of the twentieth century, scientists brought the tools of physics and chemistry to bear on studies of the Earth, transforming geology from a descriptive science to a quantitative field grounded in analysis, experiment and modeling. The geophysical and geochemical revolutions both reflected and drove a strong disciplinary emphasis on plate tectonics and planetary differentiation, not least because, for the first time, they made the Earth's interior accessible to research.

While geochemistry and geophysics occupied centre stage in the Earth sciences, another multidisciplinary transformation was taking shape nearer to the field's periphery. Paleontology had long brought a measure of biological thought to geology, in no small part because fossils provide a basis for correlating sedimentary rocks. But while it was obvious that life had evolved on the Earth, it was less clear to most Earth scientists that life had actually shaped, and been shaped, by Earth's environmental history. For example, in *Tempo and Mode in Evolution*, paleontology's key contribution to the Neodarwinian synthesis in evolutionary biology, G.G. Simpson (1944) devoted less than a page to questions of environmental interactions. As early as 1926, however, the Russian scientist Vladimir Vernadsky had published *The Biosphere*, setting forth the argument that life has shaped our planet's surface environment throughout geologic time. Vernadsky also championed the idea of a noosphere, a planet transformed by activities of human beings. A few years later, the Dutch microbiologist Lourens Baas-Becking (1934) coined the term *geobiology* to describe the interactions between organisms and environment at the chemical level. Whereas most paleontologists stressed morphology and systematics, Vernadsky and Baas-Becking focused on metabolism – and in the long run that made all the difference.

Geobiological thinking moved to centre stage in the 1970s with articulation of the Gaia Hypothesis by James Lovelock (1979). Much like Vernadsky before him, Lovelock argued that life, air, water and rocks interact in complex ways within an integrated Earth system. More controversially, he posited that organisms regulate the Earth system for their own benefit. While this latter view, sometimes called 'strong Gaia,' has found little favor with biologists or Earth scientists, most now accept the more general view that Earth surface environments cannot be understood without input from the life sciences. The seeds of these ideas may have been planted earlier, but it was Lovelock who really captured the attention of a broad scientific community.

As the twentieth century entered its final decade, interest in geobiology grew, driven by an increasing emphasis within the Earth sciences on understanding our planetary surface, and supported by accelerating research on the microbial control of elemental cycling, the ecological diversity of microbial life under even the most harsh environmental conditions (commonly referred to as *extremeophiles*), the use of microbes to ameliorate pollution (bioremediation) or recover valuable metals from mine waste (biorecovery), Earth's ancient microbial history, and efforts to understand human influences on the Earth surface system. And, in the twenty-first century, universities are increasingly supporting research and education in geobiology, international journals (e.g., *Geobiology, Biogeosciences*) have prospered, textbooks have been published (e.g., Schlesinger, 1997; Canfield *et al.*, 2005; Konhauser, 2007; Ehrlich and Newman, 2009), and conferences occur regularly. Without question, geobiology has come of age.

1.2   Life interacting with the Earth

Geobiology is predicated on the observation that biological processes interact with physical processes at and near the Earth's surface. Take, for example, carbon, the defining element of life. Within the *biosphere* – the sum of all environments that support life on Earth – carbon exists in a number of forms and in several key reservoirs. It is present as CO_2 in the atmosphere; as CO_2, HCO_3^- and CO_3^{2-} dissolved in fresh and marine waters; as carbonate minerals in soils, sediments and rocks; and as a huge variety of organic molecules in organisms, in sediments and soils, and dissolved in lakes and oceans. Physical processes move carbon from one reservoir to another; for example, volcanoes add CO_2 to the atmosphere and chemical weathering removes it. Biological processes do as well. In two notable examples, photosynthesis reduces CO_2 to sugar, and respiration oxidizes organic molecules to CO_2. Since the industrial revolution, humans have oxidized sedimentary organic matter (by burning fossil fuels) at rates much higher than those characteristic of earlier epochs, making us important participants in the Earth's carbon cycle. Given the centrality of the carbon cycle to both ecology and climate, its biological and geological components are explored in two early chapters of this book (Chapters 2 and 3) and revisited in the context of human activities in Chapter 22.

Other biologically important elements also cycle through the biosphere. Sulfur, nitrogen, and iron (Chapters 4–6) all link the physical and biological Earth, interacting with each other and, importantly, with the carbon cycle. And oxygen, key to environments that support large animals, including humans, is regulated by a complex and incompletely understood set of processes that, again, have both biological and physical components (Chapter 7).

Unlike physical processes, life evolves, and so the array of biological processes in play within the biosphere has changed through time. The state of the environment supporting biological communities has changed as well. Indeed, given the close relationship between environment and population distributions on the present day Earth, it is reasonable to hypothesize that evolving life has significantly influenced the chemical environment through time and, conversely, that environmental change has influenced the course of evolution.

While metabolism encompasses many of the biological cogs in the biosphere, other processes also play important roles. For example, many organisms precipitate minerals, either indirectly by altering local chemical environments (Chapter 8), or directly by building mineralized skeletons (Chapter 10). Today, skeletons dominate the deposition of carbonate and silica on the seafloor, although this was not true before the evolution of shells, spicules and tests. More subtly, organisms interact with clays and other minerals in a series of surface interactions that are only now beginning to be understood (Chapter 9). While much of geobiology focuses on chemical processes, organisms influence the Earth through physical activities as well – think of microbial communities that can stabilize sand beds (Chapter 16) or worms that irrigate sediments as they burrow (Chapter 11). The example of burrowing reminds us that while microorganisms garner much geobiological attention, plants and animals also act as geobiological agents, and have done so for more than 500 million years (Chapter 11).

In short, Earth surface processes once considered to be largely physical in nature – for example weathering and erosion – are now known to have key biological components (Chapter 12). Life plays a critical role in the Earth system.

1.3   Pattern and process in geobiology

Geobiologists, then, study how organisms influence the physical Earth and vice versa, and how biological and physical processes have interacted through our planet's long history. Much of this research focuses on illuminating *process*: field and experimental studies of how organisms participate in the Earth system, and what consequences these activities have for local to global environmental state. Geobiological research can be fundamental – that is, aimed at achieving a basic understanding of the Earth system and its evolution – or it can be applied. In the case of the latter, microbial populations have been deployed and even engineered to perform tasks that range from concentrating gold dispersed in the talus piles of mines, and removing arsenic from the water supply of Los Angeles, to respiring vast amounts of the petroleum that gushed into the Gulf of Mexico in 2010. Building on earlier chapters, Chapters 13–16 focus on techniques that are prominent in modern geobiological research.

Elucidating the changing role of life through Earth history, sometimes called *historical geobiology*, begins with a basic understanding of geobiological processes, but from there takes on a distinctly geological slant. We would like to interpret the geologic record in terms of active processes and chemical states, but rocks preserve only pattern. Thus, the geobiological interpretation of ancient sedimentary rocks requires that we understand how biological processes and aspects of the ambient environmental state are reflected in the geologically preservable patterns they create. For example, we can use the sulfur isotopic composition of minerals in billion-year-old shales to constrain the biological workings of the ancient sulfur cycle and sulfate abundance in ancient seawater, but can do so only in light of present day observations and experiments that show how biological and physical processes result in particular isotopic patterns.

Of course, there are at least two features that complicate this linkage of geobiological process to geologic pattern. For one, populations evolve, so biological processes observable today may not been active during the deposition of ancient sedimentary rocks. For this reason, historical geobiology has among its goals the establishment of evolutionary pattern in Earth history. The second complication is that many environmental states on the ancient Earth have no modern counterpart. Most obviously, modern surface environments are permeated with oxygen in ways unlikely to have existed during the first two billion years of our planet's development. Other differences exist, as well. Therefore, the present-day Earth system is far removed from the earliest systems where life evolved and then spread out across the planet; it represents a long accumulation of biological, physical and chemical changes through Earth history. Following a chapter on the origin of life (Chapter 17), perhaps the ultimate example of the intimate relationship between biological and physical processes, we present three chapters that outline Earth's geobiological history (Chapters 19–21). Oxygen, biological evolution and chemical change dominate these discussions, but there are other aspects to the story. For example, Chapter 18 discusses how the diversity of minerals found on Earth has expanded through time as the biosphere has changed, providing a twenty-first century account of an intriguing subject suggested long ago by Vernadsky.

Finally, there is the question of us. Either directly or indirectly, humans appropriate nearly half of the total primary production on Earth's land surface. We fix as much nitrogen as bacteria do, and shuttle phosphate from rocks to the oceans at unprecedented rates. As Vernadsky predicted in his early discussion of the noosphere, humans have become extraordinarily important agents of geobiological change. In areas that range from climate change to eutrophication, from ocean acidification to Earth's declining supplies of fossil fuels and phosphate fertilizer, the human footprint on the biosphere is large and growing. Our societal future depends in part on understanding the geobiological influences of humans and in governing the technological processes that have come to play such important roles in the modern Earth system (Chapter 22).

1.4 New horizons in geobiology

It is difficult, if not impossible, to predict the future, and while it would be fun to attempt a forecast of the status of geobiology in say 20 years, we will avoid this. Rather, we highlight that under all circumstances, geobiology will increasingly look to the heavens. *Astrobiology* can be thought of as the application of geobiological principles to the study of planets and moons beyond the Earth. At the moment, claims about life in the universe largely constitute under-constrained statistical extrapolations from our terrestrial experience: some hold that life is abundant throughout the universe, but intelligent life is rare (Ward and Brownlee, 2000), while others suggest that life is rare, but intelligence more or less inevitable wherever life occurs (Conway Morris 2004). Clearly, the way forward lies in exploration. Both remote sensing and lander operations have made remarkable strides during the past decade (e.g., Squyres and Knoll, 2006), so we can be confident that on planets and moons within our solar system, direct observation of potentially geobiological patterns will sharply constrain arguments about life in our planetary neighborhood. And arguments about life in nearby solar systems will be framed in terms of geobiological models of planetary atmospheres glimpsed by Kepler and its technological descendents (Kasting, 2010).

This book, then, is a status report. It contains detailed but accessible summaries of key issues of geobiology, hopefully capturing the state and breadth of this emerging discipline. We have tried to be inclusive in our choice of topics covered within this volume. We recognize, however, that the borders defining geobiology are fluid, and we have likely missed or underrepresented some relevant geobiological topics. We apologize in advance for this. We also hope and trust that in the future, geobiology will expand in both depth and breadth well beyond what is offered here. Our crystal ball is cloudy, but we can be certain that a similar book written twenty years from now will differ fundamentally from this one.

References

Baas-Becking LGM (1934) *Geobiologie of inleiding tot de milieukunde Diligentia Wetensch*, Serie 18/19. van Stockum's Gravenhage, The Hague.

Canfield DE, Kristensen E, Thamdrup B (2005) *Aquatic Geomicrobiology*. Elsevier, Amsterdam.

Conway Morris S (2004) *Life's Solution: Inevitable Humans in a Lonely Universe*. Cambridge University Press, Cambridge.

Ehrlich HL, Newman DK (2009) *Geomicrobiology*, 5th edn. Marcel Dekker, New York.

Kasting J (2010) *How to Find a Habitable Planet*. Princeton University Press, Princeton, NJ.

Konhauser K (2007) *Introduction to Geomicrobiology*. Blackwell Publishing, Malden, MA.

Lovelock J (1979) *Gaia: A New Look at Life on Earth*. Oxford University Press, Oxford.

Schlesinger WH (1997) *Biogeochemistry: An Analysis of Global Change*. Academic Press, San Diego, CA.

Simpson GG (1944) *Tempo and Mode in Evolution*. Columbia University Press, New York.

Squyres S, Knoll AH, eds (2006) *Sedimentary Geology at Meridiani Planum, Mars*. Elsevier Science, Amsterdam. [Also published as *Earth and Planetary Science Letters* **240**(1).]

Vernadsky VL (1926) *The Biosphere*. English translation by D.B. Langmuir, Copernicus, New York, 1998.

Ward P, Brownlee D (2000) *Rare Earth: Why Complex Life Is Uncommon in the Universe*. Springer-Verlag, Berlin.

2
THE GLOBAL CARBON CYCLE: BIOLOGICAL PROCESSES

Paul G. Falkowski

Department of Earth and Planetary Sciences and Institute of Marine
and Coastal Sciences, Rutgers University, New Brunswick, NJ 08901, USA

2.1 Introduction

Carbon is the fourth most abundant element in our solar system and its chemistry forms the basis of all life on Earth. It is used both as the fundamental building block for all structural biological molecules and as an energy carrier. However, the vast majority of carbon on the surface of this planet is covalently bound to oxygen or its hydrated equivalents, forming mineral carbonates in the lithosphere, soluble ions in the ocean, and gaseous carbon dioxide in the atmosphere. These oxidized (inorganic) forms of carbon are moved on time scales of centuries to millions of years between the lithosphere, ocean and atmosphere via tectonically driven acid-based reactions. Because these reservoirs are so vast (Table 2.1) they dominate the carbon cycle on geological time scales, but because the reactions are so slow, they are also difficult to measure directly within a human lifetime.

The 'geological' or 'slow' carbon cycle is critical for maintaining Earth as a habitable planet (Chapter 2), but entry of these oxidized forms of carbon into living matter requires the addition of hydrogen atoms. By definition, the addition of hydrogen atoms to a molecule is a chemical reduction reaction. Indeed, the addition or removal of hydrogen atoms to and from carbon atoms (i.e., 'redox' reactions), is the core chemistry of life. The processes which drive these core reactions also form a second, concurrently operating global carbon cycle which is biologically catalysed and operates millions of times faster than the geological carbon cycle (Falkowski, 2001). In this chapter, we consider the 'biological', or 'fast' carbon cycle, focusing on how it works, how it evolved, and how it is coupled to the redox chemistry of

a few other elements, especially nitrogen, oxygen, sulfur, and some selected transition metals.

2.2 A brief primer on redox reactions

When carbon is directly, covalently linked to hydrogen atoms, the resulting (reduced) molecules are called *organic*. Like acid–base reactions, all reduction reactions must be coupled to a reverse reaction in another molecule or atom; that is the reduction of carbon is coupled to the oxidation of another element or molecule. Under Earth's surface conditions, the addition of hydrogen atoms to carbon requires the addition of energy, while the oxidation of carbon-hydrogen (C–H) bonds yields energy. Indeed the oxidation of C–H bonds forms the basis of energy production for all life on Earth.

Although biologically mediated redox reactions (see Box 2.1) occur rapidly, the products are often kinetically inert. Hence, while it is relatively easy to measure the rate at which a plant converts carbon dioxide into sugars, the product, sugar, is stable. It can be purchased from a local grocery store and kept in a jar in sunlight. It does not spontaneously catch fire or explode. Yet when you eat it, your body extracts the energy from the C–H bonds, and oxidizes the sugar to CO_2 and H_2O.

2.3 Carbon as a substrate for biological reactions

Approximately 75 to 80% of the carbon on Earth is found in an oxidized, inorganic form either as the gas carbon dioxide (CO_2) or its hydrated or ionic equivalents,

Fundamentals of Geobiology, First Edition. Edited by Andrew H. Knoll, Donald E. Canfield and Kurt O. Konhauser.
© 2012 Blackwell Publishing Ltd. Published 2012 by Blackwell Publishing Ltd.

Table 2.1 Carbon Pools in the major reservoirs on Earth

Pools	Quantity (Gt carbon)
Atmosphere	835
Oceans	38,400
Total inorganic	37,400
Surface layer	670
Deep Layer	36,730
Total organic	1,000
Lithosphere	
Sedimentary carbonates	>60,000,000
Kerogens	15,000,000
Terrestrial biosphere (total)	2,000
Living biomass	600–1,000
Dead biomass	1,200
Aquatic biosphere	1–2
Fossil Fuels	4,130
Coal	3,510
Oil	230
Gas	~300
Other (peat)	250

namely bicarbonate (HCO_3^-) and carbonate (CO_3^{2-}) (see Fig. 2.1). These inorganic forms of carbon are interconvertible, depending largely on pH and pressure, and the three forms partition into the lithosphere, ocean and atmosphere[1] (see Chapter 3). Virtually all inorganic carbon in the oceans is in the form of HCO_3^- with an average concentration of about 2.5 mM. This carbon is removed in association with calcium and magnesium as carbonate minerals. Although the precipitation of carbonates is thermodynamically favourable in the contemporary ocean, it is kinetically hindered, and virtually all carbonates are formed by organisms. The biological precipitation of carbonates is *not* a result of redox reactions, but rather of acid-base reactions; hence, although virtually all carbonates are biologically derived, they remain as oxidized, inorganic carbon. The mineral phases of inorganic carbon are inaccessible to further biological reactions. The total reservoir of inorganic carbon in the ocean is approximately 50 times that of the atmosphere. Indeed, the ocean controls the concentration of CO_2 in the atmosphere on time scales of decades to millennia.

2.3.1 Carbon fixation

Entry of inorganic carbon into biological processes involves a process called carbon 'fixation', and there are only two biological mechanisms that lead to the fixation of inorganic carbon: chemoautotrophy and photoautotrophy. Before we consider these in turn, let us first examine what carbon 'fixation' is.

The term carbon 'fixation' is an anachronism that means 'to make non-volatile'. It applies when a gaseous CO_2 is biochemically converted to a solute. There are several enzymatically catalysed reactions that can lead to carbon fixation, however, by far the most important is based on the activity of ribulose 1,5-bisphosphate carboxylase/oxygenase, or Rubisco (Falkowski and Raven, 2007). This enzyme is thought to be the most abundant protein complex on Earth, and it specifically reacts with CO_2 (i.e., it does not recognize hydrated forms of the substrate). Rubisco catalyses a reaction with a 5 carbon sugar, ribulose 1,5 bisphosphate, leading to the formation of two molecules of 3-phosphoglycerate (see Figs 2.2 and 2.3). This reaction, discovered in the late 1940s and early 1950s by Melvin Calvin, Andrew Benson and Jack Bassham, forms the basis of the pathway of carbon acquisition by most photosynthetic organisms (Benson and Calvin, 1950).

It should be noted that Rubisco imprints a strong biological isotope signature that is used extensively in geochemistry. There are two stable isotopes of C in nature: ^{12}C, containing 6 protons and 6 neutrons, accounts for 98.90%, and ^{13}C, containing 6 protons and 7 neutrons, accounts for 1.10%. In the fixation of CO_2 by Rubisco, the enzyme preferentially reacts with the lighter isotope; the net result is that 3-phosphogycerate is enriched by about 25 parts per thousand in ^{12}C relative to the CO_2 in the air or water (Park, 1961). This isotopic fractionation provides a basis for understanding the impact of the biological carbon cycle over geological time (Kump and Arthur, 1999).

The fixation of CO_2 by Rubisco is *not* an oxidation/reduction reaction; the carboxylic acid group has the same oxidation state as CO_2. The biochemical

---

[1] Pure forms of carbon as (e.g.) diamond or graphite are relatively rare and do not undergo biological reaction.

Box 2.1   Redox reactions

The term *oxidation* was originally used by chemists in the latter part of the 18th century to describe reactions involving the addition of oxygen to metals, forming metallic oxides. For example:

$$3Fe + 2O_2 \rightarrow Fe_3O_4 \qquad (B2.1.1)$$

The term *reduction* was used to describe the reverse reaction, namely the removal of oxygen from a metallic oxide, for example, by heating with carbon:

$$Fe_3O_4 + 2C \rightarrow 3Fe + 2CO_2 \qquad (B2.1.2)$$

Analysis of these reactions established that the addition of oxygen is accompanied by the removal of electrons from an atom or molecule. Conversely, reduction is accompanied by the addition of electrons. In the specific case of organic reactions that involve the reduction of carbon, the addition of electrons is usually balanced by the addition of protons. For example, the reduction of carbon dioxide to formaldehyde requires the addition of four electrons *and* four H^+ – that is, the equivalent of four hydrogen atoms.

$$O=C=O + 4e^- + 4H^+ \rightarrow (CH_2O)_n + H_2O \qquad (B2.1.3)$$

Thus, from the perspective of organic chemistry, oxidation may be defined as the addition of oxygen, the loss of electrons, or the loss of hydrogen atoms (but not hydrogen ions, H^+); conversely, reduction can be defined as the removal of oxygen, the addition of electrons, or the addition of hydrogen atoms.

Oxidation–reduction reactions only occur when there are pairs of substrates, forming pairs of products:

$$A_{ox} + B_{red} \leftrightarrow A_{red} + B_{ox} \qquad (B2.1.4)$$

Photosynthesis uses energy from the sun to reduce inorganic carbon to form organic matter; i.e., photosynthesis is a biochemical reduction reaction. In oxygenic photosynthesis, CO_2 is the recipient of the electrons and protons, and thus becomes reduced (it is the A in Equation B2.1.4). Water is the electron and proton donor, and thus becomes oxidized (it is the B in Equation B2.1.4). The oxidation of two moles of water requires the addition of 495 kJ of energy. The reduction of CO_2 to the simplest organic carbon molecule, formaldehyde, requires 176 kJ of energy. The energetic efficiency of photosynthesis can be calculated by dividing the energy stored in organic matter by that required to split water into molecular hydrogen and oxygen. Thus, the maximum overall efficiency of photosynthesis, assuming no losses at any intermediate step, is 176/495 or about 36%.

reduction of 3-phosphoglycerate is the second step in the carbon fixation pathway, and leads to the formation of an aldehyde. This is the only reduction step in the so-called Calvin cycle. The rest of the pathway is primarily devoted to regenerating ribulose 1,5-bisphosphate, and leaves no discernable geochemical signal.

2.3.2   Chemoautotrophy

Chemoautotrophs (literally, 'chemical self feeders') are organisms capable of reducing sufficient inorganic carbon to grow and reproduce in the absence of light energy and without an external organic carbon source. Chemoautotrophs likely evolved very early in Earth's history and this process is exclusively carried out by prokaryotic organisms in both the domains Archaea and Bacteria (Stevens and McKinley, 1995).

Early in Earth's history, H_2 was probably an important constituent of the atmosphere a major reductant used by organisms to reduce inorganic carbon to organic biomass (Jørgensen, 2001). Although this process can still be found in marine sediments, where H_2 is produced during the anaerobic fermentation of organic matter, and in hydrothermal environments, where the gas is produced as a byproduct of serpentization, free H_2 is scarce on Earth's surface. Rather, most of the hydrogen is combined (oxidized) by microbes with other atoms, such as sulfur or oxygen, and to a much lesser extent, nitrogen. Hence, most contemporary chemoautotrophs oxidize H_2S or NH_4^+, but also other reduced compounds such as Fe^{2+}.

The driver for chemoautotrophic carbon fixation ultimately depends on a thermodynamically favourable redox gradient. For example, the oxidation of H_2S by microbes in deep sea vents is coupled to the reduction of oxygen in the surrounding water. Hence, this reaction is dependent on the chemical redox gradient between the ventilating mantle plume and the ocean interior that thermodynamically favours oxidation of the plume gases (Jannasch and Taylor, 1984) (Box 2.2).

Chemoautotrophy supplies a relatively small amount of organic carbon to the planet (probably <1%), however this mode of nutrition is critically important in sediments, anoxic basins, and in completing several elemental cycles, including that of N and S. Thus, chemoautotrophy is common in sediments and anoxic water columns where strong redox gradients develop. A classic example would be the oxygen-sulfide interface in microbial mats where sulfide-oxidizing chemoautotrophs thrive, although countless other examples could also be named (Canfield and Raiswell, 1999). Reductants for chemoautotrophs can be generated within in the Earth's crust. An important example, as mentioned above, is the hydrothermal fluids generated in mid-ocean ridge spreading centres. Here, the sulfide and ferrous iron liberated with the fluids sup-

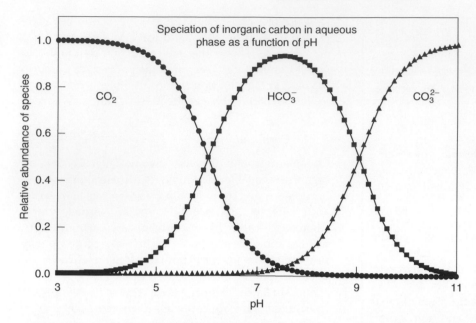

Figure 2.1 The relative distribution of the three major species of dissolved inorganic carbon in water as a function of pH. Note that at pH of seawater (~8.1), approximately 95% of the inorganic carbon is in the form of bicarbonate anion.

port the chemoautotrophic growth of sulfide- and Fe-oxidizers, which use oxygen as the oxidant.

However, early in Earth's history, due to the lack of oxygen, the redox gradients would have been small and hence there would have been no pandemic outbreak of chemoautotrophy. The vents themselves would have supplied H_2 and CO_2, for example, which could have supported the chemoautotrophic growth of methanogens (Canfield *et al.*, 2006), but this would have been on a much smaller scale than the chemoautotrophic sulfide oxidation supported in modern vents. Importantly, magma chambers, volcanism, and vent fluids are tied to either subduction or to spreading regions, which are transient features of Earth's crust and hence only temporary habitats for chemoautotrophs. In the Archean and early Proterozoic oceans, the chemoautotrophs would have had to have been dispersed throughout the oceans by physical mixing in order to colonize new vent regions (Raven and Falkowski, 1999).

2.3.3 *Photoautotrophy*

Photoautotrophy ('self feeding on light') is the biological conversion of light energy to the fixation of CO_2 in the form of organic carbon compounds. To balance the electrons, a source of reductant is also required. It should be noted that while all photoautotrophs are photosynthetic, not all photosynthetic organisms are photoautotrophs. Many organisms are capable of photosynthesis but can (and, sometimes must) supplement that metabolic strategy with the assimilation of organic carbon (Falkowski and Raven, 2007).

Photoautotrophy can be written as an oxidation–reduction reaction of the general form:

$$2H_2A + CO_2 + Light \xrightarrow{\text{\textit{Pigment}}} (CH_2O) + H_2O + 2A$$

$$(2.1)$$

In this representation, light is specified as a substrate, with some of the energy of the absorbed light stored in the products. All photoautotrophic bacteria, with the important exception of the cyanobacteria, are incapable of evolving oxygen. In these (mostly) anaerobic organisms, compound A is, for example, an atom of sulfur, and the pigments are bacteriochlorophylls (Van Niel, 1941; Blankenship *et al.*, 1995). All other photoautotrophs, including the cyanobacteria, eukaryotic algae, and higher plants, are oxygenic; that is, Equation 2.1 can be modified to:

$$Chl\ a + 2H_2O + CO_2 + Light \longrightarrow (CH_2O) + H_2O + O_2$$

$$(2.2)$$

where Chl *a* is the ubiquitous plant pigment chlorophyll *a*. Equation 2.2 implies that chlorophyll *a* catalyses a reaction or a series of reactions whereby light energy is used to oxidize water:

$$Chl\ a + 2H_2O + Light \longrightarrow 4H^+ + 4e + O_2 \qquad (2.3)$$

yielding gaseous, molecular oxygen. True photoautotrophy is restricted to the domains Bacteria and Eukarya. Although some Archaea and Bacteria use the pigment rhodopsin to harvest light, they require organic matter to fuel their metabolism (Beja *et al.*, 2000) and are not photoautotrophs.

2.4 The evolution of photosynthesis

We have discussed above the production of organic matter by photoautrophy, but photosynthesis is more

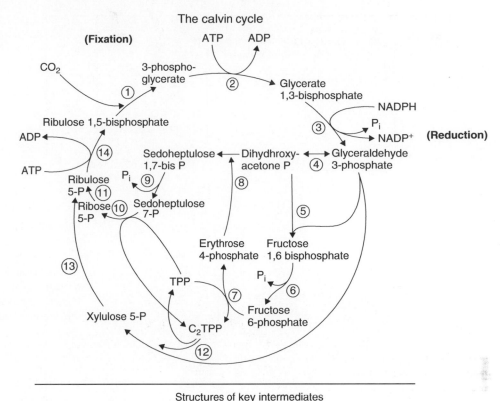

Figure 2.2 The reactions of ribulose 1,5 bisphosphate carboxylase/oxygenase. In the carboxylation process, the enzyme reacts with CO_2 to produce 2 molecules of 3-phosphogycerate, which is the first stable intermediate in the Calvin–Benson cycle. Alternatively, in the oxygenase process, the enzyme reacts with O_2, leading to the production 2 phosphoglycolate which ultimately is respired to CO_2.

broadly defined than this. Normally we consider photosynthesis to include all organisms that use light energy to synthesize new cells and this includes also photoheterotrophs which incorporate organic matter from the environment into their biomass. By far the most efficient and ubiquitous light harvesting systems for photosynthesis are based on porphyrins. The metabolic pathway for the synthesis of porphyrins is extremely old (Mauzerall,

1992); relic porphyrin molecules can be isolated from some ancient Archean (older than 2.5 billion years) rocks. It has been proposed that the porphyrin-based photosynthetic energy conversion apparatus originally arose from the need to prevent UV radiation from damaging essential macromolecules such as nucleic acids and proteins (Mulkidjanian and Junge, 1997). Indeed, most photosynthetic bacteria retain an ability to harvest UV light for

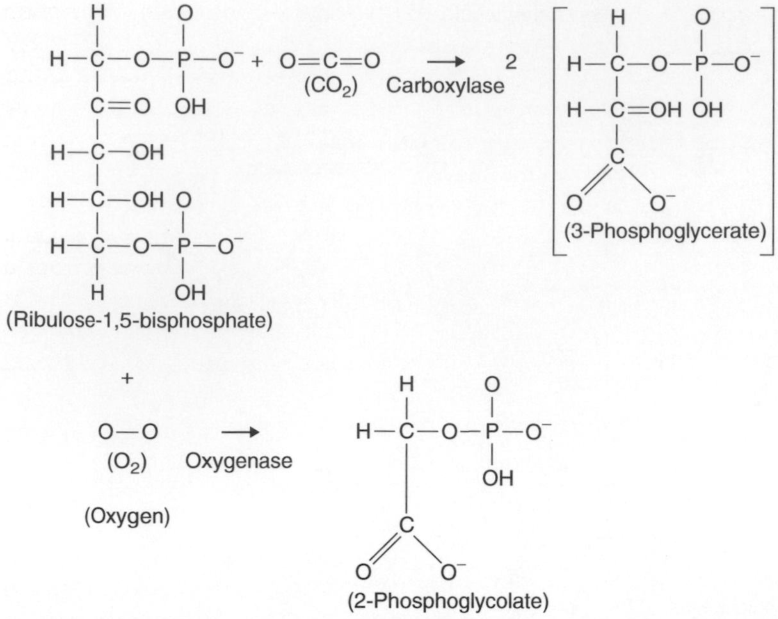

Figure 2.3 The basic Calvin–Benson cycle. CO_2 is condensed with ribulose 1,5 bisphosphate to form 3 phosphoglycerate (1), which is subsequently reduced to an aldehyde (reactions 2 and 3). The remaining reactions in the cycle are directed towards regenerating ribulose 1,5 bisphosphate; the intermediate structures are indicated.

Box 2.2 The Nernst Equation

As mentioned in the text, oxidation–reduction reactions involve the transfer of electrons. The tendency for a molecule to accept or release an electron is viewed relative to the ability of a 'standard' molecule to do the same, and this is normally the standard hydrogen electrode (SHE), which is represented by the following reaction:

$$2H^+_{(aq)} + 2e^- \rightarrow H_{2(g)} \tag{B2.2.1}$$

The SHE is defined at 25 °C, and one atmosphere of H_2 gas, and for an $H^+_{(aq)}$ activity of 1 (pH = 0). The tendency of a species to accept or liberate electrons is formally known as *electrode potential* or *redox potential*, E. The SHE is arbitrarily assigned an E of 0. The redox potential of reactions at standard state (unit activity for reactants and products), and relative to the SHE are given the designation E_0. In comparing to the SHE, reactions are always written as reduction reactions after the following general form:

$$aA_{oxid} + be^- + cH^+ \rightarrow dA_{red} - gG \tag{B2.2.2}$$

It is often more useful to define the redox potential at pH = 7, an environmentally relevant pH, and when so defined, the redox potential is denoted by the symbols E_0' or sometimes E_{m7}. The E_0' for a standard hydrogen electrode is −420 mV.

It is rare for organisms to live under standard-state chemical conditions (unit activity concentrations or reactants and products), and the electrode potential can be modified with the Nernst equation to reflect environmental conditions. With the general equation represented in B2.2.2, the Nernst equation is written as:

$$E = E_0 + 2.3(RT/nF) \log_{10} (a^a_{Aoxid} a^c_{H+} / a^d_{Ared} a^g_G) \tag{B2.2.3}$$

Where E is the redox potential (in volts) under environmental conditions, E_0 is the standard redox potential, F is Faraday's constant (= 96 485 coulombs = the electical charge in 1 mole of electrons), n is the number of moles of electrons (Faradays) transferred in the half-cell reaction, R is the Boltzmann gas constant, T is temperature in Kelvin, and a_y^x represents the activity (sometimes concentration is used, but this is not strictly correct) of species y raised to the stoichiomentric factor x in the balanced half reaction equation. The value of 2.3(RT/F) is 59 mV.

photosynthesis, but these bacteria cannot split water and evolve oxygen. These are termed 'anoxygenic' organisms, with light-mediated sulfide oxidation, as introduced above, a common (but not exclusive) metabolism.

In the ancient oceans, anoxygenic photosynthesis had profound biogeochemical consequences. It led not only to the formation of organic matter, but to the oxidation of such reductants as Fe^{2+} (to Fe^{3+}, which precipitated as Fe oxides), which was found in abundance in ancient oceans and S^{2-} (to S^0 or SO_4^{2}), which was found in ancient hydrothermal springs, in ancient oceans during some time periods (Canfield, 1998; Brocks *et al.*, 2005) and in contemporary analogs such as in Yellowstone National Park. While these reductants were ultimately resupplied via weathering in the case of Fe, or from hydrothermal fluids in the case of sulfide, the rate of supply was slow relative to the rate of at which organisms can photosynthesize. Hence there was a drive to find a reductant for photosynthesis that is virtually limitless. The obvious molecule is H_2O.

Liquid water contains ~55 kmol H_2O per m^3, and there are 10^{18} m^3 of water in the hydrosphere and cryosphere. However, the use of H_2O as a reductant for CO_2-fixation to organic matter requires a larger energy input than does the use of Fe^{2+} or S^{2-}. Indeed, for oxygenic photosynthesis to occur, several innovations on the old anoxygenic anaerobic photosynthetic machinery had to occur (Blankenship *et al.*, 2007). Among these innovations were the evolution of: (a) a new photosynthetic pigment, chlorophyll *a*, which operates at a higher energy level than bacterial chlorophylls; (b) two photochemical reaction centres that operate in series, one of which splits water, the second of which forms a biochemical reductant that is used to reduce inorganic carbon; and (c) a unique complex comprised of four Mn atoms bound to a group of proteins that forms the 'oxygen evolving complex', i.e., the site in which four electrons are sequentially extracted from two molecules of liquid water, one at a time, via the absorption of four photons.

The photochemical apparatus responsible for oxygenic photosynthesis is the most complex energy transduction system found in nature; there are well over 100 genes necessary for its synthesis (Shi and Falkowski, 2008). It appears to have arisen only once, in a single clade of bacteria (the cyanobacteria), and has never been appropriated by any other prokaryote. The origin and evolutionary trajectory of oxygenic photosynthesis remains obscure (Falkowski and Raven, 2007). It almost certainly arose sometime in the Archean Eon, although the timing is uncertain (see Chapter 7). The two photosystems appear to have different origins: the water splitting system is derived from purple photosynthetic bacteria, while the second reaction centre is derived from green sulfur bacteria. How the two reactions became incorporated into a single organism is unknown.

In order to oxidize water, the photochemical reaction must generate an oxidant with a potential of +0.8 V (Em_7) or more. This is significantly greater than is found in any extant anoxygenic photoautotroph (the highest is ca. +0.4 V). The oxidizing potential in oxygenic photosynthesis is the highest in nature, and ultimately, oxygenic photosynthesis became the primary mechanism for reducing CO_2 and forming organic carbon. Once established, it freed the microbial world from a limited supply of reductants for carbon fixation, and it decoupled the biological carbon cycle from the geological carbon cycle on time scales of millenia (Falkowski and Raven, 2007).

2.5 The evolution of oxygenic phototrophs

2.5.1 The cyanobacteria

Cyanobacteria are the only oxygenic phototrophs known to have existed before ~2 Ga. There is some suggestion that the 1.8 billion year old fossil *Grypania* may represent an eukaryotic algae (Han and Runnegar, 1992), but this has not been firmly established. Cyanobacteria numerically dominate the phototrophic community in contemporary marine ecosystems, and clearly their continued success bespeaks an extraordinary adaptive capacity.

By 2 billion years ago, cyanobacteria were probably the major primary producers (a primary producer is an organism that supplies organic matter to heterotrophs), with likely contributions from anoxygenic phototrophs and chemoautotrophs. In the contemporary ocean, the cyanobacteria fix approximately 60% of the ~45 Pg C assimilated annually by aquatic phototrophs (Falkowski and Raven, 2007). Their proportional contribution to 'local' marine primary productivity is greatest in the oligotrophic central ocean gyres that form about 70% of the surface waters of the seas. Two major groups of marine cyanobacteria can be distinguished. The phycobilin-containing *Synechococcus* are more abundant nearer the surface and the (divinyl) chlorophyll *b*-containing *Prochlorococcus* are generally more abundant at depth (Chisholm, 1992) (Table 2.2).

Some cyanobacteria not only fix inorganic carbon, but also fix N_2. Biological reduction of N_2 to NH_3 (i.e., 'fixation') is catalysed by nitrogenase, a heterodimeric enzyme that is irreversibly inhibited by O_2. Molecular phylogenetic trees suggest that N_2 fixation evolved in Bacteria prior to the evolution of oxygenic photosynthesis (Zehr *et al.*, 1997) and was acquired by cyanobacteria relatively late in their evolutionary history (Shi and Falkowski, 2008). The early evolution of nitrogenase is also indicated by the very large Fe requirement for the enzyme; the holoenzyme contains 38 iron atoms. This transition metal was much more available in the water column of the oceans in the Archean and

	Ocean NPP			Land NPP
Seasonal				
April–June	10.9			15.7
July–September	13.0			18.0
October–December	12.3			11.5
January–March	11.3			11.2
Biogeographic				
Oligotrophic	11.0		Tropical rainforests	17.8
Mesotrophic	27.4		Broadleaf deciduous forests	1.5
Eutrophic	9.1		Broadleaf and needleleaf forests	3.1
Macrophytes	1.0		Needleleaf evergreen forests	3.1
			Needleleaf deciduous forest	1.4
			Savannas	16.8
			Perennial grasslands	2.4
			Broadleaf shrubs with bare soil	1.0
			Tundra	0.8
			Desert	0.5
			Cultivation	8.0
Total	48.5			56.4

Table 2.2 Annual and seasonal net primary production (NPP) of the major units of the biosphere (after Field *et al.*, 1998)

Source: Field *et al.* (1998). All values in GtC. Ocean NPP estimates are binned into three biogeographic categories on the basis of annual average C_{sat} for each satellite pixel, such that oligotrophic = C_{sat} < 0.1 mg m^{-3}, mesotrophic = 0.1 < C_{sat} < 1 mg m^{-3}, and eutrophic = C_{sat} > 1 mg m^{-3} (Antoine *et al.*, 1996). This estimate includes a 1 GtC contribution from macroalgae (Smith, 1981). Differences in ocean NPP estimates between Behrenfeld and Falkowski (1997b) and those in the global annual NPP for the biosphere and this table result from (i) addition of Arctic and Antarctic monthly ice masks; (ii) correction of a rounding error in previous calculations of pixel area; and (iii) changes in the designation of the seasons to correspond with Falkowski *et al.* (1998). The macrophyte contribution to ocean production from the aforementioned is not included in the seasonal totals. The vegetation classes are those defined by DeFries and Townshend (DeFries and Townshend, 1994).

early Proterozoic Eons than it is today (Berman-Frank *et al.*, 2001). Indeed, iron has been suggested to limit nitrogen fixation in the contemporary ocean (Falkowski 1997).

2.5.2 The eukaryotes

Eukaryotes are unicellular or multicellular organisms whose cells possess nuclei that contain their DNA. All eukaryotes were derived from a symbiotic association between a host cell closely affiliated with the Archaea (Woese, 1998), and an organelle derived from the Bacteria. The symbiosis occurred via the engulfment of the latter cell by the former, after which most of the genes in the bacterium were either lost of transferred to the host. There appears to have been only two primary symbiotic events. The first involved a purple bacterium, which may have been photosynthetic (Woese *et al.*, 1984); this organism ultimately came to be a mitochondrion, the primary energy producing system for eukaryotes. In a second symbiotic event, a host eukaryote engulfed a cyanobacterium. The cyanobacterium would, over time, lose many of its genes and become a plastid (or,

specifically in green plants, a chloroplast) (Martin and Herrmann, 1998), leading the evolution of an oxygenic photosynthetic eukaryote. Because they lost so many genes, the endosymbiotic cyanobacteria became incapable of independent existence outside the symbiotic association; essentially they are enslaved by the host cell (Bhattacharya and Medlin, 1998). All eukaryotic photoautotrophs are oxygenic (Falkowski *et al.*, 2004).

While there are eight major phyla of photosynthetic eukaryotes, they can be broadly lumped into two major groups: a green and a red clade. The former contain, in addition to chlorophyll *a*, a second photosynthetic pigment, chlorophyll *b*. The latter does not contain chlorophyll *b*, but rather other pigments, especially chlorophyll *c*. Regardless of the evolutionary history of the host, however, the core photosynthetic machinery is highly conserved. That is, the same basic structure that evolved in cyanobacteria is found in every oxygenic photosynthetic eukaryote. Indeed, in some cases, the gene sequences are so highly conserved across the ca. 3 billion years of evolution that they have been called 'frozen metabolic accidents' – genes that essentially have ceased evolving (Shi *et al.*, 2005).

Approximately 500 million years ago, one green alga ultimately became the progenitor of all higher plants (Knoll, 1992; Kenrick and Crane, 1997). Although there are in excess of 250,000 species of extant higher plants, they are all relatively closely related to each other. Except for diatoms that live in soils, member of the red clade of photosynthetic eukaryotes successfully colonized land. These organisms, however, became extremely successful in aquatic ecosystems. All of the major photosynthetic eukaryotes in the ocean arose by *secondary* endosymbiosis, where eukaryotic phagotrophic flagellates ingested eukaryotic phototrophs (Falkowski *et al.*, 2004). This process led to the evolution of diatoms and other closely related groups (including the kelps), haptophyte algae including coccolithophores, and dinoflagellates (Delwiche, 2000). These three groups, which are very distantly related, have left fossils that allow some physical reconstruction of their evolutionary history. There is fossil evidence that a (macrophyte) red alga existed 1.2 Ga, with unicellular (cysts) and macrophytic green algae from 1 and 0.6 Ga bp respectively.

While fossils referable to heterokonts (which include diatoms) are known from the late Proterozoic, it is possible that these are not photosynthetic, granted the much later (280 Ma bp) origin of *photosynthetic* heterokonts suggested by molecular clock data calibrated from the molecular phylogeny and fossil record of diatoms (Bhattacharya *et al.*, 1992; Kooistra *et al.*, 2007). The precursors of modern dinoflagellates probably originated more than 250 million years ago (Falkowski *et al.*, 2004), while at least the coccolithophorid variants of the haptophytes are first known from fossils of the early Triassic period (de Vargas *et al.*, 2007).

This brief account of the evolution and diversification of aquatic photoautotrophs reveals that there is much more higher-taxon diversity among aquatic than terrestrial primary producers; 95% or more of species involved in terrestrial primary producers are embryophytes derived from a single class (Charophyceae) of a single Division (Chlorophyta) among the eight Divisions of eukaryotic aquatic photoautotrophs. By contrast, there is no comparable taxonomic dominance of primary producers in aquatic ecosystems. Simply put, the land is green and the oceans are red (Falkowski *et al.*, 2003).

2.6 Net primary production

Photosynthesis allows an organism to convert inorganic carbon to organic matter, yet the organism must also use some of that organic matter for its own metabolic demands. Hence, ecologists have devised another term: net primary production (NPP), defined as that fraction of photosynthetically produced carbon that is retained following all respiratory costs of the photoautotrophs

and consequently is available to the next trophic level in the ecosystem (Lindeman, 1942).

In terrestrial ecosystems, NPP is derived from empirical measurements of leaf area (i.e., the ability of a plant to absorb light), leaf nitrogen content (a surrogate for the content of the photosynthetic machinery, as leaves are the photosynthetic organ of a plant), irradiance, temperature and water availability. To a first order, the distribution of plants on land is controlled by water availability (Mooney *et al.*, 1987). When water is available, the productivity of terrestrial plants is mainly controlled by how much light is intercepted by the leaves (Nobel *et al.*, 1993; Field *et al.*, 1998). Models for NPP on land are relatively easy to verify; plants can be weighed and their carbon content can be measured.

In aquatic ecosystems, net primary production is generally estimated from incorporation rates of ^{14}C-labelled inorganic carbon (supplied as bicarbonate) into organic matter. Alternative approaches include oxygen exchange in an enclosed volume (a technique which is far less sensitive and much more time-consuming) and variable fluorescence (which embraces a variety of approaches) (Falkowski *et al.*, 1998). As for terrestrial ecosystems, extrapolation of photosynthetic rates from any technique to the water column requires a mathematical model. The commonly used strategy is to derive an empirical relationship between photosynthesis and irradiance for a given phytoplankton or macrophyte community, and to integrate that (instantaneous) relationship over time and depth in the water column. Assuming some respiratory costs during the photo-period and during the dark, this approach gives an estimate of net primary production for the water column (Behrenfeld and Falkowski, 1997).

Using satellite-based images of terrestrial vegetation and ocean water leaving radiances (i.e., the visible light scattered back to space from the ocean), it is possible to construct maps of the global distributions of NPP. Briefly, the terrestrial NPP is constructed from estimates of leaf cover and irradiance. For the oceans, phytoplankton chlorophyll is derived based on the ratio of green to blue light that is reflected back to space. In the absence of photosynthetic pigments in the upper ocean, radiances leaving water would appear blue to an observer in space. In the presence of chlorophylls and other photosynthetic pigments, some of the blue radiation is absorbed in the upwelling stream of photons, thereby reducing the overall radiance; the water becomes 'darker'. The fraction of blue light that is absorbed is proportional to the chlorophyll concentration. After accounting for scattering and absorption by the atmosphere, satellite-based maps of both terrestrial vegetation and ocean color can be used to infer the spatial and temporal variations of photosynthetic biomass from which NPP is derived (Behrenfeld and Falkowski, 1997).

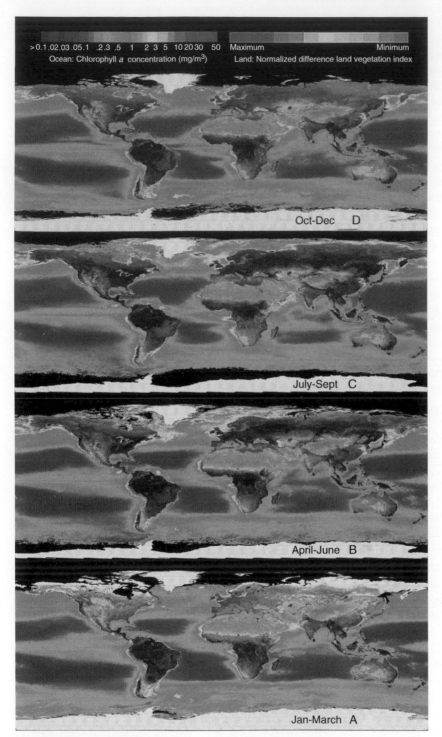

Figure 2.4 Seasonal, maps of global inventories of photosynthetic biomass in the oceans and on land as derived from satellite images. These data are used to construct global estimates of primary production (e.g. Field *et al.*, 1998). Note the strong seasonal variations in biomass, especially at high latitudes in both hemispheres (data courtesy of NASA).

Monthly, seasonal and annual maps of global ocean net primary production have been developed that incorporate a primary production model into global ocean colour images of phytoplankton chlorophyll (Fig. 2.4). Such models suggest that annual net ocean photosynthetic carbon fixation is about 45±5 GtC per annum (Behrenfeld and Falkowski, 1997a, b). This productivity is driven by a photosynthetic biomass that amounts to ca. 1 GtC. Hence, phytoplankton biomass in the oceans turns over on the order of once every two weeks; the overwhelming majority (~85%) of the productivity is consumed by heterotrophs in the upper ocean, and about 15% sinks into the ocean interior where virtually all is consumed ('remineralized') and converted back to inorganic carbon (see also Chapter 3). Only a very small fraction of the organic matter produced is buried in marine sediments. Simply put, less than 1% of the photosynthetic biomass on Earth accounts for about 50% of the net primary productivity (Field *et al.*, 1998). What about the other 50%?

Net primary production in terrestrial ecosystems accounts for approximately 52 GtC per annum, and is supported by a biomass of approximately 500 GtC (Field *et al.*, 1998). Thus, the turnover time of terrestrial plant biomass is on the order of a decade. As in the ocean, the overwhelming majority of terrestrial plant production is recycled back to inorganic carbon, primarily by fungi and microbes which degrade dead plants. In terrestrial ecosystems, virtually all of the photosynthetic biomass is consumed by heterotrophs long after it was produced. For example, the leaves on deciduous trees remain until the autumn, when they then die and fall to the ground. Over the coming years, virtually 100% of the leaf litter will be consumed by microbes and fungi, but very little will have been eaten by an animal. Terrestrial ecosystems contain large amounts of carbon associated with living structures (e.g. wood in trees), but their photosynthetic rates are not sufficiently different than aquatic ecosystems, which have a much faster rate of ecological metabolism. Thus, although the recycling time for terrestrial plant production is much longer than that in the ocean, the fates are basically the same; virtually all of the organic matter is reprocessed by heterotrophs resulting in a biological cycle that has relatively small impact on the ocean/atmosphere inventory of CO_2 on time scales of decades or centuries (Falkowski *et al.*, 2000).

2.7 What limits NPP on land and in the ocean?

All organisms are primarily composed of six major elements: H, C, N, O, P and S (Schlesinger, 1997). The ratios of these elements can vary widely, depending on the type of organism and the ecosystem; however, one or more of these elements often limits NPP. The limitations fundamentally differ between aquatic and terrestrial ecosystems.

Terrestrial plants have an absolute requirement for water, which clearly is not a problem for aquatic photoautotrophs. However, even when water is ample, carbon fixation is not necessarily maximal. One of the major limiting factors limiting NPP on land is the availability of CO_2 itself. Although Rubisco is the most abundant protein on Earth, it has a low affinity for CO_2. Saturation values for CO_2 in terrestrial ecosystems are in the order of ca. 500 parts per million, significantly above that found in the contemporary atmosphere (Mooney *et al.*, 1991). To make matters even worse, the enzyme also 'mistakes' O_2 as a substrate (hence, the origin of the term 'oxygenase' in the appellation of the enzyme). Indeed, in most terrestrial plants, approximately 35% of the time, the enzyme reacts with O_2 leading to the imme-

diate respiration of two carbon products which are never converted into biomass. This 'photorespiratory' process leads to a large loss in potential NPP when CO_2 is not saturating. Some terrestrial plants have overcome this limitation by developing a high-affinity carbon fixation process which allows them to use lower levels of CO_2; these so-called C_4 plants are primarily tropical grasses (such as cane sugar), and account for approximately 15% of global terrestrial NPP (Berry, 1999). Thus, on land, both water and CO_2 are major factors limiting NPP.

In aquatic ecosystems, water and inorganic carbon are not limiting. Although the concentration of gaseous CO_2 in the oceans is only ~10 µM, the total concentration of inorganic carbon is ~2.5 mM, of which 95% is in the form of HCO_3^-. Although HCO_3^- cannot be used by Rubisco directly, the anion can be transported into cells from the sea and converted to CO_2 by the enzyme carbonic anhydrase, which is one of the most catalytically active enzymes known. The transport and dehydration of HCO_3^- by aquatic photoautotrophs is called a 'carbon concentrating mechanism', and it virtually assures that carbon never limits NPP in aquatic ecosystems (Kaplan and Reinhold, 1999). If water or inorganic carbon are not limiting, what is?

Globally, nitrogen and phosphorus are the two elements that immediately limit NPP in both lakes and the oceans. It is frequently argued that since N_2 is abundant in both the ocean and the atmosphere, and since in principle it can be biologically reduced to the equivalent of NH_3 by N_2-fixing cyanobacteria, then nitrogen cannot be limiting on geological time scales (Redfield, 1958; Barber, 1992). Therefore, phosphorus, which is supplied to the ocean by the weathering of continental rocks, must ultimately limit biological productivity (Broecker and Peng, 1982; Tyrell, 1999). The underlying assumptions of these tenets should, however, be considered within the context of the evolution of biogeochemical cycles.

The main source of fixed inorganic nitrogen to the oceans is biological nitrogen fixation. This is strictly a prokaryotic process, and in the modern ocean, it is conducted mainly by cyanobacteria. On the early Earth, however, before the evolution of nitrogen fixers, electrical discharge or bolide impacts may have promoted NO formation from the reaction between N_2 and CO_2, but the yield for these reactions is extremely low. Moreover, any volcanogenic NH_3 in the atmosphere would have photodissociated from UV radiation (Kasting, 1990) while N_2 would have been stable (Warneck, 1988; Kasting, 1990). Biological N_2 fixation is a strictly anaerobic process (Postgate, 1971), and the genes encoding the catalytic subunits for nitrogenase are highly conserved in cyanobacteria and other prokaryotes that fix nitrogen, strongly suggesting a common ancestral origin (Zehr *et al.*, 1995). Since fixed inorganic nitrogen was likely to have been scarce prior to the evolution of diazotrophic (nitrogen fixing) organisms, there was strong evolutionary selection for nitrogen fixation on the early Earth.

The formation of nitrate from ammonium by nitrifying bacteria requires molecular oxygen (see Chapter 7); hence, nitrification must have evolved following the formation of free molecular oxygen in the oceans by oxygenic photoautotrophs. Therefore, from a geological perspective, the conversion of ammonium to nitrate probably proceeded rapidly and provided a substrate, NO_3^-, that eventually could serve both as a source of nitrogen for photoautotrophs and as an electron acceptor for a diverse group of heterotrophic, anaerobic bacteria, the denitrifiers.

In the sequence of the three major biological processes that constitute the nitrogen cycle, denitrification must have been the last to emerge. This process, which permits the reduction of NO_3^- to (ultimately) N_2, requires hypoxic or anoxic environments and is sustained by high sinking fluxes of organic matter. With the emergence of denitrification, the ratio of fixed inorganic N to dissolved inorganic phosphate in the ocean interior could only be depleted in N relative to the sinking flux of the two elements in POM. Indeed, in all of the major basins in the contemporary ocean, the N:P ratio of the dissolved inorganic nutrients in the ocean interior is conservatively estimated at 14.7 by atoms (Fanning, 1992) or less (Anderson and Sarmiento, 1994), compared to the Redfield ratio of 16/1.

There are three major conclusions that may be drawn from the foregoing discussion. First, because the sinking flux of particulate matter has an organic N to P ratio (about 16/1) which exceeds the N to P ratio of the deep ocean dissolved inorganic pool (average 14.7/1 as noted above), the average upward flux of nutrients is slightly enriched in P relative to the requirements of the photoautotrophs (Redfield, 1958; Gruber and Sarmiento, 1997). Hence, with some exceptions (Wu *et al.* 2000), dissolved inorganic fixed nitrogen generally limits primary production throughout most of the world's oceans on ecological times scales (Barber, 1992); (Falkowski, 1998). Second, the N to P ratio of the deep-ocean inorganic dissolved pool was established by biological processes, not vice versa (Redfield, 1934; Redfield *et al.*, 1963). Third, if dissolved inorganic nitrogen rather than phosphate limits productivity in the oceans, then it follows that the ratio of nitrogen fixation to denitrification plays a critical role in determining primary production and the net biologically mediated exchange of CO_2 between the atmosphere and ocean (Codispoti, 1995).

On ecological time scales, primary production is limited by nutrient supply and the efficiency of nutrient utilization in the euphotic zone. There are three major areas of the world ocean where inorganic nitrogen and phosphate are in excess throughout the year, yet the mixed layer depth appears to be shallow enough to support active primary production (if the mixed layer depth extends below the depth where sufficient light is available for primary production, production becomes limited): these are the eastern equatorial Pacific, the subarctic Pacific, and Southern (i.e., Antarctic) Oceans. For the subarctic North Pacific, Miller *et al.* (1991) suggested a tight coupling between phytoplankton production and consumption by zooplankton. This grazer-limited hypothesis was used to explain why the phytoplankton in the North Pacific do not form massive blooms in the spring and summer like their counterparts in the North Atlantic (Banse, 1992).

In the mid-1980s, however, it became increasingly clear that the concentration of trace metals, especially iron, was extremely low in all three of these regions (Martin *et al.*, 1991). Indeed, in the eastern equatorial Pacific the concentration of soluble iron in the euphotic zone is only 100 to 200 pM, about an order of magnitude lower than found in other areas of the open ocean. Although iron is the most abundant transition metal in the Earth's crust, in its most commonly occurring form, Fe^{3+}, it is virtually insoluble in oxygenated water. The major source of iron to the euphotic zone of the ocean is aeolian dust, originating from continental deserts. In the three major areas of the world oceans with high inorganic nitrogen in the surface waters and low chlorophyll concentrations, the flux of aeolian iron is extremely low (Duce and Tindale, 1991). In experiments in which iron was artificially added on a relatively large scale to the waters in the equatorial Pacific, Southern Ocean, and subarctic Pacific there were rapid and dramatic increases in photosynthetic energy conversion efficiency and phytoplankton chlorophyll concentrations (Boyd *et al.*, 2007). Beyond doubt, NPP and export production in all three regions is limited by the availability of a single micronutrient – iron.

2.8   Is NPP in balance with respiration?

On long time scales, a very small fraction of net primary production escapes respiration to be become buried in sediments and transferred to the lithosphere. This process has profound influence on planetary redox state. As organic matter, by definition, contains reducing equivalents, its burial requires that oxidizing equivalents accumulate elsewhere in the system. Indeed, the net burial of organic matter on geological time scales implies the oxidation of the atmosphere and ocean; i.e., the accumulation of free molecular oxygen requires burial and sequestration of organic carbon (see Chapter 7). How can we assess how much organic carbon is buried?

Recall that Rubisco strongly discriminates against the heavier isotope of carbon, ^{13}C. The ultimate source of CO_2 to Earth's surface is volcanism with a distinct isotopic signal. The recycling of organic matter and carbonate rocks by weathering also represents a source of IC (inorganic carbon) to the ocean-atmosphere system. Therefore volcanism plus weathering (which should on average also carry a volcanogenic isotopic signal) represent the source of IC. The two sinks are comprised of the burial of inorganic carbonates, and organic carbon in sediments. Imagine that over some long period of time, a fraction of the total carbon input to the Earth surface is buried and preserved as organic matter. This removal of ^{13}C-depleted organic carbon burial leads to an enrichment of ^{13}C in the inorganic pools. The magnitude of the ^{13}C carbonate enrichment depends on the proportion of the total IC input removed as organic carbon. The higher the proportion of organic carbon removal, the more enriched the ^{13}C carbonates (Kump and Arthur, 1999). Removal proportions of these two fractionations are the basis of the isotopic mass balance quantified as:

$$F_{in} * \delta^{13}C_{in} = F_{carb} * \delta^{13}C_{carb} + F_{org} * \delta^{13}C_{org} \qquad (2.4)$$

where F_{in} represents the total flux of IC into the ocean–atmosphere system, and $\delta^{13}C_{in}$ is its isotopic composition. Similarly, F_{carb} and F_{org} are the burial fluxes of carbonate carbon and organic carbon, while $\delta^{13}C_{carb}$ and $\delta^{13}C_{org}$ are the respective isotopic compositions of these removal fluxes. At steady state, we have the equality:

$$F_{in} = F_{carb} + F_{org} \qquad (2.5)$$

and furthermore, we can express the fractionation between inorganic carbon and organic carbon in the oceans as:

$$\Delta_{org} = \delta^{13}C_{carb} - \delta^{13}C_{or} \qquad (2.6)$$

We also define a fraction f, which describes the burial proportion of organic carbon relative to the total flux of IC into the system

$$f = F_{org} / F_{in} \qquad (2.7)$$

From these equalities, we generate the following expression for f:

$$f = (\delta^{13}C_{carb} - \delta^{13}C_{in}) / \Delta_{org} \qquad (2.8)$$

At present, the average isotopic value of input DIC is ca. −5‰, the $\delta^{13}C_{carb}$ is 0‰ and Δ_{org} is ca. −25‰. Hence, to balance the input with the output, the ratio of buried organic carbon relative to total carbon is about 0.20. That is, for every 1 atom of carbon buried as organic matter, about 5 atoms are buried as carbonate (Berner, 2004). Values for $\delta^{13}C_{carb}$ and Δ_{org} (and probably also $\delta^{13}C_{in}$) have changed through time (see Chapter 3) and hence, these serve as sedimentary archives of changes in carbon sources and sinks, thereby providing the best monitor available to reconstruct the role of the biological carbon cycle in the oxidation state of the planet's surface.

2.9 Conclusions and extensions

Fundamentally, the biological carbon cycle is driven by redox reactions far from thermodynamic equilibrium. On time scales of centuries and even millennia, the formation of organic matter, which is a reduction reaction, is closely coupled to respiration, an oxidation reaction, such that a quasi-steady state is achieved. The steady-state condition buffers the redox reactions on the planetary surface, and allows the highly efficient metabolic reactions of oxygen-dependent heterotrophy to proceed across vast stretches of geological time (Knoll, 2003). Indeed, animal life depends on this buffering capacity.

Humans extract fossil fuels from the lithosphere approximately 1 million times faster than they were formed. That is, each year of production of fossil fuel represents 1 million years of burial of organic matter by either plants on land (primarily yielding coal) or phytoplankton in the ocean (primarily yielding petroleum and natural gas). We burn these fossil fuels, thereby reoxidizing the organic matter back to inorganic carbon. The human combustion of fossil fuels is analogous to the action of volcanoes. However, even if humans burn the vast reserves of fossil fuels in the ground, it would have little impact on Earth's oxygen supply; there simply is so much oxygen in the atmosphere relative to the fuel available. However, the combustion of fossil fuels has rapidly altered the atmospheric inventory of CO_2, which, in turn, can alter Earth's climate and the pH of the ocean (Falkowski *et al.*, 2000). Can this problem be ameliorated by accelerating the biological carbon cycle?

The short answer is, no. Of the current emission of ca. 8.5 Gt C per annum from the burning of fossil fuels, approximately 54% remains in the atmosphere; the residual 46% is split, almost evenly between terrestrial and marine ecosystems. The ability of terrestrial ecosystems to absorb the anthropogenically produced CO_2 will weaken over the coming decades as the concentration of the gas begins to saturate Rubisco. Moreover, even if humans could devote large areas of land to the growth of trees, the trees would have to be harvested and preserved – they could not be allowed to enter a respiratory reaction that would oxidize the organic matter, hence the CO_2 would be liberated back to the atmosphere.

In the oceans, accelerating the export of organic matter to the ocean interior would potentially help reduce the impact of anthropogenic CO_2 emissions. However, the only reasonable mechanism to do this is to add a limiting nutrient to the ocean. One possible mechanism is based on the addition of iron to areas of the ocean that are iron limited. While such an approach may temporarily reduce the rate of accumulation of CO_2 in the atmosphere, it is a significant geoengineering problem with potentially deleterious side-effects (Martin, 1990).

Perhaps the most interesting aspect of the biological carbon cycle is that the overwhelming majority of buried organic matter is associated with rocks, especially shales and mudstones, and is not directly accessible to oxidation on ecological time scales. The interaction between the formation of organic matter and its slow leak into the lithosphere truly distinguishes Earth from all the other planets in our solar system (Falkowski and Godfrey, 2008).

References

Anderson L, Sarmiento J (1994) Redfield ratios of remineralization determined by nutrient data analysis. *Global Biogeochemical Cycles* **8**, 65–80.

Antoine D, Andre J-M, Morel A, et al. (1996) Oceanic primary production, 2: estimation at global scale from satellite (Coastal Zone Color Scanner) chlorophyll. *Global Biogeochemical Cycles* **10**(1), 57–69.

Banse K (1992) Grazing, temporal changes of phytoplankton concentrations, and the microbial loop in the open sea. In: *Primary Productivity and Biogeochemical Cycles in the Sea* (ed Falkowski PG). Plenum, New York, pp. 409–440.

Barber RT (1992) Geological and climatic time scales of nutrient variability. In: *Primary Productivity and Biogeochemical Cycles in the Sea* (ed Falkowski). Plenum, New York, pp. 89–106.

Behrenfeld M, Falkowski PG (1997a) A consumer's guide to phytoplankton productivity models. *Limnology and Oceanography* **42**, 1479–1491.

Behrenfeld MJ, Falkowski PG (1997b) Photosynthetic rates derived from satellite-based chlorophyll concentration. *Limnology and Oceanography* **42**, 1–20.

Beja OL, Aravind EV, Koonin MT, et al. (2000) Bacterial rhodopsin: evidence for a new type of phototrophy in the sea. *Science* **289**, 1902–1906.

Benson A, Calvin M (1950) Carbon dioxide fixation by green plants. *Annual Review of Plant Physiology* **1**, 25–42.

Berman-Frank I, Cullen JT, Shaked Y, Sherrell RM, Falkowski PG (2001) Iron availability, cellular iron quotas, and nitrogen fixation in *Trichodesmium*. *Limnology and Oceanography* **46**, 1249–1260.

Berner RA (2004) The Phanerozoic Carbon Cycle: CO_2 and O_2. Oxford, University Press, Oxford.

Berry J (1999) Ghosts of biospheres past. *Nature* **400**, 509–510.

Bhattacharya D, Medlin L (1998) Algal phylogeny and the origin of land plants. *Plant Physiology* **116**, 9–15.

Bhattacharya D, Medlin L, Wainwright PO, et al. (1992) Algae containing chlorophylls *a* + *c* are polyphyletic: molecular evolutionary analysis of the Chromophyta. *Evolution* **46**, 1801–1817.

Blankenship RE, Madigan MT, Bauer CE, (eds) (1995) *Anoxygenic Photosynthetic Bacteria*. Kluwer Scientific, Dordrecht.

Blankenship R, Sadekar S, Raymond J (2007) The evolutionary transition from anoxygenic to oxygenic photosynthesis. In: *Evolution of Primary Producers in the Sea* (eds Falkowski PG, Knoll AH). Academic Press, San Diego, pp. 21–35.

Boyd PW, Jickells T, Law CS, et al. (2007) Mesoscale iron enrichment experiments 1993-2005: Synthesis and future directions. *Science* **315**, 612–617.

Brocks JJ, Love GD, Summons RE, Knoll AH, Logan GA, Bowden SA (2005) Biomarker evidence for green and purple sulfur bacteria in an intensely stratified Paleoproterozoic sea. *Nature* **396**, 450–453.

Broecker WS, Peng T-H (1982) *Tracers in the Sea*. EI-digio Press, Palisades, NY.

Canfield DE (1998) A new model for Proterozoic ocean chemistry. *Nature* **396**, 450–453.

Canfield DE, Raiswell R (1999) The evolution of the sulfur cycle. *American Journal of Science* **299**, 697–723.

Canfield DE, Rosing MT, Bjerrum C (2006) Early anaerobic metabolisms. *Philosophical Transactions of the Royal Society B Biological Sciences* **361**, 1819–1834.

Chisholm SW (1992) Complementary niches of *Synechococcus* and *Prochlorococcus*. In: *Primary Productivity in the Sea* (ed Falkowski PG). Plenum Press, New York, pp. 213–237.

Codispoti L (1995) Is the ocean losing nitrate? *Nature* **376**, 724.

de Vargas C, Probert Aubry MP, Young J (2007) Origin and evolution of coccolithophores: from coastal hunters to oceanic farmers. In: *Evolution of Primary Producers in the Sea* (eds Falkowski PG, Knoll AH). Academic Press, New York.

DeFries R, Townshend J (1994) NDVI-derived land cover classification at global scales. *International Journal of Remote Sensing* **15**, 3567–3586.

Delwiche C (2000) Tracing the thread of plastid diversity through the tapestry of life. *American Naturalist* **154**, S164-S177.

Duce RA, Tindale NW (1991) Atmospheric transport of iron and its deposition in the ocean. *Limnology and Oceanography* **36**, 1715–1726.

Falkowski PG (1997) Evolution of the nitrogen cycle and its influence on the biological sequestration of CO_2 in the ocean. *Nature* **387**, 272–275.

Falkowski PG (1998) Biogeochemical controls and feedbacks on ocean primary productivity. *Science* **281**, 200–206.

Falkowski PG (2001) Biogechemical Cycles. *Encyclopedia of Biodiversity* **1**, 437–453.

Falkowski PG, Godfrey L (2008) Electrons, life, and the evolution of Earth's oxygen cycle. *Philosophical Transactions of the Royal Society B Biological Sciences* **363**, 2705–2716.

Falkowski PG, Raven JA (2007) *Aquatic Photosynthesis*. Princeton, University Press Princeton, NJ.

Falkowski P, Barber R, Smetacek V (1998) Biogeochemical controls and feedbacks on ocean primary production. *Science* **281**, 200–206.

Falkowski P, Scholes RJ, Boyle E, et al. (2000) The global carbon cycle: A test of our knowledge of earth as a system. *Science* **290**, 291–296.

Falkowski P, Katz M, van Schootenbrugge B, Schofield O, Knoll AH (2003) Why is the land green and the ocean red? In: *Coccolithophores – From Molecular Processes to Global Impact* (eds Young J, Thierstein HR). Springer-Verlag, Berlin.

Falkowski PG, Katz ME, Knoll AH, et al. (2004) The evolution of modern eukaryotic phytoplankton. *Science* **305**, 354–360.

Fanning K (1992) Nutrient provinces in the sea: concentration ratios, reaction rate ratios, and ideal covariation. *Journal of Geophysical Research* **97C**, 5693–5712.

Field C, Behrenfeld M, Randerson J, Falkowski P (1998) Primary production of the biosphere: integrating terrestrial and oceanic components. *Science* **281**, 237–240.

Gruber N, Sarmiento J (1997) Global patterns of marine nitrogen fixation and denitrification. *Global Biogeochemical Cycles* **11**, 235–266.

Han T-M, Runnegar B (1992) Megascopic eukaryotic algae from the 2.1-billion year old Neogene iron-formation. *Science* **257**, 232–235.

Jannasch HW, Taylor CD (1984) Deep-sea microbiology. *Annual Review of Microbiology* **38**, 487–514.

Jørgensen BB (2001) Space for hydrogen. *Nature* **412**, 286–289.

Kaplan A, Reinhold L (1999) CO_2 concentrating mechanisms in photosynthetic microorganisms. *Annual Review of Plant Science and Plant Molecur Biology* **50**, 539–570.

Kasting JF (1990) Bolide impacts and the oxidation state of carbon in the Earth's early atmosphere. *Origins Life and Evolution Biosphere* **20**, 199–231.

Kenrick P, Crane P (1997) The origin and early evolution of plants on land. *Nature* **389**, 33–39.

Knoll AH (1992) The early evolution of eukaryotes: a geological perspective. *Science* **256**, 622–627.

Knoll AH (2003) *Life on a Young Planet: The First Three Billion Years of Evolution on Earth.* Princeton NJ, Princeton University Press.

Kooistra W, Gersonde R, Medlin LK, Mann DG (2007) The origin and evolution of the diatoms: their adaptation to a planktonic existence. In: Evolution of Primary Producers in the Sea (eds Falkowski PG, Knoll AH). Academic Press, New York.

Kump L, Arthur A (1999) Interpreting carbon-isotope excursions: carbonates and organic matter. *Chemical Geology* **161**, 181–198.

Lindeman R (1942) The trophic-dynamic aspect of ecology. *Ecology* **23**, 399–418.

Martin JH (1990) Glacial–interglacial CO_2 change: The iron hypothesis. *Paleoceangraphy* **5**, 1–13.

Martin JH, Gordon RM, Fitzwater SE (1991) The case for iron. *Limnology and Oceanography* **36**(8, What Controls Phytoplankton Production in Nutrient-Rich Areas of the Open Sea?): 1793–1802.

Martin W, Herrmann RG (1998) Gene transfer from organelles to the nucleus: how much, what happens, and why? *Plant Physiology* **118**, 9–17.

Mauzerall D (1992) Light, iron, Sam Granick and the origin of life. *Photosynthesis Research* **33**(2), 163–170.

Mooney HA, Drake BG, Luxmoore RJ, Oechel WC, Pitelka LF (1991) Predicting ecosystem responses to elevated CO_2 concentrations. *BioScience* **41**(2), 96–104.

Mooney HA, Pearcy RW, Ehleringer J (1987) Plant physiological ecology today. *BioScience* **37**, 18–20.

Mulkidjanian A, Junge W (1997) On the origin of photosynthesis as inferred from sequence analysis – a primordial UV-protector as common ancestor of reaction centers and antenna proteins. *Photosynthesis Research* **51**(27–42).

Nobel PS, Forseth IN, Long SP (1993) Canopy structure and light interception. In: *Photosynthesis and Production in a Changing Environment: A Field and Laboratory Manual* (eds Hall DO, Scurlock JMO, Bolhàr-Nordenkampf HR, Leegood RC, Long SP). Chapman & Hall, London, pp. 79–90.

Park R (1961). Carbon isotope fractionation during photosynthesis. *Geochimica et Cosmochimica Acta* **21**, 110–126.

Postgate JR, (ed) (1971) The Chemistry and Biochemistry of Nitrogen Fixation. Plenum Press, New York.

Raven JA, Falkowski PG (1999). Oceanic sinks for atmospheric CO_2. *Plant Cell Environment* **22**(6), 741–755.

Redfield A, Ketchum B, Richards F (1963) The influence of organisms on the composition of sea-water. In: The Sea (ed Hill M). New York, *Interscience*. **2**, 26–77.

Redfield AC (1934) On the Proportions of Organic Derivatives in Sea Water and Their Relation to the Composition of Plankton. Liverpool, James Johnstone Memorial Volume.

Redfield AC (1958) The biological control of chemical factors in the environment. *American Scientist* **46**, 205–221.

Schlesinger WH (1997) Biogeochemistry: An Analysis of Global Change. Academic Press, New York.

Shi T, Bibby TS, Jiang L, Irwin AJ, Falkowski PG (2005) Protein interactions limit the rate of evolution of photosynthetic genes in Cyanobacteria. *Molecular Biology Evolution* **22**(11), 2179–2189.

Shi T, Falkowski PG (2008). Genome evolution in cyanobacteria: the stable core and the variable shell. *Proceedings of the National Academy of Sciences, USA* **105**, 2510–2515.

Smith S (1981). Marine macrophytes as a global carbon sink. *Science* **211**, 838–840.

Stevens TO, McKinley JP(1995). Lithoautotrophic microbial ecosystems in deep basalt aquifers. *Science* **270**, 450–454.

Tyrell T (1999) The relative influences of nitrogen and phosphorus on oceanic primary production. *Nature* **400**, 525–531.

Van Niel CB (1941) The bacterial photosyntheses and their imporance for the general problem of photosynthesis. *Advances in Enzymology* **1**, 263–328.

Warneck P (1988) Chemistry of the Natural Atmosphere. Academic Press, New York.

Woese C (1998) The universal ancestor. *Proceedings of National Academy of Sciences USA* **95**: 6854–6859.

Woese CR, Stackebrandt E, Weisburg WG, *et al.* (1984) The phylogeny of purple bacteria; The alpha subdivision. *Systematic and Applied Microbiology* **5**, 315–326.

Wu JF, Sunda W, Boyle EA, Karl DM (2000) Phosphate depletion in the western North Atlantic Ocean. *Science* **289**(5480), 759–762.

Zehr JP, Mellon M, Braun S, Litaker W, Steppe T, Paerl HW(1995) Diversity of heterotrophic nitrogen-fixation genes in a marine cyanobacterial mat. *Applied and Environmental Microbiology* **0061**, 02527–02532.

Zehr JP, Mellon MT, Hiorns WD (1997) Phylogeny of cyanobacterial nifH genes: Evolutionary implications and potential applications to natural assemblages. *Microbiology* **143**, 1443–1450.

3

THE GLOBAL CARBON CYCLE: GEOLOGICAL PROCESSES

Klaus Wallmann[1] and Giovanni Aloisi[1,2]

[1]Leibniz Institute for Marine Sciences (IFM-GEOMAR), Wischhofstrasse, 1–3; 24148, Kiel, Germany
[2]UMR CNRS 7159LOCEAN, Universite Pierre et Marie Curie, Paris, France

3.1 Introduction

In this chapter, we will discuss the geological carbon cycle including all processes controlling the distribution of carbon between the interior of the Earth, the lithosphere, the hydrosphere and the atmosphere (Fig. 3.1, Table 3.1). Our discussion focuses on the geochemical evolution of carbon inventories and fluxes on a multi-million year timescale. We will first address the magnitude of carbon fluxes in the pre-human Holocene and then over the last million years. We next examine the feedbacks stabilizing the distribution of carbon on our planet and present a balanced geological carbon cycle. Subsequently, we explore proxies and models to explain how carbon cycling may have changed over the Earth's geological history in parallel with its biological and geological evolution.

3.2 Organic carbon cycling

Land plants and phytoplankton use solar energy to convert atmospheric CO_2 into biomass via photosynthesis. The overall stoichiometry of photosynthesis can be expressed as:

$$CO_2 + H_2O \Rightarrow C(H_2O) + O_2 \qquad (3.1)$$

where the simple carbohydrate $C(H_2O)$ represents organic carbon formed by this process. Molecular oxygen (O_2) is released as an important by-product of this reaction (see Chapter 7). Most of the biomass formed by photosynthesis, however, is rapidly consumed by microorganisms living in soils and in the oceans. Just as humans do, they use the biomass as an energy resource and recycle CO_2 into the atmosphere (electron acceptors other than O_2 are also used to oxidize organic matter as described below):

$$C(H_2O) + O_2 \Rightarrow CO_2 + H_2O \qquad (3.2)$$

This respiration process consumes more than 99% of the biomass produced by photosynthesis both on land and in the sea (Table 3.2). On land, the remaining biomass accumulates in living vegetation, soils and detritus, but this is only transient, as most terrestrial organic matter is finally exported into the oceans by rivers in dissolved or particulate form (IPCC, 2007). Overall, approximately 4–5 Tmol yr^{-1} of terrestrial organic matter accumulates in marine sediments deposited on the continental shelf (Burdige, 2005). During certain periods of the geological past (e.g. Permian, Carboniferous) however, the accumulation of terrestrial organic carbon in swamps and wetlands contributed significantly to the global carbon burial flux (Berner and Canfield, 1989).

Ancient marine sediments represented by shales are by far the largest reservoir of organic carbon on our planet. Organic matter is liberated from these rocks, and other rocks including carbonates, by physical erosion and chemical weathering. Even though this old organic matter is much more resistant towards microbial degradation than young organic carbon in soils, studies indicate that it is oxidized by microbes in the presence of atmospheric oxygen (Petsch *et al.*, 2001). Thus, if the oxidation reactions proceed much faster than the geological processes (uplift, physical erosion, chemical weathering) that expose fossil organic matter to the atmosphere

Fundamentals of Geobiology, First Edition. Edited by Andrew H. Knoll, Donald E. Canfield and Kurt O. Konhauser.
© 2012 Blackwell Publishing Ltd. Published 2012 by Blackwell Publishing Ltd.

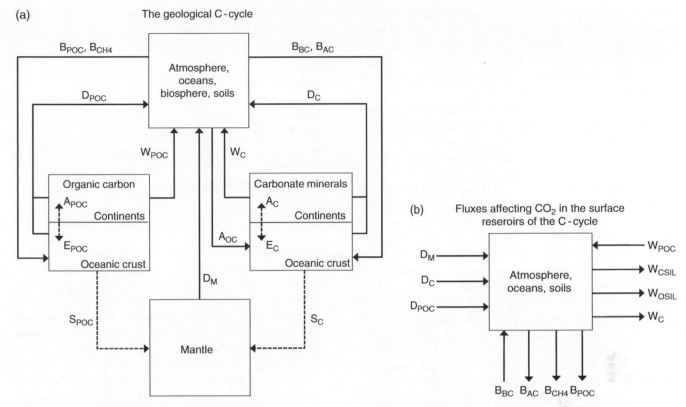

Figure 3.1 (a) The geological carbon (C) cycle; (b) fluxes affecting the concentration of carbon dioxide (CO_2) in the surface reservoir of the carbon cycle. Notes: fluxes that are discussed in detail and quantified in the present work are solid arrows: B_{POC} (burial of POC in marine sediments), B_{CH4} (burial of methane in marine sediments), B_{BC} (burial of biogenic carbonates in marine sediments), B_{AC} (burial of authigenic carbonates in marine sediments), D_{POC} (metamorphic degassing of POC), D_C (metamorphic degassing of carbonate minerals), D_M (degassing of the mantle), W_{POC} (weathering of POC on the continents), W_C (weathering of carbonates on the continents), A_{OC} (alteration of oceanic crust), W_{CSIL} (weathering of silicate minerals on the continents), W_{OSIL} (weathering of silicate minerals in marine sediments). Other carbon fluxes are (dashed arrows): E_{POC} (erosion and transport of POC to the ocean), A_{POC} (accretion of POC during continental collision), E_C (erosion and transport of carbonate minerals to the ocean), A_C (accretion of carbonate minerals during continental collision), S_{POC} (subduction of POC in the mantle), S_C (subduction of carbonate minerals in the mantle).

Table 3.1 Distribution of carbon on our planet (pre-human Holocene)

Substance	Mass (10^{18} mol)
Carbonate C in rocks	5000
Organic C in rocks	1250
Methane in gas hydrates	0.1–0.3
C in soil	0.3
Inorganic C dissolved in deep ocean (>100 m water depth)	2.7
Inorganic C dissolved in upper ocean (0–100 m water depth)	0.1
Atmospheric CO_2	0.06
Terrestrial biosphere	0.05
Marine biosphere	0.0005

Modified from Berner (2004) and Wallmann (2001).

Table 3.2 The global cycle of organic carbon (pre-human Holocene, fluxes in Tmol C yr^{-1})

	Land
CO_2-fixation in living biomass	4700
Accumulation of terrestrial organic carbon in vegetation, soil and detritus	33
Export of terrestrial organic carbon to the oceans	33
Weathering of fossil organic carbon (W_{POC})	8–16
	Oceans
CO_2-fixation in living biomass	4040
Export of marine biomass into the deep ocean	800
Rain of particulate organic matter to the seafloor	190
Burial of organic matter in surface sediments	10–65
Burial of organic matter, authigenic carbonate and methane in deep sediments ($B_{POC} + B_{AC} + B_{CH4}$)	5.4–27

Modified after (IPCC, 2007); (Sarmiento and Gruber, 2006); Burdige, 2007; Wallmann, 2001; Wallmann *et al.*, 2008. Note: symbols for fluxes correspond to those appearing in Fig. 3.1.

(Berner, 2004), which is normally true (Bolton *et al.*, 2006), then upon exposure, a significant fraction of fossil organic carbon is ultimately degraded into CO_2. The total rate of organic carbon weathering is estimated as ~8 Tmol of particulate organic carbon (POC) yr^{-1}, calculated from a rock denudation rate of 22×10^{15} g yr^{-1} (Berner and Berner, 1996) and an average POC concentration in weathering rocks of 0.45 wt-% (Lasaga and Ohmoto, 2002).

The total rate of physical erosion, however, is larger than the rate of continental denudation since much sediment eroded from upland areas is deposited in lowland areas without reaching the oceans. Thus, only about 10% of the sediments eroded in the United States reaches the oceans, whereas the remaining 90 % is re-deposited on the continents (Berner and Berner, 1996). On the other hand, physical erosion has been greatly accelerated by agriculture and other human land-uses. It is thus very difficult to constrain the modern and pre-human rates of physical erosion and organic matter weathering. Considering the available data on physical erosion and continental sediment recycling (Gaillardet *et al.*, 1999b) and the modeling results on POC weathering in shales (Bolton *et al.*, 2006), we estimate that about 8–16 Tmol of CO_2 yr^{-1} were produced by POC weathering during the pre-human Holocene.

Phytoplankton living in the surface layer of the oceans contribute significantly to the global primary production, and most of the organic matter produced (typically about 80%) is consumed by microbes and zooplankton within the upper ~ 100 m of the water column (Sarmiento and Gruber, 2006). Of the fraction exported into the deeper ocean as sinking particles (Table 3.2), most is degraded during passage to the seafloor. Furthermore, of the particulate organic matter reaching the seafloor a large proportion is consumed by microbes and other benthic organisms living in the top 10 cm of the sediment column. Only a small fraction of this organic matter is finally preserved into marine sediments accumulating at the deep-sea seafloor. However, much more organic matter is preserved in rapidly depositing shallow shelf sediments and there is a consensus that most marine organic carbon is currently buried in these sediments (Dunne *et al.*, 2007; Hedges and Keil, 1995; Middelburg *et al.*, 1993). The rate of organic carbon burial in seafloor sediments is only poorly constrained (Burdige, 2007) and falls into the broad range of 10–65 Tmol yr^{-1} (Table 3.2). Estimates generated from marine productivity and particle export data (Dunne *et al.*, 2007) are much higher than estimates based on mass balances from sediment data (Hedges and Keil, 1995).

Before final preservation, much organic matter is decomposed by anaerobic microbes, of which sulfate reducers are particularly important (see Chapter 5).

These microbes use seawater sulfate as an oxidant and convert sedimentary organic matter into bicarbonate (HCO_3^-):

$$C(H_2O) + \frac{1}{2}SO_4^{2-} \Rightarrow HCO_3^- + \frac{1}{2}H_2S \qquad (3.3)$$

Much of the sulfide (HS^- and H_2S) produced is fixed as pyrite (FeS_2) and other authigenic sulfide minerals. Further down in the sediment column, where sulfate is completely consumed, other microbes transform organic carbon into methane (CH_4) and CO_2:

$$C(H_2O) \Rightarrow \tfrac{1}{2}CO_2 + \tfrac{1}{2}CH_4 \qquad (3.4)$$

Natural gas hydrates are formed in marine sediments from methane produced in deep sediment layers by the microbial degradation of organic matter. Applying the appropriate kinetic rate laws for the microbial degradation of organic matter in marine sediments (Middelburg, 1989; Wallmann *et al.*, 2006), it is estimated that only ~20 ± 10% (~ 2–13 Tmol C yr^{-1}) of the organic matter buried in surface sediments is still preserved after 1 million years of microbial degradation. The major portion of buried organic matter (80 ± 10 %) is, thus, transformed into HCO_3^-, CO_2 and CH_4. These metabolites are either recycled into the ocean or fixed as authigenic carbonate minerals, gas hydrates and natural gas in marine sediments (Dickens, 2003; Wallmann *et al.*, 2006; Wallmann *et al.*, 2008). The burial flux of authigenic $CaCO_3$ in anoxic marine sediments has been estimated as 3.3–13.3 Tmol yr^{-1} (Wallmann *et al.*, 2008) while the accumulation rate of methane and methane hydrates in deep sediments may fall into the range of 0.1–0.6 Tmol yr^{-1} (Dickens, 2003).

When considering long-term climatic change and the geochemical evolution of the Earth on a million year time scale, the most relevant fluxes in the recycling of organic carbon are the weathering of fossil organic carbon on land (8–16 Tmol yr^{-1}) and the burial of organic matter, authigenic carbonates and methane in marine sediments (5.4–27 Tmol yr^{-1}; Table 3.2).

3.3 Carbonate cycling

Most of the carbon residing on the surface of the Earth is bound in carbonate minerals (Table 3.1), and the turnover of these carbonates has a profound effect on geological carbon cycling. Today, carbonates are mostly formed by marine organisms using calcium carbonates ($CaCO_3$, calcite or aragonite) as shell material. The overall stoichiometry of this biogenic calcification can be formulated as:

$$Ca^{2+} + 2HCO_3^- \Rightarrow CaCO_3 + CO_2 + H_2O \qquad (3.5)$$

Biogenic carbonate formation thus serves as a sink for seawater bicarbonate and as a source for atmospheric CO_2 (Table 3.3).

Calcification occurs on shallow and warm continental shelf areas where corals and other organisms produce aragonite and calcite in large scale at an estimated $10\,Tmol\ CaCO_3\ yr^{-1}$ (Kleypas, 1997). Due to smaller shelf area and a lower sea-level stand, carbonate accumulation was probably reduced to ~3 Tmol yr^{-1} during the last glacial maximum (Kleypas, 1997). The average $CaCO_3$ accumulation rate on the continental shelf over the last one million years may thus be estimated as 4–7 Tmol yr^{-1}.

Much more biogenic carbonate is formed by calcareous plankton living in the open oceans, and about $40–130\,Tmol\ CaCO_3\ yr^{-1}$ of this pelagic carbonate is exported into the deep ocean by sinking particles (Berelson *et al.*, 2007; Dunne *et al.*, 2007). Most of the exported carbonate is, however, dissolved on its way to the seafloor because deep water masses are often undersaturated with respect to aragonite and calcite. Moreover, a large fraction of $CaCO_3$ raining to the seafloor is dissolved within surface sediments by metabolic CO_2 produced during the microbial decomposition of sedimentary organic matter (Archer, 1996). The accumulation of pelagic carbonates in seafloor sediments is thus reduced to 10 Tmol yr^{-1} (Berelson *et al.*, 2007). The accumulation rate of pelagic carbonates has probably changed little through glacial/interglacial climate cycles, and the overall global accumulation rate of

biogenic carbonate over the last 1 million years may thus be estimated as 14–17 Tmol yr^{-1}.

Bicarbonate dissolved in seawater is precipitated as carbonates during the reaction of oceanic crust with seawater at moderate temperatures (0–50 °C):

$$\text{basaltic ocean crust} + HCO_3^- \\ \Rightarrow CaCO_3 + \text{altered ocean crust} \quad (3.6)$$

Because protons released from HCO_3^- during carbonate precipitation are mostly consumed by seafloor weathering reactions (Wallmann, 2001), the alteration of ocean crust serves as a sink for seawater bicarbonate but not as source for CO_2. Rates of $CaCO_3$ accumulation in oceanic crust are estimated at $1.5–2.4\,Tmol$ of $CaCO_3$ yr^{-1} (Wallmann, 2001).

Carbonates also accumulate in anoxic marine sediments because large amounts of carbonate alkalinity are formed from organic matter decomposition by microbial sulfate reduction, by the anaerobic oxidation of methane, and by the weathering of silicate minerals (Wallmann *et al.*, 2008). The burial flux of authigenic $CaCO_3$ in anoxic marine sediments has been estimated as 3.3–13.3 Tmol yr^{-1} (Wallmann *et al.*, 2008). This carbon flux contributes significantly to the burial of atmospheric CO_2 in marine sediments and may be regarded as part of the organic carbon burial flux.

Most of the bicarbonate in river water is derived from the weathering and dissolution of carbonates on land. Carbonate weathering serves as an important sink for atmospheric CO_2 and a source of seawater bicarbonate:

$$CaCO_3 + CO_2 + H_2O \Rightarrow Ca^{2+} + 2HCO_3^- \quad (3.7)$$

River chemistry data suggest that 12.3 Tmol of CO_2 yr^{-1} are presently consumed by carbonate weathering (Gaillardet *et al.*, 1999b). This flux is probably not strongly affected by human land use and other anthropogenic effects. During glacial sea-level low-stands, carbonate banks and reefs were exposed to the atmosphere. Some studies suggest that because of this, carbonate weathering was enhanced by 2–6 Tmol CO_2 yr^{-1} during the last glacial maximum (Ludwig *et al.*, 1999; Munhoven, 2002). Other studies (Lerman *et al.*, 2007) conclude that lower temperatures, diminished runoff and expanded ice area would have reduced carbonate weathering by 30%. Overall, the average rate of CO_2 consumption through carbonate weathering over the last 1 million years probably falls in the range of 10–16 Tmol yr^{-1}.

Table 3.3 Inorganic carbon cycling (pre-human Holocene, fluxes in Tmol C yr^{-1})

CO_2 sources	
Mantle degassing (D_M)	3.1–5.5
Metamorphism of carbonates (M_C)	2.0–4.0
Biogenic carbonate burial (B_{BC})	20.0
CO_2 sinks	
Silicate weathering on land (W_{CSIL})	10–11
Silicate weathering in marine sediments (W_{OSIL})	3.3–13.3
Carbonate weathering (W_C)	12.3
HCO_3^- sources	
Carbonate weathering (W_C)	24.6
Silicate weathering on land (W_{CSIL})	10–11
Silicate weathering in marine sediments (W_{OSIL})	3.3–13.3
HCO_3^- sinks	
Biogenic carbonate burial (B_{BC})	40.0
Alteration of oceanic crust (A_{OC})	1.5–2.4

Notes: symbols for fluxes correspond to those appearing in Fig. 3.1; B_{BC} and W_C differ for HCO_3^- and CO_2 because of the 2:1 stoichiometry of these carbon species in carbon precipitation–dissolution processes (Equation 3.9).

3.4 Mantle degassing

Magma is produced by the partial melting of mantle rocks at spreading centres, subduction zones, and hot spot volcanoes. Carbon is an incompatible element that is strongly enriched in the partial melt and delivered to the surface environment as lava is extruded. ^{3}He, a primordial isotope stored in the mantle, has a similar solubility in magma as CO_2 and can be used to track and trace volcanic CO_2 emissions. The largest ^{3}He fluxes are found at submarine spreading centres. The global ^{3}He flux within this tectonic setting (1000 mol ^{3}He yr^{-1}) is well constrained (Farley *et al.*, 1995), and the molar $CO_2/^3$He ratio in the volcanic volatile phase has been measured in hydrothermal fluids (Resing *et al.*, 2004) and a large number of fluid and melt inclusions (Marty and Tolstikhin, 1998). It converges towards a value of 2×10^9 (Resing *et al.*, 2004). Applying this ratio and the total ^{3}He flux, the CO_2 emissions at spreading centres become 2 Tmol yr^{-1}.

Unfortunately, it is not possible to derive the global ^{3}He flux at subduction zones from the few reliable ^{3}He flux measurements available. Thus, mantle-CO_2 fluxes have been calculated from estimates for magma production and degree of partial melting in this tectonic setting (Marty and Tolstikhin, 1998). With this approach, the release of mantle-CO_2 at subduction zones is estimated as 0.3–0.5 Tmol yr^{-1} (Marty and Tolstikhin, 1998; Sano and Williams, 1996). Additional CO_2 is also released at subduction zones by the metamorphism of subducted carbonate rocks and fossil organic carbon (see below). Hot spot and intra-plate volcanoes tap deeper into the mantle, and estimates of CO_2 emissions from these types of volcano, though poorly constrained, fall into the range of 0.8–3 Tmol yr^{-1} (Allard, 1992; Marty and Tolstikhin, 1998). The total rate of CO_2 release from the mantle is thus 3.1–5.5 Tmol CO_2 yr^{-1}.

3.5 Metamorphism

Metamorphism takes place at convergent continental margins where during subduction and associated mountain building, rocks are exposed to high temperatures and pressures. Carbonate and organic matter-bearing rocks may release CO_2 into the atmosphere under these conditions. Indeed, Becker *et al.* (2008) estimated that 0.9 Tmol yr^{-1} of metamorphic CO_2 is emitted into the atmosphere in the Himalayas. This CO_2 comes from carbonate rocks and to a smaller extent from organic carbon exposed to high temperatures (~300 °C) at a depth of ~6 km below the surface.

The ^{13}C/^{12}C and ^{3}He/CO_2 ratios in volcanic gases emitted at subduction zones indicate that most of the CO_2 comes from subducted carbonates (Sano and Williams, 1996). CO_2 measurements at fumaroles indicate that ~3.3 Tmol CO_2 yr^{-1} is emitted into the atmosphere by metamorphic decarbonation of subducting slabs (Sano and Williams, 1996). Decarbonation is promoted by the infiltration of H_2O-rich fluids into carbonate-bearing rocks. This water is mostly released from the lower oceanic crust or from serpentinite, a water-rich

mineral formed by the reaction of upper mantle rocks with seawater. Numerical modeling predicts a global CO_2 flux from subducted carbonates of 0.35–3.12 Tmol yr^{-1} (Gorman *et al.*, 2006). Considering contributions from the Himalayas and other major collision zones, the total metamorphic release of CO_2 from carbonate-bearing rocks is estimated as 2 – 4 Tmol CO_2 yr^{-1}. The CO_2 contribution from fossil organic carbon is estimated using the ^{13}C/^{12}C ratios measured at CO_2-rich fumorals and hot springs in volcanic arcs (Sano and Williams, 1996) and back-arcs of subduction zones (Seward and Kerrick, 1996) and in the Himalayas (Becker *et al.*, 2008). Overall, the metamorphic release of organic matter amounts to 0.4–0.6 Tmol of CO_2 yr^{-1}.

3.6 Silicate weathering

Chemical weathering transforms primary silicate minerals such as feldspars into clays and other particulate and dissolved products. The overall reaction may be represented as:

$$\text{primary silicates} + CO_2 \Rightarrow \text{clays} \\ + \text{dissolved metal cations} \qquad (3.8) \\ + H_4SiO_4 + HCO_3^-$$

Dissolved metal cations (Mg^{2+}, Ca^{2+}, Na^+, K^+, ...) and silica (H_4SiO_4) are released during this reaction while CO_2 is transformed into bicarbonate (HCO_3^-). The dissolved products are transported into the oceans through rivers and groundwater discharge. Silicate weathering is the most important sink for atmospheric CO_2 on geological time scales. It removes CO_2 from the atmosphere and increases the dissolved bicarbonate load of the oceans.

Studies of river chemistry indicate that about 11.7 Tmol of atmospheric CO_2 yr^{-1} are presently consumed by silicate weathering (Gaillardet *et al.*, 1999b), with young basaltic rocks and tephra the most reactive, contributing ~4.1 Tmol CO_2 yr^{-1} to this global weathering rate (Dessert *et al.*, 2003). Silicate weathering is promoted by physical erosion because the reactive surface area of silicate minerals is greatly enhanced by the grinding of rocks into finer particles (Gaillardet *et al.*, 1999a). Physical erosion has been increased by human land use since cropland, pasture, and open range export much more particulate material than natural forest and grassland (Berner and Berner, 1996). Humans have increased the sediment transport by global rivers through soil erosion by about 16% {Syvitski, 2005, p. 3949}. Considering these man-made effects on physical erosion, the pre-human rate of silicate weathering is here estimated as 10–11 Tmol CO_2 yr^{-1}.

Silicate weathering is enhanced in soils compared to bare rock. This is because soils have a higher partial pressure of CO_2 due to organic matter decompositions

(Lerman *et al.*, 2007) and because plants and microbes release organic chemicals which remove Al^{3+} and other cations from the surface of silicate minerals promoting dissolution (Berner and Berner, 1996). Rates of silicate weathering were, therefore, much lower before the advent of land plants during the Paleozoic (Berner, 2004; see chapter 11).

Most importantly, silicate weathering depends on climate and the variables of temperature, runoff, and pCO_2 (Berner, 1994). The hydrological cycle – including evaporation, precipitation, and runoff – is accelerated under warm surface conditions, while high surface temperatures are usually related to elevated pCO_2 values. This threefold link creates a strong climate sensitivity of silicate weathering. The rate of silicate weathering was probably reduced by 10–40 % during the last glacial maximum by cooler surface temperatures, lower pCO_2, diminished runoff and the sealing of land surfaces by the expansion of continental ice shields (Lerman *et al.*, 2007; Munhoven, 2002). Thus, the global rate of silicate weathering averaged over the last 1 million years is probably in the range of $6 - 10\,Tmol\ CO_2\ yr^{-1}$.

Reactive silicate minerals not weathered on land are ultimately deposited on the seafloor in continental margin sediments, and these minerals are reactive within sediments (Maher *et al.*, 2004). Silicate weathering rates are especially high in anoxic sediments rich in labile organic matter. Indeed, most of the metabolic CO_2 produced during organic matter decomposition is converted into HCO_3^- by reaction with feldspars, volcanic ash and other silicate minerals (Wallmann *et al.*, 2008). The rate of silicate weathering in marine sediments has been estimated as $5-20\,Tmol\ CO_2\ yr^{-1}$ (Wallmann *et al.*, 2008). About one third of the bicarbonate generated by marine weathering is fixed in authigenic carbonates and is treated here as part of the organic carbon cycle as it is ultimately produced by organic matter decomposition. The remaining fraction ($3.3-13.3\,Tmol\ of\ HCO_3^-\ yr^{-1}$) is recycled into the oceans. These new data suggest that the rate of marine silicate weathering may be as high as the rate of continental silicate weathering.

3.7 Feedbacks

Drastic changes in atmospheric pCO_2 and seawater bicarbonate concentration are mitigated by negative feedbacks in the geological carbon cycle. As an example, inorganic carbon is removed from the oceans and atmosphere when levels of atmospheric CO_2 or seawater bicarbonate are too high. Such negative feedbacks also work in the opposite direction, where, for example, depleted inventories of inorganic carbon in the oceans and atmosphere can be restored to pre-depletion levels.

The most important negative feedback on atmospheric pCO_2 is through the temperature sensitivity of terrestrial silicate weathering (see below). If volcanoes deliver high doses of CO_2 to the oceans and atmosphere, the concentrations of CO_2 reach only moderate levels. This is because the removal rate of CO_2 by silicate and carbonate weathering is accelerated due to higher temperatures imparted by the higher CO_2 concentrations (Walker *et al.*, 1981). Conversely, with a reduced source of CO_2, atmosphere CO_2 concentrations do not fall too low as the resulting lower temperatures reduce the removal rate of CO_2 by weathering. Therefore, the temperature sensitivity of weathering reactions acts to stabilize atmospheric pCO_2.

Bicarbonate concentrations in the oceans are stabilized by another negative feedback loop. Bicarbonate is mainly removed from the oceans by the burial of biogenic carbonates in sediments (see above). Carbonate burial depends on the saturation state of seawater with respect to calcite and aragonite. With high bicarbonate concentrations in seawater, oceans tend to be oversaturated with respect to these carbonate minerals, increasing the accumulation of carbonate in sediments and stabilizing bicarbonate concentration at a moderate level. The equilibrium reaction representing carbonate precipitation and dissolution within the oceans is shown in Equation 3.9.

$$Ca^{2+} + 2HCO_3^- \Leftrightarrow CaCO_3 + CO_2 + H_2O \qquad (3.9)$$

This reaction equation also demonstrates an additional negative feedback on CO_2 since surplus CO_2 delivered by volcanoes and other sources is removed from the oceans and atmosphere by enhanced carbonate dissolution at the seafloor.

Organic carbon burial in the seafloor is another important CO_2 sink (see above). It is ultimately regulated by the inventory of dissolved phosphate in the oceans since phosphate is the major limiting nutrient for marine productivity on geological timescales. Since organic carbon burial and phosphorus removal are enhanced when oceanic phosphate inventories are high, the phosphate inventory and marine productivity are restored to moderate levels by organic matter burial in a negative feedback loop. Phosphate is delivered to the oceans through chemical weathering on land, and since weathering rates are accelerated at high pCO_2, enhanced phosphate delivery to the oceans and elevated organic matter burial dates are promoted by high pCO_2 levels. Thus, organic matter burial provides a negative feedback for the stabilization of atmospheric pCO_2 values (Wallmann, 2001). The C:P ratio of marine phytoplankton increases at elevated pCO_2 levels (Riebesell *et al.*, 2007). Therefore, more organic carbon can be fixed and stored in marine sediments for a given oceanic phosphate inventory when surplus CO_2 is added to the atmosphere. Together, these negative feedbacks have acted to maintain the carbon inventory of oceans and atmosphere at a moderate level throughout Earth's history.

Positive feedbacks may amplify the effects of natural and anthropogenic perturbations to the carbon cycle and destabilize the carbon inventories in oceans and atmosphere. An example relates to marine productivity and organic carbon burial, with dissolved iron in seawater playing a key role in this feedback loop. Iron is another essential nutrient for marine plankton, limiting marine productivity in many areas of the modern ocean (Sarmiento and Gruber, 2006). Iron is mostly delivered to the oceans as wind-blown dust particles. At low pCO_2 values and surface temperatures, land tends to be drier and winds are stronger, resulting in a larger dust and iron flux to the oceans, which may ultimately increase CO_2 removal via ocean fertilization. Through this chain of processes, low pCO_2 values may be further diminished by enhanced marine productivity and organic carbon burial rates in a positive feedback loop. Fortunately, these positive feedbacks are usually mitigated by more powerful negative feedback loops on geological time scales. Otherwise, life on Earth could not be maintained.

3.8   Balancing the geological carbon cycle

In what follows, we apply a mass balance approach to constrain the fluxes in the geological carbon cycle. Ice core data show that atmospheric pCO_2 has ranged from 190 and 280 µatm over the last 400 kyr (Petit *et al.*, 1999) responding to glacial/interglacial cycles without a significant long-term trend. It may thus be concluded that the carbon cycle reached a pseudo-steady state during the last 1 million years with an average pCO_2 of ~230 µatm overprinted by short term oscillations. This steady state is achieved through negative feedbacks stabilizing the global CO_2 inventory in oceans and atmosphere (see above). The long-term mass balance for CO_2 may thus be formulated as:

$$CO_2 \text{ sinks} = CO_2 \text{ sources} \qquad (3.10)$$

with

CO_2 sinks = carbonate weathering (W_C) + silicate weathering on land (W_{CSIL}) + silicate weathering in marine sediments (W_{OSIL}) + burial of POC (B_{POC}), authigenic carbonates (B_{AC}), and methane (B_{CH4}) in marine sediments

CO_2 sources = mantle degassing (D_M) + metamorphic degassing of carbonates (D_C) + metamorphic degassing of POC (D_{POC}) + POC weathering (W_{POC}) + biogenic carbonate burial (B_{BC})

where the symbols for the different fluxes of carbon are the same as those which appear in Fig. 3.1. We also assume that the total carbon inventory of oceans and atmosphere reached a steady state:

$$\text{carbon sinks} = \text{carbon sources} \qquad (3.11)$$

with

carbon sinks = biogenic carbonate burial (B_{BC}) + burial of POC (B_{POC}), authigenic carbonates (B_{AC}), and methane (B_{CH4}) in marine sediments + alteration of oceanic crust (A_{OC})

carbon sources = mantle degassing (D_M) + metamorphic degassing of carbonates (D_C) + metamorphic degassing of POC (D_{POC}) + POC weathering (W_{POC}) + carbonate weathering (W_C)

Sediment core data reveal a constant base line for the carbon isotopic composition of biogenic carbonates and seawater over the last one million years (Zachos *et al.*, 2001) overprinted by glacial/interglacial fluctuations (Sarmiento and Gruber, 2006). An isotopic mass balance can thus be defined to further constrain the magnitude of geological carbon fluxes:

$$\delta^{13}C \text{ of carbon sinks} = \delta^{13}C \text{ of carbon sources} \qquad (3.12)$$

with

$\delta^{13}C$ of carbon sinks = $\delta^{13}C_{CA} \times (B_{BC} + A_{OC}) + \delta^{13}C_{PA} \times (B_{POC} + B_{AC} + B_{CH4})$

$\delta^{13}C$ of carbon sources = $\delta^{13}C_M \times D_M + \delta^{13}C_{CW} \times (D_C + W_C) + \delta^{13}C_{PW} \times (D_{POC} + W_{POC})$

The carbon isotopic compositions of the different carbon sinks and sources are given as (Wallmann, 2001): carbonate minerals accumulating at the seafloor ($\delta^{13}C_{CA} = 0$ ‰), sedimentary POC accumulating at the seafloor ($\delta^{13}C_{PA} = -24$ ‰), carbonate minerals subject to weathering and metamorphosis ($\delta^{13}C_{CW} = +2$ ‰), fossil POC subject to weathering and metamorphosis ($\delta^{13}C_{PW} = -26$ ‰) and mantle CO_2 ($\delta^{13}C_M = -5$ ‰).

The isotopic composition of carbonates and POC has changed over time (Fig. 3.2). Thus, the $\delta^{13}C$ values of the input fluxes produced by the weathering and metamorphism of ancient rocks ($\delta^{13}C_{CW}$ and $\delta^{13}C_{PW}$) differ from the modern output fluxes generated by POC and $CaCO_3$ accumulation at the seafloor ($\delta^{13}C_{CA}$ and $\delta^{13}C_{PA}$).

Balanced carbon fluxes can be calculated applying the three mass balance equations for CO_2, carbon, and the isotopic composition of the different carbon sources and sinks considering also the range of fluxes estimated above. Technically speaking, this can be achieved with constrained minimization techniques. The results of this exercise are listed in Table 3.4. They indicate that 7.3 Tmol of CO_2 yr^{-1} are released by mantle degassing and metamorphism while 10.6 Tmol of CO_2 yr^{-1} are consumed by silicate weathering on land and in marine sediments. An additional 4.3 Tmol of CO_2 yr^{-1} are released by the

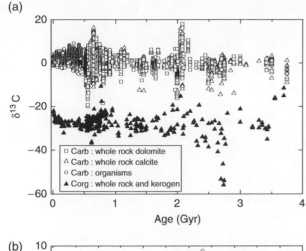

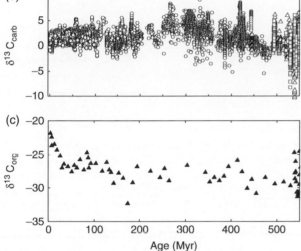

Figure 3.2 $\delta^{13}C_{carb}$ and $\delta^{13}C_{org}$ throughout Earth's history. (a) Complete record for the past 3.8 Gyr. (b) Phanerozoic $\delta^{13}C_{carb}$ record. (c) Phanerozoic $\delta^{13}C_{org}$ record. Note: data from Strauss and Moore (1992), Hayes *et al.* (1999) and Prokoph *et al.* (2008).

turnover of biogenic carbonates (burial – weathering) while 1.0 Tmol of CO_2 yr^{-1} are fixed by POC cycling (burial – weathering). Silicate weathering is, thus consuming volcanic and metamorphic CO_2 and additional CO_2 produced by an imbalance in sedimentary carbon cycling. Seawater bicarbonate is maintained at a steady level by carbonate and silicate weathering adding 23.4 and 10.6 Tmol of HCO_3^- yr^{-1} to the oceans and by the burial of biogenic carbonate and crustal alteration removing 32.0 and 2.0 Tmol of seawater HCO_3^- yr^{-1}. Volcanism is adding mantle carbon into the oceans and atmosphere which is removed by the net growth of the sedimentary carbonate and POC inventories (burial – weathering – metamorphism) and by the alteration of oceanic crust. The residence time of carbon in the oceans and atmosphere is calculated as 102 kyr considering the carbon inventory (2.86×10^6 Tmol of C, Table 3.1) and the balanced input and output fluxes (28.0 Tmol of C yr^{-1}, Table 3.4).

Table 3.4 Mean carbon fluxes over the last 1 million years (in Tmol C yr^{-1})

	Flux Range	Balanced Flux
Carbon sources		
Mantle degassing (D_M)	3.1–5.5	4.3
Metamorphic degassing of carbonate rocks (D_C)	2.0–4.0	2.5
Chemical weathering of carbonate rocks (W_C)	10–16	11.7
Metamorphic degassing of POC (D_{POC})	0.4–0.6	0.5
Chemical weathering of POC (W_{POC})	8–16	9.0
Total		*28.0*
Carbon sinks		
Burial of biogenic carbonate at the seafloor (B_{BC})	14–17	16.0
Alteration of oceanic crust (A_{OC})	1.5–2.4	2.0
Burial of POC, authigenic carbonates, and methane in marine sediments ($B_{POC} + B_{AC} + B_{CH4}$)	5.4–27	10.0
Total		*28.0*
CO_2 sources		
Mantle degassing (D_M)	3.1–5.5	4.3
Metamorphic degassing of carbonate rocks (D_C)	2.0–4.0	2.5
Metamorphic degassing of POC (D_{POC})	0.4–0.6	0.5
Chemical weathering of POC (W_{POC})	8–16	9.0
Burial of biogenic carbonate at the seafloor (B_{BC})	14–17	16.0
Total		*32.3*
CO_2 sinks		
Chemical weathering of silicate rocks on land (W_{CSIL})	6–10	7.1
Chemical weathering of silicates in marine sediments (W_{OSIL})	3.3–13.3	3.5
Chemical weathering of carbonate rocks (W_C)	10–16	11.7
Burial of POC, authigenic carbonates, and methane in marine sediments ($W_{POC} + W_{AC} + W_{CH4}$)	5.4–27	10.0
Total		*32.3*

Note: symbols for fluxes correspond to those appearing in Fig. 3.1.

3.9 Evolution of the geological carbon cycle through Earth's history: proxies and models

In what follows, we describe the evolution of the carbon cycle over the past 3.8 Gyr based on available proxy records and results from geochemical models. We begin by discussing the proxy records.

3.9.1　Carbon isotope composition of marine carbonates ($\delta^{13}C_{carb}$) and organic matter ($\delta^{13}C_{org}$)

The stable carbon isotope composition of marine sedimentary carbonates ($\delta^{13}C_{carb}$) provides a record of how the $\delta^{13}C$ of oceanic dissolved inorganic carbon (DIC = CO_3^{2-} + HCO_3^- + CO_2) has varied through time. Together with the stable carbon isotope composition of marine sedimentary organic carbon and kerogen ($\delta^{13}C_{org}$), an estimate of the isotopic fractionation accompanying the production and burial of organic carbon ($\varepsilon_{TOC} \approx \delta^{13}C_{carb} - \delta^{13}C_{org}$) can be obtained (Hayes *et al.*, 1999). Of interest to us here is the use of the ε_{TOC} record to estimate the fraction of carbon buried as organic matter in marine sediments (f_o). This provides information on the pace and mechanism of oxidation of the Earth's surface (Hayes and Waldbauer, 2006). In its simplest form, the relation between ε_{TOC} and f_o is:

$$\delta^{13}C_{carb} = \delta^{13}C_{in} + f_o \times \varepsilon_{TOC} \qquad (3.13)$$

where $\delta^{13}C_{in}$ is the stable carbon isotopic composition of the carbon input to the surficial system via volcanism, metamorphisms, and continental weathering. This simple relation is valid when the global carbon cycle is in steady state and submarine hydrothermal weathering reactions do not influence the $\delta^{13}C$ of oceanic DIC (Bjerrum and Canfield, 2004; Hayes and Waldbauer, 2006).

Figure 3.2 shows the most recent compilation of $\delta^{13}C_{carb}$ through Earth history (Prokoph *et al.*, 2008) along with a compilation of $\delta^{13}C_{org}$ data obtained by combining a dataset for the last 800 Myr (Hayes *et al.*, 1999) with the Precambrian (540–3800 Myr) dataset of Strauss and Moore (1992). The interested reader can consult the original literature for more information on which type of inorganic and organic phases are analyzed to generate these records and for potential analytical pitfalls.

3.9.2　Strontium isotope composition of marine carbonates ($^{87}Sr/^{86}Sr$)

The $^{87}Sr/^{86}Sr$ ratio in seawater is homogeneous throughout the ocean (Veizer and Compston, 1974). Thus, the $^{87}Sr/^{86}Sr$ ratio as captured in marine carbonates can be used as an indicator of the $^{87}Sr/^{86}Sr$ ratio in ancient seawater. This reflects the relative intensity of the two Sr sources to the ocean: silicate weathering, which releases Sr with high $^{87}Sr/^{86}Sr$ ratios (modern values between 0.705 and 0.735; Gaillardet *et al.*, 1999b), and seafloor spreading, which releases basaltic Sr with low $^{87}Sr/^{86}Sr$ ratios (modern average ~0.703; Teagle *et al.*, 1996). Over long time scales, the evolution of $^{87}Sr/^{86}Sr$ composition of the mantle should also be considered. Thus, mantle $^{87}Sr/^{86}Sr$ increased from ~0.700 at 4.0 Gyr to 0.703 at present due to $^{87}Rb/^{86}Sr$ decay (Shields, 2007). There might also be a temperature-dependent Sr isotope fractionation

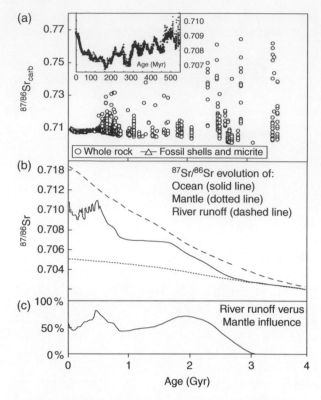

Figure 3.3 $^{87}Sr/^{86}Sr$ evolution of the ocean as an indicator of oceanic crust alteration versus continental weathering inputs of Sr. (a) $^{87}Sr/^{86}Sr$ data throughout the past 3.8 Gyr (Prokoph *et al.*, 2008). Inset: detail of the Phanerozoic $^{87}Sr/^{86}Sr$ record; (b) secular evolution of $^{87}Sr/^{86}Sr$ of the mantle (dotted line), riverine input (dashed line) and ocean (solid line); (c) relative influence of river inputs and ocean crust alteration as Sr sources to the ocean deduced from the curved entry or line in panel (b). Notes: solid line in (b) considers lower envelope of Precambrian $^{87}Sr/^{86}Sr$ values in Prokoph *et al.* (2008); see Shields (2008) for assumptions made to obtain the river runoff curve in panel b; panels (b) and (c) are redrawn from Shields (2008).

as Sr is incorporated into carbonate minerals (Fietzke and Eisenhauer, 2006), but this is usually not considered.

Figure 3.3 presents a compilation of $^{87}Sr/^{86}Sr$ of marine carbonates throughout Earth's history (Prokoph *et al.*, 2008). Analyses come from diagenetically unaltered or poorly-altered low-Mg calcite shells of marine organisms (brachiopods, belemnites, corals, inoceramids, planktic foraminifera, oysters) and the apatite shells of conodonts supplemented by whole rock samples to fill stratigraphic gaps in the fossil data and for the Precambrian. While the Phanerozoic record is considered robust, Precambrian $^{87}Sr/^{86}Sr$ values have likely been increased considerably during sedimentary diagenesis (Veizer, 1989).

3.9.3　Proxies of atmospheric pCO_2

Because of the fundamental role of atmospheric CO_2 in regulating climate (Mann *et al.*, 1998; Berner, 2004; Royer *et al.*, 2007), a number of proxies relating the geochemical characteristics of sediments to past atmospheric CO_2 concentrations have been developed:

1   The $\delta^{13}C$ of pedogenetic minerals (Cerling, 1991). The $\delta^{13}C$ of calcium carbonate or carbonate in goethite formed in soils reflects the relative contribution of biologically produced and atmospherically-derived CO_2, each with distinct $\delta^{13}C$ signatures. With constraints on the contribution from biologically derived CO_2, the $\delta^{13}C$ of pedogenetic minerals can be used to estimate past atmospheric CO_2 concentrations. This technique has been applied to soils from both the Phanerozoic (Ekart *et al.*, 1999) and the Precambrian (Rye *et al.*, 1995; Sheldon, 2006).

2   The $\delta^{13}C$ of phytoplankton (Freeman and Hayes, 1992). The concentration of dissolved CO_2 is one of the key parameters controlling the carbon isotope difference between the carbon source (dissolved CO_2) and photosynthetically-produced biomass. With careful consideration, the $\Delta^{13}C$ (= $\delta^{13}C_{carb}$ − $\delta^{13}C_{org}$) can be used to reconstruct past atmospheric pCO_2 levels, and this method has been applied to both the Phanerozoic and to certain periods in the Precambrian (Pagani, 1999a, 1999b; Kaufman and Xiao, 2003; Fletcher *et al.*, 2005).

3   Stomatal distribution (Van der Burgh *et al.*, 1993). In C3 plants, the flux of CO_2 into the leaves is regulated by pores (stomata). Since the abundance of stomata on the leaf epidermis is inversely proportional to pCO_2, the stomatal density of fossil leaves can be used to reconstruct past atmospheric pCO_2 levels. This method provides past pCO_2 estimates in the time interval 0–402 Mya (Rundgren and Beerling, 1999; McElwain and Chaloner, 1995).

4   The $\delta^{11}B$ of planktonic foraminifera (Pearson and Palmer, 2000). The $\delta^{11}B$ of trace boron incorporated in the shells of planktonic foraminifera is thought to reflect that of dissolved $B(OH)_4^-$, one of the two dissolved boron species (the other is $B(OH)_3$). Since the $\delta^{11}B$ of dissolved boron species is a function of seawater pH, the $\delta^{11}B$ of planktonic foraminifera can be used to estimate the pH of surface waters. In conjunction with knowledge of the dissolved inorganic carbon concentration, these estimates are used to calculate atmospheric pCO_2. This method has been applied in the 0–60 Myr time interval.

5   Calcified cyanobacteria (Kah and Riding, 2007). Cyanobacteria – microorganisms with a ≥2.7 Gyr evolutionary history which perform oxygenic photosynthesis – can induce the precipitation of $CaCO_3$ in the cyanobacterial sheath (the matrix of organic molecules which surrounds the cyanobacterial cell). This is thought to occur at low pCO_2 levels, concomitant to the activation of carbon concentrations mechanisms (CCMs), which make HCO_3^- available to cyanobacteria for photosynthesis in CO_2-limited environments. Because

CCMs are induced when pCO_2 is lower than 10× the present atmospheric level of CO_2 (PAL) in modern cyanobacteria, the first appearance of calcified cyanobacteria in the sedimentary record (1.2 Gyr) is taken to indicate that atmospheric pCO_2 fell below 10 PAL at that time. This method can be applied only to cyanobacteria that were supposedly calcified in direct contact with bottom ocean waters, rather than in a cyanobacterial mat where chemical conditions differ greatly from those of bottom waters (Aloisi, 2008).

6   Weathering rinds on river gravel (Hessler *et al.*, 2004). Pebbles formed 3.2 Gyr ago by continental erosion and river transport of felsic volcanic rocks have an outer layer of altered material (weathering rind) rich in Fe(II) carbonate. Thermodynamic considerations predict that Fe(II) carbonate is the most likely mineral phase to form in this ancient riverine environment if pCO_2 levels are above 2–3 PAL (at lower pCO_2 Fe(II)-layer silicates

(a)

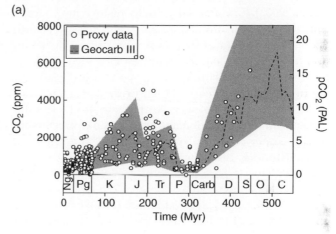

(b)

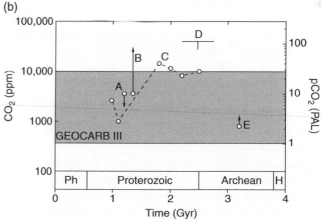

Figure 3.4 Atmospheric pCO_2 throughout Earth's history. (a) Phanerozoic record showing proxy data (circles) and results of the GEOCARB III model. (b) Precambrian proxy data obtained using calcified cyanobacteria (A; Riding and Kah, 2006), acritarchs (B; Kaufman and Xiao, 2003), Paleosols (C,D; Sheldon, 2006 and Rye *et al.*, 1995) and weathering rinds (E; Hessler *et al.*, 2004).

would form). Thus, atmospheric pCO_2 at 3.2 Gyr must have been $\geq$2–3 PAL.

Figure 3.4 summarizes the atmospheric pCO_2 proxy data and compares it with Phanerozoic atmospheric pCO_2 estimates obtained with the GEOCARB III model of the global carbon cycle (Berner and Kothavala, 2001).

3.10   The geological C cycle through time

3.10.1   Archean (3.8–2.5 Gyr)

Mantle outgassing and hydrothermal circulation were likely higher in the Archean than today due to a higher thermal flux from the Earth's interior (Sleep and Zahnle, 2001; Lowell and Keller, 2003). The rates of these processes are proportional to the rate of hydrothermal heat flow which can be reconstructed from the thermal characteristics of the oceanic lithosphere and the mantle and the growth history of the continental crust (Lowell and Keller, 2003). Continental growth history, however, is only poorly constrained (Kemp and Hawkesworth, 2005), leading to a great uncertainty of past degassing and hydrothermal circulation rates. One view – the 'big bang' model of continental growth – proposes that continental masses reached their present area very early in Earth's history (<4 Gyr; Armstrong, 1991). Another theory is that the continental growth took place mainly during super-events of crust formation at 2.7, 1.9 and 1.2 Ga (Condie, 1998). Based on an episodic continental growth history where continents grew from 10% to 80% of current area between 3.2 Gyr and 2.5 Gyr (Lowell and Keller, 2003), Hayes and Waldbauer (2006) calculate that the CO_2 degassing flux from the mantle at 3.8 Gs was about nine times higher than today.

With reduced continental surface area and enhanced thermal fluxes, alteration of oceanic crust and basalt carbonation would have probably been the largest CO_2 sink in the Archean (Walker, 1990), accounting for up to 90% of the inorganic carbon removal (Sleep and Zahnle, 2001; Godderis and Veizer, 2000). This removal is evidenced by the widespread presence of hydrothermal carbonates in greenstone belts (Veizer *et al.*, 1989; Nakamura and Kato, 2002).

Archean $^{87}Sr/^{86}Sr$ values (Fig. 3.3a) are quite variable ranging from low values of ~0.701 to much higher values of ~0.764. This variability is thought to have resulted from the addition of heavier Sr during sediment diagenesis. Therefore, one typically views the lowest measured values as representative of the $^{87}Sr/^{86}Sr$ of contemporaneous seawater (Prokoph *et al.*, 2008). Following this, the ocean had very low $^{87}Sr/^{86}Sr$ values in the Archean. This is partly due to the fact that the $^{87}Sr/^{86}Sr$ of the mantle, which is the ultimate source of Sr to the ocean, has increased though time due to radiogenic decay of ^{87}Rb

(Shields, 2007) (Fig. 3.3b). The $^{87}Sr/^{86}Sr$ record, however, departs from the mantle $^{87}Sr/^{86}Sr$ curve around 2.7–2.5 Gyr (Fig. 3.3b) suggesting that at this time the dominant source of Sr to the ocean changed from ocean ridge hydrothermal sources to continental weathering (Shields, 2007; Fig. 3.3c). A general picture emerges which broadly divides Earth's history into pre-Neoarchean times, when the exogenic system was affected principally by inputs from the mantle and post-Neoarchean where mantle fluxes decreased relative to those from continental weathering (Veizer and Compston, 1976; Rey and Coltice, 2008).

The most striking feature of the $\delta^{13}C_{carb}$ and $\delta^{13}C_{org}$ records is, despite some important deviations, the relative constancy of values throughout Earth's history (Fig. 3.2). The $\delta^{13}C$ of carbon released into oceans and atmosphere via mantle degassing, metamorphism and weathering ($\delta^{13}C_{in}$) is usually assumed to have been constant through time and equal to $\sim$−5‰ (Kump and Arthur, 1999; Bjerrum and Canfield, 2004; Hayes and Waldbauer, 2006). Supporting this is the composition of mantle-derived diamonds and carbonatites which have $\delta^{13}C \sim$ −5‰ independent of age of emplacement or location (Kyser, 1986; Mattey, 1987; Pearson *et al.*, 2004). When these average $\delta^{13}C$ values are applied to the simplified isotopic mass balance defined in equation 13, it can be concluded that ~20% of carbon was buried in organic form for the past 3.5 Gyr (see, however, Bjerrum and (Canfield, 2004) for a somewhat different view). Hayes and Waldbauer (2006) reconstructed the fraction of carbon buried in the organic form (f_0) and the net accumulation of reduced carbon in the crust in the past 3.8 Gyr using the $\delta^{13}C$ proxies introduced above in conjunction with an isotopic mass balance (similar to Equation 3.13, above) and assumptions on the history of mantle degassing and crust-mantle carbon exchanges (Fig. 3.5). These results seem to indicate that the Archean sedimentary record carries the geochemical imprint of biological carbon cycling operating in an astonishingly stable mode.

The Archean isotope records ends with a prominent negative $\delta^{13}C_{org}$ excursion ($\delta^{13}C_{org}$ down to −60‰) between 2.8 and 2.6 Gyr (Fig. 3.2). Three possible scenarios have been proposed to explain this event:

1 High ^{13}C depletions caused by high rates of methane oxidation by aerobic methanotrophic bacteria and the incorporation of strongly ^{13}C depleted bacterial biomass into the organic carbon pool. The strongly ^{13}C-depleted biomass is produced as methanotrophs oxidize highly ^{13}C depleted methane (Hayes, 1994).

2 Sulfate, instead of oxygen, was used to oxidize methane anaerobically by a microbial consortium of sulfate-reducing bacteria and methanotrophic *Archaea* (Hinrichs, 2002).

3 The ^{13}C-depleted organic matter originated as organic particles formed from a ^{13}C-depleted organic haze in the

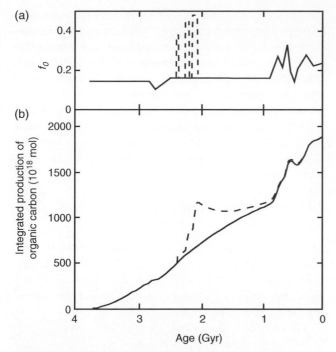

Figure 3.5 Integrated net production of organic carbon, equivalent to the net release of oxidizing power (Emol O_2), in the past 3.8 Gyr (from Hayes and Waldbauer (2006)). (a) estimated values of f_0, the portion of carbon buried in the organic form compared to total buried carbon in the past 3.8 Gyr; (b) Integrated production of organic carbon. Notes: dotted lines consider that the extreme ^{13}C-enrichment of carbonates between 2.3 and 2.0 Gyr reflects a global signal of the ocean (this idea is now considered unlikely, see Hayes and Waldbauer, 2006).

atmosphere. Modeling suggests that such a haze would form when atmospheric $CH_4/CO_2 \geq 1$ (Pavlov *et al.*, 2001). A methane-rich Archean atmosphere, and its associated greenhouse warming, is consistent with an emerging view that Archean atmospheric levels where not as high as once thought.

3.10.2 Proterozoic (2.5 Ga–540 Ma)

The Proterozoic $\delta^{13}C_{carb}$ record starts with one of the most controversial excursions of the Precambrian: a globally expressed positive $\delta^{13}C$ excursion of up to 12‰ between 2.3 and 2.0 Ga (Karhu and Holland, 1996; Melezhik *et al.*, 1999). Similar enrichments occur also in the 2.44–2.30 Ga and 1.92 – 1.97 Ga intervals but it is not known if these are global events (Melezhik *et al.*, 1999). These were initially interpreted as reflecting a nearly threefold increase in the fraction of carbon buried as organic matter (Karhu and Holland, 1996), with a corresponding huge increase in the net release of oxidizing power with consequences for the accumulation of O_2 in the atmosphere (the 'great oxidation event'; Karhu and

Holland, 1996; see dotted lines in Fig. 3.5). But a parallel increase in the $\delta^{13}C_{org}$ signal – which would result from primary producers using the ^{13}C-enriched DIC pool – has not been observed; nor have the large organic matter deposits that would result from increased production (Hayes and Waldbauer, 2006). Furthermore, such high levels of primary productivity require unlikely high supplies of phosphate to the ocean (Aharon, 2005).

In another explanation, ^{13}C-enriched carbonates can also be produced in methanogenic sediments, where fermentative processes add ^{13}C-enriched CO_2 to pore waters (Claypool and Kaplan, 1974; Hayes and Waldbauer, 2006). Such diagenetic carbonates are known to occur in modern settings where methanogenesis is intense (Matsumoto, 1989). The methane produced is ^{13}C-depleted (−60 to −100‰) and could be the source of the ^{13}C-depleted inorganic carbon and carbonates where methane is oxidized anaerobically in conjunction with sulfate reduction (Aloisi *et al.*, 2002) or aerobically with oxygen, perhaps explaining some the early Proterozoic ^{13}C-depleted carbonates. Should a diagenetic origin of the ^{13}C-rich carbonates be confirmed, it would mean that the ^{13}C-enriched carbonates from the early Proterozoic do not necessarily imply an increase in organic carbon burial and exceptional rates of oxygen production at this time (Fig. 3.5b, solid line as opposed to dashed line).

In the period from 1.8 to 1.3 Gyr, there is little change in the $\delta^{13}C$ of organic or inorganic carbon (Fig. 3.2). This could be due to stability in crustal dynamics, climate and oxidation state of the surface environment (Buick *et al.*, 1995; Braiser and Lindsay, 1998) or high oceanic DIC, which would buffer oceanic $\delta^{13}C$ with respect to changes in the carbon inputs to and outputs from the ocean (Bartley and Kah, 2004). From 1.3 Gyr to the Precambrian–Cambrian boundary (540 Myr), the variability in $\delta^{13}C_{carb}$ becomes extreme, with values ranging from −12 to 8‰ between 750 and 600 Myr (Halverson *et al.*, 2005). The events driving these extreme $\delta^{13}C$ shifts may have been truly exceptional.

The most extreme isotope variations are partly coincident with global 'snowball Earth' glaciations (Knoll, 1991). Both isotopic discrimination (ε_{TOC}) and the fraction of carbon buried in the organic form (f_0) are high just prior to the Sturtian (between 750 and 700 Myr) and Marinoan (between 663 and 636 Myr) glaciations (Halverson *et al.*, 2005). Very positive $\delta^{13}C$ carbonate values occur at the onset of global glaciations followed by negative $\delta^{13}C$ carbonate values during these global ice ages (Kaufman and Knoll, 1995). The most negative values are recorded in post-glacial 'cap' carbonates covering the glaciogenic deposits and during the so-called Shuram–Wonoka anomaly at around 580 to 550 Ma (Fike *et al.*, 2006). Rothman *et al.* (2003) have argued that these large negative isotope excursions resulted from the periodic partial oxidation of a huge Neoproterozoic marine DOC pool.

3.10.3 Phanerozoic

During most of the Cambrian to Devonian, pCO_2 values were higher than today (Berner, 1997, 2004, 2006; Farkaš *et al.*, 2007; Wallmann, 2004), likely in the range of 1000–8000 µatm (Royer, 2006; Royer *et al.*, 2007) (Fig. 3.4). These high levels were probably caused by reduced rates of silicate weathering on land. In part, high CO_2 was a consequence of lower solar luminosity. Quite simply, with a less luminous Sun, higher CO_2 concentrations are necessary to raise the Earth surface temperature to the point for sufficient CO_2 removal by weathering. Also, this time was before the advent of land plants and their enhancement of silicate weathering through rooting activity leading to soil formation, enhanced rates organic matter decay, and elevated soil pCO_2 concentrations as discussed above.

A major dip in pCO_2 occurred during the Carboniferous and Permian. The rise and spread of land plants over this period is the ultimate reason for this prominent reduction in pCO_2 (Berner, 1997). As just mentioned, vascular plants greatly promoted soil formation and the consumption of CO_2 via silicate weathering. Extensive burial of organic carbon in wetlands and swamps also added to the dramatic draw-down in pCO_2. Enhanced organic carbon burial is clearly documented by the very positive marine $\delta^{13}C$ values (Fig. 3.2) indicating the pronounced sequestration of ^{12}C in organic matter.

Atmospheric pCO_2 values recovered and reached a maximum during the Jurassic and Cretaceous. These elevated pCO_2 values, ranging between 500 and 4000 µatm, were probably related to enhanced rates of mantle degassing and volcanism. The $^{87}Sr/^{86}Sr$ of marine waters reached a prominent minimum during this period. This indicates high rates of hydrothermal activity at mid-ocean ridges (Fig. 3.3), which correlates with the high rates of seafloor spreading and mantle degassing of CO_2 (Farkaš *et al.*, 2007; Wallmann, 2004).

Over the last ~40 million years, pCO_2 values declined towards the very low Pleistocene to Holocene value of ~230 µatm. This decline was probably caused by enhanced rates of silicate weathering. The collision of India with Asia formed the Himalayas and the Tibetan Plateau increased the rate of physical erosion and CO_2 removal by silicate weathering (Gaillardet *et al.*, 1999b; Raymo, 1994). The marine $^{87}Sr/^{86}Sr$ record supports this scenario. It shows a very rapid increase in $^{87}Sr/^{86}Sr$ over the late Cenozoic indicating that Sr with very high continental-like ratios was released into the oceans (Wallmann, 2001).

During the Anthropocene, pCO_2 values may increase towards a value of ~1000 µatm due mainly to the burning of fossil fuel. Prior to the onset of industrialization, ~10.5 Tmol of CO_2 yr^{-1} were released by POC weathering and metamorphism (Table 3.4). Burning of fossil fuel increased the release of CO_2 from fossil organic carbon by almost two orders of magnitude to >600 Tmol of CO_2 yr^{-1} (IPCC, 2007). CO_2 from fossil organic sources is strongly depleted in the heavy ^{13}C isotope. Thus, the isotopic composition of the atmosphere and the surface ocean has shifted towards more negative $\delta^{13}C$ values due to anthropogenic CO_2 emissions. Models predict that anthropogenic CO_2 will be removed from the atmosphere and taken up by the oceans within the next ~100 kyr (Ridgwell and Hargreaves, 2007). The future geological record will thus document the Anthropocene as a sharp negative excursion in the $\delta^{13}C$ of carbonates and organic biomarkers.

3.11 Limitations and perspectives

The geological carbon cycle considers the long-term distribution of carbon on our planet. Thus, most short-term biological fluxes are not considered in the geological carbon cycle as they have rather little effect on the long-term evolution of the major carbon inventories. These are explored in Chapter 2.

A number of important physical processes are typically neglected by most researchers working on global carbon cycling. Thus, the dissolution of atmospheric CO_2 in the oceans, the transport of dissolved carbon compounds by ocean currents, and the out-gassing of oceanic CO_2 into the atmosphere are usually not resolved in long-term carbon cycle models. It is usually assumed that the carbon distribution between the oceans and atmosphere attains equilibrium within less than one million years. While this assumption is basically correct, it may not be strictly valid for ancient oceans with higher inventories of DIC or supporting much lower rates of biological activity.

Motivated by these problems and further important limitations arising in the traditional geological approach, a new generation of geobiologists are coordinating with scientists from other fields such as physical and biological oceanography and climatology to develop new Earth system models resolving the complex interactions between biological, physical, chemical, and geological processes. These ultimately control the pCO_2 of ancient and future atmospheres, seawater pH and the distribution of carbon on our planet.

References

Aharon P (2005) Redox stratification and anoxia of the early Precambrian oceans: implications for carbon isotope excursions and oxidation events. *Precambrian Research* **137**, 207–222.

Allard P (1992) Global emissions of helium-3 by subaereal volcanism. *Geophysical Research Letters* **19**(4), 1479–1481.

Aloisi G (2008) The calcium carbonate saturation state in cyanobacterial mats throughout Earth's history. *Geochimica et Cosmochimica Acta* **72**(24), 6037–6060.

Aloisi G, Bouloubassi I, Heijs SK, *et al.* (2002) CH$_4$-consuming microorganisms and the formation of carbonate crusts at cold seeps. *Earth and Planetary Science Letters* **203**(1), 195–203.

Archer D (1996) A data-driven model of the global calcite lysocline. *Global Biogeochemical Cycles* **10**(3), 511–526.

Armstrong RL (1991) The persistent myth of crustal growth. *Australian Journal of Earth Sciences* **38**(5), 613–630.

Bartley JK, Kah LC (2004) Marine carbon reservoir, C_{org}–C_{carb} coupling, and the evolution of the Proterozoic carbon cycle. *Geology* **32**(2), 129–132.

Becker JA, Bickle MJ, Galy A, Holland TJB (2008) Himalayan metamorphic CO_2 fluxes: quantitative constrains from hydrothermal springs. *Earth and Planetarian Science Letters* **265**, 616–629.

Berelson WE, Balch WM, Najjar R, Feely RA, Sabine C, Lee K (2007) Relating estimates of $CaCO_3$ production, export, and dissolution in the water column to measurements of $CaCO_3$ rain into sediment traps and dissolution on the sea floor: a revised global carbonate budget. *Global Biogeochemical Cycles* **21**(GB1024).

Berner RA (1994) GEOCARB II: a revised model of atmospheric CO_2 over Phanerozoic time. *American Journal of Science* **294**, 56–91.

Berner RA (1997) The rise of plants and their effect on weathering and atmospheric CO_2. *Science* **276**, 544–546.

Berner RA (2004) *The Phanerozoic Carbon Cycle: CO_2 and O_2*. Oxford University Press, Oxford.

Berner RA (2006) GEOCARBSULF: a combined model for Phanerozoic atmospheric O_2 and CO_2. *Geochimica et Cosmochimica Acta* **70**, 5653–5664.

Berner EK, Berner RA (1996) *Global Environment: Water, Air and Geochemical Cycles*. Prentice Hall, Englewood Cliffs, NJ.

Berner RA, Canfield DE (1989) A new model for atmospheric oxygen over Phanerozoic time. *American Journal of Science* **289**, 333–361.

Berner RA, Kothavala Z (2001) GEOCARB III: a revised model of atmospheric CO_2 over Phanerozoic time. *American Journal of Science* **301**, 182–204.

Bjerrum CJ, Canfield DE (2004) New insights into the burial history of organic carbon on the early Earth. *Geochemistry, Geophysics, Geosystems* **5**(8), doi:10.1029/2004GC000713.

Bolton EW, Berner RA, Petsch ST (2006) The weathering of sedimentary organic matter as a control on atmospheric O_2: II. Theoretical modeling. *American Journal of Science* **306**, 575–615.

Bowring SA, Housh T (1995) The Earth's early evolution. *Science*, **269**, 1535–1540.

Braiser MD, Lindsay JF (1998) A billion years of environmental stability and the emergence of eukaryotes: new data from northern Australia. *Geology* **23**, 555–558.

Buick R, Des Marais DJ, Knoll AH (1995) Stable isotopic compositions of carbonates from the Mesoproterozoic Bangemall Group, northwestern Australia. *Chemical Geology* **123**, 153–171.

Burdige DJ (2005) Burial of terrestrial organic matter in marine sediments: a re-assessment. *Global Biogeochemical Cycles* **19**(GB4011), doi:10.1029/2004GB002368.

Burdige DA (2007) Preservation of organic matter in marine sediments: controls, mechanisms, and an imbalance in sediment organic carbon budgets? *Chemistry Reviews* **107**, 467–485.

Claypool GE, Kaplan IR (1974) The origin and distribution of methane in marine sediments. In: *Natural Gases in Marine Sediments* (ed Kaplan IR). Plenum, New York, pp. 99–139.

Condie KC (1998) Episodic continental growth and super-continents: a mantle avalanche connection? *Earth and Planetary Science Letters* **163**, 97–108.

Dickens GR (2003) Rethinking the global carbon cycle with a large dynamic and microbially mediated gas hydrate capacitor. *Earth and Planetary Science Letters* **213**, 169–183.

Dunne JP, Sarmiento JL, Gnanadesikan A (2007) A synthesis of global particle export from the surface ocean and cycling through the ocean interior and on the seafloor. *Global Biogeochemical Cycles* **21**(GB4006), doi:10.1029/2006GB002907.

Ekart DD, Cerling TE, Montanez IP, Tabor NJ (1999) A 400 million year carbon isotope record of pedogenetic carbonate; implications for paleoatmospheric carbon dioxide. *American Journal of Science* **299**, 805–827.

Farkaš J, Böhm F, Wallmann K, *et al.* (2007) Calcium isotope budget of Phanerozoic oceans: implications for chemical evolution of seawater and its causative mechanism. *Geochimica et Cosmochimica Acta* **71**, 5117–5134.

Farley KA, Maier-Reimer E, Schlosser P, Broecker WS (1995) Constraints on mantle ^{3}He fluxes and deep-sea circulation from an oceanic general circulation model. *Journal of Geophysical Research* **100**(B3), 3829–3939.

Fietzke J, Eisenhauer A (2006) Determination of the temperature-dependent stable strontium isotope (Sr-88/Sr-86) fractionation via braketing standard MC-ICP-MS. *Geochemistry, Geophysics, Geosystems*, **7**, Q08009, doi:10.1029/2006GC001243.

Fike DA, Grotzinger JP, Pratt LM, *et al.* (2006) Oxidation of the Ediacaran Ocean. *Nature*, **444**, 744–747.

Fletcher BJ, Beerling DJ, Brentnall SJ (2005) Fossil bryophytes as recorders of ancient CO_2 levels: experimental evidence and a Cretaceous case study. *Global Biogeochemical Cycles* **19**, GB3012, doi:10.1029/2005GB002495.

Freeman KH, Hayes JM (1992) Fractionation of carbon isotopes by phytoplankton and estimates of ancient CO_2 levels. *Global Biogeochemical Cycles* **6**, 185–198.

Gaillardet J, Dupré B, Allègre CJ (1999a) Geochemistry of large river suspended sediments: Silicate weathering or recycling tracer. *Geochimica et Cosmochimica Acta* **63**(23/24), 4037–4051.

Gaillardet J, Dupré B, Louvat P, Allègre CJ (1999b) Global silicate weathering and CO_2 consumption rates deduced from the chemistry of large rivers. *Chemical Geology* **159**, 3–30.

Godderis Y, Veizer J (2000) Tectonic control of chemical and isotopic composition of ancient oceans: the impact of continental growth. *American Journal of Science* **300**, 434–461.

Gorman PJ, Kerrick DM, Connolly JAD (2006) Modeling open system metamorphisc decarbonation of subducting slaps. *Geochemistry, Geophysics, Geosystems* **7**(4), Q4007, doi:10.1029/2005GC001125.

Halverson GP, Hoffman PF, Schrag DP, Maloof AC, Rice AHN (2005) Toward a Neoproterozoic composite carbon-isotope record. *GSA Bulletin* **117**(9/10), 1181–1207.

Hayes JM (1994) Global mathanotrophy at the Archean–Proterozoic transition. In: *Early Life on Earth* (ed Bengston S). Nobel Symposium 84, Columbia University Press, New York, pp. 220–236.

Hayes JM, Waldbauer JR (2006) The carbon cycle and associated redox processes through time. *Philosophical Transactions of the Royal Society B* **361**, 931–950.

Hayes JM, Strauss H, Kaufman AJ (1999) The abundance of ^{13}C in marine organic matter and isotopic fractionation in the global biogeochemical cycle of carbon during the past 800 Ma. *Chemical Geology* **161**, 103–125.

Hedges JI, Keil RG (1995) Sedimentary organic matter preservation: an assessment and speculative synthesis. *Marine Chemistry* **49**, 81–115.

Hessler AM, Lowe DR, Jones RL, Bird DK (2004) A lower limit for atmospheric carbon dioxide levels 3.2 billion years ago. *Nature* **428**, 736–738.

Hinrichs KU (2002) Microbial fixation of methane carbon at 2.7 Ga: was an anaerobic mechanism possible? *Geochemistry, Geophysics, Geosystems* **3**, doi: 10.1029/2001GC000286.

IPCC (2007) Climate change 2007: the physical science basis. In: *Contribution of Working Group I to the Fourth Assessment Report of the Intergovernmental Panel on Climate Change.* IPCC, Geneva, 996pp.

Kah LC, Riding R (2007) Mesoproterozoic carbon dioxide levels inferred from calcified cyanobacteria. *Geology* **35**(9), 799–802.

Karhu JA, Holland HD (1996) Carbon isotopes and the rise of atmospheric oxygen. *Geology* **24**(10), 867–870.

Kaufman AJ, Knoll AH (1995) Neoproterozoic variations in the C-isotopic composition of seawater: stratigraphic and biogeochemical implications. *Precambrian Research* **73**, 27–49.

Kaufman AJ, Xiao S (2003) High CO_2 levels in the Proterozoic atmosphere estimated from analyses of individual microfossils. *Nature* **20**, 121–148.

Kemp AIS, Hawkesworth CJ (2005) Granitic perspectives on the generation and secular evolution of the continental crust. In: *The Crust*, Vol. 3 (ed Rudnick RL), *Treatise on Geochemistry* (eds Holland HD, Turekian KK). Elsevier-Pergamon, Oxford, pp. 349–410.

Kleypas JA (1997) Modeled estimates of global reef habitat and carbonate production since the last glacial maximum. *Paleoceanography* **12**(4), 533–545.

Knoll AH (1991) End of the Proterozoic Eon. *Scientific American* **265**, 64–73.

Kump LR, Arthur MA (1999) Interpreting carbon-isotope excursions: carbonates and organic matter. *Chemical Geology* **161**(1–3), 181–198.

Kyser TK (1986) Stable isotope variations in the mantle. Stable isotope variations in high temperature geological processes. *Reviews of Mineralogy* **16**, 141–164.

Lasaga A, Ohmoto H (2002) The oxygen geochemical cycle: dynamics and stability. *Geochimica et Cosmochimica Acta* **66**(3), 361–381.

Lerman A, Wu L, Mackenzie FT (2007) CO_2 and H_2SO_4 consumption in weathering and material transport to the ocean, and their role in the global carbon cycle. *Marine Chemistry* **106**, 326–350.

Lowell RP, Keller SM (2003) High-temperature seafloor hydrothermal circulation over geologic time and Archean banded iron formations. *Geophysical Research Letters* **30**(7), doi: 10.1029/2002GL016536.

Ludwig W, Amiotte-Suchet P, Probst J-L (1999) Enhanced chemical weathering of rocks during the last glacial maximum: a sink for atmospheric CO_2? *Chemical Geology* **159**, 147–161.

Maher K, DePaolo DJ, Lin JC-F (2004) Rates of silicate dissolution in deep-sea sediment: in situ measurement using

$^{234}U/^{238}U$ of pore fluids. *Geochimica et Cosmochimica Acta* **68**(22), 4629–4648.

Mann EM, Bradley RS, Hughes K (1998) Global-scale temperature patters and climate forcing over the past six centuries. *Nature* **392**, 779–787.

Marty B, Tolstikhin IN (1998) CO_2 fluxes from mid-ocean ridges, arcs and plumes. *Chemical Geology* **145**, 233–248.

Matsumoto R (1989) Isotopically heavy oxygen-containing siderite derived from the decomposition of methane hydrate. *Geology* **17**, 707–710.

Mattey DP (1987) Carbon isotopes in the mantle. *Terra Cognita* **7**, 31–37.

McElwain JC, Chaloner WG (1995) Stomatal density and index of fossil plants track atmospheric carbon dioxide in the Palaeozoic. *Annals of Botany* **76**, 389–395.

Melezhik VA, Fallick AE, Medvedev PV, Makarikhin VV (1999) Extreme $^{13}C_{carb}$ enrichment in ca. 2.0 Ga magnesite-stromatolite-dolomite-'red beds' association on a global context: a case for the world-wide signal enhanced by local environment. *Earth Science Reviews* **48**, 71–120.

Middelburg JJ (1989) A simple model for organic matter decomposition in marine sediments. *Geochimica et Cosmochimica Acta* **53**, 1577–1581.

Middelburg JJ, Vlug T, van der Nat FJWA (1993) Organic matter mineralization in marine systems. *Global and Planetary Change* **8**, 47–58.

Munhoven G (2002) Glacial-interglacial changes of continental weathering: estimates of the related CO_2 and HCO_3^- flux variations and their uncertainties. *Global and Planetary Change* **33**, 155–176.

Nakamura K, Kato Y (2002) Carbonate minerals in the Warrawoona Group, Pilbara Craton: Implications for continental crust, life, and global carbon cycle in the Early Archean. *Resource Geology* **52**(2), 91–100.

Pagani M, Arthur MA, Freeman KH (1999a) Miocene evolution of atmospheric carbon dioxide. *Paleoceanography* **14**, 273–292.

Pagani M, Freeman KH, Arthur MA (1999b) Late Miocene atmospheric CO_2 concentrations and the expansion of C4 grasses. *Science* **285**, 876–879.

Pavlov AA, Kasting JF, Eigenbrode JL, Freeman KH (2001) Organic haze in Earth's early atmosphere: source of low-C-13 Late Archean kerogens? *Geology* **29**(11), 1003–1006.

Pearson PN, Palmer MR (2000) Atmospheric carbon dioxide concentrations over the past 60 million years. *Nature* **406**, 695–699.

Pearson DG, Canil D, Shirey SB (2004) Mantle samples included in volcanic rocks: xenoliths and diamonds. In: *Treatise on Geochemistry*, Vol. 2 (eds H.D. Holland HD, Turekian KK). Elsevier, Oxford, pp. 172–221.

Petit JR, Jouzel J, Raynaud D, *et al.* (1999) Climate and atmospheric history of the past 420,000 years from the Vostok ice core, Antarctica. *Nature* **399**(6735), 429–436.

Petsch ST, Eglinton TI, Edwards KJ (2001) 14C-dead living biomass: evidence for microbial assimilation of ancient organic carbon during shale weathering. *Science* **292**, 1127–1131.

Prokoph A, Shields GA, Veizer J (2008) Compilation and time-series analysis of a marine carbonate &sbull;^{18}O, &sbull;^{13}C, $^{87}Sr/^{86}Sr$ and &sbull;^{34}S database through Earth history. *Earth-Science Reviews* **87**, 113–133.

Raymo ME (1994) The Himalayas, organic carbon burial, and climate in the Miocene. *Paleoceanography* **9**(3), 399–404.

Resing JA, Lupton JE, Feely RA, Lilley MD (2004) CO_2 and ^{3}He in hydrothermal plumes: implications for mid-ocean ridge CO_2 flux. *Earth and Planetary Science Letters* **226**, 449–464.

Rey PF, Coltice N (2008) Neoarchean lithospheric strengthening and the coupling of the Earth's geochemical reservoirs. *Earth and Planetary Science Letters* **36**(8), 635–638

Ridgwell A, Hargreaves JC (2007) Regulation of atmospheric CO_2 by deep-sea sediments in an Earth system model. *Global Biogeochemical Cycles* **21**(2), GB2008.

Riebesell U, Schulz KG, Bellerby RGJ, *et al.* (2007) Enhanced biological carbon consumption in a high CO_2 ocean. *Nature* **450**, 545-U10.

Rothman DH, Hayes JM, Summons, RE (2003) Dynamics of the Neoproterozoic carbon cycle. *Proceedings of the National Academy of Sciences,USA* **100**, 8124-8129.

Royer DL (2006) CO_2-forced climate thresholds during the Phanerozoic. *Geochimica et Cosmochimica Acta* **70**, 5665–5675.

Royer DL, Berner RA, Park J (2007) Climate sensitivity constrained by CO_2 concentrations over the past 420 million years. *Nature* **446**, 530–532.

Rundgren M, Beerling D (1999) A Holocene CO_2 record from the stomatal index of sub-fossil *Salix herbacea* L. leaves from northern Sweden. *Holocene* **9**, 509–513.

Rye R, Kuo PH, Holland HD (1995) Atmospheric carbon dioxide concentrations before 2.2 billion years ago. *Nature* **378**, 603–605.

Sano Y, Williams SN (1996) Fluxes of mantle and subducted carbon along convergent plate boundaries. *Geophysical Research Letters* **23**(20), 2749–2752.

Sarmiento JL, Gruber N (2006) *Ocean Biogeochemical Cycles*. Princeton University Press, Princeton, NJ.

Sheldon ND (2006) Precambrian paleosols and atmospheric CO_2 levels. *Precambrian Research* **147**, 148–155.

Seward TM, Kerrick DM (1996) Hydrothermal CO_2 emission from the Taupo Volcanic Zone, New Zealand. *Earth and Planetary Science Letters* **139**, 105–113.

Shields GA (2007) A normalised seawater strontium isotope curve and the Neoproterozoic-Cambrian chemical weathering event. *eEarth Discussions* **2**(69–84).

Sleep NH, Zahnle K (2001) Carbon dioxide cycling and implications for climate on ancient Earth. *Journal of Geophysical Research-Planets* **106**(E1), 1373–1399.

Strauss H, Moore TB (1992) Abundances and isotopic compositions of carbon and sulfur species in whole rock and kerogen samples. In: *The Proterozoic Biosphere* (eds Schopf JW, Klein C). Cambridge University Press, Cambridge, pp. 709–798.

Syvitski J P M, Vorosmarty C.J, Kettner A J, Green P. (2005) Impact of humans on the flux of terrestrial sediment to the global coastal ocean. *Science* **308**(5720), 376-380.

Teagle DAH, Alt JC, Bach W, Halliday AN, Erzinger J (1996) Alteration of upper ocean crust in a ridge-flank hydrothermal upflow zone: mineral, chemical and isotopic constraints from hole 896A. *Proceedings of the Ocean Drilling Program, Scientific Results* **148**, 119–150.

Van der Burgh J, Visscher H, Dilcher DL, Kürschner WM (1993) Paleoatmospheric signatures in Neogene fossil leaves. *Science* **260**, 1788–1790.

Veizer J (1989) Strontium isotopes in seawater through time. *Annual Review of Earth and Planetary Sciences* **17**, 141–167.

Veizer J, Compston W (1974) $^{87}Sr/^{86}Sr$ composition of seawater during the Phanerozoic. *Geochimica et Cosmochimica Acta* **38**, 1461–1484.

Veizer J, Compston W (1976) Sr-87–Sr-86 in Precambrian carbonates as an index of crustal evolution. *Geochimica et Cosmochimica Acta* **40**(8), 905–914.

Veizer J, Laznicka P, Jansen SL (1989) Mineralization through geologic time: recycling perspective. *American Journal of Science* **289**, 484–524.

Walker JCG (1990) Precambrian evolution of the climate system. *Global and Planetary Change* **82**(3–4), 261–289.

Walker JCG, Hays PB, Kasting JF (1981) A negative feedback mechanism for the long-term stabilization of Earth's surface temperature. *Journal of Geophysical Research* **86**, 9776–9782.

Wallmann K (2001) Controls on Cretaceous and Cenozoic evolution of seawater composition, atmospheric CO_2 and climate. *Geochimica et Cosmochimica Acta* **65**(18), 3005–3025.

Wallmann K (2004) Impact of atmospheric CO_2 and galactic cosmic radiation on Phanerozoic climate change and the marine &sbull;^{18}O record. *Geochemistry, Geophysics, Geosystems* **5**(1), doi:10.1029/2003GC000683.

Wallmann K, Aloisi G, Haeckel M, Obzhirov A, Pavlova G, Tishchenko P (2006) Kinetics of organic matter degradation, microbial methane generation, and gas hydrate formation in anoxic marine sediments. *Geochimica et Cosmochimica Acta* **70**, 3905–3927.

Wallmann K, Aloisi G, Haeckel M, *et al.* (2008) Silicate weathering in anoxic marine sediments. *Geochimica et Cosmochimica Acta* **72**, 3067–3090.

Zachos JC, Pagani M, Sloan L, Thomas E, Billups K (2001) Trends, rhythms, and aberrations in global climate 65 Ma to present. *Science* **292**, 686–693.

4

THE GLOBAL NITROGEN CYCLE

Bess Ward

Department of Geosciences, Princeton University, Princeton, NJ 08540 USA

4.1 Introduction

The geobiology of nitrogen is dominated by large inert reservoirs and small biological fluxes. The largest reservoirs are nitrogen gas (in the atmosphere and dissolved in the ocean) and sedimentary nitrogen (sequestered in continental crust). Because most organisms cannot utilize gaseous nitrogen, we distinguish between fixed nitrogen compounds (which contain no N–N bonds) that are biologically available, and the dinitrogen gases (N_2 and N_2O) that are largely inaccessible to organisms. Fluxes into and out of the fixed nitrogen pools are biologically controlled, and microbes control the rates of transformations and the distribution of nitrogen among inorganic and organic pools.

Possibly more than any other biologically important element, the global nitrogen cycle has been perturbed by anthropogenic activities. The rate of industrial nitrogen fixation now approximately equals the natural rate, resulting in a two- to threefold increase in the total inventory of fixed N on the surface of the Earth through agricultural fertilizer applications (Galloway et al., 2004). Nitrogen oxides enter the atmosphere via fossil fuel combustion and catalyse the atmospheric chemistry of ozone through pathways that were either nonexistent or insignificant prior to humans. Because of the relative biological inaccessibility of nitrogen, ecosystems have responded to the increased flux of fixed N with changes in the rates and fates of production, and in some cases, large changes in ecosystem chemistry and health.

The microbiology of nitrogen transformations has been intensely studied for over a century, but important new discoveries have been made in only the last decade.

These involve the discovery of previously suspected, but unknown processes, as well as the discovery of new and diverse microbes involved in many of the nitrogen cycle fluxes.

4.2 Geological nitrogen cycle

Nitrogen is an essential element for life on Earth and the fourth most abundant element (after C, O and H) in the biosphere (living plus dead organic matter) at about 0.3% by weight. Unlike the other major biologically important elements, nitrogen (N) is a minor constituent of the Earth's crust, averaging 50 ppm in continental crust (Wedepohl, 1995). Fresh igneous rocks contain essentially no nitrogen (e.g. fresh volcanic lava), so nitrogen in rocks is ultimately derived from the atmosphere via biotic and abiotic nitrogen fixation. The process of N sequestration in rocks involves accumulation of organic matter in low energy marine environments, where is it slowly decomposed and the ammonium that is liberated partitions into clay minerals. In high energy environments, clays and organic matter do not accumulate and the coarser grained sands do not acquire ammonium (Boyd, 2001). Nitrogen is returned to the atmosphere during metamorphism. All of the ammonium in the Earth's crust has been derived ultimately from the atmosphere through nitrogen fixation (Boyd, 2001) and this constitutes trapping of about a quarter of the original atmospheric inventory.

Although primary rocks contain essentially no nitrogen, organic N is a relatively large reservoir in the global N inventory (biosphere plus PON and DON in the ocean and much of the N in soils and sediments;

Fundamentals of Geobiology, First Edition. Edited by Andrew H. Knoll, Donald E. Canfield and Kurt O. Konhauser.
© 2012 Blackwell Publishing Ltd. Published 2012 by Blackwell Publishing Ltd.

Table 4.1 Major nitrogen reservoirs on Earth

Reservoir		Tg	Reference
Atmosphere	N_2	3.7×10^9	Sorai *et al.*, 2007
	N_2O	1.4×10^3	Sorai *et al.*, 2007
Biosphere	Marine	3.0×10^3	Lerman *et al.*, 2004
		5.0×10^2	Sorai *et al.*, 2007
	Terrestrial	5.4×10^3	Lerman *et al.*, 2004
		2.9×10^4	Chameides and Perdue, 1997
		7.7×10^3	Sorai *et al.*, 2007
Ocean	N_2	1.46×10^6	Calculated from Emerson *et al.*, 2002
	N_2O	0.34	Calculated from Emerson *et al.*, 2002
	NO_3^-	6.0×10^5	Mackenzie, 1998
	PON	9.0×10^4	Sorai *et al.*, 2007
	DON	8.1×10^5	Sorai *et al.*, 2007
Geological	Continental crust	1.3×10^9	Wedepohl, 1995
	Crustal rocks	6.4×10^8	Boyd, 2001
	Oceanic crust	8.9×10^5	Li *et al.*, 2007
	Coastal sediments	3.2×10^4	Lerman *et al.*, 2004
	Deep ocean sediments	2.0×10^9	Chameides and Perdue, 1997
	Soil	1.4×10^5	Batjes, 1996
		2.2×10^4	Lerman *et al.*, 2004

DON = dissolved organic nitrogen; PON = particulate organic nitrogen.

Table 4,1), and some of this material is preserved in sediments in the form of complex biomolecules that are studied as biomarkers of past microbial life. It has been estimated that fossil hopanoids (pentacyclic compounds similar to sterols that occur in bacterial membranes and are well preserved in sedimentary rocks) contain as much organic carbon as all living biota (~10^6 Tg C; Ourisson *et al.*, 1987). Hopanoids vary widely in structure (Romer, 1993) and some contain nitrogen. These trace levels of nitrogen contribute to the geological reservoir of N in petroleum and other fossil organic deposits, and thus to the production of N oxides during fossil fuel combustion.

By far the largest reservoir of total nitrogen on Earth is the dinitrogen gas (N_2) in the atmosphere (Table 4.1). N_2 is also the major form of nitrogen in the ocean. The most abundant form of nitrogen in soils and marine sediments is organic nitrogen, produced by biological processes. Although organic N can also be degraded by biological processes, it tends to accumulate in soils and sediments because it becomes more and more resistant to biological attack as it ages. Thus the large organic N reservoirs on land and in the ocean have very long residence times and some of this material eventually transfers to the sedimentary and rock reservoirs. The molecular composition of dead organic matter is complex and largely unresolved, but much of the residual organic matter in soils and marine sediments

may be derived from decay-resistant residues of microbial cell walls.

Transformations and transfers among the reservoirs (Table 4.2) are predominantly controlled by biological processes, so it is interesting that the amount of nitrogen in the biomass itself is vanishingly small compared to the atmospheric and geologic reservoirs. The total amount of N in land and ocean biota is variously estimated at up to 10×10^3 Tg, (10^{12} g) a million-fold less than the atmospheric N_2 reservoir.

The great size of the atmospheric and geologic reservoirs suggests that abiotic processes should also be important in controlling the fluxes between and sizes of these reservoirs. Abiotic processes such as weathering may have changed the magnitude of fluxes in geological processes and the size of the N reservoir in the atmosphere and in rocks to a greater extent than could be accomplished by biological fluxes. A combination of biological and geological processes has caused major excursions in the atmospheric oxygen content over long stretches of Earth history (see Chapter 7). How has the concentration of N_2 in the atmosphere varied as compared to the historical variation in oxygen content of the atmosphere? The geological fluxes of nitrogen can be connected to those of carbon using a C/N ratio for representative materials (Table 4.3) (Berner, 2006). There is no apparent correlation between N/C and age of the shales or their organic content. Similarly, the N/C of coals does not

Flux		Tg yr^{-1}	Reference
Inputs			
Fixation	Natural terrestrial	107	Galloway *et al.*, 2004 for early 1990s
	Natural oceanic	121	Galloway *et al.*, 2004 for early 1990s
		110	Capone *et al.*, 2001
		130–58	Deutsch *et al.*, 2007
	Leguminous crops	31.5	Cleveland *et al.*, 1999
	Chemical fertilizer	100	Galloway *et al.*, 2004 for early 1990s
	Fossil fuel combustion	24.5	Galloway *et al.*, 2004 for early 1990s
Lightning		21	Jickels, 2006
Volcanoes		5	Schumann and Huntreiser, 2007
Losses		0.04	Sano *et al.*, 2001
Denitrification	Natural terrestrial (land and rivers)	115	Galloway *et al.*, 2004 for early 1990s
		154	Lerman *et al.*, 2004
	Natural oceanic	123	Lerman *et al.*, 2004
		193	Galloway *et al.*, 2004 for early 1990s
		285	Middleburg *et al.*, 1996
		400	Codispoti, 2007
		2030	Archer *et al.*, 2002
Industrial combustion		7	Jickels, 2006
Biomass burning		41.6	Andrea and Merlet, 2001
Burial (ocean sedimentation)		25	Brandes and Devol, 2002

Table 4.2 Major fluxes in the global N cycle

Table 4.3 N content of geological reservoirs (calculated from Berner, 2006)

	% C	%N	N/C
Shales	0.013–23	0.005–0.7	0.011–0.083
Coal	57–86	1.0–2.1	0.015–0.024
Gases from volcanic/ metamorphic fumaroles	55–83.5	0.5–1.6	0.012–0.056

vary systematically with thermal grade. The overall range in N/C, a factor of 2–7, is relatively small and similar for all three reservoirs listed in Table 4.3.

With these observations and using geochemical models developed to study the carbon cycle (Berner, 2001, 2006), Berner (2006) computed that the total N_2 inventory in the atmosphere was nearly invariant over Phanerozoic time (the last 600 Mya). This is in clear contrast to the other major atmospheric component, oxygen, which has varied over the same time from as low as 10% to as high as 40% of the atmosphere (Berner *et al.*, 2003).

Before the colonization of the continents by land plants, most N fixation would have occurred in the oceans. If N-fixing microbes, such as cyanobacteria in mats, had been able to survive on land before the establishment of the ozone shield, then terrestrial N fix-

ation may have commenced long before the Ordovician/ Silurian colonization of land by chlorophyte-type plants. This scenario would be compatible with Berner's contention (Berner, 2006) that the N_2 content of the atmosphere has not varied significantly over the entire Phanerozoic Eon. Biological N_2-fixation would have been established at close to its historical average level while photosynthesis gradually increased the oxygen content of the atmosphere, such that major changes in gas ratios would have occurred before the Phanerozoic. Similar scale changes in O_2/N_2 ratios have occurred during the Phanerozoic, but these have been due to changes in O_2, mirrored by large changes in CO_2 concentration.

4.3   Components of the global nitrogen cycle

The nitrogen cycle consists of fluxes (Table 4.2) between the reservoirs (Table 4.1). Because most organisms are not capable of capturing N_2 and "fixing" it into reduced nitrogen for biological processes, the distinction between N_2 (and N_2O) and all other forms of nitrogen is a meaningful one. The most important source of fixed N is biological nitrogen fixation (Table 4.2), which is the sole domain of a few restricted groups of Bacteria and Archaea. Production of fixed N also occurs via lightning and emissions from volcanoes (about 2.8×0^9 mol yr^{-1}

including mid-ocean ridges, hot spots and subduction regions). These natural sources of fixed N are very small (Table 4.2), and are thought to be have been relatively invariant over Earth history (Sano *et al.*, 2001).

It is difficult to describe the preindustrial N cycle quantitatively. The most comprehensive and quantitative evaluation of historical, modern and future N cycling was performed by Galloway *et al.* (2004) who reconstructed the N cycle for 1860, compiled current data for 1993 and forecast future conditions for 2050. We can evaluate the early industrial N cycle by comparing the N budgets for 1860 and 1993. Human activities were already important factors in 1860. Taking the conditions of the late 1990's to represent 100% of the total anthropogenic change up to 1997, Vitusek *et al.* (1997) found that 25% of the total change in deforestation had occurred by ~1700, while 25% of the impact of industrial N fertilizer occurred by ~1975. Although only 16% of total atmospheric emissions were estimated to derive from human activities in 1860, changes in land use (deforestation, annual biomass burning) and concentrated food production had already made indelible marks on the global N cycle by that time.

Human-induced changes in the N cycle are easy to identify (Vitusek *et al.*, 1997):

1 N fertilizer: Organic fertilizers have been used since the beginning of agriculture, but they represent minor recycling of N, rather than net additions to the fixed N inventory. Preindustrial terrestrial biological N fixation (agriculture and natural) was estimated at 120 Tg N yr^{-1}, a value that had decreased about 15% by 1993 (Galloway *et al.*, 2004). Industrial N fixation via the Haber–Bosch process increased rapidly from essentially zero in the 1940s to about 100 Tg annually in 1995 and a further 20% to 121 Tg annually by 2005 (Galloway *et al.*, 2008). Most industrial fixed N (86%) is used to make fertilizer; at least half of global fertilizer applications are now in the developing countries of Asia.

2 Fossil fuel combustion continues to liberate N from the long-term geological reservoirs. Even though the N content of coal and oil is low and variable, the immense net transfer of old organic matter via combustion to the atmosphere now accounts for 24.5 Tg N yr^{-1}, or about a quarter of the total atmospheric emissions. Fossil fuels contributed ~0.3 Tg N yr^{-1} in 1860 (Galloway *et al.*, 2004).

3 Nitrogen-fixing crops: Cultivation of naturally N-fixing crops has replaced mixed vegetation in natural systems, substantially increasing the total biological N fixation by 32–53 Tg N yr^{-1} from agriculture (Galloway *et al.*, 2004). Thus total biological N fixation has increased, even though the natural rate has declined (see above).

4 N mobilization by various human activities liberates N from biological storage reservoirs: deforestation/biomass burning, conversion of forest and savannahs to croplands, drainage of wetlands and peat burning, erosion, all increase biologically available N (Vitusek *et al.*, 1997). By 1993, human activities in total (industrial Haber Bosch process, biological N fixation in agriculture, and fossil fuel burning) had increased the production of fixed N to 156 Tg yr^{-1} (Galloway *et al.*, 2004), more than doubling the natural rate of new nitrogen production.

Modern emissions of fixed N (NH_3, NO_x) to the atmosphere are dominated by anthropogenic sources, whereas in early industrial times, 1860, natural sources such as lightning and emissions from soil and vegetation dominated (Galloway *et al.*, 2004). The modern flux is dominated by reduced N (NH_3) coming from animals and agriculture and fossil fuel burning. Even in 1860, atmospheric emissions from food and energy production showed a clear fingerprint of human activity (Galloway *et al.*, 2004), and human activity increasingly controls the fixed N fluxes. Human inputs exceeded natural atmospheric emissions of oxidized (NO_x) and reduced N by about a factor of five in the early 1990s, and this dominance is expected to increase to a factor of 10 by 2050 (Galloway *et al.*, 2004).

The oceans contribute a small net flux of reduced N (about 6% of total oxidized plus reduced inorganic N atmospheric emissions) to the atmosphere, but receive a quarter of the deposition. Deposition is predominantly in the form of oxidized nitrogen (nitrate or nitric acid) because the reduced forms tend to cycle rapidly and are deposited locally. The remainder is deposited on land, which leads to low-level N enrichment of the entire Earth, even the open ocean, which is far removed from the overwhelmingly terrestrial sources. Duce *et al.* (2008) concluded that anthropogenic N deposition in the open ocean could be responsible for ~3% of annual new primary production. The effects are much stronger on land and in estuaries and coastal oceans, where total N deposition is completely dominated by anthropogenic sources. The majority of these N additions cannot be directly accounted for and are assumed to have accumulated in soils, vegetation or groundwater, or been denitrified (Galloway *et al.*, 2008). Excess N loading from atmospheric deposition has been linked to eutrophication in temperate aquatic systems and to loss of biodiversity in temperate grasslands (Stevens *et al.*, 2004). Vitousek *et al.* (1997) concluded on the basis of an extensive literature review that human alterations of the N cycle had contributed substantially to the acidification of surface waters and accelerated the losses of diversity among terrestrial plants and animals, as well as contributing to long-term changes in estuarine and coastal ecosystems leading to fisheries declines.

Table 4.4 Oxidation states of biogeochemically important N compounds (adapted from Chamiedes and Perdue, 1997)

Compound	Name	N oxidation state
NH_2, NH_4^+	Ammonia, ammonium	−3
R-NH_3	Amino acids, organic nitrogen polymers	−3
NH_2OH	Hydroxylamine	−1
N_2	Dinitrogen gas	0
N_2O	Nitrous oxide	+1
NO	Nitric oxide	+2
HONO, NO_2^-	Nitrous acid, nitrite	+3
NO_2	Nitrogen dioxide	+4
N_2O_5, HNO_3, NO_3^-	Dinitrogen pentoxide, nitric acid, nitrate	+5

4.4 Nitrogen redox chemistry

With an atomic number of 7 and an atomic weight of 14, nitrogen has five valence electrons and occurs in oxidation states ranging from −3 to +5. Nitrogen occurs in nature in six of its eight possible oxidation states (Table 4.4), and the odd oxidation states are most common, except for the ground state N_2 with an oxidation state of zero.

$$NO_3^- \leftrightarrow NO_2(g) \leftrightarrow NO_2^- \leftrightarrow NO(g) \leftrightarrow N_2O(g) \leftrightarrow N_2 \leftrightarrow NH_4^+$$

[+5] [+4] [+3] [+2] [+1] [0] [−3]

Nitrogen in the biosphere is usually in the -3 oxidation state as the amino group of biomolecules, and this form of nitrogen is most abundant in rocks and fossil organic matter. There are no N-containing primary minerals of any significance in the lithosphere, and, other than N_2 in the atmosphere and dissolved in natural waters, inorganic nitrogen is most abundant as nitrate in seawater and other aquatic systems. The most extreme oxidation states are those that are stable in the most oxidizing (e.g. ocean water) and most reducing (e.g. anoxic sediments) environments.

4.5 Biological reactions of the nitrogen cycle

Many organisms produce and obtain energy from the oxidation-reduction reactions that convert nitrogen among its stable states, and in so doing, control the distribution of nitrogen on Earth. These processes are described below, and they make up the biological nitrogen cycle, which is shown in Fig. 4.1. Most steps in the biological N cycle are uniquely the domain of microbes. The main role for macroorganisms is in the transfer of ammonium among various diverse organic nitrogen compounds, but the physiology of eukaryotes does very little to influence the nitrogen budget on either a local or global basis. Some eukaryotes, such as humans, do have a huge effect on the N budget, however, as explained above, but this is not directly related to their metabolism or biochemical peculiarities.

4.5.1 Nitrogen fixation

N_2 is a relatively inert gas, due to the strength of the triple bond that joins the two atoms. The use of molecular nitrogen by organisms thus requires substantial energy expenditure, usually estimated at the cost of 16 ATP (and eight electrons) per molecule N_2 reduced. The only known biological mechanism of N fixation involves the enzyme nitrogenase, which is composed of two multisubunit proteins. The most common form of Component I, which donates electrons to N_2, contains both iron and molybdenum at the active site. Alternative forms contain either iron and vanadium or iron alone. Component I proteins (nitrogenase, the protein that actually splits nitrogen) are coordinated with those of Component II (azoferredoxin, also known as dinitrogenase reductase, the protein that donates electrons to nitrogenase). Thus nitrogen fixation not only requires energy but also a high iron quota. The nitrogenase protein complex is highly conserved, suggesting an ancient origin. Some degree of lateral gene transfer has no doubt occurred and accounts for the wide phylogenetic distribution of nitrogenase, but a high degree of coherence between the phylogenies inferred from 16S rRNA genes and nitrogenase genes supports early evolution from a common ancestor (Raymond *et al.*, 2004).

The first step in nitrogen fixation is the nitrogenase-catalysed reduction of N_2 to NH_3 within the cells of a restricted group of microbes.

$$N_2 + 5H_2O \rightarrow 2NH_4^+ + 2OH^- + 1.5O_2$$

The best known systems involve up to 20 proteins that must be coregulated, and the system is inducible, usually expressed only under nitrogen limitation. Recent reports of high nitrogen fixation rates in the presence of high external ammonium concentrations imply more complex regulation.

As for many transformations in the nitrogen cycle, study of nitrogen fixation has benefitted from the use of specific molecular assays to detect the genes that encode the key enzymes. Identification of nitrogen fixing microbes in the environment has focused on *nifH*, the gene that encodes azoferredoxin. Great diversity among

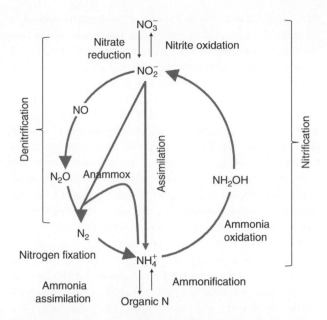

Figure 4.1 Diagram of the biological nitrogen cycle showing the main inorganic forms in which nitrogen occurs in natural and anthropogenically influenced environments.

nifH genes in a range of natural habitats has been reported, and the degree of diversity is not obviously correlated with the degree of nitrogen limitation or the observed rates of N fixation in the environment (Zehr *et al.*, 2003). Numerous and diverse nitrogenase genes have been detected in microbes living in environments where nitrogen is rarely if ever limiting, and where nitrogen fixation is rarely detected, suggesting (1) that organisms maintain the genetic capability for this energy-expensive process even when it is not used for long periods or (2) the presence of the genetic capability is not indicative of the actual process occurring in a particular place because microbes are motile and their distribution is determined by dispersal, depending on the physics of the system.

Most environmentally important nitrogen fixers are identified as (1) endosymbionts with leguminous plants (including crops and trees), (2) cyanobacteria, especially filamentous types in the ocean, microbial mats and lakes, (3) anaerobes, including methanogens, sulfate reducers and fermenters, and (4) free living soil bacteria such as *Azotobacter* (Zehr *et al.*, 2003). Thus, nitrogen fixers are distinguished mainly by metabolic attributes not associated with nitrogen fixation, and they all have close relatives with similar metabolic characteristics that are unable to fix nitrogen. The ecological advantage of nitrogen fixation enables growth under conditions that would be prohibitive for non-fixers, but comes at the cost of a high demand for reducing power.

The triple bond of N_2 can also be broken by abiotic processes, including lightning and combustion of fossil fuels and forest fires. In this case, the resulting fixed N is in the form of NO gas.

$$N_2 + O_2 \rightarrow 2NO$$

Using very high temperatures and pressures, the industrial fixation of N_2 to NH_3 by the Haber–Bosch process has doubled the global rate of atmospheric N_2 fixation since 1950, a major factor in the historical changes to the global N cycle (see above).

4.5.2 Ammonium assimilation

Ammonium is assimilated into organic matter by microbes during both autotrophic and heterotrophic growth. Several enzymes are involved in the process, and they all result in the assimilation of ammonium into amino acids, which are then incorporated into proteins. Proteins are the basis of enzymes, which do the catalytic work of biological processes, and they are the main reservoir of N in living biomass. N is also an important component of the nucleotides that comprise DNA and RNA.

Although it is dangerous to make wide generalizations, it is probably safe to say that the vast majority of, if not all, microbes are capable of ammonium assimilation. For many organisms, it is the preferred (or only) inorganic nitrogen source, probably because ammonium is at the same oxidation state as amino acids, and thus little metabolic energy is required for its incorporation.

4.5.3 Ammonification or mineralization

The nitrogen in biomass is returned to the ecosystem by degradation of the macromolecules assembled from proteins and nucleotides. This is referred to as mineralization because it returns the organic N to the inorganic, or mineral, form. The essential step is deamination, removal of the amino group from proteins and amino acids, which results in the release of ammonium. This process is excretion, which occurs during recycling of macromolecules during growth and metabolism of organisms. Ammonium is also returned to the environment when organic matter is degraded by heterotrophic microbes. Because it is in the reduced form required for incorporation into biomass with minimal energy expenditure, ammonium rarely accumulates in the environment, but is rapidly recycled into biomass. An exception occurs in anoxic sediments, where ammonium accumulates to high levels, due to the absence of oxidants needed to support microbial utilization of the ammonium.

4.5.4 Nitrification

For a few specialized microbes, ammonium serves as a source of reducing power and not simply a source of nitrogen for building biomass. Nitrifying bacteria oxidize ammonium to nitrite, and then to nitrate. The two steps are performed by two different groups of organisms; no single organism is known to catalyse the complete conversion of ammonium to nitrate. All of the bacteria that perform nitrification are primarily autotrophic, utilizing the reducing power of ammonium or nitrite to fix CO_2 by the Calvin cycle. While slow and inefficient growth results, the process has an important effect on the N cycle, by catalyzing the net conversion of ammonium to nitrate. Recently, some Archaea have been shown to be able to oxidize ammonium. The pathway of archaeal ammonium oxidation is still unknown, but it appears to be quite different from the well known pathway in bacteria. Archaeal ammonia oxidizers are also autotrophic, fixing CO_2 by the 3-hydroxypropionate pathway, but they may also be mixotrophic, relying on organic carbon assimilation to an unknown degree. The reactions of bacterial nitrification are shown below.

Ammonium oxidation
$$2H^+ + NH_3 + 2e^- + O_2 \rightarrow NH_2OH + H_2O$$

$$NH_2OH + H_2O \rightarrow HONO + 4e^- + 4H^+$$

$$2H^+ + .5O_2 + 2e^- \rightarrow H_2O$$
Net reaction: $NH_3 + 1.5O_2 \rightarrow HONO + H_2O$

Nitrite oxidation
$$HNO_2 + H_2O \rightarrow HNO_3 + 2e^- + 2H^+$$

In oxic environments, nitrate is the stable form, and it accumulates up to tens or hundreds of micromolar concentrations in seawater and lakes. Although not shown in the reactions above, ammonia-oxidizing bacteria also produce nitrous oxide via two pathways. One is a reductive pathway that is analogous to the nitrite reduction of denitrifiers (see next section) and the other is a less well characterized pathway thought to involve decomposition of hydroxylamine. The proportion of ammonium that is released as N_2O vs. NO_2^- is negatively correlated with oxygen concentration at high cell densities in culture. It has proven very difficult, however, to determine what controls N_2O production in nitrifiers and the correlation with oxygen concentration may not apply to naturally occurring population densities. N_2O production has been detected in cultivated Archaeal ammonia (Santoro et al., 2011). Archaeal ammonia oxidizers may thus play an important role in producing N_2O in aquatic and terrestrial environments.

The key enzyme in ammonium oxidation by both Bacteria and Archaea is the first enzyme in the process, ammonia monooxygenase, encoded by the *amoABC* genes. *amoA* has been widely used as a genetic marker for the organisms responsible for the process (Ward and

O'Mullan, 2005). As is usual for molecular studies, a broader diversity of *amoA* genes has been discovered than was known from cultivated microbes. The ammonia-oxidizing Archaea were first identified from their genes, and only later cultivated (Konneke et al., 2005). They appear to be very abundant in both aquatic and terrestrial environments, and in some environments, much more abundant than their better known bacterial counterparts (Wuchter et al., 2006, Mincer et al., 2007). Cultivated aerobic nitrifiers have relatively long generation times, on the order of a day, which is consistent with the low energy yield of ammonium and nitrite oxidation, and the energetic costs of CO_2 fixation.

Nitrosomonas and *Nitrobacter* have long been identified as the typical, and presumably most important, genera of ammonia-oxidizing and nitrite-oxidizing bacteria, respectively. These are well characterized from laboratory studies as capable of true chemolithoautotrophy – obtaining all reducing power from either ammonium or nitrite, and all carbon from CO_2. Genetic investigations, however, have shown that the cultivated types of ammonia oxidizers are not representative of those in the environment. Among the ammonia-oxidizing bacteria, the most prevalent in the ocean is a *Nitrosospira* type (Bano and Hollibaugh, 2000; O'Mullan and Ward, 2005) that is not present in culture collections and has proven resistant to cultivation. This lack of cultivation might be because its lifestyle and metabolic requirements differ from the cultured forms and researchers have been unable to determine their requirements.

In addition to *Nitrobacter*, a more recently discovered and cultivated genus known as *Nitrospira* is also abundant in the environment. A relative of this genus appears to be the most abundant nitrite-oxidizing bacterium in the ocean (Mincer et al., 2007)

The genomes of several cultivated nitrifying bacteria, both ammonia and nitrite oxidizers, have been completely sequenced. As expected, the complete pathways that allow chemolithoautotrophic growth on either ammonium or nitrite and CO_2 are represented in the genomes. Although they also possess a complete tricarboxylic acid cycle, which is usually involved in degradation of carbon substrates for heterotrophic growth, the nitrifiers appear to favour an autotrophic lifestyle due to a dearth of transporters for organic carbon molecules. Even the nitrite oxidizers that were suspected of mixotrophic growth are not capable of growth on 6-carbon compounds and appear to have very limited heterotrophic capabilities. The heterotrophic capabilities of the ammonia oxidizing Archaea are uncertain at present.

4.5.5   Denitrification

The nitrate formed by nitrification, or resulting from the oxidation of N in the atmosphere, can be utilized by a wide range of bacteria as a respiratory substrate. Most denitrifying organisms are bacteria, and most are facultative aerobes. They respire oxygen when oxygen is present, but when oxygen is depleted, denitrifiers switch to an anaerobic respiration in which NO_3^- is the initial electron acceptor. Complete denitrifiers reduce nitrate sequentially to nitrite, then to nitric oxide, nitrous oxide and finally to dinitrogen gas:

$$NO_3^- \rightarrow NO_2^- \rightarrow NO \rightarrow N_2O \rightarrow N_2$$

Each intermediate is used as a respiratory substrate, but growth is best on nitrate. The strong bonds connecting two N atoms are established at the N_2O step; N_2O is a linear molecule in which the nitrogen atoms are linked by a double bond. Dinitrogen gas is the final product, thus completing the N cycle and returning N_2 gas to the atmosphere and removing it from biological availability. Denitrification (and anammox, see below) is thus the budgetary balance for nitrogen fixation. The balance between nitrogen fixation and denitrification determines the total inventory of fixed N on Earth. Whether the nitrogen budget is in balance or not is the subject of much current debate (see below).

N_2O is an obligate intermediate in the complete denitrification pathway, and it is assumed that incomplete denitrification is responsible for the accumulations of N_2O that occur in anoxic environments. Because the enzymatic pathway by which N_2O is produced in denitrifiers is homologous with the reductive pathway in bacterial nitrifiers, however, it is difficult to determine which organisms are responsible for N_2O production in stratified environments.

Most denitrifiers are heterotrophs and thus utilize organic carbon for energy and to build biomass. The complete degradation of organic matter by denitrification results in the net production of ammonium (ammonification from nitrogen-bearing organic materials). A few kinds of denitrifying bacteria couple nitrate respiration to the oxidation of an inorganic substrate and use the reducing power from that substrate to fix CO_2, thus achieving a lithoautotrophic lifestyle. Reduced sulfur is usually the lithotrophic substrate, and these organisms are well suited for life at the anoxic interface of calm water and sediments.

The diversity and distribution of denitrifying bacteria have been investigated from a genetic approach, focusing on genes in the nitrate reduction pathway. Three genes have been widely applied for this purpose. Two of them encode the reduction of nitrite to nitric oxide; this is the only step in the pathway for which two different enzymes are known. The genes that encode these enzymes are known as *nir*, and genetic investigations suggest that there are thousands of different versions, as proxies for thousands of different kinds of bacteria capable of denitrification, in the environment. The third signature gene is *nosZ*, which encodes the nitrous oxide reduction enzyme. It is also diverse and widely distributed, although not all denitrifiers are capable of reducing N_2O.

Bacteria capable of denitrification are very common in aquatic and terrestrial environments, especially as many of them grow well by aerobic respiration. In oxygen depleted environments, such as found in hemipelagic marine sediments, and the three major oxygen minimum zones of the world ocean, denitrifiers can be a significant if not dominant portion of the total microbial community.

A related metabolism is that of dissimilatory nitrate reduction to ammonium (DNRA) in which the first step is nitrate respiration as in denitrifiers. The nitrite is then reduced directly to ammonium, without the intermediates of denitrification. Organisms that carry out DNRA perform the process as a respiratory mechanism, and can couple it to heterotrophic or autotrophic growth based on sulfur oxidation.

4.5.6 Anammox

Anaerobic ammonium oxidation (anammox), in which ammonium is oxidized microbially using nitrite as an oxidant, is a thermodynamically favourable reaction, but until 1995, organisms responsible for it had never been detected or identified.

$$NH_4^+ + NO_2^- \rightarrow N_2 + 2H_2O$$

Biogeochemical distributions in sediments and across oxic–anoxic interfaces in some aquatic environments had suggested that such a process was occurring (Richards, 1965; Bender *et al.*, 1977). Organisms capable of the process were first identified in wastewater treatment systems (Mulder *et al.*, 1995; van de Graaf *et al.*, 1995), and have now been found in marine, terrestrial and freshwater environments as well. The anammox organisms are all members of a previously obscure phylum of bacteria known as the Planctomycetes. Typical of Planctomycetes, the anammox organisms possess unique intracellular membrane bound compartments, and in the case of anammox, these appear to be linked to their unique metabolism. The oxidation of ammonium with nitrite involves hydrazine as an intermediate, an explosive known more commonly as rocket fuel. The oxidation reactions take place inside the anammoxosome (van Niftrik *et al.*, 2004), a membrane-bound compartment. The membrane lipids of the anammox cell contain novel lipids known as ladderanes, from their ladder like structure. These have now been detected in ancient sediments.

Four different genera of anaerobic ammonia-oxidizing Planctomycetes are known, but it appears that only one of them is common in natural environments. Those in culture grow very slowly, with minimum generation times of several days to a few weeks. This is much slower than most other bacteria under optimal conditions, and it remains to be determined whether anammox bacteria in nature grow at a similarly low rate.

Anammox bacteria from wastewater systems live in apparently obligate consortia with other organisms, usually denitrifiers or aerobic ammonia-oxidizing bacteria (Sliekers *et al.*, 2002, Third *et al.*, 2001). Wastewater influent contains very high ammonium levels and trace levels of oxygen. The ammonia-oxidizing bacteria consume the oxygen in the process of oxidizing some of the ammonium to nitrite. The anammox bacteria combine the nitrite with ammonium to produce N_2 gas at the very low levels of oxygen maintained by the ammonia-oxidizing bacteria. Thus, anammox performs the same role in the nitrogen budget as denitrification under conditions usually associated with denitrification.

When it became clear that the anammox process occurs in nature, it was also recognized that previous estimates of denitrification rates had probably also included anammox. In many ways, it probably does not matter whether denitrifiers or anammox bacteria are responsible for the loss of fixed nitrogen. But because the two groups of organisms have such different lifestyles and metabolic constraints, it might make a difference to environmental regulation and response to changing environmental conditions. Both groups perform their N_2 production under very low or zero oxygen conditions, but denitrifiers usually have an alternative metabolism and can live quite well as facultative aerobes; not so for anammox bacteria, which appear to live only anaerobically (Kartal *et al.*, 2008). Denitrifiers are versatile heterotrophs, capable of utilizing a wide variety of organic carbon compounds; anammox bacteria are obligate autotrophs. Thus denitrification has an important role in the carbon cycle, which is the degradation of organic matter under oxygen poor conditions. In contrast, the slow, energy-limited growth of anammox bacteria is likely to have a negligible effect on the carbon cycle directly. Denitrifiers produce and consume N_2O, while it does not appear that N_2O is involved in the nitrogen transformations of anammox. All of these differences suggest that denitrification and anammox respond differently to environmental conditions such as supply and concentration of oxygen, organic carbon, etc. and leave a different imprint on the chemistry of their environment.

Despite the complications of a consortium, the genome of an anammox bacterium has been published (Strous *et al.*, 2006). The pathways by which the anammox bacteria make a living were not immediately obvious, even from the complete genome. The

organisms grow as obligate autotrophs, fixing CO_2, although they are capable of metabolizing a few simple organic acids. Unlike the aerobic nitrifiers who utilize the Calvin cycle for CO_2 fixation, anammox uses the acetyl CoA pathway, although the genes in this pathway were not highly expressed during growth. Similarly, the enzymes responsible for the oxidation of hydrazine are still unclear.

Because the key enzymes and genes involved in the nitrogen metabolism of anammox have only recently been discovered, molecular detection of anammox organisms in the environment has relied on their 16S rRNA genes, depending on the assumption that only the few Planctomycetes genera, which can be defined by their 16S rRNA genes, are involved. On the basis of 16S rRNA genes, only the genus *Scalindua* appears to be abundant in the environment (Schmid *et al.*, 2007). Ladderane lipids are also used as a unique biomarker for anammox, but these cannot distinguish between the genera.

4.5.7 Linked reactions in the biological nitrogen cycle

From this discussion of the biological N cycle, it is clear that several of the steps are tightly coupled and directly dependent upon each other. Some of these interactions are interesting because they lead to counterintuitive distributions of N compounds and concentrations that seem to contrast with the magnitude of the fluxes. We explore some of the specific coupled reactions below (Fig. 4.2).

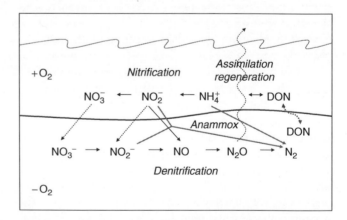

Figure 4.2 Transformations in the nitrogen cycle emphasizing the linkages between processes that occur in aquatic environments. Denitrification and anammox both occur in anoxic conditions, but are linked ultimately to nitrification (favoured under oxic conditions) for supply of substrates. Ammonium, the initial substrate for nitrification, is regenerated by mineralization of organic matter, but is also assimilated as the preferred N source by many organisms. Solid arrows signify microbial transformations while dotted arrows imply diffusion.

4.5.7.1 Coupled nitrification/denitrification

The oxidation of ammonium to nitrite and nitrate produces substrates that can be used for respiration by conventional denitrifying bacteria as well as by anammox bacteria. In order to be coupled, however, the organisms must be situated in a gradient such that oxygen supply is sufficient to allow aerobic nitrification but not high enough to inhibit the anaerobic processes of denitrification and anammox. Stable gradients that allow this coupling occur at sediment–water interfaces in the coastal ocean and lakes, in the stratified water column of oxygen minimum zones in the oceans, in stratified basins and lakes, and in the interstices of soils. The result of coupling these processes across diffusional gradients is the net consumption of ammonium and the net production of N_2 gas. While this represents a net removal of fixed nitrogen, this is sometimes desirable in estuarine and coastal environments where excess N loading can lead to increased primary production and eutrophication. In N limited systems, as N input increases, organic production increases, leading to increased oxygen demand for its decomposition. Under aerobic conditions, decomposition produces the ammonium that drives the coupled nitrification/denitrification cycle. If organic loading leads to an oxygen demand in excess of supply by ventilation or mixing, the nitrification part of the couple will be inhibited, and the ammonium will tend to accumulate in the system. Fixed N will then be maintained in the system instead of being removed as N_2 via denitrification or anammox. Under anoxic conditions, it should be possible for anammox to oxidize this ammonium anaerobically, but many anoxic systems accumulate high levels of ammonium, suggesting that anammox is limited by the absence of an oxidized N compound.

4.5.7.2 Denitrification/anammox

While denitrification and anammox have the same net effect on the N cycle (i.e., the loss of fixed nitrogen through N_2 production), their relative contributions to this flux are poorly known and probably quite variable among ecosystems. Denitrification depends on substrates produced by aerobic nitrification, while anammox depends on substrates that, under anoxic conditions, could be supplied by denitrification. Ammonium is produced by ammonification of organic compounds during anaerobic growth of denitrifers, and the nitrite required for anammox is an intermediate in denitrifier respiration. Anammox apparently must be coupled to nitrification, dissimilatory nitrate reduction to ammonium or denitrification for supply of its N substrates, and the natural lifestyle of anammox organisms is thus not surprisingly as members of consortia.

4.5.7.3 Ammonium assimilation/regeneration

Ammonium has already been identified as a key substrate and product in the coupled N cycle reactions described above. In both aerobic and anaerobic environments, ammonium cycles rapidly between biomass and the inorganic pool without being involved in redox chemistry. As the most favourable N compound for assimilation into biomass, as well as the first product of organic nitrogen degradation, ammonium is usually transferred rapidly by assimilation and regeneration. Thus its concentration in the environment is usually very low under oxic conditions, despite its involvement in very large fluxes. The rates of ammonium oxidation and assimilation may therefore be more correlated with ammonium flux than with actual concentrations.

4.6 Atmospheric nitrogen chemistry

The nitrogen cycle in aquatic and terrestrial environments is dominated by the biological processes described above. The nitrogen cycle of the atmosphere, however, is dominated by abiotic interactions, that are nevertheless closely coupled to biological processes in the water and on land. The inorganic nitrogen gases NO, NO_2 and N_2O are all produced as intermediates or side reactions in nitrification and denitrification, and ammonia volatilization adds another reactive N molecule to the atmosphere. For an excellent discussion and quantitative treatment of the effect of biological and human processes on atmospheric nitrogen chemistry, see Chamiedes and Perdue (1997).

Nitrous oxide is the most stable of the nitrogen gases other than N_2, and it is relatively unreactive in the troposphere. It is emitted to the atmosphere at the Earth's surface through biological and abiotic processes (fossil fuel burning) and is destroyed only after being transported to the stratosphere, where it is destroyed by photochemical reactions. In the troposphere, most N_2O is converted to N_2, either by direct photolysis by UV light, or by reaction with singlet oxygen, an interaction that can also yield NO:

$$N_2O + uv \rightarrow N_2 + O^*$$

$$N_2O + O^* \rightarrow N_2 + O_2$$

$$N_2O + O^* \rightarrow 2NO$$

The abundance of N_2O in the stratosphere is of interest for two reasons. First, N_2O itself is a powerful greenhouse gas with a life time of about 150 years, comparable to that of CO_2 in the atmosphere, but with a radiative forcing equivalent to ~200-fold that of CO_2 on a per molecule basis. The concentration of CO_2 is about 1000-fold higher than that of N_2O, but the total radiative forcing due to N_2O is almost 20% of that attributed to CO_2 (IPCC, 2007).

The concentration of N_2O in the atmosphere has increased approximately 16%, from about 275 parts per billion by volume (ppbv) before the industrial revolution to about 320 ppbv today. Some of this increase is due to the oxidative liberation of trace N in fossil fuels during energy generation. Some N_2O production is also due to agricultural practices, and results ultimately from increased fertilizer additions to partially water-logged soils, where coupled nitrification/denitrification under low oxygen conditions results in the net release of N_2O.

The second reason that even trace levels of N_2O in the atmosphere are significant is that NO produced by interaction with oxygen atoms exerts a controlling influence over the concentration of ozone in the stratosphere. In the troposphere and lower stratosphere, where ozone concentrations are relatively low, NO contributes to ozone production by interaction with hydroperoxy radicals:

$$NO + HO_2 \rightarrow NO_2 + OH$$

$$NO_2 + uv \rightarrow NO + O^*$$

$$O^* + O_2 \rightarrow O_3$$

Net reaction: $HO_2 + O_2 \rightarrow OH + O_3$

NO also causes the catalytic destruction of ozone in the stratosphere.

$$NO + O_3 \rightarrow NO_2 + O_2$$

$$O_3 + uv \rightarrow O_2 + O$$

$$NO_2 + O \rightarrow NO + O_2$$

Net reaction: $2O_3 + uv \rightarrow 3O_2$

Atmospheric chemists tend to consider all the nitrogen gases that do not include N-N bonds as NO_y, which includes NO_x (NO and NO_2) and NO_z (NO_3, N_2O_5, HONO, HNO_3, RNO_3, etc). NO_y gases are very reactive and are involved in a complex suite of interactions that include the oxidation of volatile organic carbon compounds and CO, and the generation of ozone and hydrogen peroxide. NO_y compounds are involved in the production of photochemical smog in the troposphere.

4.7   Summary and areas for future research

The global nitrogen cycle includes important geological and biological components, in which relatively small biological fluxes control the availability of nitrogen between large reservoirs. The versatile redox chemistry of nitrogen means that it occurs in a wide range of compounds, whose concentrations are mostly controlled by microbial transformations.

As mentioned in the discussion of individual processes in the biological N cycle above (and described in more detail in Chapter 13), nature is provided with an astounding diversity of microbes that are involved in every step of the N cycle. Does each variety of, for example, denitrifier or nitrogen fixer occupy a unique niche, the dimensions of which we are not yet capable of identifying? Does this immense diversity provide resilience for ecosystem function, such that overall rates of denitrification or nitrogen fixation are not seriously perturbed when the environment changes, although the suite of organisms performing the reactions changes to those more suitable for the new environment? Among the diverse organisms with the potential for each of the N cycle reactions, it often appears that just a few are dominant in a particular environment. Are these key players important in similar ecosystems around the world? Are the key players interchangeable such that dominance results from chance interactions between the diverse assemblage and some key environmental factor that may change at any time?

The inventory of fixed nitrogen on Earth is controlled by the balance between nitrogen fixation and denitrification/anammox. There is no *a priori* reason to think that the two processes should always be equivalent, but the ramifications of life on Earth of a long-term imbalance would be substantial. Already one of the least abundant elements in the lithosphere, in contrast to its relatively high requirement for life, further depletion of nitrogen by excess denitrification would constrain production on a global basis. As the macronutrient most limiting for life, any sustained additions through excess nitrogen fixation would likely have large effects on the composition and growth rates of natural communities. This could lead to increased CO_2 consumption and to limitation by other, more trace level, elements (Chapter 6).

Nitrous oxide is produced and consumed in both nitrification and denitrification, and although the pathways are fairly well known in cultivated organisms, it is still not clear to what degree the two processes are involved in N_2O production in nature. In the ocean, for example, highest N_2O concentrations are found in oxygen-limited environments. These are the same locations where coupled nitrification/denitrification occurs, so it is not obvious whether both or one of the processes is responsible for net N_2O production. If excess N loading to the coastal or even open ocean were to lead to increased production of organic matter, which subsequently resulted in increased oxygen demand and the expansion of anoxic environments, would this lead to an increase in N_2O release to the atmosphere (Duce *et al.*, 2008)?

Through net production of fixed nitrogen and net fertilization of natural and managed ecosystems, humans have already perturbed the natural nitrogen cycle. Even beyond changes to the biological N cycle, these changes have the potential to influence atmospheric chemistry far into the future as well. Although the NO_y compounds are highly reactive and short lived, N_2O has very long residence time in the atmosphere. Because it is the source of some of the NO_y compounds via its photochemistry, the long-term increase in N_2O already documented has probably already affected the concentration of trace NO_y compounds in the atmosphere, and thus their reaction rates. Global warming aside, what will be the long-term effects of the anthropogenic nitrogen perturbations on the atmosphere and the quality of the environment?

References

Andrea MO, Merlet P (2001) Emission of trace gases and aerosols from biomass burning. *Global Biogeochemical Cycles* **15**, 955–966.

Archer DA, Morford JL, Emerson SR (2002) A model of suboxic sedimentary diagenesis suitable for automatic tuning and gridded global domains. *Global Biogeochemical Cycles* **16**, doi:10.1029/2000GB001288.

Bano N, Hollibaugh JT (2000) Diversity and distribution of DNA sequences with affinity to ammonia-oxidizing bacteria of the beta subdivision of the class Proteobacteria in the Arctic Ocean. *Applied and Environmental Microbiology* **66**, 1960–1969.

Batjes NH (1996) Total carbon and nitrogen in the soils of the world. *European Journal of Soil Science* **47**, 151–163.

Bender ML, Fanning KA, Froelich PN, Heath GR, Maynard V (1977) Interstitial nitrate profiles and oxidation of sedimentary organic matter in the eastern equatorial Atlantic. *Science* **195**, 605–609.

Berner RA, Beerling DJ, Dudley R, Robinson MG, Wildman, RA Jr, (2003) Phaerozoic atmospheric oxygen. *Annual Review of Earth and Planetary Science* **31**, 105–134.

Berner RA (2001) Modeling atmospheric O_2 over Phanerozoic time. *Geochimica et Cosmochemica Acta* **65**, 685–694.

Berner RA (2006) Geological nitrogen cycle and atmospheric N_2 over Phanerozoic time. *Geology* **34**, 413–415

Boyd SR (2001) Nitrogen in future biosphere studies. *Chemical Geology* **176**, 1–30.

Brandes JA, Devol AH (2002) A global marine fixed-nitrogen isotopic budget: implications for Holocene nitrogen cycling. *Global Biogeochemical Cycles* **4**, 1120–1134.

Capone DG (2001) Marine nitrogen fixation: what's the fuss? *Current Opinion in Microbiology* **4**, 241–348.

Chamiedes WL, Perdue EM (1997) Biogeochemical Cycles: A computer-interactive study of earth system science and global change. Oxford University Press, Oxford

Cleveland CC, Townsend AR, Schimel DS, *et al.* (1999) Global patterns of terrestrial biological nitrogen (N_2) fixation in natural ecosystems. *Global Biogeochemical Cycles* **13**, 623–645.

Codispoti LA (2007) An oceanic fixed nitrogen sink exceeding 400 Tg N a^{-1} vs the concept of homeostasis in the fixed-nitrogen inventory. *Biogeosciences* **4**, 233–253.

Duce RA, *et al* (2008) Impacts of atmospheric anthropogenic nitrogen on the open ocean. *Science* **320**, 893–897.

Deutsch C, Sarmiento JL, Sigman DM, Gruber N, Dunne JP (2007) Spatial coupling of nitrogen inputs and losses in the ocean. *Nature* **445**, 163–167.

Emerson S, Stump C, Johnson B, Karl DM (2002) In situ determination of oxygen and nitrogen dynamics in the upper ocean. *Deep-Sea Research I* **49**, 941–952.

Galloway JN, Dentener FJ, Capone DB, *et al.* (2004) Nitrogen cycles: past, present and future. *Biogeochemistry* **70**, 153–226.

Galloway, JN, Townsend AR, Erisman JW, *et al.* (2008) Transformation of the nitrogen cycle: Recent trends, questions, and potential solutions. *Science* **320**, 889–892.

IPCC (2007) Climate Change 2007: Synthesis Report. Contribution of Working Groups I, II and III to the Fourth Assessment Report of the Intergovernmental Panel on Climate Change [Core Writing Team, Pachauri, R.K and Reisinger, A. (eds.)]. IPCC, Geneva, Switzerland, 104 pp.

Jickels T (2006) The role of air-sea exchange in the marine nitrogen cycle. *Biogeosciences* **3**, 271–280.

Kartal G, Keltjens KT, Jetten MSM (2008) The metabolism of anammox. In: Encyclopedia of Life Sciences. John Wiley & Sons, Ltd, Chichester. doi: 10.1002/9780470015902. a0021315.

Konneke M, Bernhard AE, de la Torre JR, Walker CB, Waterbury JB, Stahl DA (2005) Isolation of an autotrophic ammonia-oxidizing marine archaeon. *Nature* **437**, 543–546.

Lerman A, Mackenzie FT, Ver LM (2004) Coupling of the perturbed C-N-P cycles in industrial time. *Aquatic Geochemistry* **10**, 3–32.

Li L, Bebout GE, Idleman BD (2007) Nitrogen concentration $\delta^{15}N$ of altered oceanic crust obtained on ODP Legs 129 and 185: Insights into alteration-related nitrogen enrichment and the nitrogen subduction budget. *Geochimica et Cosmochimica Acta* **71**, 2344–2360.

Mackenzie FT (1998) Our changing planet: An introduction to Earth system science and environmental change, 2nd edn. Prentice-Hall, Upper Saddle River, NJ.

Middleburg JJ, Soetaert K, Herman PMJ, Heip CHR (1996) Denitrification in marine sediments: A model study. *Global Biogeochemical Cycles* **10**, 661–673.

Mincer TJ, Church MJ, Taylor LT, Preston C, Karl DM, Delong EF (2007) Quantitative distribution of presumptive archaeal and bacterial nitrifiers in Monterey Bay and the north Pacific subtropical gyre. *Environmental Microbiology* **9**, 1162–1175.

Mulder, A, van de Graaf AA, Robertson LA, Kuenen JG (1995) Anaerobic ammonium oxidation discovered in a denitrifying fluidized bed reactor. *FEMS Microbiology Ecology* **16**, 177–184.

O'Mullan GD, Ward, BB (2005) Relationship of temporal and spatial variabilities of ammonia-oxidizing bacteria to nitrification rates in Monterey Bay, CA. *Applied and Environmental Microbiology* **71**, 697–705.

Ourisson G, Rohmer M, Poralla K (1987) Prokaryotic hopanoids and other polyterpenoid sterol surrogates. *Annual Reviews of Microbiology* **41**, 301–333.

Raymond J, Siefert JL, Staples CR, Blankship, R E (2004) The natural history of nitrogen fixation. *Molecular Biology and Evolution* **21**, 541–554.

Richards FA (1965) Anoxic basins and fjords. In: Chemical Oceanography (Eds. Riley JP and Skirrow G). Academic Press, London, pp. 611–645.

Romer M (1993) The biosynthesis of triterpenoids of the hopane series in the Eubacteria: a mine of new enzyme reactions. *Pure and Applied Chemistry* **65**, 1293–1298.

Sano Y, Takahata N, Nishio Y, Fischer TP, Williams SN (2001) Volcanic flux of nitrogen from the Earth. *Chemical Geology* **171**, 263–271.

Santoro AE, Buchwald C, McIlvin MR, Casciotti KL (2011) Isotopic signature of N_2O produced by marine ammonia-oxidizing Archaea. *Science* **333**, 1282–1285.

Schmid MC, Risgaard-Petersen N, *et al.* (2007) Anaerobic ammonium-oxidizing bacteria in marine environments: widespread occurrence but low diversity. *Environmental Microbiology* **9**, 1476–1484.

Schumann U, Huntreiser H (2007) The global lightning-induced nitrogen oxides source. *Atmospheric Chemistry and Physics* **7**, 3823–2907.

Sliekers AL, Derwort N, Gomez JLC, Strous M, Kuenen JG, Jetten MSM (2002) Completely autotrophic nitrogen removal over nitrite in one single reactor, *Water Research* **36**, 2475–2482.

Sorai M, Yoshida N, Ishikawa M (2007) Biogeochemical simulation of nitrous oxide cycle based on the major nitrogen processes. *Journal of Geophysical Research* **112**, doi: 10.1029/2005JG000109.

Stevens CJ, Dise, NB, Mountford JO, Gowing DJ (2004) Impact of nitrogen deposition on the species richness of grasslands. *Science* **303**(5665), 1876–1879.

Strous M, Pelletier E, Mangenot S, *et al.* (2006) Deciphering the evolution and metabolism of an anammox bacterium from a community genome. *Nature* **440**, 790–794.

Third KA, Sliekers AO, Kuenen JG, Jetten MSM (2001) The CANON system (completely autotrophic nitrogen-removal over nitrite) under ammonium limitation: Interaction and competition between three groups of bacteria. *Systematic and Applied Microbiology* **24**, 588–596.

van de Graaf AA, deBruijn P, Mulder, A, *et al.* (1995) Anaerobic ammonium oxidation is a biologically mediated process. *Applied and Environmental Microbiology* **61**, 1246–1251.

van Niftrik LA, Fuerst JA, Sinninghe Damste JS, Kuenen, JG, Jetten MSM, Strous M (2004) The anammoxosome: an intra-cytoplasmic compartment in anammox bacteria. *FEMS Microbiological Letters* **233**, 7–13.

Vitusek PM, Howarth RW, Likens GE, *et al.* (1997) Human alteration of the global nitrogen cycle: causes and consequences *Issues in Ecology* **1**, 1–17.

Ward BB, O'Mullan GD (2005) Community level analysis: Genetic and biogeochemcial approaches to investigate community composition and function in aerobic ammonia oxidation. In: *Methods in Enzymology* **397**, 395–413.

Wedepohl KH (1995) The composition of the continental crust. *Geochimca et Cosmochimica Acta* **59**, 1217–1232.

Wuchter C, Abbas B, Coolen MJL, *et al.* (2006) Archaeal nitrification in the ocean. *Proceedings of the National Academy of Sciences, USA* **103**, 12317–12322.

Zehr JP, Jenkins BD, Short SN, Steward GF (2003) Nitrogenase gene diversity and microbial community structure: a cross-system comparison. *Environmental Microbiology* **5**, 539–554.

5

THE GLOBAL
SULFUR CYCLE

Donald E. Canfield[1] and James Farquhar[2]

[1]Institute of Biology and Nordic Center for Earth Evolution (NordCEE), University of
Southern Denmark Campusvej 55 5230 Odense M, Denmark
[2]Department of Geology and ESSIC, University of Maryland, College Park, Maryland 20742, USA

5.1   Introduction

From a geobiological perspective, the sulfur cycle is rich indeed. With valence states ranging from +6 to –2, sulfur compounds are used as electron donors and electron acceptors in numerous microbial metabolisms. Some of these are quite ancient and are among the earliest identifiable metabolisms in the geologic record. Others play important roles in modern geochemical systems. They influence the oxidation state of the system, and they interact with the biogeochemical cycling of other elements such as C, O, N and Fe in the hydrosphere, atmosphere and other components of the surface Earth environment. Processes driven by volcanism and tectonics link the Earth surface sulfur reservoir with the mantle. The nature of this exchange has varied through time depending, among other things, on the chemistry of the ocean and on the intensities of volcanism as well as hydrothermal exchange with the ocean crust and transport of sulfur into the mantle by subduction. The size of the surface sulfur reservoir has likely responded to these dynamics. Surface sulfur is also connected to the atmospheric sulfur cycle, and characteristic isotope signatures produced by atmospheric reactions are transferred to surface sulfur pools providing new ways to evaluate the role of sulfur biology early in Earth's history. This chapter explores the geobiology of the sulfur cycle. We look at the interface between sulfur-based microbial metabolism and Earth surface chemistry through time, and we explore how geological factors such as tectonics and volcanism have influenced these interactions.

5.2   The global sulfur cycle from two perspectives

There are two very different ways one can look at the sulfur cycle. One way is geological in nature and explores the vectors of sulfur exchange between the surface reservoirs of sulfur and between these reservoirs and the mantle; this view allows one to explore the dynamics of the Earth surface sulfur reservoir. The second view is biological in nature and explores the numerous ways in which organisms use and transform and redistribute sulfur compounds. A geobiological understanding of the sulfur cycle benefits from both approaches.

5.2.1   Geological perspective

We begin by looking at the sulfur cycle from a geological perspective (Fig. 5.1; see also Canfield, 2004). The three most significant long-term pathways by which sulfur is transferred from the mantle into the surface environment, and eventually the oceans, are associated with generation of oceanic crust (see Canfield, 2004 for review). These include: (1) volcanic outgassing of SO_2 and H_2S, (2) release of H_2S during hydrothermal circulation, and (3) the weathering of igneous sulfides during the hydrothermal circulation of oxic seawater. There are also less significant contributions associated with oxidative weathering of sulfide in juvenile continental crust and arc volcanism, but a significant part of this sulfur does not come from pristine mantle but rather Earth's crust. The surface reservoir consists of sulfur mostly as sulfate or sulfide minerals in various continental rocks.

Fundamentals of Geobiology, First Edition. Edited by Andrew H. Knoll, Donald E. Canfield and Kurt O. Konhauser.

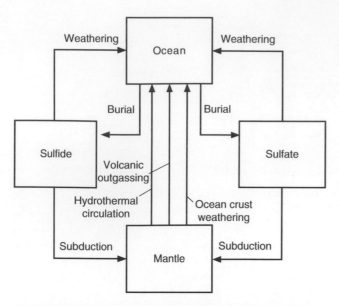

Figure 5.1 Sulfur cycle from a geological perspective. Sulfide and sulfate minerals are weathered on land, delivering sulfate to the oceans. Marine sulfate may be formed into sulfide during sulfate reduction, returning sulfur back to the sulfide pool, or sulfate may be precipitated in evaporite basins. The mantle also represents a source of sulfur to the Earth's surface. Mantle sources include volcanic outgassing of mainly SO_2 gas, the delivery of sulfide to the oceans through hydrothermal circulation at mid ocean ridges and the weathering of sulfide in ocean crust. The subduction of sediments containing sulfide and sulfate represents return fluxes of sulfur back into the Mantle. Modified from Canfield (2004).

When these rocks are brought to the surface weathering environment, sulfides are oxidized to sulfate in the presence of oxygen, and this sulfate is transported by rivers to the oceans. Sulfate accumulates in the oceans where its concentration reflects a balance between these sources, and sinks including the burial of reduced sulfur in sediments as mineral sulfides (mostly pyrite, FeS_2) and the deposition of sulfate evaporate deposits. Some sulfate is also taken up by ocean crust during hydrothermal circulation (principally as sulfide formed by thermochemical or biological sulfate reduction and also as sulfate minerals; Alt, 1995; Shanks, 2001).

Sedimentary sulfides are formed from the sulfide produced by microbial sulfate reduction. The distribution of sulfide in marine sediments is not uniform, and this distribution ultimately determines whether sulfur is transferred back to the mantle or whether it is recycled to continental weathering environments. Pyrite in passive margin (continental shelf) sediments forms the largest sedimentary sulfide sink, and these rocks may be brought back by tectonic processes into the weathering environment. Some sulfides, such as those formed in deep-sea sediments or in the ocean crust, may be lost to the mantle during subduction. Some of the earliest

evidence for recycling of sulfur to the mantle is found as Archean-aged diamond sulfide inclusions (Farquhar *et al.*, 2002; Chaussidon *et al.*, 1987). Overall, the balance of these vectors transporting sulfur between the surface environment and the mantle fix the size of the surface sulfur inventory. Their relative rates and the resulting size of the surface sulfur reservoir have likely changed with time (Canfield, 2004).

5.2.2   *Biological perspective*

The biological sulfur cycle (presented in Fig. 5.2) not only moderates the transformation of sulfur between its different oxidation states, but it also leaves isotopic signatures of these transformations in the geologic record. Each of the compounds listed in Fig. 5.2 is potentially available for microbial sulfur metabolism, and the arrows point to the metabolic products. Table 5.1 lists in equation form a number of the microbial metabolisms. We begin with sulfate reduction, which is the ultimate source of biologically-produced sulfide in the environment.

Sulfate reduction may be either assimilatory, where the product sulfide is used for anabolic needs (such as to build sulfur-containing amino acids), or dissimilatory, where the process is catabolic and used for energy gain. Dissimilatory sulfate reduction is a strictly prokaryotic process broadly distributed within the Bacteria (Fig. 5.3), with, in addition, the hyperthermophilic *Archaeoglobus* within the Archaea. Most known sulfate reducers are members of the δ-proteobacteria. A majority of sulfate reducers use organic compounds as an electron donor, although in some species, H_2 utilization is also possible (Rabus *et al.*, 2002). In most cases, sulfate reducers coexist with other organisms that hydrolyse and ferment complex molecules into smaller moieties (e.g. see review in Canfield, 2005) and therefore depend on a complex anaerobic microbial food chain to deliver the substrates they can use. Sulfate reduction is the starting point for the biological part of the sulfur cycle.

In the presence of light, a large number of so-called anoxygenic phototrophic bacteria (these organisms live by photosynthesis but do not produce oxygen) can oxidize free sulfide to elemental sulfur or sulfate (minor thiosulfate is also possible; see Pfenig, 1975). Sulfide-oxidizing anoxygenic phototrophics have been informally classified into four major groups: the purple sulfur bacteria, which are all found within the γ proteobacteria, the purple non-sulfur bacteria which are distributed among the α and β proteobacteria, and the green sulfur and green non-sulfur bacteria, each with their own lineage. If one were to generalize, the 'sulfur' bacterial designation includes those phototrophs which tend to oxidize sulfide as autotrophs, whereas the 'non-sulfur' bacteria are generalists, oxidizing sulfide, but also oxidizing and

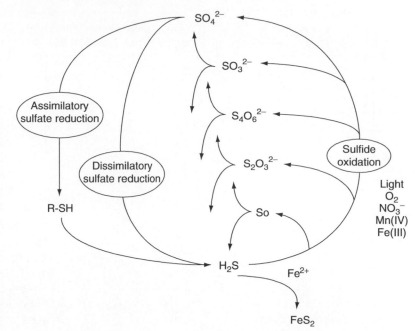

Figure 5.2 Sulfur cycle from a microbial perspective. Ultimately, sulfide is formed from sulfate either by assimilatory sulfate reduction or through dissimilatory sulfate reduction. Most sulfide formed in nature is reoxidized again through light-mediated photosynthetic pathways, biologically mediated non-photosynthetic pathways or through inorganic reactions. Numerous sulfur-intermediate compounds may be formed, and these may be oxidized, reduced, or in many cases disproportionated to sulfide and sulfate. Important reactions are listed in Table 5.1. Figure modified from Canfield (2001).

Table 5.1 Some important S metabolisms

$SO_4^{2-} + 2CH_2O \rightarrow H_2S + 2HCO_3^-$	Sulfate reduction
$2SO_3^{2-} + 3CH_2O + H^+ \rightarrow 2H_2S + 3HCO_3^-$	Sulfite reduction
$4S^o + 4H_2O \rightarrow 3H_2S + SO_4^{2-} + 2H^+$	Sulfur disproportionation
$4SO_3^{2-} + 2H^+ \rightarrow H_2S + 3SO_4^{2-}$	Sulfite disproportionation
$S_2O_3^{2-} + H_2O \rightarrow H_2S + SO_4^{2-}$	Thiosulfate disproportionation
$H_2S + CO_2 \rightarrow CH_2O + H_2O + 2S^o$	Phototrophic sulfide oxidation to sulfur
$H_2S + 2CO_2 + H_2O \rightarrow 2CH_2O + SO_4^{2-} + 2H^+$	Phototrophic sulfide oxidation to sulfate
$2S^o + 3CO_2 + 5H_2O \rightarrow 3CH_2O + 2SO_4^{2-} + 4H^+$	Phototrophic sulfur oxidation to sulfate

incorporating organic compounds as photoheterotrophs. Sulfide-oxidizing anoxygenic photosynthetic bacteria are common in nature, and they can be found in sulfidic marine and freshwater bodies (e.g., Fig. 5.4), in microbial mats, stromatolites, and in tidal sands.

In the absence of light, sulfide may be oxidized by non-phototrophic bacteria (often called colourless sulfur bacteria) using O_2 or nitrate (see below). The ability to oxidize sulfide is widely distributed among the proteobacteria, and it is also found within the Archaea (see Canfield, 2005 for a review). Many sulfide oxidizers oxidize sulfide to sulfate through elemental sulfur, which they store in their cells and oxidize when sulfide is lacking. Many sulfide oxidizers also display fascinating behaviour including gliding and swarming to maximize their sulfide oxidation rates (see review in Canfield *et al.*, 2005), and some also, as discussed more fully below, concentrate nitrate to high concentrations in a central vacuole to maximize sulfide oxidation rates.

A variety of sulfide oxidation products are formed during microbial and abiological (by reaction with oxygen, and Fe and Mn oxides, for example) sulfide oxidation including elemental sulfur (S^0), thiosulfate ($S_2O_3^{2-}$), and sulfite (SO_3^{2-}). These all can be microbially oxidized and reduced, and they also can be disproportionated to sulfate and sulfide. The disproportionation of elemental sulfur is perhaps the best studied of the disproportionation pathways, and it likely has great significance in the environment (e.g., Habicht and Canfield, 2001). Most sulfur disproportionators are found in the δ subdivision of the proteobacteria (Finster, 2008), although a representative from the γ subdivision has also been described (Obraztsova *et al.*, 2002). Many disproportionators are also sulfate reducers, although obligate disproportionators are also found (Finster *et al.*, 1998).

The microbial sulfur cycle is more complex than described above, but we highlight here, dissimilatory sulfate reduction, anoxygenic phototrophic sulfide oxidation, elemental sulfur disproportionation, and nonphototrophic sulfide oxidation to sulfate as the main sulfur metabolic pathways.

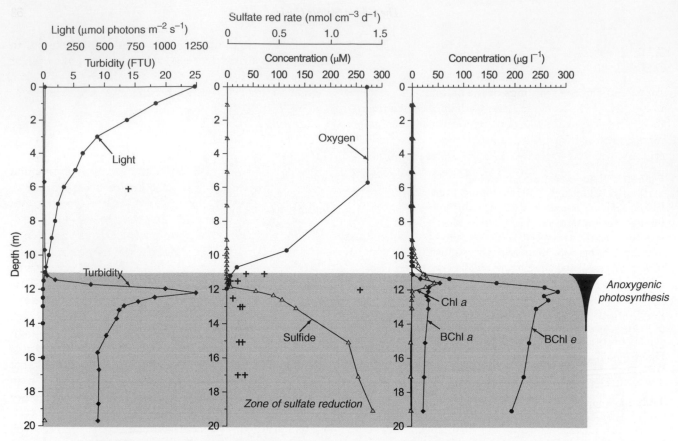

Figure 5.3. Important features of the biology and chemistry of meromictic Lake Cadagno in Switzerland. Figure modified from Gregersen *et al.* (2009) with sulfate reduction data (crosses) from Canfield *et al.* (2010).

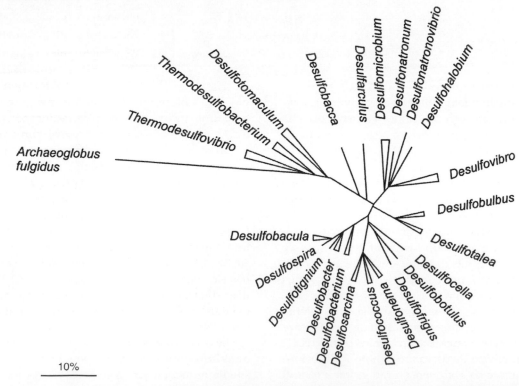

Figure 5.4 The phylogeny of major groups of sulfate-reducers as determined from 16S rRNA sequence comparisons. Most of these organisms are from the δ-subdivision of proteobacteria. Exceptions include the *Desulfotomaculum* group, represented in the firmicutes, the deep-branching bacterial *Thermodesulfobacterterium* and *Thermodesulfovibrio* groups and the archaeal *Archaeoglobus*.

5.2.3   *Illustration of the S cycle in a modern environment*

An example of the workings of the biological S cycle is presented for meromictic Lake Cadagno, Switzerland, in Fig. 5.4. Here, a weak density stratification inhibits mixing of oxygen into the deep portion the lake (Tonolla *et al.*, 2005; Del Don *et al.*, 2001), and sulfide accumulates in the denser anoxic zone due to sulfate reduction (Tonolla *et al.*, 2005; Del Don *et al.*, 2001). Light penetrates to the top of the anoxic zone, fueling two different populations of anoxygenic phototrophs; one is a mixed population of purple sulfur bacteria, which contains BChl *a*, and the other is a green sulfur bacterial population (*Chlorobium* sp.) using BChl *e* (Tonolla *et al.*, 1999). Rates of sulfate reduction are highest at the chemocline, reflecting the high rates of photosynthetic carbon production in this zone. One would expect non-phototrophic sulfide oxidizers at or near the O_2–sulfide interface, and the formation of elemental sulfur and other S intermediate compounds during phototrophic and non-phototrophic sulfide oxidation (see review in Canfield *et al.*, 2005; Pfennig, 1975). Indeed, elemental sulfur concentrations are typically highest at the oxic-anoxic interface in sulfidic water columns (e.g. Zopfi *et al.*, 2001; Luther *et al.*, 1991), although they have not been measured in Lake Cadagno. Non-photosynthetic sulfide oxidizers have also yet to be identified in the lake, but sulfate reducers capable of sulfur disproportionation have been found in intimate association with the purple-sulfur bacteria (Peduzzi *et al.*, 2003). The importance of these disproportionators in the sulfur cycle of the lake, however, is unclear (Canfield *et al.*, 2010).

With this general introduction, we will explore the evolution of the sulfur cycle through time, and then look at how the S cycle interfaces with other element cycles.

5.3   The evolution of S metabolisms

All of the sulfur metabolic pathways represented in Fig. 5.2 could be supported with an appropriate source of sulfide or sulfate and with the availability of light. These conditions would likely have been met on the very early Earth. Sulfide would have been available from volcanism and hydrothermal discharge (e.g., Holland, 1984; Canfield, 2004), and sulfide would have intercepted light in terrestrial hot spring settings such as found in modern-day Yellowstone Park, Iceland, and the North Island of New Zealand, providing conditions conducive for phototrophic oxidation. Sulfate and elemental sulfur would have been produced through various atmospheric pathways (Walker and Brimblecombe, 1985; Farquhar *et al.*, 2001; Pavlov and Kasting, 2002), providing the ingredients for reductive and disproportionative metabolisms. Therefore, we shift the focus from whether the metabolisms could have existed, to whether their existence is supported by biological or geological evidence, and we also raise the question, when did the specific metabolic pathways first evolve?

5.3.1   *Molecular evidence for the relative antiquity of S metabolisms*

Molecular phylogenies are evolutionary histories that are constructed from the sequence comparisons of genetic material. Those phylogenies constructed from the 16S rRNA gene suggest an early emergence of sulfate reducers within the bacterial domain of the Tree of Life, in particular, members of the family *Thermodesulfobacteriaceae* (Klein *et al.*, 2001). However, sequence comparisons of the key genes in the sulfate reduction process, in particular dissimilatory sulfite reductase (*dsr*), demonstrate that members of the *Thermodesulfobacteriaceae* obtained their *dsr* genes by lateral gene transfer from later-evolved sulfate reducers of the δ proteobacteria (Klein *et al.*, 2001). This means that while members of the *Thermodesulfobacteriaceae* may have emerged early in microbial evolution, their ability to reduce sulfate may have come much later.

The *dsr* genes in *Thermodesulfovibrio* species as well as some Gram-positive sulfate reducers (Firmicutes) are apparently more deeply rooted than in *Thermodesulfobacterium* species or other members of the δ proteobacteria suggesting that sulfate reduction originated in other groups before it was established among the proteobacteria (Loy *et al.*, 2008). Loy *et al.* (2008), further propose that the *dsr* gene originated early in microbial evolution, before the independent diversification of the domains Bacteria and Archaea, before the evolution of sulfide-oxidizing bacteria, and possible with organisms originally reducing sulfite rather than sulfate. This analysis suggests that a number of sulfur metabolisms are apparently ancient, though it is vague on the order of emergence (see also Xiong, 2006 for a similar conclusion).

This simple analysis is supported by a more complex whole-genome approach where individual gene's phylogenies are resolved against organismal phylogenies (generated from housekeeping genes, which are not believed to transfer between organisms). To do so algorithms are constructed to model gene birth, gene loss, gene duplication and horizontal transfer (David and Alm, 2011). When these analyses are calibrated against the best-known timing of important biological innovations, a time-resolved picture of metabolic innovation emerges. This type of complex analysis is in it infancy, but results point to a very early emergence of a broad spectrum of sulfur metabolisms, including sulfate reduction, but with a particular importance for sulfite and thiosulfate transformations. We eagerly await further insight with this powerful analytical method.

5.3.2 Geological evidence for the antiquity of S metabolisms

It is well known that several microbial metabolisms fractionate sulfur isotopes (e.g., Kaplan and Rittenberg, 1964; Canfield, 2001; Johnston *et al.*, 2008). The patterns and magnitude of these fractionations can be used to deduce the evolutionary history of sulfur metabolisms, and when in Earth history these metabolisms gained environmental significance. The best studied fractionating process is sulfate reduction, where the product sulfide is depleted in the minor sulfur isotopes (^{33}S, ^{34}S and ^{36}S) compared to the major isotope ^{32}S. The range of fractionations observed in experiments with pure cultures and natural populations of sulfate reducers extend from no fractionation to fractionations of many tens of per mil (e.g., Canfield, 2001). The earliest rocks revealing the presence of large fractionations characteristic of ancient sulfur metabolisms are those from 3.49 billion year old (Ga) quartz-barite deposits of the Dresser Formation, North Pole, Western Australia. The geological history of these rocks is complex, but the chert–barite units are believed to have formed in a restricted, sulfate-rich, shallow-water marine environment where the barite was originally precipitated as gypsum (Buick and Dunlop, 1990). Relatively high fractionations between sulfate and sulfide of up to 21 ‰ have been found (Shen *et al.*, 2001), and these contrast with small fractionations typical of contemporaneous sedimentary rocks. The high fractionations in the Dresser Formation have been interpreted to suggest the activities of sulfate-reducers in this locally sulfate-rich environment instead of the more common Archean sulfate-poor conditions where fractionations are not expressed (Habicht *et al.*, 2002; see below).

Several recent studies have revisited these rocks and employed multiple sulfur isotope analysis of the sulfides and sulfate, which include ^{33}S and in some cases ^{36}S, and inferences made using the context of mass-independent isotope effects (Philippot *et al.*, 2007; Shen *et al*, 2009; see Box 5.1). Philippot *et al.* (2007) documented populations of microscopic pyrites with ^{34}S-depletions and ^{33}S enrichments, which could not have formed from direct reduction of the sulfate present in the deposits. A different source for the sulfide seems likely. Indeed, the isotopic signature of these microscopic sulfides points to an origin from the microbial disproportionation of elemental sulfur, which is inferred to have formed from the photolysis of SO_2 in the atmosphere by mass-independent pathways. This observation does not rule out microbial sulfate reduction, and indeed, another study documented a relationship among all four sulfur isotopes suggesting that a population of pyrites is related to sulfate through microbial sulfate reduction (Shen *et al.*, 2009). It therefore seems likely that both sulfate reducers

Box 5.1

The past 10 years of research have documented an unusual type of isotope signature that is common in the most ancient terrestrial sedimentary rock samples. These isotopic signatures are mass-independent, meaning that they preserve isotope fractionations that are different from those predicted by the mass difference between the isotopes. The most common solid and liquid phase transformations yield mass-dependent isotope fractionations because the differences in the strengths of chemical bonds (and chemical potential energy) involving different isotopes are proportional to the mass-difference between the different isotopes. Since biology is linked to chemistry through chemical potential energy, it is also strongly characterized by these mass-dependent isotope effects. Isotopic fractionations for some reactions, however, can follow complicated reaction pathways, and factors other than mass may come into play. This yields reaction rates for some isotopic species that are different from those predicted solely on the basis of mass differences between the isotopes. One way to describe these effects is to use the Δ^{33}S, which expresses the difference between the expected isotope composition of δ^{33}S from a standard fractionation and the measured δ^{33}S. Formally, $\Delta^{33}S = \delta^{33}S - 1000(1 + \delta^{34}S/1000)^{0.515}$. Δ^{33}S values of close to zero indicate mass-dependent effects, and significant non-zero Δ^{33}S is taken as an indication of mass-independent effects.

and sulfur disproportionators were present in this environment, and indeed other marine environments in the Paleoarchean (Wacey *et al.*, 2010).

The next evidence for microbial sulfur disproportionation is not seen until the early Mesoproterozoic (about 1.3 Ga), where ^{33}S again demonstrates the activities of sulfur disproportionating organisms in the environment, although in a different way (Johnston *et al.*, 2005a). This evidence is not related to mass-independent sulfur isotope signatures in the record but to smaller variations in ^{33}S that are produced because of differences in the metabolic pathways of sulfur disproportionation and sulfate reduction (Johnston *et al.*, 2005b; Box 5.1). Johnston *et al.* (2005a) argued that the onset of a disproportionation isotope signal after 1.3 Ga may have reflected a higher degree of oxygenation of the surface environment and more active operation of the oxidative sulfur cycle including those sedimentary processes generating elemental sulfur. A similar argument was made by Canfield and Teske (1996) based on the magnitude of ^{34}S fractionations, but these signals are not time coincident, and appear to be providing different types of information about the role of dispropotionators in Proterozoic environments. In particular, the ^{33}S approach is probably more sensitive to the influence of disproportionation, while the ^{34}S approach provides information on both sulfate availability for sulfate reduction and the overprint of disproportionation reactions. The lack of evidence for disproportionators in sequences with ages between the Dresser Formation and 1.3 Ga does not mean that they were absent, only that their metabolism did not leave a discernible isotopic signal.

From our discussion above, we might expect very early geologic evidence for anoxygenic sulfide-oxidizing organisms, as based on at least some molecular phylogenic arguments, these organisms are seemingly quite ancient. Unfortunately, sulfur isotope fractionations during sulfide oxidation are small, and thus far, undiagnostic (e.g., Canfield, 2001). Anoxygenic phototrophs do leave distinct organic biomarkers, but even so, the earliest report of anoxygenic phototrophic biomarkers (chlorobactane, a biomarker for green sulfur bacteria, and okenone, a biomarker for purple sulfur bacteria) is from the Barney Creek Formation from northern Australia dated at 1.64 Ga (Brocks *et al.*, 2005). Despite the lack of earlier geologic evidence, phylogenic arguments (as discussed above), coupled with the available geologic evidence for the timing of the emergence of sulfur disproportionation, would place the evolution of sulfide-oxidizing anoxygenic phototrophs well into the early Archean.

5.4 The interaction of S with other biogeochemical cycles

We now turn to the interactions between the sulfur cycle and the element cycles of carbon, oxygen, nitrogen, and iron.

5.4.1 Connections with the carbon cycle

The sulfur and carbon cycles are linked through common metabolic pathways and shared environments where they interact. Three of the fundamental connections are described below.

5.4.1.1 Sulfate reduction

Sulfate reduction is an extremely important pathway of organic carbon mineralization in anoxic sulfate-rich environments. For example, in coastal marine sediments, sulfate reduction typically accounts for one half of the total carbon oxidation (Jørgensen, 1982; Canfield, 1993). In low-sulfate environments, like anoxic lake sediments, sulfate reduction is less important (Holmer and Storkholm, 2001; Canfield *et al.*, 2005). In a modeling study, Habicht *et al.* (2002; Fig. 5.5) showed that, in the

absence of bioturbation, sulfate concentrations must drop to around 1 mM (present seawater is 28 mM) before sulfate reduction rates are significantly reduced, which carries important implications for studies of sulfur and carbon in the geologic past (see below). This relationship desperately needs field verification and further modeling studies.

5.4.1.2 Anaerobic oxidation of methane

Anaerobic oxidation of methane (AOM) was identified from geochemical observations nearly 40 years ago (e.g., Reeburgh, 1969), and for many years, it was rejected by a vociferous fraction of the microbial ecological community. However, recent studies have demonstrated that the process is mediated by a microbial consortium consisting of an archaeal methanotroph (ostensibly a methanogen) that apparently conducts reverse methanogenesis, and a sulfate reducer that utilizes the products from the methanogen (Hinrichs *et al.*, 1999; Boetius *et al.*, 2000). The methanotroph and sulfate reducer are often found physically associated in various arrangements of up to several hundreds of cells (Boetius *et al.*, 2000; Orphan *et al.*, 2001). It was traditionally thought that the methanotroph transferred H_2 to the sulfate reducer (Hoeler *et al.*, 1994), but experiments suggest that this is probably not true (Nauhaus *et al.*, 2002; Moran *et al.*, 2008). Recent work points to methyl sulfide (H_3C-SH) as a likely intermediate compound, being produced by the methanotroph and subsequently utilized by the sulfate reducer (Moran *et al.*, 2008).

AOM interfaces the carbon and sulfur cycles by regulating the flux of methane to the atmosphere. Currently, marine sediments produce about 1.3×10^{13} mol yr^{-1} methane. This estimate is determined by combining the carbon oxidation rate by sulfate reduction in sediments (1.3×10^{14} mol yr^{-1}; Canfield *et al.*, 2005) with the observation that rates of methane production are usually about 10 times lower (Canfield, 1993). Most of this methane is currently oxidized in sediments by AOM, but lower marine sulfate concentrations on the early Earth should have allowed more methane to flux to the atmosphere (e.g., Catling *et al.*, 2007). The effect of low sulfate concentration would have been twofold. First, among lower sulfate would have limited sulfate reduction, leaving more organic carbon for methanogens. Second, very low sulfate concentrations would have supported less AOM, allowing a higher proportion of the methane to escape the sediment. For these reasons, methane may have been an important greenhouse gas early in Earth history, perhaps dominating over CO_2 at some times (Domagal-Goldman *et al.*, 2008; Kasting, 2005; Pavlov *et al.*, 2000; see chapter 3). Interesting feedbacks between AOM, sulfate, and oxygen may have also influenced the history of atmospheric oxygen levels. Catling *et al.* (2007) argued

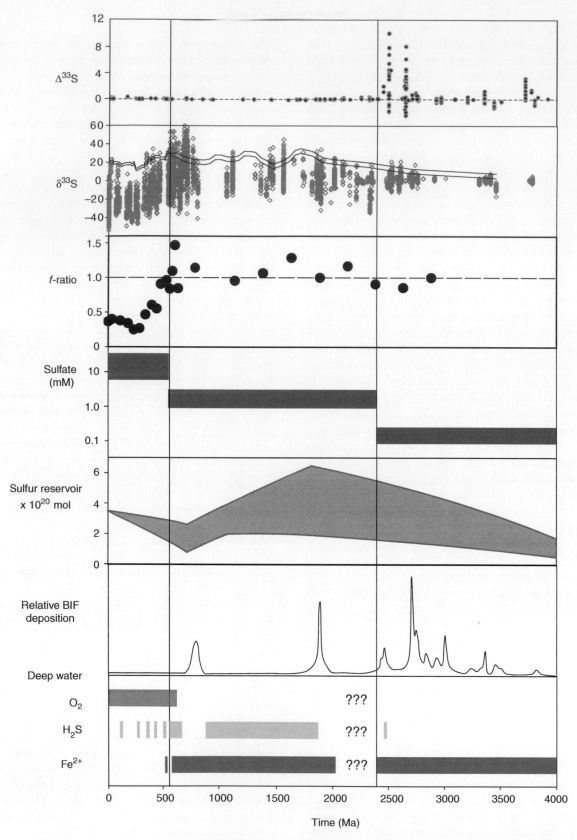

Figure 5.5. A compilation of data relevant to the evolution of the sulfur cycle through time, as well as the evolution of marine chemistry. Vertical lines represent the Cambrian-Ediacaran boundary at 542 Ma and the 'Great Oxidation Event' at 2400 Ma. The term 'deep water' chemistry refers to chemistry below the mixed upper layer and into the deep sea including oxygen-minimum zone settings. Question marks indicate suspected chemistry, but lack of data either direct or indirect. Two or more deep water chemistries at the same time could indicate either spatial or vertical structure in the chemistry of the water column. See text for further details.

that as oxygen rose, sulfate levels in the oceans increased due to the oxidative weathering of sulfate on land. This, in turn, increased sediment AOM reducing the flux of methane in the atmosphere. With reduced methane oxidation in the atmosphere, oxygen could rise to higher levels than would have occurred in absence of AOM.

5.4.1.3 Anoxygenic photosynthesis

Anoxygenic phototrophs interface with the carbon cycle when they build organic biomass through their phototrophic activities and most importantly when they fix CO_2. The largest modern marine basin housing anoxygenic phototrophs is the Black Sea. Here, a significant population of so-called 'brown' strains of green-sulfur bacteria is identified through its diagnostic pigment Bchl *e*. These organisms are extremely low-light adapted (Overmann *et al.* 1992) as necessitated by the low light available (around 10^{-3} µmol quanta m^{-2} s^{-1}; Manske *et al.*, 2005) where they live. Probably due to the low light, the importance of these bacteria to sulfide oxidation and carbon fixation in the Black Sea is very small, but it is thought to have varied over time as the depth to the chemocline has changed (Jørgensen *et al.*, 1991; Manske *et al.*, 2005). In other environments where light is more available, anoxygenic phototrophs can be very important, contributing to 83% of the total productivity in Fayettville Green Lake and up to 50% or more in many other sulfidic lakes (see Canfield *et al.*, 2005 for a summary). Such high amounts of carbon fixation by anoxygenic phototrophs require intense carbon, sulfur and nutrient recycling within the anoxic sulfidic waters. As mentioned above, biomarker evidence also demonstrates the presence of anoxygenic phototrophs in ancient marine basins, but their importance in carbon cycling in these basins remains unclear.

5.4.2 *Connections with the oxygen cycle*

The biogeochemical cycles of oxygen and sulfur interact in ways both biological and distinctly geological. These are outlined below.

5.4.2.1 Sulfide oxidation

As mentioned above, microbes from both the Bacteria and the Archaea engage in sulfide oxidation. Most of these organisms act as autotrophs (there is a link to the carbon cycle here), or mixotrophs. Aerobic sulfide oxidizers must compete with abiological reactions between oxygen and sulfide. Canfield *et al.* (2005) calculated that biological-mediated aerobic sulfide oxidation can outcompete biological sulfide oxidation when cell densities are on the order of 10^4 to 10^5 cells ml^{-1}, oxidizing sulfide at rates of 10 to 100 nM h^{-1}. The sulfide oxidation rates

(and presumably also cell densities) in active systems like microbial mats are orders of magnitude faster than this (Canfield *et al.*, 2005).

Some sulfide oxidizers such as *Thiovulum* sp. display fascinating coordinated behaviour. These organisms attach their cells to solid surfaces with slime threads. They spin around these threads generating currents which enhance the flux of oxygen to the organisms and increase their metabolic rate (Fenchel and Glud, 1998). As discussed below, nitrate-utilizing sulfide oxidizers also demonstrate fascinating behaviour.

5.4.2.2 Oxygen production

Somewhat counter to intuition, the burial of pyrite in sediments represents a net source of oxygen to the atmosphere. Indeed, this relationship was identified over 160 years ago by Jacques Joseph Ebelman (see Berner and Maasch, 1996), but rediscovered in modern times by Garrels and Perry (1974). The connections between oxygen and sulfur are made through a series of reactions beginning with the production of oxygen and organic carbon through photosynthesis (Equation 5.1), the oxidation of organic carbon through sulfate reduction (Equation 5.2) and the formation of pyrite (Equation 5.3). The individual reactions are as follows (calcite formation is used to balance the excess production of bicarbonate, Equation 5.4), with the overall reaction presented in Equation 5.5:

$$16CO_2 + 16H_2O \rightarrow 16CH_2O + 16O_2 \tag{5.1}$$

$$8SO_4^{2-} + 16CH_2O \rightarrow 8H_2S + 16HCO_3^- \tag{5.2}$$

$$2Fe_2O_3 + 8H_2S + O_2 \rightarrow 4FeS_2 + 8H_2O \tag{5.3}$$

$$16Ca^{2+} + 32HCO_3^- \rightarrow 16CaCO_3 + 16H_2O + 16CO_2 \tag{5.4}$$

$$2Fe_2O_3 + 16Ca^{2+} + 16HCO_3^- + 8SO_4^{2-} \rightarrow \\ 4FeS_2 + 16CaCO_3 + 8H_2O + 15O_2 \tag{5.5}$$

Through the Phanerozoic, pyrite formation typically accounts for less than 25% of the oxygen production, with organic carbon burial providing the rest (e.g., Garrels and Lerman, 1981). However, in the Neoproterozoic, and through much of the Proterozoic, pyrite burial appears to be of equal importance to carbon burial in terms of oxygen production (Canfield, 2005). This has largely been overlooked in discussions of the Precambrian history of atmospheric oxygen but illustrates a fundamental aspect of the connections between sulfur and oxygen.

5.4.2.3 Sulfide weathering

Just as pyrite burial plays a role in oxygen production, sulfide weathering plays a pivotal role in oxygen consumption. Evidence from detrital pyrites in 2.7 to 3.0 Ga Archean river deposits suggests that there was insufficient oxygen early in Earth's history to completely oxidize sulfides (Rasmussen and Buick, 1999; England *et al.*, 2002). Limiting the oxidative weathering of sulfide would limit the flux of sulfate to the oceans and contribute to low seawater sulfate concentrations. Pyrite weathering is very oxygen sensitive, and models suggest that a 100 μm pyrite grain would oxidize completely in 20 000 years in water saturated with around 10^{-5} present atmospheric oxygen levels (PAL) (Canfield *et al.*, 2000; Anbar *et al.*, 2007). However, it may be that higher oxygen levels would be required to have an impact on sulfide oxidation as such low concentrations of oxygen would quickly be depleted as a result of the weathering process (e.g., Bolton *et al*, 2006).This problem would benefit from further careful dedicated modeling.

5.4.2.4   Recorder of oxygen levels

Sulfur not only acts as a source and sink for oxygen, but its isotopic composition provides insights into past atmospheric oxygen concentrations. This comes from carefully considering the isotopic composition of the minor sulfur isotopes (see Box 5.1). Indeed, mass-independent sulfur isotope fractionations (seen as non-zero $\Delta^{33}S$) are found in sedimentary rocks older than 2.4 billion years old, but in younger rocks, fractionations are mass-dependent (Fig. 5.5). The preferred explanation for these sulfur isotope variations is that before 2.4 Ga, atmospheric sulfur species such as sulfur dioxide (and possibly sulfur monoxide) were photolysed with deep UV radiation (Farquhar and Wing, 2003). Experiments with sulfur dioxide in closed photocells with deep UV have produced a strong mass-independent isotope signal (Farquhar *et al.* 2001). The penetration of deep UV through Earth's atmosphere implies the absence of an ozone shield and low atmospheric oxygen levels, as an ozone shield disappears below an oxygen threshold of ~10^{-3} of present atmospheric levels (PAL) (Kasting and Donahue, 1980; Levine, 1985).

Another link between atmospheric sulfur and oxygen was made by Pavlov and Kasting (2002), who noted that the network of reactions describing atmospheric sulfur chemistry behaves in different ways at different oxygen levels. These workers argued that at oxygen levels greater than a threshold of 10^{-5} PAL, almost all sulfur exits the atmosphere as sulfate, whereas at lower oxygen concentrations sulfur also exits as elemental sulfur. Therefore, the 'bottleneck' provided by one exit channel at higher oxygen concentrations will homogenize isotope signals, and the opening of a second exit channel at lower oxygen concentrations is a requirement for more efficient transfer of the mass-independent signal to the surface reservoirs.

A third connection to oxygen relates to the amount of cycling that sulfur underwent. The preservation of a mass-independent sulfur isotope signal in the ocean requires limited cycling between sulfate and sulfide. Low oxygen would limit oxidative weathering and as a result may have limited this cycling (Farquhar *et al.*, 2000). Low sulfate concentrations in the oceans may have also allowed spatial variations in isotopic composition because of the non-conservative behaviour of sulfate in low sulfate oceans (e.g., Kaufman *et al.*, 2007).

5.4.3   Connections with the nitrogen cycle

The interface between the sulfur and nitrogen cycles is microbiological. There are many sulfide-oxidizing bacteria that use nitrate as an electron acceptor, oxidizing sulfide to elemental sulfur or sulfate and producing either N_2 gas or NH_4^+ (see Shultz and Jørgensen, 2001; review in Canfield *et al.*, 2005). These organisms are active in sulfidic sediments, water columns and microbial mats. This type of metabolism is also associated with bacterial gigantism where some sulfide oxidizers concentrate nitrate into a central vacuole. This concentration process can be quite spectacular, with nitrate reaching internal cellular concentrations of up to 500 mM (e.g. Fossing *et al.*, 1995). For most bacteria that obtain their nutrition from outside the cell, small size benefits nutrient acquisition. In nitrate-concentrating organisms, the cytoplasm forms a thin layer around the central vacuole and gigantism increases the ratio of vacuole volume to cytoplasm volume. This is a decided advantage when the electron donor (nitrate in this case) is delivered from within the cell.

Nitrate-concentrating sulfide-oxidizing bacteria demonstrate some fascinating microbial behaviour. For example, sulfide oxidizers of the genus *Thioploca* live as multicellular filaments within a common sheath in which the trichomes can actively glide. The filaments collect and concentrate nitrate from the overlying water, showing a strong positive tactic response to nitrate, and they then migrate into the sediment to oxidize sulfide (Fossing *et al.*, 1995). In another case, the giant sulfide oxidizer *Thiomargarita namibiensis*, which grows up to 0.5 mm in diameter, concentrates nitrate into a huge central vacuole (Schulz *et al.*, 1999). This organism is shaped like a ball, is non-motile, and apparently recharges its nitrate tank when periodically re-suspended into the water column.

5.4.4   Connections with the iron cycle

The balance between sulfur and iron in the present oceans is described in Table 5.2. Overall, S and Fe (the reactive Fe fraction, not total Fe) enter the oceans with a S/Fe ratio of about 0.5. Because of considerable Fe

Table 5.2 FeHR and S budgets for the modern ocean

	×10¹² mol yr⁻¹	
	FeHR	S
Inputs		
Terriginous sources[a]	6.5 (1.7)	2.6 (0.6)
Hydrothermal	0.25 (0.09)	0.5 (0.4)
Volcanic	–	0.2 (0.1)
Total	6.75 (1.7)	3.3 (0.7)
Outputs		
River proximal areas	5.5 (0.6)	1.6 (0.5)
Shelf sediments	0.73 (0.25)	0.22 (0.07)
Slope/deep sea	0.53 (0.17)	1.47 (0.8)
S/FeHR deep sea	2.8 (1.5)	

[a]Includes aeolian and glacial sources.
Uncertainties are in brackets.

removal in areas proximal to rivers and continental shelves, the S and Fe is available to the continental slope and deep sea in an S/Fe flux ratio of 2.8 ± 1.5, or a modest excess of sulfur over Fe when compared to the stoichiometric constraints of pyrite formation (FeS_2). Today, much sulfide is consumed by reaction with oxygen to reform sulfate and to form other sulfur intermediate compounds, but this combination of sinks has not always been the same. At times in the past when oxygen was low, the balance of iron to sulfur played a much more important role in controlling deep ocean chemistry.

Our entrance point to these interactions is to recognize that anoxic marine water will be dominated by either Fe^{2+} (henceforth referred to as a ferruginous state) or sulfide (henceforth referred to as a sulfidic state). The balance between a sulfidic or ferruginous states depends on the relative fluxes of sulfide or Fe^{2+} to the deep ocean such that if the H_2S/Fe flux exceeds 2 (imposed by pyrite stoichiometry), sulfidic waters will develop, but if the flux is less than 2, the waters will be ferruginous (Canfield, 1998; Canfield *et al.*, 2008). Iron comes from a combination of sources including hydrothermal input and terrestrial particulates, a portion of which (typically about 25%) is sulfide reactive, forming iron sulfide minerals (Raiswell and Canfield, 1998; Raiswell *et al.*, 2006). Of the total reactive iron entering the oceans, about 10% is deposited in deep sea settings (Table 5.2; Raiswell *et al.*, 2006). Sulfide also comes from hydrothermal sources, but the amount of sulfide is ultimately controlled by the amount of sulfate entering the oceans and the sulfate reduction rates.

Many factors can influence the ratio of sulfide to reactive iron and enact potential shifts between sulfidic and ferruginous states. If the riverine flux of sulfate to the oceans is reduced relative to Fe, such as is inferred for the Archean, the oceans will tend toward a ferruginous state (Fig. 5.5). If the total surface inventory of sulfur were

lower as a result of the long-term subduction of pyrite-rich sediments, the production of sulfate during weathering might be reduced and the H_2S/Fe would also favour ferruginous oceans. Also, sulfate influences the high temperature redox chemistry in mid ocean ridge hydrothermal systems. Low marine sulfate concentrations favour more reducing conditions and Fe^{2+}-enriched vent fluids (Kump and Seyfried, 2005). These latter mechanisms have been suggested to explain the reappearance of ferruginous oceans in the latest NeoproterozoicEon (Canfield *et al.*, 2008; discussed later in this chapter).

5.5 The evolution of the S cycle

We use the above discussion to frame further reflections on the evolution of the sulfur cycle and how this evolution has shaped the evolving nature of Earth surface chemistry.

5.5.1 *Evolution of sulfate concentrations*

We begin by considering how the biogeochemical environment has influenced the history of seawater sulfate concentrations. The hypothesis of limited free oxygen in the early atmosphere implies a limited source of sulfate to the oceans. This is supported by the observation that the ³⁴S fractionations between sulfide and sulfate are typically small before 2.4 Ga (Fig. 5.5). Indeed, sulfate reducers are known to produce low fractionations when reducing low concentrations of sulfate (Habicht *et al.*, 2002); less than 5 ‰ fractionation is generated when sulfate reducers use <200 µM sulfate (Fig. 5.6).

After about 2.4 Ga, mass-independent fractionations all but disappear, signaling an increase in atmospheric

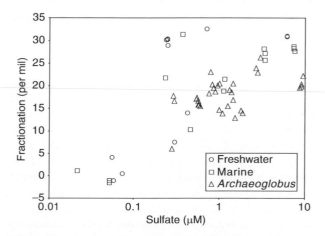

Figure 5.6 The relationship between sulfate concentration and isotope fractionation during sulfate reduction by natural populations of marine and freshwater sulfate reducers, as well as the archaeal sulfate reducer *Archaeoglobus fulgidus*. The figure is modified from Habicht *et al.* (2002).

oxygen (Fig. 5.5). Following the reasoning of Habicht *et al.* (2002), a near concomitant increase in ^{34}S fractionations (Fig. 5.5) signals a rise in seawater sulfate concentrations. This is consistent with an influx of sulfate from the oxidative weathering of sulfides in a newly oxygen-enriched atmosphere. But what levels did sulfate attain? Mass balance calculations reveal that through the remainder of the Precambrian, most sulfur was removed from the oceans as pyrite (Canfield, 2004; Fig. 5.5), which would be consistent with the geologic observations for limited sulfate evaporite deposition (Grotzinger and Kasting, 1993; Canfield and Farquhar, 2009) and relatively low sulfate levels. Holland (1984) has calculated that for sulfate to precipitate in association with halite evaporite deposition, the ion product of dissolved sulfate and calcium in normal seawater must exceed 28 mM ($[SO_4^{2-}][Ca^{2+}]$ > 28 mM). Proterozoic calcium concentrations are not well constrained, but if they encompass the Phanerozoic range of 10 to 40 mM (Horita *et al.*, 2002), then the lack of significant sulfate evaporites implies sulfate concentrations below 0.7 to 2.8 mM. These values are much lower than the present 28 mM and generally consistent with other, though rather imprecise, Proterozoic sulfate concentration estimates (Shen *et al.*, 2002; Kah *et al.*, 2004; Canfield, 2004).

Another approach to constraining the concentration of sulfate in the ocean has been introduced by Johnston *et al.* (2006) who suggested that sulfate will behave non-conservatively in the oceans at sulfate concentrations less than 1–4 mM, and that evidence for non-conservative behaviour is provided by the style of covariation between ^{33}S and ^{34}S in sequences like the 1780–1835 Ma Rove Formation.

There was a rise in sulfate concentrations in the latest NeoproterozoicEon as evidenced by the massive deposition of sulfate evaporites in the middle East and Asia as well as from sulfate concentration estimates from fluid inclusions within bedded halites (Horita *et al.*, 2002) and sulfur isotope arguments (Fike *et al.*, 2006; Canfield *et al.*, 2007). From fluid inclusion studies, late Neoproterozoic sulfate concentrations may have reached 15 mM, although they apparently dropped rapidly again into the early Cambrian to values of 5 mM or less (Horita *et al.*, 2002; Petrychenko *et al.*, 2005). Through the remainder of the Phanerozoic, sulfate cycled between 5 and 28 mM (see summary in Berner, 2004) and therefore was considerably elevated in concentration compared to Proterozoic values. Elevated Phanerozoic sulfate levels are also consistent with mass balance calculations showing substantial sulfate evaporite deposition in the Phanerozoic (Canfield, 2004; Canfield and Farquhar, 2008; Fig. 5.5) and the common occurrence of Phanerozoic sulfate evaporites (Grotzinger and Kasting, 1993). A cartoon displaying the history of seawater sulfate levels is shown in Fig. 5.5.

5.5.2 Evolution of ocean redox chemistry

The interplay between the cycles of S, C and Fe defines the redox history of the oceans. We begin by considering the history of banded iron formation (BIF) deposition which helps up to identify the nature of deep ocean chemistry early (and also later) in Earth history. Banded iron formations are rather common sedimentary rocks before 2.4 Ga, with further episodes of BIF deposition around 1.9 to 2.0 Ga and again episodes of BIF deposition later in the NeoproterozoicEon (Fig. 5.5). The deposition of BIF requires a supply of Fe^{2+} from ferruginous marine bottom waters (e.g., Holland, 1984). The widespread occurrence of BIF in certain time windows suggests that the ferruginous deep water conditions during these times were also widespread and possibly global. As described above, ferruginous conditions develop when the flux ratio of H_2S/Fe to deep ocean waters is <2. In the case of sediments depositing before 2.4 Ga, a low H_2S/Fe flux ratio would have resulted, at least in part, from limited pyrite weathering in a very low oxygen environment, resulting in a low flux of riverine sulfate to the oceans. The resulting low sulfate concentrations would have influenced the chemistry of mid-ocean ridge hydrothermal fluids, enriching them in Fe^{2+} compared to fluids venting today.

Most BIFs date from before 2.4 Ga, and therefore, they were deposited predominantly under nearly anoxic atmospheric conditions. For this reason, Holland (1984, 2006) has advanced the idea that BIF deposition ceased as the deep oceans became flushed with oxygenated bottom waters in response to rising atmospheric oxygen levels. Canfield (1998), however, noted that relatively high oxygen levels of about 40 to 50% PAL would be required to oxygenate the deep ocean, and such high levels so early in Earth's history would be inconsistent with a later Neoproterozoic rise in oxygen levels as suggested from the sulfur isotope record (Canfield and Teske, 1996). Alternatively, Canfield (1998) noted that BIF deposition ceased as the ^{34}S fractionation dramatically increased (see Figure 5). The increase in fractionation is consistent with an increase in marine sulfate levels in response to the oxidative weathering of sulfides enabled by increased atmospheric oxygen levels as discussed above. Therefore, Canfield (1998) argued that increased sulfate availability to the oceans increased the H_2S/Fe flux ratio to the deep oceans to >2; in this model the ocean became sulfidic.

The timing for the initiation of these conditions is unclear as there is return to significant BIF deposition around 1.9 Ga, and the nature of water chemistry between 2.4 and 2.0 Ga is uncertain (see Canfield, 2005). However, the 1.9 Ga Gunflint Iron Formation clearly gave way to anoxic sulfidic conditions with the deposition of the overlying Rove Formation (Poulton *et al.*,

2004). A variety of persistent sulfidic basins are found from 1.4 to 1.6 Ga in Australia (Shen *et al.*, 2002, 2003), and Mo isotopes suggest substantial removal of Mo from the global ocean under sulfidic conditions at 1.6 Ga (Arnold *et al.*, 2004). Also, Mo concentrations through Earth's history demonstrate that seawater was Mo-starved during the Mesoproterozoic and up to the late Neoproterozoic (Scott *et al.*, 2008). These data are consistent with substantial Mo removal under sulfidic conditions (Scott *et al.*, 2008), and are generally in line with the Canfield (1998) model.

This model, however, has also evolved. Canfield (1998) originally called for pervasive deep-water sulfidic conditions. Recent studies, however, suggest that sulfidic conditions were not evenly distributed throughout the deep-ocean and they may have been restricted to regions of the outer shelf and slope, analogous to an expanded sulfidic oxygen-minimum zone (Lyons *et al.*, 2009; Poulton *et al.*, 2010; Johnston *et al.*, 2010). This was shown in a detailed study of the Animikie Basin where the original Gunflint to sulfidic Rove transition was first described (Poulton *et al.* 2004). Indeed, when looked at on a regional scale extending over 100 km from the seashore into the basin, geochemical evidence points to ferruginous waters distal to the sulfidic wedge defining the Rove Formation (Poulton *et al.*, 2010). In this case, sulfide was preferentially removed as pyrite in these upper slope regions, possibly due to a high coastal organic carbon flux, driving high enough rates of sulfide production by sulfate reduction to counter the reactive iron flux from settling particles on the shelf and the advective ferrous iron flux from the deeper ocean. (Johnston *et al.*, 2010; Poulton *et al.*, 2010; Poulton and Canfield, 2011). Newer studies have also documented ferruginous conditions in some Mesoproterozoic basins and possible some open outer shelf settings (Planovsky *et al.*, 2011).

Moving forward into the Neoptoterozoic, Canfield *et al.* (2008) concluded from the analysis of sediments from 35 different geologic formations, that there is evidence for sulfidic deep waters around 0.75 Ga, but afterwards, and until the Cambrian-Precambrian boundary, ferruginous deep waters were common. This would be consistent with the identification of Banded Iron Formations associated with a snowball Earth glaciations, but the new results demonstrate that ferruginous conditions were a general feature of later Neoproterozoic water chemistry and not just restricted to glacial intervals. Although sulfidic condition were apparently rare in the interval from 0.75 Ga to the Cambrian/Precambrian border, Li *et al.* (2010) identified a wedge of sulfidic water in the Ediacaran-aged (0.635 to 0.542 Ga) Doushantuo Formation in South China.

Starting at 0.58 Ga there is the first evidence for deep ocean oxygenation (Canfield *et al.*, 2007; Fike *et al.*, 2006; Shen *et al.*, 2008), which was apparently widespread (Canfield *et al.*, 2008). Still, oxygenation was not global as there is evidence for contemporaneous, anoxic, ferruginous, marine deep-water (Canfield *et al.*, 2008).

One can reasonably ask, why was there a return to ferruginous deepwater chemistry in the Neoproterozoic? The answer is not completely clear, but Canfield *et al.* (2008) argued that these conditions might have resulted, in part, from a decreased size of the surface sulfur reservoir. This argument follows an earlier modeling study by Canfield (2004) who reasoned that an important consequence of sulfidic deepwater chemistry was the subduction of pyritic sediments into the mantle (see Fig. 5.1). If operating over a long period of time, this process could have significantly reduced the size of the surface sulfur reservoir (see Fig. 5.5) by the NeoproterozoicEon. Also, the S isotope record suggests a return to low sulfate concentrations between the Sturtian (about 700 Ma) and Marinoan glaciations (about 630 Ma) (Canfield, 2004). This low sulfate concentration may have increased the flux of Fe^{2+} during mid ocean ridge hydrothermal circulation (Kump and Seyfried, 2005). Either individually or in combination, these factors may have tipped the balance of the deep water H_2S/Fe flux ratio towards ferruginous conditions.

There was a return to apparently widespread water-column sulfidic conditions near the Cambrian-Precambrian boundary (see Canfield *et al.*, 2008), and at least periodic and even sometimes widespread euxinia during Paleozoic and in the Cretaceous as well (Berry and Wilde, 1976; Werne *et al.*, 2002; Jenkyns, 1980; Berner and Raiswell, 1981; Gill *et al.*, 2011). The importance of water-column euxinia, however, seemed to decrease dramatically by the early Devonian (Dahl *et al.*, 2010). Despite the significance of sulfidic conditions, oxic bottom waters were probably a normal feature of deep-ocean chemistry through most of the Phanerozoic (e.g. Dahl *et al.*, 2010). A summary of deep water chemistry through time is shown in Fig. 5.5.

5.6 Closing remarks

Organisms involved in the biogeochemical cycling of sulfur compounds evolved early in Earth history, and their activities have left an indelible mark in the geological record. This is in large part because the sulfur cycle interacts intimately with the cycling of many other elements of biogeochemical interest like oxygen, carbon, iron, and nitrogen. The sulfur cycle has also responded to changes in the ecology of the Earth's surface environment as well as to changes brought about by fundamentally geological processes such as volcanic outgassing, hydrothermal interactions and subduction. Overall, the sulfur cycle has heavily influenced the evolving chemistry of the Earth surface. Indeed, one can argue that, to a great extent, the history of Earth surface chemistry has been defined by the dynamics of the sulfur cycle.

Acknowledgements

We wish to acknowledge the helpful comments of Dave Johnston and discussions with Bo Thamdrup. Financial support from Danmarks Grundforskningsfond, a Guggenheim fellowship to JF, NASA and NSF.

References

Alt JC (1995) Sulfur isotopic profile through the oceanic-crust – sulfur mobility and seawater-crustal sulfur exchange during hydrothermal alteration. *Geology* **23**, 585–588.

Anbar AD, Duan Y, Lyons TW, *et al.* (2007) A whiff of oxygen before the Great Oxidation Event? *Science* **317**, 1903–1906.

Arnold GL, Anbar AD, Barling J. Lyons TW (2004) Molybdenum isotope evidence for widespread anoxia in mid-Proterozoic oceans. *Science* **304**, 87–90.

Berner RA, Raiswell R (1983) Burials organic-carbon and pyrite sulfur in sediments over Phanerozoic time-a new theory. *Geochemica et Cosmochimica Acta* **47**, 855–862.

Berner RA (2004) *The Phanerozoic Carbon Cycle: CO₂ and O₂*. Oxford University Press, Oxford.

Berner RA, Maasch KA (1996) Chemical weathering and controls on atmospheric O₂ and CO₂: Fundamental principles were enunciated by J.J. Ebelman in 1845. *Geochimica et Cosmochimica Acta* **60**, 1633–1637.

Berry WBN, Wilde P (1978) Progressive ventilation of the oceans- an explanation for the distribution of the lower paleozoic black shales. *American Journal of Science* **278**, 257–275.

Boetius A, Ravenschlag K, Schubert CJ, *et al.* (2000) A marine microbial consortium apparently mediating anaerobic oxidation of methane. *Nature* **407**, 623–626.

Bolton EW, Berner RA, Petsch ST (2006) The weathering of sedimentary organic matter as a control on atmospheric O-2: II. Theoretical modeling. *American Journal of Science* **306**, 575–615.

Brocks JJ, Love GD, Summons RE, Knoll AH, Logan GA, Bowden SA (2005) Biomarker evidence for green and purple sulphur bacteria in a stratified Palaeoproterozoic sea. *Nature* **437**, 866–870.

Buick R, Dunlop JSR (1990) Evaporitic sediments of early Archaean age from the Warrawoona Group, North Pole, Western Australia. *Sedimentology* **37**, 247–277.

Canfield DE (1993) Organic matter oxidation in marine sediments. In *Interactions of C, N, P and S Biogeochemical Cycles and Global Change* (eds Wollast R, Mackenzie FT, Chou L). Springer, Berlin, pp. 333–363.

Canfield DE (1998) A new model for Proterozoic ocean chemistry. *Nature* **396**, 450–453.

Canfield DE (2001) Biogeochemistry of sulfur isotopes. In *Reviews in Mineralogy and Geochemistry* (eds Valley JW, Cole DR). Mineralogical Society of America, Blacksburg, VA, pp. 607–636.

Canfield DE (2004) The evolution of the Earth surface sulfur reservoir. *American Journal of Science* **304**, 839–861.

Canfield DE (2005) The early history of atmospheric oxygen: Homage to Robert M. Garrels. *Annual Review of Earth and Planetary Sciences* **33**, 1–36

Canfield DE, Farquhar J (2009) Animal evolution, bioturbation, and the sulfate concentration of the oceans. *PNAS* **106**, 8123–8127.

Canfield DE, Teske A (1996) Late Proterozoic rise in atmospheric oxygen concentration inferred from phylogenetic and sulphur-isotope studies. *Nature* **382**, 127–132.

Canfield DE, Habicht KS, Thamdrup B (2000) The Archean sulfur cycle and the early history of atmospheric oxygen. *Science* **288**, 658–661.

Canfield DE, Thamdrup B, Kristensen E (2005) Systematics and Phylogeny. *Aquatic Geomicrobiology* **48**, 1–21.

Canfield DE, Poulton SW, Narbonne GM (2007) Late-Neoproterozoic deep-ocean oxygenation and the rise of animal life. *Science* **315**, 92–95.

Canfield DE, Poulton SW, Knoll AH, *et al.* (2008) Ferruginous conditions dominated later Neoproterozoic deep-water chemistry. *Science* **321**, 949–952.

Canfield DE, Farquhar J, Zerkle AL (2010) High isotope fractionation during sulfate reduction in a low-sulfate euxinic ocean analog. *Geology* **38**, 415–418.

Catling DC, Claire MW, Zahnle KJ (2007) Archean methane, oxygen and sulfur. *Geochimica Et Cosmochimica Acta* **71**, A151–A151.

Chaussidon M, Albarede F, Sheppard SMF (1987) Sulfur isotope heterogeneity in the mantle from ion microprobe measurements of sulfide inclusions in diamonds. *Nature* **330**, 242–244.

Dahl TW, Hammarlund EU, Anbar AD, *et al.* (2010) Devonian rising atmospheric oxygen correlated to the radiations of terrestrial plants and large predatory fish. *PNAS* **107**, 17911–17915.

David LA, Alm EJ (2011) Rapid evolutionary innovation during an Archaean genetic expansion. *Nature* **469**, 93–96.

Del Don C, Hanselmann KW, Peduzzi R, Bachofen R (2001) The meromictic alpine Lake Cadagno: Orographical and biogeochemical description. *Aquatic Sciences* **63**, 70–90.

Domagal-Goldman S, Kasting J, Johnston DT, Farquhar J (2008) Organic haze, glaciations and multiple sulfur isotopes in the Mid-Archean Era. *Earth and Planetary Science Letters* **269**, 2369–2372.

England GL, Rasmussen B, Krapez B, Groves DI (2002) Palaeoenvironmental significance of rounded pyrite in siliciclastic sequences of the Late Archaean Witwatersrand Basin: oxygen-deficient atmosphere or hydrothermal alteration? *Sedimentology* **49**, 1133–1156.

Farquhar J, Wing BA (2003) Multiple sulfur isotopes and the evolution of the atmosphere. *Earth and Planetary Science Letters* **213**, 1–13.

Farquhar J, Bao HM, Thiemens M (2000) Atmospheric influence of Earth's earliest sulfur cycle. *Science* **289**, 756–758.

Farquhar J, Savarino J, Airieau S, Thiemens MH (2001) Observation of the wavelength-sensitive mass-dependent sulfur isotope effects during SO₂ photolysis: Implications for the early atmosphere. *Journal of Geophysical Research* **106**, 32829–32839.

Farquhar J, Wing BA, McKeegan KD, Harris JW, Cartigny P, Thiemens MH (2002) Mass-independent sulfur of inclusions in diamond and sulfur recycling on early Earth. *Science* **298**, 2369–2372.

Fenchel T, Glud RN (1998) Veil architecture in a sulphide-oxidizing bacterium enhances countercurrent flux. *Nature* **394**, 367–369.

Fike DA, Grotzinger JP, Pratt LM, Summons RE (2006) Oxidation of the Ediacaran Ocean. *Nature* **444**, 744–747.

Finster K (2008) Microbial disproportionation of inorganic sulfur compounds. *Journal of Sulfur Chemistry* **29**, 281–292.

Finster K, Liesack W, Thamdrup B (1998) Elemental sulfur and thiosulfate disproportion by *Desulfocapsa sulfoexigens* sp nov, a new anaerobic bacterium isolated from marine surface sediment. *Applied and Environmental Microbiology* **64**, 119–125.

Fossing H, Gallardo VA, Jørgensen BB, *et al.* (1995) Concentration and transport of nitrate by the mat-forming sulphur bacterium *Thioploca*. *Nature* **374**, 713–715.

Garrels RM, Lerman A (1981) Phanerozoic cycles of sedimentary carbon and sulfur. *Proceedings of the National Academy of Sciences of the United States of America* **78**, 4652–4656.

Garrels RM, Perry Jr. EA (1974) Cycling of carbon, sulfur, and oxygen through geologic time. In *The Sea* (ed, Goldberg ED). John Wiley & Sons, Inc., New York, pp. 303–336.

Gill BC, Lyons TW, Young SA, Kump LR, Knoll AH, Saltzman (2011) Geochemical evidence for widespread euxinia in the later Cambrian ocean. *Nature* **469**, 80–83.

Gregersen LH, Habicht KS, Peduzzi S, *et al.* (2009) Dominance of a clonal green sulfur bacterial population in a stratified lake. *FEMS Microbiology Ecology* **70**, 30–41.

Grotzinger JP, Kasting JF (1993) New constraints on Precambrian ocean composition. *Journal of Geology* **101**, 235–243.

Habicht KS, Canfield DE (2001) Isotope fractionation by sulfate-reducing natural populations and the isotopic composition of sulfide in marine sediments. *Geology* **29**, 555–558.

Habicht KS, Gade M, Thamdrup B, Berg P, Canfield DE (2002) Calibration of sulfate levels in the Archean ocean. *Science* **298**, 2372–2374.

Hinrichs K-U, Hayes JM, Sylva SP, Brewer PG, DeLong EF (1999) Methane-consuming archaebacteria in marine sediments. *Nature* **398**, 802–805.

Hoehler TM, Alperin MJ, Albert DB, Martens CS (1994) Field and laboratory studies of methane oxidation in an anoxic marine sediment: Evidence for a methanogen-sulfate reducer consortium. *Global Biogeochemical Cycles* **8**, 451–463.

Holland HD (1984) *The Chemical Evolution of the Atmosphere and Oceans*. Princeton University Press, Princeton, NJ.

Holland HD (2006) The oxygenation of the atmosphere and oceans. *Philosophical Transactions of the Royal Society, London B.* **361**, 903–15.

Holmer M, Storkholm P (2001) Sulphate reduction and sulphur cycling in lake sediments: a review. *Freshwater Biology* **46**, 431–451.

Horita J, Zimmermann H, Holland HD (2002) Chemical evolution of seawater during the Phanerozoic: Implications from the record of marine evaporites. *Geochimica et Cosmochimica Acta* **66**, 3733–3756.

Jenkyns HC (1980) Cretaceous anoxic events – from continents to oceans. *Journal of the Geological Society* **137**, 171–188.

Johnston DT, Wing BA, Farquhar J, *et al.* (2005a) Active microbial sulfur disproportionation in the Mesoproterozoic. *Science* **310**, 1477–1479.

Johnston DT, Farquhar J, Wing BA, Kaufman A, Canfield DE, Habicht KS (2005b) Multiple sulfur isotope fractionations in biological systems: A case study with sulfate reducers and sulfur disproportionators. *American Journal of Science* **305**, 645–660.

Johnston DT, Poulton SW, Fralick PW, Wing BA, Canfield DE, Farquhar J (2006) Evolution of the oceanic sulfur cycle at the end of the Paleoproterozoic. *Geochimica et Cosmochimica Acta* **70**, 5723–5739.

Johnston DT, Farquhar J, Habicht KS, Canfield DE (2008) Sulphur isotopes and the search for life: strategies for identifying sulphur metabolisms in the rock record and beyond. *Geobiology* **6**, 425–435.

Johnston DT, Poulton SW, Dehler C, *et al.* (2010) An emerging picture of Neoproterozoic ocean chemistry: Insights from the Chuar Group, Grand Canyon, USA. *Earth and Planetary Science Letters*, **290**, 64–73.

Jørgensen BB (1982) Mineralization of organic matter in the sea bed - the role of sulfate reduction. *Nature* **296**, 643–645.

Jørgensen BB, Fossing H, Wirsen CO, Jannasch HW (1991) Sulfide oxidation in the anoxic Black Sea chemocline. *Deep-Sea Research* **38**, S1083–S1103.

Jørgensen BB, Weber A, Zopfi J (2001) Sulfate reduction and anaerobic methane oxidation in Black sea sediments. *Deep-Sea Research I* **48**, 2097–2120.

Kah LC, Lyons TW, Frank TD (2004) Low marine sulphate and protracted oxygenation of the Proterozoic biosphere. *Nature* **431**, 834–838.

Kaplan IR, Rittenberg SC (1964) Microbiological fractionation of sulphur isotopes. *Journal of General Microbiology* **34**, 195–212.

Kasting JF (2005) Methane and climate during the Precambrian era. *Precambrian Research* **137**, 119–129.

Kasting JF, Donahue TM (1980) The evolution of atmospheric ozone. *Journal of Geophysical Research-Oceans and Atmospheres* **85**, 3255–3263.

Kaufman AJ, Johnston DT, Farquhar J, *et al.* (2007) Late Archean biospheric oxygenation and atmospheric evolution. *Science* **317**, 1900–1903.

Klein M, Friedrich M, Roger AJ, *et al.* (2001) Multiple lateral transfers of dissimilatory sulfite reductase genes between major lineages of sulfate-reducing prokaryotes. *Journal of Bacteriology* **183**, 6028–6035.

Kump LR, Seyfried Jr. WE (2005) Hydrothermal Fe fluxes during the Precambrian: Effect of low oceanic sulfate concentration and low hydrostatic pressure on the composition of black smokers. *Earth and Planetary Science Letters* **235**, 654–662.

Levine JS (1985) *The Photochemistry of Atmospheres: Earth, the other Planets, and Comets*. Academic Press, San Diego, CA.

Li C, Love GD, Lyons TW, Fike DA, Sessions AL, Chu X (2010) A stratified redox model for the Ediacaran ocean. *Science* **328**, 80–83.

Loy A, Duller S, Wagner M (2008) Evolution and ecology of microbes dissimilating sulfur compounds: Insights from siriheme sulfite reductases. In *Microbial Sulfur Metabolism* (eds, Dahl C, Friedrich CG) Springer, Amsterdam, pp 46–59.

Luther GW, III, Church T, Powell D (1991) Sulfur speciation and sulfide oxidation in the water column of the Black Sea. *Deep-Sea Research* **38**, S1121–S1137.

Lyons TW, Reinhard CT, Scott C (2009) Redox redux. *Geobiology* **7**, 489–494.

Manske AK, Glaeser J, Kuypers MAM, Overmann J (2005) Physiology and phylogeny of green sulfur bacteria forming a monospecific phototrophic assemblage at a depth of 100 meters in the Black Sea. *Applied and Environmental Microbiology* **71**, 8049–8060.

Moran JJ, Beal EJ, Vrentas, JM, Orphan VJ, Freeman KH, House CH (2008) Methyl sulfides as intermediates in the anaerobic oxidation of methane. *Environmental Microbiology* **10**, 162–173.

Nauhaus K, Boetius A, Krüger M, Widdel F (2002) *In vitro* demonstration of anaerobic oxidation of methane coupled to sulphate reduction in sediment from a marine gas hydrate area. *Environmental Microbiology* **4**, 296–305.

Obraztsova AY, Francis CA, Tebo BM (2002) Sulfur disproportionation by the facultative anaerobe Pantoea agglomerans SP1 as a mechanism for chromium(VI) reduction. *Geomicrobiology Journal* **19**, 121–132.

Orphan VJ, House CH, Hinrichs K-U, McKeegan KD, DeLong EF (2001) Methane-consuming Archaea revealed by directly coupled isotopic and phylogenetic analysis. *Science* **293**, 484–487.

Overmann J, Cypionka H, Pfennig N (1992) An extremely low-light-adapted phototrophic sulfur bacterium from the Black Sea. *Limnology and Oceanography* **37**, 150–155.

Pavlov AA, Kasting JF (2002) Mass-independent fractionation of sulfur isotopes in Archean sediments: Strong evidence for an anoxic Archean atmosphere. *Astrobiology* **2**, 27–41.

Pavlov AA, Kasting JF, Brown LL, Rages KA, Freedman R (2000) Greenhouse warming by CH_4 in the atmosphere of early Earth. *Journal of geophysical research* **105**, 11981–11990.

Peduzzi S, Tonolla M, Hahn D (2003) Isolation and characterization of aggregate-forming sulfate-reducing and purple sulfur bacteria from the chemocline of meromictic Lake Cadagno, Switzerland. *FEMS Microbiology Ecology* **45**, 29–37.

Petrychenko OY, Peryt TM, Chechel EL (2005) Early Cambrian seawater chemistry from fluid inclusions in halaite from Siberian evaporites. *Chemical Geology* **219**, 149–161.

Pfennig N (1975) The phototrophic bacteria and their role in the sulfur cycle. *Plant Soil* **43**, 1–16.

Philippot P, Van Zuilen M, Lepot K, Thomazo C, Farquhar J, Van Kranendonk A (2007) Elemental-sulfur reducing or disproportionating organisms in a similar to 3,5 Myr-old seafloor setting. *Geochimica et Cosmochimica Acta* **71**, A786–A786.

Planavsky NJ, McGoldrick P, Scott CT, Li C, Reinhard CT, Kelly AE, Chu X, Bekker A, Love GD, Lyons TW (2011) Widespread iron-rich conditions in the mid-Proterozoic ocean. *Nature* **477**, 448–451.

Poulton SW, Canfield DE, Fralick P (2004) The transition to a sulfidic ocean ~1.84 billion years ago. *Nature* **431**, 173–177.

Poulton SW, Fralick PW, Canfield DE (2010) Spatial variability in oceanic redox structure 1.8 billion years ago. *Nature Geoscience* **3**, 486–490.

Poulton SW, Canfield DE, (2011) Ferruginous conditions: a dominant feature of the ocean through Earth's history. *Elements* 7, 107–112.

Rabus R, Brüchert V, Amann J, Könneke M (2002) Physiological response to temperature changes of the marine, sulfate-reducing bacterium *Desulfobacterium autotrophicum*. *FEMS Microbiology Ecology* **42**, 409–417.

Raiswell R, Canfield DE (1998) Sources of iron for pyrite formation in marine sediments. *American Journal of Science* **298**, 219–245.

Raiswell R, Tranter M, Benning LG, *et al.* (2006) Contributions from glacially derived sediment to the global iron(oxyhydr)

oxide cycle: Implications for iron delivery to the oceans. *Geochimica et Cosmochimica Acta* **70**, 2765–2780.

Rasmussen B, Buick R (1999) Redox state of the Archean atmosphere: Evidence from detrital heavy minerals in ca. 3250–2750 Ma sandstones from the Pilbara Craton, Australia. *Geology* **27**, 115–118.

Reeburgh WS (1969) Observations of gases in Chesapeake Bay sediments. *Limnology and Oceanography* **14**, 368–375.

Schulz HN, Brinkhoff T, Ferdelman TG, Marine MH, Teske A. Jørgensen BB (1999) Dense populations of a giant sulfur bacterium in Namibian shelf sediments. *Science* **284**, 493–495.

Schulz HN, Jorgensen BB (2001) Big bacteria. *Annual Review of Microbiology* **55**, 105–137.

Scott C, Lyons TW, Bekker A, Shen Y, Poulton SW, Chu X, Anbar AD (2008) Tracing the stepwise oxygenation of the Proterozoic ocean. *Nature* **452**, 456-U5.

Shanks WC (2001) Stable isotopes in seafloor hydrothermal systems: vent fluids, hydrothermal deposits, hydrothermal alteration, and microbial processes. *Stable Isotope Geochemistry* **43**, 469–525.

Shen Y, Buick R, Canfield DE (2001) Isotopic evidence for microbial sulphate reduction in the early Archean era. *Nature* **410**, 77–81.

Shen Y, Canfield DE, Knoll AH (2002) Middle Proterozoic ocean chemistry: Evidence from the McArthur Basin, Northern Australia. *American Journal of Science* **302**, 81–109.

Shen Y, Knoll AH, Walter MR (2003) Evidence for low sulphate and anoxia in a mid-Proterozoic marine basin. *Nature* **423**, 632–635.

Shen Y, Zhang T, Hoffman PF (2008) On the coevolution of Ediacaran oceans and animals. *Proceedings of the National Academy of Sciences of the United States of America* **105**, 7376–7381.

Shen Y, Farquhar J, Masterson A, Kaufman AJ, and Buick R (2009) Evaluating the role of microbial sulfate reduction in the early Archean using quadruple isotope systematics. *Earth and Planetary Science Letters* **279**(3–4), 383–391.

Tonolla M, Peduzzi R, Hahn D (2005) Long-term population dynamics of phototrophic sulfur bacteria in the chemocline of Lake Cadagno, Switzerland. *Applied and Environmental Microbiology* **71**, 3544–3550.

Tonolla M, Demarta A, Peduzzi R, Hahn D (1999) In situ analysis of phototrophic sulfur bacteria in the chemocline of meromictic Lake Cadagno (Switzerland). *Applied and Environmental Microbiology* **65**, 1325–1330.

Wacey D, McLoughlin N, Whitehouse MJ, Kilburn MR (2010) Two coexisting sulfur metabolisms in ca. 3400 Ma sandstones. *Geology* **38**, 1115–1118.

Walker JCG, Brimblecombe P (1985) Iron and sulfur in the pre-biologic ocean. *Precambrian Research* **28**, 205–222.

Werne JP, Sageman BB, Lyons TW, Hollander DJ (2002) An integrated assessment of a 'type euxinic' deposit: Evidence for multiple controls on black shale deposition in the Middle Devonian Oatka Creek Formation. *American Journal of Science* **302**, 110–143.

Xiong J (2006) Photosynthesis: what color was its origin? *Genome Biology* **7**, article 245.5

Zopfi J, Kjær T, Nielsen LP, Jørgensen BB (2001) Ecology of *Thioploca* spp.: nitrate and sulfate storage in relation to chemical microgradients and influence of *Thioploca* spp. on the sedimentary nitrogen cycle. *Applied and Environmental Microbiology* **67**, 5530–5537.

6

THE GLOBAL IRON CYCLE

Brian Kendall[1], Ariel D. Anbar[1,2], Andreas Kappler[3] and Kurt O. Konhauser[4]

[1]School of Earth and Space Exploration, Arizona State University, Tempe, Arizona, 85287, USA
[2]Department of Chemistry and Biochemistry, Arizona State University, Tempe, Arizona, 85287, USA
[3]Geomicrobiology, Center for Applied Geosciences, University of Tübingen, Sigwartstrasse 10, 72076, Tübingen, Germany
[4]Department of Earth and Atmospheric Sciences, University of Alberta, Edmonton, Alberta, T6G 2E3, Canada

6.1 Overview

It should come as no surprise that iron, the fourth most abundant element in the Earth's crust (Taylor and McLennan, 1985), is essential in biology. Yet, in today's oceans, iron is a vanishingly rare element (Fig. 6.1). Its concentration – typically <1 nM (Johnson et al., 1997; Boye et al., 2001; Cullen et al., 2006) – is so low that iron scarcity limits biological productivity across large areas of the Earth's surface (Martin and Fitzwater, 1988). This peculiar situation is a consequence of the chemical behaviour of iron on an oxygenated Earth. In the presence of abundant O_2, the element is found primarily in the Fe(III) oxidation state, which forms poorly soluble oxyhydroxides. Why, then, is iron required by biology? Most likely, this is a legacy of early evolution when iron was ubiquitous on land and in the sea. It also helped that iron binds strongly to a variety of anionic ligands (involving oxygen, nitrogen, and sulfur) and could readily have been incorporated into biological compounds such as enzymes.

The story of iron geobiology is therefore a story in which the evolution of one geochemical cycle – that of oxygen – wreaked havoc with another – iron – that is essential to life's distribution on Earth. As O_2 levels rose, ocean iron abundances fell (Fig. 6.2). The acquisition of iron became more difficult. Microbes that depended on iron redox cycling for their metabolisms were driven from the ocean and land surface, which they once ruled, to the ocean depths. Eventually, they were confined to eking out a grubby living in ocean and lake sediments, where they dominate the biogeochemical cycling of iron to this day.

To understand this story and its implications, it is necessary to review the inorganic geochemistry of iron, its uses in biology, and the ways in which geochemistry and biology intersect in modern environments. These topics are addressed in Sections 6.2 and 6.3 below. In Section 6.4, we return to the evolutionary tale.

6.2 The inorganic geochemistry of iron: redox and reservoirs

Iron is the final product of nuclear fusion in stars because of its high binding energy per nucleon. It is therefore a relatively abundant element in the cosmos, and a major constituent of rocky planets. During early planetary differentiation, the high density of metallic iron relative to silicates caused iron to sink to the interior, so that it became the dominant constituent of the Earth's core. This is the Earth's major iron reservoir (Fig. 6.3). As a result of this partitioning, as well as later differentiation during partial melting of the mantle, the abundance of iron in the Earth's continental crust (~7 wt%; McLennan, 2001) is significantly less than in undifferentiated meteorites (~18 wt% in CI chondrites; Palme and Jones, 2003).

The redox state of the Earth's crust is such that iron in igneous crustal rocks occurs primarily in the Fe(II) oxidation state ('ferrous' iron), and also in the Fe(III) oxidation state ('ferric' iron), rather than as iron metal. Iron is present in igneous rocks in a wide range of minerals. It is a major constituent of common minerals such as the olivine mineral fayalite (Fe_2SiO_4), magnetite (Fe_3O_4) and pyrite (FeS_2), but is also found in a host of mineral classes such as pyroxenes, amphiboles and phyllosilicates.

Fundamentals of Geobiology, First Edition. Edited by Andrew H. Knoll, Donald E. Canfield and Kurt O. Konhauser.
© 2012 Blackwell Publishing Ltd. Published 2012 by Blackwell Publishing Ltd.

Fundamentals of Geobiology

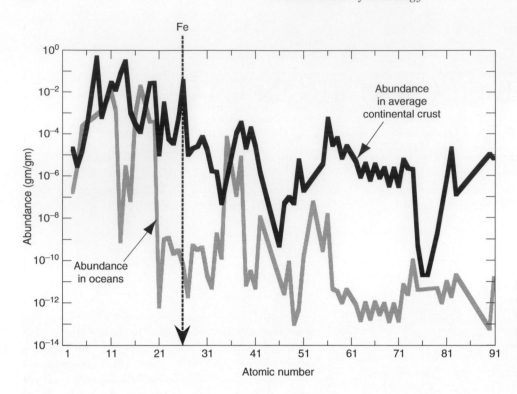

Figure 6.1 Abundance of the elements in the continental crust and oceans. Despite being one of the most abundant elements in the continental crust, iron has a very low concentration in the ocean.

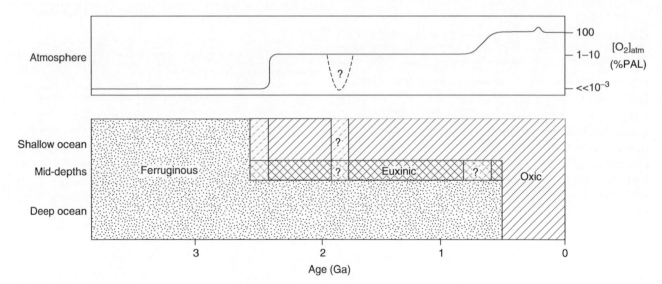

Figure 6.2 Redox conditions in the atmosphere and oceans over time. During the Archean when the atmosphere was essentially devoid of O_2, the oceans were predominantly anoxic and Fe(II)-rich (ferruginous) and sulfate-poor. Near the close of the Archean eon, the evolution of oxygenic photosynthesis led to limited O_2 accumulation along ocean margins. The first major increase in atmospheric O_2 at ~2.4 Ga was accompanied by widespread surface ocean oxygenation. Increased sulfate inputs to the oceans (from the oxidative weathering of crustal sulfide minerals) boosted rates of microbial hydrogen sulfide production along productive ocean margins (sites of high organic carbon export), leading to local expressions of anoxic and sulfidic (euxinic) conditions at mid-depths. The deep oceans remained ferruginous. This ocean redox structure likely held sway for the next ~2 Gyr, albeit with some spatiotemporal variation. Significant changes during this time include a possible decline in the atmosphere and ocean O_2 content at ~1.9 Ga, an expansion of euxinic and possibly low-O_2 conditions at 1.85 Ga that terminated the deposition of large iron formations, and a return to predominantly ferruginous oceans during the widespread, low-latitude Neoproterozoic glaciations. The onset of substantial deep-ocean oxygenation may have been delayed until the Ediacaran Period following a second major increase in atmospheric O_2 levels. Predominantly Oxygenated, iron-scarce oceans may not have been fully established until the Paleozoic. Modified from Lyons *et al.* (2009b).

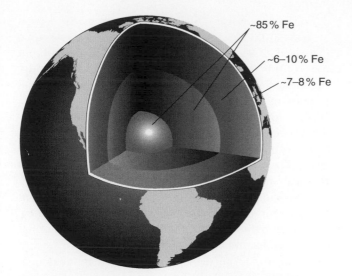

Figure 6.3 Iron distribution in the Earth. Most of the iron resides in the Earth's core as a result of early planetary differentiation that caused the dense iron metal to sink to the interior. Ocean crust (~8 wt%) is slightly enriched in iron relative to continental crust (~7 wt%). Classical estimates for the average mantle iron content were ~6 wt%. However, recent estimates suggest values closer to ~10 wt% for the lower mantle. Sources of data: crust – Hofmann (1988) and McLennan (2001); mantle – Palme and O'Neill (2003), Khan *et al.* (2008); Verhoeven *et al.* (2009) and Javoy *et al.* (2010); core – McDonough (2003) and Javoy *et al.* (2010).

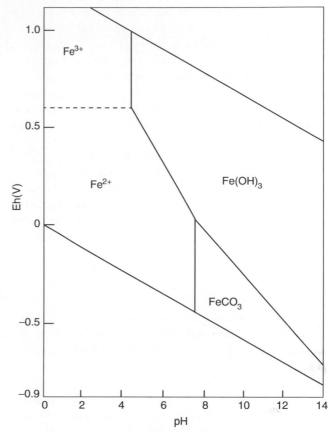

Figure 6.4 Eh-pH diagram for the Fe-O_2-H_2O-CO_2 system at 25 °C. The boundary between aqueous and solid phases is based on a dissolved iron concentration of 10^{-5} mol kg^{-1}. Modified from Langmuir (1997).

From a geobiological perspective, the iron cycle becomes interesting when chemical and biological weathering breaks down these iron-bearing minerals, releasing the element into aqueous solution. The transport and distribution of iron then depend strongly on pH, Eh (redox) and the presence or absence of other dissolved constituents that coordinate with Fe(II) or Fe(III) to form dissolved complexes, colloids or poorly soluble mineral phases. Biology strongly affects these parameters just as the availability of iron (or its absence) affects biological activity. These interactions give rise to complex and dynamic biogeochemical cycling of iron (Boyd and Ellwood, 2010; Konhauser *et al.*, 2011a; Radic *et al.*, 2011; Raiswell, 2011), as will be discussed in later sections.

The behaviour of iron is most strongly shaped by redox conditions because of the different chemical bonding affinities of Fe(II) and Fe(III). A chemical bond of particular importance to the transport and distribution of iron is that between Fe(III) and OH$^-$ (the 'hydroxyl' ion). The equilibrium constant for the formation of FeOH^{2+} is large ($K \approx 10^2$; Millero and Hawke, 1992; Millero *et al.*, 1995; Stefánsson, 2007). The addition of further OH$^-$ groups to Fe(III) is even more strongly favoured, leading to the formation of a host of neutral hydrolysis species such as FeOOH and Fe(OH)$_3$, and other hydrated Fe(III) oxyhydroxides such as disor-

dered ferrihydrite (Fe$_{8.2}$O$_{8.5}$[OH]$_{7.4}$ + 3H$_2$O), which ages to ferrihydrite (5Fe$_2$O$_3$•9H$_2$O) (Michel *et al.*, 2010). These species are only sparingly soluble, causing Fe to be removed from solution. Hence, the amount of iron dissolved in groundwaters, rivers and seawater decreases sharply with increasing Eh and pH (Fig. 6.4).

Iron can also be effectively removed from solution under the opposite condition of low Eh if there is an abundance of dissolved sulfide (HS$^-$ and H$_2$S). Under these conditions, Fe(II) reacts readily to form mono- and disulfide species, leading ultimately to the production of insoluble iron sulfide (pyrite) via a series of intermediate iron–sulfur species (Luther, 1991; Schoonen and Barnes, 1991). This is a major pathway for the immobilization of iron in anoxic environments, such as in sulfide-rich lakes and ocean basins, and the pore fluids of marine sediments and soils (Berner, 1970, 1984). This chemistry couples the biogeochemistry of iron to that of sulfur. When Fe(II) is in excess of HS$^-$, iron can be removed from solution in the presence of bicarbonate and phosphate, leading to the formation of siderite and vivianite, respectively (Krom and Berner, 1980; Coleman, 1985).

A major consequence of these chemical characteristics is that iron is typically scarce in modern natural waters, at least in comparison to its abundance in rocks. During oxidative weathering, Fe(III) is produced and immediately immobilized, generating iron oxide residues and leaving average river water with a typical dissolved iron concentration of only ~40 nM (higher concentrations can be found in rivers draining peatlands); a large fraction of the iron load in rivers is colloidal or suspended (Dai and Martin, 1995; Krachler *et al.*, 2005, 2010). Colloidal iron is largely removed from solution during mixing of freshwaters with seawater in estuaries. There, the high ionic strength of seawater neutralizes surface charges on colloidal particles, allowing them to coagulate and precipitate (Gustafsson *et al.*, 2000; Krachler *et al.*, 2010). Far from shore, then, other sources may be more important (Fig. 6.5). Most notable are dust particles, which are believed to be the primary source of iron to the open oceans (Jickells *et al.*, 2005). In glaciated regions, the flux of bioavailable iron supplied by melting glaciers and icebergs can be similar to the aeolian flux, as shown for the Southern Ocean (Lannuzel *et al.*, 2008; Raiswell *et al.*, 2008). However, as on land, chemical decomposition of these particles rapidly results in the production of ferric oxyhydroxide particulates. A major challenge for marine ecosystems, then, is to acquire this iron before particulates settle to the seafloor. In general, the transport of iron from the continents to the open ocean is thought to be dominated by the formation of nanoparticulate oxyhydroxides. Stabilization ('ageing') of these nanoparticles is suggested to permit long-distance transport to sites in the open ocean where they can be converted ('rejuvenated') to more bioavailable forms (Raiswell, 2011).

Iron derived from high-temperature hydrothermal systems in the deep sea may constitute another source (Chu *et al.*, 2006; Bennett *et al.*, 2008). It was thought that most of this iron is removed either as pyrite in sulfide-rich vent fluids or as ferrihydrite when hydrothermal plumes mix with oxygenated seawater (Lilley *et al.*, 2004). However, recent spectromicroscopic measurements of carbon and iron in particles from hydrothermal plumes on the East Pacific Rise suggests that some Fe(II) is stabilized by organic complexation, preventing its removal into insoluble minerals. Such particles may then provide a source of bioavailable iron to environments outside of the mid-ocean ridge (Toner *et al.*, 2009). In oxygen-deficient Precambrian oceans, hydrothermal sources were probably an important source of iron to seawater, and enabled the formation of massive deposits of chemical sediment known as iron formation (see Chapter 8).

An additional, but poorly quantified, source of ferrous iron is dissolved in sedimentary pore fluids that are anoxic, or nearly so, but not sulfidic. Such pore fluids occur on continental margins, where high biological productivity in overlying seawater yields a high flux of organic carbon to the sediments on continental shelves. This influx of reduced carbon generates anoxia, allows microbial Fe(III) reduction to take place, and hence results in the presence of dissolved Fe(II), the concentration of which may approach ppm levels in pore fluids. Benthic iron fluxes from river-dominated continental margins could potentially be orders of magnitude greater than non-river dominated shelves. Some of this iron may escape to seawater and hence provide a critical source of iron to near-shore ecosystems (Severmann *et al.*, 2010). This process is also known to occur on the

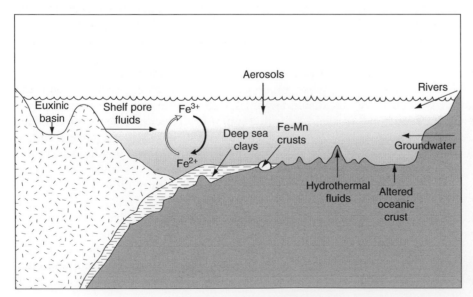

Figure 6.5 Schematic diagram illustrating the sources and sinks of iron in modern seawater. Dissolved and particulate iron is supplied to seawater by rivers, groundwaters, aeolian dust, hydrothermal vents and alteration of oceanic crust, and anoxic pore waters along continental margins. In glaciated regions, iron would also be supplied by melting glaciers and icebergs. In oxygenated seawater, iron is oxidized and precipitated as poorly soluble iron oxyhydroxides that are removed into sediments. Iron is also removed as insoluble iron sulfide minerals in restricted euxinic basins such as the Black Sea. Modified from Anbar and Rouxel (2007).

shelves of basins containing anoxic and sulfidic deep waters, such as the Black Sea. Systematic studies of sedimentary iron speciation and iron isotopes in such basins demonstrate that pore fluid-derived iron migrates along the chemocline to the deep basins, where it is immobilized as pyrite (Lyons and Severmann, 2006; Severmann *et al.*, 2008; Fehr *et al.*, 2010).

Unsurprisingly in view of these considerations, iron has a very short residence time of only up to a few hundred years in the modern oceans (Johnson *et al.*, 1997). As a result, iron is ubiquitous in marine sediments. Outside of sulfidic basins, it is delivered to the seafloor as ferric oxyhydroxide precipitates. However, as these precipitates are buried, bacterial sulfate reduction generates H_2S or HS^- (depending on pore fluid pH) in sedimentary pore fluids, converting a large fraction of these ferric oxides to ferrous sulfides.

6.3 Iron in modern biology and biogeochemical cycles

6.3.1 Fe as a micronutrient

Iron is essential to most organisms. It is generally found in the centre of metalloproteins that mediate redox reactions. Some of these proteins serve as enzymes that facilitate the transfer of electrons used to generate chemical energy for the cell. Some major Fe-containing enzymes include hydrogenases, iron–sulfur proteins and cytochromes. Hydrogenases are proteins that catalyse the reduction of a substrate by adding H_2, which can be obtained from either various intracellular respiratory processes, or extracellularly from aqueous solution. Iron–sulphur proteins are electron carriers that range from simple molecules containing one Fe–S centre to complexes containing multiple types of Fe–S clusters. Fe_2S_2 (ferredoxin) and Fe_4S_4 are the most common. Each Fe–S centre has at least two redox states, a reduced ferrous form and an oxidized ferric form, and each centre carries only one electron at a time. Cytochromes are proteins that have an iron-containing porphyrin ring (known as heme) that is capable of alternating between Fe(II) and Fe(III). There are a number of different cytochromes based on differences in the side groups of the porphyrin ring (heme *a*, *O*, *b*, *c* and *d*), each with a different electrode potential, and hence each occurs in a different location in the electron transport chain. Some serve as the terminal reductases in metabolic pathways, passing the electrons onto a terminal electron acceptor, whereas other cytochromes specifically facilitate the transfer of electrons from the external environment into the transport chain (i.e. those that oxidize Fe(II), H_2, H_2S, and $S_2O_3^{2-}$).

Other Fe-containing enzymes (nitrogenase) are found in organisms that fix atmospheric nitrogen gas (N_2).

Nitrogenase catalyses the breaking of the triple bonds between each nitrogen atom, and then bonds the nitrogen to hydrogen atoms via the reaction: $N_2 + 3H_2 \rightarrow 2NH_3$. All nitrogenases have an iron- and sulfur-containing cofactor that facilitates the electron transfers. Due to the oxidative properties of O_2 on the Fe–S cofactors, most nitrogenases are irreversibly inhibited by the presence of O_2. Thus, nitrogen fixing organisms utilize mechanisms to exclude O_2 – a particular challenge for cyanobacteria that produce O_2 via photosynthesis. Some cyanobacteria cope by expressing specialized nonphotosynthetic cells within their filaments (called *heterocysts*) that serve as O_2-free microenvironments for nitrogen fixation (Fay *et al.*, 1968). Other cyanobacteria photosynthesize strictly during daylight and fix nitrogen at night (Bebout *et al.*, 1993).

6.3.2 Fe in redox reactions

6.3.2.1 Aerobic Fe(II) oxidation

The occurrence of bacteria that gain energy from the oxidation of Fe(II) to Fe(III) is generally limited by the availability of dissolved Fe^{2+}. This is not an insignificant problem because at neutral pH and under fully aerated conditions, Fe(II) rapidly oxidizes chemically to Fe(III), which is then hydrolysed to ferrihydrite. The kinetic relationship that describes chemical Fe(II) oxidation at circumneutral pH values is:

$$\frac{d[Fe(II)]}{dt} = k[Fe(II)][OH^-]pO_2$$

where $k = 8(\pm 2.5) \times 10^{13}$ min^{-1} atm^{-1} mol^{-2} l^{-2} at 25 °C (Singer and Stumm, 1970). As is evident from the equation, pH and oxygen availability have strong influences on the reaction rate, which explains why at low pH or low oxygen concentrations, dissolved Fe^{2+} is quite stable (e.g. Liang *et al.*, 1993). Accordingly, the most efficient way for a microorganism to survive on Fe(II) is to either grow under acidic conditions (as an acidophile) or under low-O_2 conditions at circumneutral pH (as a microaerophile) because in both cases, the chemical reaction kinetics are sufficiently diminished that microorganisms can harness Fe(II) oxidation for growth.

There are a number of acidophilic Fe(II)-oxidizing bacteria that grow autotrophically on Fe(II), using O_2 as its terminal electron acceptor (Blake and Johnson, 2000):

$$2Fe^{2+} + 0.5O_2 + 2H^+ \rightarrow 2Fe^{3+} + H_2O$$

The best characterized acidophiles are *Acidothiobacillus ferrooxidans* and *Leptospirillum ferrooxidans*. They grow well at mine waste disposal sites where reduced sources of iron are continuously regenerated during acid mine drainage. Another iron-oxidizing bacterium is *Sulfolobus acidocaldarius* that lives in hot, acidic springs at temperatures near boiling. All of the Fe(II)-oxidizing bacteria use ferrous iron for both the generation of energy (in the form of ATP [adenosine triphosphate]) and reducing power to convert CO_2 into organic carbon. Since it takes on average 50 mol of Fe^{2+} to assimilate 1 mol of carbon (Silverman and Lundgren, 1959), cells such as *A. ferrooxidans* must oxidize a large amount of ferrous iron in order to grow. Consequently, even a small number of bacteria can be responsible for generating significant concentrations of Fe^{3+}.

Under neutral pH, but with O_2 levels below 1.0 mg l⁻¹ and redox conditions about 200–300 mV lower than typical surface waters (characteristics of some iron-rich springs, stratified bodies of water and hydrothermal vent systems), microaerophilic bacteria, such as *Gallionella ferruginea*, play an important role in Fe(II) oxidation. *Gallionella*-type oxidizers are bean-shaped cells that grow at the terminus of a helical structure called a stalk which is composed largely of polysaccharides frequently encrusted by ferrihydrite (Hanert, 1992). Unlike the acidophiles, the neutrophiles can harness much more energy because at pH 7, the electrode potential of the couple $Fe(OH)_3/Fe^{2+}$ ($E^{0'} \approx 0$ V; Thamdrup, 2000) is substantially lower than the redox couple of O_2/H_2O ($E^{0'} = 0.81$ V). This indicates that Fe(II) oxidation can generate significant energy at circumneutral pH to support ATP production. Although *G. ferruginea* grows chemotrophically at a pH just below 7 on a medium with Fe(II) salts and fixes all of its carbon autotrophically from CO_2 (Hallbeck and Pedersen, 1991), there is at present no conclusive evidence that they actually derive energy from Fe(II) oxidation. Interestingly, *G. ferruginea* does not form a stalk at a pH < 6 or under very micro-oxic conditions, where O_2 is present but the redox potential is –40 mV (Hallbeck and

Pederson, 1990). This suggests that the stalk represents an organic surface upon which ferrihydrite can precipitate and, in doing so, protect the cell itself from becoming mineralized. In a similar manner, it has been suggested that another bacterium, *Leptothrix ochracea*, induces ferrihydrite precipitation on its sheath as a means to detoxify the presence of any free oxygen in their environment (Nealson, 1982). These examples certainly imply that Fe(II) oxidation need not be directly tied to energy production.

6.3.2.2 Anaerobic chemolithoautotrophic Fe(II) oxidation

Ferrous iron has also been observed to undergo microbial oxidation under anoxic conditions thus closing the iron redox cycle even in O_2-free environmental systems (Fig. 6.6). In anoxic environments, Fe(II) is relatively stable since neither nitrate nor sulfate react chemically with Fe(II) at appreciable rates at low temperature. Only Mn(IV) and high concentrations of nitrite have been shown to be relevant abiotic chemical oxidants for Fe(II) (Buresh and Moraghan 1976; Rakshit *et al.* 2008). Biological oxidation of Fe^{2+} in the absence of oxygen can occur via photoferrotrophy (discussed below) and chemoautotrophy. During the latter process, oxidation of Fe^{2+} occurs in the absence of light with nitrate as the electron acceptor according to the following equation (Straub *et al.*, 1996):

$$10Fe^{2+} + 2NO_3^- + 24H_2O \rightarrow 10\ Fe(OH)_3 + N_2 + 18H^+$$

Nitrate-dependent Fe(II) oxidation has been shown to be widespread in sediments (Straub and Buchholz-Cleven, 1998). Most of the described nitrate-reducing, Fe(II)-oxidizing strains depend on an organic

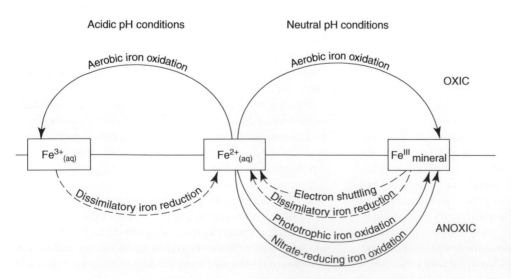

Figure 6.6 Schematic diagram summarizing the microbial iron redox reactions under conditions of acidic and neutral pH.

co-substrate (e.g. acetate, Kappler *et al.*, 2005a) and truly lithoautotrophic strains have not been isolated in pure culture. Weber *et al.* (2006) isolated an Fe(II)-oxidizing bacterium that was suggested to be able to oxidize Fe(II) autotrophically; however, this strain could not be transferred continuously in lithoautotrophic culture. Similarly, some strains of nitrate-dependent bacteria oxidizing Fe(II) in the absence of an organic co-substrate were isolated from the deep oceans (Edwards *et al.*, 2003), but it is unclear whether these strains can be cultivated for successive generations with Fe(II) as the sole electron donor. So far, the chemolithoautotrophic enrichment culture described by Straub *et al.* (1996) is the only culture oxidizing Fe(II) with nitrate autotrophically (without the addition of any organic substrate). From gene analysis it is known that this culture consists of four organisms, including three chemoheterotrophic nitrate-reducing bacteria (*Parvibaculum lavamentivorans, Rhodanobacter thiooxidans* and *Comamonas badia*), plus a fourth organism related to the chemolithoautotrophic Fe(II)-oxidizing bacterium *Sideroxydans lithotrophicus* (Blothe and Roden, 2009). The complexity of this culture potentially suggests that a consortium of organisms is needed for autotrophic Fe(II) oxidation coupled to nitrate-reduction.

In contrast to the microaerophilic strains discussed above, at least one mixotrophic nitrate-reducing Fe(II)-oxidizing strain (*Acidovorax* sp. BoFeN1) was shown to benefit directly from Fe(II) oxidation. Higher cell numbers were observed when oxidizing Fe^{2+} and the organic co-substrate compared to oxidation of the organic compounds alone (Muehe *et al.*, 2009). During Fe(II) oxidation, these organisms precipitate goethite, lepidocrocite, ferrihydrite or poorly crystalline Fe(III) phosphates depending on the geochemical conditions (Miot *et al.*, 2009; Larese-Casanova *et al.*, 2010). They precipitate iron minerals distant to the cells but also form mineral crusts at the cell surface and in the space between the outer and inner cell membranes, known as the periplasm (Miot *et al.*, 2009). These electron microscopical and synchrotron-based spectro-microscopical studies, in combination with iron isotope analysis (Kappler *et al.*, 2010), suggest that Fe(II) oxidation takes place at least to some extent in the periplasm.

6.3.2.3 Photosynthetic Fe(II) oxidation

The existence of anoxygenic photosynthetic Fe(II) oxidation (photoferrotrophy) was suggested nearly 20 years before the discovery of the first microorganisms catalysing this reaction. Garrels *et al.* (1973) and Hartmann (1984) suggested photoferrotrophy as a deposition mechanism for iron formation under O_2-free conditions in the Precambrian. Light instead of O_2 could have facilitated Fe(II) oxidation, via photosynthesis that used Fe(II) rather than H_2O as an electron donor, producing Fe(III) instead of O_2. Two decades later this hypothesis was validated by the discovery of the first photoferrotrophic microorganisms (Widdel *et al.*, 1993). Currently, a variety of phylogenetically diverse strains of anoxygenic Fe(II)-oxidizing phototrophs including purple sulfur, purple non-sulfur, and green sulfur bacteria are known to catalyse oxidation of Fe(II) to Fe(III) according to the following reaction (Hegler *et al.*, 2008):

$$4\,Fe^{2+} + HCO_3^- + 10H_2O + h\nu \rightarrow 4\,Fe(OH)_3 + (CH_2O) + 7H^+$$

where $h\nu$ is a quantum of light.

Both the aerobic and anaerobic Fe(II)-oxidizing bacteria face the problem of limited availability of dissolved Fe(II) and the possible inhibitory effect of the very poor solubility of the ferric oxyhydroxide end products of their metabolism. The formed particles are positively charged due to their high points of net zero charge (ZPC); e.g. pH ≈ 8 for ferrihydrite (Posth *et al.*, 2010). Therefore, in the proximity of cells, Fe(III) cations, colloids and particles are expected to adsorb to cell surfaces that are in general negatively charged due to a high content of carboxylic, phosphoryl and/or hydroxyl groups. On the one hand, Fe(III) encrustation can lead to the accumulation of trace metals and nutrients that naturally adsorb onto such particles. Other advantages include protection from dehydration, while the protons released during Fe(III) mineral precipitation near the cell surface could also increase the energy yield of iron oxidation by increasing the pH-gradient utilized in the proton motive force (Chan *et al.*, 2004). The downside of iron encrustation comes from the potential reduction in the diffusion and uptake of substrates and nutrients to the cell, leading to the stagnation of cell metabolism and eventually to cell death. In terms of light availability, iron encrustation can have opposing effects; it may limit absorption of key wavelengths of light, but the minerals may serve as a UV-shield to protect against damage by high radiation (Phoenix *et al.*, 2001).

For the stalk- or sheath-forming aerobic Fe(II)-oxidizing bacterial genera *Gallionella* and *Leptothrix*, it was suggested that the microbially produced and excreted organic matrices are used for extracellular capture of the Fe(III) minerals produced. However, for these strains it remains unanswered where the Fe(II) is oxidized (i.e. in the cytoplasm, in the periplasm, or at the cell surface). In terms of the photoferrotrophs, it has recently been proposed that Fe(II) oxidation happens in the periplasm of the cells (Croal *et al.*, 2004; Jiao and Newman, 2007). Since cell encrustation is not observed this further raises the question of how Fe(III) is transported out of the cytoplasm to the cell exterior and then away from the cells without sorbing to the organic ligands. An acidic pH microenvironment, the use of organic ligands to keep the Fe(III) in solution in close proximity to the cell, or the shedding off of organic-mineral aggregates from the cell surface have all been suggested as plausible methods used by the bacteria (Kappler and Newman 2004; Schädler *et al.*, 2009; Hegler *et al.*, 2010).

6.3.2.4 Anaerobic Fe(III) reduction

In addition to primary producers, there are a variety of other species that heterotrophically break down existing organic carbon to either carbon dioxide or methane gas. The type of heterotrophic metabolism that occurs in nature depends on what oxidants are available and, in the situation when multiple electron acceptors are present (as in the uppermost sediment layers), on the free energy yield of the specific reaction. Thus, the decomposition of freshly deposited organic material in sediments proceeds in a continuous sequence of redox reactions, with the most electropositive oxidants (such as O_2 and $NO_3^-)^-$ consumed at or near the surface, and progressively poorer oxidants (Mn(IV), Fe(III), SO_4^{2-}, and CO_2) consumed at depth until the labile organic fraction is exhausted and the deeper sediments are left with a composition very different from the sediments originally deposited (Froehlich *et al.*, 1979).

Below the zone of dissimilatory Mn(IV) reduction, and at the depth of complete nitrate removal from pore waters, is where Fe(III) reduction takes place. Dissimilatory iron reduction is broadly distributed amongst several known bacterial genera. *G. metallireducens* and *S. putrefaciens* were among the first bacteria studied in pure culture that could gain energy from coupling Fe(III) reduction to the oxidation of H_2 and/or simple fermentation products, including short- and long-chain fatty acids, alcohols and various monoaromatic compounds:

$$CH_3COO^- + 8Fe(OH)_3 \rightarrow 8Fe^{2+} + 2HCO_3^- + 15OH^- + 5H_2O$$

Since then, many more species, including a number of hyperthermophilic Archaea, some sulfate- and nitrate-reducers, and methanogens have shown the capacity for reducing ferric iron minerals (Lovley *et al.*, 2004).

Ferric iron minerals occur in soils and sediment in a wide variety of forms, ranging from amorphous to crystalline phases. The amorphous to poorly-ordered iron oxyhydroxides, such as ferrihydrite or goethite, are the preferred sources of solid-phase ferric iron for Fe(III)-reducing bacteria (Lovely and Phillips, 1987). More crystalline Fe(III) oxides (e.g. hematite and magnetite) and Fe-rich clays (e.g. smectite) are also microbially reducible, and some experimental observations suggest that these minerals may provide

energy for cellular growth comparable to that derived from the poorly crystalline phases (e.g. Roden and Zachara, 1996; Kostka *et al.*, 2002), although this may only be the case for optimal experimental conditions. Variations in Fe(III) reduction rates are related to a number of factors, including the amount of surface area exposure, crystal morphology, particle aggregation, composition of the aqueous solution in which the microorganisms grow, and the amount of Fe^{2+} sorbed to the oxide surface (e.g. Urrutia *et al.*, 1998). Importantly, with such great heterogeneity in reactivity towards microbial reduction, it is not surprising that Fe(III) can represent a long-term electron acceptor for organic matter oxidation, even at sediment depths where other anaerobic respiratory processes are thermodynamically predicted to dominate (Roden, 2003).

Until recently, it was believed that the reduction of ferric iron-containing minerals (dissolved Fe^{3+} dominates at pH < 4) necessitated direct contact of the microorganism with the mineral surface. Once in contact with the surface, the Fe(III)-reducers are faced with the problem of how to effectively access an electron acceptor that cannot diffuse into the cell. These criteria thus require that Fe(III)-reducing bacteria must not only be able to actively recognize an iron mineral surface and attach to it, but that they must also be capable of activating or producing proteins that specifically interact with that mineral surface. Work with species such as *G. metallireducens* has shown that they are chemotactic and are able to sense the gradient of reduced metal ions emanating from the dissolution of the oxyhydroxide phases under anoxic conditions. After moving towards the solid phases, the bacteria specifically express flagella and pili that help them adhere to the Fe(III) oxyhydroxides (Childers *et al.*, 2002). Alternatively, *Shewanella algae* relies on the production of hydrophobic surface proteins that facilitate greater cell adhesion (Caccavo *et al.*, 1997). Once the bacteria attach to the mineral surface they begin shuttling electrons from a reduced source within the cytoplasm, across the plasma membrane and periplasmic space, to the outer membrane. Located there are iron reductase enzymes that transfer those electrons directly to the Fe(III) mineral surface, causing a weakness in the Fe–O bond and invariably its reductive dissolution (Lower *et al.*, 2001). Electron transfer via direct contact between cells and the mineral surface has been suggested to be mediated also via conductive pili (Reguera *et al.*, 2005; Gorby *et al.*, 2006; El-Naggar *et al.*, 2010).

More recently, it was discovered that other Fe(III)-reducing bacteria, such as *Shewanella* species, overcome the solubility problem by utilizing dissolved or even non-dissolved organic compounds as electron shuttles between the cell surface and the Fe(III) oxides, which may be located at some distance away from the cell. One example is the quinone moieties in exogenous humic compounds which bacteria can reduce to the semi-quinone and hydroquinone oxidation state via the oxidation of acetate or lactate (Lovley *et al.*, 1996; Jiang and Kappler, 2008; Roden *et al.*, 2010). The reduced humics subsequently transfer electrons abiotically to Fe(III), producing Fe^{2+}, and in doing so, regenerate the oxidized form of the humic compound for another cycle. The extent of electron transfer from reduced humics to Fe(III) was shown to depend on the redox potential of the Fe(III) species and thus on the type of Fe(III) minerals present (Bauer and Kappler, 2009). Some *Shewanella* species (*S. oneidensis*) have also been shown to produce and excrete their own quinone compounds that function in a similar manner to natural humics (Marsili *et al.* 2008; von Canstein *et al.*, 2008), while the closely related *S. algae* produces soluble melanin which might serve as another type of electron conduit for Fe(III) oxyhydroxide reduction (Turick *et al.*, 2002). Significantly, Fe(III) reduction rates are faster in the presence of organic electron shuttles than in their absence because they are likely to be more accessible for microbial reduction than poorly soluble Fe(III) oxyhydroxides (Nevin and Lovley, 2000).

6.3.3   Fe acquisition by siderophores

The poor solubility of Fe(III) at circumneutral pH and its correspondingly low concentrations in solution (~10^{-10} mol l^{-1}) means that it is often the limiting nutrient for growth. Many bacteria and fungi get around this impasse by excreting low molecular weight, Fe(III)-specific ligands known as siderophores (Neilands, 1989). In soils and seawater, siderophores or their breakdown products can be so abundant that they dominate ferric iron (e.g. Wilhelm and Trick, 1994). Siderophores have several properties that make them ideal Fe(III) chelators. They contain a general preponderance of oxygen atoms, are soluble, and provide bi- and multidente ligands that can form multiple coordinative positions around the central Fe^{3+} cation. Significantly, they form especially strong 1:1 surface complexes, and their association constants for Fe(III) can be several times higher than common soil organic acids (e.g. oxalic acid). This is an important property because it maintains dissolved iron in a soluble form that minimizes its loss from the aqueous environment by the precipitation of solid-phase ferric hydroxide (Hider, 1984).

The biosynthesis of siderophores is tightly controlled by iron levels. When soluble Fe^{3+} concentrations are low, siderophore production becomes activated by the presence of ferric iron-containing minerals, with higher levels of siderophores produced in response to increasingly insoluble iron sources (e.g. Hersman *et al.*, 2000). Other studies have documented that some species generate different types and amounts of siderophores depending on the type of iron mineral present.

Importantly, it appears that different siderophores are required to sequester Fe(III) from different iron minerals, and that changing the iron mineralogy can elicit a specific response from the same microorganism. In any event siderophores represent an extremely successful solution to the problem of obtaining dissolved iron from stable iron solid phases.

On a much larger scale, recent iron enrichment experiments in the equatorial Pacific have demonstrated that with the addition of soluble iron, a threefold increase in the concentration of Fe-binding organic ligands occurred, leading to a concomitant increase in microbial biomass production (Hutchins and Bruland, 1998). Interestingly, many species produce siderophores in great excess of their requirements (because many are lost via diffusion and advection), yet when levels of iron become sufficiently high (i.e. an order of magnitude above micromolar levels), their production is repressed and the cells meet their iron needs via low-affinity iron uptake systems (Page, 1993).

6.4   Iron through time

6.4.1   Evidence for major changes in biogeochemical cycles

6.4.4.1   Introduction

Iron is a redox-sensitive element. It forms poorly soluble iron oxyhydroxides in the presence of O_2, poorly soluble iron sulfides in the presence of dissolved sulfide, but soluble Fe(II) complexes in anoxic and sulfide-free environments. Hence, the biogeochemistry of iron, along with that of carbon and sulfur, is linked to the history of Earth surface oxygenation (Canfield, 2005; Lyons *et al.*, 2009a; Poulton and Canfield, 2011). Ideally, the evolution of the iron biogeochemical cycle can be reconstructed from changes in the iron concentration of the oceans over geological time. Unfortunately, the rock record does not directly preserve samples of ancient seawater. Instead, we must infer the major changes in iron biogeochemical cycling through the petrological and geochemical characteristics of ancient sedimentary rocks such as iron formations (e.g. Konhauser *et al.*, 2009; Bekker *et al.*, 2010; Planavsky *et al.*, 2010a) and fine-grained mudrocks (e.g. Canfield *et al.*, 2008; Scott *et al.*, 2008; Poulton *et al.*, 2010; Planavsky *et al.*, 2011). Even then, an additional complication arises from the fact that we cannot sample most of the sedimentary rocks deposited from pre-Jurassic open ocean seafloor because this material has since been recycled into the mantle by subduction at convergent plate margins. We have little choice but to rely upon the fragmentary sedimentary rock record, preserved along continental margins and in intracratonic basins, to draw our inferences on the evolution of the iron biogeochemical cycle.

In addition to episodes of iron formation deposition, a sizable fraction of our evidence for the spatiotemporal distribution of Fe(II) in the oceans comes from sedimentary iron speciation analyses of fine-grained sedimentary rocks. Briefly, as reviewed in Lyons and Severmann (2006), three basic parameters are employed: (1) the ratio of total iron to aluminum (Fe_T/Al), (2) the ratio of highly reactive iron to total iron (Fe_{HR}/Fe_T), and (3) the ratio of pyrite iron to highly reactive iron (Fe_{PY}/Fe_{HR}) (an older related term, the degree of pyritization [DOP], is a more conservative estimate of the degree to which Fe_{HR} has been converted to pyrite). As its name implies, Fe_{HR} comprises biogeochemically reactive iron, specifically pyrite plus other iron minerals (e.g. ferric oxides, magnetite, and iron-rich carbonates) that can react with sulfide in the water column or in sediments during early diagenesis (Poulton *et al.*, 2004). Modern sediments deposited from locally anoxic bottom waters have ratios of Fe_{HR}/Fe_T that are higher (typically >0.38) compared to modern sediments deposited from oxygenated bottom waters (average = 0.26 ± 0.08; Raiswell and Canfield, 1998) and Phanerozoic oxic sediments (average = 0.14 ± 0.08; Poulton and Raiswell, 2002). Elevated ratios of Fe_{HR}/Fe_T (and Fe_T/Al) reflect Fe_{HR} transport, scavenging, and enrichment in sediments relative to background siliciclastic sources either because of pyrite formation in an anoxic and sulfidic (euxinic) water column or because of iron-rich mineral formation

in an anoxic and Fe(II)-rich (ferruginous) water column (Poulton *et al.*, 2004; Lyons and Severmann, 2006). The extent to which Fe_{HR} has been converted to pyrite is then used to determine whether the local water column was euxinic ($Fe_{PY}/Fe_{HR} > 0.8$) or ferruginous ($Fe_{PY}/Fe_{HR} < 0.8$) (Poulton *et al.*, 2004).

6.4.4.2 Iron reigns supreme: the Archean oceans

Several lines of evidence point to an anoxic Archean atmosphere, including the preservation of mass-independent fractionation (MIF) of sulfur isotopes in sedimentary rocks ($pO_2 < 0.001\%$ of present atmospheric levels [PAL]; Farquhar *et al.*, 2000, 2007; Pavlov and Kasting, 2002; Farquhar and Wing, 2003), the presence of detrital uraninite, pyrite and siderite in fluvial deposits (Fleet, 1998; Rasmussen and Buick, 1999; England *et al.*, 2002; Hofmann *et al.*, 2009), low Fe^{3+} to Fe_T ratios of spinels in Archean impact-produced spherules (Krull-Davatzes *et al.*, 2010), and evidence for substantial iron mobilization and loss from paleosols during weathering (Rye and Holland, 1998; Sugimori *et al.*, 2009). The Archean oceans were probably also anoxic and contained abundant iron in the form of dissolved Fe(II) complexes. The most obvious evidence for this are the iron formations – the major source of industrial iron ore (Fig. 6.7). A uniquely Precambrian rock type, many of these chemical sedimentary rocks comprise alternating

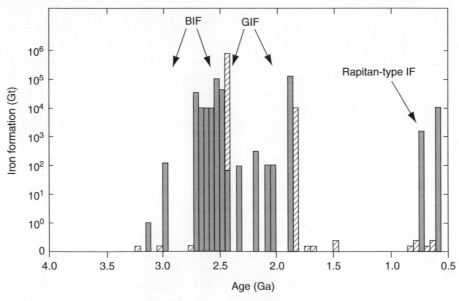

Figure 6.7 Distribution of iron formation deposits over geological time (plotted as time bins of 50 Myr). The diagonal bars indicate significant uncertainty in the age of the iron formation. The Archean and earliest Paleoproterozoic are dominated by deeper-water banded iron formation (BIF), whereas the rest of the Paleoproterozoic is commonly characterized by the deposition of shallower-water granular iron formation (GIF). The change in textural

style of iron formation deposition may relate to the first major increase of atmospheric O_2, but is still poorly understood. Neoproterozoic iron formations, often referred to as Rapitan-type after the type locality in northwestern Canada, may be a product of ferruginous oceans during widespread, low-latitude glaciations and/or enhanced hydrothermal Fe(II) inputs to rift basins. Modified from Bekker *et al.* (2010).

(metre- to sub-millimetre-thick) layers of iron-rich minerals and silicate/carbonate minerals, with a typical bulk chemical composition of ~20–40 wt% Fe and ~40–60 wt% SiO_2. Known specifically as banded iron formation (or BIF) because of their layering, some of these BIF (such as those of the Hamersley Group of Western Australia and Transvaal Supergroup of South Africa) that were deposited at the Archean–Paleoproterozoic transition are hundreds of meters thick and formed over vast depositional areas of ~10^5 km^2 (Morris, 1993; Trendall, 2002; Klein, 2005).

It is now widely believed that the ultimate source of the iron in iron formation is Fe(II) from hydrothermal systems (Jacobsen and Pimentel-Klose, 1988; Derry and Jacobsen, 1990; Bau and Möller, 1993) that were located either distally (i.e. from mid-ocean ridges) or proximally (i.e. from shallow submarine volcanoes) to the site of iron formation deposition. Given a distal source, it was suggested that upwelling anoxic waters had Fe(II) concentrations of 40–120 μM, assuming equilibrium with siderite and calcite (Holland, 1984; Canfield, 2005). However, the upwelling rate required to account for iron formation sedimentation rates is approximately an order of magnitude higher than maximum sedimentation rates in modern coastal environments (Konhauser et al., 2007a). An origin proximal to hydrothermal plumes is suggested by a correlation between mantle plume activity and iron formation deposition between 3.8 and 1.85 Ga (Isley, 1995; Barley et al., 1997, 2005; Isley and Abbott, 1999; Bekker et al., 2010).

Anoxic deep oceans would have facilitated transport of Fe(II) from deeper to shallower waters (Cloud, 1968; Holland, 1973). However, the next step – oxidation of Fe(II) to Fe(III) and the precipitation of ferric oxyhydroxides – is where the picture becomes blurry (Fig. 6.8). Three main hypotheses have been proposed for the

(a)

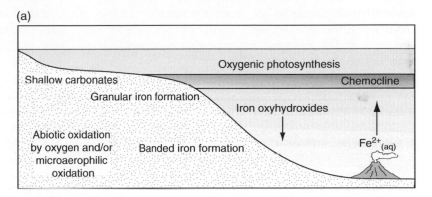

(b)

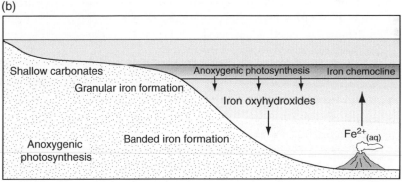

Figure 6.8 Main hypotheses for the mechanism of Fe(II) oxidation necessary for iron formation deposition. (a) Abiotic or microbially catalysed reaction of Fe(II) with dissolved O_2 released during cyanobacterial oxygenic photosynthesis. (b) Direct microbial oxidation during anoxygenic photosynthesis (photoferrotrophy). (c) Abiotic photo-oxidation of dissolved Fe(II) by ultraviolet light. In the vicinity of hydrothermal plumes from shallow submarine volcanoes, photo-oxidation of Fe(II) would likely have been insignificant relative to the formation of ferrous silicate and/or ferrous carbonate minerals. Anoxygenic photosynthesis is suggested to be the dominant oxidation mechanism in the Archean oceans, but the abiotic and/or microbially catalysed oxidation of Fe(II) by O_2 ultimately took on a major role following the evolution of oxygenic photosynthesis. Modified from Bekker et al. (2010).

(c)

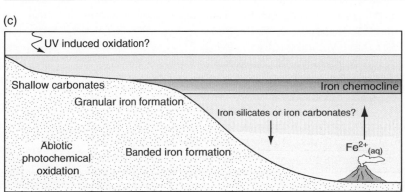

mechanism of Fe(II) oxidation: (1) abiotic or microbially catalysed reaction of Fe(II) with dissolved O_2 released during cyanobacterial oxygenic photosynthesis (Cloud, 1968, 1973); (2) direct microbial oxidation during anoxygenic photosynthesis (photoferrotrophy; Garrels *et al.*, 1973; Hartman, 1984; Widdel *et al.*, 1993; Konhauser *et al.*, 2002; Kappler *et al.*, 2005b; Crowe *et al.*, 2008), and (3) abiotic photo-oxidation of dissolved Fe(II) by ultraviolet light (Cairns-Smith, 1978; Braterman *et al.*, 1983, 1984; François, 1986). Early laboratory experiments demonstrated the plausibility of photochemical Fe(II) oxidation (Anbar and Holland, 1992). Subsequently, experiments by Konhauser *et al.* (2007a) showed that photo-oxidation of Fe(II) in an anoxic Precambrian surface ocean was likely negligible in close proximity to hydrothermal plumes from shallow submarine volcanoes. In these environments, photosynthesis is the leading candidate for providing the oxidant. It is possible that prior to ocean oxygenation, photoferrotrophy was generally the mechanism of choice for producing ferric oxyhydroxides, but at some stage, oxygen played a greater role. The timing of this transition is unresolved.

It has long been postulated that oxygen oases – local regions in the surface ocean where rates of cyanobacterial oxygenic photosynthesis were high enough to permit O_2 accumulation – might have existed before the first major increase in atmospheric O_2 (Kasting, 1993). Hydrocarbon biomarker evidence (2-methylhopanes) from 2.7–2.5 Ga black shales may suggest cyanobacteria had evolved by the Late Archean (Brocks *et al.*, 1999, 2003; Summons *et al.*, 1999; Eigenbrode *et al.*, 2008; Waldbauer *et al.*, 2009). However, an indigenous origin for these molecules has been challenged (Brocks *et al.*, 2008; Rasmussen *et al.*, 2008; Brocks, 2011) and the same biomarkers can be synthesized by anoxygenic photoautotrophs (Rashby *et al.*, 2007; Welander *et al.*, 2010). Morphological characteristics of some Late Archean stromatolites may suggest that cyanobacteria evolved by 2.7 Ga (Buick, 1992, 2008; Bosak *et al.*, 2009). Sterane biomarkers, considered a diagnostic hallmark of O_2-dependent eukaryotes (Summons *et al.*, 2006), are also known from Late Archean black shales (Brocks *et al.*, 1999, 2003; Waldbauer *et al.*, 2009) but the concerns about younger geological and anthropogenic contaminants (Brocks *et al.*, 2008; Rasmussen *et al.*, 2008; Brocks, 2011) have yet to be fully resolved.

Geochemical signatures in black shales appear to provide a clearer picture of environmental O_2 levels near the end of the Archean Eon. In the Late Archean Mt. McRae Shale (Hamersley Basin, Western Australia) and Ghaap Group (Griqualand West Basin, South Africa), molybdenum and rhenium enrichments together with sulfur, nitrogen, and molybdenum isotope signatures point to the presence of surface ocean O_2 beneath a low-O_2 atmosphere (<0.001% PAL based on MIF of sulfur

isotopes; Anbar *et al.*, 2007; Kaufman *et al.*, 2007; Wille *et al.*, 2007; Garvin *et al.*, 2009; Godfrey and Falkowski, 2009; Reinhard *et al.*, 2009; Duan *et al.*, 2010a; Kendall *et al.*, 2010). Black shales deposited on the slope of the Campbellrand–Malmani carbonate platform in the Griqualand West Basin contain high rhenium and low molybdenum enrichments which, along with iron speciation data, indicate the presence of dissolved O_2 in bottom waters beneath the photic zone. These same redox proxies show that the mildly oxygenated surface waters gave way to an anoxic deeper ocean (Kendall *et al.*, 2010). Collectively, these observations imply that nutrient-rich regions along Late Archean ocean margins were sites of significant O_2 accumulation more than 100 million years before O_2 began to appreciably accumulate in the atmosphere.

Hence, photosynthetic O_2 could have contributed to the precipitation of ferric oxyhydroxides as early as 2.7 Ga (other controversial geological and geochemical evidence may suggest an even earlier origin for oxygenic photosynthesis; e.g. Rosing and Frei, 2004; Hoashi *et al.*, 2009; Kato *et al.*, 2009; Kerrich and Said, 2011). Arguably, the strongest peak in iron formation deposition occurred at 2.7–2.45 Ga. Although these iron formations have been interpreted as the products of elevated mantle plume activity and hence an increased Fe(II) supply to seawater (e.g. Barley *et al.*, 2005; Bekker *et al.*, 2010), it is also possible there is a link between iron formation deposition and oxygenic photosynthesis. Such a link would be implausible if during periods of iron formation deposition, photosynthetic O_2 production was retarded by the widespread adsorption of the nutrient phosphorus onto sedimenting ferric oxyhydroxides (Bjerrum and Canfield, 2002). However, Konhauser *et al.* (2007b) subsequently showed that such particles would not constitute a significant phosphorus sink in the silica-rich Archean oceans. Photosynthetic O_2 merits consideration as an oxidant for the precipitation of ferric oxyhydroxides in Late Archean and earliest Paleoproterozoic oceans, though further research, particularly on the spatiotemporal distribution of O_2, is required to elucidate its significance relative to anoxygenic photosynthesis. Rare earth element (REE) data from Archean iron formations suggest that abiotic oxidation of Fe(II) by free oxygen was limited, but permits the possibility that microaerophilic Fe(II) oxidation was an important oxidation mechanism (along with anoxygenic photosynthesis) in the Late Archean oceans (Planavsky *et al.*, 2010b).

6.4.4.3 The Paleoproterozoic Great Oxidation Event and its impact on the iron cycle

The Paleoproterozoic Earth witnessed the first major increase in atmospheric O_2 (the Great Oxidation Event), as marked by several lines of evidence, including the disappearance of MIF of S isotopes between 2.45 and 2.32 Ga (Farquhar *et al.*, 2000, 2011; Bekker *et al.*, 2004; Guo *et al.*, 2009; Johnston, 2011), a significant increase in Cr abundances in iron formations after 2.48 Ga (Konhauser *et al.*, 2011b), the appearance of red beds, sediment-hosted stratiform copper deposits, phosphorites, manganese deposits and increasingly abundant oxidized Fe(III) in paleosols between 2.5 and 2.0 Ga (Cloud, 1968; Chandler, 1980; Eriksson and Cheney, 1992; Rye and Holland, 1998; Bekker *et al.*, 2004; Canfield, 2005; Holland, 2006; Farquhar *et al.*, 2011; Murakami *et al.*, 2011; Sekine *et al.*, 2011), and a contemporaneous growth in the diversity of minerals driven by the availability of elements in their oxidized forms (Hazen *et al.*, 2009; Sverjensky and Lee, 2010). In addition, the appearance of $CaSO_4$-rich evaporites (Chandler, 1988; El Tabakh *et al.*, 1999; Melezhik *et al.*, 2005; Schröder *et al.*, 2008), enhanced expressions of mass-dependent fractionation of sulfur isotopes (e.g. Cameron, 1982; Canfield and Raiswell, 1999; Canfield *et al.*, 2000; Bekker *et al.*, 2004; Guo *et al.*, 2009), and increased molybdenum abundances in euxinic black shales (Scott *et al.*, 2008) point to a rise in seawater sulfate and molybdenum concentrations. The most likely explanation is an increase in the oxidative weathering of crustal sulfide minerals and hence the riverine transport of SO_4^{2-} and MoO_4^{2-} to the oceans. Atmospheric O_2 levels rose above 0.001% PAL (required to eliminate the MIF of sulfur isotopes; Pavlov and Kasting, 2002; Farquhar and Wing, 2003) but likely remained at least an order of magnitude below the present level (Canfield, 2005; Holland, 2006).

What was the impact of the Great Oxidation Event on the iron biogeochemical cycle? From the rock record, it appears that deposition of iron formation was limited between ~ 2.4 and 1.9 Ga and large deposits (≥10 000 Gt) are not known from the rock record (Isley and Abbott, 1999; Bekker *et al.*, 2010). The connection between the end of large iron formation deposition and the Great Oxidation Event raises an obvious possibility – the rising atmospheric O_2 levels ventilated the deep oceans, leading to a major reduction in the oceanic iron reservoir via the formation of poorly soluble ferric oxyhydroxides (Holland, 2006). However, other explanations can be envisioned that involve the continuation of deep ocean anoxia. The distribution of U–Pb ages from detrital zircons and subduction-related granitoids, together with a paucity of greenstones, tonalite–trondhjemite–granodiorite (TTG) suites, and large igneous provinces (LIPs), points to a widespread slowdown of magmatic activity between 2.45 and 2.20 Ga (Condie *et al.*, 2009). Hence, the general absence of iron formation deposition at this time could reflect low hydrothermal Fe(II) inputs to an otherwise anoxic deep ocean. Another possibility is that increased sulfate fluxes to the oceans stimulated

larger rates of microbial sulfide production, leading to an expansion of euxinic waters along productive ocean margins (regions with high organic carbon fluxes) and the removal of Fe(II) into insoluble iron sulfide minerals (Canfield, 2005). The global 2.2–2.1 Ga Lomagundi positive carbon isotope excursion is thought to reflect extensive burial of organic carbon and the release of oxidizing power to the environment (Karhu and Holland, 1996). A more oxic surface environment may have substantially increased seawater sulfate concentrations, and hence the rate of microbial hydrogen sulfide production, although pyrite burial and a return to lower oceanic redox conditions could have resulted in a lower seawater sulfate concentration after the Lomagundi event (Schröder *et al.*, 2008). Carbon isotope compositions and iron speciation data from Lomagundi-age sedimentary rocks are consistent with ocean stratification, including water column euxinia (Bekker *et al.*, 2008; Scott *et al.*, 2008).

A final widespread episode of large iron formation deposition at 1.88–1.85 Ga (predominantly as shallow-water granular iron formation, also known as GIF; Bekker *et al.*, 2010) indicates a return to Fe(II)-rich deep oceans. This change may be associated with an increase in mantle plume activity (Bekker *et al.*, 2010). An alternative (though not mutually exclusive) hypothesis is a decline in atmospheric and hence oceanic O_2 levels. Support for this comes from the occurrences of ferric oxyhydroxide precipitation in high-energy, shallow-water environments, which requires Fe(II) transport from anoxic deeper waters into shallow continental shelves where reaction with photosynthetic O_2 can occur (Canfield, 2005 and references therein). Furthermore, chromium isotope compositions in the 1.9 Gyr-old Gunflint Formation (Ontario, Canada) are not fractionated relative to igneous rocks, suggesting minimal oxidative mobilization of chromium from the upper crust because of a low-O_2 atmosphere (Frei *et al.*, 2009).

6.4.4.4 Iron's fall and redemption: The end of large iron formation deposition and the nature of Middle Proterozoic ocean chemistry

With the exception of occurrences commonly associated with low-latitude Neoproterozoic glacial deposits (Hoffman and Schrag, 2002) and sporadic, generally small Middle Proterozoic deposits (Bekker *et al.*, 2010), iron formations disappear from the rock record after ~1.85 Ga. This major change in the iron biogeochemical cycle has attracted substantial interest among biogeochemists for the past dozen years. Originally, the disappearance of the large iron formations was attributed to a major rise in atmospheric O_2 concentrations that led to the development of mildly oxygenated and iron-scarce deep oceans (Holland, 1984,

2006). However, Canfield (1998) suggested instead that a smaller atmospheric O_2 increase would allow the persistence of anoxic deep oceans while simultaneously increasing seawater sulfate concentrations to the point where elevated rates of microbial hydrogen sulfide production led to widespread ocean euxinia and the removal of dissolved Fe(II) into iron sulfide minerals.

Several lines of geochemical evidence, particularly iron speciation data, has since been advanced in support of some variant of the 'Canfield' ocean, the most compelling of which was the apparent capture of the transition from ferruginous to euxinic deep ocean conditions at ~1.84 Ga in the Animikie Basin (Lake Superior region; Poulton *et al.*, 2004). At least three episodes of deep ocean euxinia were also documented in the 1.8–1.4 Ga McArthur Basin of northern Australia (Shen *et al.*, 2002, 2003), including an example of shallow photic zone euxinia indicated by hydrocarbon biomarkers of green and purple sulfur bacteria in the 1.64 Ga Barney Creek Formation (Brocks *et al.*, 2005; Brocks and Schaeffer, 2008). Consistent with an expansion of euxinic environments, the end of large iron formation deposition is approximately contemporaneous with the first appearance of exhalative (SEDEX) lead–zinc–sulfide mineralization in the rock record (Lyons *et al.*, 2006). Low seawater sulfate concentrations (perhaps ≤1 mM; Canfield *et al.*, 2010), inferred from ^{34}S-rich pyrites and rapid variations in the seawater sulfate isotope composition (Lyons *et al.*, 2009a and references therein), may fingerprint widespread pyrite burial and the permanent removal of sulfur from the Earth's surface via subduction of the oceanic crust and its sedimentary cover. The return of ferruginous oceans and the increase in iron formation deposition in the Neoproterozoic (Canfield *et al.*, 2008) may then reflect the product of extreme sulfate limitation (and hence low rates of microbial hydrogen sulfide production) in a Canfield Ocean that lasted over 1 Gyr (Canfield, 2004).

Was a global expansion of ocean euxinia truly the end of iron's dominance in the deep oceans? Lyons *et al.* (2009a, b) pointed out several problematic issues with the Canfield ocean hypothesis. For example, it is difficult to sustain global euxinia because such conditions should result in widespread depletion of bioessential, redox-sensitive metals (e.g. molybdenum, copper), thereby eradicating the high rates of primary production (which enables high organic carbon export fluxes) required to sustain euxinia. Further, the idea of global deep ocean euxinia is at odds with the molybdenum abundances and isotope compositions of Proterozoic euxinic shales. Because molybdenum burial rates in euxinic environments are high, the molybdenum seawater concentration is sensitive to the extent of euxinic seafloor. Hence, global deep ocean euxinia should easily strip the ocean of nearly all dissolved Mo. However,

Proterozoic euxinic shales contain molybdenum enrichments intermediate between that of Archean and Phanerozoic shales, indicating that wholesale drawdown of the Proterozoic oceanic Mo inventory did not occur (Scott *et al.*, 2008). Molybdenum isotope data from euxinic shales further supports this contention (Arnold *et al.*, 2004; Kendall *et al.*, 2009, 2011). The existence of large expanses of oxic or ferruginous seafloor is implied.

If the Middle Proterozoic deep ocean was not globally euxinic, then it must have either contained dissolved O_2 or was still ferruginous. Slack and Cannon (2009) advanced the possibility that a large bolide impact at 1.85 Ga mixed oxic surface waters with anoxic deep waters on a global scale, leading to a new low-O_2 state (~1 µM O_2) for the deep ocean. However, this is a geologically instantaneous event and is unlikely to explain a permanent change in ocean redox chemistry. Independent of the bolide hypothesis, the existence of some low-O_2 regions (<5 µM O_2) in the deep oceans after 1.85 Ga is supported by mineralogical and geochemical data (cerium anomalies) from 1.74–1.71 Ga seafloor-hydrothermal Si–Fe–Mn sedimentary rocks deposited in association with volcanogenic massive sulfides (VMS) (Slack *et al.*, 2007, 2009). Sporadic occurrences of oxide-facies, VMS-related exhalites (Slack *et al.*, 2007, 2009; Bekker *et al.*, 2010) and the absence of Middle Proterozoic marine manganese deposits (anoxic deep-sea conditions are required to permit accumulation of soluble Mn(II); Holland, 2006) are also consistent with a weakly oxygenated deep ocean. Such conditions, if widespread, would have effectively terminated the deposition of large iron formations.

However, recent studies have the potential to bring about a major paradigm shift in our thinking of Middle Proterozoic deep ocean chemistry. By expanding their 2004 study of the 1.9–1.8 Ga Animikie Basin from a single drill core to multiple localities, Poulton *et al.* (2010) sought to elucidate for the first time the paleobathymetric variations in Late Paleoproterozoic ocean chemistry from shallow to deeper waters along a continental margin. Their target: the <1.84 Ga Rove and Virginia Formations which are thought to be deposited after the global cessation of large iron formation deposition (represented in part by the underlying Gunflint Formation). Iron speciation analyses revealed a spatial transition from oxic surface to mid-depth euxinic waters, which in turn gave way to ferruginous deeper waters. Poulton *et al.* (2010) argued that although the deep oceans remained ferruginous, the expansion of water column euxinia along continental margins (perhaps further aided by a diminished ratio of iron to hydrogen sulfide in hydrothermal fluids because of a higher oceanic sulfate concentration; Kump and Seyfried, 2005), was sufficiently widespread to end iron formation deposition. Evidence for ferruginous deep oceans some tens of millions of years after the end of large iron formation deposition comes in the form of small iron formation deposits, carbonate and clay minerals whose compositions are indicative of iron-rich bottom waters, and Gunflint-type microfossils in the Ashburton Basin of Western Australia (Wilson *et al.*, 2010). The lone example of a large deposit after 1.85 Ga is represented by the GIF of the ca. 1.8 Ga Frere Formation (Earaheedy Basin, Western Australia; Pirajno *et al.*, 2009).

Younger Middle Proterozoic iron speciation evidence for water column euxinia mostly come from the intracratonic McArthur Basin whose connection to the global ocean was probably restricted to some degree and hence the data from this region may not be representative of open ocean conditions (Lyons *et al.*, 2009a). To address this major gap in our understanding of Middle Proterozoic ocean chemistry, Planavsky *et al.* (2011) obtained iron speciation data from mudrocks ranging in age between 1.7 and 1.2 Ga. In all cases, they found abundant evidence for ferruginous deep-ocean conditions in a diverse range of paleogeographic settings, including a passive margin (1.7 Ga Chuanlinggou Formation, northern China), a passive margin that evolved into a foredeep setting (1.2 Ga Borden Basin, Arctic Canada), a restricted extensional setting (1.45 Ga Belt Supergroup, north-central USA), and a continental back-arc environment (1.64 Ga Barney Creek and Lady Loretta formations, northern Australia). A prevalence of ferruginous deep oceans is most harmonious with the Middle Proterozoic occurrences of small iron formation deposits (Bekker *et al.*, 2010), available geochemical data (including the inferred molybdenum budget; Lyons *et al.*, 2010), and the growing evidence for similar conditions in the Neoproterozoic (Planavsky *et al.*, 2011), which we discuss next.

6.4.4.5 Iron's persistent march: the ferruginous Neoproterozoic deep oceans

As mentioned previously, the Neoproterozoic Earth witnessed a return to significant iron formation deposition. The close association of iron formations with low-latitude Neoproterozoic glacial deposits is considered a logical consequence of a global, multi-million-year-long 'Snowball Earth' glaciation because ocean stagnation should lead to a build-up of dissolved Fe(II) in the ice-covered oceans. This idea is further supported by REE signatures, cerium anomalies, and enrichments in iron, manganese, and other redox-sensitive trace elements in post-glacial cap carbonates (e.g., Huang *et al.*, 2011). At the end of the glaciation when ocean circulation is re-established, the upwelling of dissolved Fe(II) into oxic shallower waters would drive ferric oxyhydroxide precipitation. This could also

occur during the glaciation where ice cover was sufficiently thin to allow oxygenic photosynthesis (Hoffman and Schrag, 2002). Widespread glaciation would also facilitate iron formation deposition by cutting off the supply of riverine sulfate to the oceans and by increasing the iron-to-hydrogen sulfide ratio in hydrothermal fluids via the lowered hydrostatic pressure that would accompany a significant drop in global sea level (Kump and Seyfried, 2005). This latter explanation is also consistent with less severe 'Slushball Earth' glaciations (ice-free equatorial oceans; Fairchild and Kennedy, 2007). The mineralogical simplicity of the iron-bearing phases (predominantly hematite) in Neoproterozoic compared to Archean–Paleoproterozoic iron formations may arise from limited organic carbon delivery to the glaciated oceans, which would have retarded the formation of reduced iron phases during diagenesis (Halverson *et al.*, 2011). Others advocate that iron formation deposition was partly or primarily related to enhanced hydrothermal fluxes in restricted rift basins during the breakup of the supercontinent Rodinia (Young, 2002; Eyles and Januszczak, 2004; Bekker *et al.*, 2010; Basta *et al.*, 2011).

Recently, Canfield *et al.* (2008) presented an impressive compilation of more than 700 iron speciation analyses from 34 different formations and concluded that the Neoproterozoic deep oceans were predominantly ferruginous between ca. 750 and 530 Ma. A similar conclusion was reached for the ca. 800–742 Ma Chuar Group (Johnston *et al.*, 2010). Li *et al.* (2010) then provided a detailed picture of Ediacaran ocean chemistry through iron speciation analyses on the 635–551 Ma Doushantuo Formation. They showed that mid-depth euxinic waters were sandwiched within ferruginous deep waters in the Nanhua Basin, South China. Other examples of water column euxinia from the Neoproterozoic were previously noted (Canfield *et al.*, 2008; Scott *et al.*, 2008; Johnston *et al.*, 2010), but Li *et al.* (2010) were the first to clearly delineate the paleobathymetric distribution of euxinic and ferruginous conditions beneath oxic surface waters for a Neoproterozoic continental margin.

Most delightfully, the proposed redox structure for the Middle Proterozoic and Neoproterozoic oceans – oxic surface waters, mid-depth euxinic waters in regions of elevated organic carbon export along productive ocean margins, and ferruginous deeper waters – is very similar (Fig. 6.9; Li *et al.*, 2010; Poulton *et al.*, 2010; Planavsky *et al.*, 2011). Furthermore, this redox structure probably had its roots along Late Archean ocean margins prior to the first major rise in atmospheric O_2 (Reinhard *et al.*, 2009; Kendall *et al.*, 2010; Scott *et al.*, 2011). These recent observations raise the tantalizing possibility that the same basic ocean redox structure has held sway for ~2 Gyr of Earth's middle age (Planavsky *et al.*, 2011; Poulton and Canfield, 2011).

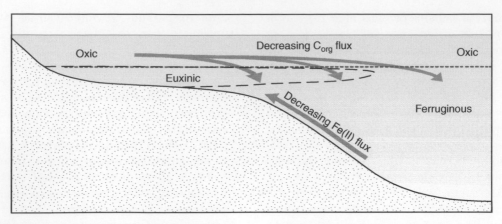

Figure 6.9 Redox conditions in the Proterozoic oceans based on Poulton *et al.* (2010), Li *et al.* (2010) and Planavsky *et al.* (2011). The surface oceans were mildly oxygenated after the initial rise of atmospheric O_2 but the deep oceans were anoxic. At mid-water depths, locally high organic carbon (C_{org}) and sulfate fluxes stimulated extensive microbial hydrogen sulfide production. Anoxic and sulfidic (euxinic) conditions developed when sulfide became sufficiently abundant to quantitatively titrate Fe(II) from the water column through sedimentary pyrite formation. These euxinic waters gave way at depth to anoxic and iron-rich (ferruginous) conditions where Fe(II) still remained in solution after the removal of all sulfide as pyrite. A similar redox structure has been proposed for the Late Archean ocean, with the exception that oxygenated surface waters were confined to the ocean margins. Modified from Poulton *et al.* (2010).

6.4.4.6 Iron dethroned: the transition to fully oxygenated Phanerozoic oceans

Abundant geochemical data point to a second major increase in atmospheric O_2 levels during the late Neoproterozoic (e.g. Des Marais *et al.*, 1992; Canfield and Teske, 1996; Hurtgen *et al.*, 2005; Fike *et al.*, 2006; McFadden *et al.*, 2008; Scott *et al.*, 2008; Knauth and Kennedy, 2009), but this O_2 rise did not immediately spell the end of iron's reign in the deep oceans. Indeed, it was not until approximately 580 Ma that the first glimmerings of substantial, regionally stable deep-ocean oxygenation appear in the geochemical record. In the Avalon Peninsula of Newfoundland (Canada), iron speciation data appear to capture a transition from ferruginous deep waters before and during the ca. 580 Ma Gaskiers glaciation to oxygenated deep waters that persisted for at least 15 Myr afterwards. The minimum atmospheric O_2 level required to enable this expansion of deep ocean O_2 was calculated to be 15% PAL assuming waters supplying the deep ocean had an oxygen content of at least $50\,\mu M$ (to account for deep-sea oxygen deficits and aerobic respiration by Ediacaran metazoans; Canfield *et al.*, 2007). Similar iron speciation evidence of deep ocean oxygenation was found in late Ediacaran sedimentary rocks of the Windermere Supergroup, Western Canada (Canfield *et al.*, 2008; Shen *et al.*, 2008). Molybdenum abundances in euxinic black shales increase to Phanerozoic-like levels between 663 and 551 Ma, pointing to a larger dissolved Mo inventory in more extensively oxygenated oceans (Scott *et al.*, 2008).

However, the iron speciation evidence for ferruginous deep oceans from other late Ediacaran sections indicates that ocean oxygenation was not global at this time (Canfield *et al.*, 2008; Li *et al.*, 2010). In fact, geochemical evidence indicates that non-trivial expanses of seafloor remained anoxic to the Precambrian–Cambrian boundary (Canfield *et al.*, 2008; Wille *et al.*, 2008; Ries *et al.*, 2009) and perhaps well into the early Phanerozoic (Goldberg *et al.*, 2007; Dahl *et al.*, 2010; Gill *et al.*, 2011). Elucidating the nature and timing of the transition to predominantly oxygenated and iron-scarce oceans and its significance for metazoan evolution remains a high priority for the biogeochemistry community. Other than enhanced iron transport and scavenging during oceanic anoxic events (Meyer and Kump, 2008) and in restricted anoxic basins (like the modern Black Sea; Lyons and Severmann, 2006; Severmann *et al.*, 2008), iron biogeochemical cycling in the near-pervasively oxygenated Phanerozoic oceans was limited primarily to anoxic sediments, particularly the benthic iron flux on continental margins (e.g. Homoky *et al.*, 2009; Severmann *et al.*, 2010).

6.4.2 Consequences for biology

Iron has likely played a significant role in cellular metabolism since the very first microorganisms appeared on Earth. It has been argued that the very first prebiotic metabolisms occurred in the vicinity of deep-sea hydrothermal vents and coupled the reduction of CO_2 into an anionic carboxylate group using energy from the reaction between FeS (mackinawite) with dissolved H_2S at temperatures of 100°C to form pyrite:

$$FeS + H_2S \rightarrow FeS_2 + H_2 \quad \Delta G^{0'} = -38.4\,kJ\,mol^{-1}$$

Importantly, the reaction is exergonic, yielding sufficient free energy for the formation of adenosine triphosphate (ATP) from adenosine diphosphate (ADP), which requires 31.8 kJ mol^{-1} (Drobner *et al.*, 1990). Under slightly acidic conditions, and in the presence of dissolved Fe^{2+}, pyrite has a positively-charged surface that would adsorb inorganic anions (i.e. carbonate, sulfide, phosphate) and negatively-charged organic molecules (e.g. Russell and Hall, 1997; Bebié and Schoonen, 1999). These products could then accumulate and polymerize to more complex compounds directly on the pyrite. The central role of iron in those early biochemical reactions may also explain their later incorporation into a number of enzymes, such as Fe–S proteins and cytochromes (Wächtershäuser, 1988).

The generation of H$_2$ as a by-product from the above reaction would have proven fortuitous because it could eventually have been used as an electron donor in subsequent metabolic reactions, gradually supplementing and then replacing the original energy source when the first primitive hydrogenase activity had evolved (Kandler, 1994). Simultaneously, the electrons would have needed disposing of, and the possible presence of ferric iron on the seafloor due to surface photochemical reactions in the absence of an effective UV screen would have served this purpose nicely:

$$0.5H_2 + Fe(OH)_3 \rightarrow Fe^{2+} + 2OH^- + H_2O$$

There is certainly evidence to suggest that ferric iron was available in the early Archean – the presence of iron formations in the 3.8–3.7 Ga Isua Greenstone Belt in Greenland (Dauphas *et al.* 2004) and the Nuvvuagittuq Supracrustal Belt in northern Québec (Dauphas *et al.*, 2007). What is less clear is when Fe(III)-reducing bacteria evolved and took advantage of the ferric iron as an electron donor. Based on the hyperthermophilic lifestyle of some bacteria that are deeply rooted in the phylogenetic tree, it has been suggested that Fe(III) reducers are very old indeed (Vargas *et al.*, 1998; Kashefi and Lovley, 2000). Some supporting evidence for an ancient Fe(III) reduction pathway comes from highly negative δ^{56}Fe values in 2.9 Gyr old organic carbon- and magnetite-rich shales (Rietkuil Formation, South Africa; Yamaguchi *et al.*, 2005) and 2.7 Gyr old pyrites (Manjeri Formation, Zimbabwe; Archer and Vance, 2006). These δ^{56}Fe values are comparable with the negative fractionations observed from cultures of dissimilatory Fe(III)-reducing (DIR) bacteria (Johnson *et al.*, 2005) and in modern ferric oxyhydroxide-rich chemical sediments (Tangalos *et al.*, 2010) and the oxic–anoxic boundary of modern ferruginous lakes (Teutsch *et al.*, 2009) where DIR is taking place. Recent iron, carbon, and oxygen isotope studies of Late Archean to earliest Paleoproterozoic (2.7–2.45 Ga) sedimentary rocks from

South Africa and Western Australia also support a prominent role for DIR (Czaja *et al.*, 2010; Heimann *et al.*, 2010). Iron and carbon isotope signatures in metacarbonates from iron formation of the Isua Greenstone Belt are similar to those of the Late Archean iron formations, suggesting that DIR is as ancient as the oldest known sedimentary rocks on Earth (Craddock and Dauphas, 2011). Coupling the reduction of Fe(III) minerals to the oxidation of organic matter also explains the low content of organic carbon in iron formations (<0.5%; Gole and Klein, 1981), as well as the abundance of light carbon isotope signatures associated with the interlayered carbonate minerals (Perry *et al.*, 1973; Walker, 1984; Baur *et al.*, 1985).

The lightest δ^{56}Fe values are observed in sedimentary pyrites and black shales between ~2.7 and 2.5 Ga (Johnson *et al.*, 2008). This distinctive isotopic transition may reflect a radiation of DIR (Fig. 6.10). It has been hypothesized that the increased expression of DIR may be coupled to the evolution of oxygenic photosynthesis, which would provide an abundant supply of organic carbon and Fe(III). After the Great Oxidation Event, δ^{56}Fe variability is attenuated, consistent with a decline in open-ocean DIR arising from lower reactive Fe(III) availability in response to the expansion of oxygenated and sulfidic (from increased sulfate availability) waters that remove Fe from solution (Johnson *et al.*, 2008).

Alternatively, the light δ^{56}Fe values could reflect the preferential sequestration of heavy iron isotopes onto iron oxyhydroxides during episodes of large iron formation deposition. This would leave seawater with a pool of isotopically light iron, which could subsequently be incorporated into pyrite (Rouxel *et al.*, 2005; Anbar and Rouxel, 2007). However, for this process to explain the lightest δ^{56}Fe signatures (−3.5‰) alone, it would require ~90% of the dissolved iron pool to be removed as iron oxyhydroxides, at least episodically. It is likely that both abiotic and DIR-driven iron isotope fractionation played an important role in the Late Archean oceans.

Severmann *et al.* (2008) proposed that a benthic iron flux from continental shelves, made isotopically light by the combined effects of Fe(II) oxidation and DIR, supplied isotopically light iron to seawater. Additional iron isotope fractionation, favouring the uptake of lighter isotopes, could have occurred during pyrite formation in deeper euxinic waters if iron was not quantitatively removed from solution (likely the case because Late Archean oceans were Fe(II)-rich and sulfate-poor). Hence, the light δ^{56}Fe from 2.7–2.5-Gyr-old black shales and pyrites may reflect an increase in the extent of iron redox cycling in response to O$_2$ accumulation along ocean margins. A fingerprint of DIR's influence on this shelf-to-basin iron 'shuttling' is seen in the distinctive pattern of δ^{56}Fe vs. Fe/Al in recent Black Sea sediments (Severmann *et al.*, 2008) and in some Devonian black

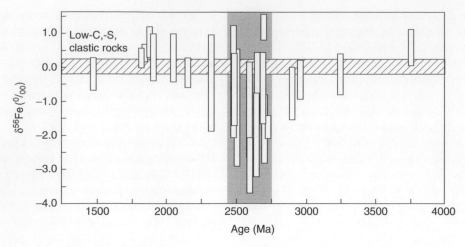

Figure 6.10 Iron isotope compositions of Archean to Middle Proterozoic sedimentary rocks (iron formations and carbon-, sulfur-, and/or iron-rich shales) and sedimentary sulfide minerals. The horizontal bar reflects the typical range in iron isotope composition for organic carbon- and sulfur-poor sedimentary rocks of Archean to Phanerozoic age. Light iron isotope compositions of less than $-2‰$ are exclusive to the Late Archean (shaded region). Iron isotope compositions are reported as per mil deviations from the average of igneous rocks: $\delta^{56}Fe\ (‰) = [(^{56}Fe/^{54}Fe)_{sample}/(^{56}Fe/^{54}Fe)_{Igneous\ Rocks} - 1] \times 1000$. Modified from Johnson *et al.* (2008).

shales (Duan *et al.*, 2010b). A similar pattern is observed in black shales of the 2.5 Ga Mt. McRae Shale (Duan *et al.*, 2007). Czaja *et al.* (2010) suggested that in the Late Archean Hamersley Basin, DIR played an important role in providing isotopically light iron to deeper water environments (Jeerinah Formation euxinic shales and Carawine Dolomite carbonates/shales). In contrast, they show that light $\delta^{56}Fe$ signatures of shallow-water carbonates (Wittenoom Formation) were likely a result of Fe(II) oxidation in the photic zone (cf. Rouxel *et al.*, 2005; Anbar and Rouxel, 2007). Hence, iron oxyhydroxides fueling DIR could have been isotopically light and together with DIR-related isotope fractionation would have resulted in an amplified light $\delta^{56}Fe$ signature relative to that achieved by either process alone (Severmann *et al.*, 2008).

One problem is that a complementary marine sink with heavy $\delta^{56}Fe$ to balance the predominantly light $\delta^{56}Fe$ signatures in Late Archaean sedimentary materials has not been identified. Such a sink could be represented by pelagic sediments, which are not preserved in the geological record (Steinhoefel et al., 2010). However, Guilbaud *et al.* (2011) argued that the need for a heavy $\delta^{56}Fe$ marine sink and a large oceanic pool with light $\delta^{56}Fe$ could be eliminated if the light pyrite $\delta^{56}Fe$ arises from isotope fractionation during abiotic pyrite formation rather than DIR. This hypothesis proposes that a small degree of Fe(II) utilization in iron-rich oceans enabled the full expression of iron isotope fractionation between Fe^{2+}, mackinawite, and pyrite, with minimal impact on the isotopic composition of the remaining oceanic Fe(II) pool. When a substantial proportion of dissolved Fe(II) was depleted via pyrite or oxide mineral formation, as in the Proterozoic and Phanerozoic oceans, then the expression of isotope fractionation became muted, leading to a narrow range of heavier pyrite $\delta^{56}Fe$ (Guilbaud *et al.*, 2011).

Nevertheless, microbial processes likely played an important role in the formation of the iron- and silica-rich layers in BIFs. Anoxygenic phototrophic Fe(II) oxidation and microaerophilic Fe(II) oxidation are temperature-dependent microbial processes. Seawater temperature fluctuations in the photic zone may have resulted in repeated cycles of microbially catalysed iron oxyhydroxide deposition and abiotic silica precipitation (Posth *et al.*, 2008). Others argue instead for a prominent role of DIR in generating the alkalinity necessary for the precipitation of siderite and for concentrating silica, which was then precipitated as diagenetic minerals (e.g. Fischer and Knoll, 2009; Heimann *et al.*, 2010).

With the evolution of photoferrotrophy, biological Fe(II) oxidation would have superseded photochemical oxidation because the bacteria could grow deeper in the water column where UV radiation would be effectively absorbed (Kappler *et al.*, 2005b). As long as a source of ferrous iron and nutrients were available, photoferrotrophy could have contributed to ferric iron deposition onto vast areas of seafloor, some of which became manifest as iron formation. What is poorly constrained, however, is when did Fe(II)-based anoxygenic photosynthesis first take place, and how did the photoferrotrophs respond to the evolution and diversification of cyanobacteria in the water column? At present, there is no actual physical or chemical evidence for photoferrotrophs in the rock record, but a number of independent lines of evidence do suggest their presence on the early Earth. First and foremost, molecular phylogenetic analysis of a number of enzymes involved in (bacterio-)chlorophyll biosynthesis suggests that anoxygenic photosynthetic lineages are almost certain to be more deeply rooted than the oxygenic cyanobacterial lineages (Xiong, 2006). Second, modern anoxygenic phototrophs, including photoferrotrophs, display the ability to utilize multiple substrates such as H_2S, H_2 and Fe(II) (Croal *et al.*, 2009). Yet, in the Archean oceans, dissolved sulfide would have been removed from seawater by reacting with the abundant Fe(II) in the deep sea, while available H_2 would have been consumed at depth by methanogenic bacteria (Konhauser *et al.*, 2005). By contrast, the input of dissolved Fe(II) from mid-ocean ridges was almost certainly greater during the Archean. Third, the recovery of 2α-methylhopanes from bitumens in the 2.6 Ga Marra Mamba Iron Formation and the 2.5 Ga Mt. McRae Shale (Hamersley Basin) initially led researchers to conclude that oxygenic photosynthesis was already in existence at that time because those biomarkers were considered unique to cyanobacteria (Brocks *et al.*, 1999; Summons *et al.*, 1999). However, most recently it has been demonstrated that an anoxygenic Fe(II)-oxidizing phototroph, *Rhodopseudomonas palustris*, generates substantial quantities of 2-methylhopanoids in the absence of oxygen (Rashby *et al.*, 2007), making the case for Fe(II)-oxidizing phototrophs at 2.6 Ga just as plausible as that for cyanobacteria.

When cyanobacteria eventually did begin to dominate the ocean's photic zones, the oxygen they produced would have allowed other bacteria to begin elaborating on their electron transport chain to include special terminal reductase enzymes that made it possible to pass electrons directly onto O_2. The benefit for those cells was that they could now harness more energy from the inorganic and organic substrates they oxidized. In the case of Fe(II) oxidation, chemolithoautotrophs, such as *Gallionella*, may even have thrived under early low oxygen conditions as they would have enjoyed a kinetic advantage over inorganic or photosynthetic reactions (Holm, 1989). At 1.9–1.8 Ga, Fe(II)-oxidizing bacteria appear to have colonized large tracts of the shallow marine environment where iron-rich deep waters were brought into contact with fully oxygenated surface waters (Planavsky *et al.*, 2009; Wilson *et al.*, 2010).

6.5 Summary

The central role of iron in modern biology is probably a legacy of the early Earth, when the Archean atmosphere was essentially devoid of O_2 and the anoxic oceans contained abundant dissolved, and thus, bioavailable Fe(II).

Iron was likely to have been involved in the first prebiotic metabolic reactions near deep-sea hydrothermal systems. Dissimilatory Fe(III)-reducing bacteria may have had their origins in the Early Archean. Anoxygenic photosynthesis (photoferrotrophy) would likely have been an important metabolic process in the iron-rich oceans and its origin is thought to predate cyanobacterial oxygenic photosynthesis.

Archean and earliest Paleoproterozoic iron formations were probably the direct and indirect products of microbial processes. When O_2 was scarce, anoxygenic photosynthesis may have driven Fe(II) oxidation. With the advent of cyanobacterial O_2 production and accumulation in surface waters, Fe(II)-oxidizing bacteria is likely to have become an important component of microbial ecosystems at the interface between oxic and ferruginous water masses. Hence, microaerophilic Fe(II) oxidation and the abiotic oxidation of Fe(II) by photosynthetically produced O_2 likely played an increasingly important role in the Late Archean oceans. Ultimately, Fe(II) oxidation by O_2 became the principle metabolic process driving the formation of Paleoproterozoic iron formations after the Great Oxidation Event.

Conventional wisdom held that the ferruginous oceans disappeared along with the large iron formations after 1.85 Ga because of the establishment of oxic or euxinic deep oceans in response to rising atmospheric O_2 levels. Recent geochemical studies, however, point to a complex stratification of the oceans during Earth's middle age, with oxic surface waters underlain by mid-depth euxinic waters along productive ocean margins, which in turn gave way to predominantly ferruginous deep oceans. The end of large iron formation deposition at 1.85 Ga may then reflect an expansion of euxinic water masses along ocean margins, which led to the titration of upwelling, dissolved Fe(II) as insoluble sedimentary sulfides. The development of low-O_2 conditions in some regions may also have removed some Fe(II) from solution as iron oxyhydroxides. Nevertheless, ferruginous deep oceans likely continued throughout the rest of the Proterozoic and perhaps into the early Phanerozoic.

Seawater iron concentrations declined dramatically upon the development of the largely oxygenated Phanerozoic oceans because of the poor solubility of iron in the presence of O_2 at the circumneutral pH of seawater. Consequently, the residence time of iron in the oceans became very low, and the biogeochemical cycling of iron was restricted primarily to anoxic sediments. Exceptions to this rule include oceanic anoxic events and restricted anoxic basins. Despite the fact that iron is the limiting nutrient for biological productivity over large parts of the modern ocean, it continues to play a prominent role as a micronutrient and in microbially mediated redox reactions. This is facilitated by the development of siderophores, which play an important role as Fe(III) chelators that enable bacteria and fungi to satisfy their metabolic requirements when iron is scarce in the environment.

As is the case for other redox-sensitive elements, the story of iron geobiology is one that reflects the planetary co-evolution of life and its environment. Bioavailable iron was plentiful on the anoxic early Earth and may have played an important role in the first metabolisms. Following the advent of cyanobacterial oxygenic photosynthesis, the biogeochemical cycles of iron and oxygen clashed. For much of Earth's middle age, oxygen dominated the surface environments whereas iron continued to play an important role in the ocean depths. Finally, the expansion of O_2 throughout most of the Phanerozoic oceans limited iron biogeochemical cycling to ocean floor sediments, as is observed today.

Acknowledgements

This work was supported by the National Science Foundation, the German Research Foundation (DFG), the Natural Sciences and Engineering Research Council of Canada, the NASA Astrobiology Institute, and the Agouron Institute. Noah Planavsky is thanked for insightful comments that improved the manuscript. We are grateful to Sue Selkirk for drafting the figures.

References

Archer C, Vance D (2006) Coupled Fe and S isotope evidence for Archean microbial Fe(III) and sulfate reduction. *Geology* **34**, 153–156.

Anbar AD, Holland HD (1992) The photochemistry of manganese and the origin of banded iron formations. *Geochimica et Cosmochimica Acta* **56**, 2595–2603.

Anbar A, Rouxel O (2007) Metal stable isotopes in paleoceanography. *Annual Review of Earth and Planetary Sciences* **35**, 717–746.

Anbar AD, Duan Y, Lyons TW, *et al.* (2007) A whiff of oxygen before the Great Oxidation Event? *Science* **317**, 1903–1906.

Arnold GL, Anbar AD, Barling J, Lyons TW (2004) Molybdenum isotope evidence for widespread anoxia in Mid-Proterozoic oceans. *Science* **304**, 87–90.

Barley ME, Pickard AL, Sylvester PJ (1997) Emplacement of a large igneous province as a possible cause of banded iron formation 2.45 billion years ago. *Nature* **385**, 55–58.

Barley ME, Bekker A, Krapež B (2005) Late Archean to Early Paleoproterozoic global tectonics, environmental change and the rise of atmospheric oxygen. *Earth and Planetary Science Letters* **238**, 156–171.

Basta FF, Maurice AE, Fontboté L, Favarger PY (2011) Petrology and geochemistry of the banded iron formation (BIF) of Wadi Karim and Um Anab, Eastern Desert, Egypt: implications for the origin of Neoproterozoic BIF. *Precambrian Research* **187**, 277–292.

Bau M, Möller P (1993) Rare earth element systematics of the chemically precipitated component in Early Precambrian iron formations and the evolution of the terrestrial atmosphere-hydrosphere-lithosphere system. *Geochimica et Cosmochimica Acta* **57**, 2239–2249.

Bauer I, Kappler A (2009) Rates and extent of reduction of Fe(III) compounds and O_2 by humic substances. *Environmental Science & Technology* **43**, 4902–4908.

Baur ME, Hayes JM, Studley SA, Walter MR (1985) Millimeter-scale variations of stable isotope abundances in carbonates from banded iron-formations in the Hamersley Group of Western Australia. *Economic Geology* **80**, 270–282.

Bebié J, Schoonen MAA (1999) Pyrite and phosphate in anoxia and an origin-of-life hypothesis. *Earth and Planetary Science Letters* **171**, 1–5.

Bebout BM, Fitzpatrick MW, Paerl HW (1993) Identification of the sources of energy for nitrogen fixation and physiological characterization of nitrogen-fixing members of a marine microbial mat community. *Applied and Environmental Microbiology* **59**, 1495–1503.

Bekker A, Holland HD, Wang PL, *et al.* (2004) Dating the rise of atmospheric oxygen. *Nature* **427**, 117–120.

Bekker A, Holmden C, Beukes NJ, Kenig F, Eglinton B, Patterson WP (2008) Fractionation between inorganic and organic carbon during the Lomagundi (2.22–2.1 Ga) carbon isotope excursion. *Earth and Planetary Science Letters* **271**, 278–291.

Bekker A, Slack JF, Planavsky N, *et al.* (2010) Iron formation: the sedimentary product of a complex interplay among mantle, tectonic, oceanic, and biospheric processes. *Economic Geology* **105**, 467–508.

Bennett SA, Achterberg EP, Connelly DP, Statham PJ, Fones GR, German CR (2008) The distribution and stabilisation of dissolved Fe in deep-sea hydrothermal plumes. *Earth and Planetary Science Letters* **270**, 157–167.

Berner RA (1970) Sedimentary pyrite formation. *American Journal of Science* **268**, 1–23.

Berner RA (1984) Sedimentary pyrite formation: an update. *Geochimica et Cosmochimica Acta* **48**, 605–615.

Bjerrum CJ, Canfield DE (2002) Ocean productivity before about 1.9 Gyr ago limited by phosphorus adsorption onto iron oxides. *Nature* **417**, 159–162.

Blake R, Johnson DB (2000) Phylogenetic and biochemical diversity among acidophilic bacteria that respire on iron. In: *Environmental Microbe–Metal Interactions* (ed Lovley D). American Society of Microbiology Press, Washington, DC, pp. 53–78.

Blöthe M, Roden EE (2009) Composition and activity of an autotrophic Fe(II)-oxidizing, nitrate-reducing enrichment culture. *Applied and Environmental Microbiology* **75**, 6937–6940.

Bosak T, Liang B, Sim MS, Petroff AP (2009) Morphological record of oxygenic photosynthesis in conical stromatolites. *Proceedings of the National Academy of Sciences* **106**, 10939–10943.

Boyd PW, Ellwood MJ (2010) The biogeochemical cycle of iron in the ocean. *Nature Geoscience* **3**, 675–682.

Boye M, van den Berg CMG, de Jong JTM, Leach H, Croot P, de Baar HJW (2001) Organic complexation of iron in the Southern Ocean. *Deep-Sea Research I* **48**, 1477–1497.

Braterman PS, Cairns-Smith AG, Sloper RW (1983) Photo-oxidation of hydrated Fe^{2+}-significance for banded iron formations. *Nature* **303**, 163–164.

Braterman PS, Cairns-Smith AG, Sloper RW (1984) Photo-oxidation of iron(II) in water between pH 7.4 and 4.0. *Journal of the Chemical Society, Dalton Transactions: Inorganic Chemistry*, 1441–1445.

Brocks JJ (2011) Millimeter-scale concentration gradients of hydrocarbons in Archean shales: live-oil escape or fingerprint of contamination? *Geochimica et Cosmochimica Acta* **75**, 3196–3213.

Brocks JJ, Schaeffer P (2008) Okenane, a biomarker for purple sulfur bacteria (Chromatiaceae), and other new carotenoid derivatives from the 1640 Ma Barney Creek Formation. *Geochimica et Cosmochimica Acta* **72**, 1396–1414.

Brocks JJ, Logan GA, Buick R, Summons RE (1999) Archean molecular fossils and the early rise of eukaryotes. *Science* **285**, 1033–1036.

Brocks JJ, Buick R, Summons RE, Logan GA (2003) A reconstruction of Archean biological diversity based on molecular fossils from the 2.78 to 2.45 billion-year old Mount Bruce Supergroup, Hamersley Basin, Western Australia. *Geochimica et Cosmochimica Acta* **67**, 4321–4335.

Brocks JJ, Love GD, Summons RE, Knoll AH, Logan GA, Bowden SA (2005) Biomarker evidence for green and purple sulphur bacteria in a stratified Palaeoproterozoic sea. *Nature* **437**, 866–870.

Brocks JJ, Grosjean E, Logan GA (2008) Assessing biomarker syngeneity using branched alkanes with quaternary carbon (BAQCs) and other plastic contaminants. *Geochimica et Cosmochimica Acta* **72**, 871–888.

Buick R (1992) The antiquity of oxygenic photosynthesis: evidence from stromatolites in sulphate-deficient Archaean lakes. *Nature* **255**, 74–77.

Buick R (2008) When did oxygenic photosynthesis evolve? *Philosophical Transactions of the Royal Society B* **363**, 2731–2743.

Buresh RJ, Moraghan JT (1976) Chemical reduction of nitrate by ferrous iron. *Journal of Environmental Quality* **5**, 320–325.

Caccavo F Jr, Schamberger PC, Keiding K, Nielsen PH (1997) Role of hydrophobicity in adhesion of the dissimilatory Fe(III)-reducing bacterium *Shewanella alga* to amorphous Fe(III) oxide. *Applied and Environmental Microbiology* **63**, 3837–3843.

Cairns-Smith AG (1978) Precambrian solution photochemistry, inverse segregation, and banded iron formations. *Nature* **276**, 807–808.

Cameron EM (1982) Sulphate and sulphate reduction in early Precambrian oceans. *Nature* **296**, 145–148.

Canfield DE (1998) A new model for Proterozoic ocean chemistry. *Nature* **396**, 450–453.

Canfield DE (2004) The evolution of the Earth surface sulfur reservoir. *American Journal of Science* **304**, 839–861.

Canfield DE (2005) The early history of atmospheric oxygen. *Annual Review of Earth and Planetary Sciences* **33**, 1–36.

Canfield DE, Raiswell R (1999) The evolution of the sulfur cycle. *American Journal of Science* **299**, 697–723.

Canfield DE, Teske A (1996) Late Proterozoic rise in atmospheric oxygen concentration inferred from phylogenetic and sulphur-isotope studies. *Nature* **382**, 127–132.

Canfield DE, Habicht KS, Thamdrup B (2000) The Archean sulfur cycle and the early history of atmospheric oxygen. *Science* **288**, 658–661.

Canfield DE, Poulton SW, Narbonne GM (2007) Late-Neoproterozoic deep-ocean oxygenation and the rise of animal life. *Science* **315**, 92–95.

Canfield DE, Poulton SW, Knoll AH, *et al.* (2008) Ferruginous conditions dominated later Neoproterozoic deep-water chemistry. *Science* **321**, 949–952.

Canfield DE, Farquhar J, Zerkle AL (2010) High isotope fractionations during sulfate reduction in a low-sulfate euxinic ocean analog. *Geology* **38**, 415–418.

Chan CS, De Stasio G, Welch SA, *et al.* (2004) Microbial polysaccharides template assembly of nanocrystal fibers. *Science* **303**, 1656–1658.

Chandler FW (1980) Proterozoic red bed sequences of Canada. *Geological Survey of Canada Bulletin* **311**, 1–53.

Chandler FW (1988) Diagenesis of sabkha-related, sulphate nodules in the early Proterozoic Gordon Lake Formation, Ontario, Canada. *Carbonates and Evaporites* **3**, 75–94.

Childers SE, Ciufo S, Lovley DR (2002) *Geobacter metallireducens* accesses insoluble Fe(III) oxide by chemotaxis. *Nature* **416**, 767–769.

Chu NC, Johnson CM, Beard, BL, *et al.* (2006) Evidence for hydrothermal venting in Fe isotope compositions of the deep Pacific Ocean through time. *Earth and Planetary Science Letters* **245**, 202–217.

Cloud PE (1968) Atmospheric and hydrospheric evolution on the primitive Earth. *Science* **160**, 729–736.

Cloud PE (1973) Paleoecological significance of the banded iron-formation. *Economic Geology and the Bulletin of the Society of Economic Geologists* **68**, 1135–1143.

Coleman ML (1985) Geochemistry of diagenetic non-silicate minerals: kinetic considerations. *Philosophical Transactions of the Royal Society of London A* **315**, 39–56.

Condie KC, O'Neill C, Aster RC (2009) Evidence and implications for a widespread magmatic shutdown for 250 Myr on Earth. *Earth and Planetary Science Letters* **282**, 294–298.

Craddock PR, Dauphas N (2011) Iron and carbon isotope evidence for microbial iron respiration throughout the Archean. *Earth and Planetary Science Letters* **303**, 121–132.

Croal LR, Johnson CM, Beard BL, Newman DK (2004) Iron isotope fractionation by Fe(II)-oxidizing photoautotrophic bacteria. *Geochimica et Cosmochimica Acta* **68**, 1227–1242.

Croal LR, Jiao Y, Kappler A, Newman DK (2009) Phototrophic Fe(II) oxidation in the presence of H$_2$: implications for banded iron formations. *Geobiology* **7**, 21–24.

Crowe SA, Jones C, Katsev S, *et al.* (2008) Photoferrotrophs thrive in an Archean Ocean analogue. *Proceedings of the National Academy of Sciences* **105**, 15938–15943.

Cullen JT, Bergquist BA, Moffett JW (2006) Thermodynamic characterization of the partitioning of iron between soluble and colloidal species in the Atlantic Ocean. *Marine Chemistry* **98**, 295–303.

Czaja AD, Johnson CM, Beard BL, Eigenbrode JL, Freeman KH, Yamaguchi KE (2010) Iron and carbon isotope evidence for ecosystem and environmental diversity in the ~2.7 to 2.5 Ga Hamersley Province, Western Australia. *Earth and Planetary Science Letters* **292**, 170–180.

Dahl TW, Hammarlund EU, Anbar AD, Bond DPG, Gill BC, Gordon GW, Knoll AH, Nielsen AT, Schovsbo NH, Canfield DE (2010). Devonian rise in atmospheric oxygen correlated to the radiations of terrestrial plants and large predatory fish. *Proceedings of the National Academy of Sciences* **107**, 17911–17915.

Dai MH, Martin JM (1995) First data on trace metal level and behaviour in two major Arctic river-estuarine systems (Ob and Yenisey) and in the adjacent Kara Sea, Russia. *Earth and Planetary Science Letters* **131**, 127–141.

Dauphas N, van Zuilen M, Wadhwa M, Davis AM, Marty B, Janney PE (2004) Clues from Fe isotope variations on the origin of Early Archean BIFs from Greenland. *Science* **306**, 2077–2080.

Dauphas N, Cates NL, Mojzsis SJ, Busigny V (2007) Identification of chemical sedimentary protoliths using iron isotopes in the >3750 Ma Nuvvuagittuq supracrustal belt, Canada. *Earth and Planetary Science Letters* **254**, 358–376.

Derry LA, Jacobsen SB (1990) The chemical evolution of Precambrian seawater: evidence from REEs in banded iron formation. *Geochimica et Cosmochimica Acta* **54**, 2965–2977.

Des Marais DJ, Strauss H, Summons RE, Hayes JM (1992) Carbon isotope evidence for the stepwise oxidation of the Proterozoic environment. *Nature* **359**, 605–609.

Drobner E, Huber H, Stetter KO (1990) *Thiobacillus ferrooxidans*, a facultative hydrogen oxidizer. *Applied and Environmental Microbiology* **56**, 2922–2923.

Duan Y, Anbar AD, Arnold GL, Gordon GW, Severmann S, Lyons TW (2007) The iron isotope variations in the ~2.5 Ga Mt. McRae Shale. *Geological Society of America Abstracts with Programs* **39**, 449.

Duan Y, Anbar AD, Arnold GL, Lyons TW, Gordon GW, Kendall B (2010a) Molybdenum isotope evidence for mild environmental oxygenation before the Great Oxidation Event. *Geochimica et Cosmochimica Acta* **74**, 6655–6668.

Duan Y, Severmann S, Anbar AD, Lyons TW, Gordon GW, Sageman BB (2010b) Isotopic evidence for Fe cycling and repartitioning in ancient oxygen-deficient settings: examples from black shales of the mid- to late Devonian Appalachian basin. *Earth and Planetary Science Letters* **290**, 244–253.

Edwards KJ, Rogers DR, Wirsen CO, McCollom TM (2003) Isolation and characterization of novel psychrophilic, neutrophilic, Fe-oxidizing, chemolithoautotrophic alpha- and, gamma-Proteobacteria from the deep sea. *Applied and Environmental Microbiology* **69**, 2906-2913.

Eigenbrode JL, Freeman KH, Summons RE (2008) Methylhopane hydrocarbon biomarkers in Hamersley Province sediments provide evidence for Neoarchean aerobiosis. *Earth and Planetary Science Letters* **273**, 323–331.

El-Nagger MY, Wanger G, Leung KM, et al. (2010) Electrical transport along bacterial nanowires from *Shewanella oneidensis* MR-1. *Proceedings of the National Academy of Sciences* **107**, 18127–18131.

El Tabakh M, Grey K, Pirajno F, Schreiber BC (1999) Pseudomorphs after evaporitic minerals interbedded with 2.2 Ga stromatolites of the Yerriba basin, Western Australia: origin and significance. *Geology* **27**, 871–874.

England GL, Rasmussen B, Krapez B, Groves DI (2002) Palaeoenvironmental significance of rounded pyrite in siliciclastic sequences of the Late Archean Witwatersrand Basin: oxygen-deficient atmosphere or hydrothermal alteration? *Sedimentology* **49**, 1133–1156.

Eriksson PG, Cheney ES (1992) Evidence for the transition to an oxygen-rich atmosphere during the evolution of red beds in the Lower Proterozoic sequences of southern Africa. *Precambrian Research* **54**, 257–269.

Eyles N, Januszczak N (2004) 'Zipper-rift': a tectonic model for Neoproterozoic glaciations during the breakup of Rodinia after 750 Ma. *Earth-Science Reviews* **65**, 1–73.

Fairchild IJ, Kennedy MJ (2007) Neoproterozoic glaciation in the Earth system. *Journal of the Geological Society, London* **164**, 895–921.

Farquhar J, Wing BA (2003) Multiple sulfur isotopes and the evolution of the atmosphere. *Earth and Planetary Science Letters* **213**, 1–13.

Farquhar J, Bao H, Thiemens M (2000) Atmospheric influence of Earth's earliest sulfur cycle. *Science* **289**, 756–758.

Farquhar J, Peters M, Johnston DT, et al. (2007) Isotopic evidence for Mesoarchean anoxia and changing atmospheric sulphur chemistry. *Nature* **449**, 706–709.

Farquhar J, Zerkle AL, Bekker A (2011) Geological constraints on the origin of oxygenic photosynthesis. *Photosynthesis Research* **107**, 11–36.

Fay P, Stewart WDP, Walsby AE, Fogg GE (1968) Is the heterocyst the site of nitrogen fixation in blue-green algae? *Nature* **220**, 810–812.

Fehr MA, Andersson PS, Hålenius U, Gustafsson Ö, Mörth CM (2010) Iron enrichments and Fe isotope compositions of surface sediments from the Gotland Deep, Baltic Sea. *Chemical Geology* **277**, 310–322.

Fike DA, Grotzinger JP, Pratt LM, Summons RE (2006) Oxidation of the Ediacaran ocean. *Nature* **444**, 744–747.

Fischer WW, Knoll AH (2009) An iron shuttle for deepwater silica in Late Archean and early Paleoproterozoic iron formation. *Geological Society of America Bulletin* **121**, 222–235.

Fleet ME (1998) Detrital pyrite in Witwatersrand gold reefs: x-ray diffraction evidence and implications for atmospheric evolution. *Terra Nova* **10**, 302–306.

François LM (1986) Extensive deposition of banded iron formations was possible without photosynthesis. *Nature* **320**, 352–354.

Frei R, Gaucher C, Poulton SW, Canfield DE (2009) Fluctuations in Precambrian atmospheric oxygenation recorded by chromium isotopes. *Nature* **461**, 250–253.

Froelich PN, Klinkhammer GP, Bender ML, et al. (1979) Early oxidation of organic matter in pelagic sediments of the eastern equatorial Atlantic: suboxic diagenesis. *Geochimica et Cosmochimica Acta* **43**, 1075–1090.

Garrels RM, Perry EA Jr, MacKenzie FT (1973) Genesis of Precambrian iron-formations and the development of atmospheric oxygen. *Economic Geology* **68**, 1173–1179.

Garvin J, Buick R, Anbar AD, Arnold GL, Kaufman AJ (2009) Isotopic evidence for an aerobic nitrogen cycle in the latest Archean. *Science* **323**, 1045–1048.

Gill BC, Lyons TW, Young SA, Kump LR, Knoll AH, Saltzman MR (2011) Geochemical evidence for widespread euxinia in the Later Cambrian ocean. *Nature* **469**, 80–83.

Godfrey LV, Falkowski PG (2009) The cycling and redox state of nitrogen in the Archaean ocean. *Nature Geoscience* **2**, 725–729.

Goldberg T, Strauss H, Guo Q, Liu C (2007) Reconstructing marine redox conditions for the early Cambrian Yangtze platform: evidence from biogenic sulphur and organic carbon isotopes. *Palaeogeography, Palaeoclimatology, Palaeoecology* **254**, 175–193.

Gole MJ, Klein C (1981) Banded iron-formations through much of Precambrian time. *Journal of Geology* **89**, 169–183.

Gorby YA, Yanina S, McLean JS, et al. (2006) Electrically conductive bacterial nanowires produced by *Shewanella oneiden-*

sis strain MR-1 and other microorganisms. *Proceedings of the National Academy of Sciences* **103**, 11358–11363.

Guilbaud R, Butler IB, Ellam RM (2011) Abiotic pyrite formation produces a large Fe isotope fractionation. *Science* **332**, 1548–1551.

Guo Q, Strauss H, Kaufman AJ, *et al* (2009) Reconstructing Earth's surface oxidation across the Archean-Proterozoic transition. *Geology* **37**, 399–402.

Gustafsson O, Widerlund A, Andersson PS, Ingri J, Roos P, Ledin A (2000) Colloid dynamics transport of major elements through a boreal river-brackish bay mixing zone. *Marine Chemistry* **71**, 1–21.

Hallbeck L, Pedersen K (1990) Culture parameters regulating stalk formation and growth rate of *Gallionella ferruginea*. *Journal of General Microbiology* **136**, 1675–1680.

Hallbeck L, Pedersen K (1991) Autotrophic and mixotrophic growth of *Gallionella ferruginea*. *Journal of General Microbiology* **137**, 2657–2661.

Halverson GP, Poitrasson F, Hoffman PF, *et al.* (2011) Fe isotope and trace element geochemistry of the Neoproterozoic synglacial Rapitan iron formation. *Earth and Planetary Science Letters* **309**, 100–112.

Hanert HH (1992) The genus *Gallionella*. In: *The Prokaryotes* (ed Balows A), 2nd edn. Springer, Berlin, pp. 4082–4088.

Hartman H (1984) The evolution of photosynthesis and microbial mats: a speculation on banded iron formations. In: *Microbial Mats: Stromatolites* (eds Cohen Y, Castenholz RW, Halvorson HO). Alan Liss, New York, pp. 451–453.

Hazen RM, Ewing RC, Sverjensky DA (2009) Evolution of uranium and thorium minerals. *American Mineralogist* **94**, 1293–1311.

Hegler F, Posth NR, Jiang J, Kappler A (2008) Physiology of phototrophic iron(II)-oxidizing bacteria – implications for modern and ancient environments. *FEMS Microbiology Ecology* **66**, 250–260.

Hegler F, Schmidt C, Schwarz H, Kappler A (2010) Does a low pH-microenvironment around phototrophic Fe^{II}-oxidizing bacteria prevent cell encrustation by Fe^{III} minerals? *FEMS Microbiology Ecology* **74**, 592–600.

Heimann A, Johnson CM, Beard BL, *et al.* (2010) Fe, C, and O isotope compositions of banded iron formation carbonates demonstrate a major role for dissimilatory iron reduction in ~2.5 Ga marine environments. *Earth and Planetary Science Letters* **294**, 8–18.

Hersman LE, Huang A, Maurice PA, Forsythe JH (2000) Siderophore production and iron reduction by *Pseudomonas mendocina* in response to iron deprivation. *Geomicrobiology Journal* **17**, 261–273.

Hider RC (1984) Siderophore mediated absorption of iron. *Structure and Bonding* **58**, 25–87.

Hoashi M, Bevacqua DC, Otake T, *et al.* (2009) Primary haematite formation in an oxygenated sea 3.46 billion years ago. *Nature Geoscience* **2**, 301–306.

Hofmann AW (1988) Chemical differentiation of the Earth: the relationship between mantle, continental crust, and oceanic crust. *Earth and Planetary Science Letters* **90**, 297–314.

Hoffman PF, Schrag DP (2002) The snowball Earth hypothesis: testing the limits of global change. *Terra Nova* **14**, 129–155.

Hofmann A, Bekker A, Rouxel O, Rumble D, Master S (2009) Multiple sulphur and iron isotope composition of detrital pyrite in Archaean sedimentary rocks: a new tool for provenance analysis. *Earth and Planetary Science Letters* **286**, 436–445.

Holland HD (1973) The oceans: a possible source of iron in iron formations. *Economic Geology and the Bulletin of the Society of Economic Geologists* **68**, 1169–1172.

Holland HD (1984) *The Chemical Evolution of the Atmosphere and Oceans*. Princeton University Press, Princeton, NJ.

Holland HD (2006) The oxygenation of the atmosphere and oceans. *Philosophical Transactions of the Royal Society B* **361**, 903–915.

Holm NG (1989) The ^{13}C/^{12}C ratios of siderite and organic matter of a modern metalliferous hydrothermal sediment and their implications for banded iron formations. *Chemical Geology* **77**, 41–45.

Homoky WB, Severmann S, Mills RA, Statham PJ, Fones GR (2009) Pore-fluid Fe isotopes reflect the extent of benthic Fe redox cycling: evidence from continental shelf and deep-sea sediments. *Geology* **37**, 751–754.

Huang J, Chu X, Jiang G, Feng L, Chang H (2011) Hydrothermal origin of elevated iron, manganese and redox-sensitive trace elements in the *c.* 635 Ma Doushantuo cap carbonate. *Journal of the Geological Society, London* **168**, 805–815.

Hurtgen MT, Arthur MA, Halverson GP (2005) Neoproterozoic sulfur isotopes, the evolution of microbial sulfur species, and the burial efficiency of sulfide as sedimentary pyrite. *Geology* **33**, 41–44.

Hutchins DA, Bruland KW (1998) Iron-limited diatom growth and Si:N uptake ratios in a coastal upwelling regime. *Nature* **393**, 561–564.

Isley AE (1995) Hydrothermal plumes and the delivery of iron to banded iron formation. *Journal of Geology* **103**, 169–185.

Isley AE, Abbott DH (1999) Plume-related mafic volcanism and the deposition of banded iron formation. *Journal of Geophysical Research* **104**, 15461–15477.

Jacobsen SB, Pimentel-Klose MR (1988) A Nd isotopic study of the Hamersley and Michipicoten banded iron formations: the source of REE and Fe in Archean oceans. *Earth and Planetary Science Letters* **87**, 29–44.

Javoy M, Kaminski E, Guyot F, *et al.* (2010) The chemical composition of the Earth: enstatite chondrite models. *Earth and Planetary Science Letters* **293**, 259–268.

Jiang J, Kappler A (2008) Kinetics and thermodynamics of microbial and chemical reduction of humic substances: implications for electron shuttling in natural environments. *Environmental Science & Technology* **42**, 3563–3569.

Jiao Y, Newman, DK (2007) The pio operon is essential for phototrophic Fe(II) oxidation in *Rhodopseudomonas palustris* TIE-1. *Journal of Bacteriology* **189**, 1765–1773.

Jickells TD, An ZS, Andersen KK, *et al.* (2005) Global iron connections between desert dust, ocean biogeochemistry, and climate. *Science* **308**, 67–71.

Johnson KS, Gordon RM, Coale KH (1997) What controls dissolved iron concentrations in the world ocean? *Marine Chemistry* **57**, 137–161.

Johnson CM, Roden EE, Welch SA, Beard BL (2005) Experimental constraints on Fe isotope fractionation during magnetite and Fe carbonate formation coupled to dissimilatory hydrous ferric oxide reduction. *Geochimica et Cosmochimica Acta* **69**, 963–993.

Johnson CM, Beard BL, Roden EE (2008) The iron isotope fingerprints of redox and biogeochemical cycling in modern and ancient Earth. *Annual Review of Earth and Planetary Sciences* **36**, 457–493.

Johnston DT (2011) Multiple sulfur isotopes and the evolution of Earth's surface sulfur cycle. *Earth-Science Reviews* **106**, 161–183.

Johnston DT, Poulton SW, Dehler C, *et al.* (2010) An emerging picture of Neoproterozoic ocean chemistry: insights from the Chuar Group, Grand Canyon, USA. *Earth and Planetary Science Letters* **290**, 64–73.

Kandler O (1994) The early diversification of life. In: *Early Life on Earth* (ed Bengtson S). Nobel Symposium No. 84. Columbia University Press, New York, pp. 152–161.

Kappler A, Newman DK (2004) Formation of Fe(III)-minerals by Fe(II)-oxidizing photoautotrophic bacteria. *Geochimica et Cosmochimica Acta* **68**, 1217–1226.

Kappler A, Schink B, Newman DK (2005a) Fe(III)-mineral formation and cell encrustation by the nitrate-dependent Fe(II)-oxidizer strain BoFeN1. *Geobiology* **3**, 235–245.

Kappler A, Pasquero C, Konhauser KO, Newman DK (2005b) Deposition of banded iron formations by anoxygenic phototrophic Fe(II)-oxidizing bacteria. *Geology* **33**, 865–868.

Kappler A, Johnson CM, Crosby HA, Beard BL, Newman DK (2010) Evidence for equilibrium iron isotope fractionation by nitrate-reducing iron(II)-oxidizing bacteria. *Geochimica et Cosmochimica Acta* **74**, 2826–2842.

Karhu JA, Holland HD (1996) Carbon isotopes and the rise of atmospheric oxygen. *Geology* **24**, 867–870.

Kashefi K, Lovley DR (2000) Reduction of Fe(III), Mn(IV), and toxic metals at 100°C by *Pyrobaculum islandicum*. *Applied and Environmental Microbiology* **66**, 1050–1056.

Kasting JF (1993) Earth's early atmosphere. *Science* **259**, 920–926.

Kato Y, Suzuki K, Nakamura K, *et al.* (2009) Hematite formation by oxygenated groundwater more than 2.76 billion years ago. *Earth and Planetary Science Letters* **278**, 40–49.

Kaufman AJ, Johnston DT, Farquhar J, *et al.* (2007) Late Archean biospheric oxygenation and atmospheric evolution. *Science* **317**, 1900–1903.

Kendall B, Creaser RA, Gordon GW, Anbar AD (2009) Re-Os and Mo isotope systematics of black shales from the Middle Proterozoic Velkerri and Wollogorang Formations, McArthur Basin, northern Australia. *Geochimica et Cosmochimica Acta* **73**, 2534–2558.

Kendall B, Reinhard CT, Lyons TW, Kaufman AJ, Poulton SW, Anbar AD (2010) Pervasive oxygenation along late Archean ocean margins. *Nature Geoscience* **3**, 647–652.

Kendall B, Gordon GW, Poulton SW, Anbar AD (2011) Molybdenum isotope constraints on the extent of late Paleoproterozoic ocean euxinia. *Earth and Planetary Science Letters* **307**, 450–460.

Kerrich R, Said N (2011) Extreme positive Ce-anomalies in a 3.0 Ga submarine volcanic sequence, Murchison Province: oxygenated marine bottom waters. *Chemical Geology* **280**, 232–241.

Khan A, Connolly JAD, Taylor SR (2008) Inversion of seismic and geodetic data for the major element chemistry and temperature of the Earth's mantle. *Journal of Geophysical Research* **113**, B09308.

Klein C (2005) Some Precambrian banded iron formations (BIFs) from around the world: their age, geologic setting, mineralogy, metamorphism, geochemistry, and origin. *American Mineralogist* **90**, 1473–1499.

Knauth LP, Kennedy MJ (2009) The late Precambrian greening of the Earth. *Nature* **460**, 728–732.

Konhauser KO, Hamade T, Raiswell R, *et al.* (2002) Could bacteria have formed the Precambrian banded iron formations? *Geology* **30**, 1079–1082.

Konhauser KO, Newman DK, Kappler A (2005) The potential significance of microbial Fe(III)-reduction during Precambrian banded iron formations. *Geobiology* **3**, 167–177.

Konhauser KO, Amskold L, Lalonde SV, Posth NR, Kappler A, Anbar A (2007a) Decoupling photochemical Fe(II) oxidation from shallow-water BIF deposition. *Earth and Planetary Science Letters* **258**, 87–100.

Konhauser KO, Lalonde SV, Amskold L, Holland HD (2007b). Was there really an Archean phosphate crisis? *Science* **315**, 1234.

Konhauser KO, Pecoits E, Lalonde SV, *et al.* (2009) Oceanic nickel depletion and a methanogen famine before the Great Oxidation Event. *Nature* **458**, 750–753.

Konhauser KO, Kappler A, Roden EE (2011a) Iron in microbial metabolisms. *Elements* **7**, 89–93.

Konhauser KO, Lalonde SV, Planavsky NJ, *et al.* (2011b) Aerobic bacterial pyrite oxidation and acid rock drainage during the Great Oxidation Event. *Nature* **478**, 369–373.

Kostka JE, Dalton DD, Skelton H, Dollhopf S, Stucki JW (2002) Growth of Fe(III)-reducing bacteria on clay minerals as the sole electron acceptor and comparison of growth yields on a variety of oxidized iron forms. *Applied and Environmental Microbiology* **68**, 6256–6262.

Krachler R, Jirsa F, Ayromlou S (2005) Factors influencing the dissolved iron input by river water to the open ocean. *Biogeosciences* **2**, 311–315.

Krachler R, Krachler RF, von der Kammer F, *et al.* (2010) Relevance of peat-draining rivers for the riverine input of dissolved iron into the ocean. *Science of the Total Environment* **408**, 2402–2408.

Krom MD, Berner RA (1980) Adsorption of phosphate in anoxic marine sediments. *Limnology and Oceanography* **25**, 797–806.

Krull-Davatzes AE, Byerly GR, Lowe DR (2010) Evidence for a low-O_2 Archean atmosphere from nickel-rich chrome spinels in 3.24 Ga impact spherules, Barberton greenstone belt, South Africa. *Earth and Planetary Science Letters* **296**, 319–328.

Kump LR, Seyfried WE Jr (2005) Hydrothermal Fe fluxes during the Precambrian: effect of low oceanic sulfate concentrations and low hydrostatic pressure on the composition of black smokers. *Earth and Planetary Science Letters* **235**, 654–662.

Langmuir D (1997) *Aqueous Environmental Geochemistry*. Prentice-Hall, Inc., Englewood Cliffs, NJ.

Lannuzel D, Schoemann V, de Long J, Chou L, Delille B, Becquevort S, Tison JL (2008) Iron study during a time series in the western Weddell pack ice. *Marine Chemistry* **108**, 85–95.

Larese-Casanova P, Haderlein SB, Kappler A (2010) Biomineralization of lepidocrocite and goethite by nitrate-reducing Fe(II)-oxidizing bacteria: effect of pH, bicarbonate, phosphate and humic acids. *Geochimica et Cosmochimica Acta* **74**, 3721–3734.

Li C, Love GD, Lyons TW, Fike DA, Sessions AL, Chu X (2010) A stratified redox model for the Ediacaran ocean. *Science* 328, 80–83.

Liang L, McCarthy JF, Jolley LW, McNabb JA, Mehlhorn TL (1993) Iron dynamics: transformation of Fe(II)/Fe(III) during injection of natural organic matter in a sandy aquifer. *Geochimica et Cosmochimica Acta* 57, 1987–1999.

Lilley MD, Feely RA, Trefry JH (2004) Chemical and biochemical transformations in hydrothermal plumes. In: *Seafloor Hydrothermal Systems: Physical, Chemical, Biological, and Geological Interactions* (eds Humphris SE, Zierenberg RA, Mullineaux LS, Thomson RE). American Geophysical Union Monograph 91, Washington, DC, pp. 369–391.

Lovley DR, Coates JD, Blunt-Harris EL, Phillips EJP, Woodward JC (1996) Humic substances as electron acceptors for microbial respiration. *Nature* 382, 445–448.

Lovley DR, Holmes DE, Nevin KP (2004) Dissimilatory Fe(III) and Mn(IV) reduction. *Advances in Microbial Physiology* 49, 219–286.

Lower SK, Hochella MF Jr, Beveridge TJ (2001) Bacterial recognition of mineral surfaces: nanoscale interactions between Shewanella and a-FeOOH. *Science* 292, 1360–1363.

Luther GW III (1991) Pyrite synthesis via polysulfide compounds. *Geochimica et Cosmochimica Acta* 55, 2839–2849.

Lyons TW, Severmann S (2006) A critical look at iron paleoredox proxies: new insights from modern euxinic marine basins. *Geochimica et Cosmochimica Acta* 70, 5698–5722.

Lyons TW, Gellatly AM, McGoldrick PJ, Kah LC (2006) Proterozoic sedimentary exhalative (SEDEX) deposits and links to evolving global ocean chemistry. *Geological Society of America Memoirs* 198, 169–184.

Lyons TW, Anbar AD, Severmann S, Scott C, Gill BC (2009a) Tracking euxinia in the ancient ocean: a multiproxy perspective and Proterozoic case study. *Annual Review of Earth and Planetary Sciences* 37, 507–534.

Lyons TW, Reinhard CT, Scott C (2009b) Redox redux. *Geobiology* 7, 489–494.

Lyons TW, Anbar AD, Bekker A, *et al.* (2010) New view of the old ocean: a prevalence of deep iron and marginalized sulfide from the Late Archean through the Proterozoic. *Geological Society of America Abstracts with Programs* 42, 560.

Marsili E, Baron DB, Shikhare ID, Coursolle D, Gralnick JA, Bond DR (2008) Shewanella secretes flavins that mediate extracellular electron transfer. *Proceedings of the National Academy of Sciences* 105, 3968–3973.

Martin JM, Fitzwater SE (1988) Iron deficiency limits phytoplankton growth in the north-east Pacific subarctic. *Nature* 331, 341–343.

McDonough WF (2003) Compositional model for the Earth's core. In: *Treatise on Geochemistry* (eds Holland HD, Turekian KK), Vol. 2: *The Mantle and Core* (ed Carlson RW). Elsevier, Oxford, pp. 547–568.

McFadden KA, Huang J, Chu X, *et al.* (2008) Pulsed oxidation and biological evolution in the Ediacaran Doushantuo Formation. *Proceedings of the National Academy of Sciences* 105, 3197–3202.

McLennan SM (2001) Relationships between the trace element composition of sedimentary rocks and upper continental crust. *Geochemistry, Geophysics, Geosystems* 2, 1021, Paper number 2000GC000109.

Melezhik VA, Fallick AE, Rychanchik DV, Kuznetsov AB (2005) Palaeoproterozoic evaporites in Fennoscandia: implications for seawater sulphate, d13C excursions and the rise of atmospheric oxygen. *Terra Nova* 17, 141–148.

Meyer KM, Kump LR (2008) Oceanic euxinia in Earth history: causes and consequences. *Annual Review of Earth and Planetary Sciences* 36, 251–288.

Michel FM, Barrón V, Torrent J, *et al.* (2010) Ordered ferrimagnetic form of ferrihydrite reveals links among structure, composition, and magnetism. *Proceedings of the National Academy of Sciences* 107, 2787–2792.

Millero FJ, Hawke DJ (1992) Ionic interactions of divalent metals in natural waters. *Marine Chemistry* 40, 19–48.

Millero FJ, Yao W, Aicher J (1995) The speciation of Fe(II) and Fe(III) in natural waters. *Marine Chemistry* 50, 21–39.

Miot J, Benzerara K, Morin M, *et al.* (2009) Iron biomineralization by neutrophilic iron-oxidizing bacteria. *Geochimica et Cosmochimica Acta* 73, 696–711.

Morris RC (1993) Genetic modelling for banded iron-formation of the Hamersley Group, Pilbara Craton, Western Australia. *Precambrian Research* 60, 243–286.

Muehe M, Gerhardt S, Schink B, Kappler A (2009) Ecophysiology and energetic benefit of mixotrophic Fe(II)-oxidation by nitrate-reducing bacteria. *FEMS Microbiology Ecology* 70, 335–343.

Murakami T, Sreenivas B, Sharma SD, Sugimori H (2011) Quantification of atmospheric oxygen levels during the Paleoproterozoic using paleosol compositions and iron oxidation kinetics. *Geochimica et Cosmochimica Acta* 75, 3982–4004.

Nealson KH (1982) Microbiological oxidation and reduction of iron. In: *Mineral Deposits and the Evolution of the Biosphere* (eds Holland HD, Schidlowski M). Springer-Verlag, New York, pp. 51–66.

Neilands JB (1989) Siderophore systems of bacteria and fungi. In: *Metal Ions and Bacteria* (ed Doyle RJ). John Wiley & Sons, New York, pp. 141–163.

Nevin K, Lovley D (2000) Lack of production of electron-shuttling compounds or solubilization of Fe(III) during reduction of insoluble Fe(III) oxide by Geobacter metallireducens. *Applied and Environmental Microbiology* 66, 2248–2251.

Page WJ (1993) Growth conditions for the demonstration of siderophores and iron-repressible outer membrane proteins in soil bacteria, with an emphasis on free-living diazotrophs. In: *Iron Chelation in Plants and Soil Microorganisms* (eds Barton LL, Hemming BC). Academic Press, Inc., New York, pp. 75–110.

Palme H, Jones A (2003) Solar system abundances of the elements. In: *Treatise on Geochemistry* (eds Holland HD, Turekian KK), Vol. 1: *Meteorites, Comets, and Planets* (ed Davis AM). Elsevier, Oxford, UK, pp. 41–61.

Palme H, O'Neill, HSC (2003) Cosmochemical estimates of mantle composition. In: *Treatise on Geochemistry* (eds Holland HD, Turekian KK), Vol. 2: *The Mantle and Core* (ed Carlson RW). Elsevier, Oxford, UK, pp. 1–38.

Pavlov AA, Kasting JF (2002) Mass-independent fractionation of sulfur isotopes in Archean sediments: strong evidence for an anoxic Archean atmosphere. *Astrobiology* 2, 27–41.

Perry EC, Tan FC, Morey GB (1973) Geology and stable isotope geochemistry of the Biwabik Iron Formation, northern Minnesota. *Economic Geology* 68, 1110–1125.

Pirajno F, Hocking RM, Reddy SM, Jones AJ (2009) A review of the geology and geodynamic evolution of the Palaeoproterozoic Earaheedy Basin, Western Australia. *Earth-Science Reviews* **94**, 39–77.

Phoenix VR, Konhauser KO, Adams DG, Bottrell SH (2001) Role of biomineralization as an ultraviolet shield: implications for Archean life. *Geology* **29**, 823–826.

Planavsky N, Rouxel O, Bekker A, Shapiro R, Fralick P, Knudsen A (2009) Iron-oxidizing microbial ecosystems thrived in late Paleoproterozoic redox-stratified oceans. *Earth and Planetary Science Letters* **286**, 230–242.

Planavsky NJ, Rouxel O, Bekker A, *et al.* (2010a) The evolution of the marine phosphate reservoir. *Nature* **467**, 1088–1090.

Planavsky NJ, Bekker A, Rouxel OJ, *et al.* (2010b) Rare Earth Element and yttrium compositions of Archean and Paleoproterozoic Fe formations revisited: new perspectives on the significance and mechanisms of deposition. *Geochimica et Cosmochimica Acta* **74**, 6387–6405.

Planavsky NJ, McGoldrick P, Scott CT, *et al.* (2011) Widespread iron-rich conditions in the mid-Proterozoic ocean. *Nature* **477**, 448–451.

Posth NR, Hegler F, Konhauser KO, Kappler A (2008) Alternating Si and Fe deposition caused by temperature fluctuations in Precambrian oceans. *Nature Geoscience* **1**, 703–708.

Posth NR, Huelin S, Konhauser KO, Kappler A (2010) Size, density and mineralogy of cell-mineral aggregates formed during anoxygenic phototrophic Fe(II) oxidation. *Geochimica et Cosmochimica Acta* **74**, 3476–3493.

Poulton SW, Canfield DE (2011) Ferruginous conditions: a dominant feature of the ocean through Earth's history. *Elements* **7**, 107–112.

Poulton SW, Raiswell R (2002) The low-temperature geochemical cycle of iron: from continental fluxes to marine sediment deposition. *American Journal of Science* **302**, 774–805.

Poulton SW, Fralick PW, Canfield DE (2004) The transition to a sulphidic ocean ~1.84 billion years ago. *Nature* **431**, 173–177.

Poulton SW, Fralick PW, Canfield DE (2010) Spatial variability in oceanic redox structure 1.8 billion years ago. *Nature Geoscience* **3**, 486–490.

Radic A, Lacan F, Murray JW (2011) Iron isotopes in the seawater of the equatorial Pacific Ocean: new constraints for the oceanic iron cycle. *Earth and Planetary Science Letters* **306**, 1–10.

Raiswell R (2011) Iron transport from the continents to the open ocean: the aging-rejuvenation cycle. *Elements* **7**, 101–106.

Raiswell R, Canfield DE (1998) Sources of iron for pyrite formation in marine sediments. *American Journal of Science* **298**, 219–245.

Raiswell R, Benning LG, Tranter M, Tulaczyk S (2008) Bioavailable iron in the Southern Ocean: the significance of the iceberg conveyor belt. *Geochemical Transactions* **9**, 7.

Rakshit S, Matocha CJ, Coyne MS (2008) Nitrite reduction by siderite. *Soil Science Society of America Journal* **72**, 1070–1077.

Rashby SE, Sessions AL, Summons RE, Newman DK (2007) Biosynthesis of 2-methylbacteriohopanepolyols by an anoxygenic phototroph. *Proceedings of the National Academy of Sciences* **104**, 15099–15104.

Rasmussen B, Buick R (1999) Redox state of the Archean atmosphere: evidence from detrital heavy minerals in ca.

3250–2750 Ma sandstones from the Pilbara Craton, Australia. *Geology* **27**, 115–118.

Rasmussen B, Fletcher IR, Brocks JJ, Kilburn MR (2008) Reassessing the first appearance of eukaryotes and cyanobacteria. *Nature* **455**, 1101–1104.

Reguera G, McCarthy KD, Mehta T, Nicoll JS, Tuominen MT, Lovley DR (2005) Extracellular electron transfer via microbial nanowires. *Nature* **435**, 1098–1101.

Reinhard CT, Raiswell R, Scott C, Anbar AD, Lyons TW (2009) A Late Archean sulfidic sea stimulated by early oxidative weathering of the continents. *Science* **326**, 713–716.

Ries JB, Fike DA, Pratt LM, Lyons TW, Grotzinger JP (2009) Superheavy pyrite (d34Spyr > d34SCAS) in the terminal Proterozoic Nama Group, southern Namibia: a consequence of low seawater sulfate at the dawn of animal life. *Geology* **37**, 743–746.

Roden EE (2003) Fe(III) oxide reactivity toward biological versus chemical reduction. *Environmental Science & Technology* **37**, 1319–1324.

Roden EE, Zachara JM (1996) Microbial reduction of crystalline iron(III) oxides: influence of oxide surface area and potential for cell growth. *Environmental Science & Technology* **30**, 1618–1628.

Roden EE, Kappler A, Bauer I, *et al.* (2010) Extracellular electron transfer through microbial reduction of solid-phase humic substances. *Nature Geoscience* **3**, 417–421.

Rosing MT, Frei R (2004) U-rich Archaean sea-floor sediments from Greenland – indications of >3700 Ma oxygenic photosynthesis. *Earth and Planetary Science Letters* **217**, 237–244.

Rouxel OJ, Bekker A, Edwards KJ (2005) Iron isotope constraints on the Archean and Paleoproterozoic ocean redox state. *Science* **307**, 1088–1091.

Russell MJ, Hall AJ (1997) The emergence of life from iron monosulphide bubbles at a submarine hydrothermal redox and pH front. *Journal of the Geological Society of London* **154**, 377–402.

Rye R, Holland HD (1998) Paleosols and the evolution of atmospheric oxygen: a critical review. *American Journal of Science* **298**, 621–672.

Schädler S, Burkhardt C, Hegler F, *et al.* (2009) Formation of cell-iron-mineral aggregates by phototrophic and nitrate-reducing anaerobic Fe(II)-oxidizing bacteria. *Geomicrobiology Journal* **26**, 93–103.

Schoonen MAA, Barnes HL (1991) Reactions forming pyrite and marcasite from solution: II. Via FeS precursors below 100°C. *Geochimica et Cosmochimica Acta* **55**, 1505–1514.

Schröder S, Bekker A, Beukes NJ, Strauss H, van Niekerk HS (2008) Rise in seawater sulphate concentration associated with the Paleoproterozoic positive carbon isotope excursion: evidence from sulphate evaporites in the ~2.2–2.1 Gyr shallow-marine Lucknow Formation, South Africa. *Terra Nova* **20**, 108–117.

Scott C, Lyons TW, Bekker A, *et al.* (2008) Tracing the stepwise oxygenation of the Proterozoic ocean. *Nature* **452**, 456–459.

Scott CT, Bekker A, Reinhard CT, *et al.* (2011) Late Archean euxinic conditions before the rise of atmospheric oxygen. *Geology* **39**, 119–122.

Sekine Y, Tajika E, Tada R, *et al.* (2011) Manganese enrichment in the Gowganda Formation of the Huronian Supergroup: a highly oxidizing shallow-marine environment after the last

Huronian glaciation. *Earth and Planetary Science Letters* **307**, 201–210.

Severmann S, Lyons TW, Anbar A, McManus J, Gordon G (2008) Modern iron isotope perspective on the benthic iron shuttle and the redox evolution of ancient oceans. *Geology* **36**, 487–490.

Severmann S, McManus J, Berelson WM, Hammond DE (2010) The continental shelf benthic iron flux and its isotope composition. *Geochimica et Cosmochimica Acta* **74**, 3984–4004.

Shen Y, Canfield DE, Knoll AH (2002) Middle Proterozoic ocean chemistry: evidence from the McArthur Basin, northern Australia. *American Journal of Science* **302**, 81–109.

Shen Y, Knoll AH, Walter MR (2003) Evidence for low sulphate and anoxia in a mid-Proterozoic marine basin. *Nature* **423**, 632–635.

Shen Y, Zhang T, Hoffman PF (2008) On the coevolution of Ediacaran oceans and animals. *Proceedings of the National Academy of Sciences* **105**, 7376–7381.

Silverman MP, Lundgren DG (1959) Studies on the chemoautotrophic iron bacterium Ferrobacillus ferrooxidans II. Manometric studies. *Journal of Bacteriology* **78**, 326–331.

Singer PC, Stumm W (1970) Acidic mine drainage: the rate-determining step. *Science* **167**, 1121–1123.

Slack JF, Cannon WF (2009) Extraterrestrial demise of banded iron formations 1.85 billion years ago. *Geology* **37**, 1011–1014.

Slack JF, Grenne T, Bekker A, Rouxel OJ, Lindberg PA (2007) Suboxic deep seawater in the late Paleoproterozoic: evidence from hematitic chert and iron formation related to seafloor-hydrothermal sulfide deposits, central Arizona, USA. *Earth and Planetary Science Letters* **255**, 243–256.

Slack JF, Grenne T, Bekker A (2009) Seafloor-hydrothermal Si-Fe-Mn exhalites in the Pecos greenstone belt, New Mexico, and the redox state of ca. 1720 Ma deep seawater. *Geosphere* **5**, 302–314.

Stefánsson A (2007) Iron(III) hydrolysis and solubility at 25°C. *Environmental Science & Technology* **41**, 6117–6123.

Steinhoefel G, von Blanckenburg F, Horn I, *et al.* (2010) Deciphering formation processes of banded iron formations from the Transvaal and the Hamersley successions by combined Si and Fe isotope analysis using UV femtosecond laser ablation. *Geochimica et Cosmochimica Acta* **74**, 2677–2696.

Straub KL, Benz M, Schink B, Widdel F (1996) Anaerobic, nitrate-dependent microbial oxidation of ferrous iron. *Applied and Environmental Microbiology* **62**, 1458-1460.

Straub KL, Buchholz-Cleven BEE (1998) Enumeration and detection of anaerobic ferrous iron-oxidizing, nitrate-reducing bacteria from diverse European sediments. *Applied and Environmental Microbiology* **64**, 4846-4856.

Sugimori H, Yokoyama T, Murakami T (2009) Kinetics of biotite dissolution and Fe behavior under low O_2 conditions and their implications for Precambrian weathering. *Geochimica et Cosmochimica Acta* **73**, 3767–3781.

Summons RE, Jahnke LL, Hope JM, Logan GM (1999) 2-Methylhopanoids as biomarkers for cyanobacterial oxygenic photosynthesis. *Nature* **400**, 554–557.

Summons RE, Bradley AS, Jahnke LL, Waldbauer JR (2006) Steroids, triterpenoids and molecular oxygen. *Philosophical Transactions of the Royal Society B* **361**, 951–968.

Sverjensky DA, Lee N (2010) The Great Oxidation Event and mineral diversification. *Elements* **6**, 31–36.

Tangalos GE, Beard BL, Johnson CM, *et al.* (2010) Microbial production of isotopically light iron(II) in a modern chemically precipitated sediment and implications for isotopic variations in ancient rocks. *Geobiology* **8**, 197–208.

Taylor SR, McLennan SM (1985) *The Continental Crust: Its Composition and Evolution*. Blackwell, Malden, MA.

Teutsch N, Schmid M, Müller B, Halliday AN, Bürgmann H, Wehrli B (2009) Large iron isotope fractionation at the oxic-anoxic boundary in Lake Nyos. *Earth and Planetary Science Letters* **285**, 52–60.

Thamdrup B (2000) Bacterial manganese and iron reduction in aquatic sediments. In: *Advances in Microbial Ecology*, Vol. 16 (ed Schink B). Kluwer Academic/Plenum Publishers, New York, pp. 41–84.

Toner BM, Fakra SC, Manganini SJ, *et al.* (2009) Preservation of iron(II) by carbon-rich matrices in a hydrothermal plume. *Nature Geoscience* **2**, 197–201.

Trendall AF (2002) The significance of iron-formation in the Precambrian stratigraphic record. *International Association of Sedimentologists Special Publication* **33**, 33–66.

Turick CE, Tisa LS, Caccavo F Jr (2002) Melanin production and use as a soluble electron shuttle for Fe(III) oxide reduction and as a terminal electron acceptor by Shewanella algae BrY. *Applied and Environmental Microbiology* **68**, 2436–2444.

Urrutia MM, Roden EE, Fredrickson JK, Zachara JM (1998) Microbial and surface chemistry controls on the reduction of synthetic Fe(III) oxide minerals by the dissimilatory iron-reducing bacterium *Shewanella alga*. *Geomicrobiology* **15**, 269–291.

Vargas M, Kashefi K, Blunt-Harris EL, Lovley DR (1998) Microbiological evidence for Fe(III) reduction on early Earth. *Nature* **395**, 65–67.

Verhoeven O, Mocquet A, Vacher P, *et al.* (2009) Constraints on thermal state and composition of the Earth's lower mantle from electromagnetic impedances and seismic data. *Journal of Geophysical Research* **114**, B03302.

Von Canstein H, Ogawa J, Shimizu S, Lloyd JR (2008) Secretion of flavins by Shewanella species and their role in extracellular electron transfer. *Applied and Environmental Microbiology* **74**, 615–623.

Wächtershäuser G (1988) Before enzymes and templates: theory of surface metabolism. *Microbiological Reviews* **52**, 452–484.

Waldbauer JR, Sherman LS, Sumner DY, Summons RE (2009) Late Archean molecular fossils from the Transvaal Supergroup record the antiquity of microbial diversity and aerobiosis. *Precambrian Research* **169**, 28–47.

Walker JCG (1984) Suboxic diagenesis in banded iron formations. *Nature* **309**, 340–342.

Weber KA, Pollock J, Cole KA, O'Connor SM, Achenbach LA, Coates JD (2006) Anaerobic nitrate-dependent iron(II) bio-oxidation by a novel lithoautotrophic betaproteobacterium, strain 2002. *Applied and Environmental Microbiology* **72**, 686-694.

Welander PV, Coleman ML, Sessions AL, Summons RE, Newman DK (2010) Identification of a methylase required for 2-methylhopanoid production and implications for the interpretation of sedimentary hopanes. *Proceedings of the National Academy of Sciences* **107**, 8537–8542.

Widdel F, Schnell S, Heising S, Ehrenreich A, Assmus B, Schink B (1993) Ferrous iron oxidation by anoxygenic phototrophic bacteria. *Nature* **362**, 834–836.

Wilhelm SW, Trick CG (1994) Iron-limited growth of cyanobacteria: multiple siderophore production is a common response. *Limnology and Oceanography* **39**, 1979–1984.

Wille M, Kramers JD, Nägler TF, *et al.* (2007) Evidence for a gradual rise of oxygen between 2.6 and 2.5 Ga from Mo isotopes and Re-PGE signatures in shales. *Geochimica et Cosmochimica Acta* **71**, 2417–2435.

Wille M, Nägler TF, Lehmann B, Schröder S, Kramers JD (2008) Hydrogen sulphide release to surface waters at the Precambrian/Cambrian boundary. *Nature* **453**, 767–769.

Wilson JP, Fischer WW, Johnston DT, *et al.* (2010) Geobiology of the late Paleoproterozoic Duck Creek Formation, Western Australia. *Precambrian Research* **179**, 135–149.

Xiong J (2006) Photosynthesis: what color was its origin? *Genome Biology* **7**, 245.

Yamaguchi KE, Johnson CM, Beard BL, Ohmoto H (2005) Biogeochemical cycling of iron in the Archean-Paleoproterozoic Earth: constraints from iron isotope variations in sedimentary rocks from the Kaapvaal and Pilbara Cratons. *Chemical Geology* **218**, 135–169.

Young GM (2002) Stratigraphic and tectonic settings of Proterozoic glaciogenic rocks and banded iron-formations: relevance to the snowball Earth debate. *Journal of African Earth Sciences* **35**, 451–466.

7

THE GLOBAL OXYGEN CYCLE

James F. Kasting[1] and Donald E. Canfield[2]

[1]Department of Geosciences, Penn State University, University Park, PA 16802, USA
[2]Danish Center for Earth System Science (DCESS) and Institute of Biology,
University of Southern Denmark, Campusvej 55, 5230 Odense M, Denmark

7.1 Introduction

Oxygen itself is not of the greatest interest to geobiologists. Although it is the most abundant element on Earth, most oxygen is bound tightly in silicate minerals, which comprise a large fraction of Earth's mantle and crust. Some geochemical cycling of the oxygen from silicates does take place, particularly at the midocean ridges (Muehlenbacks and Clayton, 1976), but that story is of less interest to geobiologists than the story of dioxygen, O_2.

This chapter concerns the role played by O_2 (and H_2, as we will see) in redox (reduction/oxidation) reactions, and it has been guided by a recent review by Sleep (2005) as well as by books and papers by Holland (1978, 1984, 2002), Catling et al. (2001), Canfield (2005), Canfield et al. (2006), Hayes and Waldbauer (2006), Goldblatt et al. (2006), and Claire et al. (2006). Sleep emphasizes the importance of midocean ridge processes as a sink for O_2, and We strongly concur with this. The midocean ridges are the locations where most of Earth's volcanism occurs and where the connection between Earth's crustal environment and the mantle are the strongest. Furthermore, midocean ridges may have been even more important as O_2 sinks in the distant past (Kump and Barley, 2007). We extend this line of reasoning and discuss its possible significance for the timing of the rise of atmospheric O_2.

7.2 The chemistry and biochemistry of oxygen

Redox reactions are reactions in which electrons are transferred from one atom or molecule to another. An oxygen atom has six electrons in its outer shell, which means that it needs two additional electrons to form a complete shell. This makes it a powerful *electron acceptor*. Oxygen will therefore 'borrow' electrons from whatever other element, or compound, that it can. For example, in the previous chapter we learned that ferrous iron, Fe^{2+}, can be *oxidized* to ferric iron, Fe^{3+}, by O_2. One can write this reaction compactly by expressing iron in the form of oxides:

$$2\,FeO + \tfrac{1}{2}O_2 \rightarrow Fe_2O_3 \qquad (7.1)$$

Here, FeO represents ferrous iron bound in silicates, and Fe_2O_3 is the mineral hematite. When this reaction occurs, one electron from each of the two ferrous iron atoms is partially transferred to the oxygen atom that came from O_2. The iron has been oxidized, because it reacted with O_2, but the key point is that it gave up, or donated, one of its valence electrons.

Note that oxygen is written as O_2 in Equation 7.1. That is because free oxygen atoms combine with each other to form molecular oxygen, O_2, and in the process two pairs of electrons are shared between the two atoms, giving each a complete outer shell. O_2 makes up 21% of Earth's atmosphere, and we will talk more about why that is so below.

Oxygen is of great interest to biologists because it plays a crucial role in metabolism. Aerobic organisms, including nearly all eukaryotes (organisms that have a membrane-bound cell nucleus), use O_2 for respiration. Similarly, when organisms die, aerobic bacteria can use O_2 to decompose the dead organic matter. If one follows geochemists' notation and uses CH_2O to represent more

Fundamentals of Geobiology, First Edition. Edited by Andrew H. Knoll, Donald E. Canfield and Kurt O. Konhauser.
© 2012 Blackwell Publishing Ltd. Published 2012 by Blackwell Publishing Ltd.

complex forms of organic matter, then both respiration and decay can be written as

$$CH_2O + O_2 \rightarrow CO_2 + H_2O \qquad (7.2)$$

Here, the two oxygen atoms from O_2 have borrowed four electrons from the carbon atom, increasing its valence state from 0 to +4. In doing so, they have formed carbon dioxide (CO_2) and water (H_2O). As part of this process, oxygen decreased its valence state from 0 (in O_2) to −2 (in both CO_2 and H_2O). Hydrogen in CH_2O and H_2O is assigned a valence state of +1.

If equation 2 is flipped around, it becomes the geochemists' representation of *oxygenic photosynthesis*:

$$CO_2 + H_2O \rightarrow CH_2O + O_2 \qquad (7.3)$$

Equation 7.3 is carried out by green plants and algae, both of which are eukaryotes, and also by cyanobacteria, which are prokaryotes (organisms which lack a membrane-bound cell nucleus). As we learned already in Chapter 2, this reaction is responsible for producing nearly all the O_2 in Earth's atmosphere. But there are subtleties to this argument, as we saw in Chapter 3. One needs to bury the organic carbon in sediments in order to create a net source for atmospheric O_2. Otherwise, Equation 7.3 will be quickly reversed by respiration. We will put some numbers into this argument later in this chapter.

In chemical terms, the CO_2 in Equation 7.3 is reduced to organic carbon. By this, we mean that the carbon atom gained, or accepted, electrons from the water molecule, while the water molecule served as the reductant, or *electron donor*. This took energy which was provided by sunlight.

CO_2 can also be reduced by other means. For example, one form of *anoxygenic photosynthesis* uses molecular hydrogen (H_2) as the reductant

$$CO_2 + 2H_2 \rightarrow CH_2O + H_2O \qquad (7.4)$$

However, H_2 has probably always been scarcer than H_2O in Earth's surface environment, so organisms that learned how to carry out Equation 7.3 eventually dominated the surface biosphere. And that, at the most basic level, is why our atmosphere is rich in O_2.

7.3   The concept of redox balance

When one compound in a reaction is oxidized, another compound must be reduced. Hence, redox balance is nothing more than conservation of electrons. In this chapter, we will apply the principle of redox balance to the combined atmosphere–ocean system and calculate a *global redox budget*. The redox budget of the Earth is balanced, but the oxidation state is not because, as we will see, the Earth continually loses hydrogen to space and is

becoming more oxidized with time. This rate is slow today, but could have been significant in the past. The atmosphere–ocean system also exchanges oxidized and reduced species with the crust and the mantle. The redox state of the mantle is evidently well buffered (see below) and is probably little affected by this exchange, but the crustal redox state could have been highly influenced, as we shall explore below.

Interestingly, it does not matter precisely how one expresses redox balance, as long as one is self-consistent. Typically, global redox balance today is expressed in terms of an O_2 budget. O_2 enters the atmosphere–ocean system when reduced compounds (mainly organic carbon and pyrite, FeS_2) are buried in sediments; it is consumed when atmospheric O_2 reacts with reduced minerals during weathering and with reduced gases released from volcanoes. As we will discuss further below, the lifetime of O_2 relative to these geologic processes is long – roughly two million years – and, hence, it makes sense to consider the atmosphere and ocean as a single system.

We could just as well keep track of redox balance in terms of H_2, rather than O_2. After all, the largest available reservoir of both oxygen and hydrogen is seawater, for which one can write the equilibrium reaction

$$H_2 + \tfrac{1}{2}O_2 \leftrightarrow H_2O \qquad (7.5)$$

For every mole of water that is split apart by photosynthesis or by UV photolysis, one mole of H_2 and one-half mole of O_2 are produced. Similarly, if one-half mole of O_2 is consumed during weathering, then the equivalent of one mole of H_2 will be left behind in the atmosphere. So, balancing the H_2 budget should give precisely the same answer as balancing the O_2 budget, as long as one flips the sign of each term and multiplies by a factor of 2 to account for the stoichiometry of Equation 7.5.

Because of its low present atmospheric concentration (0.55 ppm), it makes little sense to consider the present Earth surface redox balance in terms of H_2. But, on the early Earth H_2 should have been long-lived, whereas O_2 would have had a low concentration and an atmospheric lifetime ranging from seconds to hours. Thus, in the second half of this chapter, when we discuss the Archean Eon, we will recast atmospheric redox balance in terms of H_2, rather than O_2.

7.4   The modern O_2 cycle

7.4.1   *The biological O_2 cycle*

As discussed above (see also Chapter 2), O_2 is produced by oxygenic photosynthesis, and it is consumed by respiration and decay. These processes are extremely fast by geological standards. The rate of photosynthesis on land is about $60\,Gt\ C\ yr^{-1}$ (Prentice *et al.*, 2001;

see Chapter 2); which is equivalent to a net O_2 production rate of 5×10^{15} mol yr^{-1}, according to Equation 7.3. Net photosynthesis here is the rate of plant growth. The gross photosynthetic rate is higher, but some of the plant biomass produced is respired again, particularly at night. The rate of marine photosynthesis, or net primary productivity (NPP), is about 45Gt C yr^{-1}, or 3.8×10^{15} mol O_2 yr^{-1} (Prentice *et al.*, 2001). Both of these production processes for O_2 are nearly balanced by decay. The marine carbon cycle, though, has a leak (to sediments) of the order of a few tenths of a percent of NPP (Holland, 1978). We shall return to this below, but for now let us assume that the biological organic carbon cycle (the topic of this section) is exactly balanced.

The atmosphere contains about 3.6×10^{19} mol O_2. The residence time of O_2 in the atmosphere is equal to the reservoir size divided by the output flux, that is, the rate of consumption by respiration and decay. As we have just seen, this rate is approximately equal to the production rate from photosynthesis. Hence, the residence time for O_2 relative to the terrestrial organic carbon cycle is 3.6×10^{19} mol$/5 \times 10^{15}$ mol $yr^{-1} \cong 7000$ yr. In other words, each O_2 molecule in the atmosphere cycles through the terrestrial biosphere once every 7000 yr, on average. By comparison, the CO_2 concentration of the preindustrial atmosphere, prior to 1800 CE, was about 280 ppm. This is lower than the O_2 concentration by a factor of 750; hence, the lifetime of atmospheric CO_2 relative to the terrestrial organic carbon cycle is about 7000 yr$/750 \cong 10$ yr.

7.4.2 The geologic O_2 cycle

7.4.2.1 Sources of O_2

Atmospheric O_2 levels are controlled by the small imbalances in the biological oxygen cycle. Thus, the major source for O_2, in a geological sense, comes from the leak in the marine organic carbon cycle. According to Holland (2002), the burial rate of organic carbon in sediments is $(10 \pm 3.3) \times 10^{12}$ mol yr^{-1}. This amounts to a little over 0.25% of marine NPP (Table 7.1). Berner (2004) estimates an organic C burial rate of 5.3×10^{12} mol yr^{-1}, or about half Holland's value. We use Holland's value for this and many other fluxes in Table 7.1 because the values are self-consistent; however, it should be borne in mind that all these fluxes are probably uncertain by a factor of at least 2.

An additional large source of O_2 comes from the burial of pyrite (FeS_2) in sediments. Pyrite is produced from the sulfide generated from bacterial sulfate reduction in sediments (see Chapter 5). Chemically, we can represent this as

$$2Fe_2O_3 + 8H_2SO_4 + 15CH_2O \rightarrow 4FeS_2 + 15CO_2 + 23H_2O$$

$$(7.6)$$

Table 7.1 The modern geologic O_2 budget

O_2 production	
From burial in sediments	Effect on O_2 $(10^{12}$ mol $yr^{-1})$
Burial of C^0	10 ± 3.3
Burial of FeS_2	7.0 ± 3.6
Burial of FeO	0.9 ± 0.4
Loss of hydrogen to space	0.024
Total	17.9
O_2 loss	
Oxidation of C^0 during weathering	7.5 ± 2.5
Oxidation of FeS_2 during weathering	7.0 ± 3.6
Oxidation of FeO during weathering	1.0 ± 0.6
Surface outgassing of H_2	0.1
Surface outgassing of SO_2 and H_2S	0.2
Submarine outgassing of H_2S	1.0
Submarine H_2 from serpentinization	0.2
Total	17.0

This reaction illustrates some of the complexities of redox chemistry. Here, eight S atoms are being reduced from a valence state of +6 in sulfate (represented here as sulfuric acid, H_2SO_4) to –1 in pyrite, requiring a net addition of 56 electrons. At the same time, four Fe atoms are reduced from the +3 state in hematite to +2 in pyrite, which takes an additional 4 electrons. The 60 electrons required to perform both reductions are derived from oxidation of 15 carbon atoms from the 0 valence state in CH_2O to +4 in CO_2.

Now, O_2 does not appear explicitly Equation 7.6 the way it has been written. But the 4 moles of pyrite that are formed, and subsequently buried, are equivalent to the burial of 15 moles of organic carbon; hence, they contribute 15 moles of O_2 to the atmosphere-ocean system (Garrels and Perry, 1974). The net rate of O_2 addition from pyrite burial, according to Holland (2002), is $(7.0 \pm 3.6) \times 10^{12}$ mol yr^{-1}. This is nearly as large as the direct production of O_2 from organic carbon burial itself. The burial of 'FeO' in sediments also contributes to oxygen production at a rate of 0.9×10^{12} mol yr^{-1}. If we add these three terms (burial of organic carbon, pyrite, and FeO), the net production of O_2 is of the order of 18×10^{12} mol yr^{-1}, with an uncertainty of the order of 50%.

7.4.2.2 Sinks for O_2 (surface processes)

This geological production of O_2 must be balanced by a corresponding geological O_2 sink; otherwise, atmospheric O_2 would change. Again, following Holland (2002), the greatest consumption of atmospheric O_2 occurs during the continental weathering of organic carbon, principally kerogen (Equation 7.7), pyrite, and FeO from sedimentary (and to some extent igneous) rocks.

$$C^0 + O_2 \rightarrow CO_2 \qquad (7.7)$$

Equation 7.7 is similar to Equation 7.2 except that CH_2O has been replaced by C^0. This reflects the fact that kerogen has lost much of its hydrogen and oxygen and, hence, is closer to graphite in chemical composition than to fresh organic matter. The rate of consumption of O_2 during organic carbon weathering is $(7.5 \pm 2.5) \times 10^{12}$ mol yr^{-1} while pyrite weathering consumes an additional $(7.0 \pm 3.6) \times 10^{12}$ mol yr^{-1} O_2, and FeO oxidation contributes $(1.0 \pm 0.6) \times 10^{12}$ mol yr^{-1} to O_2 consumption. Summing these terms, the total rate at which O_2 is consumed during weathering is about 15.5×10^{12} mol yr^{-1}, again with an uncertainty of the order of 50%.

The difference between the calculated rate of O_2 production $(17.9 \times 10^{12}$ mol $yr^{-1})$ and O_2 consumption during weathering $(15.5 \times 10^{12}$ mol $yr^{-1})$ is 2.4×10^{12} mol yr^{-1}. This is well within the errors of each estimate, and so one could say that the geological O_2 cycle is effectively balanced. However, there are additional O_2 sinks that we have not yet considered which may be important on long time scales. These include the oxidation of reduced gases (and fluids) emanating from surface, or subaerial, volcanoes and those coming from deep sea hydrothermal vents. We analyze these terms separately below.

Volcanic gases released subaerially consist mostly of H_2O and CO_2, with smaller amounts of H_2, CO, and reduced sulfur gases (SO_2 and H_2S). The concentrations of H_2 and CO are related to those of H_2O and CO_2 by way of the oxygen fugacity of the magmas from which they are released. The magmas are relatively reduced; hence, a small percentage (~1%) of the hydrogen is released as H_2, and a slightly larger percentage (~2%) of the carbon is released as CO. (See discussion in Holland, 1984, p. 40ff.). Outgassing of CO is equivalent, from a redox standpoint, to outgassing of H_2 because of the reaction

$$CO + H_2O \rightarrow CO_2 + H_2 \qquad (7.8)$$

If carbon is injected into the atmosphere-ocean system as CO and leaves the system as CO_2 (in carbonates), then an equivalent amount of H_2 has been generated.

The release rate of hydrogen from surface volcanoes has been a topic of recent debate. Holland (2002) estimates an H_2 flux of 1.4×10^{12} mol yr^{-1} by scaling the H_2 outgassing rate to the CO_2 outgassing flux (Holland lists an even larger H_2 input, ~5×10^{12} mol yr^{-1}, but this number includes both volcanic and submarine fluxes.) This H_2 flux corresponds to an O_2 sink of about 7×10^{11} mol yr^{-1} (see Equation 7.5). This is about one-third of the difference between O_2 production and O_2 consumption rates calculated above, suggesting that release of H_2 by surface volcanoes could make a significant contribution to the net O_2 sink.

Other authors, though, have argued that the H_2 flux from subaerial volcanoes is significantly lower. For example, Canfield *et al.* (2006) estimate a volcanic H_2 flux of $(1.8–5.0) \times 10^{11}$ mol yr^{-1}. They derive this number by comparing the observed amount of H_2 in volcanic gases with that of SO_2, and then scaling by the (better known) volcanic flux of SO_2, $(2.3–3.1) \times 10^{11}$ mol yr^{-1} (Halmer *et al.*, 2002). This procedure seems reasonably well constrained. Hayes and Waldbauer (2006) also get a lower number: they estimate the H_2 source from arc volcanism to be 1.7×10^{11} mol yr^{-1}. We will adopt a conservative estimate of 2×10^{11} mol H_2 yr^{-1} from surface volcanism here. This corresponds to an O_2 sink of 1×10^{11} mol yr^{-1}.

The sulfurous gases emanating from surface volcanoes also form an O_2 sink. Thus, volcanic SO_2 is oxidized to sulfuric acid (or sulfate) by the reaction

$$SO_2 + H_2O + \tfrac{1}{2} O_2 \rightarrow H_2SO_4 \qquad (7.9)$$

Some H_2S is also emitted from volcanoes and from low-temperature vents such as those within Yellowstone National Park in the USA. The H_2S flux is roughly one-tenth that of SO_2 (Berresheim and Jaeschke, 1983), but oxidizing H_2S to sulfate requires an eight-electron transfer, compared to only two electrons for Reaction 7.9, and so the O_2 requirement is about 50% that of SO_2. Hence, the net O_2 sink from subaerial volcanic SO_2 plus H_2S combined is approximately 2×10^{11} mol yr^{-1}, or about twice estimate for the sink provided by subaerial H_2.

7.4.2.3 Sinks for O_2 (submarine processes)

Large quantities of reduced gases also emanate from midocean ridge hydrothermal vents. Holland (2002) estimated average dissolved H_2S and H_2 concentrations of 7 mmol kg^{-1} and 2 mmol kg^{-1}, respectively, implying a total O_2 demand of 15 mmol kg^{-1}. The H_2S concentration is highly variable and can be anywhere within the range of 3–80 mmol kg^{-1} at different vent sites (Von Damm, 1995, 2000). Exactly where the H_2S is coming from is unclear. Isotopic studies suggest that much of it is magmatic (i.e. it is leached from the surrounding basalts; Alt, 1995), as its $\delta^{34}S$ values are typically closer to that of mantle sulfides (0‰) than to seawater sulfate (21‰). However, some H_2S could also be derived directly from seawater sulfate if the reduction process occurs at temperatures below 300°C (Arthur, 2000). The observed preponderance of H_2S over Fe^{+2} in vent fluids, as compared to the 12:1 ratio of Fe:S in basalts (Walker and Brimblecombe, 1985), suggests that reduction of seawater sulfate is an important source for the H_2S. Both magmatic sulfide and ferrous iron are probably important O_2 sinks. One can see this by writing the redox reaction (Lecuyer and Ricard, 1999; Sleep, 2005).

$$H_2SO_4 + 12FeO \rightarrow 4Fe_3O_4 + H_2S \qquad (7.10)$$

Fe_3O_4, *magnetite*, is an intermediate oxidation state of iron that can be thought of as one part FeO and one part Fe_2O_3; hence, two iron atoms have been oxidized to ferric iron. The 12:1 ratio between FeO and H_2SO_4 in equation 7.10 effectively makes up for the 12:1 ratio of Fe:S in the basalts, suggesting that magmatic sulfide and ferrous iron provide comparable O_2 sinks.

We will follow Sleep (2005) and assume for the sake of budgeting that the sulfate is reduced by reaction with ferrous iron. To estimate the net oxygen sink, the O_2 demand of 15 mmol kg^{-1} should be multiplied by the flux of hot seawater coming from the vents, about 5×10^{13} kg yr^{-1} (Wolery and Sleep, 1976; Mottl and Wheat, 1994). This yields an O_2 sink of $\sim 0.75 \times 10^{12}$ mol yr^{-1}. Lecuyer and Ricard (1999) and Sleep (2005) derive slightly larger numbers by analyzing the ferric iron content of seafloor and combining this with seafloor production rates. So, we round this number up to 1×10^{12} mol yr^{-1}, recognizing that, like other fluxes in our redox budget, it is highly uncertain.

An additional chemical redox process, termed *serpentinization*, also takes place within the modern midocean ridge systems (Sleep, 2005). Some of the rocks deep within the ridge systems are *ultramafic*, meaning rich in iron and magnesium. Peridotite, which contains the minerals olivine and pyroxene, is the most common example. Especially along slow-spreading ridges like the Mid-Atlantic Ridge, deep fractures allow warm seawater to interact with these ultramafic rocks at depth. Unlike normal seafloor basalts, these ultramafic rocks generate hydrogen as they are altered to form various *serpentine minerals*. We can write this reaction as

$$3FeO + H_2O \rightarrow Fe_3O_4 + H_2 \qquad (7.11)$$

If the water contains dissolved CO_2, then some of the CO_2 can be reduced to CH_4, as witnessed at the Lost City ventfield along the Mid-Atlantic Ridge (Kelley *et al.*, 2005; Proskurowski *et al.*, 2008).

The O_2 sink from serpentinization has been estimated by Sleep (2005) at 0.2×10^{12} mol yr^{-1} – about one-fifth the O_2 sink from oxidation of the seafloor by sulfate. Thus, this process is of relatively minor importance to the modern geologic O_2 budget, but on the early Earth, when O_2 and sulfate were less abundant, it may well have been the dominant O_2 sink.

If we add the midocean ridge O_2 sink to that from subaerial volcanism, we compute a net 'volcanic' O_2 sink of about 1.5×10^{12} mol yr^{-1}, with about 80% of this coming from the midocean ridges. This is a little over half the 2.4×10^{12} mol yr^{-1} needed to bring net O_2 production and consumption into balance, according to Holland (2002). Given the large uncertainties in all of these numbers, this level of agreement is probably as good as can be expected.

We finish this discussion by calculating the geologic lifetime of O_2. The amount of O_2 in the atmosphere is 3.6×10^{19} mol. The estimated rate of O_2 consumption by weathering and by oxidation of volcanic gases is about 18×10^{12} mol yr^{-1}. Hence, the geologic lifetime of O_2 is 3.6×10^{19} mol$/1.8 \times 10^{13}$ mol yr^{-1} = 2 million years (my). This then is the time scale over which atmospheric O_2 concentrations could fluctuate significantly if inputs and outputs to the system were to become unbalanced. This time scale also shows that atmospheric O_2 levels are essentially fixed on time scales of immediate concern to us.

7.4.3 O_2 over Phanerozoic time

Actually, atmospheric O_2 concentrations have probably been relatively high throughout the entire Phanerozoic, the last 542 million years. The persistence of animal life through the Phanerozoic indicates that oxygen did not dip below about 10% of present levels. Also, experiments on the oxygen sensitivity of terrestrial plant materials to burning suggest that Phanerozoic oxygen probably stayed within about 50% of present levels through most of this time (Wildman *et al.*, 2004). At higher oxygen levels, terrestrial vegetation would have been severely impacted by fires. Models have been constructed to attempt to hindcast the Phanerozoic history of atmospheric oxygen. These models are based on the principles outlined above, where the burial of organic carbon and pyrite sulfur represent oxygen sources and the weathering of sedimentary rocks and volcanic and metamorphic gases represent oxygen sinks (Hansen and Wallmann, 2003; Berner, 2004; Bergman *et al.*, 2004). These models also attempt to parameterize how such factors as sea level changes, temperature, seafloor spreading rates and the evolution of land plants influence weathering. Some models also consider terrestrial and marine nutrient cycles and how these influence primary production in the oceans (Hansen and Wallmann, 2003; Bergman *et al.*, 2004).

Each of these models includes a variety of feedbacks to dampen oxygen fluctuations. In some cases, an oxygen dependency on weathering is included (e.g. Bergmann *et al.*, 2004); an inverse dependency of oxygen concentration on pyrite burial has also been used (e.g. Berner and Canfield, 1989). An important modeling concept has been the inclusion of rapid recycling. This means that the material most likely to be weathered is that which was most recently deposited. Rapid recycling acts as a negative feedback, dampening oxygen fluctuations. Recent model results are shown in Fig. 7.1.

7.4.4 Loss of hydrogen to space

In the previous section, we discussed the global redox budget. In doing so, we took into account the O_2 sink provided by the mantle, but we did not account for processes occurring at the top boundary of the system. Hydrogen, being a light gas, can be lost to space, whereas

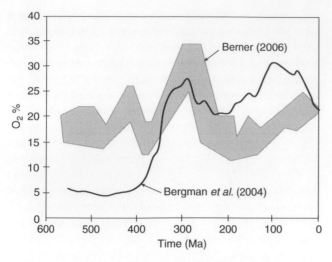

Figure 7.1 Estimates of Phanerozoic O_2 concentrations from two different models.

oxygen, which is much heavier, cannot. Because Earth's hydrogen occurs mostly as H_2O, loss of hydrogen to space leaves O_2 behind, thereby contributing to net O_2 production.

Hydrogen escape to space is a complex process, as it occurs by way of a combination of thermal and nonthermal escape mechanisms that operate near the *exobase*, around 500 km altitude (Walker, 1977; Chamberlain and Hunten, 1987). (The exobase is the altitude above which the atmosphere becomes collisionless.) Fortunately, for the present Earth at least, we do not need to worry about the details of these processes because the escape rate itself is limited by a different factor, namely, the rate at which hydrogen can diffuse upwards through the *homopause*. The homopause occurs near 90 km altitude and is that region of the atmosphere above which mixing of gases by winds and by turbulent eddies is superseded by mixing from molecular diffusion. Or, to put it another way, it is the altitude above which light gases begin to separate out from heavier ones.

In a classic paper, Hunten (1973) showed that the *diffusion-limited escape rate* is proportional to the *total hydrogen mixing ratio*, $f_{tot}(H)$, at the homopause level. A more accessible discussion is given by Walker (1977, p. 157 ff.). Mathematically, the escape rate can be approximated by the formula

$$\Phi_{esc}(H) \cong \frac{b f_{tot}(H)}{H_a} \qquad (7.12)$$

Here, b is a weighted average of the binary diffusion coefficients of H and H_2 in air, and H_a is the atmospheric pressure scale height (= RT/g)[1]*. The quantity $f_{tot}(H)$ will be defined below. Numerically, $b \cong 1.5 \times 10^{19}$ cm^{-1} s^{-1}, and

$H_a \cong 6 \times 10^5$ cm (i.e. 6 km), so $b/H_a \cong 2.5 \times 10^{13}$ cm^{-2} s^{-1}. For redox budgeting, it is convenient to write the escape rate in terms of H_2 molecules instead of H atoms. So, we can rewrite Equation 7.12 as

$$\Phi_{esc}(H_2) \cong 2.5 \times 10^{13} f_{tot}(H_2) \text{ molec cm}^{-2} \text{ s}^{-1} \qquad (7.13)$$

When expressed in this way, the total hydrogen mixing ratio can be written as

$$f_{tot}(H_2) = f(H_2) + \tfrac{1}{2} f(H) + f(H_2O) + 2f(CH_4) + \dots \qquad (7.14)$$

i.e. it is the sum of the volume mixing ratios[2]* of all of the hydrogen-bearing species, weighted (in this case) by the number of H_2 molecules that they contain.

At the homopause in today's atmosphere, most of the hydrogen is present either as H or H_2. But the total hydrogen mixing ratio is approximately constant throughout the stratosphere and mesosphere, above the region where H_2O can condense. Hence, $f_{tot}(H_2)$ can be evaluated at the tropopause, which occurs at about 10–12 km altitude over much of the Earth's surface. There, the dominant forms of hydrogen-bearing gases are H_2O (~4 ppm) and CH_4 (~1.6 ppm). Adding these two species together gives: $f_{tot}(H_2) \cong 4 \times 10^{-6} + 2(1.6 \times 10^{-6}) = 7.2 \times 10^{-6}$. Plugging this value back into Equation 7.13 then yields: $\Phi_{esc}(H_2) \cong 1.8 \times 10^8$ H$_2$ molec cm^{-2} s^{-1}. Finally, we convert this into total flux with the conversion factor: 1 molec cm^{-2} s$^{-1} \cong 268$ mol yr^{-1}. Hence, we get $\Phi_{esc}(H_2) \cong 4.8 \times 10^{10}$ mol yr^{-1}.

Each mole of H_2 is stoichiometrically equivalent to one-half mole of O_2 (reaction 7.5). Thus, the net O_2 production is $\tfrac{1}{2}(4.8 \times 10^{10}$ mol yr$^{-1}) = 2.4 \times 10^{10}$ mol yr^{-1}. This can be compared with a total geological O_2 production rate (from above) of 1.8×10^{13} mol yr^{-1}. Hence, hydrogen escape to space is completely negligible! Why take all this trouble to estimate an insignificant term? The answer is that hydrogen escape is negligible today only because Earth's atmosphere is rich in O_2. In a more reduced early atmosphere, both CH_4 and H_2 could have been much more abundant, and so H escape to space could well have been an important term in the global redox budget.

7.5 Cycling of O_2 and H_2 on the early Earth

7.5.1 *Timing and cause of the rise of atmospheric O_2*

We now turn our attention to the early Earth. As was made clear in Chapter 5, there is now convincing evidence from mass-independently fractionated sulfur isotopes that atmospheric O_2 levels were low prior to

---

[1] Here, R is the universal gas constant, T is temperature in Kelvin, and g is Earth's gravitational acceleration.

[2] The term volume mixing ratio is defined as the number density of atoms of a given species divided by the total number density of air molecules. Hence, a concentration of 1 part per million by volume corresponds to a volume mixing ratio of 1×10^{-6}.

about 2.4 Ga. The idea that atmospheric O_2 first rose at about this time is not new. It was proposed originally by Preston Cloud (Cloud, 1972) on the basis of other geologic O_2 indicators (Cloud actually put the transition at ~2.0 Ga based on less accurate radiometric age dates available at that time.) Heinrich Holland spent a good portion of his career further documenting this evidence (see, e.g. Holland, 1984, 1994, 2006). James Walker (1977) and Donald Canfield (Canfield *et al.*, 2000, 2006, 2007) have also worked on various aspects of this problem. At this point, nearly all workers agree that Archean O_2 concentrations were low, with the notable exception of Hiroshi Ohmoto and his colleagues (Ohmoto, 1996; Ohmoto *et al.*, 2004, 2006).

Just as we understand the timing of the initial O_2 rise, we also understand at least something about its cause. All workers, including Ohmoto, agree that O_2 concentrations should have remained low until the evolution of oxygenic photosynthesis by cyanobacteria. However, there is currently little agreement about the relative timing of the origin of cyanobacteria. Schopf and Packer (1987) and Schopf (1993) suggested that cyanobacteria-like microfossils were present in the Apex Chert in the 3.5 by-old Warrawoona Formation of Western Australia, but the biogenic origin has been questioned (Brasier *et al.*, 2002). Brocks *et al.* (1999) suggested that the presence of 2α-methylhopanes and steranes in 2.7-by-old sedimentary rocks from Western Australia demonstrated the presence of cyanobacteria and (O_2-dependent) eukaryotes at that time. However, Rasmussen *et al.*, (2008) have recently argued that these biomarkers are not indigenous to the sediments; furthermore, methyl-hopanes are now known to be produced by anoxygenic phototrophic bacteria as well (Rashby *et al.*, 2007). Clearly, microfossils and biomarkers provide equivocal evidence for the timing of cyanobacterial evolution (Kopp *et al.*, 2005; Kirschvink, 2006).

Additional evidence for the existence of oxygenic photosynthesis at 2.7 Ga is provided by isotopically light bulk kerogen–up to 60‰ depleted in ^{13}C compared to carbonates in sedimentary rocks of this age. Hayes (1983, 1994) has argued that this organic carbon was formed during two separate steps of biological fractionation: a first step in which CO_2 is reduced to organic matter through photosynthesis, and a second in which isotopically light CH_4 produced from anaerobic decay of this organic matter was incorporated into cell biomass by aerobic methanotrophs (organisms that combine CH_4 with O_2). Alternatively, so-called reverse methanogens living in consortia with sulfate-reducing bacteria could have produced this high degree of fractionation (Hinrichs, 2002). This latter mechanism requires a substantial source for sulfate, and that in turn requires free O_2; hence, it also supports the idea that oxygenic photosynthesis had been invented by this time.

Evidence for oxygenic photosynthesis at ~2.5 Ga is provided by molybdenum (Wille *et al.*, 2007; Anbar *et al.*, 2007) and sulfur (Kaufmann *et al.*, 2007) isotopes in ancient black shales. On the whole, the available evidence suggests that oxygenic photosynthesis had already been invented by at least 2.5 Ga and perhaps by 2.7 Ga – some 300 My prior to the time when atmospheric O_2 concentrations first rose – but the evidence is not strong. If so, then what caused the long delay between the origin of photosynthesis and the rise of atmospheric O_2? And what was the atmosphere like prior to the oxic–anoxic transition? Below, we offer some ideas on how to approach these questions.

7.5.2 Effect of low atmospheric O_2 on the global redox budget

Let us first consider how the global redox budget would have worked prior to the rise of atmospheric O_2. To keep things as simple as possible, we assume a completely abiotic Earth. We further assume that the abiotic production rate of organic carbon was negligible and that the sulfur cycle was extremely slow. The latter assumption may not seem intuitive, but oceanic sulfate concentrations would have been low, probably below 0.2 mM (Habicht *et al.*, 2002), as compared to 28 mM today. Therefore, less sulfate would have entered the midocean ridge vent systems, and hence less H_2S would have been formed by its reduction. Magmatic sulfide would similarly have become less important as an O_2 sink. Because the deep oceans were filled with ferrous iron (Holland, 1984), the H_2S given off by vents would have left the system almost immediately as pyrite, with little net effect on global redox state. Similarly, with low oceanic sulfate concentrations, less sulfur would have subducted in sediments (or as anhydrite in the oceanic crust), liberating less SO_2 from arc volcanoes. SO_2 entering the atmosphere would have left the system as a combination of SO_2, H_2S, and H_2SO_4 (Pavlov and Kasting, 2002). But, because of sulfur's reduced mobility, its concentration in volcanic gases should have been lower than that of H_2, and so its contribution to the global redox budget was probably small.

Other terms in the global O_2 budget should have been absent as well on the prebiotic Earth. Specifically, oxidative weathering of the continents should have been negligible because of vanishingly small atmospheric O_2 concentrations. This leaves surface and submarine outgassing of H_2 as the dominant O_2 sinks. If we use modern values for these numbers, without scaling them to the higher heat flux of Archean, the combined O_2 sink would have been of the order of 3×10^{11} mol yr^{-1}.

The list of O_2 sources on the prebiotic Earth is short. Escape of hydrogen to space, for which the modern rate is 2.4×10^{10} mol yr^{-1}, is the only real source. Indeed, the rate of H_2 escape overestimates the rate of abiotic O_2 production, as half of the hydrogen that escapes today comes from CH_4, and CH_4 itself comes mostly from the decomposition of organic matter, a process that by presumption did not exist on the prebiotic Earth. The actual net source for O_2 on the prebiotic Earth would have been the escaping hydrogen that was generated by H_2O photolysis, about 1.3×10^{10} mol yr^{-1}. [One can see this by computing the contribution of H_2O to $f_{tot}(H_2)$ in Equation 7.14]. This term is so low because H_2O is effectively condensed out of the atmosphere at the *tropopause cold trap*, around 10–12 km. Climate models predict that an Archean Earth with a CO_2-rich atmosphere warmed by a dimmer young Sun would have had an even more effective cold trap, and thus an even drier stratosphere (see, e.g. Kasting *et al.*, 1984; Kasting and Ackerman, 1986). Hence, we conclude that the abiotic O_2 source would have been much smaller than the potential O_2 sink from reduced volcanic gases.

The global redox budget must still have been balanced, however. It does not make sense to compute redox balance with O_2 on an O_2-poor Archean Earth, so instead we balance production and loss of H_2 (Walker, 1977). The net source of H_2 from surface volcanism and from serpentinization should be twice the corresponding O_2 sink listed in Table 7.1, or 6×10^{11} mol yr^{-1}. To convert back to photochemists' units, we can divide this number by 268, yielding 2.2×10^9 H_2 molec cm^{-2} s^{-1}. This H_2 production rate must have been balanced by the loss of hydrogen to space, which is given by Equation 7.13. Solving for the total hydrogen mixing ratio yields: $f_{tot}(H_2) \cong 10^{-4}$. H_2O makes a negligible contribution to f_{tot}, and other H-bearing gases should have been scarce; hence, the atmospheric H_2 mixing ratio should also have been of the order of 10^{-4}, or 100 ppmv. This value is near the lower end of the range of H_2 mixing ratios considered in various published models of prebiotic atmospheric composition (e.g. Kasting *et al.*, 1984; Kasting, 1993).

In reality, the source of hydrogen on the early Earth was almost certainly higher than the estimate just given, for two reasons. First, the geothermal heat flow was higher as a consequence of greater heat generation by radioactive decay and left over heat from planetary formation. An increase in geothermal heat flow by a factor of 2–3 at 4.0 Ga relative to today is not unreasonable (McGovern and Schubert, 1989; Sleep, 2007). If this excess heat resulted in thicker oceanic crust (Moores, 2002; Sleep, 2007), then the volume of oceanic crust produced at the ridges should have been proportional to heat flow squared, yielding an increase at 4.0 Ga by a factor of 4–9. But this may well underestimate the rate of serpentinization, and associated H_2 production, because the crust may also have been more mafic as a consequence of higher upper mantle temperatures and a corresponding greater degree of partial melting. So, let us choose

10 as a plausible value for the enhancement in H_2 outgassing at 4.0 Ga. The same balancing procedure used above would then predict an atmospheric H_2 mixing ratio of 10^{-3}, rather than 10^{-4}. That is a more typical value assumed in early atmosphere models.

7.5.3 Prebiotic O_2 concentrations

With estimates of the H_2 concentration in the prebiotic atmosphere, it is straightforward (but tedious) to compute the concentration of atmospheric O_2. This is done with a *photochemical model*. In such a model, one divides the atmosphere up into a number of layers, calculates the solar UV radiation in each layer, and uses this information to compute photochemical reaction rates between different atmospheric gases. Modeling of this sort shows that O_2 would have been a short-lived species within such an atmosphere. Near the surface, it would have been consumed in a matter of seconds by reaction with H_2 (reaction 7.5 going to the right), with the reaction catalyzed by the byproducts of H_2O photolysis. But, O_2 would also have been produced higher up in the atmosphere by CO_2 photolysis, followed by recombination of O atoms to form O_2

$$CO_2 + h\nu \rightarrow CO + O \tag{7.15}$$

$$O + O + M \rightarrow O_2 + M \tag{7.16}$$

Here, 'M' represents a third molecule, necessary to carry off the excess energy of the collision.

By putting all of this together in a computer model, one derives estimates for early atmospheric composition similar to that shown in Fig. 7.2. The assumed surface pressure of this atmosphere is 10^5 Pa (1 bar), close to the value 1.103×10^5 Pa (1.013 bar) for the modern Earth. In this calculation, the atmospheric H_2 mixing is close to 10^{-3}, the value derived above, N_2 is present at roughly

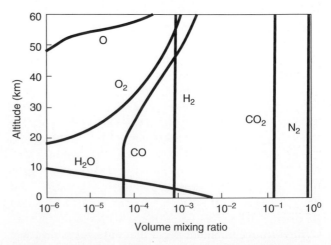

Figure 7.2 Vertical profiles of major atmospheric gases in a simulated prebiotic Earth atmosphere (from Kasting, 1993).

the same concentration as today (~0.8 bar) and the CO_2 mixing ratio is 0.2, about 500 times higher than at present. This assumed high value is necessary to provide enough greenhouse effect to compensate for 25–30% reduced solar luminosity early in Earth history (Gough, 1981; Kasting, 1993).

Our calculation demonstrates that O_2 would indeed have been a trace species in a prebiotic atmosphere. Its predicted mixing ratio in the upper stratosphere is ~10^{-3}, comparable to that of H_2, but near the surface its mixing ratio drops to a value of ~10^{-13}, indicating that it is essentially absent. Indeed, that is just what one needs to get life started, as O_2 would have been a poison both to prebiotic chemistry and to unprotected early organisms.

7.5.4 Oxidation of the mantle and crust

7.5.4.1 Oxidation of the mantle

An early Earth H_2-rich atmosphere could have led indirectly to substantial oxidation of both the crust and mantle. This counterintuitive result is a direct consequence of the escape of hydrogen to space. Recall that most of Earth's hydrogen arrived originally in the form of H_2O; therefore, when H_2 is lost, O_2 must accumulate somewhere. We know that O_2 did not build up in the atmosphere until ~2.4 Ga; hence, it must have reacted into either the crust, the mantle, or both. We saw earlier that serpentinization reactions may have proceeded at rates considerably faster than today. If we scale the Archean serpentinization rate up by a factor of 10, then O_2 went into the seafloor as ferric iron at a rate of 2×10^{12} mol yr^{-1}, roughly equal to the present rate of subduction of ferric iron into the mantle, ~1.7×10^{12} mol yr^{-1} (O_2 equivalent) (Lecuyer and Ricard, 1999). This accidental equivalence results from cancellation of competing effects: subduction has likely been slowing down over time, but the crust that is subducted today is more highly oxidized because of its interaction with seawater sulfate.

If ferric iron was subducted at this rate during the first 2 By of Earth's history, the equivalent amount of O_2 taken down into the mantle would have been 4×10^{21} moles. This is a large amount of O_2—about 100 times the amount present in today's atmosphere, or one-tenth of the oxygen in the ocean as water. That said, it has evidently had little effect on mantle oxidation state over time (Delano, 2001; Canil, 2002; Li and Lee, 2004), presumably because the mantle is well buffered by both iron and carbon.

7.5.4.2 Oxidation of the crust

O_2 has also accumulated in the Earth's continental crust. Continental rocks contain an estimated 8×10^{21} mol of ferric iron (Lecuyer and Ricard, 1999). Most of this iron should have been in the ferrous state originally, so oxidizing it would have required the addition of 2×10^{21} mol of O_2, according to equation 1. This O_2 could have come from photosynthesis followed by organic carbon burial (Equation 7.3). However, the total amount of organic carbon in the crust is only about 1.3×10^{21} mol (Garrels and Lerman, 1981). Furthermore, much of the ferric iron in older, Archean cratonic rocks appears to have been in place since soon after the rocks formed, that is, oxidation did not wait for the oxidation of the atmosphere at 2.4 Ga and the subsequent initiation of the oxidative weathering of continental rocks. This suggests that escape of hydrogen to space may have played a role in oxidizing the crust, as well (Catling *et al.*, 2001). If we scale the modern H_2 surface outgassing rate up by a factor of 10, and assume that all of this H_2 was lost to space, then the O_2 equivalent production rate would have been 1×10^{12} mol yr^{-1}. Over 2 billion years this would amount to a net O_2 production of 2×10^{21} mol – just the amount needed to oxidize the crust.

Indeed, Catling *et al.* (2001) and Claire *et al.* (2006) suggested that progressive oxidation of the crust delayed the rise of atmospheric O_2 from 2.7 Ga until 2.4 Ga. In their view, reduced gases were released from crustal metamorphism at a greater rate in the past, and these gases retarded the O_2 rise. (The term 'metamorphism' refers to relatively low temperature processes, such as mountain building, in which rocks do not fully melt.) The details of how crustal oxidation actually proceeded were not spelled out in their paper. However, the Archean crust, much of which is characterized by the presence of *greenstone belts* (Holland, 1984), was also more mafic than the modern crust and, hence, may have been more prone to oxidative processes such as serpentinization. So, progressive crustal oxidation could well have played a role in delaying the rise of atmospheric O_2.

7.5.5 The post-biotic Archean atmosphere

Thus far, we have focused on abiotic processes. Geobiologists suspect, however, that life has been present since at least 3.5 Ga, based on the occurrence of both microfossils and stromatolites, as well as isotopically light organic carbon and sulfur. Life should have modified the Archean atmosphere in a variety of different ways, even prior to the origin of cyanobacteria. As originally pointed out by Walker (1977), the most significant of these was probably the generation of methane, CH_4, by *methanogenic bacteria*, or *methanogens*. Methanogens would have operated initially on abiotically produced H_2 and on organic matter produced by carbon-fixing organisms. They would have likely converted much of the available atmospheric H_2 into CH_4 (Walker, 1977; Kharecha *et al.*, 2005).

High concentrations of atmospheric CH_4 would have had a number of consequences that would have been important for biology. CH_4 could have contributed to the greenhouse effect, thereby helping to compensate for reduced solar luminosity (Pavlov *et al.*, 2000; Haqq-Misra *et al.*, 2008). More importantly for our purposes here, production of CH_4 could also have contributed significantly to global redox balance, especially after the origin of oxygenic photosynthesis. Because the hydrogen required to produce organic matter photosynthetically was coming from H_2O, the rate of production of organic matter, and of methane, could have far exceeded the rate that was possible beforehand. If the escape of hydrogen to space remained diffusion-limited, then the net rate of O_2 production could have risen, possibly contributing to the eventual rise of O_2 (Catling *et al.*, 2001).

7.6 Synthesis: speculations about the timing and cause of the rise of atmospheric O_2

Let us return to two issues raised earlier: Why did atmospheric O_2 rise at about 2.4 Ga, and what exactly caused this to happen? We agreed that cyanobacteria produced the O_2, but whether they originated at or before 2.4 Ga remains controversial. If they arose at 2.4 Ga, then this question is moot, but if they arose before this time then further explanation is needed.

One can imagine a number of reasons why the rise of atmospheric O_2 might have been delayed. A potentially promising explanation – although it has yet to be supported by data – is that early cyanobacteria had not yet developed mechanisms for protecting their *nitrogenase*. As discussed in Chapter 4, nitrogenase is the enzyme used to convert N_2 to fixed nitrogen, and it is poisoned by O_2. Hence, most of the organisms that can fix nitrogen are anaerobic prokaryotes. Cyanobacteria are an exception, however, as they have evolved a variety of mechanisms for keeping their nitrogenase separate from O_2. Some filamentous cyanobacteria fix nitrogen in specialized cells called *heterocysts*; others employ circadian rhythms to separate O_2 production and nitrogen fixation in time. If primitive cyanobacteria lacked these (complex) nitrogenase protection mechanisms, then their ability to spread globally would have been limited because they would have put other anaerobic N-fixers out of business due to their oxygen production, limiting the availability of N to the biosphere.

Other, more geochemical explanations for the delay in the rise of O_2 have been suggested. Holland (2002) has argued that if volcanic gases emitted during the Archean had a lower $S:H_2O$ ratio, this would have made them a larger net O_2 sink, because less outgassed SO_2 would have been buried as pyrite. In a more recent analysis Holland (2009) argues that increased rates of recycling of both sulfur and carbon with time led to increased O_2

production. In making this link, he assumes that the ratio of carbonate to organic carbon in sediments has remained at 4:1 throughout most of Earth's history. This argument is based on the relative constancy of the $\delta^{13}C$ values of carbonate rocks over time (Schidlowski *et al.*, 1983; Hayes and Waldbauer, 2006). Some authors (DesMarais *et al.*, 1992; Bjerrum and Canfield, 2002) have suggested that organic carbon burial rates may increase with time, but their alternative interpretations are themselves disputed (Claire *et al.*, 2006). The isotopic data are probably too spotty for the Archean to provide conclusive backing for any of these ideas.

Other authors have suggested that the abiotic production rate of hydrogen has declined with time (Kasting *et al.*, 1993; Kump *et al.*, 2001; Catling *et al.*, 2001; Holland, 2002; Claire *et al.*, 2006; Kump and Barley, 2007). As discussed earlier, the progressive mantle oxidation theories have now been ruled out, but the Catling/Claire mechanism of progressive continental oxidation remains viable. As discussed in this chapter, a gradually decreasing O_2 sink from seafloor hydrothermal processes may also be part of the explanation. Because volcanic gases released at depth are more reduced than those released at the surface, this could reflect a switch from dominantly submarine volcanism to a mix of submarine plus subaerial volcanism (Kump and Barley, 2007). Or, as discussed in this chapter, it may also reflect a gradual change in the composition of seafloor, from ultramafic to mafic, accompanied by decreased rates of hydrothermal serpentinization.

All current explanations for the rise of O_2 are speculative, of course, and are likely to remain so in the near future. Better data on H_2 fluxes from surface volcanoes and from submarine hydrothermal systems are needed, along with better models for Earth's tectonic evolution. Other parts of this chapter are less likely to change, however. Redox balance is a requirement, not a hypothesis, and any explanation for the rise of atmospheric O_2 must necessarily account for it. Hopefully, the discussion offered here will help others to use the concept of global redox balance to better understand Earth's history.

References

Alt JC (1995) Subseafloor processes in mid-ocean ridge hydrothermal systems. In: *Seafloor Hydrothermal Systems* (eds Humphris SE, Zierenberg RA, Mullineaux LS, Thomson RE). American Geophysical Union, Washington, DC, pp. 85–114.

Anbar AD, Duan Y, Lyons TW, *et al.* (2007) A whiff of oxygen before the great oxidation event? *Science* **317**, 1903–1906.

Arthur MA (2000) Volcanic contributions to the carbon and sulfur geochemical cycles and global change. In: *Encyclopedia of Volcanoes* (ed Sigurdsson S). Academic Press, San Diego, CA, pp. 1045–1056.

Bergman NM, Lenton TM, Watson AJ (2004) COPSE: A new model of biogeochemical cycling over Phanerozoic time. *American Journal of Science* **304**, 397–437.

Berner RA (2004) *The Phanerozoic Carbon Cycle: CO₂ and O₂.* Oxford University Press, Oxford.

Berner RA (2006) GEOCARBSULF: A combined model for Phanerozoic atmospheric O₂ and CO₂. *Geochimica et Cosmochimica Acta* **70**, 5653–5664.

Berresheim H, Jaeschke W (1983) The contribution of volcanoes to the global atmospheric sulfur budget. *Journal of Geophysical Research* **88**, 3732–3740.

Bjerrum CJ, Canfield DE (2002) Ocean productivity before about 1.9 Gyr ago limited by phosphorus adsorption onto iron oxides. *Nature* **417**, 159–162.

Brasier MD, Green OR, Jephcoat AP, *et al.* (2002) Questioning the evidence for the Earth's oldest fossils. *Nature* **416**, 76–81.

Brocks JJ, Logan GA, Buick R, Summons RE (1999) Archean molecular fossils and the early rise of eukaryotes. *Science* **285**, 1033–1036.

Canfield DE (2005) The early history of atmospheric oxygen: Homage to Robert M. Garrels. *Annual Review of Earth and Planetary Science* **33**, 1–36.

Canfield DE, Habicht KS, Thamdrup B (2000) The Archean sulfur cycle and the early history of atmospheric oxygen. *Science* **288**, 658–661.

Canfield D, Rosing MT, Bjerrum C (2006) Early anaerobic metabolisms. *Philosophical Transactions of the Royal Society B* **361**, 1819–1836.

Canfield D, Poulton SW, Narbonne GM (2007) Late-Neoproterozoic deep-ocean oxygenation and the rise of animal life. *Science* **315**, 92–95.

Canil D (2002) Vanadium in peridotites, mantle redox and tectonic environments: Archean to present. *Earth and Planetary Science Letters* **195**, 75–90.

Catling DC, Zahnle KJ, McKay CP (2001) Biogenic methane, hydrogen escape, and the irreversible oxidation of early Earth. *Science* **293**, 839–843.

Chamberlain JW, Hunten DM (1987) *Theory of Planetary Atmospheres.* Academic Press, Orlando, FL.

Claire MW, Catling DC, Zahnle KJ (2006) Biogeochemical modelling of the rise in atmospheric oxygen. *Geobiology* **4**, 239–269.

Cloud PE (1972) A working model of the primitive Earth. *American Journal of Science* **272**, 537–548.

Delano JW (2001) Redox history of the Earth's interior: Implications for the origin of life. *Origins of Life and the Evolution of the Biosphere* **31**, 311–341.

DesMarais DJ, Strauss H, Summons RE, Hayes JM (1992) Carbon isotope evidence for the stepwise oxidation of the Proterozoic environment. *Nature* **359**, 605–609.

Garrels RM, Lerman A (1981) Phanerozoic cycles of sedimentary carbon and sulfur. *Proceedings of the National Academy of Sciences, USA* **78**, 4652–4656.

Garrels RM, Perry EA (1974) Cycling of carbon, sulfur and oxygen through geological time. In: *The Sea* (ed Goldberg ED). John Wiley & Sons, Inc., New York, pp. 303–316.

Goldblatt C, Lenton TM, Watson AJ (2006) Bistability of atmospheric oxygen and the great oxidation. *Nature* **443**, 683–686.

Gough DO (1981) Solar interior structure and luminosity variations. *Solar Physics* **74**, 21–34.

Habicht KS, Gade M, Thamdrup B, Berg P, Canfield DE (2002) Calibration of sulfate levels in the Archean ocean. *Science* **298**, 2372–2374.

Halmer MM, Schmincke H-U, Graf H-F (2002) The annual volcanic gas input into the atmosphere, in particular into the stratosphere: A global data set for the past 100 years. *Journal of Volcanology and Geothermal Research* **115**, 511–528.

Haqq-Misra JD, Goldman SD, Kasting PJ, Kasting JF (2008) A revised, hazy methane greenhouse for the early Earth. *Astrobiology* **8**, 1127–1137.

Hayes JM (1983) Geochemical evidence bearing on the origin of aerobiosis, a speculative hypothesis. In: *Earth's Earliest Biosphere: Its Origin and Evolution* (ed Schopf JW). Princeton University Press, Princeton, NJ, pp. 291–301.

Hayes JM (1994) Global methanotrophy at the Archean-Preoterozoic transition. In: *Early Life on Earth* (ed Bengtson S). Columbia University Press, New York, pp. 220–236.

Hayes JM and Waldbauer JR (2006) The carbon cycle and associated redox processes through time. *Philosophical Transactions of the Royal Society of London B* **361**, 931–950.

Hinrichs KU (2002) Microbial fixation of methane carbon at 2.7 Ga: Was an anaerobic mechanism responsible? *Geology, Geochemistry, Geophysics* **3**, #1042, doi:10.1029/2001GC000286.

Holland HD (1978) *The Chemistry of the Atmosphere and Oceans.* John Wiley & Sons, Inc., New York.

Holland HD (1984) *The Chemical Evolution of the Atmosphere and Oceans.* Princeton University Press, Princeton, NJ.

Holland HD (1994) Early Proterozoic atmospheric change. In: *Early Life on Earth* (ed Bengtson S). Columbia University Press, New York, pp. 237–244.

Holland HD (2002) Volcanic gases, black smokers, and the great oxidation event. *Geochimica et Cosmochimica Acta* **66**, 3811–3826.

Holland HD (2006) The oxygenation of the atmosphere and oceans. *Philosophical Transactions of the Royal Society of London B* **361**, 903–915.

Holland HD (2009) Why the atmosphere became oxygenated: A proposal. *Geochimica et Cosmochimica Acta* **73**, 5241–5255.

Hunten DM (1973) The escape of light gases from planetary atmospheres. *Journal of Atmospheric Science* **30**, 1481–1494.

Kasting JF (1993) Earth's early atmosphere. *Science* **259**, 920–926.

Kasting JF, Ackerman TP (1986) Climatic consequences of very high CO₂ levels in the earth's early atmosphere. *Science* **234**, 1383–1385.

Kasting JF, Pollack JB, Crisp D (1984) Effects of high CO₂ levels on surface temperature and atmospheric oxidation state of the early earth. *Journal of Atmospheric Chemistry* **1**, 403–428.

Kasting JF, Eggler DH, Raeburn SP (1993) Mantle redox evolution and the oxidation state of the Archean atmosphere. *Journal of Geology* **101**, 245–257.

Kaufman AJ, Johnston DT, Farquhar J, *et al.* (2007) Late Archean biospheric oxygenation and atmospheric evolution. *Science* **317**, 1900–1903.

Kelley DS, Karson JA, Fruh-Green GL, *et al.* (2005) A serpentinite-hosted ecosystem: The Lost City hydrothermal vent field. *Science* **307**, 1428–1434.

Kharecha P, Kasting JF, Siefert JL (2005) A coupled atmosphere-ecosystem model of the early Archean Earth. *Geobiology* **3**, 53–76.

Kirschvink JL (2006) Archean sterol biomarkers do not prove oxygenic photosynthesis. *Geochimica et Cosmochimica Acta* **70**, A320–A320.

Kopp RE, Kirschvink JL, Hilburn I, Nash CZ (2005) The Paleoproterozoic snowball Earth: A climate disaster triggered by the evolution of photosynthesis. *Proceedings of the National Academy of Sciences, USA* **102**, 11,131–111,136.

Kump LR, Barley ME (2007) Increased subaerial volcanism and the rise of atmospheric oxygen 2.5 billion years ago. *Nature* **448**, 1033–1036.

Kump LR, Kasting JF, Barley ME (2001) The rise of atmospheric oxygen and the 'upside-down' Archean mantle. *Geology, Geochemistry, Geophysics* (online) **2**, paper number 2000GC000114.

Lecuyer C, Ricard Y (1999) Long-term fluxes and budget of ferric iron: Implication for the redox states of the Earth's mantle and atmosphere. *Earth and Planetary Science Letters* **165**, 197–211.

Li ZXA, Lee CTA (2004) The constancy of upper mantle fO_2 through time inferred from V/Sc ratios in basalts. *Earth and Planetary Science Letters* **228**, 483–493.

McGovern PJ, Schubert G (1989) Thermal evolution of the Earth: Effects of volatile exchange between atmosphere and interior. *Earth and Planetary Science Letters* **96**, 27–37.

Moores EM (2002) Pre-1 Ga (pre-Rodinian) ophiolites: Their tectonic and environmental implications. *GSA Bulletin* **114**, 80–95.

Mottl MJ, Wheat CG (1994) Hydrothermal circulation through mid-ocean ridge flanks: Fluxes of heat and magnesium. *Geochimica et Cosmochimica Acta* **58**, 2225–2237.

Muehlenbachs K, Clayton RN (1976) Oxygen isotope composition of the oceanic crust and its bearing on seawater. *Journal of Geophysical Research* **81**, 4365–4369.

Ohmoto H (1996) Evidence in pre-2.2 Ga paleosols for the early evolution of atmospheric oxygen and terrestrial biota. *Geology* **24**, 1135–1138.

Ohmoto H, Watanabe Y, Kumazawa K (2004) Evidence from massive siderite beds for a CO_2-rich atmosphere before ~1.8 billion years ago. *Nature* **429**, 395–399.

Ohmoto H, Watanabe Y, Ikemi H, Poulson SR, Taylor BE (2006) Sulphur isotope evidence for an oxic Archaean atmosphere. *Nature* **442**, 908–911.

Pavlov AA, Kasting JF (2002) Mass-independent fractionation of sulfur isotopes in Archean sediments: Strong evidence for an anoxic Archean atmosphere. *Astrobiology* **2**, 27–41.

Pavlov AA, Kasting JF, Brown LL, Rages KA, Freedman R (2000) Greenhouse warming by CH_4 in the atmosphere of early Earth. *Journal of Geophysical Research* **105**, 11,981–911,990.

Prentice IC, Farquhar GD, Fasham MJR, *et al.* (2001) The carbon cycle and atmospheric carbon dioxide. In: *Climate Change 2001: the Scientific Basis* (eds Houghton JT, Ding Y, Griggs DJ *et al.*). Contribution of Working Group I to the Third Assessment Report of the Intergovernmental Panel on Climate Change. Cambridge University Press, New York, pp. 183–238.

Proskurowski G, Lilley MD, Seewald JS, *et al.* (2008) Abiogenic hydrocarbon production at Lost City hydrothermal field. *Science* **319**, 604–607.

Rashby SE, Sessions AL, Summons RE, Newman DK (2007) Biosynthesis of 2-methylbacteriohopanepolyols by an anoxygenic phototroph. *Proceedings of the National Academy of Sciences, USA* **104**, 15099–15104.

Rasmussen B, Fletcher IR, Brocks JJ, Kilburn MR (2008) Reassessing the first appearance of eukaryotes and cyanobacteria. *Nature* **455**, 1101–1104.

Schidlowski M, Hayes JM, Kaplan IR, Schopf JW (1983) Isotopic inferences of ancient biochemistries: Carbon, sulfur, hydrogen, and nitrogen. In: *Earth's Earliest Biosphere: Its Origin and Evolution* (ed Schopf JW). Princeton University Press, Princeton, NJ, pp. 149–186.

Schopf JW (1993) Microfossils of the Early Archean Apex chert: New evidence for the antiquity of life. *Science* **260**, 640–646.

Schopf JW, Packer BM (1987) Early Archean (3.3 billion to 3.5 billion-year-old) microfossils from Warrawoona Group, Australia. *Science* **237**, 70–73.

Sleep NH (2005) Dioxygen over geologic time. *Metal Ions in Biological Systems* **43**, 49–73.

Sleep NH (2007) Plate tectonics through time. In: *Treatise on Geophysics* (ed Schubert G), Vol. 9. Elsevier, Amsterdam, pp. 145–169.

Sleep NH, Zahnle K (2001) Carbon dioxide cycling and implications for climate on ancient Earth. *Journal of Geophysical Research* **106**, 1373–1399.

Summons JR, Jahnke LL, Hope JM, Logan GA (1999) Methylhopanoids as biomarkers for cyanobacterial oxygenic photosynthesis. *Nature* **400**, 554–557.

Von Damm KL (1995) Controls on the chemistry and temporal variability of seafloor hydrothermal fluids. In: *Seafloor Hydrothermal Systems: Physical, Chemical, Biological, and Geological Interactions* (eds Humphris SE, Zierenberg RA, Mullineaux LS, Thomson RE). American Geophysical Union, Washington, DC, pp. 222–247.

Von Damm KL (2000) Chemistry of hydrothermal vent fluids from 9 degrees-10 degrees N, East Pacific Rise: 'Time zero,' the immediate posteruptive period. *Journal of Geophysical Research* **105**, 11,203–11,222.

Walker JCG (1977) *Evolution of the Atmosphere*. Macmillan, New York.

Walker JCG, Brimblecombe P (1985) Iron and sulfur in the prebiologic ocean. *Precambrian Research* **28**, 205–222.

Wille M, Kramers JD, Nagler TF, *et al.* (2007) Evidence for a gradual rise of oxygen between 2.6 and 2.5 Ga from Mo isotopes and Re-PGE signatures in shales. *Geochimica et Cosmochimica Acta* **71**, 2417–2435.

Wolery TJ, Sleep NH (1976) Hydrothermal circulation and geochemical flux at midocean ridges. *Journal of Geology* **84**, 249–275.

8

BACTERIAL BIOMINERALIZATION

Kurt Konhauser[1] and Robert Riding[2]

[1]Department of Earth & Atmospheric Sciences, University of Alberta, Edmonton, AB, T6G 2E3, Canada
[2]Department of Earth & Planetary Sciences, University of Tennessee, Knoxville, TN 37996, USA

8.1 Introduction

A number of living organisms form mineral phases through a process termed biomineralization. Two end member mechanisms exist depending on the level of biological involvement. The first involves mineral formation without any apparent regulatory control. Termed 'biologically induced biomineralization' by Lowenstam (1981), biominerals form as incidental by-products of interactions between the organisms and their immediate environment. The minerals that form through this passive process have crystal habits and chemical compositions similar to those produced by precipitation under inorganic conditions. By contrast, 'biologically controlled biomineralization', the subject of Chapter 10, is much more closely regulated, and organisms precipitate minerals that serve physiological and structural roles. This process can include the development of intracellular or epicellular organic matrices into which specific ions are actively introduced and their concentrations regulated such that appropriate mineral saturation states are achieved. Accordingly, minerals can be formed within the organism even when conditions in the bulk solution are thermodynamically unfavourable. In this chapter we focus on the role of bacteria. Specifically, we examine the formation of iron oxyhydroxides and calcium carbonates throughout geological time, and explore how our understanding of modern biomineralization processes is shedding new insights into the evolution of the Earth's hydrosphere–atmosphere–biosphere over long time scales.

8.2 Mineral nucleation and growth

The thermodynamic principles underpinning biological mineral formation, irrespective of whether it is induced or controlled, are the same as those involved in inorganic mineral formation. In all cases, before any solid can precipitate, a certain amount of energy has to be invested to form a new interface between the prospective mineral nucleus and both the aqueous solution and the underlying substrate upon which it is formed. The amount of energy required to do this can be viewed as an activation energy barrier. The standard free energy (G^0) of a solid is lower than that of its ionic constituents in solution, and if the activation energy barrier can be overcome, the reaction proceeds towards mineral precipitation. On the other hand, if the activation energy barrier is prohibitively high, metastable solutions persist until either the barrier is reduced or the concentration of ions are diminished.

Mineral nucleation involves the spontaneous growth of a number of nuclei that are large enough to resist rapid dissolution. Formation of these 'critical nuclei' requires a certain degree supersaturation wherein the concentration of ions in solution exceeds the solubility product of the mineral phase (see Stumm and Morgan, 1996, for details). Nucleation is termed homogeneous when critical nuclei form simply by random collisions of ions or atoms in a supersaturated solution. Such 'pure' solutions, however, rarely exist; in nature, most solutions contain a wide variety of competing solid and dissolved phases. In this regard, heterogeneous nucleation occurs when critical nuclei form on those solid phases.

Fundamentals of Geobiology, First Edition. Edited by Andrew H. Knoll, Donald E. Canfield and Kurt O. Konhauser.

After critical nuclei are formed, continued addition of ions is accompanied by a decrease in free energy, resulting in mineral growth. This process goes on spontaneously until the system reaches equilibrium.

Mineral growth typically favours the initial formation of amorphous solid phases that are characterized by their high degree of hydration and solubility, and lack of intrinsic form (Nielson and Söhnel, 1971). Accordingly, minerals such as amorphous silica ($SiO_2 \bullet nH_2O$), hydrated carbonate ($CaCO_3 \bullet H_2O$), or ferric hydroxide [$Fe(OH)_3$] will nucleate readily if the solution composition exceeds their solubility. In contrast, their respective crystalline equivalents, quartz (SiO_2), calcite ($CaCO_3$), and hematite (Fe_2O_3), have higher interfacial free energies. Therefore, they nucleate slowly at ambient temperatures, even in the presence of substrates favouring heterogeneous nucleation. Often the transition between amorphous and crystalline phases involves the precipitation of metastable phases.

The nucleation rate also affects the size of the critical nuclei formed. At high ion activities, above the critical supersaturation value, new surface area is created mainly by the nucleation of many small grains characterized by high surface area to mass ratios, a regime referred to as nucleation-controlled. At lower activities, surface area generation is crystal-growth controlled, with surface area increasing by the accretion of additional ions or atoms to existing grains. In a nucleation-controlled regime, the generation of new surfaces by nucleation occurs rapidly and causes the solution supersaturation to drop below the critical value needed for nucleation. This means that in nature, supersaturation above the critical value does not typically occur for lengthy periods of time. For silica precipitation as an example, if a concentrated silica solution were emitted from a hot spring vent, it would thermodynamically be supersaturated with regard to all silica phases, but because amorphous silica has the lower interfacial free energy it nucleates first despite quartz being the more stable phase with lower solubility. Then as amorphous silica nucleates it drives the dissolved silica concentration down to its critical value below which quartz nuclei formation is prohibitively slow.

Crystalline minerals that would otherwise be difficult or impossible to nucleate directly at low temperatures can circumvent activation energy barriers by making use of the more soluble precursors as templates for their own growth. Once they begin to grow, the crystals increase in surface area, and in doing so, they drive the proximal free ion concentration below the solubility of the amorphous precursor, causing it to dissolve (Steefel and Van Cappellen, 1990). However, despite thermodynamics predicting the transformation sequence based on energetics, they say nothing about the kinetics. Sometimes the reactions are relatively quick, such as the formation of magnetite on ferric hydroxide. At other times, the reaction rates are immeasurably slow and the amorphous or metastable phases persist in sediments despite supersaturation with respect to more thermodynamically stable minerals. Those phases can show little discernible alteration for tens of millions of years until pressure–temperature changes associated with burial cause the reaction sequence to advance to the next stage (Morse and Casey, 1988). For example, amorphous silica skeletons (such as diatoms) deposited onto the seafloor slowly dissolve at shallow depths and re-precipitate as cristobalite, which remains stable to depths of hundreds of meters until it too transforms into quartz.

8.3 How bacteria facilitate biomineralization

Bacteria contribute significantly to the development of extremely fine-grained (often <1μm in diameter) mineral precipitates. All major mineral groups, whether metal oxyhydroxides, silicates, carbonates, phosphates, sulfates, sulfides, and even native metals, have been shown to precipitate as a consequence of bacterial activity. Although not directly associated with biologically controlled biomineralization, bacteria may influence the initial stages of mineralization in two significant ways:

8.3.1 Development of an ionized cell surface

Bacterial surfaces are highly variable, but commonly they have a cell wall that is overlain by additional organic layers, such as extracellular polymeric substances (EPS), sheaths and S-layers, which differ in terms of their hydration, composition and structure. In all cases, bacterial surfaces act as highly reactive interfaces. Organic ligands within the bacterial surface (such as carboxyl, hydroxyl, phosphoryl, sulfur and amine functional groups) deprotonate with increasing pH and thus impart the cells with a net negative surface charge (see Konhauser, 2007 for details). In its most simplistic form, deprotonation can be expressed by the following equilibrium reaction:

$$B - AH \leftrightarrow B - A^{-1} + H^{+1} \tag{8.1}$$

where B denotes a bacterium to which a protonated ligand type, A, is attached. The distribution of protonated and deprotonated sites can be quantified with the corresponding mass action equation:

$$K_a = \frac{\left[B - A^{-1} \right]\left[H^{+1} \right]}{\left[B - AH \right]} \tag{8.2}$$

where K_a is the dissociation constant; [$B-A^{-1}$] and [B-AH] represent the respective concentrations of exposed deprotonated and protonated ligands on the bacterium (in mol l^{-1}); and [H^{+1}] represents the activity of protons

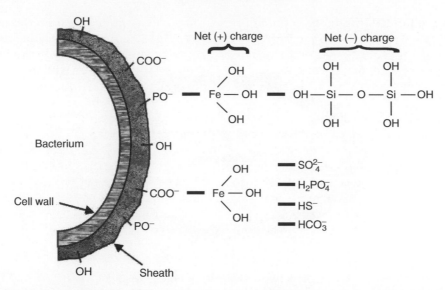

Figure 8.1 Schematic diagram of bacterial surface showing how the deprotonation of exposed functional groups leads to metal cation adsorption. Once bound, those cations can then react inorganically with dissolved anions in the external environment, such as silica, sulfate, phosphate, sulfide and bicarbonate. Depending on the anion available, different minerals may form on the bacterial surface.

in solution. Each functional group has its own K_a, and based on equation [8.2], the pH at which [B-A^{-1}] and [B-AH] are equivalent is known as the pK_a value, where $pK_a = -\log_{10}K_a$. At pH < pK_a the functional groups are predominantly protonated, and at pH > pK_a, they are predominantly deprotonated.

Due to these protonation–deprotonation reactions, the bacterial surface develops a negative surface charge at pH values characteristic of most natural environments, and in doing so, will become reactive towards charged cations and a number of mineral surfaces (Fig. 8.1). Some cations preferentially bind to different sites on the cell surface, and crucially, they are not equally exchangeable. For instance, trivalent (e.g. La^{3+}, Fe^{3+}) and divalent (e.g. Ca^{2+}, Mg^{2+}) metal cations are strongly bound to the cell wall of various bacteria, while monovalent cations (e.g. Na$^+$, K$^+$) are easily lost in competition with those metals for binding sites (Beveridge and Murray, 1976). This aspect is important to keep in mind – in natural systems, a multitude of cations and anions compete with one another for surface adsorption sites. Some of the most important factors that influence ion binding to cells are: (i) the composition of the solution, including pH and the activity of all ions; (ii) ligand spacing and their stereochemistry; (iii) ligand composition; and (iv) the balance between the initial electrostatic attractions between a soluble ion and the organic ligands, and the subsequent covalent forces that arise from electron sharing across a ion-ligand molecular orbital (see Williams, 1981).

One consequence of metal sorption is that the bound cation subsequently lowers the interfacial energy for heterogeneous nucleation of solid phases while simultaneously decreasing the surface area of the nucleus that is in contact with the bulk solution. In this manner the bacterium catalyses mineral formation simply because it has bound cations on its outer surface. Those cations react with more ions, potentially leading to biomineralization. With that said, bacteria only serve to enhance the precipitation kinetics in supersaturated solutions; they neither increase the extent of precipitation nor facilitate precipitation in undersaturated solutions (e.g. Fowle and Fein, 2001). The size of the mineral precipitate depends on a number of variables, including the concentration of ions and the amount of time through which the reactions proceed. The end result could be a bacterial wall that contains copious amounts of mineral precipitate, often approaching, or exceeding, the mass of the microorganism itself (Beveridge, 1984).

8.3.2 Metabolic processing

Each bacterium has a biogeochemical lifestyle that is optimally suited to its particular environmental conditions. And, all have two common objectives: to obtain energy and carbon. Energy can be captured from sunlight (phototrophy) or through the transfer of electrons from a reductant to an electron acceptor (chemotrophy). Carbon can be fixed through the reduction of CO_2 (autotrophy) or the consumption of pre-existing organic materials (heterotrophy). In all cases, the metabolic processes employed directly influence the chemistry and distribution of a wide range of elements. From a biomineralization standpoint this is important because metabolism affects the redox and saturation states of the fluids around the living

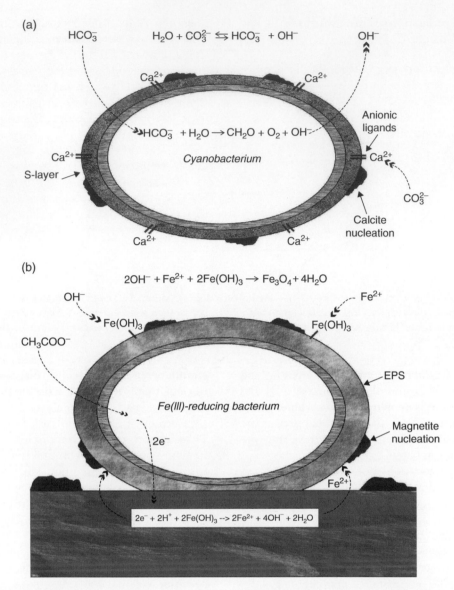

$$H_2O + CO_3^{2-} \leftrightarrows HCO_3^- + OH^-$$

(a) Cyanobacterium

$$\text{»}HCO_3^- + H_2O \rightarrow CH_2O + O_2 + OH^-$$

S-layer, Ca^{2+}, HCO$_3^-$, OH$^-$, Anionic ligands, Ca^{2+}, CO$_3^{2-}$, Calcite nucleation

(b)

$$2OH^- + Fe^{2+} + 2Fe(OH)_3 \rightarrow Fe_3O_4 + 4H_2O$$

OH$^-$, Fe(OH)$_3$, Fe^{2+}, Fe(OH)$_3$

CH$_3$COO$^-$

Fe(III)-reducing bacterium

EPS, Magnetite nucleation, 2e$^-$, Fe^{2+}

$$2e^- + 2H^+ + 2Fe(OH)_3 \text{-->} 2Fe^{2+} + 4OH^- + 2H_2O$$

Figure 8.2 Schematic of metabolically-induced biomineralization in (a) cyanobacteria and (b) Fe(III)-reducing heterotrophs. In the cyanobacteria, uptake of the bicarbonate anion leads to excretion of OH$^-$, which in turn changes the alkalinity and inorganic carbon speciation proximal to the cell surface. The generation of carbonate anions and the pre-adsorption of calcium cations to the cell's sheath can then induce calcification. In the Fe(III)-reducing bacterium, release of Fe(II) from the ferric hydroxide substratum can promote magnetite formation both on the pre-existing mineral surfaces, but also on the cell surface if deprotonated ligands had adsorbed ferric iron. In both examples, the secondary minerals form as by-products of microbial metabolism – the cell itself did not control the mineralization process.

cells (Fig. 8.2). In this regard, the microenvironment surrounding each bacterium can be quite different from the bulk aqueous environment, and as a result, mineral phases form that would not normally be predicted from the geochemistry of the bulk fluid. It is important at this stage to point out that although these mineral phases are passively formed, there exists a grey zone between what constitutes minerals intimately associated with microbial biomass and the broader association of minerals formed by microbial activities changing local fluid chemistry.

8.3.2.1　Photoautotrophs

Phototrophic organisms use pigments to absorb sunlight and generate chemical energy. Two different processes of photosynthesis are; (1) anoxygenic photosynthesis, used by green bacteria, purple bacteria and heliobacteria; and (2) oxygenic photosynthesis, used by cyanobacteria, algae and plants. Anoxygenic species utilize hydrogen gas (H$_2$), hydrogen sulfide [reaction 8.3], elemental sulfur [reaction 8.4] and dissolved ferrous iron [reaction 8.5]. In oxygenic photosynthesis, the electrons for CO_2 reduction come from splitting water into O_2 and electrons [reaction 8.6].

$$3H_2S + 6CO_2 + 6H_2O \rightarrow C_6H_{12}O_6 + 3SO_4^{2-} + 6H^+ \qquad (8.3)$$

$$12S^0 + 18CO_2 + 30H_2O \rightarrow 3C_6H_{12}O_6 + 12SO_4^{2-} + 24H^+ \qquad (8.4)$$

$$24Fe^{2+} + 6CO_2 + 66H_2O \rightarrow C_6H_{12}O_6 + 24Fe(OH)_3 + 48H^+ \qquad (8.5)$$

$$6H_2O + 6CO_2 \rightarrow C_6H_{12}O_6 + 6O_2 \qquad (8.6)$$

In reactions 8.3 and 8.4, the production of sulfate can affect the saturation state of sulfate minerals, and in some experimental studies, photoautotrophic bacteria can become encrusted by gypsum ($CaSO_4$), barite ($BrSO_4$) or celestite ($SrSO_4$) depending on the cations available to them in culture (e.g. Schultze-Lam and Beveridge, 1994). In reaction 8.5, and as discussed below in more detail, the oxidation of Fe(II) leads to the formation of dissolved Fe(III), which spontaneously hydrolyses at circumneutral pH values to form ferric hydroxide. This solid iron phase generally forms outside of cells where it can create difficulties for the bacterium because it acts as a physical barrier that inhibits diffusion of nutrients into the cell and of waste materials out of the cell. In fact, growth experiments with *Rhodomicrobium vannielli* demonstrated that the ferric hydroxide crusts formed on the cell surface impeded further Fe(II) oxidation after two to three generations (Heising and Schink, 1998). Research is currently being conducted on various species of Fe(II)-oxidizing photoautotrophs to try to understand how they survive the effects of their own metabolism (e.g. Kappler and Newman, 2004; Hegler *et al.*, 2008).

Of all the reactions above, the one which arguably has had the greatest effect on life's evolution and diversification has been the generation of O_2 via oxygenic photosynthesis [reaction 8.6]. The ability of cyanobacteria to strip electrons from the virtually unlimited supply of water meant that they were not limited by reductants, and as a consequence, they were able to rapidly colonize the Earth's surface wherever sunlight was sufficiently present. As discussed in Chapter 7 (the Global Oxygen Cycle), the oxygen they produced gradually transformed the atmosphere and surface ocean from an anoxic to oxic state. This fuelled an oxygenated biosphere that profoundly altered the direction of evolution. This also significantly affected biomineralization (see Chapter 18 on mineral evolution) because the oceans eventually attained sufficient concentrations of dissolved O_2 to oxidize reduced metals in the water column and top layers of sediment. For bacteria that had reduced metals

adsorbed to their outer surfaces, this meant that they now had to contend with the potential for metal oxidation and hydrolysis, leading to encrustation.

In addition, all phototrophs consume CO_2, or more precisely, HCO_3^- – the predominant form of dissolved inorganic carbon in neutral to mildly alkaline waters (reaction 8.7). Excretion of the unconsumed products of this reaction, hydroxyl ions, into the external environment creates localized alkalinization around the cell surfaces (see Section 8.5.1.1). This in turn induces a change in the carbonate speciation towards the carbonate (CO_3^{2-}) anion (reaction 8.8), which ultimately may lead to the precipitation of calcium carbonate (Thompson and Ferris, 1990).

$$HCO_3^- \leftrightarrow CO_2 + OH^- \qquad (8.7)$$

$$HCO_3^- + OH^- \leftrightarrow CO_3^{2-} + H_2O \qquad (8.8)$$

Cyanobacteria also provide reactive ligands to metal cations. Once bound, the cations can react with the dissolved carbonate to form mineral phases, such as aragonite or calcite (see Calcium Carbonates, below):

$$CO_3^{2-} + Ca^{2+} \leftrightarrow CaCO_3 \qquad (8.9)$$

8.3.2.2 Chemolithoautotrophs

Similar to the photoautotrophs, chemolithoautotrophs use CO_2 as their source of carbon, however, sunlight is not their source of energy. Instead, they oxidize reduced inorganic substrates coupled to a terminal electron acceptor, usually O_2. For biomineralization, the most important reactions include the oxidation of Fe(II) (reaction 8.10), Mn(II) (reaction 8.11) and CH_4 (reaction 8.12).

$$2Fe^{2+} + 0.5O_2 + 2H^+ \rightarrow 2Fe^{3+} + H_2O \qquad (8.10)$$

$$Mn^{2+} + 0.5O_2 + H_2O \rightarrow MnO_2 + 2H^+ \qquad (8.11)$$

$$CH_4 + SO_4^{2-} \rightarrow HCO_3^- + HS^- + H_2O \qquad (8.12)$$

Ferrous iron oxidation is discussed in detail below. The development of manganese oxides occurs in the same types of present-day oxic–anoxic interfacial environments where ferric hydroxide forms, but because Mn(II) is not subject to such rapid oxidation as Fe(II), at circumneutral pH most Mn(II) oxidation results from microbial catalysis (Ehrlich, 2002). At marine hydrothermal sites, MnO_2 forms as crusts around seafloor vents and where hot fluids percolate up through the sediment (Mandernack and Tebo, 1993). Analyses of the hydrothermal plumes emanating from the southern Juan de

Fuca Ridge have also shown that the particulate fraction is largely composed of bacterial cells encrusted in iron and manganese, with Fe-rich particles near the vent and Mn-rich particles further off-axis (Cowen *et al.*, 1986).

One type of bacteria, the methanotrophs, makes a living from oxidizing methane to CO_2. They are widespread in marine sediment where methane from underlying anoxic zones diffuses upwards into the overlying oxic zone, allowing them to use O_2 to oxidize methane. There is, however, another group of methanotrophs that lives in anoxic sediments and, in conjunction with other bacteria, facilitates the oxidation of methane via concomitant reduction of sulfate (reaction 8.12). This reaction is significant in limiting the release of methane to the atmosphere, and in some marine sediment it can account for nearly 100% of the downward SO_4^{2-} flux (D'Hondt *et al.*, 2002). The oxidation of methane (and other hydrocarbon gases) under anoxic conditions leads to an increase in alkalinity and the subsequent precipitation of aragonite and Mg-calcites as cements that line cavities in the sediment, and as carbonate nodules that tend to show repeated zonation with framboidal pyrite (FeS_2). The precipitation of carbonates can even produce seafloor topographic features with up to several meters of relief (Michaelis *et al.*, 2002). These precipitates serve as a major inorganic carbon sink, and help stabilize the microbial community by cementing the soft sediment and forming a hard, solid substratum that can be used for attachment by symbiotic macrofauna, such as tubeworm, clam and mussel colonies. Methane oxidation coupled to sulfate reduction further generates HS^- that not only provides an inorganic energy source for sulfur-oxidizing bacteria, such as *Beggiatoa*, *Thiothrix* and *Thioploca*, but it also reacts with Fe(II) to precipitate as iron monosulfide (reaction 8.13), and it is not uncommon to observe pyritized remains of bacteria (Sassen *et al.*, 2004).

$$Fe^{2+} + HS^- \rightarrow FeS + H^+ \tag{8.13}$$

8.3.2.3   Chemoheterotrophs

There are several chemoheterotrophic reactions that play a profound role in sediment diagenesis, including reduction of dissolved (O_2, NO_3^-, SO_4^{2-}, CO_2) and solid (MnO_2 and ferric oxyhydroxides) substrates. Of those, the reduction of nitrate (reaction 8.14), manganese(IV) (reaction 8.15), ferric iron (reaction 8.16) and sulfate (reaction 8.17) have substantial biomineralization potential because they increase pore water alkalinity, raising the saturation state for calcium carbonate minerals (Canfield and Raiswell 1991a). The net effect is precipitation of early diagenetic carbonate minerals that are relatively stable once formed, and not subject to rapid

recycling by redox reactions in the same way as sulfides and oxides (Irwin *et al.*, 1977).

$$2.5C_6H_{12}O_6 + 12NO_3^- \rightarrow 6N_2 + 15CO_2 \\ + 12OH^- + 9H_2O \tag{8.14}$$

$$CH_3COO^- + 4MnO_2 + 3H_2O \rightarrow 4Mn^{2+} + 2HCO_3^- \\ + 7OH^- \tag{8.15}$$

$$CH_3COO^- + 8Fe(OH)_3 \rightarrow 8Fe^{2+} + 2HCO_3^- \\ + 15OH^- + 5H_2O \tag{8.16}$$

$$CH_3COO^- + SO_4^{2-} \rightarrow HS^- + 2HCO_3^- \tag{8.17}$$

The reduction of ferric iron minerals also produces an increase in the concentration of Fe^{2+} in suboxic sediment pore waters, with a peak at the boundary between the Fe(III) and sulfate reduction zones. Some of this ferrous iron may diffuse upwards to be re-oxidized to ferric hydroxide inorganically by either NO_3^- or MnO_2 (e.g. Myers and Nealson, 1988). However, in marine sediment most ferrous iron is removed from solution by reaction with hydrogen sulfide produced by the underlying populations of sulfate reducing bacteria (reaction 8.17). This forms metastable iron monosulfide minerals, such as mackinawite, that are precursors to pyrite (Berner, 1984). When abundant ferric iron is available (enough to keep HS^- concentrations down) other ferrous iron-containing minerals can precipitate. One such mineral is magnetite (Fe_3O_4) that can occur as tiny (10–50 nm), rounded, poorly crystalline particles on cell surfaces. Although the actual role that Fe(III)-reducing bacteria play in magnetite formation remains unresolved, the adsorption of Fe^{2+} (produced during ferric iron reduction) to the remaining ferric hydroxide appears to be the key step in the precipitation process (reaction 8.18) (Lovley, 1990).

$$2OH^- + Fe^{2+} + 2Fe(OH)_3 \rightarrow Fe_3O_4 + 4H_2O \tag{8.18}$$

Other minerals that form in cultures of Fe(III)-reducing bacteria are siderite ($FeCO_3$) and vivianite ($Fe_3[PO_4]_2$). Marine siderite has been linked to the activity of Fe(III)-reducing bacteria, since they generate both ferrous iron and bicarbonate. By contrast, when sulfate reduction rates exceed Fe(III) reduction rates, enough HS^- is produced to precipitate iron monosulfide minerals instead (FeS has lower solubility than $FeCO_3$). Therefore, the precise spatial distribution of Fe(III) reduction and sulfate reduction in marine sediments controls whether siderite and/or pyrite form (e.g. Coleman, 1985). The formation of vivianite in many ways resembles that of siderite in that Fe^{2+} concentrations must exceed those of HS^-. It also requires that soluble phosphate is made available through oxidation of organic matter, the dissolution of phosphorous-bearing solid phases or through reduction of phosphorous-adsorbing Fe(III) oxides (Krom and Berner, 1980).

8.3.3 Microbial interactions in nature

A wide variety of microbial 'biominerals' exist in sedimentary environments (see Konhauser, 2007, chapter 4). Those formed passively are entirely dependent upon the chemical composition of the fluids in which they are growing; a mineral phase will form only if the requisite solutes are immediately available. Conversely the same bacterium in a different environment could form a different mineral phase altogether. For example, it is well known that the anionic ligands comprising a cell's surface can form covalent bonds with dissolved ferric iron species, which in turn can lead to charge reversal at the cell surface. Invariably, this positive charge will attract anionic counter-ions from solution (recall Fig. 8.1). So, in the sediment, iron adsorption to a bacterium may lead to the precipitation of an iron sulfate precipitate in the oxic zone, whereas another bacterium may instead form an iron sulfide at depth, when conditions become reducing.

It is also important to stress that in any given environment different bacteria co-exist, often utilizing different metabolic pathways. Consequently, one species may contribute to the formation of a mineral phase, while others in close proximity may have a completely different impact. Using iron as an example, bacteria in the oxic zone may oxidize dissolved Fe(II) leading to ferric hydroxide precipitation, while bacteria growing in suboxic layers may be involved in the reduction of that ferric mineral, facilitating Fe(II) release, and potentially the formation of mixed ferrous-ferric minerals, such as magnetite, or completely reduced ferrous mineral phases, such as siderite, pyrite or vivianite.

The impact that biomineralization has on elemental cycling in aqueous and sedimentary environments cannot be overstated. Present-day C, Ca, Fe, Mg, Mn, P, S and Si cycles are all strongly affected by biomineralizing processes. Although individual biomineral grains are micrometres in scale, if one adds the total amount of biomineralizing biomass, it is not difficult to imagine how they can be significant in partitioning metals from the hydrosphere into the sedimentary system. The precipitation of iron oxides and carbonate minerals by bacteria are especially relevant throughout Earth's history. For example, the extensive record of banded iron formation (BIF), from 3.8 to 0.5 billion years ago, testifies to the enormous magnitude of ferric iron deposition to the seafloor. Given that they are chemical precipitates, BIF have been used as proxies for paleo-oceanic conditions (e.g. Bjerrum and Canfield, 2002; Konhauser *et al.*, 2009; Planavsky *et al.*, 2010). On a similar, or even larger, scale calcium carbonate minerals represent the final products in the weathering of silicate minerals, and a long-term sink for atmospheric carbon dioxide. In the following sections, we explore the importance of bacterial biomineralization of both iron oxyhydroxides and calcium carbonates.

8.4 Iron oxyhydroxides

The most widespread iron biomineral is ferric hydroxide (also loosely referred to as ferrihydrite). It forms in any environment where Fe(II)-bearing waters come into contact with O_2. This includes mine wastes, springs and seeps, freshwater and marine sediment, soils and subsurface fractured rock, hydrothermal vents, and water distribution systems, to name just a few (see Konhauser, 1998). Fossils that resemble modern iron depositing bacteria have been found in laminated black cherts and hematite-bearing jaspillites from several Early Proterozoic banded iron formations (Robbins *et al.*, 1987). There is also circumstantial evidence suggesting that microbial activity was directly involved in the initial deposition of iron sediment, which later consolidated to make BIF (e.g. Konhauser *et al.*, 2002; Kappler *et al.*, 2005; Posth *et al.*, 2008).

The role bacteria play in ferric hydroxide formation can range from completely passive to that actively facilitated. Yet, by current definitions, this process is not considered biologically controlled because the bacterium does not manage all aspects of the mineralization process. In the most passive of examples, ferrous iron transported into an oxygenated environment at circumneutral pH spontaneously reacts with dissolved oxygen to precipitate inorganically as ferric hydroxide on available nucleation sites. Bacteria provide such sites by their mere presence, and over a short period of time the submerged communities can become completely encrusted in amorphous iron hydroxide as abiological surface catalysis accelerates the rate of mineral precipitation. Initial microscopic observations of such samples often indicate a paucity of bacteria, but staining the iron-rich sediment with fluorescent dyes for nucleic acids (e.g. acridine orange) often reveals high bacterial densities closely associated with the iron precipitates (e.g. Emerson and Revsbech, 1994a). In other cases, where bacteria are more actively involved, ferric hydroxide can form through the oxidation and hydrolysis of cell-bound ferrous iron (Fe^{2+}) and dissolved ferric ion species [e.g. Fe^{3+}, $Fe(OH)^{2+}$; $Fe(OH)_2^+$; $Fe(OH)_3$], or when local pH and redox conditions around cells are altered by their metabolic activity. In fact, the iron-coating on cells grown in Fe-rich cultures can be sufficiently dense to visualize the bacteria under the transmission electron microscope (TEM) without the standard use of metal stains (Fig. 8.3a).

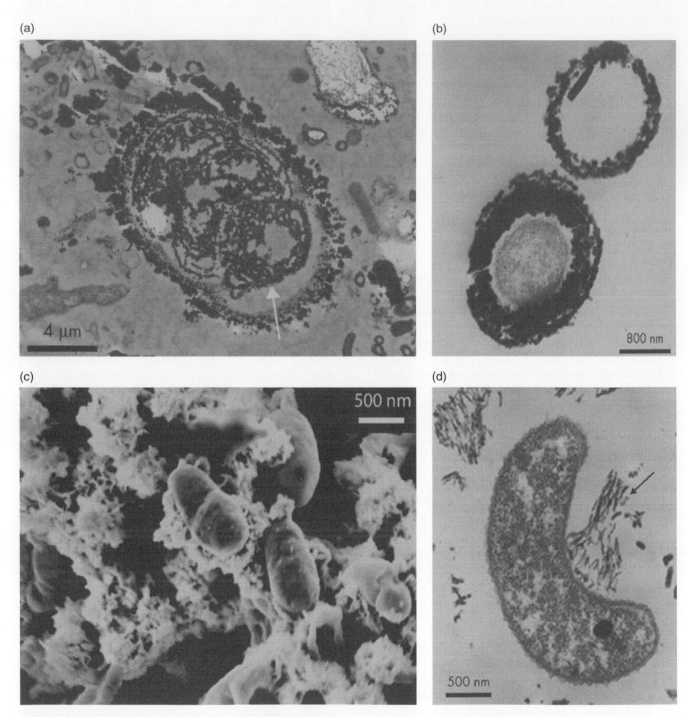

Figure 8.3 Transmission electron micrographs (TEM) and scanning electron micrographs (SEM) of iron hydroxide-cell assemblages. (a) a lysed bacterial cell with remnants of organic material retained within an iron hydroxide coating (from Konhauser, 2007). (b) two ferric hydroxide encrusted *Leptothrix ochracea* cells from an iron seep in Denmark. The cross section shows one ensheathed cell and one abandoned sheath (from Emerson, 2000). (c) the phototrophic Fe(II)-oxidizing *Rhodobacter ferrooxidans* attached to ferric hydroxide particles it generated during anoxygenic photosynthesis using dissolved Fe(II) (from Konhauser *et al.*, 2005). (d) image of *Gallionella ferruginea*, showing a portion of its stalk (arrow) attached to the cell (from Emerson, 2000).

Because of the ubiquity of iron biomineralization in nature, it has been suggested that under circumneutral conditions any microorganism that produces anionic ligands will non-specifically adsorb cationic iron species or fine-grained iron hydroxides from the surrounding waters (Ghiorse, 1984; Glasauer *et al.*, 2001). This is not unexpected given that the zero point of charge of pure ferric hydroxide is 8–9. Ferric hydroxide also develops on the organic remains of dead cells, implying that iron biomineralization can occur independently of cell physiological state.

8.4.1 Chemoheterotrophic iron mineralization

A number of microorganisms, the so-called iron depositing bacteria, facilitate iron mineralization by having surface ligands that promote Fe(II) oxidation, although it is not believed that they gain energy from the process (see Emerson, 2000 for review). The most common visible inhabitant of many freshwater, low-oxygenated iron seeps is *Leptothrix ochracea*. This chemoheterotroph frequently forms thick filamentous layers comprising a mass of tubular sheaths encrusted in iron. In an iron seep in Denmark, cell counts of 10^8–10^9 cells cm^{-3} promote iron accumulation rates of 3 mm day^{-1} (Emerson and Revsbech, 1994b), an extremely rapid sedimentation rate compared to the deposition of clastic sediments in marginal marine environments.

An interesting observation is that it is rare to find intact filaments of *L. ochracea* cells inside their sheaths (e.g. Fig. 8.3b). This correlates well with experiments showing that *Leptothrix* continuously abandons its sheath at a rate of 1–2 μm min^{-1}, leaving behind sheaths 1–10 cells in length that continue to deposit ferric hydroxide. This would seem to indicate that the bacteria use sheath secretion to avoid becoming permanently fixed in the mineral matrix (van Veen *et al.*, 1978). Other heterotrophic bacteria, such as filamentous species of *Sphaerotilus*, *Crenothrix*, *Clonothrix* and *Metallogenium*, as well as unicellular cocci of the *Siderocapsaceae* family, can induce ferric hydroxide precipitation through oxidation of organic iron chelates. In this manner, they use the organic carbon of such ligands as an energy source, resulting in the liberation and hydrolysis of ferric iron (Ghiorse and Ehrlich, 1992).

8.4.2 Photoautotrophic iron mineralization (photoferrotrophy)

As discussed above, and in detail in chapter 6, some anoxygenic photosynthetic bacteria are capable of oxidizing Fe(II) to Fe(OH)$_3$, using light for energy for CO$_2$ fixation (recall reaction 8.5). This process could be described as 'facilitated biomineralization' because ferric iron precipitates as a direct result of the metabolic activity of the microorganisms (Fig. 8.3c). These bacteria are phylogenetically diverse and include green sulfur bacteria (e.g. *Chlorobium ferrooxidans*), purple non-sulfur bacteria (e.g. *Rhodobacter ferrooxidans*) and purple sulfur bacteria (e.g. *Thiodictyon* sp.). All strains are mesophilic, with maximum growth rates at 23 °C, but are able to survive temperatures as low as 4 °C, while their growth is optimal at pH values around 7 (Hegler *et al.*, 2008). All experimental strains are also able to oxidize Fe(II) at very low light levels (down to 50 lux), which means that they can grow in light regimes befitting the photic zone of ocean water to depths of 100 m, or greater (Kappler *et al.*, 2005).

Ferrous iron can be used as an electron donor by these bacteria because the standard electrode potential for Fe^{2+}/Fe^{3+} (+0.77 V) is applicable only at very acidic pH, whereas at more neutral pH, the potential shifts to less positive values due to the low solubility of Fe^{3+} (Ehrenreich and Widdel, 1994). Photoferrotrophic growth can also be sustained by the presence of soluble ferrous iron minerals, such as siderite and iron monosulfide, but not insoluble minerals, such as vivianite, magnetite or pyrite (Kappler and Newman, 2004).

A current area of interest involves understanding how photoferrotrophs survive the mineralization process. Some have speculated that the cells prevent encrustation by creating a localized pH gradient outside the cell so that the Fe(III) remains soluble and can diffuse away from the cell until it reaches a less acidic environment (Kappler and Newman, 2004). Others have suggested that organic iron-chelating molecules may be involved in Fe(III) release by Fe(II)-oxidizing phototrophs (Straub *et al.*, 2001). The story has been complicated by the recent observation that Fe(II) is oxidized in the periplasm of some photoferrotrophs, which not only means that dissolved Fe(II) is brought into the cell, but that some dissolved form of Fe(III) must be excreted or else the periplasm would fill with ferric hydroxide (Jiao and Newman, 2007).

8.4.3 Chemolithoautotrophic iron mineralization

The formation of iron hydroxides may also stem from the ability of some chemolithoautotrophic bacteria to oxidize ferrous iron as an energy source (recall reaction 8.10). Although most enzymatic oxidation of Fe(II) occurs at extremely low pH, such as in acid mine drainage environments, the activity of *Acidithiobacillus ferrooxidans* or *Leptospirillum ferrooxidans* generally does not promote *in situ* ferric hydroxide precipitation because the Fe^{3+} formed remains soluble until more alkaline pH conditions ensue. However, at neutral pH, and under low oxygen concentrations, chemolithoautotrophic Fe(II) oxidation (e.g. by *Gallionella ferruginea*) does lead to high rates of iron mineralization (e.g. Søgaard *et al.*, 2000). In fact, the extracellular stalk can become so heavily encrusted with amorphous ferric hydroxide that the majority of its dry weight is iron (Fig. 8.3d).

Similar to *L. ochracea, Gallionella* species are common inhabitants of iron springs, and where they are abundant, the stalk material appears to form the substratum upon which subsequent Fe(II) oxidation occurs. Nonetheless, actively growing *Gallionella* and *Leptothrix* populations appear to occupy separate microniches within the same iron seep environments; the former preferring areas of sediment with lower oxygen concentrations (Emerson and Revsbech, 1994a). In wells, water pipes and field drains of water distribution systems, the large amount of iron precipitated by *Gallionella* has long been recognized as a cause of serious clogging problems (e.g. Ivarson and Sojak, 1978).

In anoxic environments, Fe(II) oxidation has also been shown to proceed with nitrate as the electron acceptor (Straub *et al.*, 1996). What was intriguing about the Straub study was the observation that nitrate-reducing species, that had not previously been grown in iron media, exhibited the capacity for ferrous iron oxidation, implying that this form of microbial oxidation may be commonplace. Indeed, bacteria that reduce nitrate with ferrous iron have now been recovered from wetlands (Weber *et al.*, 2006); shallow marine hydrothermal systems (e.g. Hafenbrandl *et al.*, 1996); and even deep sea sediments (e.g. Edwards *et al.*, 2003).

One particularly interesting feature about this form of metabolism is that most of the described nitrate-dependent Fe(II)-oxidizing strains depend on an organic co-substrate (e.g. acetate) and so far, only one truly lithoautotrophic nitrate-reducing strains has been isolated in pure culture. 16S rRNA gene analyses from the Straub study showed that this culture consists of four organisms, including three chemoheterotrophic nitrate-reducing bacteria (*Parvibaculum lavamentivorans, Rhodanobacter thiooxidans* and *Comamonas badia*), plus a fourth organism related to the chemolithoautotrophic Fe(II)-oxidizing bacterium *Sideroxydans lithotrophicus* (Blothe and Roden, 2009). The complexity of this culture potentially suggests that a consortium of organisms is needed for autotrophic Fe(II) oxidation coupled to nitrate-reduction.

8.4.4 *Hydrothermal ferric hydroxide deposits*

Perhaps the most striking example of ferric hydroxide biomineralization is at marine hydrothermal settings. Ferric iron minerals commonly precipitate directly on the seafloor, where subsurface mixing of hydrothermal fluids with infiltrating seawater produce metal-rich solutions that range in temperature from near ambient deep sea (~2 °C) to around 50 °C. The deposits themselves range from centimetre-thick oxide coatings formed by diffuse venting through underlying basalts or solid-phases sulfides to more voluminous oxide mud deposits (Juniper and Tebo, 1995).

Extensive deposits of ferric hydroxide-rich muds have been described from a number of submarine hydrothermal sites, including; (1) the shallow waters around the island of Santorini, (2) the Red and Larson Seamounts on the East Pacific Rise, and (3) the Loihi Seamount of the Hawaiian archipelago. The Santorini site is likely the most unambiguous example of ferric hydroxide precipitation because the mineralized stalks of *Gallionella ferruginea* occur so abundantly in the bottom sediment (Holm, 1987). The Red Seamount is characterized by an abundance of bacterial filaments, some of which have morphologies reminiscent of Fe(II)-oxidizing bacteria, e.g. twisted ribbons like *Gallionella ferruginea*, twisted spirals like *Leptospirillum ferrooxidans* and straight sheaths similar to *Leptothrix ochracea*. Although there is no direct evidence supporting enzymatic Fe(II) oxidation, indications of a microbial role in mineralization comes from the fact that the hydrothermal waters in which these deposits form are not conducive to rapid ferrous iron oxidation (Alt, 1988). The Loihi Seamount is the newest shield volcano at Hawaii. The impact of high Fe(II)/low H_2S emissions is apparent from the extensive deposits of ferric hydroxide that encircle the vent orifices, where anoxic hydrothermal waters mix with oxygenated seawater, and in the peripheral regions where visible mats are present. Initial microscopic analyses revealed that both deposits were rich in Fe-encrusted sheaths, similar in appearance to those of *Leptothrix ochracea* (Karl *et al.*, 1988). Since then, a number of studies have shown that the iron deposits have abundant microbial populations associated with them, up to 10^8 cells mg^{-1} (wet weight) of mat material, and that many of the cells are novel microaerophilic Fe(II)-oxidizers (Emerson *et al.*, 2010). Two interesting findings that have recently come from studies at Loihi are: (1) that the majority of ferric iron deposited around the vents is directly or indirectly attributable to bacterial activity (Emerson and Moyer, 2002), and (2) the bulk of the energy utilized by the seafloor chemolithoautotrophic communities derives from the iron released via hydrothermal emissions, and not oxidative weathering of the underlying basalts (Templeton *et al.*, 2009).

8.4.5 *Geobiological implications*

8.4.5.1 Evidence for ancient Fe(III)-precipitating microbes

For much of Earth's history the oceans were iron-rich (see Chapter 6 on the Global Iron Cycle), with dissolved Fe(II) concentrations estimated around 0.02 mM

(Holland, 1973; Morris, 1993); some 1,000 to 10 000 times modern seawater values. The ferruginous nature of the Precambrian oceans is manifested by the presence of BIF, chemical sedimentary rocks that were precipitated throughout the Archean and Paleoproterozoic. They then disappeared for a billion years before reappearing in association with episodes of global glaciation during the Neoproterozoic (Bekker *et al.*, 2010). BIF are characteristically laminated, with alternating Fe-rich (hematite, magnetite, and siderite) and silicate/carbonate (chert, jasper, dolomite and ankerite) minerals. Banding can often be observed on a wide range of scales, from coarse macrobands (meters in thickness) to mesobands (centimetre-thick units) to millimetre and submillimetre layers. Some deposits, such as in the 2.5 Ga Dales Gorge Member, in the Hamersley Group of Western Australia, show laterally contiguous layers up to a hundred in extent, suggesting that BIF were deposited uniformly over areas of tens of thousands of square kilometers (Trendall, 2002). The layering in BIF has been attributed to seasonal or decadal episodic hydrothermal pulsation and/or upwelling of anoxic, Fe-rich waters into semi-restricted basins already saturated with dissolved silica (Morris, 1993; Posth *et al.*, 2008).

Despite the lack of direct evidence, it is becoming increasingly accepted that microorganisms were involved in the primary oxidation of Fe(II) to Fe(III) in BIF. Two possible roles for bacteria are envisioned. The first is based on the production of O_2 by cyanobacteria, or their predecessors, as first proposed by Cloud (1965). In this model, these microorganisms would have flourished when Fe(II) and nutrients were available and passively induced the precipitation of ferric hydroxide through their metabolic activity and/or by sorbing aqueous Fe species to the anionic ligands exposed on their cell surfaces (Konhauser, 2000). Under an anoxic atmosphere, this O_2 could have been confined to local 'oxygen oases' associated with cyanobacterial blooms in coastal settings (Cloud, 1973). Once oxygen was present, chemolithoautotrophic Fe(II) oxidizers (using O_2 as the oxidant) could also have contributed to biogenic Fe mineral precipitation. For instance, Holm (1989) speculated that oxidation of dissolved Fe(II) by ancient forms of *Gallionella ferruginea* would have been kinetically favoured in an ocean with limited free oxygen because inorganic rates of Fe(II) oxidation at circumneutral pH are sluggish under microaerobic conditions (e.g. Liang *et al.*, 1993).

The second hypothesis, first elaborated on by Hartman (1984), is that anoxygenic photosynthetic bacteria may have coupled the C and Fe cycles prior to the evolution of oxygenic photosynthesis. The ability of primitive anoxygenic bacteria to oxidize Fe(II) is supported by the isolation of various marine and freshwater purple and green phototrophic bacteria that can use Fe(II) as a reductant for CO_2 fixation (recall discussion above). Models based on the Fe(II) oxidation rates of these strains under varying environmental conditions further demonstrate the plausibility of photoferrotrophy as a means for primary BIF precipitation (Konhauser *et al.*, 2002; Kappler *et al.*, 2005). Indeed, the ferric iron minerals that these strains produce are comparable with the Fe(III) precipitates that probably were deposited as primary BIF minerals. In particular, photoferrotrophs produce poorly crystalline ferric hydroxide, which carries a net positive charge (Kappler and Newman, 2004). Such biogenic minerals are expected to bind to organic carbon (e.g. cells), leading to deposition of cell–mineral aggregates on the seafloor. Metabolically-driven synsedimentary reactions and subsequent burial diagenesis would then transform the aggregates. Ferric hydroxide would either dehydrate to hematite or be reduced to magnetite and/or siderite, while organic carbon would be oxidized to CO_2. These reactions have been used to account for the lack of organic carbon in BIF deposits and the isotopic composition of the secondary minerals in BIF (Walker, 1984; Konhauser *et al.*, 2005).

One issue that any biological model for BIF must consider is when did cyanobacteria and photoferrotrophs evolve? Although the timing of cyanobacterial evolution is widely debated, the general consensus leans towards 2.8 to 2.7 Ga based on several lines of evidence from the rock record. For instance, based on geochemical constraints, it has been proposed that stromatolites from the 2.7 Gyr old Tumbiana Formation in Western Australia were formed by phototrophic bacteria that used oxygenic photosynthesis (Buick, 1992). This view is bolstered by the earliest recognized microfossil assemblage of colonies of filamentous and coccoid cells, from the 2.6 Gyr Campbell Group, South Africa, that are similar in appearance to modern oscillatoriacean cyanobacterial genera, such as *Phormidium* or *Lyngbya* (Altermann and Schopf, 1995). Bitumens from the 2.6 Gyr Marra Mamba Iron Formation of the Hamersley Group in Western Australia have yielded abundant amounts of 2α-methylhopanes, derivatives of prominent lipids in cyanobacteria (methyl-bacteriohopanepolyols), which imply cyanobacteria existed in the oceans at that time (Brocks *et al.*, 1999; Summons *et al.*, 1999), while extremely ^{12}C-rich kerogens from marine sediments aged 2.7–2.6 Ga have been attributed to the metabolic coupling of methane generation and its subsequent oxidation with O_2 by a group of aerobically respiring bacteria called methanotrophs (Hayes, 1983; Eigenbrode and Freeman, 2006). Most recently, trace metal speciation in 2.6 Ga black shales points towards oxygen availability in marine waters possibly several hundred meters deep (Kendall *et al.*, 2010).

In terms of anoxygenic photosynthetic bacteria, molecular phylogenetic analyses of a number of enzymes involved in (bacterio)-chlorophyll biosynthesis suggest that anoxygenic photosynthetic lineages are almost certain to be more deeply rooted than oxygenic cyanobacterial ones (Xiong, 2006). Recent studies also indicate that anoxygenic phototrophs represent a considerable fraction of biomass in modern stromatolite communities, and the recently demonstrated construction of stromatolite-like structures by the anoxygenic phototroph *Rhodopseudomonas palustris*, challenges the notion that stromatolites only indicate cyanobacteria in the geologic record (e.g. Bosak *et al.*, 2007). Furthermore, unique biomarkers for anoxygenic phototrophs, remnants of specific light-harvesting pigments, have recently been found in Paleoproterozoic strata, offering intriguing evidence for their presence on ancient Earth (Brocks *et al.*, 2005). It is interesting to note that 2-methylbacteiohopanes have also been identified in significant quantities in modern anoxygenic phototrophic Fe(II)-oxidizers (Rashby *et al.*, 2007), making the case for Fe(II)-oxidizing phototrophs at 2.7 Ga just as plausible as that for O_2-producing cyanobacteria.

8.4.5.2 What BIF tell us about ancient seawater

In natural systems where trace element sequestration by ferric hydroxide particles results from a continuum of adsorption and co-precipitation reactions (Ferris *et al.*, 1999), lumped-process distribution coefficient models can be used to relate the concentration of an element in the ferric hydroxide precipitate to the dissolved concentration present at the time of precipitation (Dzombak and Morel, 1990). This predictive aspect of the sorption reactions has subsequently been exploited to better understand the BIF record with respect to ancient seawater composition, and ultimately, nutrient availability for the ancient marine biosphere. Two examples have recently been addressed in the literature.

The first used BIF compositional data and phosphate partitioning coefficients between modern seawater and hydrothermal iron hydroxide precipitates to hypothesize that primary productivity in the Archean oceans may have been limited during BIF deposition by the strong adsorption of phosphate to the iron particles precipitating in the photic zone (Bjerrum and Canfield, 2002). The starvation of cyanobacteria would have had a knock-on effect by limiting oxygen production, which in turn, helped delay the eventual oxygenation of the atmosphere. While this study was quite revolutionary in its linking of seawater geochemistry to the productivity of the biosphere via the BIF record, its interpretation was likely complicated by the fact that the Archean oceans were also highly siliceous. Dissolved silica has a higher propensity for sorbing onto ferric hydroxides than dissolved phosphate, which means that in an amorphous silica-saturated ocean, phosphate levels may well have been sufficient to support plankton growth (Konhauser *et al.*, 2007). A recent compilation of P content in BIF and other iron-rich sediments through time has added an extra dimension to the phosphate story. Based on a five-fold increase in the phosphate concentrations in BIF from 750–630 Myr ago, a time coincident with two possible global 'Snowball Earth' glaciations, Planavsky *et al.* (2010) suggested that enhanced chemical weathering during deglaciation led to increased phosphate supply to the oceans, which in turn, led to greater cyanobacterial productivity, increased organic carbon burial, and eventually a transition to more oxidizing conditions in the oceans and atmosphere. This sequence of steps could then have facilitated the rise of animal life.

The second example used molar Ni/Fe ratios preserved in BIF over time to demonstrate the dynamic links between the physical evolution of the planet to changes in the biosphere and atmosphere. Konhauser *et al.* (2009) observed that the nickel content of BIF changed dramatically over time; from high Ni content between 3.8 and 2.7 billion years ago, dropping to about half that value between 2.7 and 2.5 billion years ago, and then slowly approaching modern values by 0.55 billion years ago. The large drop in Ni availability reflected a progressively cooling mantle and the eruption of less Ni-rich volcanic rocks (e.g. komatiites). Decreased komatiite production meant less Ni was dissolved into the oceans to become incorporated into BIF. The drop in seawater Ni at 2.7 Ga would have had profound consequences for microorganisms that depended on it, that being methane-producing bacteria, the methanogens. These bacteria have a unique Ni requirement for many of their essential enzymes, and a deficiency in the metal would have decreased their metabolism, resulting in less methane production. Crucially, methanogens have been implicated in controlling oxygen levels on the ancient Earth because the methane they produced was reactive with oxygen and kept oxygen levels low. So as long as methane was abundant, oxygen could not accumulate in the atmosphere, and indeed, it is believed that methane production must have dropped to enable the rise of atmospheric oxygen some 2.4 billion years ago, the so-called 'Great Oxidation Event' (Zahnle *et al.*, 2006).

8.5 Calcium carbonates

Bacterial carbonates are long-lived and locally abundant sediments that record both bacterial growth and the environmental factors that promote calcite/aragonite ($CaCO_3$) and dolomite [$CaMg(CO_3)_2$] precipitation in, on and around bacteria and the organic matter that they produce. These sediments include biogenic water

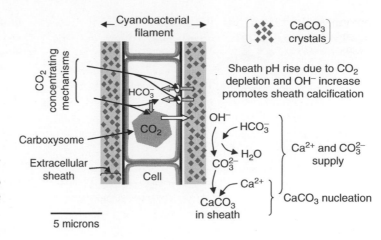

Figure 8.4 Schematic of *in vivo* cyanobacterial sheath calcification driven by CCM-enhanced photosynthesis (from Riding 2006, based on information from Miller and Colman, 1980; Thompson and Ferris, 1990; Merz, 1992; Price *et al.*, 1998; Kaplan and Reinhold, 1999; Badger and Price, 2003). Cyanobacterial CCMs include active uptake of HCO_3^- and its conversion to CO_2 for fixation by Rubisco in the carboxysome. This releases OH^-, which elevates the pH of fluids within the sheath pH, and ultimately promotes calcification.

column precipitates (whitings) that produce carbonate mud deposits on lake and sea floors, and a variety of *in situ* benthic deposits, most notably stromatolites. Present-day bacterial carbonates are deposited in diverse environments, but their geological history is best recorded in marine sedimentary rocks and extends through at least the past 2.5 Gyr, and possibly 1 Gyr more.

Calcification ($CaCO_3$ precipitation) by bacteria is not obligatory and it shows close dependence on environmental conditions, particularly the carbonate saturation state of ambient waters. It is therefore a good example of 'induced', as opposed to 'controlled' biomineralization. This reliance can be used to interpret changes in past environmental conditions, including the concentrations of dissolved inorganic carbon (e.g. CO_2, HCO_3^-) and sulfate that are required by key bacterial metabolic processes such as photosynthesis and sulfate reduction, respectively. Thus, secular patterns of bacterially induced calcification provide insights into long-term changes in Earth's surface environments that include seawater and atmospheric composition.

8.5.1 Cyanobacteria

Cyanobacterial calcification occurs when $CaCO_3$ crystals nucleate within the extracellular sheath or on the cell wall and EPS. These distinct styles of calcification have differing paleobiological and sedimentary consequences. Sheath impregnation by $CaCO_3$ crystals can create discrete microfossils such as *Girvanella* that are known since the mid-Proterozoic. By contrast, crystals formed at outer cell surfaces of bacteria that lack sheaths can subsequently be released into the water column as small loose particles or aggregates. These form milk-like, fine-grained suspensions (whitings) that settle out to accumulate as carbonate mud on lake and sea floors. However, they are not known to contain features that

reflect a specifically cyanobacterial origin and so they are difficult to discriminate from carbonate mud formed in other ways, such as by breakdown of algal and invertebrate skeletons.

8.5.1.1 Calcification mechanisms

Despite considerable progress, much remains to be understood concerning the factors influencing cyanobacterial calcification. Variation in calcification between strains within the same environment suggest intrinsic effects involving, for example, metabolic rates and mechanisms, growth stages, cell structure, type of EPS, and surface charge characteristics (Golubic, 1973). Nonetheless, three critically important factors have been recognized: (1) photosynthetic uptake of bicarbonate ions, (2) adsorption of Ca^{2+} to cell surfaces, and (3) ambient saturation state with respect to calcium carbonate.

As discussed in Section 8.3.2.1, photosynthetic uptake of bicarbonate results in pH increase near the cell (recall reaction 8.8) (Miller and Colman, 1980). Bicarbonate uptake is regarded as a response to low CO_2 availability and can be experimentally induced in cyanobacteria at atmospheric CO_2 levels below ~0.36% (Badger *et al.*, 2002). The bicarbonate is transformed into CO_2 within the cell and this process is termed a CO_2-concentrating mechanism (CCM). It involves a series of steps: active HCO_3^- transport into the cell, its intracellular conversion to CO_2, and the resulting external release of OH^- ions. OH^- release is the critical process driving calcification because it raises pH, potentially promoting $CaCO_3$ nucleation when ambient carbonate saturation is already elevated (reaction 8.19). Thus, pH increase near cyanobacterial cells resulting from CCM induction can stimulate both whiting production and sheath calcification (Riding, 2006) (Figs 8.4 and 8.5).

$$2HCO_3^- + Ca^{2+} \rightarrow CH_2O + CaCO_3 + O_2 \qquad (8.19)$$

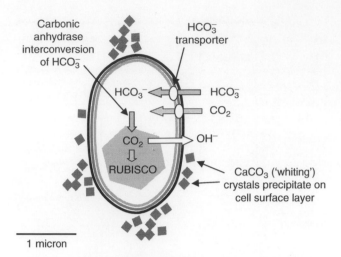

Figure 8.5 Where a sheath is lacking, CCMs promote CaCO$_3$ precipitation on the surface layer of individual picoplanktic cyanobacterial cells (see Thompson and Ferris, 1990, Fig. 3; Thompson 2000, Fig. 3). These external crystallites can be sloughed off or sedimented together with dead cells.

Evidence that Ca^{2+} adsorption is a key step in calcification comes from experimental studies that observed cyanobacteria to precipitate combinations of strontionite (SrCO$_3$), magnesite (MgCO$_3$) and mixed calcium–strontionite carbonates when grown in the presence of various combinations of Sr^{2+}, Mg^{2+} or Ca^{2+} (Schultze-Lam and Beveridge, 1994). In general, cyanobacteria are equally capable of incorporating Ca^{2+} or Sr^{2+} into carbonate mineral formation, yet Mg-containing precipitates are easily inhibited by the preferential binding of the other two cations. Other studies have documented that cyanobacteria can partition up to 1.0 wt% strontium in calcite (Ferris *et al.*, 1995). However, our understanding of calcium adsorption to cyanobacteria is complicated by the fact that they can present differing surfaces to the external environment. For instance, cyanobacteria with sheaths generally precipitate more calcium carbonate than those without sheaths. The mucilaginous sheath that envelops benthic calcified cyanobacteria is a structured form of EPS that can provide support, stability, and protection against physical damage, including ultraviolet radiation, dehydration and grazers. It also binds solutes and reduces the diffusion of carbonate ions (formed during photosynthesis) to the bulk milieu. Calcium carbonate may nucleate on the sheath surface and grow radially outwards, or it may nucleate within the sheath, leading to the sheath's impregnation with mineral material (Riding, 1977). In present-day examples sheath calcification ranges from isolated crystals (Pentecost, 1987, Fig. 1d), through a crystalline network (e.g. Friedmann 1979, Fig. 9), to a relatively solid tube of closely juxtaposed crystals (e.g. Couté and Bury, 1988, plate 2). These differences in

degree of calcification appear to relate to both environmental and biological controls. Post-mortem sheath degradation by bacteria has also been suggested to lead to calcification, but degraded sheaths are irregular in form and encrusted by carbonate to varying degrees. In contrast, fossils such as *Girvanella* exhibit regular tube morphology in which wall-thickness remains constant in individual specimens, suggesting *in vivo* sheath impregnation (Riding, 1977, 2006).

The overriding environmental control on cyanobacterial calcification appears to be aquatic saturation with respect to CaCO$_3$ minerals (Kempe and Kazmierczak, 1994). At the present-day this is reflected in the limitation of cyanobacterial sheath calcification to calcareous lakes and freshwater streams. A striking additional observation is that calcified cyanobacteria appear to be very rare in present-day marine environments, whereas sheath calcified microfossils were common in ancient shallow marine environments, especially during the Paleozoic and Mesozoic (Riding, 1982), suggesting a significant reduction in carbonate saturation. Similarly, biogenic whitings are well-known in some present-day calcareous lakes, and are inferred to have contributed a significant proportion of marine carbonate mud from the mid-Proterozoic to late Mesozoic; but their development in present-day marine environments, although strongly suspected, is not confirmed.

8.5.1.2   Sedimentary products

Biogenic whitings   Small cyanobacteria (<2 μm in size), known as picoplankton, have long been linked to fine-grained calcium carbonate precipitation in the water column. These whiting events can occur in lakes during times of seasonal blooms and locally are responsible for the bulk of the sedimentary carbonate deposits. One well described site is Fayetteville Green Lake in New York State, where active growth of the unicellular cyanobacterium *Synechococcus* sp. during the summer months leads to calcite formation on the extracellular S-layers (Thompson *et al.*, 1990). Consequently, a light rain of calcite-encrusted plankton falls to the lake bottom each summer, contributing to unconsolidated carbonate mud deposition. Interestingly, during the cold winter months, when the *Synechococcus* cells become dormant, the non-metabolizing cells develop abundant gypsum (CaSO$_4$•2H$_2$O) crystals on their S-layers instead of calcite – the formation of gypsum is the thermodynamically predictable phase given the high dissolved sulfate concentrations in the lake. Importantly, this ability to switch biominerals testifies to the lack of control exerted by the cyanobacteria in this biomineralization process. The carbonate crystals are sedimented through the water column individually, or as poorly structured aggregates along with organic cells, to accumulate as layers of carbonate mud on the lake bed, often in seasonally varved deposits (Kelts and Hsü, 1978).

There has been considerable debate regarding the origins of present-day marine 'whitings' on the Bahaman Banks. Satellite imagery has shown some whitings to cover as much as $200\,km^2$ during the summer (Shinn *et al.*, 1989) and it has been suggested that cyanobacterial whitings could account for much of the late Holocene bank-top lime muds on the Great Bahamas Bank (Robbins *et al.*, 1997). Field studies provided putative evidence in support of a microbial origin for the whitings due to the presence of 25% by weight organic matter in the solid whiting material and electron microscopy (SEM/TEM) images that showed individual whiting spheres embedded in an organic matrix, along with the presence of $CaCO_3$ crystals on cyanobacterial surfaces (Robbins and Blackwelder, 1992). However, whitings generally occur in very shallow water, and it is thus possible that they may also include a component of resuspended mud. This interpretation is supported by whiting $CaCO_3$ having a $^{14}C/^{12}C$ ratio different to that of inorganic carbon in the surrounding water (Broecker *et al.*, 2000). Consequently, re-suspension of sediment could be the dominant process involved in marine whitings on the Bahaman Banks (Broecker *et al.*, 2001).

Calcified sheaths At the present-day, calcified cyanobacteria are rare in normal marine environments, although they were common at many times in the past. This secular distribution most likely reflects the relatively low carbonate saturation state of present-day oceans. Consequently, information concerning calcified cyanobacteria has to be obtained from non-marine environments, and they are most common as tufa deposits in hardwater streams fed by springs in limestone areas. As it emerges, groundwater loses CO_2 to the atmosphere, increasing carbonate saturation, resulting in the precipitation of calcium carbonate on cyanobacterial sheaths. In fast flowing streams cyanobacterial photosynthesis may utilize CO_2 and calcification appears largely as a surface crust on the sheaths. In slower flowing streams and lakes, in contrast, the sheaths are impregnated by carbonate crystals, probably reflecting localized pH increase due to bicarbonate uptake (Merz-Preiß and Riding, 1999). These contrasting styles of calcification emphasize the large differences that can occur even within so-called 'biologically induced calcification'. Importantly, it is this sheath-induced calcification that creates microfossils, recognizable in marine environments back to at least the Mesoproterozoic (Kah and Riding, 2007). When well-preserved, these are morphologically simple but nonetheless distinctive. They include delicate tubes (e.g. *Girvanella* and *Ortonella*) and dendritic shrub-like masses (e.g. *Angusticellularia*) that

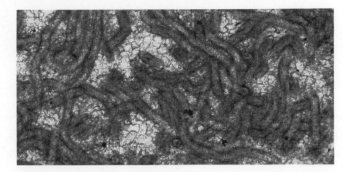

Figure 8.6 $CaCO_3$-impregnated cyanobacterial sheath preserved as the calcified microfossil *Girvanella*, early Mid-Ordovician, Lunnan, Tarim, China. Width of view 1 mm.

have present-day analogues in calcified filamentous cyanobacteria. In marine environments they typically accumulated *in situ* as components of stromatolites, oncoids, thrombolites, dendrolites, reef crusts, and other microbial mat deposits. Occasionally, however, they occurred as loosely tangled planktic or semi-planktic filaments as in some *Girvanella* (Fig. 8.6).

Tufa and travertine Aquatic mosses are often abundant and heavily calcified in tufa streams that channel the water flow. As in cyanobacteria, in these fast-flowing conditions it appears that calcification is not directly related to moss photosynthesis but instead is associated with a microbial biofilm growing on the moss that includes cyanobacteria and small algae (Freytet and Verrecchia, 1999; Pentecost, 2005). Moss and cyanobacterial tufas can create substantial barriers that block rivers, creating lakes between which water cascades over tufa dams. This localized turbulence and evaporation promotes degassing of CO_2 that enhances carbonate precipitation, providing a positive feedback to further tufa formation. Ultimately such barriers can fail or the water is diverted to other routes, but while they exist they can create spectacular natural water gardens, as at Plitvice in Croatia (Golubić *et al.*, 2008). In these environments the link between biofilms and calcification is suggested by experiments in which copper substrates toxic to microbes remain uncalcified (Srdoč *et al.*, 1985), whereas biofilms form on other substrates and calcify.

The organisms that create substrates for carbonate precipitation in ambient-water streams are responsible for the typically highly porous structure of tufa. In contrast, the chemistry and temperature of hot water springs promote even more rapid carbonate precipitation but tend to inhibit growth of all but the most hardy microorganisms. As a result, organic substrates are often inconspicuous and the resulting travertine deposit may be sinter-like (the siliceous precipitates associated with hot springs). There are complex intergradations between tufa and travertine, but hot spring water is generally subject to such rapid increase in saturation that much of the carbonate is deposited very close to the springs in ridges or steep terraces such as Mammoth Hot Springs, in Yellowstone National Park, USA (Fouke *et al.*, 2000). These deposits appear to be essentially abiogenic, and the effect of microorganisms on their fabric is less than in the tufa of cool water streams (Pentecost, 2005). However, some distinctive travertine deposits such as dendritic shrub-like fabrics, have been regarded as the calcification products of bacteria other than cyanobacteria due to the presence of numerous micron-sized bodies interpreted as the remains of bacterial cells (Chafetz and Folk, 1984).

8.5.2 Bacteria

Bacteria other than cyanobacteria have long been suggested to contribute to the formation of a variety of non-skeletal sedimentary carbonates, including carbonate mud, ooids, clotted micrite fabrics, stromatolites, peloids, and peloidal cements. Drew (1913) proposed that abundant lime mud accumulating west of Andros Island on the Bahaman Banks, and in parts of the Florida Keys, is a bacterial precipitate, and also suggested that oolitic limestone may owe its origin to 'some diagenetic change in the precipitate of finely divided calcium carbonate particles produced in this way by bacterial action'. He and other early researchers (see Ehrlich, 2002) were impressed by the abundance of bacteria in carbonate sediments. They found that when grown under laboratory conditions, a wide variety of bacteria were able to precipitate carbonate minerals. The drawback to this approach lies in the differences between laboratory and field conditions, particularly with regard to aqueous chemistry, and growth media composition and consistency. Efforts continue to elucidate the effects of specific organic functional groups on carbonate morphology and mineralogy, e.g. spherule vs. euhedral calcite or calcite vs. aragonite (e.g. Braissant *et al.*, 2004), and also to explore links between bacterial activity and the formation of carbonate grains such as mud (e.g. Thompson, 2000) and ooids (e.g. Plée *et al.*, 2008). Particularly close attention has been paid to the development of stromatolitic microfabrics, such as clotted and peloidal micrite, in microbial mats.

8.5.2.1 Calcification mechanisms in microbial mats

Microbial mats are complex, densely populated algal–bacterial communities that develop on illuminated sediment surfaces that are intermittently or permanently water-covered (e.g. Revsbech *et al.*, 1983; van Gemerden, 1993). They have long been regarded as a key to understanding stromatolites (e.g. Walcott, 1914; Monty, 1976).

In shallow water, mat life is based on energy cycling between 'primary producers' that photosynthetically fix inorganic carbon, and 'decomposers' that efficiently recycle these organic products. In present-day marine mats the main primary producers of organic matter are cyanobacteria, whereas decomposers release energy from the organic matter by transferring electrons from the reduced carbon to available oxidants such as O_2, NO_3^-, SO_4^{2-}, and CO_2 (see Konhauser, 2007, chapter 6). These processes can significantly change alkalinity and pH (Baas-Becking *et al.*, 1960), and thus, they affect carbonate precipitation and dissolution (Canfield and Raiswell, 1991b). For example, nitrate reduction 'recall reaction 8.14' and sulfate reduction (recall reaction 8.17) contribute to localized increases in HCO_3^- level, and therefore raised alkalinity, that can promote supersaturation with respect to calcium carbonate minerals (Visscher and Stolz, 2005). This, in turn, can lead to the nucleation of fine-grained carbonates that entomb living and dead microorganisms in a lithified matrix (e.g. Castanier *et al.*, 2000). In an opposing manner, aerobic respiration in the surface mat promotes carbonate dissolution through the production of dissolved CO_2

$$CH_2O + CaCO_3 + O_2 \rightarrow 2HCO_3^- + Ca^{2+} \tag{8.20}$$

Calcium-binding by EPS and its release during degradation, as well as the spatial separation of biological processes that affect carbonate dissolution and precipitation, contribute to significant local variations in carbonate precipitation within mats (Dupraz and Visscher, 2005).

8.5.2.2 Sedimentary products: stromatolites

The characteristic feature of stromatolites is lamination that reflects episodic but relatively evenly distributed accretion associated with microbial growth and calcification, and locally also with abiogenic precipitation and grain trapping. Their record commences 3.5 Ga. By contrast, thrombolites are microbial carbonates with a macroscopically clotted fabric (Aitken, 1967). They first appeared in abundance in the mid-Proterozoic, concurrent with the presence of recognizable calcified cyanobacterial filaments in the rock record, and were particularly common in the Cambrian and Early Ordovician. Present-day thrombolites that are strikingly similar in appearance to Early Palaeozoic examples form through cyanobacterial calcification in the marginal marine Lake Clifton in Western Australia (Moore and Burne, 1994), but thrombolitic fabrics are also associated with laminated fabrics in Shark Bay (Aitken, 1967) and Lee Stocking columns (Feldmann and McKenzie, 1998) that mainly form by trapping and binding sandy carbonate sediment, together with episodic lithification (Reid *et al.*, 2000). Thrombolites therefore appear to have a variety of origins, and – unlike stromatolites – have the additional distinction that their characteristic clotted appearance can be significantly enhanced by diagenetic alteration.

Although some stromatolite-like deposits may be essentially abiogenic seafloor crusts (e.g. Grotzinger and Rothman, 1996; Riding, 2008), many stromatolites, especially Phanerozoic examples, can confidently be interpreted as lithified microbial mats (Chapter 16). Present-day calcified shallow-water mats, formed by communities of cyanobacteria and other bacteria, have complex fabrics comparable with those of many Phanerozoic and also Proterozoic stromatolites. They can include well-defined microfossils such as calcified cyanobacterial sheaths, but these are generally volumetrically subordinate to fine-grained clotted and peloidal carbonate microfabrics that mostly represent the synsedimentary calcification of bacterial cells and cell products. The peloids are irregular micritic aggregates commonly 20–60 microns in size. Both clotted texture and associated peloids have been linked to bacterial calcification in general, and sulfate reduction in particular (Pigott and Land, 1986; Chafetz 1986; Heindel *et al.*, 2010). Formation of these microfabrics during degradation of organic matter, that includes bacterial cell wall material, EPS and other organic macromolecules, helps to explain their complexity (see Riding, 2000; Dupraz and Visscher, 2005).

8.5.2.3 Dolomitization

The experimental abiotic formation of dolomite [$CaMg(CO_3)_2$] at low temperatures is difficult, if not impossible, even from supersaturated solutions. In contrast, in nature dolomite commonly has been reported as an early replacement mineral of calcite and/or aragonite. Again, it has been realized that bacterial sulfate reduction can be directly involved because it overcomes the kinetic barrier to dolomite formation by increasing the pH and alkalinity, and removes sulfate which is a known inhibitor to dolomite formation (Baker and Kastner, 1981; Lyons *et al.*, 1984; van Lith *et al.*, 2003a). The environments from which bacterial dolomite has been reported range from deep marine (Friedman and Murata, 1979) to shallow hypersaline lagoons (Gunatilaka *et al.*, 1984). Since sulfate occurs in seawater as a magnesium-sulfate ion pair, its removal also increases magnesium availability for dolomite precipitation in the microenvironment around the cell (Van Lith *et al.*, 2003b). Furthermore, sulfate-reducing bacteria have been shown to experimentally precipitate dolomite crystals identical in composition and morphology to those found in the natural systems where the bacteria were isolated (Warthmann *et al.*, 2000).

Dolomite has also been shown to form in basalt-hosted aquifers, in association with methanogenic bacteria (Roberts *et al.*, 2004). Dissolution of basalt yields elevated pore-water concentrations of dissolved Ca^{2+} and Mg^{2+}, which can adsorb onto cell surfaces. Together with the methanogenic consumption of CO_2 that generates alkalinity, this locally increases carbonate saturation and results in the formation of small dolomite grains (10s of nanometers in diameter) directly on the cell surfaces, and at times, completely encrusting them (Kenward *et al.*, 2009).

8.5.3. *Geological implications*

It can be expected that bacterial calcification should have responded to long-term variations in atmosphere-hydrosphere composition because these global changes would have affected carbonate saturation and bacterial metabolism (e.g. sulfate reduction, photosynthesis). Indeed, there is evidence that these changes may be reflected in the secular abundance of bacterial carbonates, and in the fabrics developed by stromatolites and other microbial carbonates (Fig. 8.7).

8.5.3.1 Stromatolites

Microbial mat fabrics developed in association with precipitated abiogenic crusts during the Archean and early-mid Proterozoic (Sumner, 1997). This is also suggested by the regular and even laminations common in these stromatolites, and by the alternating dark (fine-grained) and light (sparry) layers of which some of them consist (Riding, 2008). As abiogenic precipitation declined and lithified microbial mats increased over time, marine stromatolites developed the uneven layering and dominantly fine-grained, clotted-peloidal fabrics that characterize many Phanerozoic examples. During the Proterozoic, reduction in abiogenic precipitation probably reflects a general decline in seawater carbonate saturation (Grotzinger, 1990) and the development of clotted-peloidal fabrics may relate to increased levels of seawater sulfate that promoted bacterial sulfate reduction. The mid-Proterozoic appearance of calcification of cyanobacterial sheaths could indicate that decline in atmospheric CO_2 had reached a critical threshold stimulating cyanobacteria to develop CCMs (Riding, 2006). In this view, the development of the clotted/peloidal (spongiostrome) and tubiform (porostromate) microfabrics that are widespread in lithified microbial mats during the Phanerozoic were the respective products of bacterial sulfate reduction and cyanobacterial CCM development. In addition, it seems likely that calcified cyanobacteria complicated and coarsened microbial carbonate macrofabrics sufficiently to give them a dominantly clotted rather than laminated appearance, and thrombolites were born.

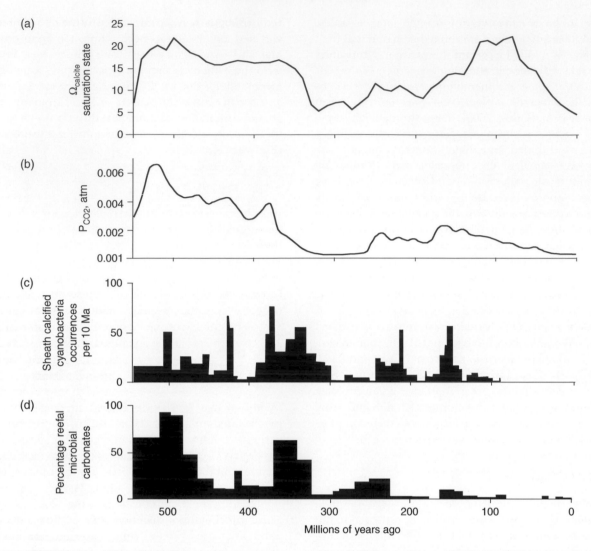

Figure 8.7 Phanerozoic trends of (a) seawater calcite saturation state ($\Omega_{calcite}$) (Riding and Liang 2005b, Fig. 5), (b) atmospheric CO_2 (Berner and Kothavala, 2001, Fig. 13 GEOCARB III), (c) marine calcified cyanobacteria occurrences (Arp *et al.*, 2001, Fig. 3d), (d) reefal microbial abundance (Kiessling, 2002, Fig. 16).

Long-term decline in the abundance of microbial carbonates, such as stromatolites, has often been attributed to both eukaryote competition and reduction in carbonate saturation state (Fischer, 1965), although the inception of this decline before the appearance of metazoans suggests the latter as the major influence (Grotzinger, 1990). This, together with a lower rate of microbial growth, could also have affected stromatolite morphology, since the external shapes of stromatolites are significantly determined by the accretion rate relative to adjacent sediment (Riding, 1993). Low relative accretion rate results in low relief, making stromatolites more prone to lateral incursion by sediment, and fostering complex shapes, such as digitate forms. In contrast, high relative accretion rates promote high relief and simple shapes, such as domes and cones. Consequently, although mid-Proterozoic increase in

morphotypic diversity (e.g. in branched stromatolites), has been regarded as a proxy for abundance, it could be that complex shape reflects low synoptic relief due to reduced relative accretion rate. If so, then it might be a sign that stromatolite growth was in decline. The Phanerozoic is characterized by episodic development of dendrolites and thrombolites, for example in the Cambro-Ordovician and Late Devonian, and also by changes in microbial carbonate abundance in general. These patterns may reflect broad positive correspondence between reefal microbial carbonate abundance (Kiessling, 2002, see their Fig. 16) and calculated seawater saturation state for $CaCO_3$ minerals during much of the Palaeozoic and Mesozoic (Riding and Liang, 2005a) (Fig. 8.7a,d). Breakdown in this pattern ~120–80 Ma ago (when the calculated saturation ratio was high but microbial carbonate abundance was low) could reflect removal of carbonate deposition by pelagic plankton that significantly reduced the actual saturation state.

Nonetheless, the important influence of saturation state on stromatolite accretion and preservation does not preclude significant additional long-term effects on stromatolite and thrombolite development as a result of eukaryote evolution. It is debatable whether metazoan grazing significantly affected stromatolite development so long as carbonate saturation was high enough to ensure microbial mat lithification, but overgrowth by skeletonized algae and invertebrates must have inhibited the formation of well-defined microbial structures from the mid-Ordovician onwards. As a result, microbial carbonates were subsumed within complex reef structures where, lacking space to form domes and columns, they developed as patchy and irregular crusts on and around skeletal organisms. This made them appear much less abundant than they really were (Pratt, 1982). Exceptions to this pattern could have occurred in the immediate aftermaths of mass extinction events that killed skeletal reef builders, and also in ecological refuges such as at present-day Shark Bay (Logan, 1961) and Lee Stocking Island (Dill *et al.*, 1986). In these situations, unconstrained by algal and invertebrate reef organisms, microbial mats were once again able to develop distinctive morphologies, including metric scale steep sided columns.

Uncalcified algae also transformed mat communities. For example, diatoms and green algae are abundant in Shark Bay (Awramik and Riding, 1988) and Lee Stocking (Riding *et al.*, 1991) mats and confer added ability to trap sediment that can be much larger (sand, gravel) than in most ancient stromatolites. This creates coarse and crudely layered fabrics that are locally thrombolitic. Thus, these examples of present-day columns reflect the combination of at least two favourable factors: the trapping abilities of both cyanobacteria and algae, and salinity and current stressed conditions that inhibit overgrowth by reef organisms. The first of these promotes rapid accretion, while the second hinders disruption of the microbial mats. Overall, however, marine microbial carbonates are scarcer today than at almost any other time during the Phanerozoic. Presumably this reflects at least two present-day conditions: the abundance of skeletal reef builders, and relatively low carbonate saturation state.

8.5.3.2 Calcified cyanobacteria

The earliest record of cyanobacterial calcification is from the early Proterozoic (Klein *et al.*, 1987), although sheath-calcified cyanobacteria seem to be remarkably scarce until ~1.2 Ga (Kah and Riding, 2007) and they only became widespread in the Neoproterozoic. They remained locally very common in shallow marine carbonates throughout much of the Palaeozoic and

Mesozoic, but then became scarce by the mid-Cretaceous and are rare in present-day normal marine environments (Riding, 1982; Arp *et al.*, 2001). This unusual secular distribution is hypothesized to be due to the combined effects of CCM induction (and therefore changes in atmospheric CO_2) and fluctuations in carbonate saturation state (Riding, 2006).

Badger *et al.* (2002) proposed that cyanobacterial CCMs developed in the Late Palaeozoic in response to both CO_2 decline and O_2 increase. However, similarly large changes in CO_2 levels are thought to have occurred in the mid-Proterozoic (Kasting, 1987), and it has been suggested that cyanobacterial CCMs first developed then, coincident with the appearance of calcified sheath microfossils (Riding, 2006; Kah and Riding, 2007). The reasoning is that under high early Proterozoic CO_2 levels, cyanobacterial photosynthesis would have relied on CO_2 diffusion that would not have significantly altered pH near the cells. As atmospheric CO_2 level fell in the mid-Proterozoic, cyanobacteria began to pump in bicarbonate to maintain photosynthesis, and the conversion of bicarbonate to CO_2 resulted in OH^- production that raised pH, stimulating calcification (see Section 8.5.1.1). Thus, cyanobacterial CCM development promoted sheath calcification and whiting production. Appearance of sheath-calcified microfossils ~1.2 Ga ago could therefore reflect reduction in atmospheric CO_2 level below the threshold of ~0.36% (~10 times the present preindustrial atmospheric level) at which cyanobacteria induce CCMs in experiments (see Badger *et al.*, 2002; Riding, 2006).

During much of the Palaeozoic and Mesozoic, fluctuations in calcified sheath abundance (Arp *et al.*, 2001) and also reefal microbial carbonate abundance (Kiessling, 2002, Fig. 16) broadly track some estimates of seawater carbonate saturation (Riding and Liang, 2005b) (Figure 8.8). The abundance of sheath calcified cyanobacteria when CO_2 levels are thought to have been high (see Berner and Kothavala, 2001), e.g. during the early-mid Palaeozoic, suggests that CCMs continued to be induced even when p_{CO2} substantially exceeded 10 PAL. This could be because calcified sheaths developed in microbial mat and reef environments where cyanobacteria experienced localized carbon limitation that necessitated CCM activation (Riding, 2006). The broader influence of p_{CO2} on CCMs may be reflected in the Mississippian, when calcified sheath abundance increased as p_{CO2} declined (Riding, 2006, 2009).

As noted with regard to reefal microbial carbonates, reduction in sheath calcified cyanobacteria ~120–80 Ma ago could reflect reduction in actual (as opposed to calculated) saturation state resulting from the burial of abundant carbonate precipitated as calcified skeletons by plankton such as coccolithophore algae and globigerine foraminifers (Riding and Liang, 2005a), and this effect may have continued into the Palaeogene. Since the

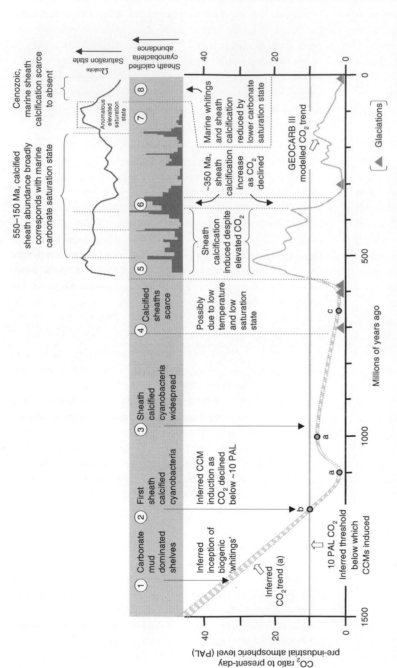

Figure 8.8 Conjectural history of cyanobacterial marine sheath calcification and picoplanktic 'whiting' precipitation. The Proterozoic inferred CO_2 trend is based on (a) Sheldon (2006), (b) Kah and Riding (2007), (c) Hyde et al. (2000) and (d) Ridgwell et al. (2003); the continuous trend line is from Berner & Kothavala (2001, fig. 13); the Neoproterozoic glaciations is from Walter et al. (2000); the occurrences of marine sheath calcified cyanobacteria is from Arp et al. (2001, fig. 3d); the calculated carbonate saturation states are from Arp et al. (2001, fig. 5); and the threshold below which CCMs are induced is based on Badger et al. (2002). Several key developments can be inferred from the figure. (1) Photosynthetic 'whitings', as reflected by widespread carbonate mud sedimentation, may have been triggered as CO_2 reduced pH buffering (see Arp et al., 2001; Riding, 2006). (2) A further decline below ~10 PAL CO_2 induced CCM development and sheath calcification at ~1200 Ma (Kah and Riding, 2007). (3) Calcified sheaths were widespread in the early Neoproterozoic (see references in Knoll and Semikhatov 1998), but (4) became scarce during 'Snowball' glaciations, possibly due to reduction in CCM development as low temperatures favoured diffusive entry of CO_2 into cells, and due to lower seawater saturation state reflecting reduction in both temperature and p_{CO_2}. (5) Sheath calcification was common in marine environments during the early-mid Palaeozoic despite elevated CO_2 suggesting that once CCMs had developed they were readily induced where carbon limitation developed, such as microbial mats. Throughout much of the Palaeozoic and early Mesozoic, calcified sheath abundance appears to vary with carbonate saturation state. (6) As CO_2 declined in the Late Devonian-Early Mississippian, calcified sheath abundance temporarily increased, possibly reflecting enhanced CCM induction, but then declined as the saturation state dropped in the Mississippian-Pennsylvanian (Riding, 2009). (7) Despite a high calculated saturation state in the Late Cretaceous-Palaeogene, planktic calcifiers probably reduced the actual saturation state sufficiently to inhibit cyanobacterial calcification. Thus, calcified sheaths were scarce in marine environments. (8) Since the Eocene, low carbonate saturation – due to low levels of both Ca ions and pCO_2 is reflected in extreme scarcity in sheath calcification in marine environments.

Eocene, under the influence of low levels of Ca^{2+} ions and p_{CO2}, seawater carbonate saturation was reduced to a Phanerozoic low only matched in the Late Palaeozoic (Riding and Liang, 2005a; see their Fig. 5a). Consequently, despite low p_{CO2} values during the Cenozoic (see Berner and Kothavala, 2001) that would have induced cyanobacterial CCMs, sheath calcification has not been observed in marine cyanobacteria, and this presumably reflects low carbonate saturation.

8.5.3.3 Biogenic whitings

It has been widely supposed that a significant proportion of Proterozoic carbonate mud may have derived from 'whitings' stimulated by photosynthetic CO_2 uptake by photosynthetic phytoplankton (Grotzinger, 1989, 1990). Through much of the Proterozoic, carbonate mud abundance appears to exhibit a first order trend of increase with decreasing age. Absence of carbonate mud in the Late Archaean (Sumner and Grotzinger, 2004) and scarcity in the Palaeoproterozoic (e.g. Kah and Grotzinger, 1992), was followed by an increase that transformed cement-rimmed platform margins into muddy carbonate ramps ~1.4–1.3Ga (Sherman *et al.*, 2000, p. 290). By the early Neoproterozoic, both calcified cyanobacteria (Swett and Knoll 1985; Knoll and Semikhatov, 1998) and carbonate mud accumulation (Herrington and Fairchild, 1989) were widespread. It can be postulated that this transition to carbonate mud dominated platforms reflects inception of whiting precipitation as p_{CO2} declined and CCMs developed in cyanobacteria (Riding, 2006).

Since CCMs promote bloom conditions by conferring ability to overcome carbon limitation (Rost *et al.*, 2003), CCM-stimulated picoplanktic mud production on carbonate shelves may have continued to exert a major influence on subsequent carbonate sedimentation whenever carbonate saturation state was sufficiently elevated, for example, during the early-mid Palaeozoic and parts of the Mesozoic. These biogenic whitings should have been further intensified whenever p_{CO2} levels fell to levels near or below ~10 present atmospheric levels. Eventually, as seawater carbonate saturation levels declined in the Cenozoic this process should have slowed, despite falling p_{CO2}. Thus, as noted above (Section 8.5.1.2), it has been suggested that present-day marine 'whitings' on the Bahama Banks could be biogenic (Robbins and Blackwelder, 1992), but may dominantly be formed by sediment resuspension (Broecker *et al.*, 2000). If present-day biogenic marine whitings produced by cyanobacteria are in fact scarce, it most likely reflects relatively low present-day levels of seawater carbonate saturation.

Bacterial calcification, strongly susceptible to environmental influence, is therefore a sensitive indicator of changes in atmospheric and seawater composition over long geological time-scales. The fabrics of benthic bacterial carbonates such as stromatolites and thrombolites, and also their abundance together with that of bacterial whitings, have altered significantly during the Precambrian and Phanerozoic. Relating these developments to both environmental change and bacterial metabolic evolution offers fruitful areas for further research.

Acknowledgements

We express our appreciation to Tanja Bosak and Linda Kah for their detailed and useful suggestions.

References

Aitken JD (1967) Classification and environmental significance of cryptalgal limestones and dolomites, with illustrations from the Cambrian and Ordovician of southwestern Alberta. *Journal of Sedimentary Petrology* **37**, 1163–1178.

Alt JC (1988) Hydrothermal oxide and nontronite deposits on seamounts in the Eastern Pacific. *Marine Geology* **81**, 227–239.

Altermann W, Schopf JW (1995) Microfossils from the Neoarchean Campbell Group, Griqualand West Sequence of the Transvaal Supergroup, and their palaeoenvironmental and evolutionary implications. *Precambrian Research* **75**, 65–90.

Arp G, Reimer A, Reitner J (2001) Photosynthesis-induced biofilm calcification and calcium concentrations in Phanerozoic oceans. *Science* **292**, 1701–1704.

Awramik SM, Riding R (1988) Role of algal eukaryotes in subtidal columnar stromatolite formation. *Proceedings of the National Academy of Sciences, USA* **85**, 1327–1329.

Baas-Becking LGM, Kaplan IR, Moore D (1960) Limits of the natural environment in terms of pH and oxidation-reduction potentials. *Journal of Geology* **68**, 243–284.

Badger M, Price D (2003) CO_2 concentrating mechanisms in cyanobacteria: molecular components, their diversity and evolution. *Journal of Experimental Botany* **54**, 609–622.

Badger MR, Hanson D, Price GD (2002) Evolution and diversity of CO_2 concentrating mechanisms in cyanobacteria. *Functional Plant Biology* **29**, 161–173.

Baker PA, Kastner M (1981) Constraints on the formation of sedimentary dolomite. *Science* **213**, 214–216.

Bekker A, Slack JF, Planavsky N, *et al.* (2010) Iron formation: a sedimentary product of the complex interplay among mantle, tectonic, and biospheric processes. *Economic Geology*, in press.

Berner RA (1984) Sedimentary pyrite formation: an update. *Geochimica et Cosmochimica Acta* **48**, 605–615.

Berner RA, Kothavala Z (2001) GEOCARB III. A revised model of atmospheric CO_2 over Phanerozoic time. *American Journal of Science* **301**, 182–204.

Beveridge TJ (1984) Mechanisms of the binding of metallic ions to bacterial walls and the possible impact on microbial ecology. In: *Current Perspectives in Microbial Ecology* (eds Reddy CA, Klug MJ). American Society for Microbiology, Washington, DC, pp. 601–607.

Beveridge TJ, Murray RGE (1976) Uptake and retention of metals by cell walls of *Bacillus subtilis*. *Journal of Bacteriology* **127**, 1502–1518.

Bjerrum CJ, Canfield DE (2002) Ocean productivity before about 1.9 Gyr limited by phosphorous adsorption onto iron oxides. *Nature* **417**, 159–162.

Blothe M, Roden, EE (2009) Composition and activity of an autotrophic Fe(II)-oxidizing, nitrate-reducing enrichment culture. *Applied and Environmental Microbiology* **75**, 6937–6940.

Braissant O, Cailleau G, Aragno M, Verrecchia EP (2004) Biologically induced mineralization in the tree *Milicia excelsa* (Moraceae): its causes and consequences to the environment. *Geobiology* **2**, 59–66.

Brocks JJ, Logan GA, Buick R, Summons RE (1999) Archean molecular fossils and the early rise of eukaryotes. *Science* **285**, 1033–1036.

Broecker WS, Sanyal A, Takahashi T (2000) The origin of Bahamian whitings revisited. *Geophysical Research Letters* **27**, 3759–3760.

Broecker WS, Langdon C, Takahashi T, Peng TH (2001) Factors controlling the rate of $CaCO_3$ precipitation on the Great Bahama Bank. *Global Biogeochemical Cycles* **15**, 589–596.

Bosak T, Greene SE, Newman DK (2007) A possible role for anoxygenic photosynthetic microbes in the formation of ancient stromatolites. *Geobiology* **5**, 119–126.

Buick R (1992) The antiquity of oxygenic photosynthesis: evidence for stromatolites in sulphate-deficient Archaean lakes. *Science* **255**, 74–77.

Canfield DE, Raiswell R (1991a) Pyrite formation and fossil preservation. In: *Taphonomy. Releasing the Data Locked in the Fossil Record* (eds Allison PA, Briggs DEG). Plenum Press, New York, pp. 337–387.

Canfield DE, Raiswell R (1991b) Carbonate precipitation and dissolution. In: *Taphonomy: Releasing the Data Locked in the Fossil Record* (eds Allison PSA, Briggs DEG). Plenum Press, New York, pp. 411–453.

Castanier S, Le Métayer-Levrel G, Perthuisot J-P (2000) Bacterial roles in the precipitation of carbonate minerals. In: *Microbial Sediments* (eds Riding R, Awramik SM). Springer-Verlag, Berlin, Germany, pp. 32–39.

Chafetz HS (1986) Marine peloids: a product of bacterially induced precipitation of calcite. *Journal of Sedimentary Petrology* **56**, 812–817.

Chafetz HS, Folk RL (1984) Travertines: depositional morphology and the bacterially constructed constituents. *Journal of Sedimentary Petrology* **54**, 289–316.

Cloud PE (1965) Significance of the Gunflint (Precambrian) microflora. *Science* **148**, 27–35.

Cloud PE (1973) Paleoecological significance of the banded iron-formation. *Economic Geology* **68**, 1135–1143.

Coleman ML (1985) Geochemistry of diagenetic non-silicate minerals. Kinetic considerations. *Philosophical Transactions of the Royal Society of London* **315**, 39–56.

Couté A, Bury E (1988) Ultrastructure d'une Cyanophycée aérienne calcifère cavernicole: scytonema julianum Frank (Richter) (Hormogonophicidae, Nostocales, Scytonemataceae). *Hydrobiologia* **160**, 219–239.

Cowen JP, Massoth GJ, Baker ET (1986) Bacterial scavenging of Mn and Fe in a mid- to far-field hydrothermal particle plume. *Nature* **322**, 169–171.

Dill RF, Shinn EA, Jones AT, Kelly K, Steinen RP (1986) Giant subtidal stromatolites forming in normal salinity waters. *Nature* **324**, 55–58.

Drew GH (1913) On the precipitation of calcium carbonate in the sea by marine bacteria, and on the action of denitrifying bacteria in tropical and temperate seas. *Journal of the Marine Biological Association* **9**, 479–524.

D'Hondt S, Rutherford S, Spivack AJ (2002) Metabolic activity of subsurface life in deep-sea sediments. *Science* **295**, 2067–2070.

Dupraz C, Visscher PT (2005) Microbial lithification in marine stromatolites and hypersaline mats. *Trends in Microbiology* **13**, 429–438.

Dzombak DA, Morel FMM (1990) *Surface Complexation Modeling. Hydrous Ferric Oxide*. John Wiley & Sons, Inc., New York.

Edwards KJ, Rogers DR, Wirsen CO, *et al.* (2003) Isolation and characterization of novel psychrophilic, neutrophilic, Fe-oxidizing, chemolithoautotrophic – Proteobacteria from the deep sea. *Applied and Environmental Microbiology* **69**, 2906–2913.

Ehrenreich A, Widdel F (1994) Anaerobic oxidation of ferrous iron by purple bacteria, a new type of phototrophic metabolism. *Applied and Environmental Microbiology*, **60**, 4517–4526.

Ehrlich HL (2002) *Geomicrobiology*, 4th edn. Marcel Dekker, New York, 768pp.

Eigenbrode JL, Freeman KH (2006) Late Archean rise of aerobic microbial ecosystems. *Proceedings of the National Academy of Sciences, USA*, **103**, 15759–15764.

Emerson D (2000) Microbial oxidation of Fe(II) and Mn(II) at circumneutral pH. In: *Environmental Microbe–Metal Interactions* (ed Lovley DR). ASM Press, Washington, DC, pp. 31–52.

Emerson D, Moyer CL (2002) Neutrophilic Fe-oxidizing bacteria are abundant at the Loihi Seamount hydrothermal vents and play a major role in Fe oxide deposition. *Applied and Environmental Microbiology* **68**, 3085–3093.

Emerson D, Revsbech NP (1994a) Investigation of an iron-oxidizing microbial mat community located near Aarhus, Denmark. Field studies. *Applied and Environmental Microbiology* **60**, 4022–4031.

Emerson D, Revsbech NP (1994b) Investigation of an iron-oxidizing microbial mat community located near Aarhus, Denmark: laboratory studies. *Applied and Environmental Microbiology* **60**, 4032–4038.

Emerson D, Fleming EJ, McBeth JM (2010) Iron-oxidizing bacteria: an environmental and genomic perspective. *Annual Reviews in Microbiology* **64**, 561–583.

Feldmann M, McKenzie JA (1998) Stromatolite-thrombolite associations in a modern environment, Lee Stocking Island, Bahamas. *Palaios* **13**, 201–212.

Ferris FG, Fratton CM, Gerits JP, Schultze-Lam S, Sherwood Lollar B (1995) Microbial precipitation of a strontium carbonate phase at a groundwater discharge zone near Rock Creek, British Columbia, Canada. *Geomicrobiology Journal* **13**, 57–67.

Ferris FG, Konhauser KO, Lyvén B, Pedersen K (1999) Accumulation of metals by bacteriogenic iron oxides in a subterranean environment. *Geomicrobiology Journal* **16**, 181–192.

Fischer AG (1965) Fossils, early life, and atmospheric history. *Proceedings of the National Academy of Sciences, USA* **53**, 1205–1215.

Fouke BW, Farmer JD, Des Marais DJ, *et al.* (2000) Depositional facies and aqueous-solid geochemistry of travertine-depositing hot springs (Angel Terrace, Mammoth Hot Springs, Yellowstone National Park, USA). *Journal of Sedimentary Research* 70, 565–585.

Fowle DA, Fein JB (2001) Quantifying the effects of *Bacillus subtilis* cell walls on the precipitation of copper hydroxide from aqueous solution. *Geomicrobiology Journal* 18, 77–91.

Freytet P, Verrecchia EP (1999) Calcitic radial palisadic fabric in freshwater stromatolites: diagenetic and recrystallized feature or physicochemical sinter crust? *Sedimentary Geology* 126, 97–102.

Friedmann EI (1979) The genus Geitleria (Cyanophyceae or Cyanobacteria): distribution of *G. calcarea* and *G. floridana* n. sp. *Plant Systematics and Evolution* 131, 169–178.

Friedman I, Murata KJ (1979) Origin of dolomite in Miocene Monterey Shale and related formations in the Temblor Range, California. *Geochimica et Cosmochimica Acta* 43, 1357–1365.

Ghiorse WC (1984) Biology of iron- and manganese-depositing bacteria. *Annual Reviews in Microbiology* 38, 515–550.

Ghiorse WC, Ehrlich HL (1992) Microbial biomineralization of iron and manganese. In: *Biomineralization Processes* (eds Skinner HCW, Fitzpatrick RW). Catena Supplement, Cremlingen, Germany, pp. 75–99.

Glasauer S, Langley S, Beveridge TJ (2001) Sorption of Fe hydr(oxides) to the surface of *Shewanella putrefaciens*: cell-bound fine-grained minerals are not always formed de novo. *Applied and Environmental Microbiology* 67, 5544–5550.

Golubić S (1973) The relationship between blue-green algae and carbonate deposits. In: *The Biology of Blue-Green Algae* (eds Carr NG, Whitton BA). Botanical Monographs 9, Blackwell, Oxford, pp. 434–472.

Golubić, S., Violante, C, Plenković-Moraj, A, and Grgasović, T, (2008). Travertines and calcareous tufa deposits: an insight into diagenesis. *Geologia Croatica* 61, 363–378.

Grotzinger JP (1989) Facies and evolution of Precambrian carbonate depositional systems. The emergence of the modern platform archetype. In: *Controls on Carbonate Platform and Basin Development* (eds Crevello PD, Wilson JL, Sarg JF). SEPM Special Publication, No. 44, Tulsa, OK, pp. 79–106.

Grotzinger JP (1990) Geochemical model for Proterozoic stromatolite decline. *American Journal of Science* 290, 80–103.

Grotzinger JP, Rothman DR (1996) An abiotic model for stromatolite morphogenesis. *Nature* 383, 423–425.

Gunatilaka A, Saleh A, Al-Temeemi A, Nassar N (1984) Occurrence of subtidal dolomite in a hypersaline lagoon, Kuwait. *Nature* 311, 450–452.

Hafenbradl D, Keller M, Dirmeier R, *et al.* (1996) Ferroglobus placidus gen. nov., sp. nov., a novel hyperthermophilic archaeum that oxidizes Fe²⁺ at neutral pH under anoxic conditions. *Archives of Microbiology* 166, 308–314.

Hartman H (1984) The evolution of photosynthesis and microbial mats: a speculation on banded iron formations. In: *Microbial Mats: Stromatolites* (eds Cohen Y, Castenholz RW, Halvorson HO). Alan R. Liss, New York, pp. 451–453.

Hayes JM (1983) Geochemical evidence bearing on the origin of aerobiosis, a speculative hypothesis. In: *Earth's Earliest Biosphere, Its Origin and Evolution* (ed Schopf JW). Princeton University Press, Princeton, NJ, pp. 291–301.

Hegler F, Posth NR, Jiang J, Kappler A (2008) Physiology of phototrophic iron(II)-oxidizing bacteria-implications for modern and ancient environments. *FEMS Microbiology Ecology* 66, 250–269.

Heindel K, Birgel D, Peckmann J, Kuhnert H, Westphal H (2010) Formation of deglacial microbialites in coral reefs off Tahiti (IODP 310) involving sulfate-reducing bacteria. *Palaios* 25, 618–635.

Heising S, Schink B (1998) Phototrophic oxidation of ferrous iron by a *Rhodomicrobium vannielii* strain. *Microbiology* 144, 2263–2269.

Herrington PM, Fairchild IJ (1989) Carbonate shelf and slope facies evolution prior to Vendian glaciation, central East Greenland. In: *The Caledonide Geology of Scandinavia* (ed. Gayer RA). Graham and Trotman, London, UK, pp. 263–273.

Holland HD (1973) The oceans: a possible source of iron in iron-formations. *Economic Geology* 68, 1169–1172.

Holm NG (1987) Possible biological origin of banded iron-formations from hydrothermal solutions. *Origin of Life* 17, 229–250.

Holm NG (1989) The 13C/12C ratios of siderite and organic matter of a modern metalliferous hydrothermal sediment and their implications for banded iron formations. *Chemical Geology* 77, 41–45.

Hyde WT, Crowley TJ, Baum SK, Peltier WR (2000) Neoproterozoic 'snowball Earth' simulations with a coupled climate/ice-sheet model. *Nature* 405, 425–429.

Irwin H, Curtis C, Coleman M (1977) Isotopic evidence for source of diagenetic carbonates formed during burial of organic-rich sediments. *Nature* 269, 209–213.

Ivarson KC, Sojak M (1978) Microorganisms and ochre deposits in field drains of Ontario. *Canadian Journal of Soil Science* 58, 1–17.

Jiao Y, Newman DK (2007) The pio operon is essential for phototrophic Fe(II) oxidation 521 in *Rhodopseudomonas palustris* TIE-1. *Journal of Bacteriology* 189, 1765–1773.

Juniper SK, Tebo BM (1995) Microbe–metal interactions and mineral deposition at hydrothermal vents. In: *The Microbiology of Deep-Sea Hydrothermal Vents* (ed. Karl DM). CRC Press, Boca Raton, FL, pp. 219–253.

Kah LC, Riding R (2007) Mesoproterozoic carbon dioxide levels inferred from calcified cyanobacteria. *Geology* 35, 799–802.

Kah LC, Grotzinger JP (1992) Early Proterozoic (1.9 Ga) thrombolites of the Rocknest formation, Northwest Territories, Canada. *Palaios* 7, 305–315.

Kaplan A, Reinhold L (1999) The CO₂-concentrating mechanism of photosynthetic microorganisms. *Annual Review of Plant Physiology and Plant Molecular Biology* 50, 539–570.

Kappler A, Newman DK (2004) Formation of Fe(III)-minerals by Fe(II)-oxidizing photoautotrophic bacteria. *Geochimica et Cosmochimica Acta* 68, 1217–1226.

Kappler A, Pasquero C, Konhauser KO, Newman DK (2005) Deposition of banded iron formations by anoxygenic phototrophic Fe(II)-oxidizing bacteria. *Geology* 33, 865–868.

Karl DM, McMurtry GM, Malahoff A, Garcia MO (1988) Loihi Seamount, Hawaii: a mid-plate volcano with a distinctive hydrothermal system. *Nature* 335, 532–535.

Kasting JF (1987) Theoretical constraints on oxygen and carbon dioxide concentrations in the Precambrian atmosphere. *Precambrian Research* 34, 205–229.

Kelts K, Hsü KJ (1978) Freshwater carbonate sedimentation. In: *Lakes Chemistry, Geology, Physics* (ed. Lerman A). Springer, New York, pp. 295–323.

Kempe S, Kazmierczak J (1994) The role of alkalinity in the evolution of ocean chemistry, organization of living systems, and biocalcification processes. *Bulletin de la Institut Océanographique (Monaco)* **13**, 61–117.

Kendall B, Reinhard CT, Lyons TW, Kaufman AJ, Poulton SW, Anbar AD (2010) Pervasive oxygenation along late Archean ocean margins. *Nature Geoscience* **3**, 647–652.

Kenward PA, Goldstein RH, González LA, Roberts JA (2009) Precipitation of low-temperature dolomite from an anaerobic microbial consortium: the role of methanogenic Archaea. *Geobiology* **7**, 556–565.

Kiessling W (2002) Secular variations in the Phanerozoic reef ecosystem. In: *Phanerozoic Reef Patterns* (eds Kiessling, W, Flügel E, Golonka J). SEPM Special Publication, No. 72, Tulsa, OK, pp. 625–690.

Klein C, Beukes NJ, Schopf JW (1987) Filamentous microfossils in the early Proterozoic Transvaal Supergroup: their morphology, significance, and paleoenvironmental setting. *Precambrian Research* **36**, 81–94.

Knoll AH, Semikhatov MA (1998) The genesis and time distribution of two distinctive Proterozoic stromatolite microstructures. *Palaios* **13**, 408–422.

Konhauser KO (1998) Diversity of bacterial iron mineralization. *Earth-Science Reviews* **43**, 91–121.

Konhauser KO (2000) Hydrothermal bacterial biomineralization: potential modern-day analogues for Precambrian banded iron formation. In: *Marine Authigenesis: From Global to Microbial* (eds Glenn CR, Lucas J, Prévôt L). SEPM Special Publication No. 66, Tulsa, OK, pp. 133–145.

Konhauser KO (2007) *Introduction to Geomicrobiology*. Blackwell, Oxford, 425pp.

Konhauser KO, Hamade T, Morris RC, *et al.* (2002) Did bacteria form Precambrian banded iron formations? *Geology* **30**, 1079–1082.

Konhauser KO, Newman DK, Kappler A (2005) The potential significance of microbial Fe(III) reduction during deposition of Precambrian banded iron formations. *Geobiology* **3**, 167–177.

Konhauser KO, Lalonde SV, Amskold L, Holland H (2007) Was there really an Archean phosphate crisis? *Science* **315**, 1234.

Konhauser KO, Pecoits E, Lalonde SV, *et al.* (2009) Oceanic nickel depletion and a methanogen famine before the Great Oxidation Event. *Nature* **458**, 750–753.

Krom MD, Berner RA (1980) Adsorption of phosphate in anoxic marine sediments. *Limnology and Oceanography* **25**, 797–806.

Liang L, McNabb JA, Paulk JM, Gu B, McCarthy JF (1993) Kinetics of Fe(II) oxygenation at low partial pressure of oxygen in the presence of natural organic matter. *Environmental Science and Technology* **27**, 1864–1870.

Logan BW (1961) Cryptozoon and associated stromatolites from the Recent, Shark Bay, Western Australia. *Journal of Geology* **69**, 517–533.

Lovley DR (1990) Magnetite formation during microbial dissimilatory iron reduction. In: *Iron Biominerals* (eds Frankel RB, Blakemore RP). Plenum Press, New York, pp. 151–166.

Lowenstam HA (1981) Minerals formed by organisms. *Science* **211**, 1126–1131.

Lyons WMB, Long DT, Hines ME, Gaudette HE, Armstrong PB (1984) Calcification of cyanobacterial mats in Solar Lake, Sinai. *Geology* **12**, 623–626.

Mandernack KW, Tebo BM (1993) Manganese scavenging and oxidation at hydrothermal vents and vent plumes. *Geochimica et Cosmochimica Acta* **57**, 3907–3923.

Merz MUE (1992) The biology of carbonate precipitation by cyanobacteria. *Facies* **26**, 81–102.

Merz-Preiß M, Riding R (1999) Cyanobacterial tufa calcification in two freshwater streams: ambient environment, chemical thresholds and biological processes. *Sedimentary Geology* **126**, 103–124.

Michaelis W, Seifert R, Nauhaus K, *et al.* (2002) Microbial reefs in the Black Sea fuelled by anaerobic oxidation of methane. *Science* **297**, 1013–1015.

Miller AG, Colman B (1980) Evidence for HCO_3^- transport in the blue-green alga (cyanobacterium) *Coccochloris peniocystis*. *Plant Physiology* **65**, 397–402.

Monty CLV (1976) The origin and development of cryptalgal fabrics. In: *Stromatolites* (ed. Walter MR), *Developments in Sedimentology*, Vol. 20, Elsevier, Amsterdam, pp. 193–249.

Moore LS, Burne RV (1994) The modern thrombolites of Lake Clifton, Western Australia. In: *Phanerozoic Stromatolites II* (eds. Bertrand-Sarfati J, Monty C). Kluwer, Dordrecht, pp. 3–29.

Morris RC (1993) Genetic modelling for banded iron-formation of the Hamersley Group, Pilbara Craton, Western Australia. *Precambrian Research* **60**, 243–286.

Morse JW, Casey WH (1988) Ostwald processes and mineral paragenesis in sediments. *American Journal of Science* **288**, 537–560.

Myers KH, Nealson KH (1988) Bacterial manganese reduction and growth with manganese oxide as the sole electron donor. *Science* **240**, 1319–1321.

Nielson AE, Söhnel O (1971) Interfacial tensions electrolyte crystal-aqueous solutions, from nucleation data. *Journal of Crystal Growth* **11**, 233–242.

Pentecost A (1987) Growth and calcification of the freshwater cyanobacterium *Rivularia haematites*. *Proceedings of the Royal Society of London, Series B* **232**, 125–136.

Pentecost A (2005) *Travertine*. Springer, Berlin, 445pp.

Pigott JD, Land LS (1986) Interstitial water chemistry of Jamaican reef sediment: sulfate reduction and submarine cementation. *Marine Chemistry* **19**, 355–378.

Planavsky NJ, Rouxel O, Bekker A, Lalonde SV, Konhauser KO, Reinhard CT, Lyons T (2010) The evolution of the marine phosphate reservoir. *Nature* **467**, 1088–1090.

Plée K, Ariztegui D, Martini R, Davaud E (2008) Unravelling the microbial role in ooid formation – results of an in situ experiment in modern freshwater Lake Geneva in Switzerland. *Geobiology* **6**, 341–350.

Posth NR, Hegler F, Konhauser KO, Kappler A (2008) Alternating Si and Fe deposition caused by temperature fluctuations in Precambrian oceans. *Nature Geoscience* **1**, 703–707.

Pratt BR (1982) Stromatolite decline – a reconsideration. *Geology* **10**, 512–515.

Price GD, Sültemeyer D, Klughammer B, Ludwig M, Badger MR (1998) The functioning of the CO_2 concentrating mechanism in several cyanobacterial strains: a review of general physiological characteristics, genes, proteins and recent advances. *Canadian Journal of Botany* **76**, 973–1002.

Rashby SE, Sessions AL, Summons RE, Newman DK (2007) Biosynthesis of 2-methylbacteriohopanepolyols by an anoxygenic phototroph. *Proceedings of the National Academy of Sciences, USA* **104**, 15099–15104.

Reid RP, Visscher PT, Decho AW, *et al.* (2000) *Nature* **406**, 989–992.

Revsbech NP, Jørgensen BB, Blackburn TH, Cohen Y (1983) Microelectrode studies of the photosynthesis and O2, H2S and pH profiles of a microbial mat. *Limnology and Oceanography* **28**, 1062–1074.

Ridgwell AJ, Kennedy MJ, Caldeira K (2003) Carbonate deposition, climate stability, and Neoproterozoic ice ages. *Science* **302**, 859–862.

Riding R (1977) Calcified Plectonema (blue-green algae), a recent example of Girvanella from Aldabra Atoll. *Palaeontology* **20**, 33–46.

Riding R (1982) Cyanophyte calcification and changes in ocean chemistry. *Nature* **299**, 814–815.

Riding R (1993) Interaction between accretionary process, relative relief, and external shape in stromatolites and related microbial deposits. *International Alpine Algae Symposium and Field-Meeting*, Munich–Vienna, August 29–September 5, 1993, Abstracts.

Riding R (2000) Microbial carbonates: the geological record of calcified bacterial-algal mats and biofilms. *Sedimentology*, **47**, 179–214.

Riding R (2006) Microbial carbonate abundance compared with fluctuations in metazoan diversity over geological time. *Sedimentary Geology* **185**, 229–238.

Riding R (2008) Abiogenic, microbial and hybrid authigenic carbonate crusts: components of Precambrian stromatolites. *Geologia Croatica* **61/2–3**, 73–103.

Riding R (2009) An atmospheric stimulus for cyanobacterial-bioinduced calcification ca. 350 million years ago? *Palaios* **24**, 685–696.

Riding R, Liang L (2005a) Geobiology of microbial carbonates: metazoan and seawater saturation state influences on secular trends during the Phanerozoic. *Palaeogeography, Palaeoclimatology, Palaeoecology* **219**, 101–115.

Riding R, Liang L (2005b) Seawater chemistry control of marine limestone accumulation over the past 550 million years. *Revista Española de Micropaleontología* **37**, 1–11.

Riding R, Awramik SM, Winsborough BM, Griffin KM, Dill RF (1991) Bahamian giant stromatolites: microbial composition of surface mats. *Geological Magazine* **128**, 227–234.

Robbins LL, Blackwelder P (1992) Biochemical and ultrastructural evidence for the origin of whitings. A biologically induced calcium carbonate precipitation mechanism. *Geology* **20** 464–468.

Robbins EI, LaBerge GL, Schmidt RG (1987) A model for the biological precipitation of Precambrian iron formations-B: morphological evidence and modern analogs. In: *Precambrian Iron-Formations* (eds Appel PWU, LaBerge GL). Theophrastus, Athens, Greece, pp. 97–139.

Robbins LL, Tao Y, Evans CA (1997) Temporal and spatial distribution of whitings on Great Bahama Bank and a new lime mud budget. *Geology* **25**, 947–950.

Roberts JA, Bennett PC, González LA, Macpherson GL, Milliken KL (2004) Microbial precipitation of dolomite in methanogenic groundwater. *Geology* **32**, 277–280.

Rost B, Riebesell U, Burkhardt S, Sültemeyer D (2003) Carbon acquisition of bloom-forming marine phytoplankton. *Limnology and Oceanography* **48**, 55–67.

Sassen R, Roberts HH, Carney R, *et al.* (2004) Free hydrocarbon gas, gas hydrate, and authigenic minerals in chemosynthetic communities of the northern Gulf of Mexico continental slope: relation to microbial processes. *Chemical Geology* **205**, 195–217.

Sheldon ND (2006) Precambrian paleosols and atmospheric CO_2 levels. *Precambrian Research* **147**, 148–155.

Schultze-Lam S, Beveridge TJ (1994) Nucleation of celestite and strontionite on a cyanobacterial S-layer. *Applied and Environmental Microbiology* **60**, 447–453.

Sherman AG, James NP, Narbonne GM (2000) Sedimentology of a late Mesoproterozoic muddy carbonate ramp, northern Baffin Island, Arctic Canada. In: *Carbonate Sedimentation and Diagenesis in the Evolving Precambrian World* (eds Grotzinger JP, James NP). SEPM Special Publication, No. 67, Tulsa, OK, pp. 275–294.

Shinn EA, Steinen RP, Lidz BH, Swart PK (1989) Perspectives. Whitings, a sedimentologic dilemma. *Journal of Sedimentary Petrology* **59**, 147–161.

Søgaard EG, Medenwaldt R, Abraham-Peskir JV (2000) Conditions and rates of biotic and abiotic iron precipitation in selected Danish freshwater plants and microscopic analysis of precipitate morphology. *Water Research* **34**, 2675–2682.

Srdoć D, Horvatinčić N, Obelić B, Krajcar I, Sliepčević A (1985) Calcite deposition processes in karst waters with special emphasis on the Plitvice Lakes Yugoslavia. *Carsus Iugoslaviae* **11**, 101–204.

Steefel CI, Van Cappellen P (1990) A new kinetic approach to modeling water-rock interaction: the role of nucleation, precursors, and Ostwald ripening. *Geochimica et. Cosmochimica Acta*, **54**, 2657–2677.

Straub KL, Benz M, Schink B, Widdel F (1996) Anaerobic, nitrate-dependent microbial oxidation of ferrous iron. *Applied Environmental Microbiology* **62**, 1458–1460.

Straub KL, Benz M, Schink B (2001) Iron metabolism in anoxic environments at near neutral pH. *FEMS Microbial Ecology* **34**, 181–186.

Stumm W, Morgan JJ (1996) *Aquatic Chemistry*, 3rd edn. John Wiley & Sons, Inc., New York.

Summons RE, Jahnke LL, Hope JM, Logan GA (1999) 2-Methylhopanoids as biomarkers for cyanobacterial oxygenic photosynthesis. *Nature* **400**, 554–557.

Sumner DY (1997) Late Archean calcite-microbe interactions: two morphologically distinct microbial communities that affected calcite nucleation differently. *Palaios*, **12**, 302–318.

Sumner DY, Grotzinger JP (2004) Implications for Neoarchaean ocean chemistry from primary carbonate mineralogy of the Campbellrand-Malmani Platform, South Africa. *Sedimentology* **51**, 1273–1299.

Swett K, Knoll AH (1985) Stromatolitic bioherms and microphytolites from the Late Proterozoic Draken Conglomerate Formation, Spitsbergen. *Precambrian Research* **28**, 327–347.

Templeton AS, Knowles EJ, Eldridge DL, *et al.* (2009) A seafloor microbial biome hosted within incipient ferromanganese crusts. *Nature Geoscience* **2**, 872–876.

Thompson JB (2000) Microbial whitings. In: *Microbial Sediments* (eds Riding R, Awramik SM). Springer-Verlag, Berlin, pp. 250–260.

Thompson JB, Ferris FG (1990) Cyanobacterial precipitation of gypsum, calcite, and magnesite from natural alkaline lake water. *Geology* **18**, 995–998.

Thompson JB, Ferris FG, Smith DA (1990) Geomicrobiology and sedimentology of the mixolimnion and chemocline in Fayetteville Green Lake, New York. *Palaios* **5**, 52–75.

Trendall AF (2002) The significance of iron-formation in the Precambrian stratigraphic record. *Special Publication from the International Association of Sedimentology* **33**, 33–66.

Trichet J, Défarge C (1995) Non-biologically supported organomineralization. *Bulletin de l'Institut Océanographique de Monaco* **14**, 203–236.

van Gemerden H (1993) Microbial mats: a joint venture. *Marine Geology* **113**, 3–25.

Van Lith Y, Warthmann R, Vasconcelos C, McKenzie JA (2003a) Sulfate-reducing bacteria induce low-temperature Ca-dolomite and high Mg-calcite formation. *Geobiology* **1**, 71–80.

Van Lith Y, Warthmann R, Vasconcelos C, McKenzie JA (2003b) Microbial fossilization in carbonate sediments: a result of the bacterial surface involvement in dolomite precipitation. *Sedimentology* **50**, 237–245.

van Veen WL, Mulder EG, Deinema MH (1978) The Sphaerotilus-Leptothrix group of bacteria. *Microbiology Reviews* **42**, 329–356.

Visscher PT, Stolz JF (2005) Microbial mats as bioreactors: populations, processes, and products. *Palaeogeography, Palaeoclimatology, Palaeoecology* **219**, 87–100.

Walcott CD (1914) Cambrian geology and paleontology III: Precambrian Algonkian algal flora. *Smithsonian Miscellaneous Collection* **64**, 77–156.

Walker JCG (1984) Suboxic diagenesis in banded iron formations. *Nature* **309**, 340–342.

Walter MR, Veevers JJ, Calver CR, Gorjan P, Hill AC (2000) Dating the 840–544 Ma Neoproterozoic interval by isotopes of strontium, carbon, and sulfur in seawater, and some interpretative models. *Precambrian Research* **100**, 371–433.

Warthmann R, Van Lith Y, Vasconcelos C, McKenzie JA, Karpoff A-M (2000) Bacterially induced dolomite precipitation in anoxic culture experiments. *Geology* **28**, 1091–1094.

Weber KA, Churchill PF, Urrutia MM, Kukkadapu RK, Roden EE (2006b) Anaerobic redox cycling of iron by wetland sediment microorganisms. *Environmental Microbiology* **8**, 100–113.

Williams RJP (1981) Physico-chemical aspects of inorganic element transfer through membranes. *Philosophical Transactions of the Royal Society of London* **B294**, 57–74.

Xiong J (2006) Photosynthesis: what color was its origin? *Genome Biology* **7**, 2451–2455.

Zahnle KJ, Claire MW, Catling DC (2006) The loss of mass-independent fractionation of sulfur due to a Paleoproterozoic collapse of atmospheric methane. *Geobiology* **4**, 271–283.

9

MINERAL–ORGANIC–MICROBE INTERFACIAL CHEMISTRY

David J. Vaughan and Jonathan R. Lloyd

Williamson Research Centre for Molecular Environmental Science and School of Earth, Atmospheric and Environmental Science, University of Manchester, Manchester M13 9PL, UK

9.1 Introduction

Over the past two decades there has been rapid growth in experimental and theoretical (computational) studies of mineral surfaces at the molecular scale. This has extended to investigations of the interactions of such surfaces with the gases of the air and with aqueous fluids, and latterly with organic molecules and with a range of microorganisms. Interest in such geochemical and geobiological systems has grown with appreciation of the interplay that takes place among minerals, fluids, organic molecules and microbes at, or very near, the surface of the Earth, in the so-called 'critical zone' that is essential for sustaining life. The geochemical and biogeochemical reactions and processes in this zone include both the dissolution and precipitation of minerals (the latter often producing highly reactive nanoparticulate phases), redox transformations, and sorption/desorption of solution species (metals, anions, organic molecules) at mineral surfaces. Microbes play a key role in these processes, generating organic conditioning films and biofilms on mineral surfaces, promoting electron transfer (redox) reactions as part of their metabolism, and sorbing metals or other solution species onto their surfaces.

The mineral–organic–microbe interactions, which are central to these essentially chemical process, take place at surfaces and interfaces can only be properly understood through experimentation (and modelling) at the molecular scale. Their importance, however, extends from the molecular to the regional and, ultimately, to the global scale. They are central to the metabolisms of simple organisms and impact upon larger biogeochemical cycles, and play a key role in the geochemical cycling of the elements.

In this chapter, we briefly describe the tools that are available to study mineral surfaces and interfaces and their interactions with organic molecules and microbes, particularly at the molecular scale. We also provide examples of some key findings from these studies, with emphasis on our own work, and suggest possible directions for future research. In a chapter such as this, it is clearly impossible to comprehensively review this large field, but an attempt is made to cover key points and to direct readers to the relevant literature. General references concerning mineral surface chemistry include Hochella and White (1990), Stumm (1992) and Vaughan and Pattrick (1995), and about mineral-microbe interactions include Banfield and Nealson (1997), with several of the chapters in the *Manual of Environmental Microbiology* (Hurst, 2007) being relevant.

9.2 The mineral surface (and mineral–bio interface) and techniques for its study

In order to have a proper understanding of mineral surfaces, interfaces, and interfacial processes, we need to address a number of questions. These include: whether and how a pristine mineral surface (growth or cleavage) differs from a simple truncation of the bulk crystal structure in the arrangement of its atoms; the roughness of the surface down to atomic level and the presence of steps, kinks, vacancies and other defects; the chemical composition of the fresh surface, before and after interaction with air, aqueous fluid, organic

Fundamentals of Geobiology, First Edition. Edited by Andrew H. Knoll, Donald E. Canfield and Kurt O. Konhauser.
© 2012 Blackwell Publishing Ltd. Published 2012 by Blackwell Publishing Ltd.

species or microbe, and the nature of the actual interface between mineral and fluid, organic or microbe. Experimental techniques for the study of surfaces and interfaces can be generally categorized into those involving diffraction, imaging or spectroscopy. Some methods can only be performed in a vacuum or ultra-high vacuum (UHV), whereas others permit *in situ* observations and measurements of surfaces in contact with fluid, organic matter or microbe. In what follows, the main techniques are described in the context of addressing the questions outlined above. In addition to the experimental methods, there is a special role in all such studies for computer simulations. Advances in theory and in code development mean that, in addressing some questions about surfaces and interfaces, computer simulations now rival experiment in their accuracy, although most theoretical studies involve close comparison with experiment in order to validate the former and aid the interpretation of the latter. Excellent examples of the power of such approaches include pioneering work on the surfaces of important minerals for 'geobiologists' such as hematite (Becker *et al.*, 1996) and pyrite (Rosso *et al.*, 1999). It is not possible, here, to review this area properly, but details of the theoretical background and many of the applications of molecular modelling approaches to mineral surfaces and other geochemical problems are given in the books by Tossell and Vaughan (1992) and Cygan and Kubicki (2001).

9.2.1 Mineral surface structure and topography

The scanning electron microscope (SEM) is a long-established standard instrument for topographical imaging of minerals and associated microbes, but one where there have been very important technique developments in recent years. In an SEM, on bombardment by an electron beam, secondary electrons are emitted from the exterior of a specimen and are used to image its surface. Modern instruments, using field-emission electron guns and low-voltage systems, now allow sub-nanometer resolution on some samples. Also, the X-rays emitted when the sample is excited by the electron beam can be detected, and their energies measured by energy dispersive spectroscopy (EDS). As these energies are characteristic of the elements present in the sample, this provides a means of determining chemical composition. Such analysis is only semiquantitative due to problems of calibration and the exact conditions of electron excitation whereby there is spreading of X-ray emission sites beyond the zone of electron irradiation. A new generation of so-called 'environmental' SEM instruments (or ESEMs) allows samples to be viewed under high humidity, so as to maintain partial hydration, which can be extremely important for observing microbial materials at surfaces. Related developments

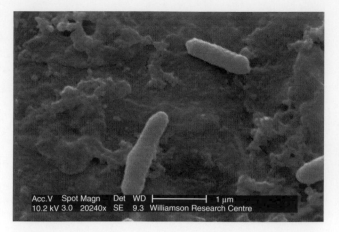

Figure 9.1 Environmental scanning electron microscope (ESEM) image showing cells of *Geobacter sulfurreducens* growing on an iron oxide substrate (after Wilkins *et al.*, 2007b).

also enable samples to be studied at very low temperatures, where biomaterials can be preserved in a 'frozen' state. An example of an ESEM image is shown in Fig. 9.1, where cells of *Geobacter sulfurreducens* are pictured growing attached to an iron oxide substrate. A comprehensive review of scanning electron microscopy and associated X-ray microanalysis is provided by Goldstein *et al.* (2003).

Transmission electron microscopy (TEM) is another very long-established technique where instrumentation and sample preparation developments have provided important new advances. TEM relies on the interpretation of images and electron diffraction patterns obtained when a focused beam of electrons passes through a very thin specimen (usually ~50 nm thick). Although not a 'surface technique' as such, resolution on the order of a few angstroms can be obtained using modern instruments, so that surface and interface regions (or the cores and rims of 'nanoscale' particles; see Fig. 9.2) can be studied. Thus, electron diffraction patterns from very small areas (so-called selected area electron diffraction, SAED) can provide structural information for crystalline mineral phases, and the combination with EDS can also provide semi-quantitative compositional data for those very small regions (McLaren, 1991; White, 1985). Another example of the power of TEM (combined with EDS) is shown in Fig. 9.3 where *Geobacter* has facilitated the reduction of U(VI) in solution to U(IV) which has precipitated both outside the cell and in the periplasmic region of the organism. Electron energy loss spectroscopy (EELS) is another TEM analytical technique and is particularly sensitive for lower atomic number elements (as well as some metals). Here, a proportion of the transmitted electrons that have passed through the specimen are inelastically scattered and excite either valence or core level electrons. Core level EELS is elementally sensitive and the fine structure at the onset of the spectrum

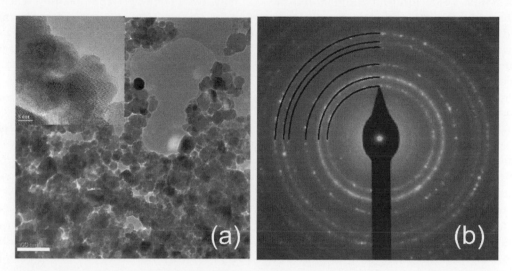

Figure 9.2 Transmission electron microscope (TEM) data for iron oxide nanoparticles produced by bacterial reduction (with *Geobacter sulfurreducens*) of inorganically synthesized feroxyhyte:(a) image of the nanoparticles showing grain sizes and morphologies (note the inset at higher magnification showing differences between rims and cores); (b) selected area electron diffraction (SAED) pattern for the sample showing that it has been transformed to magnetite (after Cutting *et al.*, 2009).

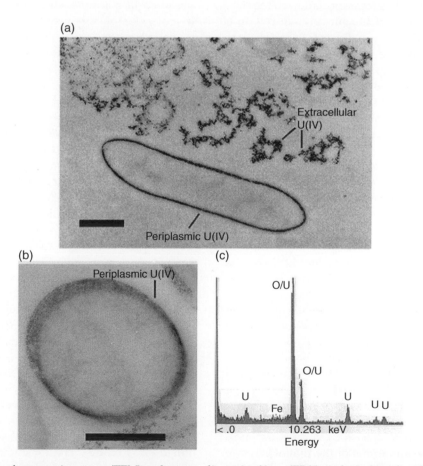

Figure 9.3 Transmission electron microscope (TEM) and energy dispersive X-ray (EDX) analysis data showing *Geobacter* with precipitated extracellular U(IV) and periplasmic U(IV) (Scale bars: in (a) 0.5 μm; in (b) 0.5 μm) (after Lloyd *et al.*, 2002).

provides information about local bonding around the core excited atom (similar to that from the X-ray absorption spectra discussed later in this article). So-called 'electron-filtering' TEM instruments have the ability to selectively filter the inelastic electrons of the transmitted beam, so that angstrom-resolution images of the distribution of elements can be produced; for example, the iron and oxygen distribution can be seen on extremely fine-grain hematite. This is called electron spectroscopic imaging (ESI). As with the SEM, important advances have been made in sample preparation for TEM studies by using low temperatures (down to −196 °C) to preserve microbial cells and other biomaterials, and then keeping the sample at these temperatures in the (cryo) TEM. These developments are discussed in detail by Beveridge *et al.* (2007).

The actual arrangement of atoms at the surface of a mineral (monolayer) can be studied using low energy electron diffraction (LEED). In this method, which requires the sample to be in UHV (and therefore dry and clean) and to be a single crystal with a well-ordered surface structure, a beam of electrons (of typically 20–200 eV energy) is fired at the surface and generates a back-scattered electron diffraction pattern. This pattern of spots may be analysed to give information on the size, symmetry and orientation of the surface unit cell. In a more sophisticated analysis, spot intensities can be used to provide accurate information on the positions of atoms at the surface of the crystal (or of an ordered overlayer on that surface). A detailed account is provided in the book by Van Hove *et al.* (1986).

The indirect observations at 'atomic resolution' made using LEED can be contrasted with the direct measurements which can be made using atomic force microscopy and scanning tunneling microscopy (AFM, STM). These are the most important members of a family of methods ('scanning probe' microscopies, SPM) first developed about twenty years ago and which have revolutionised surface science (see Dufrene, 2004, 2007). STM functions by using a piezoelectric device to scan a very fine stylus-like tip across the flat surface of a material (which must be electrically conducting) whilst applying an electrical potential. Monitoring the variations in the current passing between sample surface and tip at constant potential during scanning (or in the potential at a constant current) enables an image of the surface to be acquired. In the very clean environment of a UHV chamber, this image of the electron density distribution can be at atomic resolution. In Fig. 9.4, atomic resolution images of the (1 1 1) crystal face of magnetite (Fe_3O_4) are shown (after Cutting *et al.*, 2006). These STM data are interpreted as different terminations of the magnetite arising from different level 'slices' through the bulk structure which is also shown in Fig. 9.4. Thus, there

may be unexpected complexities when considering only a single 'face' of a particular mineral.

The AFM technique involves the same kind of scanner system, but in this case the very small tip is mounted on the end of a flexible arm (cantilever) and movement of the tip towards or away from the surface during scanning is measured by reflecting a laser beam off the end of the arm. Here, minute forces of attraction and repulsion are detected and used to construct a topographical profile of the specimen in x, y and z dimensions (Dufrene, 2007). As the tip can be made almost atomically sharp and since movement is precisely controlled by piezoceramic voltage input, molecular resolution is possible in UHV environments. Somewhat lower resolution studies (nanometer to micron) can be conducted in air, or with the surface beneath a layer of fluid. In work of microbiological relevance, surfaces with attached microbes can be imaged and measurements made of forces of adhesion as well as surface topography (see Fig. 9.5a). It has even been possible to study mineralization processes at the surface of a living bacterium, as in the work of Obst *et al.* (2006) where calcium carbonate nucleation on immobilized cells of a cyanobacterium (*Synechococcus leopoliensis*) was observed in real time, showing changes in the surface microtopography of the organism (see Fig. 9.5b). A particular strength of the SPM methods is that three-dimensional information is obtained about surfaces; a weakness is that they do not directly provide any chemical (elemental or species analysis) information.

The other area of development which has revolutionized studies of mineral surfaces concerns the use of synchrotron radiation. In a synchrotron, a packet of electrons is accelerated around a large diameter (~20–100 m) storage ring and tangentially emits intense 'white' radiation (mostly X-rays) which can be used for a wide variety of diffraction, scattering, spectroscopic and imaging experiments, as well as experiments that combine techniques, such as spectral-imaging and diffraction-imaging. Experiments involving X-ray spectroscopy to determine surface composition are discussed in the following section. In the context of studies of surface structure and topography, measuring the changes in intensity of a beam of X-rays reflected at glancing incidence by a flat mineral surface (in X-ray reflectivity studies) can provide data on surface roughness at nanometre scale, and of the development of layers of surface alteration produced by reaction with air, water, or other gases and fluids (Wogelius and Vaughan, 2000). Important developments relevant to the subject matter of this review have also been associated with synchrotron techniques employing 'soft' X-rays (ie in the energy range ~75–2150 eV). The review by Bluhm *et al.* (2006) discusses several of these developments in the context of the facilities at the Advanced Light Source (Berkeley,

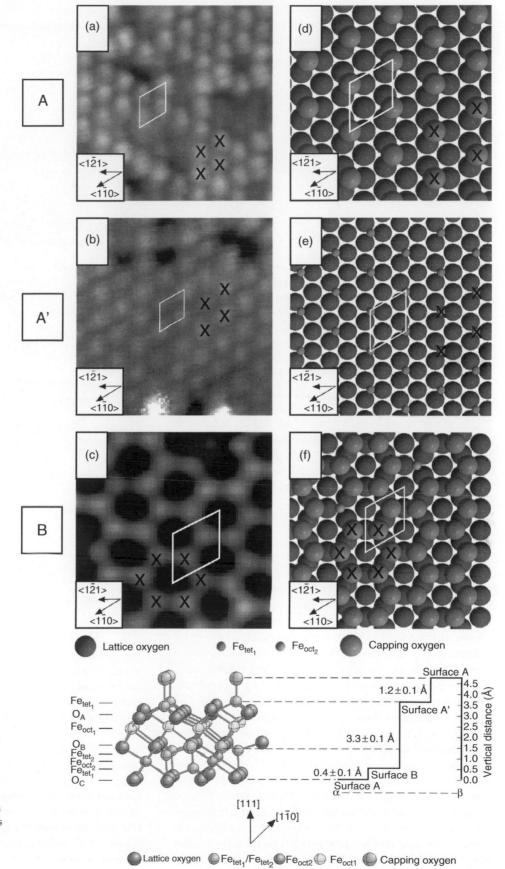

Figure 9.4 Scanning tunneling microscope (STM) images of the (111) surface of magnetite obtained under UHV conditions: (a) $10 \times 10 \, nm^2$ ($100 \times 100 \, Å^2$) image of the A surface; (b) $12 \times 12 \, nm^2$ ($120 \times 120 \, Å^2$) image of the A' surface; (c) $1.6 \times 1.6 \, nm^2$ ($16 \times 16 \, Å^2$) image of the B surface, with corresponding models of the surfaces (d,e and f, respectively) which are interpreted as arising from different level slices through the bulk structure of magnetite, as illustrated on the right hand side of the figure (redrawn after Cutting *et al.*, 2006).

(a)

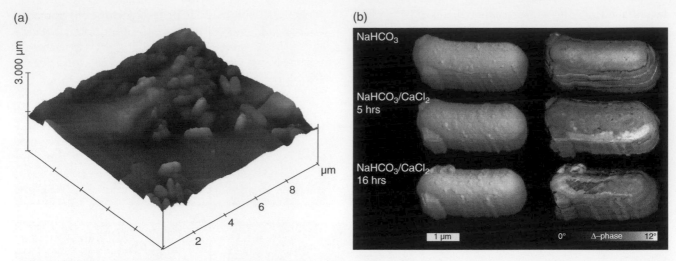

Figure 9.5 (a) An atomic force microscope (AFM) image showing cells of *Geobacter sulfurreducens* on a stepped surface of goethite (FeOOH); note that quantitative data are obtained in all three dimensions (Wilkins, Wincott, Vaughan and Lloyd, unpublished data). (b) AFM images showing the growth with time of protruberances at the surface of a cyanobacterium exposed to a solution of NaHCO₃ to which Ca²⁺ was added. The two sets of images were obtained under different conditions (after Obst *et al.*, 2006).

California, USA). A particularly good example is the application of the recently developed technique of scanning transmission X-ray microscopy (STXM) to map the spatial distribution of iron species throughout *Pseudomonas aeruginosa* biofilms and, hence, assess the influence of chemical heterogeneity on biomineralization. This work by Hunter *et al.* (2008) revealed that Fe(II) and Fe(III) occur in localized microenvironments, with Fe(III) being mainly associated with cell surfaces and Fe(II) in extracellular space.

9.2.2 Mineral surface composition

Elemental and chemical species analysis of the mineral surface in its initial state and following reaction is clearly central to any investigation. X-ray photoelectron spectroscopy (XPS) is the 'workhorse' method of surface chemical analysis. In this technique, a sample contained within a UHV chamber is bombarded with monoenergetic photons (usually $MgK\alpha$ or $AlK\alpha$ X-rays) causing the ejection of electrons from core and valence orbitals (Fig. 9.6; see also Fig. 9.12 for examples of XPS data). The kinetic energies of these electrons are determined in an energy analyser to yield an X-ray photoelectron spectrum. In the related technique of Auger electron spectroscopy (which can commonly be performed in the same instrument as XPS), the energy of a secondary electron (a so-called Auger electron; see Fig. 9.6) is measured. This is an electron ejected during the relaxation of the system, following ejection of a photoelectron, when an electron in a higher energy orbital drops down to fill a hole in an inner shell. This process is an alternative to emission of X-rays of characteristic energy (as are measured in X-ray

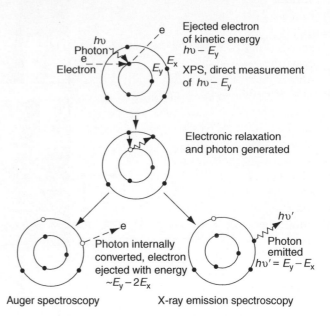

Figure 9.6 Diagram to show the principles of X-ray photoelectron spectroscopy (XPS), auger electron spectroscopy (AES), and X-ray emission spectroscopy, which is also the principle of the electron microprobe and of energy dispersive (X-ray) spectroscopy (EDS) analysis used in electron microscopes.

emission spectroscopy techniques such as electron probe microanalysis). In XPS, the electron kinetic energies that are measured are directly related to the binding energies of electrons in the material being studied, and hence to the specific elements present. Peak positions provide a means of identifying the elements, peak intensities provide a semi-quantitative measure of their amounts. Because electrons can only escape from the topmost layers (a few nanometers) of a material, this offers a sensitive method of surface analysis. In favourable cases, precise peak position also varies with element speciation, enabling further information to be gained (e.g. not only identifying sulphur but distinguishing sulfide vs. sulfate, or ferric from ferrous iron). Rastering of the incident beam also enables mapping of surface chemistry at micron level. A disadvantage of conventional XPS measurements is that they have to be made in a UHV environment, so that wet or volatile materials cannot be studied. Recently, however, photoelectron spectroscopy under ambient pressure and temperature conditions, and hence of wet surfaces, has become possible at certain synchrotron facilities (e.g. ALS Berkeley, BESSY Berlin). These so-called ambient pressure photoemission spectroscopy (APPES) instruments (see Ogletree *et al.*, 2009; Bluhm *et al.*, 2006) are set to make important fundamental contributions to chemical studies of the mineral–organic–microbe interface.

The other long established technique employed in surface analysis is secondary ion mass spectrometry (SIMS). Of course, mass spectrometry, the identification and analysis of quantities and ratios of the isotopes of a specific element or of molecular entities by discrimination on the basis of atomic or molecular mass, has numerous applications in the natural sciences. The SIMS methods involve bombarding the surface of a sample with a beam of ions, thus causing the ejection of secondary ions from surface layers. These secondary ions are taken in to a spectrometer where determination of their masses enables identification of atomic or molecular species and, from the intensities of peaks in the mass spectrum, their percentage contributions. In this way, the chemical composition of the surface can be determined. Because the bombardment of the surface removes material from deeper layers as the process continues, the SIMS technique can provide a profile of compositional changes as a function of depth beneath the surface (generally on scales from nm to μm). As with XPS and AES, rastering of the beam can be employed to image areas of a surface as well as producing point analyses. A disadvantage of SIMS has been that the incident beam may cause changes in the chemistry of the sample, particularly if it comprises organics or biomaterials, or may disrupt the surface so as to compromise any attempts at 'depth profiling'. However, recent developments involving new types of incident beam to bombard the sample (e.g. a beam of C^{60} molecules) appear set to revolutionize this technique, and to enable delicate samples to be analysed at nanometer resolution (hence the term 'NanoSIMS'). A relevant example is shown in Fig. 9.7 and this is further discussed below. Details concerning the SIMS technique in general can be found in Macrae (1995), McMahon and Cabri (1998) and references therein.

The availability of intense (synchrotron) radiation sources has also made possible the development of techniques based on the absorption of photons by a sample, particularly X-rays as in X-ray absorption spectroscopy (XAS). Here, the absorption of X-rays over a relatively small energy range as a result of the electrons in a particular shell (e.g. the K shell) of an element of interest being excited to higher energy state, or to the continuum, is a powerful probe of the local environment around an atom. This is because the fine structure of the absorption spectrum both in the region of the edge – the X-ray absorption near edge structure (XANES), and above the absorption edge – the extended X-ray absorption fine structure (EXAFS), is sensitive to the numbers, types and distances of the shells of atoms surrounding the absorber element. These methods also have the advantage of working whether that absorber element is contained in a crystalline or amorphous solid, on a surface, or in a liquid or glass. Enhanced surface sensitivity can be achieved by reflecting the incident X-ray beam off a flat mineral surface at grazing incidence (ReflEXAFS). The XAS methods are particularly powerful in probing the local environment of species formed when metals (or other, often toxic, elements or molecules) in aqueous solution interact with the surfaces of minerals, whether these are nanoparticulate mineral phases such as is commonly the case with iron (oxyhydr)oxides or iron sulfides, or well developed crystals. Such interactions may give rise to sorption of the species from solution onto the mineral surface, or some kind of partial or total replacement of atoms of the substrate, or the precipitation of a new phase. In contrast to replacement or precipitation, sorption processes imply an uptake of the solution species that is limited by the availability of sites at the surface where sorption may occur. An example of an XAS study of sorption is illustrated in Fig. 9.8. Here the mineral substrate is the tetragonal form of FeS (mackinawite), commonly formed from the activity of sulfate-reducing bacteria in anaerobic sediments, and the species in solution is the highly toxic actinide element, neptunium. The EXAFS data show almost no variation over a wide range of Np concentrations (see Fig. 9.8) and are consistent with formation of an inner sphere complex (i.e. where the Np bonds directly to atoms at the FeS surface with coordination and bond lengths as shown in Fig. 9.8).

9.2.3   *The mineral–(fluid) organic–microbe interface*

Although some of the techniques discussed above require the sample to be in a vacuum and hence be removed from contact with a fluid or biomaterial, others are powerful methods for the *direct* study of the mineral–(fluid)organic-microbe interface. This is true of the SPM techniques and synchrotron methods such as XAS.

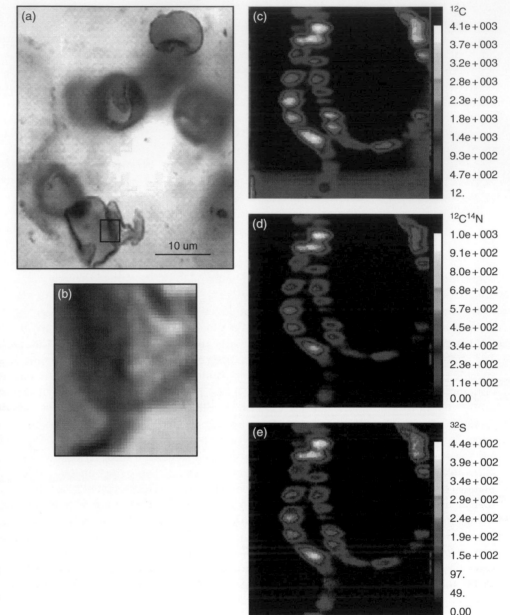

Figure 9.7 NanoSIMS of Late
Precambrian microfossils from
the Bitter Springs Formation of
Australia: (a, b) are transmitted
light optical photomicrographs
of a polished thin section of
chert; (b) is the area within the
rectangle marked on (a) where
two cells are in contact; (c, d, e)
are NanoSIMS elements maps
of ^{12}C, $^{12}C^{14}N$, and ^{32}S,
respectively, of the cell contact
imaged in (b). The maps show
a one-to-one correspondence
between these isotopes. This
correspondence, along with the
globular aligned character, are
strong indicators of biogenicity
(after Oehler *et al.*, 2006, 2009).

Others like (E)SEM and TEM offer the possibility of retaining much of the information lost when samples are studied *in vacuo* by using cryo techniques or through the use of a partial vacuum.

AFM is not just restricted to imaging because the use of AFM tips on easily deformed cantilevers can provide an exquisite method for measuring forces of adhesion. Indeed, this AFM measurement of force (or adhesive energies) is, possibly, as important as AFM imaging. For example, methods have been worked out for attaching a bacterium to an AFM tip so that the cell can be slowly touched to a mineral surface, and the (initial) repulsive and (eventual) adhesive forces

measured (Lower *et al.*, 2001, 2005). It is also possible to measure the deformability or the elasticity of microbial surfaces (see, for example, Yao *et al.*, 1999). These adhesion and elasticity forces approach the limits of AFM detection (i.e. tens of pN), but they can be measured. Another exciting possibility with AFM is the measurement of ionized sites (or electrostatics) on the surface of a single cell (Sokolov *et al.*, 2001). Here, small changes in the amperage (microamperes to picoamperes) of the tip are monitored as it travels over the surface and extremely small current alterations give an indication of exposed reactive sites. The applications of AFM in the life sciences and possible future

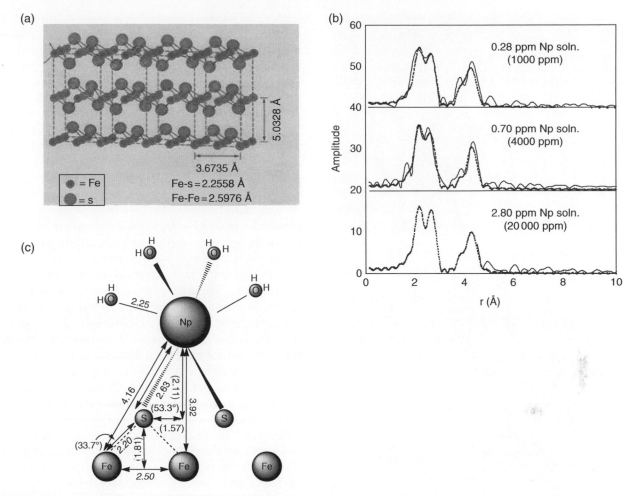

Figure 9.8 Interaction of Np in aqueous solution with the surface of mackinawite (FeS): (a) the structure of mackinawite; (b) Fourier transform of the EXAFS spectrum of Np at different solution concentrations 'reacted' with FeS; (c) model derived from the EXAFS data of Np atom bonding to the FeS surface as an inner sphere complex (after Moyes *et al.*, 2002).

developments have been reviewed by Parot *et al.* (2007) in an introductory article of a special journal issue devoted to the technique.

There will always be a role for both transmitted and reflected light optical microscopy (practical maximum resolution ~1μm) in studying mineral surfaces and interfaces and the associated microbial material. Biomaterials can be further characterized in terms of their components using fluorescent markers. When these fluorescent markers are used with confocal scanning laser microscopes (CSLM) an even wider world of microscopy opens up. The use of a laser provides a coherent beam of light for better resolving power and focussing accuracy; now, high resolution optical slices of a sample can be obtained and, with suitable computing power and optical reconstruction software, three-dimensional imaging can be achieved (Lawrence and Neu, 2006). An example is shown in Fig. 9.9, taken from the work of Brydie *et al.* (2005, 2009). Here, as further discussed below, CSLM has been used to image a

biofilm grown in a simulated fracture between two quartz glass surfaces. With new ratiometric fluorescent probes it is possible to probe such biofilms for some of the geochemical parameters (e.g. pH) that control mineralization processes (Hunter and Beveridge, 2005). Furthermore, advances in multiple-photon excitation CSLMs can, through light activation of highly specific fluorescent ligands, pinpoint the distances between such small cellular components as molecules. Recent advances in optical methods are now overcoming the diffraction imposed limits to resolution of focusing light microscopes (180 nm in the focal plane and 500 nm along the optic axis). As outlined by Hell (2003), techniques such as stimulated emission depletion microscopy (STED) and photoactivated localization microscopy (PALM) offer nanoscale imaging with focused light. These techniques have been used to image intracellular fluorescent proteins at nanometer resolution (Betzig *et al.*, 2006) and even to video record fast physiological phenomena at the nanoscale (Westphal *et al.*, 2008).

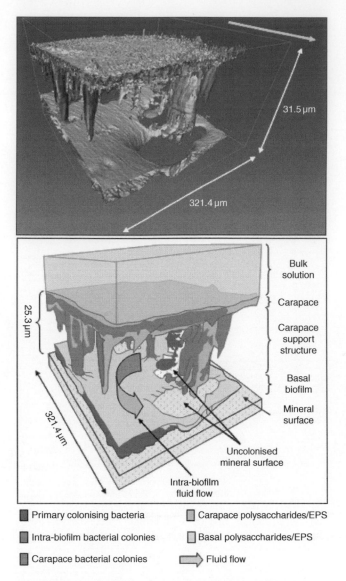

Figure 9.9 Confocal scanning laser microscope (CSLM) image of a biofilm grown between two quartz glass plates by introducing a nutrient solution and inoculating with *Pseudomonas aeruginosa*. Below the image is a schematic diagram showing the various components of the biofilm (after Brydie *et al.*, 2004, 2009).

If we extend from the visible region of the electromagnetic spectrum to the infrared, a range of other techniques present important opportunities for research (see McMillan and Hofmeister, 1988, for a mineralogically oriented overview). This is because energy in the infrared (~0.75–100 μm) region of the electromagnetic spectrum can be absorbed by material through vibrational processes such as the bending and stretching of the bonds between atoms, and this can be observed in gaseous, liquid or solid phases. In the simplest case, measurement of the positions and intensities of absorption peaks as a function of energy can provide a 'fingerprint' identification of a particular molecule or

crystalline material present in a sample. Infrared (IR) spectra can be measured in absorption mode, allowing the beam to pass through the sample, or reflection mode by reflecting off the surface of a solid crystal or compressed powder. Surface coatings can also be studied in reflection mode. More detailed analysis of an infrared spectrum involves associating individual peaks with the bending or stretching of particular bonds in the material under investigation. This can provide information on molecular or crystalline structure, or on chemical changes taking place in a system with time or with changing conditions. Recent decades have seen the development of equipment enabling imaging of samples in the infrared to facilitate the mapping of particular components in complex samples, and the availability of intense IR light at synchrotron facilities offers the prospect of important advances in improved spacial resolution and detection limits. Vibrational spectra can also be studied using Raman spectroscopy. Here, monochromatic light from a suitable source such as a powerful laser is directed through a translucent sample and the light scattered at 90° or 180° is recorded. The scattered light is of two kinds, Rayleigh, where no frequency change occurs, and Raman which involves a change of frequencies arising from vibrational effects. As with IR spectroscopy, Raman spectra can be used to provide 'fingerprint' identification of particular materials, or more detailed information on molecular or crystal structure, elastic or thermodynamic properties. It is another method which has been refined over the years; for example, so as to study spectral signals from very small areas using the 'Raman microprobe'. Raman spectroscopy is one of the large number of techniques discussed by Geesey *et al.* (2002) in reviewing spectroscopic methods for characterizing microbial transformations of minerals, whereas Swain and Stevens (2007) outline the use of Raman microspectroscopy for the non-invasive biochemical analysis of single cells.

9.3 Mineral-organic-microbe interfacial processes: some key examples

9.3.1 *Minerals and organic molecules*

At the most fundamental level, the interaction between a mineral surface and a microbe is that between the surface and an organic molecule. Molecular scale studies of such interactions are now beginning; one example is an atomic resolution STM study of interactions of the magnetite (111) surface with three contrasting organic molecules: formic acid, pyridine and carbon tetrachloride (Cutting *et al.*, 2006). The complexities of this mineral surface have been discussed above and illustrated in Fig. 9.4. Figure 9.10 shows an STM image of the so-called A' surface of the mineral after exposure to a small

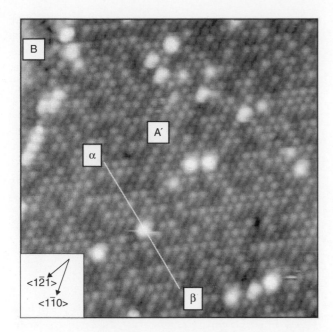

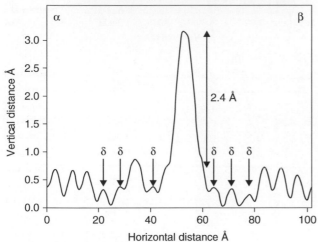

Figure 9.10 Scanning tunneling microscope (STM) image (200 Å by 200 Å) of the (1 1 1) surface of magnetite obtained under UHV conditions. The 'spheres' imaged in the background (at 0.6 nm (6 Å) separation) are individual tetrahedral site iron atoms at the mineral surface. The sample has been exposed to a small amount of the organic molecule pyridine (C_5H_5N); the bright spots are interpreted as individual pyridine molecules bonded to the surface via interaction of N with the tetrahedral Fe atoms. The cross section (α-β) shows surface topography, and the height of the sorbed pyridine molecule matches this interpretation (after Cutting *et al.*, 2006).

amount of pyridine. Individual molecules of pyridine give rise to the bright features in the image, and the topography determined by the cross section labeled α-β is consistent with interaction of the N atoms of pyridine with tetrahedral site Fe atoms. More extensive coating of the mineral surface by appropriate organics is clearly the first stage in biofilm formation.

9.3.2 Biofilms and interfacial processes

Traditional microbiological approaches have focused on studying planktonic cell cultures grown under well-characterized, highly mixed growth regimes. However, in most natural environments, microbial cells grow on solid substrates as 'biofilms'; hence the mineral substrate is crucial. The traditional approaches have tended to limit many geomicrobiological studies, because the latter form of cell growth has a major impact on the host cell physiology, and also on the physical and chemical environment associated with the biofilm (Stoodley *et al.*, 2002). A useful definition of 'biofilms' is that given by Characklis and Marshall (1990) in stating that microbial cells attach to solids (e.g. minerals) and 'immobilized cells grow, reproduce and produce extracellular polymers which frequently extend from the cell forming a tangled matrix of fibers, which provide structure to the assemblage termed a biofilm'. As noted above, it is well known that biofilms at a solid–fluid interface can maintain conditions which are very different to those in the associated bulk fluid (e.g. in terms of pH, pO_2 etc). In the geological context, three key points are illustrated by work in our laboratories (Brydie *et al.*, 2005, 2009). First, in experiments involving the growth of biofilms in a fracture simulated using two silica glass surfaces ~30 μm apart, CSLM imaging (see Fig. 9.9) of a mature biofilm showed complex columnar structures. Crucially, however, substantial areas of the 'mineral' surface remained free from biofilm material and able to interact with the fluid. Second, in subsequent experiments, very low concentrations of iron in solution were introduced into the fracture containing the (*Pseudomonas aeruginosa*) biofilm; this led to formation of iron oxyhydroxide precipitates (lepidocrocite) under conditions where such precipitation would not have occurred without the presence of such a biofilm. Third, the development of biofilms can have an important impact on the flow of fluid through fractures or through porous media. Figure 9.11 illustrates the results of experiments involving the flow of fluid through a column packed with pure quartz sand (Brydie *et al.* 2005). Following inoculation with *Pseudomonas* and biofilm growth, the hydraulic conductivity of the sand column drops by several orders of magnitude. Also shown in this figure are ESEM images of biofilm grown between the quartz grains in the column under realistic concentrations of nutrients.

9.3.3 Mineral–microbe interactions and redox processes

The above examples illustrate how microbial colonization can have indirect impacts on the hydrology and chemistry of geological systems, through increases in biomass and the production of associated metabolites and exopolymers. However, redox transformations of

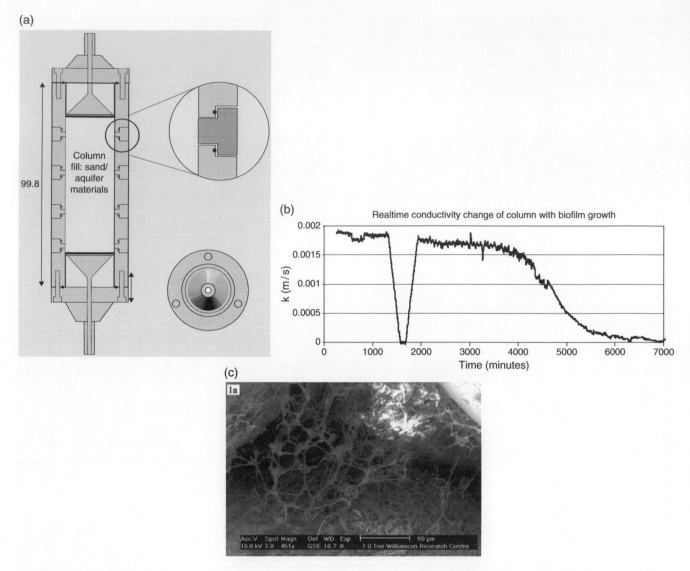

Figure 9.11 Diagram to illustrate column experiments on the nature of biofilm growth in porous media and its impact on fluid flow: (a) cylindrical column used in experiments – 76 × 22 mm internal dimensions; (b) decrease in hydraulic conductivity over time following inoculation with *Pseudomonas aeruginosa* and growth of biofilm; (c) ESEM image of biofilm occupying pore space between quartz grains (after Brydie *et al.*, 2004).

many mineral phases can also be linked directly to microbial respiration. These direct processes often involve highly specialized microorganisms that use the mineral surface as a source of electrons for the respiration (or reduction) of oxygen or nitrate, or couple the oxidation of organic matter (or another suitable electron donor) to the reduction of the mineral phase itself, normally in the absence of oxygen as an electron acceptor. The range of elements that can be harnessed for respiratory electron flow during these transformations is large, with redox changes that are quite complex and potentially difficult to dissect in experimental systems. For this reason, the microbial redox cycle for iron is a useful model system to examine. Not only is iron the

fourth most abundant element, and one that has a key role in many biological systems, it is dominated by only two oxidation states (Fe(II) or ferrous iron, and Fe(III) or ferric iron) which are involved in many environmentally important oxidation and reduction processes. In addition, there is a range of mineralogical techniques that are ideally suited to the study of microbial interactions with iron minerals, and the phases laid down via their metabolism by specialist microorganisms. As well as the methods discussed above, these include techniques related to the magnetic properties of iron compounds, such as the synchrotron technique XMCD (see below) and methods for the measurement of bulk magnetic properties (magnetometry); also the technique involving resonant absorption of gamma rays known as Mössbauer spectroscopy. These methods can provide insights into the oxidation states, spin states, magnetic ordering and crystallographic site distributions of iron in minerals and related materials.

9.3.3.1 Microbial oxidation of Fe(II)

Focusing first on the use of Fe(II) minerals as electron donors for microbial respiration, geomicrobiological studies have shown that there are at least three contrasting mechanisms for Fe(II) oxidation. Under acidic conditions, Fe(II) can be oxidized aerobically (Johnson, 1998), while under anaerobic conditions at neutral pH, Fe(II) oxidation by phototrophs or denitrifying microorganisms has been recognized (Weber *et al.* 2006; Kappler and Newman, 2004). The latter two processes, although environmentally important (e.g. potentially linked to processes as diverse as controlling metal/radionuclide solubility through to the precipitation of banded iron formation; Weber *et al.*, 2006; Kappler *et al.*, 2005), are relatively newly characterized and have not been studied in the same detail as the oxidation of Fe(II) at low pH. For this reason, we will now focus on this latter process for the first of our case studies. Aqueous Fe(II) is very unstable in the presence of oxygen at neutral pH; however, at low pH (<4.0; Stumm and Morgan, 1981), ferrous iron is oxidized very slowly by oxygen and is an excellent source of electrons for aerobic 'acidophilic' prokaryotes. Although the free energy associated with the oxidation of Fe(II) is quite small, the relatively large amounts of ferrous iron in many acidic environments means that this process can play a defining role in controlling the chemistry of such sites. This is especially true where pyrite (FeS$_2$) is present, for example at mine-waste sites, as when this mineral is exposed to moisture and air it oxidizes spontaneously with either molecular oxygen or ferric iron acting as the oxidant. A typical overall reaction can be written:

$$FeS_2 + 6Fe^{3+} + 3H_2O \rightarrow 7Fe^{2+} + S_2O_3^{2-} + 6H^+$$

although, as noted by Rimstidt and Vaughan (2003), the detailed reactions involved are more complex, involving a whole series of electron transfer steps. Critically, however, this process is accelerated dramatically by the presence of Fe(II)-oxidizing bacteria including *Acidithiobacillus* and *Leptospirillum* species (Hallberg and Johnson, 2001) that effectively regenerate highly corrosive ferric iron, leading to widespread sulfide mineral dissolution and the formation of conditions associated with 'acid mine drainage'. The closely related phase arsenopyrite (FeAsS) is particularly environmentally hazardous, being a source of arsenic pollution as well as acid mine waters. Recent work involving both abiotic

and microbial (*Leptospirillum ferrooxidans*) oxidative dissolution experiments, with mineral surface characterisation using ESEM, XPS and AES, has demonstrated the effectiveness of microbial involvement in accelerating the arsenopyrite breakdown process (Corkhill *et al.*, 2008; see Fig. 9.12).

9.3.3.2 Microbial reduction of Fe(III)

In the absence of oxygen in the subsurface, Fe(III) minerals can also make excellent electron acceptors for the anaerobic growth of specialist dissimilatory Fe(III)-reducing bacteria, including a very wide range of Archaea and Bacteria (Lovley *et al.*, 2004). The environmental relevance of microbial Fe(III) reduction has been well documented (Thamdrup, 2000, Lovley, 1991), with Fe(III) being the dominant electron acceptor in many subsurface environments (Lovley and Chapelle, 1995). As such, Fe(III)-reducing communities can be responsible for the majority of the organic matter oxidized in such environments (Lovley, 1993), while dramatically influencing the mineralogy of sediments. For example, the reductive dissolution of insoluble Fe(III) oxides can result in the release of potentially toxic levels of Fe(II), and also trace metals that were bound by the Fe(III) minerals. Depending on the chemistry of the water, a range of reduced minerals can also be formed including magnetite (Fe$_3$O$_4$), siderite (FeCO$_3$) and vivianite resulting in a change in structure of the sediments. Finally, Fe(III)- and Mn(IV)-reducing microorganisms can also impact on the fate of other high valence contaminant metals through direct enzymatic reduction, and also via indirect reduction catalysed by biogenic Fe(II). Metals reduced by these mechanisms include U(VI), Cr(VI) and Tc(VII) and, in such cases, metal reduction results in immobilization of these potentially toxic and mobile metals in sediments (Lloyd, 2003). The mechanisms of Fe(III)-reduction have been studied in most detail in two model Fe(III)-reducing bacteria; *Shewanella oneidensis* and *Geobacter sulfurreducens*. Indeed, research on these organisms has been given added impetus through the availability of the genome sequences, and suitable genetic systems for the generation of deletion mutants for both of these organisms (Coppi *et al.*, 2001; Myers *et al.*, 2000). Although the terminal reductase has yet to be identified unequivocally in either organism, the involvement of *c*-type cytochromes is implicated in electron transport to Fe(III) in several studies (Gaspard *et al.*, 1998; Lloyd *et al.*, 2003; Magnuson *et al.*, 2000; Myers and Myers, 1993, 1997; Beliaev *et al.*, 2001). In some examples, activities have been localized to the outer membrane or surface of the cell, consistent with a role in direct transfer of electrons to Fe(III) oxide that are highly insoluble at circumneutral pH (Gaspard *et al.*, 1998; DiChristina *et al.*, 2002; Lloyd *et al.*, 2002; Myers and Myers, 1992,

(a)

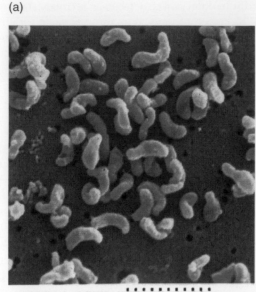

(b)

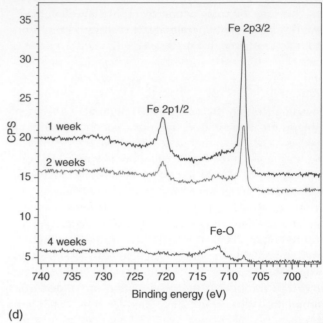

(c)

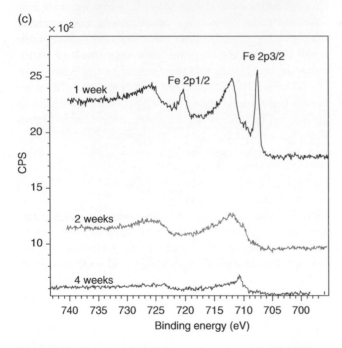

(d)

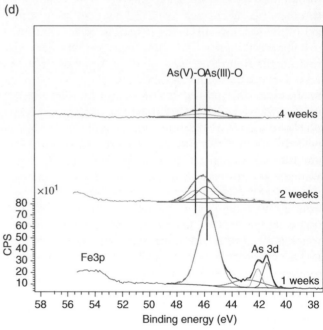

Figure 9.12 Oxidative breakdown of arsenopyrite (FeAsS) involving the organism *Leptospirillum ferrooxidans*: (a) ESEM image of *Leptospirillum*; (b) Fe2p XPS data for the arsenopyrite surface abiotically oxidized for periods of 1, 2 and 4 weeks; (c) the same experiment in the presence of *Leptospirillum*; (d) As 3D XPS data for the arsenopyrite surface oxidized in the presence of *Leptospirillum* for 1, 2 and 4 weeks. The Fe2p spectra show development of Fe^{3+} and Fe oxide peaks over time with much more rapid oxidation in the biotic system. The As3d spectra show evidence for the later onset of As^{3+} to As^{5+} oxidation (after Corkhill *et al.*, 2008).

2001). In addition to the proposed direct transfer of electrons to Fe(III), soluble 'electron shuttles' are also able to transfer electrons between metal-reducing prokaryotes and the mineral surface. This mechanism alleviates the requirement for direct contact between the microorganism and mineral. For example, humics and other extra-cellular quinones are utilized as electron acceptors by Fe(III)-reducing bacteria (Lovley *et al.*, 1996), and the reduced hydroquinone moieties are able to abiotically transfer electrons to Fe(III) minerals. The oxidized humic is then available for reduction by the microorganism, leading to further rounds of electron shuttling to the insoluble mineral (Nevin and Lovley, 2002). Very low concentrations of an electron shuttle (e.g. 100 nM) of the humic analog anthraquinone-2,6-disulfonate (AQDS) can rapidly accelerate the reduction of Fe(III) oxides (Lloyd *et al.*, 1999). Recent studies with *Shewanella* species have also shown that flavin molecules (FMN and riboflavin) are secreted to act as endogenous electron shuttles in this organism (von Canstein *et al.*, 2008; Marsili *et al.*, 2008). The environmental significance of such processes, however, remains to be confirmed. The addition of small amounts of electron shuttle to cultures of Fe(III)-reducing bacteria is, however, a very useful strategy to accelerate dramatically the rate of Fe(III) reduction in laboratory studies. A final mechanism of electron transfer to Fe(III) minerals that has also received much attention is the formation of conductive nanowires projecting from the surface of *Geobacter* and *Shewanella* species (Gorby *et al.*, 2006; Reguera *et al.*, 2005). Here, STM has proved particularly useful in confirming that the proteinaceous structures formed by these bacteria are indeed conductive. Again, the environmental relevance of these potentially important extracellular structures remains to be confirmed.

In addition to STM, several other techniques are useful to image the interactions of Fe(III)-reducing bacteria with the mineral surfaces that they respire. For example, hydrated films of poorly crystalline Fe(III) oxides offer good models for the highly bioavailable mineral phases that are the preferred substrates for these organisms in natural environments. Here, environmental scanning electron microscopes are particularly useful in visualizing the interactions of the microbial cells with the hydrated samples (Wilkins *et al.*, 2007b), showing clearly the impact of electron transfer at the cell surface on the Fe(III) mineral phase (see Fig. 9.1). Interestingly, probing the mineral surface with XPS in this study failed to identify Fe(II), confirmed by challenging the samples with redox active chemical probes highly sensitive to exposed Fe(II) (Wilkins *et al.*, 2007b). These observations suggest electron transfer is possible into the mineral substrate, effectively increasing the amount of electron acceptor available to the organism, which in turn has important implications for the growth yield of these organisms in the subsurface. For higher resolution imaging of the microbe–mineral interface, TEM is a very useful technique, offering near atomic resolution at the price of working in a vacuum with dried samples. Here, detailed characterization of the starting and end products of Fe(III) reduction has confirmed the recalcitrance of crystalline phases such as hematite to reduction by *Geobacter sulfurreducens*, in stark contrast to highly bioavailable amorphous phases such as ferrihydrite or feroxyhyte, which are transformed rapidly to magnetite (Cutting *et al.*, 2009; see Fig. 9.2). Colorimetric (ferrozine) assays for Fe(II), in combination with XPS

analyses confirmed a lack of ferrous iron associated with the crystalline mineral phases after incubation with the Fe(III)-reducing bacterium. A major conclusion from these data is that surface area alone is not the critical factor controlling the rate and extent of Fe(III) reduction in the subsurface.

Of course, in natural environments Fe(III)-reducing bacteria will not interact with Fe(III) phases in isolation, and sorbed trace elements can play a role in controlling both the rate and the end-point of microbial reduction (Zachara *et al.*, 2001). For mixed metal systems, additional techniques are useful to monitor the structural evolution of the end products of metal bioreduction, including the synchrotron technique X-ray magnetic circular dichroism (XMCD). XMCD provides unique information on magnetic minerals as it is element specific and can quantify the amount and the valence state of Fe on the lattice sites within the structure of ferrite spinels such as magnetite. In our recent papers, this technique has been used, in combination with XAS analyses, to characterize a range of biogenic magnetites from cultures of Fe(III)-reducing and magnetotactic bacteria (Coker *et al.*, 2007), and also biospinels substituted with a range of transition metals including Co, Ni and Zn (Coker *et al.* 2008). This technique is now being applied in mechanistic studies to look at the changes in biomagnetite structure when the Fe(II)-bearing biomineral is used as a potent reductant to treat a range of organic and inorganic contaminants.

9.3.3.3 Ancient biogeochemical processes and biogenicity

The examples of mineral-organic–microbe interfacial processes given above all concern present-day living systems, whereas the rapidly developing field of geobiology also includes the study of ancient (fossil) microbial systems. A key question here, particularly when examining the oldest rocks, is whether or not observed features interpreted as 'fossils' are truly biogenic in origin (see Chapter 16). Two approaches to answering this question have been described in recent publications. The first involves the use of the NanoSIMS technique (described above) to map the sub-micron scale chemistry of putative microfossils. For example, Oehler *et al.* (2006, 2009) mapped C, N, S, Si and O distributions in microfossils from the ~850 million year old Bitter Springs Formation of Australia (see Fig. 9.7). The data obtained on the chemistry of particular structures were consistent with a biological origin, and provided information on the silicification (fossilization) process and the biosignatures of specific microorganisms and microbial communities. These studies also led the authors to propose new criteria for assessing the biogenicity of problematic kerogenous materials.

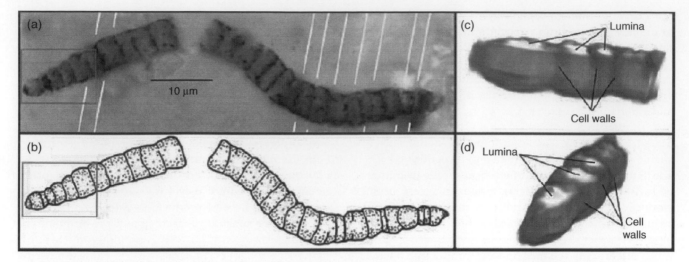

Figure 9.13 Fossil cyanobacterium trichome (*Cephalophytarion laticellulosum*) from the Precambrian (~850 Ma old) Bitter Springs Formation (Central Australia): (a) and (b) photomicrograph of a thin section with derived interpretive drawing: (c) and (d) images from 3-D Raman spectroscopy of the region outlined by a box in (a) and (b) (after Schopf and Kudryavtsev, 2005).

The second approach has involved obtaining much improved data on the morphology and structure of a 'microfossil' by using some form of three-dimensional imaging. For example, Schopf *et al.* (2006) used confocal scanning laser microscopy to image Precambrian microscopic organisms. As a non-intrusive and non-destructive technique capable of submicron-scale resolution, CSLM can be used to provide three-dimensional information on morphology, taphonomy and fidelity of preservation of fossil microbes and related microorganisms. In a related development, Schopf and Kudryavtsev (2005) used three-dimensional Raman imaging of Precambrian microorganisms to obtain similar information and probe their molecular structural composition. In Fig. 9.13, a fossil cyanobacterium from ~850-million-year-old rocks is shown using traditional optical imaging in thin section and using three-dimensional Raman spectroscopy, with the latter clearly revealing much new information. The impact of such studies is illustrated by publications combining the evidence from imaging with that from carbon isotope data, and which make a case for microbial life extending back at least ~3500 million years into the Archean (Schopf *et al.*, 2007).

9.3.4 Mineral–organic–microbe interactions: past, present, future

The processes taking place at the mineral–microbe interface were surely amongst the very first associated with the emergence of life on Earth. They have played a key role in the biogeochemical cycling of the elements since that time, and the part played by them in the past, in processes such as the formation of the geologically ancient banded iron formations (currently the dominant world suppliers of iron), is only now being fully appreciated.

Recent years have seen rapid growth in the understanding of mineral-microbe interactions, prompted by both advances in methods for studying minerals and their surfaces at the molecular scale, and by developments in microbiology, not least those involving molecular techniques. Only a few examples of the systems already studied just in our laboratories have been discussed in the present article; we have also recently investigated mineral–microbe interaction processes in systems involving a range of metals including Cr, Hg, Tc, U, Pu and As (see, for example, Islam *et al.*, 2004; Burke *et al.*, 2006; McBeth *et al.*, 2007; Wilkins *et al.*, 2007a). Such studies are of relevance to the mobility of these toxic metals in the environment, to strategies limiting their environmental impact, and to the development of technologies for the bioremediation of contaminated sites. For example, in the case of arsenic contamination of drinking water over large areas of the Indian subcontinent and of SE Asia, it has been shown that a significant role is played by the microbial reduction of arsenic, leading to its mobilization. The functional gene involved in this process has been identified, allowing the identification of active organisms (Héry *et al.*, 2009). Future research in mineral–organic–microbe interactions will surely see benefits to our understanding of biogeochemical cycles at the molecular level, helping in the quantitative prediction of migration of contaminants in the 'critical zone'. It should also bring a more fundamental understanding of how particular microorganisms function, including those capable of flourishing under extreme conditions of pH, temperature or pressure. The practical applications of such work are relevant to many of the key problems now facing society, including the safe containment of hazardous (including nuclear) wastes, the cleanup of contaminated soils and waters, and development of efficient (clean) technologies for processing of minerals or recovery of metals and toxic materials from wastes.

Acknowledgements

The authors acknowledge the funding received from NERC, BBSRC and EPSRC in support of much of the work described in this article. The contributions made by numerous colleagues and students, in particular J. Brydie, J. Charnock, C. Corkhill, R. Cutting, M. Wilkins, P. Wincott are also gratefully acknowledged.

References

Banfield JF, Nealson KH (eds) (1997) *Geomicrobiology: Interactions between Microbes and Minerals*. Reviews in Mineralogy and Geochemistry, **35**, 448pp.

Becker U, Hochella MF Jr, Apra E (1996) The electronic structure of hematite {001} surfaces: applications to the interpretation of STM images and heterogeneous surface reactions. *American Mineralogist* **81**, 1301–1314.

Beliaev AS, Safarini DA, McLaughlin JL, Hunnicutt D (2001) MtrC, an outer membrane decahaem c cytochrome required for metal reduction in *Shewanella putrefaciens* MR-1. *Molecular Microbiology* **39**, 722–730.

Betzig E, Patterson GH, Sougrat R, *et al.* (2006) Imaging intracellular fluorescent proteins at nanometer resolution. *Science* **313**, 1642–1645.

Beveridge TJ, Moyles D, Harris B (2007) Electron microscopy. In: *Methods for General and Molecular Microbiology* (eds Reddy *et al.*). ASM Press, Washington DC.

Bluhm H (and 28 other authors) (2006) Soft X-ray microscopy and spectroscopy at the molecular environmental science beamline at the Advanced Light Source. *Journal of Electron Spectroscopy and Related Phenomena* **150**, 86–104.

Brydie JR, Wogelius RA, Merrifield CM, *et al.* (2004) The μ2M project on quantifying the effects of biofilm growth on hydraulic properties of natural porous media and on sorption equilibria: an overview. In: *Understanding the Micro to Macro Behaviour of Rock-Fluid Systems* (ed Shaw RP). Geol. Soc. London Spec Pubn. **249**, 131–144.

Brydie JR, Wogelius RA, Boult S, Merrifield CM, Vaughan DJ (2009) Model system studies of the influence of bacterial biofilm formation on mineral surface reactivity. *Biofouling* **25**, 463–472.

Burke IT, Boothman C, Lloyd JR, *et al.* (2006) Redoxidation behaviour of technetium, iron and sulfur in estuarine sediments. *Environmental Science and Technology* **40**, 3529–3535.

Characklis WG, Marshall KC (eds) (1990) *Biofilms*. John Wiley & Sons, New York.

Coker VS, Pearce CI, Lang C, *et al.* (2007) Cation site occupancy of biogenic magnetite compared to polygenic ferrite spinels determined by X-ray magnetic circular dichroism *European Journal of Mineralogy* **19**, 707–716.

Coker VS, Pearce CI, Pattrick RAD, *et al.* (2008) Probing the site occupancies of Co-, Ni-, and Mn-substituted biogenic magnetite using XAS and XMCD. *American Mineralogist* **93**, 1119–1132.

Coppi MV, Leang C, Sandler SJ, Lovley DR (2001) Development of genetic system for *Geobacter sulfurreducens*. *Applied and Environmental Microbiology* **67**, 3180–3187.

Corkhill CL, Wincott PL, Lloyd JR, Vaughan DJ (2008) The oxidative dissolution of arsenopyrite (FeAsS) and enargite (Cu_3AsS_4) by *Leptospirillum ferrooxidans*. *Geochimica et Cosmochimica Acta* **72**, 5616–5633.

Cutting RS, Muryn CA, Thornton G, Vaughan DJ (2006) Molecular scale investigations of the reactivity of magnetite Geochimica et Cosmochimica with formic acid, pyridine, and carbon tetrachloride. *Acta Applied and Environmental Microbiology* **70**, 3593–3612.

Cutting RS, Coker VS, Fellowes JW, Lloyd JR, Vaughan DJ (2009) Mineralogical and morphological constraints on the reduction of Fe(III) minerals by *Geobacter sulfurreducens*. *Geochmica et Cosmochimica Acta* **73**, 4004–4022.

Cygan RT, Kubicki JD (eds) (2001) *Molecular Modelling Theory: Applications in the Geosciences*. Reviews in Mineralogy and Geochemistry, **42**, 531pp.

DiChristina TJ, Moore CM, Haller CA (2002) Dissimilatory Fe(III) and Mn(IV) reduction by *Shewanella putrefaciens* requires ferE, a homolog of the pulE (gspE) type II protein secretion gene. *Journal of bacteriology* **184**, 142–151.

Dufrene YF (2004) Using nanotechniques to explore microbial surfaces. *Nature Reviews in Microbiology* **2**, 451–560.

Dufrene YF (2007) Atomic force microscopy. In: *Methods for General and Molecular Microbiology* (eds Reddy *et al.*). ASM Press, Washington DC.

Gaspard S, Vazquez F, Holliger C (1998) Localization and solubilization of the iron(III) reductase of *Geobacter sulfurreducens* **64**, 3188–3194.

Geesey GG, Neal AL, Suci PA, Peyton BM (2002) A review of spectroscopic methods for characterizing microbial transformations of minerals. *Journal of Microbiological Methods* **512**, 125–139.

Goldstein J, Newbury DE, Echlin P, *et al.* (2003) *Scanning electron microscopy and X-ray microanalysis*, 3rd edn. Springer, New York, 689pp.

Gorby YA, Yanina S, McLean JS, *et al.* (2006).Electrically conductive bacterial nanowires produced by *Shewanella oneidensis* strain MR-1 and other microorganisms. *Proceedings of the National Academy of Sciences, USA* **103**, 11358–11363.

Hallberg KB, Johnson DB (2001). Biodiversity of acidophilic prokaryotes. *Advances in Applied Microbiology* **49**, 37–84.

Hell SW (2003) Toward fluorescence nanoscopy. *Nature Biotechnology* **21**, 1347–1355.

Héry M, Gault AG, Rowland HAL, Lear G, Polya DA, Lloyd JR (2009) Molecular and cultivation-dependent analysis of metal-reducing bacteria implicated in arsenic mobilisation in South East Asian aquifers. *Applied Geochemistry*, **23**, 3215–3223.

Hochella MF Jr, White AF (eds) (1990) *Mineral-Water Interface Geochemistry*. Reviews in Mineralogy and Geochemistry, **23**, 603pp.

Hunter RC, Beveridge TJ (2005)Application of a pH-sensitive fluoroprobe (C-SNARF-4) for pH microenvironmental analysis in *Pseudomonas aeruginosa* biofilms. *Applied and Environmental Microbiology,* **71**, 2501–2510.

Hunter RC, Hitchcock AP, Dyness JJ, Obst M, Beveridge TJ (2008) Mapping the speciation of iron in *Pseudomonas aeruginosa* biofilms using scanning transmission X-ray microscopy. *Environmental Science and Technology* **42**, 8766–8772.

Hurst CJ (ed.) (2007) *Manual of Environmental Microbiology,* 3rd edn. ASM Press, Washington DC, 1293pp.

Islam FS, Gault AG, Boothman C, *et al.* (2004) Role of metal-reducing bacteria in arsenic release from Bengal Delta sediments. *Nature* **430**, 68–71.

Johnson DB (1998) Biodiversity and ecology of acidophilic microorganisms. *FEMS Microbiology Ecology* **27**, 307–317.

Kappler A, Newman DK (2004) Formation of Fe(III)-minerals by Fe(II)-oxidizing photoautotrophic bacteria. *Geochimica et Cosmochimica Acta,* **68**, 1217–1226.

Kappler A, Pasquero C, Konhauser KO, Newman DK (2005) Deposition of banded iron formations by anoxygenic phototrophic Fe(II)-oxidizing bacteria. *Geology* **33**,865–868.

Lawrence JR, Neu TR (2006) Laser scanning microscopy. In: *Methods for General and Molecular Microbiology* (eds Reddy *et al.*). ASM Press, Washington DC.

Lloyd JR (2003) Microbial reduction of metals and radionuclides. *FEMS Microbiology Reviews* **27**, 411–425.

Lloyd JR, Blunt-Harris EL, Lovley DR (1999) The periplasmic 9.6 kDa c-type cytochrome of *Geobacter sulfurreducens* is not an electron shuttle to Fe(III). *Journal of Bacteriology* **181**, 7647–7649.

Lloyd JR, Chesnes J, Glasauer S, Bunker DJ, Livens FR, Lovley DR (2002) Reduction of actinides and fission products by Fe(III)-reducing bacteria. *Geomicrobiological Journal* **19**, 103–120.

Lloyd JR, Leang C, Hodges Myerson AL, *et al.* (2003) Biochemical and genetic characterization of PpcA, periplasmic cytochrome in *Geobacter sulfurreducens. Biochemical Journal* **369**, 153–161.

Lovley DR (1991) Dissimilatory Fe(III) and Mn(IV) reduction. *Microbiological Reviews* **55**, 259–287.

Lovley, DR (1993) Dissimilatory metal reduction. *Annual Review of Microbiology* **47**, 263–290.

Lovley DR, Chapelle FH (1995) Deep subsurface microbial processes. *Reviews in Geophysics* **33**, 365–381.

Lovley DR, Coates JD, Blunt-Harris EL, Phillips EJP, Woodward JC (1996) Humic substances as electron acceptors for microbial respiration. *Nature* **382**, 445–448.

Lovley DR, Holmes DE, Nevin KP (2004). Dissimilatory Fe(III) and Mn(IV) Reduction. *Advances in Microbial Physiology* **49**, 219–286.

Lower SK, Tadanier CJ, Hochella MF Jr (2001) Dynamics of the mineral-microbe interface: use of biological force microscopy in biogeochemistry and geomicrobiology. *Geomicrobiology Journal,* **18**, 63–76.

Lower BH, Yongsunthon R, Vellano FP III, Lower SK (2005) Simultaneous force and fluorescence measurements of a protein that forms a bond between a living bacterium and a solid surface. *Journal of Bacteriology* **187**, 2127–2137.

Macrae ND (1995) Secondary ion mass spectrometry and geology. *Canadian Mineralogist* **33**, 219–236.

Magnuson TS, Hodges-Myerson AL, Lovley DR (2000) Characterization of a membrane-bound NADH-dependent Fe(3+) reductase from the dissimilatory Fe(3+)-reducing bacterium *Geobacter sulfurreducens. FEMS Microbiology Letters* **185**, 205–211.

Marsili E, Baron DB, Shikhare ID, Coursolle D, Gralnick JA, Bond DR (2008) *Shewanella* secretes flavins that mediate extracellular electron transfer. *Proceedings of the National Academy of Sciences, USA* **105**, 3968–3973.

McBeth JM, Lear G, Morris K, Burke IT, Livens FR, Lloyd JR (2007). Technetium reduction and reoxidation in aquifer sediments. *Geomicrobiology Journal* **24**,189–197.

McLaren AC (1991) *Transmission Electron Microscopy of Minerals and Rocks.* Cambridge University Press, Cambridge.

McMillan PF, Hofmeister AM (1988) Infrared and Raman spectroscopy. In: *Spectroscopic Methods in Mineralogy and Geology* (Hawthorne FC, ed.). Reviews in Mineralogy, **18**, 99–160.

McMahon G, Cabri LJ (1998) The SIMS technique in ore mineralogy. In: *Modern Approaches to Ore and Environmental Mineralogy* (eds Cabri, Vaughan), Mineralogical Association of Canada Short Course Series, vol. **27**, 199–240.

Myers CR, Myers JM (1992) Localization of cytochromes to the outer membrane of anaerobically grown *Shewanella putrefaciens* MR-1. *Journal of Bacteriology* **174**, 3429–3438.

Myers CR, Myers JM (1993) Ferric reductase is associated with the membranes of anaerobically grown *Shewanella putrefaciens* MR-1. *FEMS Microbiology Letters* **108**, 15–21.

Myers CR, Myers JM (1997) Cloning and sequence of cymA, a gene encoding a tetraheme cytochrome c required for reduction of iron(III), fumarate and nitrate by *Shewanella putrefaciens* MR-1. *Journal of Bacteriology* **179**, 1143–1152.

Myers JM, Myers CR (2001) Role of outer membrane cytochromes OmcA and OmcB of *Shewanella putrefaciens* MR-1 in reduction of manganese dioxide. *Applied and Environmental Microbiology* **67**, 260–269.

Myers JM, Myers M, Myers CR (2000) Role of tetraheme cytochrome CymA in anaerobic electron transport in cells of *Shewanella putrefaciens* MR-1 with normal levels of menaquinone. *Journal of Bacteriology* **182**, 67–75.

Nevin KP, Lovley DR (2002) Mechanisms for Fe(III) oxide reduction in sedimentary environments. *Geomicrobiology Journal* **19**, 141–159.

Obst M, Dittrich M, Kuehn H (2006) Calcium adsorption and changes of the surface microtopography of cyanobacteria studied by AFM, CFM, and TEM with respect to biogenic calcite nucleation. *Geochemistry Geophysics Geosystems* **7**, 1–15.

Oehler DZ, Robert F, Mostefaoui S, Meibom A, Selo M, McKay DS (2006) Chemical mapping of Proterozoic organic matter at sub-micron spatial resolution. *Astrobiology* **6**, 838–850.

Oehler DZ, Robert F, Walter MR, *et al.* (2009) NANOSIMS: insights to biogenicity and syngeneity of Archaean carbonaceous structures. *Precambrian Research* **173**, 70–78.

Ogletree DF, Bluhm H, Hebenstreit ED, Salmeron M (2009) Photoelectron spectroscopy under ambient pressure and temperature conditions. *Nuclear Instruments and Methods in Physics Research* A **601**, 151–160.

Parot P, Dufrene YF, Hinterdorfer P, *et al.* (2007) Past, present and future of atomic force microscopy in life sciences and medicine. *Journal of Molecular Recognition* **20**, 418–431.

Reguera G, McCarthy KD, Mehta T, Nicoll JS, Tuominen MT, Lovley DR (2005) Extracellular electron transfer via microbial nanowires. *Nature* **435**, 1098–1101.

Rimstidt JD, Vaughan DJ (2003) Pyrite oxidation: a state-of-the-art assessment of the reaction mechanism. *Geochimica et Cosmochimica Acta*, **67**, 873–880.

Rosso KM, Becker U, Hochella MF JR (1999) Atomically resolved electronic structure of pyrite{100} surfaces: an experimental and theoretical investigation with implications for reactivity. *American Mineralogist* **84**, 1535–1548.

Schopf JW, Kudryavtsev AB (2005) Three-dimensional Raman imagery of Precambrian microscopic organisms. *Geobiology* **3**, 1–12.

Schopf JW, Tripathi AB, Kudryavtsev B (2006) Three-dimensional confocal optical imagery of Precambrian microscopic organisms. *Astrobiology* **6**, 1–16.

Schopf JW, Kudryavtsev AB, Czaja AD, Tripathi AB (2007) Evidence of Archean life: stromatolites and microfossils. *Precambrian Research* **158**, 141–155.

Sokolov I, Smith DS, Henderson GS, Gorby YA, Ferris FG (2001) Cell surface electrochemical heterogeneity of the Fe(III) reducing bacteria *Shewanella putrefaciens*. *Environmental Science and Technology*, **35**, 341–347.

Stoodley P, Sauer K, Davies DG, Costerton JW (2002) Biofilms as complex differentiated communities. *Annual Review of Microbiology*, **56**, 187–209.

Stumm W (1992) *Chemistry of the Solid–Water Interface*. Wiley-Interscience, New York (N.Y.), 448pp.

Stumm W, Morgan JJ (1981) *Aquatic Chemistry : an introduction emphasizing chemical equilibria in natural waters*. John Wiley & Sons, Inc., New York.

Swain RJ, Stevens MM (2007) Raman microspectroscopy for non-invasive biochemical analysis of single cells. *Biochemical Society Transactions* **35**, 544–549.

Thamdrup B (2000) Bacterial manganese and iron reduction in aquatic sediments. *Advances in Microbiology and Ecology* **16**, 41–84.

Tossell JA, Vaughan DJ (1992) *Theoretical Geochemistry: Applications of Quantum Mechanics in the Earth and Mineral Sciences*. Oxford University Press, New York, 528pp.

Van Hove MA, Weinberg WH, Chan C-M (1986) *Low-Energy Electron Diffraction*. Springer Verlag, Berlin.

Vaughan DJ, Pattrick RAD (eds) (1995) *Mineral Surfaces*. Mineralogical Society, Chapman & Hall, London, 384pp.

von Canstein H, Ogawa J, Shimizu S, Lloyd JR (2008). Flavin secretion by *Shewanella* species and their role as extracellar redox mediators. *Applied and Environmental Microbiology* **74**, 615–623.

Weber, KA, Achenbach LA, Coates JD (2006). Microorganisms pumping iron: anaerobic microbial iron oxidation and reduction. *Nature Reviews Microbiology*, **4**, 752–764.

Westphal V, Rizzoli SO, Lauterbach MA, Kamin D, Jahn R, Hell SW (2008) Video-rate far-field optical nanoscopy dissects synaptic vesicle movement. *Science* **320**, 246–249.

White JC (ed.) (1985) *Application of Electron Microscopy in the Earth Sciences*. Mineralogical Association of Canada Short Course Series vol. 11, Ottawa, Canada.

Wilkins MJ, Livens FR, Vaughan DJ, Beadle I, Lloyd JR (2007a) The influence of microbial redox cycling on radionuclide mobility in the subsurface at a low-level radioactive waste storage site. *Geobiology* **5**, 293–301.

Wilkins MJ, Wincott PL, Vaughan DJ, Livens FR, Lloyd JR (2007b). Growth of *Geobacter sulfurreducens* on poorly crystalline Fe(III) oxyhydroxide coatings: an integrated study combining light and electron microscopy with X-ray photoelectron spectroscopy. *Geomicrobiology Journal* **24**, 199–204.

Wogelius RA, Vaughan DJ (2000) Analytical, experimental and computational methods in environmental mineralogy. In: *Environmental Mineralogy* (Vaughan DJ, Wogelius RA, eds), European Mineralogical Union Notes in Mineralogy, vol. 2, 7–88, Etvos University Press, Budapest.

Yao X, Jericho M, Pink D, Beveridge T (1999) Thickness and elasticity of gram-negative murein sacculi measured by atomic force microscopy. *Journal of Bacteriology* **181**, 6865–6875.

Zachara JM, Fredrickson JK, Smith SC, Gassman PL (2001) Solubilization of Fe(III) oxide-bound trace metals by a dissimilatory Fe(III) reducing bacterium. *Geochimica et Cosmochimica Acta*, **65**, 75–93.

10
EUKARYOTIC SKELETAL FORMATION

Adam F. Wallace[1], Dongbo Wang[2], Laura M. Hamm[2], Andrew H. Knoll[3] and Patricia M. Dove[2]

[1] Earth Sciences Division, Lawrence Berkeley National Laboratory, Berkeley CA 94720, USA
[2] Department of Geosciences, Virginia Polytechnic Institute and State University, Blacksburg VA 24061, USA
[3] Department of Organismic and Evolutionary Biology, Harvard University, Cambridge MA 02138, USA

10.1 Introduction

Biomineralization is the field of study dedicated to understanding the processes by which living beings cause minerals to form. Here, we focus on biologically *controlled* or *directed* routes to the formation of mineralized skeletons by eukaryotic organisms. Some prokaryotes also demonstrate a remarkable level of control over the mineralization process (notably, magnetotactic bacteria), but passive, or induced modes of mineral deposition prevail in bacteria and are discussed in Chapter 8. Traditionally, frontier research on controlled biomineralization has been dispersed across multiple scientific and medical disciplines, with relatively little cross-pollination of ideas between fields. However, modern approaches to biomineralization are promoting extensive communication, and these increased interactions are facilitated by a new urgency to understand the history and future of Earth and its inhabitants. Adding to this renaissance is the availability of many new analytical and culturing techniques in the biological and chemical sciences, as well as expanding trends toward true interdisciplinary communication and collaboration.

As molecular biologists decipher the structure, function, and genetic expression of biomolecular species involved in all aspects of biomineral formation, the materials community is seeking to harness Nature's ability to construct a wide variety of morphologically complex, organic-mineral composites for technological applications. Meanwhile, evolutionary biologists and paleoceanographers are using the abundance, distribution, and chemical composition of skeletal fossils to investigate the history of life. Skeletal chemistry is commonly used to infer past environmental conditions, including climate history; however, with increasing evidence that the chemical nature of an organism's internal crystallization environment may differ substantially from its external surroundings, we recognize the need to understand how mineralization pathways (either inorganic or biologically-controlled) influence the impurity content and isotopic fractionation of elements in biomineral phases. Current investigations within the crystal growth community are addressing these issues by determining the roles of organic macromolecules (peptides, proteins, carbohydrates and other biomolecules) in mediating crystal growth, morphology, and impurity contents, in both classical and non-classical crystallization systems. As each community continues to advance, a complete and inclusive physical model of biomineralization processes may develop. In this chapter, we discuss specific well-studied organisms that, in our opinion, best showcase several forefront issues currently facing the biomineralization community. We attempt to synthesize the enormous topic of eukaryote skeletal formation in a way that highlights areas where continued or renewed efforts may yield considerable insights into the nature of controlled mineralization processes. The nature and breadth of this subject prevents equal treatment of all mineralizing groups in a single chapter. We refer readers with deeper interests in specific organisms to several previous volumes dedicated entirely to biomineralization (Crick 1989; Leadbeater and Riding 1986; Mann et al., 1989; Simkiss and Wilbur 1989; Mann 2001; Dove et al., 2003).

Fundamentals of Geobiology, First Edition. Edited by Andrew H. Knoll, Donald E. Canfield and Kurt O. Konhauser.
© 2012 Blackwell Publishing Ltd. Published 2012 by Blackwell Publishing Ltd.

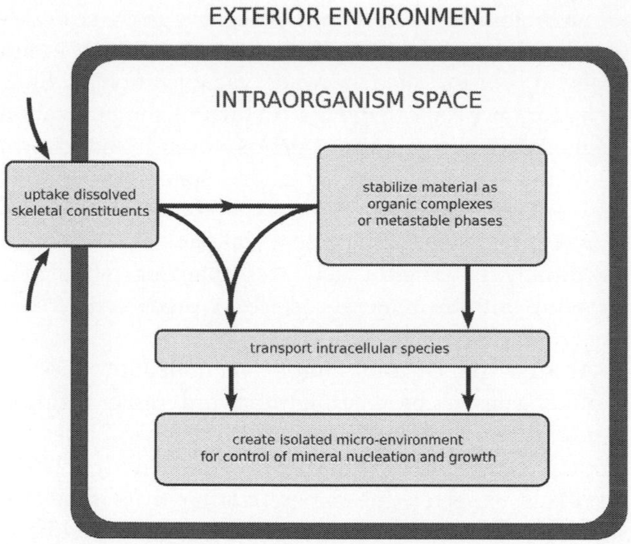

Figure 10.1 Elements of active mineralization strategies employed by eukaryotic organisms. Elucidating the molecular level details of these processes, the interplays amongst them, and their interactions with the surrounding genetically controlled environment are the main objectives of biomineral research.

We focus on the calcium carbonate and silica mineralizers because a number of organisms within these groups are both prevalent in the fossil record and among the most thoroughly investigated biomineralization systems. Fascinating new frontiers are emerging from research on phosphate biomineralization, particularly of vertebrate bone and teeth. Space does not allow us to develop this topic, but for those interested in gaining a perspective on current research in phosphatization, we refer readers to Weiner (2006) and Olstza *et al.* (2007).

Our discussion is organized around the single versus multicellular character of the organisms rather than by skeletal composition. While the details of the mineralization processes differ among the selected representatives of these groups, they all share certain common elements (Fig. 10.1). Broadly speaking, mineralization occurs by concentrating dissolved skeletal constituents from the external environment, stabilizing or storing these constituents as intermediates, utilizing various mechanisms for intracellular transport, and creating highly-regulated micro-environments for mineral deposition. The first section considers three unicellular organisms. The Foraminifera were chosen because their tests are widely used in paleotemperature reconstructions. For a current discussion of coccolithophorid mineralization see Mackinder *et al.* (2010). Radiolarians are also presented for their growing use in paleoenvironmental studies, and to demonstrate the limitations of traditional macroscopic approaches to biomineralization. The diatoms are discussed in detail because they are one of the most dominant members of planktonic

assemblages in modern oceans, and as such, are currently among the most extensively investigated silica mineralizing organisms. Frustule formation is an example of a template-mediated mineralization process in which non-enzymatic macromolecules facilitate biosilica formation.

In the second section of the chapter, we consider mineralization strategies employed by multicellular organisms. First, we discuss recent findings which suggest that mineralization of amorphous silica by sponges is regulated through enzymatic processes rather than by template-mediated reactions as in the unicellular diatoms. We then discuss molluscs, which fabricate complex composite structures of multiple calcium carbonate polymorphs by utilizing biochemical templates to determine the nucleating phase and proteins to direct the mineralization process. Non-classical crystallization processes involving an amorphous phase also appear to also be involved. To address this emerging aspect of calcification processes, we then explore mineralization by the sea urchin, which uses amorphous calcium carbonate (ACC) as a precursor to crystalline $CaCO_3$ throughout its embryonic development. In particular, the urchins provide an example of how biochemical and genetic approaches can be used to identify and assign functions to individual proteins and also to provide considerable *in vivo* insights into the mineralization process.

The final section briefly discusses the evolutionary history of skeleton-forming eukaryotes, as reconstructed from phylogeny and the fossil record. Throughout the chapter we attempt to note emerging research areas and questions, but also recognize that some of the recent ideas in the literature are provocative and still await careful testing and validation. This is, however, part of the energy that comes from a rapidly advancing scientific arena. We think you will agree that our understanding of skeletal mineralization is poised for tremendous strides in the coming decade.

10.2 Mineralization by unicellular organisms

10.2.1 The Foraminifera: chemical signatures as paleoenvironmental indicators

Foramanifera comprise one of the most diverse groups of microscopic protists in the modern ocean. Many of these protozoans, members of the eukaryotic superkingdom Rhizaria, develop a variety of carbonate biominerals as shells, or tests. Of the 15 foraminiferal orders recognized by the micropaleontological community, seven are capable of secreting calcite exoskeletons (Sen Gupta, 1999; Armstrong and Brasier, 2006). Others mineralize aragonite and opaline silica, and two groups can assemble foreign particles into tests using carbonate or organic matter as cements. Early foraminiferans did not form mineralized skeletons and their lack of preservation has likely skewed older models of how these organisms evolved over deep time. However, with the emergence of techniques that combine molecular and fossil data, new studies suggest a significant radiation of non-skeletonized forms that eventually gave rise to the wide variety of forms recognized today (Pawlowski *et al.*, 2003). Rhizarian microfossils in 750 million year old rocks support the phylogenetic inference that forams began to diversify well before their appearance as fossils in Cambrian age rocks (Porter *et al.*, 2003).

Because they combine large population sizes, excellent preservation, well-understood environmental distributions, and rapid evolutionary turnover, Foraminifera have long been the focus of paleoenvironmental research, especially for Cenozoic oceans (e.g. Sen Gupta, 1999). These remarkable organisms occur in both planktonic and benthic settings, enabling them to document conditions throughout the water column in ancient oceans. Moreover, aspects of their test chemistry record ambient environments at the time and place of mineral precipitation. Over the last decade, an ever-increasing variety of compositional signatures have emerged that can be correlated with or calibrated to the formation environments of forams and other calcifiers (e.g. Table 10.1). In particular, a significant effort has been directed at using the tests of calcifying forams to quantify relationships between stable isotopes and/or trace elements and the paleochemistry of marine water masses. For example, O^{18} and Mg contents have been used to determine simultaneously the temperature and salinity of past oceans (Lear *et al.*, 2000; Lea, 2006). A number of field and laboratory studies show that the Mg/Ca ratio in foram tests can faithfully record paleotemperatures of the oceans (Nurnberg *et al.*, 1996; Rosenthal *et al.*, 1997; Lea *et al.*, 1999; Lear *et al.*, 2000, 2002; Toyofuku *et al.*, 2000; Anand *et al.*, 2003; Lea, 2006). As discussed later, however, there are significant caveats to making the recognized proxy models more robust and to testing the many new proxies that are being introduced in current literature.

10.2.1.1　Mineralization of calcareous tests

Extensive *in vivo* observations suggest that the mineralization strategies employed by foraminiferans should allow these organisms to serve as faithful indicators of the ambient conditions during test formation. In particular, some forams undergo mineralization by processes that, at least in part, (i) utilize classical nucleation and growth processes, and (ii) use captured seawater as starting material for the calcification process (Erez, 2003).

From biological studies, Erez and colleagues have developed a model for the stepwise mineralization of foram tests – a process called 'seawater vacuolization' (J. Erez, pers. comm.). As seen in Fig. 10.2 (a, b, c), mineralization begins when the organism extends a pseudopod to form a delimited or compartmentalized space that engulfs ambient seawater. From this packet of solution, calcification occurs on the compartment walls as a thin layer that eventually becomes a new test chamber while also growing onto the mineral surfaces of preexisting chambers (Erez, 2003).

At first glance, this simple vacuolization process would seem to be a straightforward mineralization model for eukaryotes. However, numerous species are known to produce tests that are depleted in Mg relative to levels expected for mineralization from seawater and there is significant variation amongst individuals within a single species (Segev and Erez, 2006). This variability suggests that biological factors must be active and provides evidence that mineralization cannot simply be a passive reflection of seawater chemistry at the time of entrapment. Clearly, there is an ongoing need to assess whether/when forams begin to modify the composition of the seawater packet after the compartment is formed. For example, the organism could modify: (i) local chemical conditions at the site of mineralization (local pH, supersaturation, rate of precipitation); (ii) the supply or removal of ions via channels and pumps; (iii) availability of organic macromolecules, including proteins, polysaccharides, or other organic matrix components; or (iv) mineralogy, texture and/or preferred orientation of crystals in the developing mineralized structure. The simple model in Fig. 10.3 (Erez, 2003) illustrates how biological processes can significantly modulate local chemistry. In particular, acidic vacuoles supply CO_2 to the symbionts found in many foram species and to the basic vacuoles contained within the organism, thereby influencing pH and supersaturation that, in turn, could affect trace element contents. Moreover, assuming that the system is a semi-closed environment, this results in a CO_2 competition that could also influence isotopic fractionation. Investigation of the *in vivo* environment is a critical research area that is poised to advance rapidly with new microelectrode techniques to measure pH and ion activities within these microcompartments. Coupled biological–chemical models such as Fig. 10.3 suggest that key variables such as pH and ion activities likely vary significantly over length scales of just a few microns or less. Given that microelectrode measurements were used to constrain this physical picture, it would seem this model merits further testing. This kind of knowledge will be essential to establish a robust understanding of mineralization processes *in vivo*.

Table 10.1 Partial list of tracers and proxies being used in paleoceanography based upon chemical signatures in carbonate biominerals

Tracer	Proxy	Reference
Mg/Ca	Temperature, Mg concentration	Nurnberg *et al.* (1996)
		Lea *et al.* (1999)
		Lear *et al.* (2000)
		Rosenthal *et al.* (2002)
		Dickson (2002)
		Dekens *et al.* (2002)
		Billups and Schrag (2002)
		Anand *et al.* (2003)
		Eggins *et al.* (2003)
		Russell *et al.* (2004)
		Ries (2004)
		Cronin *et al.* (2005)
		Rathmann & Kuhnert (2007)
		Borremans *et al.* (2009)
		Healey *et al.* (2008)
Sr/Ca	Temperature	Chivas *et al.* (1986)
		Beck *et al.* (1992)
		Stoll *et al.* (2002)
		Wanamaker *et al.* (2008)
		Borremans *et al.* (2009)
Cd/Ca	Phosphate, circulation	Boyle *et al.* (1995)
		Ripperger *et al.* (2008)
Ba/Ca	Alkalinity, circulation, meltwater discharge	Lea and Spero (1994)
		Hall and Chan (2004)
Zn/Ca	CO_3^{2-}	Marchitto *et al.* (2000)
		Henderson (2002)
U/Ca	Redox potential, productivity, CO_3^{2-}	Becker *et al.* (2001)
		Chase *et al.* (2001)
		Shen and Dunbar (1995)
		Russell *et al.* (2004)
V/Ca	Redox potential	Hastings *et al.* (1996)
		Tribovillard *et al.* (2006)
B/Ca	pH	Yu *et al.* (2007)
Mg/Li	Temperature	Bryan and Marchitto (2008)
$\delta^{11}B$	pH	Zeebe (2005)
$\delta^{13}C$	Productivity, circulation	Boyle and Keigwin (1985)
		Klein *et al.* (1996)
		Bemis *et al.* (2000)
		Basack *et al.* (2009)
		Henderson (2002)
$\delta^{18}O$	Isotopic composition, temperature	Jones and Quitmyer (1989)
		Spero and Lea (1996)
		Lear *et al.* (2000)
		Elderfield and Ganssen (2000)
		Billups and Schrag (2002)
		Pearson *et al.* (2001)
		Ortiz *et al.* (1996)
		Beltran *et al.* (2007)
		Rollion-Bard *et al.* (2008)
		Puceat *et al.* (2007)
$\delta^{44}Ca$	Calcium concentration, temperature, possibly biomineralization pathway	Sime *et al.* (2007)
		Griffith *et al.* (2008)
		Kozdon *et al.* (2009)
$\delta^{24,25,26}Mg$	Paleotemperature	Chang *et al.* (2004)
$\delta^{88}Sr$	Salinity, chemical weathering rates, etc.	Veizer (1989)
		Ruggeberg *et al.* (2008)

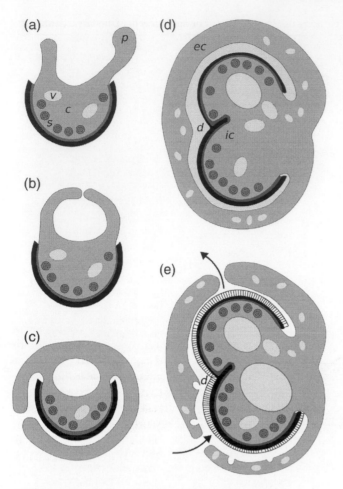

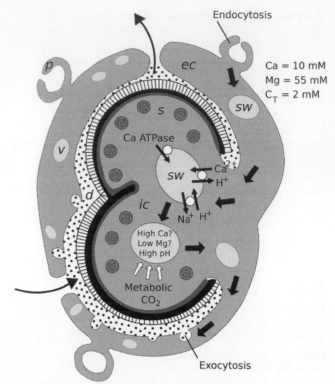

Figure 10.2 Stepwise model for calcification by perforate foraminifera. (a) The process begins when pseudopods extend into the surrounding seawater, creating (b) an enclosed compartment of seawater for mineralization. (c, d) Portions of the calcified test can also act as templates for mineral deposition through a similar process whereby pseudopods form seawater filled compartments around preexisting mineral chambers. Evidence suggests mineralization within these compartments proceeds by accumulation of an amorphous $CaCO_3$ phase that subsequently transforms to calcite; (e) seawater flux through a compartment creates chemical conditions to thicken overgrowths as a secondary layer, likely by classical mineralization process (with permission from Erez, 2003). The symbols are: v - vacuole, p - pseudopodia, s - symbiotic space, d - delimited biomineralization space, c - cytoplasm, ic - intralocullar cytoplasm, ec - extralocullar cytoplasm.

10.2.1.2 Decoupling vital effects from true environmental signatures

New studies of foram mineralization processes have rejuvenated interest in the long-recognized problem of 'vital effects' and their influence on trace element and isotopic signatures. The concept of the vital effect was first introduced by Urey *et al.* (1951) to explain why oxygen isotopes measured in echinoderm shells depart from temperature curves calibrated from inorganically precip-

Figure 10.3 Simplified model of ion fluxes between the mineralization environment and seawater in perforate foraminifera (with permission from Erez, 2003). Initials are the same as described in Figure 10.2 except sw - seawater, and C_T - total inorganic carbon. Future studies will likely show how the organic molecules within the deposition environment impose another layer onto this construct to influence the formation of specific $CaCO_3$ polymorphs and their compositions.

itated carbonates. Urey and colleagues described these deviations as 'physiological offsets' that likely had both 'kinetic' and 'taxonomic' origins (McCrea, 1950; Epstein and Lowenstam, 1954; Weber and Raup, 1966). Despite the fact that these offsets were found in many subsequent studies of isotopic and elemental signatures, the difficulty of interpreting this complicating factor has been largely 'swept under the rug.' By necessity, researchers have selected species that were the most faithful recorders of the environments in which they formed or, worse, proposed new proxies that simply ignore the potential for significant biological overprinting (Weiner and Dove, 2003). Over the last few years, efforts have become refocused toward understanding how mineralization processes occur and how the inherent biochemical/biological bias in measured chemical signatures originates. This holds the long-term promise of converting vital effects from nuisances to sharper tools in paleoenvironmental reconstruction (Weiner and Dove, 2003). Here we highlight examples from recent studies of foram mineralization that are leading the way toward sorting out the physiological processes that give rise to vital effects.

As noted above, Foraminifera display a large variability in the Mg content of their tests. Species can contain from near-zero to almost 20 mol % $MgCO_3$, with significant deviations from Mg/Ca ratios expected for inorganically formed calcite (Bentov and Erez, 2006; Segev and Erez, 2006). These variations are reported for different species from the same environment as well as for individual shells that are themselves highly heterogeneous. Until recently, only a few studies (Delaney *et al.*, 1985) had unambiguously determined if this variability could be attributed solely to species-specific differences or if it was a true an indicator of the ambient chemical and thermal conditions. Recent culturing studies (Russell *et al.*, 2004; Hintz *et al.*, 2006) have been particularly useful in assessing this kind of question for Mg as well as other elemental and isotopic signals. Similar questions extend to whether latitudinal differences could also influence Mg/Ca dependencies for paleotemperature reconstructions. For example, von Langen *et al.* (2005) showed that temperate and polar foraminera have similar Mg/Ca temperature calibration equations but with a considerable offset in the pre-exponential constant. Similarly, parallel ship-board studies are critical to testing and validating assumptions that underlie reconstruction models of sea surface temperatures to constrain models for climate periodicities (Lear *et al.*, 2002; Medina-Elizalde and Lea, 2005; Lea *et al.*, 2005; Waldeab *et al.*, 2007; Medina-Elizalde *et al.*, 2008) and habitat migration (Eggins *et al.*, 2003).

With so many factors at play, a concerted effort to discern first and second order controls must include novel approaches to studying foram mineralization. At this writing, new *in vivo* studies are revealing mechanism-based explanations for Mg signatures to be complex. Even for the relatively simple forams, multiple mineralization processes may be involved. For example, Bentov and Erez (2005) showed that tests of the perforate foram *Amphistegina lobifera* are a composite of two types of calcite: a thin high-Mg primary layer (1–10 μm) and thicker low-Mg layers (3–35 μm) that cover both sides of the primary layer. The high Mg layer (up to 20 mol % $MgCO_3$) is associated with an organic matrix, consistent with the idea that this type of calcite is formed via an amorphous intermediate (Raz *et al.*, 2000). The idea that forams may use multiple pathways to calcite crystallization in such close temporal and physical proximity raises the question of whether there is a general relationship between high and low Mg-calcites and nonclassical versus classical mineralization processes, respectively. Different processes could explain aspects of the reported heterogeneity in foram shells, their different temperature dependencies, and possibly aspects of the vital effect.

Bentov and Erez (2006) have assembled what may be the most comprehensive roadmap available to uncover the factors that regulate foram mineralization and the

Table 10.2 Summary of possible mechanisms for controlling Mg:Ca ratio in foram tests (see Bentov and Erez, 2006)

Cellular controls on Mg composition by regulating chemistry of the parent solution
- *Transport systems that allow passive Mg diffusion*
- *Cellular buffering by ATP binding*
- *Sequestration within cellular organelles, mitochondria*

Possible influence of the organic matrix on the mineralization process
- *Polymorph selection*
- *Incorporation of impurities and trace elements*
- *Selective development of crystal facets*

Skeleton growth by classical and nonclassical crystallization pathways
- *Deposition of amorphous transient phase and transformation process favors high Mg*
- *Ion by ion growth may limit maximum Mg uptake*
- *Proportions of these processes may affect intrashell variability*

origins of vital effects. Recognizing the significance of their finding that forams employ the multiple mineralization processes discussed above, Bentov and Erez synthesized available data to emphasize influences on Mg signatures that merit further investigation (Table 10.2). Electron microprobe analyses of bulk tests from *Amphistegina lobifera* and *A. lessonii* support the idea that a physiological mechanism must exist to exclude or reduce the activity of Mg at the site of calcification or within the compartment that is created by their pseudopods (Segev and Erez, 2006). Support for this physiological influence is found in measurements showing that biogenic calcites do not incorporate Mg in similar proportion to inorganically grown calcite crystals (Wasylenki *et al.*, 2005). Perhaps these foram species optimize skeletal growth at the lower Mg/Ca values characteristic of seawater for much of the Cenozoic Era (Hardie, 1996) and have evolved pathways to remove Mg from the engulfed packet of solution. Possible processes include passive membrane transport systems for Mg diffusion (Preston, 1998), chemical buffering with ATP binding reactions (Romani and Maguire, 2002), and sequestration by nearby organelles (Bentov and Erez, 2006). A second mode of controlling Mg levels in foram shells is through the activity of an organic matrix. Macromolecules of polysaccharides, carboxyl-rich peptides and proteins are known to influence carbonate polymorph selection (Wada *et al.*, 1999; Levi *et al.*, 1998; Kotachi *et al.*, 2006; Xu *et al.*, 2008), mineralization processes (Raz *et al.*, 2000) and Mg partitioning into crystallographically unique faces (Lowenstam and Weiner, 1989; Davis *et al.*, 2004; Wasylenki *et al.*, 2005; Stephenson *et al.*, 2008). Again, the highest Mg calcites (before diagenesis) are intimately associated with the organic matrix, particularly the first layer of a new chamber which may begin as an transient amorphous phase (Fig. 10.2d). Indeed, a recent study (Wang *et al.*, 2009) demonstrates that poly-carboxylic acids can dramatically increase the Mg content of amorphous calcium carbonates. Retention of Mg during recrystallization of these disordered phases may provide a plausible mechanism for high-Mg biogenic carbonates (Cheng *et al.*, 2007).

Mention of an amorphous precursor brings us to the third aspect of understanding how organisms influence elemental and isotopic signatures. While biological mineralization through a transient intermediate is considered in greater detail later (see discussion of echinoderms and molluscs), it is important to note that this alternative pathway has implications for understanding chemical signatures. After all, if the bulk shell composition indeed reflects multiple crystallization processes, then it follows that interpretations of composition must make assumptions about the signals contained in the carbonate produced by each process. Table 10.2 shows how variations in the relative quantities of each type of calcite produced for an individual test chamber can account for some of the reported intrashell variability. To answer these questions, we point again to the need to understand underlying chemical processes. Finally, the less glamorous but nonetheless important roles of *in vivo* pH and the coupled aqueous carbonate chemistry are critical to determining the chemical driving force for mineralization within the organism. While pH control on mineralization is known, some evidence also suggests Mg levels are influenced by pH (Russell *et al.*, 2004). At the time of writing, it is clear that we have only begun to appreciate the influence of biomolecules on the formation and chemical properties of the inorganic components of foram shells. The new *in vivo* evidence, however, suggests that we are entering an exciting time of rapid advancement.

10.2.1.3 Mineralization from seawater: Insights for elemental signatures in corals?

Space limitations prohibit us from developing a discussion of coral mineralization, but we would be remiss not to mention these important organisms. In some parts of the world, corals have a significant impact on carbon chemistry. Approximately 40% of net oceanic $CaCO_3$ precipitation occurs in tropical areas along coastal zones where coral reefs predominate (Gattuso *et al.*, 1998; Milliman, 1993). Corals probably account for about half of this amount, but in recent years, ocean acidification has been severely decreasing their productivity (Bellwood *et al.*, 2004; Milliman and Droxler, 1996). Unlike planktonic organisms whose carbonate products are largely recycled on short time scales, the corals sequester carbonate for long periods. Corals are also used to reconstruct chemical conditions in the ocean because their calcification process is based upon

capturing seawater and bringing it to the site of mineralization (Silverman *et al.*, 2007a, b).

Corals and forams share a number of common features in calcification. Analogous to the pseudopods of forams, corals are believed to utilize a tissue pumping mechanism to capture seawater and create a locally supersaturated area while possibly modifying the compositions of intracellular fluids. In particular, recent evidence suggests that, like forams, corals shift the pH of captured seawater to values above 8 (Schneider and Erez, 2006; Al-Horani *et al.*, 2003; Cohen and McConnaughey, 2003). By creating a local calcification environment analogous to the semi-closed system utilized by forams, it is possible that corals also influence chemical signatures of the resulting minerals (Raz-Bahat *et al.*, 2006; Cohen and McConnaughey, 2003). Finally, corals produce biomineral structures with different levels of trace elements suggesting that, again, multiple crystallization pathways may be involved. Meibom *et al.* (2006, 2008) have shown that isotopic and trace element (including Mg and Sr) abundances differ between the centre of calcification (the earliest aragonite, precipitated on an organic template) and an enveloping coat of fibrous aragonite. As in the case of forams, this suggests that bulk chemical analyses may record a mixing line between two distinct mineralization processes. And, again, as in forams, the intertwined roles of symbionts are part of the picture, at least for reef-forming hermatypic corals. *In vivo* approaches that include tracer and microelectrode methods are certain to shift the emphasis of coral research from a descriptive to a mechanism-based understanding.

10.2.2 The radiolarians: limitations of macroscopic approaches to biomineralization

Although forams precipitate calcite, the Rhizaria are more broadly characterized by silica precipitation, as tests or scales (see below). Ecologically, paleontologically and bigeochemically, the most prominent silica-precipitating rhizarians are the Radiolaria. Traditionally, radiolarians were divided into three groups: the Polycystina, the Phaeodaria and the strontianite-precipitating Acantharia. Molecular data now indicate that the polycystines (which make up nearly all of the radiolarian fossil record) and acantharians are closely related, but that the phaeodarians represent an independent origin of siliceous tests within the superkingdom (Cavalier-Smith and Chao, 2003; Danelian and Moreira, 2004; Kunitomo *et al.*, 2006). Polycystine radiolaria (De Wever *et al.*, 2001) live as unicells or colonies and are characterized by a well-developed internal skeleton, complex cytoplasm-encapsulated microtubule arrays (axopodia), and a capsular membrane dividing the endoplasmic and ectoplasmic regions of the cell (Fig. 10.4a). Polycystine radiolarians that live in

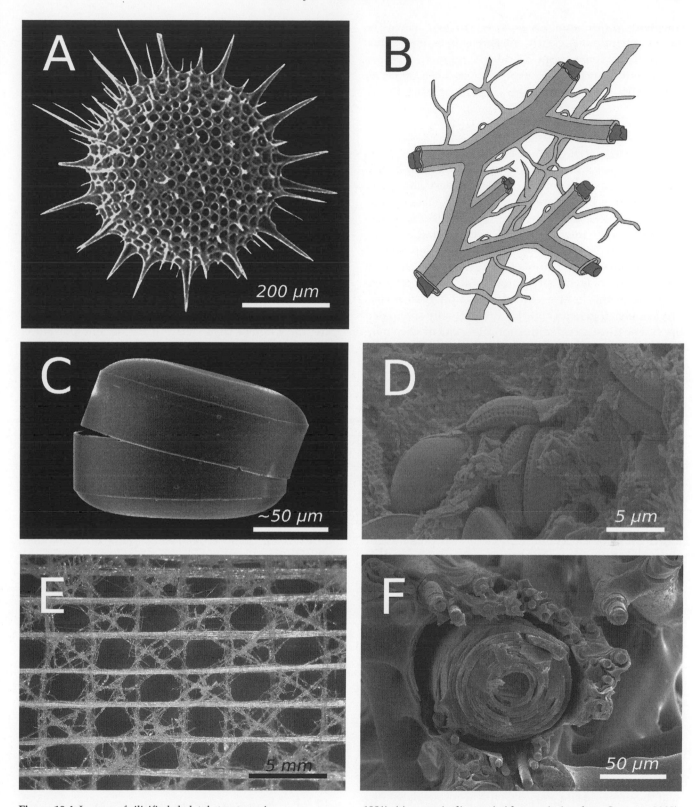

Figure 10.4 Images of silicified skeletal structures in radiolarians, diatoms, and sponges. (a) radiolarian (with permission from Perry, 2003); (b) cartoon of the relationship between cytoplasmic membranes (light grey) and skeletal elements (dark grey) in spongiose radiolaria (after Anderson, 1981); (c) a centric diatom (with permission from Sumper, 2002); (d) pennate diatoms (courtesy of J.D. Schiffbauer); (e) the glass sponge *Euplectella* sp.; and (f) a cross-section through one of its spicules showing the presence of several lamellar structures (e and f reproduced with permission from Weaver *et al.*, 2007).

the photic zone commonly possess algal symbionts. The earliest microfossils clearly interpretable as polycystine radiolarians come from Middle Cambrian rocks in Queensland, Australia (Won and Below, 1999; De Wever *et al.*, 2001).

10.2.2.1 Test formation

As one of the earliest silica-mineralizing protists with representatives still living in oceans, radiolarians are a natural choice of model organism for investigating processes of silica biomineralization. Indeed, silica deposition in radiolarians has been extensively studied with the aid of light microscopy for over a hundred years (Haeckel, 1887). However, early investigations were quite limited by the technology of the age, and although some cellular level structures and processes were observable, the nature of processes and mechanisms driving skeletogenesis could only be broadly speculated upon. The most detailed theoretical model of silica mineralization to emerge from this era was asserted by Thompson (1942), who argued that the close-packing of alveoli (gas filled vesicles) within the ectoplasm created a network of interconnected void spaces, whose surfaces promoted silica deposition by locally reducing the thermodynamic barrier to nucleation. Initially, the elegance and simplicity of this model appealed to a wide scientific base – so much so that elements of it have since been applied to a number of biomineralization systems, and bio-inspired materials syntheses (Oliver *et al.*, 1995; Zhu *et al.*, 1998; Volkmer *et al.*, 2003; Shi *et al.*, 2006; Wang *et al.*, 2007). One of the most recent and high-profile incarnations of the Thompson model was presented by Sumper (2002), who sought to describe the mechanism of polyamine mediated frustule formation in centric diatoms. Despite its popularity, Thompson's description of radiolarian skeletogenesis is invalidated by a number of experimental observations, as described by Anderson (1983):

> Not all species possess closely packed alveoli yet exhibit rather complex skeletons. Some skeletons, moreover, possess long spicules, sometimes elegantly twisted or ornamented with little *resemblance* to interfaces among close-packed spherical or spheroidal symplastic surfaces.

Advances in light and electron microscopy also revealed that the growing siliceous skeleton is encapsulated by a specialized cytoplasmic membrane, or cytokalymma (Fig. 10.4b), which expands to match the rate of silica growth within (Anderson, 1976, 1980, 1981). Moreover, the interstitial spaces between alveoli, when present, do not act as templates for silica nucleation. The cytokalymma is involved in controlling silica formation within

its interior (the cisterna), and presumably does so by regulating the intravesicle concentration of dissolved Si species and organic macromolecules, hydrogen ion activity, and other as yet undetermined environmental parameters.

Unfortunately, little is known about the chemical composition of the cytokalymma, its internal chemistry, or the nature of its macromolecular contents. Assays of the amino acid content of radiolarian skeletons taken from recent sediment cores reported the following relative abundances: glycine > aspartic acid > glutamic acid > alanine > valine > serine > leucine > threonine (King, 1974, 1975, 1977). This sequence differs considerably from comparable work on diatom frustules which found high abundances of hydroxylated amino acids (e.g. serine and threonine) and lesser amounts of exotic derivatives of proline and hydroxylysine (Hecky *et al.*, 1973; Volcani, 1981; Swift and Wheeler, 1990, 1992). The significance of these differences is debatable, as it is unclear whether characterizing the organic portion of siliceous tests in this way is meaningful, since the bulk of silica-associated organic material may not belong to molecular species that are involved in the mineralization process.

Significant efforts made in recent years to identify and characterize molecules directly involved in the mineralization of diatoms and siliceous sponges are paying off with new insights into the mechanisms of biologically-controlled silicification. Comparable advances in our understanding of radiolarian skeletogenesis at the molecular level have been hindered by difficulties in culturing these organisms in the laboratory (Kouduka *et al.*, 2006). Major advances towards understanding silica mineralization in radiolarians may be obtained through targeted research aimed at extracting and characterizing those macromolecular species from the skeletons of recently living specimens that are involved in silicification. Further work is also required to understand forefront questions related to silicon uptake, metabolism and intracellular transport in these organisms. Some of the first steps are currently being taken in this area as a certain fluorescent molecule was found to be an effective tracer of silicate species *in vivo* (Shimizu *et al.*, 2001) enabling for the first time accurate spatially and temporally resolved information about silicate transport and deposition in polycystine radiolaria (Ogane *et al.*, 2010).

10.2.3 *The diatoms: non-enzymatic routes to biosilica formation*

Diatoms are unicellular algae that belong to the superkingdom Chromalveolata. Among the most diverse of all known protistan clades, the diatoms are ecologically important in both marine and freshwater ecosystems, as

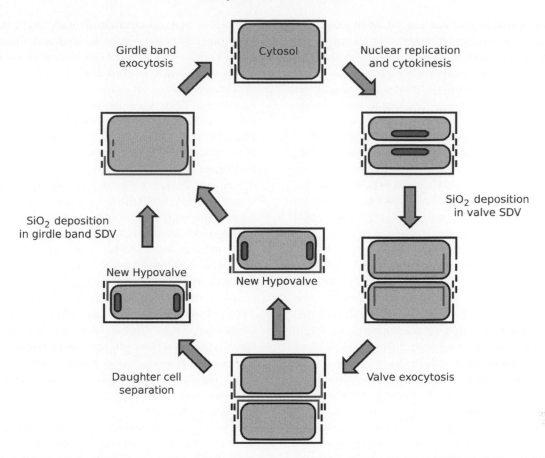

Figure 10.5 Overview of the diatom's asexual, vegetative life cycle. Silica mineralization occurs within specialized silica deposition vesicles (SDV). Valve formation proceeds rapidly in parallel with cell division. Mineralization of girdle bands is not strictly coupled with cell division, and may occur at different rates, depending on the species, and environmental conditions. Figure after Kröger and Sumper, 2000.

well as in some soil and aerial habitats (Werner, 1977; Volcani, 1981; Round, 1981). The silicified portion of the cell is termed the 'frustule,' and is composed of two valves, the epitheca and hypotheca, which may be partially obscured by a set of siliceous structures called girdle bands. Diatom species have traditionally been divided into two broad morphological subgroups, the radially symmetric *centrics* (Fig. 10.4c) and bilaterally symmetric *pennates* (Fig. 10.4d); molecular phylogeny makes it clear that pennate diatoms form a monophyletic group nested within the paraphyletic centrics (Kooistra *et al.*, 2007). Many pennate varieties also have a prominent, but discontinuous slit, or raphe, that spans the entire length of the valve, and again, raphide pennates constitute a monophyletic group nested within paraphyletic non-raphide pennates (Kooistra *et al.*, 2007; Volcani, 1981; Crawford, 1981). Since their first appearance in the fossil record (~140 Ma), diatoms have assumed important regulatory roles over the global geochemical cycles of both silicon and carbon (Nelson *et al.*, 1995; Smetacek, 1999; Ragueneau *et al.*, 2000; Van

Cappellen, 2003). The rapid diversification of the diatoms following the K/T extinction event (~65 Ma) elevated these algae to their current position as the most prolific silicifying organisms in modern seas, where they sequester gigatonnes of biosilica annually, and as a result, suppress dissolved silica levels in the surface ocean to the current low levels (Racki and Cordey, 2000).

10.2.3.1 The cell cycle

The normal means of reproduction is asexual cell division (mitosis), in which two daughter cells are produced from a single parent cell (Fig. 10.5). Because the cell wall is silicified and rigid, formation of the daughter cells necessarily occurs within the parent frustule (Scala and Bowler, 2001; Zurzolo and Bowler, 2001). During the first stage of cell division, the hypotheca and epitheca begin to separate, allowing the cytoplasm to expand. This movement is facilitated by the formation of a second set of girdle bands attached to the hypotheca. As the cytoplasm expands, the nucleus moves into position alongside the girdle bands. For some diatom species, the nucleus is usually positioned near the girdle bands during interphase, and therefore no initial migration is necessary; however, the nucleus may also reside at the centre of cell, or against the inside wall of the epitheca, and must migrate before cell division can proceed (Volcani, 1981). Nuclear motion is assisted by the presence of strong, hollow, protein filaments called microtubules (MT), and associated microtubule centres (MC), which form an extensive cytoskeletal network during the premiotic stage (Pickett-Heaps, 1991). The MT/MC cytoskeleton also plays a probable, but as yet undetermined role in valve morphogenesis (Volcani, 1981; Pickett-Heaps, 1991). Once the nucleus is in place, it replicates and divides. The details of nuclear replication in diatoms are reviewed by Pickett-Heaps (1991). As mitosis completes, the cytoplasm begins to undergo cytokinesis or division. During cytokinesis the cell membrane (plasmalemma) develops a small invagination near the girdle bands that continues to grow until the two identical nuclei produced during mitosis reside in separate plasmalemmas within the original frustule.

Following cytokinesis, each daughter cell possesses only one valve, which originally belonged to the parent cell. In the newly formed cells, the inherited valve assumes the role of the epitheca (Zurzolo and Bowler, 2001), and a new hypotheca is made from scratch. Synthesis of the new hypovalve occurs within a specialized compartment termed the silica deposition vesicle (SDV), which appears in the daughter cell shortly after cytokinesis terminates. The origin of the SDV is unknown, but its initial appearance in the cell is spatially correlated with the position of the nucleus, and a nucleus-associated microtubule centre (Pickett-Heaps, 1991; Zurzolo and Bowler, 2001). Silicon, in an as yet undetermined form, is transported by an unknown mechanism, across the SDV plasma membrane (the silicalemma), where it interacts with the organic contents of SDV. The silicalemma must progressively expand during silica deposition in order to accommodate the developing hypovalve inside the SDV. When the new hypotheca is complete, it is coated with a thin, but robust carbohydrate layer that helps protect against dissolution. The silicalemma then merges with the plasmalemma, and the valve is excised from the cell. When both daughter cells have synthesized and excised new hypovalves, they separate from one another, and the process of cell division begins anew in each of the fledgling cells.

10.2.3.2   Silicon uptake

The surface ocean is slightly alkaline (pH = 7.5–8.5) and typically contains 1–100 micromoles of Si per liter (with surface waters where diatoms thrive skewed toward the low end of that range) (Tréguer *et al.*, 1995). Under these conditions Si is present as neutral and ionized monomeric species (Iler, 1979). $H_4SiO_4°$ comprises about 97% of the total soluble Si pool, and is the form of dissolved Si utilized by most marine species (Del Amo and Brzezinski, 1999; Wischmeyer *et al.*, 2003); There is some evidence, however, suggesting that certain species (e.g. *Phaeodactylum tricornutum*) may be able to use both $H_4SiO_4°$ and $H_3SiO_4^-$ (Riedel and Nelson, 1985; Del Amo and Brzezinski, 1999).

Active transport of silicon across the diatom cell membrane is coupled 1:1 with the transport of sodium ions in marine species (Bhattacharyya and Volcani, 1980), and is strongly dependent upon the concentration of silicic acid in the extracellular environment. As $H_4SiO_4^0$ increases, the uptake rate asymptotically approaches a maximum value, which indicates that the transport process is carrier mediated (Paasche, 1973a, b; Azam *et al.*, 1974; Martin-Jézéquel *et al.*, 2000; Hildebrand and Wetherbee, 2003). This behaviour has been described by many authors within the context of hyperbolic saturation models (e.g. Mechaelis–Menten type enzyme kinetics) (Sullivan, 1976; Conway and Harrison, 1977; Conway *et al.*, 1976; Riedel and Nelson, 1985; Del Amo and Brzezinski, 1999).

A family of five silicon transporter (SIT) genes has been isolated from *Cylindrotheca fusiformis* (Hildebrand *et al.*, 1997, 1998). The first of these genes to be discovered (*SIT1*) encodes a protein composed of 548 amino acids that contains as many as 10 hydrophobic transmembrane sequences, a long hydrophilic carboxyterminus, and a signature amino acid sequence of sodium symporters (Hildebrand *et al.*, 1997; Hildebrand, 2000). Hildebrand *et al.* (1997) demonstrated that silicic acid transport protein encoded by *SIT1* has characteristics consistent with the results of whole cell uptake experiments. The amino acid sequences of the hydrophobic regions are highly conserved among the five proteins (~90%); however, the carboxy-terminal regions are far more variable (~45–57%), which prompts speculation that this portion of the protein interacts with other proteins and may be involved in the regulation and localization of these proteins within the cell (Hildebrand, 2000; Hildebrand *et al.*, 1997, 1998). More recent investigations show that silicon transport is highly regulated at the cellular level, since the temporal expression of individual SIT genes is correlated with the various phases of the diatom cell cycle (Thamatrakoln and Hildebrand, 2007). A comparison of SIT proteins from a number of centric and pennate diatom species revealed the presence of two highly conserved amino acid sequences (glutamate–glycine–X–glutamine and glycine–arginine–glutamine) that may participate in the reversible binding of silicic acid during transport (Thamatrakoln *et al.*, 2006).

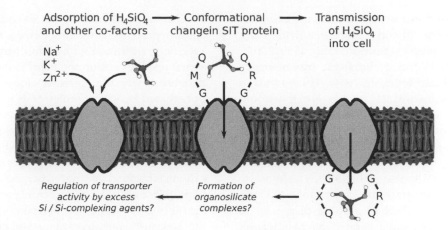

Figure 10.6 The silicon transport pathway as proposed by Thamatrakoln *et al.*, 2006. Adsorption of silicic acid and enzyme co-factors induces a conformational shift in the SIT protein resulting in the influx of H_4SiO_4 into the cell. In the cytoplasm, high levels of dissolved silicon may be stabilized through complexation with organic silicic acid binding agents. The activity of the Si uptake system may be regulated by the abundance of excess silicic acid complexing agents, which may bind to the SIT protein and induce a conformational change that blocks Si uptake. Q, M, and G represent the amino acids glutamine, methionine, and glycine respectively. X is an additional amino acid whose identity is not conserved within the SIT gene family. Modeled after Thamatrakoln *et al.*, 2006.

The regulatory mechanism underlying the activity of silicon transport proteins remains somewhat speculative. The main assumption in the literature is that the binding of Na$^+$ (or K$^+$) and $H_4SiO_4^0$ to the SIT surface causes a conformational change in the protein (Thamatrakoln *et al.*, 2006), allowing both species to enter the cell (Fig. 10.6). However, silicic acid uptake is also inhibited in zinc-deficient environments (De La Rocha *et al.*, 2000), which suggests that the activity of silicon transport proteins may be at least partially dependent upon the concentration of Zn^{2+} ions. The recent characterization of silicon transport proteins in the centric marine species *Chaetoceros muelleri* has revealed a highly conserved amino acid motif (cysteine–methionine–leucine–aspartate; abbreviated CMLD), which may act as a binding site for Zn^{2+} ions; experiments with model peptides have demonstrated that the CMLD motif has a high affinity for zinc (Grachev *et al.*, 2005; Sherbakova *et al.*, 2005).

10.2.3.3 Intracellular transport and storage of silicon

Intracellular pools of soluble silicon were first identified by Werner (1966). Subsequent studies showed that under certain conditions, the soluble pool may account for as much as 50% of the total cellular silicon content (Azam *et al.*, 1974; Chisholm *et al.*, 1978; Sullivan, 1979; Binder and Chisholm, 1980; Taylor, 1985; Claquin and Martin-Jézéquel, 2005). Recent measurements, which assume that Si is distributed evenly throughout the cell as monosilicic acid, indicate that for a range of species, the internal pool may be 19–340 mM $H_4SiO_4^0$,

(Hildebrand, 2000). The size of the Si pool is not constant, and varies systematically over the course of the cell cycle (Hildebrand and Wetherbee, 2003). Although precise determination of the intracellular silicon concentration is difficult, all studies indicate that diatoms are capable of maintaining sizeable internal Si pools (Chisholm *et al.*, 1978; Sullivan, 1979; Binder and Chisholm, 1980; Hildebrand, 2000), which exceed the solubility of amorphous silica (Iler, 1979; Harrison and Loton, 1995).

In the cell, uncomplexed polysilicic acids are unlikely to exist for any appreciable length of time because they condense rapidly to form colloidal silica. Silicon is probably not stored in particulate form either, because large abundances of colloidal material could potentially compromise the integrity of cellular membranes (Hildebrand, 2000). Therefore, intracellular silicon most likely exists as monomeric or low molecular weight oligomeric species, which are complexed, or reversibly sequestered by organic binding components (Werner, 1966; Azam *et al.*, 1974; Schmid and Schulz, 1979; Sullivan, 1986). Bhattacharyya and Volcani (1983) isolated an unusual class of ionophores (specific for both sodium and silicon) from the lipid membranes of *Nitzschia alba* and suggested that intracellular transport of silicon could occur by ionophore-mediated diffusion. This hypothesis is consistent with the nearly ubiquitous existence of silicon in the cell; however, these species have not yet been characterized, and the precise role that ionophores might play in maintaining intracellular silicon pools remains speculative (Hildebrand, 2000).

The potential of Si to form complexes with other intracellular organics is largely unexplored; however, several ^{29}Si NMR studies now document the existence of stable hypervalent silicon complexes with open chain sugar-acids (polyols) (Kinrade *et al.*, 1999, 2001a, b; Benner *et al.*, 2003) and cyclic sugar monomers in aqueous solution (Lambert *et al.*, 2004). These five and six-coordinated organosilicate anions are the dominant species in basic solutions, and silicon concentrations of up to 3.0 mol l^{-1} have been obtained without gelling, even in the presence of alkali metal cations (Kinrade *et al.*, 1999). Kinrade *et al.* (2001b) showed that these complexes are also stable at physiologically realistic silicon concentrations (1.4 mmol l^{-1}), and account for as much as 30% of soluble Si species at neutral pH. These data have been used to interpret recent spectroscopic information, which suggests that hypervalent Si complexes may exist as a transient species during cell wall synthesis in *Navicula pelliculosa* (Kinrade *et al.*, 2002).

10.2.3.4 Apparent roles of biomolecules in the mineralization of diatom cell walls

Studies of diatom cell division (Pickett-Heaps, 1991) show that silica deposition occurs within specialized silica deposition vesicles. However, because the appearance of the SDV is coupled with silica deposition, very little is known about the composition, constituent molecules, or internal chemistry of these vesicles. Vrieling *et al.* (1999) used a pH sensitive fluorescent dye to show that the SDV has an internal pH near 5. Coombs and Volcani (1968) showed that protein synthesis increased markedly during silica deposition. Subsequent investigations suggested that the proteinaceous component of diatom cell walls was enriched in glycine, and hydroxy-functional amino acid residues (Nakajima and Volcani, 1969, 1970; Hecky *et al.*, 1973; Swift and Wheeler, 1990, 1992). Nakajima and Volcani (1969, 1970) also identified three unusual amino acid residues (ε-*N,N,N*-trimethyl-δ-hydroxylysine, its phosphorylated derivative, and 3,4-dihydroxyproline) that were not detected by subsequent investigations (Hecky *et al.*, 1973; Swift and Wheeler, 1990, 1992). Unfortunately, these studies were unable to isolate the cell wall-associated proteins, and functional studies could not be performed. Based on these data, Hecky *et al.* (1973) proposed that hydroxyl-rich proteins in the SDV might form a template upon which orthosilicic acid molecules could condense.

More recent attempts to characterize cell wall-associated macromolecules have identified three novel protein families (frustulins, pleuralins, and silaffins/silacidins) and long chain polyamines (LCPA) as major cell wall constituents (Kröger *et al.*, 1994;, 1997, 1999, 2000, 2001, 2002; Kröger and Sumper, 1998; Wenzl *et al.*, 2008). The silaffins, silicidins, and LCPA influence silica formation *in vitro*, and therefore have probable, but as yet incompletely understood roles in silica deposition. Silaffin proteins are characterized by the presence of specific post-translational modifications to lysine and serine residues within the peptide backbone. The modified lysine residues possess elongated alkane chains, punctuated by amine moieties in varying states of protonation, while serines are typically phosphorylated, and negatively charged under physiological conditions (Fig. 10.7a). The silacidins and LCPA express chemical motifs that are similar to the post-translational modifications observed in the structure of silaffin proteins (Fig. 10.7b,c). Silacidins are acidic proteins, which resemble the highly phosphorylated silaffin peptide backbone, but lack polyaminated lysine derivatives (Wenzl *et al.*, 2008). Conversely, LCPA are nearly identical to the polyamine-modified lysine side-chains in silaffins but they are not associated with a peptide backbone (Kröger and Sumper, 2000).

Native silaffin-1A and 1B promote the condensation of silicic acid and aid in the flocculation of silica particles *in vitro*. LCPA exhibit similar behaviour, but only in the presence of phosphate, and other polyvalent anions like sulfate, citrate, and silacidins (Kröger *et al.*, 2000; Kröger and Sumper, 2000; Wenzl *et al.*, 2008). The final size of the colloidal particles is related to the precise mixture of silaffins and LCPA used, as well as the solution pH, and the molecular weight of the polyamine fraction (Kröger *et al.*, 1999, 2000, 2001, 2002). Native silaffin-2 has no intrinsic ability to nucleate silica, but does appear to serve in some sort of regulatory capacity, since it specifically inhibits the activity of natSil-1A, and promotes LCPA dependent silica formation (Kröger and Sumper, 2000).

In vivo, the numerous phosphate and amine functionalities on silaffins/silacidins and LCPA are likely to exhibit opposite charges. Kröger *et al.* (2002) showed that electrostatic interactions between these groups causes silaffin and polyamine mixtures to self-assemble into a relatively dense 'silaffin matrix,' which may catalyse the condensation of silicic acid and serve as a template for silica nucleation. Although most species probably contain a mixture of species-specific silaffins/silacidins and polyamines, some lack silaffins altogether, and long chain polyamines constitute the majority of biosilica-associated organic material. Sumper (2002) proposed a model for LCPA-mediated cell wall formation for species in the genus *Coscinodiscus*, in which he argued that the contact points between hexagonally close packed micro-emulsions of LCPA might serve as nucleation sites for silica. Further, co-precipitation of LCPA with the silica phase would result in an overall decrease in the size of the emulsions with time, ultimately leading to a series of honeycomb-like silica layers with diminishing pore size. This model successfully reproduces the highly symmetric patterns exhibited by

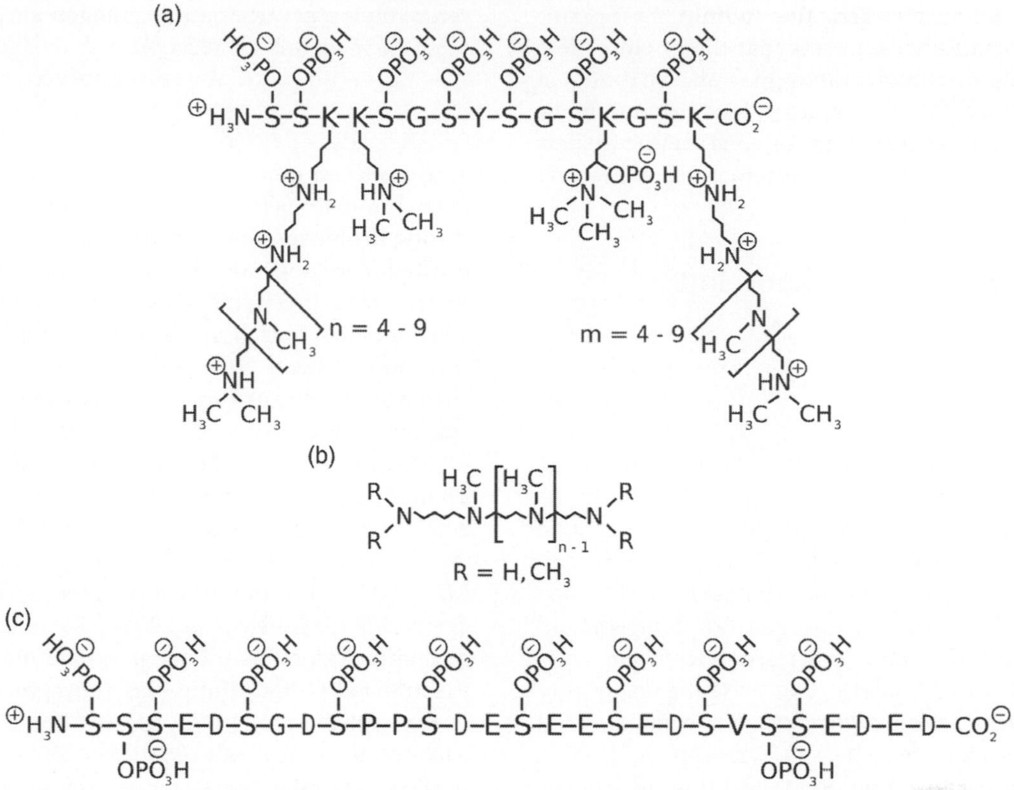

Figure 10.7 Structural representations of macromolecular species implicated in the mineralization of siliceous diatom cell walls (frustules). (a) Silaffin 1A$_1$ (after Kröger *et al.*, 1999, 2001, 2002); (b) long-chain polyamines (after Kröger *et al.*, 1999); and (c) silacidin A (after Wenzl *et al.*, 2008).

members of *Coscinodiscus* and other morphologically simple genera, but cannot wholly explain how more complex asymmetric patterns are generated by other species (Sumper, 2002). More complex patterns are probably generated by temporal and spatial cooperation of silaffins, silacidins, polyamines, and other perhaps unknown intracellular components.

The nature of the interaction between polymerized silicic acid species and polyamines, both natural and synthetic, has been extensively investigated (Behrens *et al.*, 2007; Brunner *et al.*, 2004; Lutz *et al.*, 2005; Sumper, 2004; Sumper and Kröger, 2004; Sumper and Lehmann, 2006; Sumper *et al.*, 2003). In the absence of anionic species, polyaminated species apparently stabilize sols of nanopartiticulate amorphous silica (Sumper *et al.*, 2003). The introduction of anionic species (e.g. phosphate, sulfate, citrate, silacidins) induces the separation of the polyamines from the aqueous phase, forming a porous matrix where preformed silica nanoparticles grow to their final size, which varies with the molecular weight of the polyamine utilized and the concentration of anionic species (Sumper *et al.*, 2003; Kröger *et al.*, 1999). Recent observations show that silica deposition during diatom cell wall synthesis consists of an initial mineralization step, where a thin version of the new valve is

rapidly formed, followed by a second stage where the valve structure thickens. This suggests that the aggregation processes investigated by Sumper and co-workers is most applicable to later stages of frustule development. The organic-silica interface in living diatom cells, as revealed by atomic force microscopy (Crawford *et al.*, 2001), is surprisingly smooth, and devoid of discernable boundaries between silica nanospheres. The absence of texture implies that the initial stage of silica deposition is dominated by the surface-directed nucleation of amorphous silica on the organic component of the cell wall, rather than colloidal aggregation and adsorption. A recent study on the kinetics of silica nucleation on model anionic and cationic biosurfaces provides some insight into which molecular contacts on the organic matrix promote silica nucleation during the early stages of valve formation (Wallace *et al.*, 2009). Under the conditions of the study (SiO$_2$(aq) $\leq$ 1000 ppm and pH = 5.0) negatively charged surface species (carboxyl groups) facilitated silica nucleation, while positively charged surfaces (amine-terminated) were resistant to silica deposition. However, the greatest rates of silica mineralization were measured on surfaces containing both anionic and cationic species, suggesting that silica formation may be mediated by cooperative interactions between oppositely charged sites within the organic matrix. This result also suggests that the origin of fine scale patterning on diatom valves may also be rooted in these interactions. That is, silica deposition might be preferentially concentrated in regions of the matrix where favourable amine-anion interactions are locally abundant.

10.3 Mineralization by multicellular organisms

10.3.1 The sponges: enzyme-catalysed formation of biosilica

Members of the phylum Porifera (the sponges) are known to utilize a number of inorganic materials in the construction of mineralized tissues, including iron oxides (Towe and Rutzler, 1968; Garrone, 1978), calcium and magnesium carbonates (Jones, 1970; Hartman and Goreau, 1970), and amorphous silica (Simpson and Volcani, 1981). Silica, and to a lesser extent the polymorphs of calcium carbonate, comprise the major mineral components of skeletons within the three extant sponge classes: Calcarea ($CaCO_3$), Demospongiae (SiO_2 and $CaCO_3$) and Hexactinellida (SiO_2) (Fig. 10.4e). The oldest fossils reliably interpreted as mineralized sponge spicules occur in strata just below the Proterozoic-Cambrian boundary, perhaps 545–543 Ma (Brasier et al., 1997), but macrofossil impressions and biomarker molecules indicate an earlier (>635 Ma; Gehling and Rigby, 1996; Love et al., 2008) origin for the group. The apparent maximum in abundance and geographic extent of reef building glass sponges occurred during the Jurassic when extensive deep-water sponge reefs formed a ~7000 km long discontinuous belt along the northern shelf of the Tethys and its proximal proto-Atlantic basins (Wendt et al., 1989; Krautter et al., 2001). Since that time, siliceous sponges have not been important as reef builders, although modern hexactinellid build-ups are known from ca. 200 m depths off the coast of British Columbia (Krautter et al., 2001, 2006) and Washington State. Although sponges are widely regarded as the phylogenetically oldest metazoans (Xiao et al., 2005), recent comparative genomics work suggests that the Ctenophores (typically grouped within Cnidaria) may reside closer to the base of the metazoan branch of the tree of life than the Porifera (Dunn et al., 2008); however, additional research is needed to confirm these findings.

Recent advances in sponge biomineralization have been focused upon specific enzymes involved in silica spiculogenesis, rather than $CaCO_3$ deposition. Moreover, certain aspects of sponge calcification strategies, and the molecules involved therein, bear striking similarities to approaches employed by the calcifiers discussed later in this chapter. Therefore, the remainder of this discussion is concerned only with silicification processes. For a recent review of carbonate spiculogenesis in sponges we refer the reader to Uriz (2006).

10.3.1.1 Silica spiculogenesis

The concentration of silica in modern seawater (1–100 μmol l^{-1}) is far lower than when sponges evolved during the Neoproterozoic Era (Siever, 1992). As significantly higher aqueous Si concentrations are required to induce silica precipitation in vivo (>2 mM l^{-1} at 25 °C), silica-mineralizing organisms must employ silicic acid transport systems that both harvest and concentrate silicon from the surrounding environment. Silicic acid transport proteins are now documented for a number of diatom species (Hildebrand et al., 1997, 1998; Thamatrakoln et al., 2006; Thamatrakoln and Hildebrand, 2007), higher plants (Liang et al., 2005; Mitani and Ma, 2005; Ma et al., 2006, 2007; Rains et al., 2006; Yamaji and Ma, 2007), and the siliceous demosponge Suberites domuncula (Schröder et al., 2004). As in diatoms, silicic acid transport in this sponge is apparently coupled with the transport of sodium ions; however, comparative analysis of the protein amino acid sequences indicates a lack of homology between diatom and sponge Si transporters. Investigations with complementary DNA sequences (cDNA) demonstrate that the gene that codes for the sponge Si transporter is up-regulated in sclerocyte cells located adjacent to developing silica spicules (Schröder et al., 2004).

The present indication is that silica spiculogenesis in hexactinellids and demosponges proceeds by a similar, multi-stage process (Müller et al., 2006, 2008a), in which lightly-silicified and highly-structured proteinaceous axial filaments (initially localized inside sclerocytes), grow into mature spicules within a semi-isolated extracellular compartment defined by several closely associated sclerocytes and a matrix of collagen, chitin, and glycoprotein (e.g. lectin, galectin) (Schröder et al., 2006; Müller et al., 2008c). Precisely how sufficiently-high Si levels are maintained within this intercellular environment during spicule growth remains unknown; however, there is at least some evidence to suggest that Si is stored in a pre-condensed state (e.g. as a silica gel) in specialized vesicles (termed silicasomes) within the sclerocytes, and that it can be rapidly transported in this form to the surface of the growing spicule, where enzymes located in the extracellular compartment and on the active growth surface simultaneously accelerate the dissolution of the unstable silica gel (silicase) and the re-precipitation of amorphous silica (silicatein) (Schröder et al., 2007b, 2008). The resulting spicules exhibit multiple, regularly-spaced lamellae that are concentric about the axial filament and are separated by thin layers of organic material (Fig. 10.4f) (Garrone et al., 1981).

The organic material localized within the axial filament and at the interface between lamellar structures consists largely of silicatein-type proteins (Cha et al.,

1999; Weaver and Morse, 2003). In *ex vivo* silica precipitation assays, silicateins accelerate both the hydrolysis of organosilicate precursors to silicic acid, and the subsequent condensation of dissolved silicate species to SiO_2 (Shimizu *et al.*, 1998; Cha *et al.*, 1999; Müller *et al.*, 2008b). Comparison of silicatein amino acid sequences with highly homologous members of the cathepsin L family of proteases suggests that the silicatein active site contains serine and histidine residues that are responsible for the enzyme's observed catalytic activity (Shimizu *et al.*, 1998). This result is now prompting some speculation on the actual mechanisms of organosilicate hydrolysis (Cha *et al.*, 1999) and silicic acid condensation (Fairhead *et al.*, 2008) as promoted by silicateins (Fig. 10.8a,b). However, the proposed reaction pathways are not yet fully supported by experimental measurements or molecular models. Moreover, while it is interesting that the same enzyme (and presumably the same active site) seems capable of catalyzing two distinct reactions that are potentially related to silica biomineralization, there is no experimental evidence confirming the presence of organosilicate complexes *in vivo* (Birchall, 1995); therefore, the catalytic activity of silicateins in sponges may be restricted to the silicic acid condensation reaction *in vivo* (Fairhead *et al.*, 2008).

Silicase, a member of the carbonic anhydrase protein family (enzymes that hydrate carbon dioxide), promotes the hydrolysis of Si–O–Si linkages in *ex vivo* assays, and is closely associated with silicatein proteins during spiculogenesis (Schröder *et al.*, 2007a). The silicase active site contains Zn^{2+}, and the proposed mechanism of Si–O hydrolysis is similar to that suggested by a recent computational study of electrolyte promoted silica dissolution (Wallace *et al*, 2010). However, as with silicatein, the proposed reaction mechanism is speculative (Fig. 10.8c), and assumes that the active sites for CO_2 hydration and Si–O bond hydrolysis are the same. It is not yet clear why the expression of silicatein and silicase proteins is so highly coupled with silica formation, as they exhibit seemingly opposite functions with regard to silica mineralization; however, there are two plausible mechanisms by which silicase may work in tandem with silicatein during spiculogenesis. The first, which has already been alluded to, requires silicase to act upon pre-condensed silica species in silicasomes; afterward, the liberated silicic acid might be re-precipitated in a controlled fashion by neighboring silicateins. Alternatively, silicase may utilize its carbonic anhydrase activity to modulate the pH of the intercellular region surrounding the developing spicule, thereby creating more favourable conditions for silica deposition.

Chitin and collagen are also closely associated with biogenic silica in sponges, but play questionable roles in spiculogenesis; nonetheless, there is an emerging literature concerned with the novel properties of silica-based composite materials involving these biomolecules (Ehrlich *et al.*, 2008a,b). Of primary interest to silica

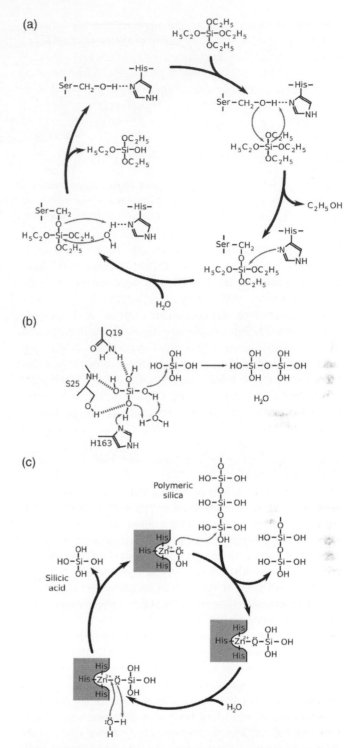

Figure 10.8 Proposed reaction mechanisms for sponge proteins silicatein and silicase. Small arrows represent the rearrangement of electrons during an individual reaction step at the enzyme active site (a) silicatein-catalysed hydrolysis of Si-alkoxides (Cha *et al.*, 1999) ; (b) silicic acid condensation via the silicatein-catalysed proton shuttling mechanism proposed by Fairhead *et al.* (2008). Binding of silicic acid monomers at the protein active site results in the formation of ephemeral ionized silicate species that rapidly react with solution borne silicic acid to form small oligomers; (c) silicase-catalysed hydrolysis of Si– O bonds. Figures used with permission from Schröder *et al.* (2008) and Fairhead *et al.* (2008).

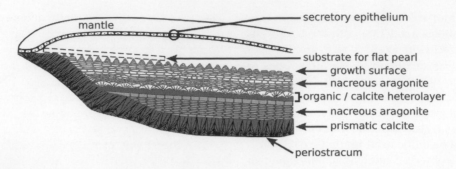

Figure 10.9 Schematic of a vertical section of the outer edge of the shell and mantle of the mollusc *Haliotis rufescens* showing the nacreous aragonite and prismatic calcite layers. The size of the extrapallial space is enlarged for clarity. Note that the secretory epithelium of the mantle is in direct contact with the biomineral growth surface (used with permission, Zaremba *et al.*, 1996).

biomineralization is the recent discovery of silica-associated long-chain polyamines (similar to those that direct silica deposition in diatoms) in the marine sponge *Axinyssa aculeata* (Matsunaga *et al.*, 2007). This discovery implies that silica deposition in diatoms and sponges may be more similar than previously thought, and hints at the existence of a common molecular machinery underlying at least some aspects of biosilicification. However, further investigation is required to establish a mechanistic link between the silica mineralizing strategies employed by these and other silicifying organisms.

10.3.2 The molluscs: model for roles of matrix proteins in biomineral formation

Molluscs use an elaborate organic matrix to exercise strict control over CaCO$_3$ nucleation and growth. A look through previous sections suggests controlled biomineral formation is frequently, if not always, intimately linked to organic components produced and assembled by specialized cells or organelles within the organism. The fundamental importance of macromolecules was first recognized when early researchers found that key chemical aspects of macromolecules are shared by a number of calcifying organisms; namely, the presence of acidic carboxyl moieties and, to a lesser extent, phosphate and sulfate groups (Weiner *et al.*, 1983). Since that time, proteins exhibiting high concentrations of these species have been recognized or implicated in a number of roles related to biogenic mineral formation, including ion transport, polymorph selection, crystal nucleation, and growth (Addadi *et al.*, 2001). While advances have been made, the specific and individual roles of organic components remain elusive.

10.3.2.1 Shells as composites with remarkable fracture toughness

The mollusc shell is a composite material with extraordinary strength/weight properties that are often noted by materials scientists. To construct this type of material, many mollusc species secrete phase controlling proteins

from mantle epithelial cells (Marin, 2004) to selectively deposit calcite or aragonite (Belcher *et al.*, 1996; Addadi *et al.*, 2006) in close association with an organic matrix (Fig. 10.9 and Fig. 10.10a). These biomolecule components can impart favourable mechanical properties unto the mineral constituents of the shell. In addition to the crossed lamellar structure displayed by most mollusc shells, the nacreous (Fig. 10.10b) and prismatic mineralized layers (Fig. 10.10c) confer a considerable amount of strength to the shell. The formation of these layers is strongly mediated by the organic matrix.

Nacreous layer- Mollusc nacre consists of uniform layers of aragonite single crystal tablets (Fig. 10.10b) interspaced by layers of organic material (Nudelman *et al.*, 2008). This network of organic and interlocking crystals has been noted as a model for designing bioinspired materials with high fracture toughness. The current model for its formation is that the polysaccharide (β-chitin) and mildly hydrophobic silk-like proteins (rich in Ala and Gly) comprise the framework portion of the organic matrix while the soluble fraction is an assemblage of hydrophilic, highly acidic glycoproteins in β-sheet conformation (Nudelman *et al.*, 2006; Addadi *et al.*, 2006). It is the highly ordered nature of the chitin that apparently orients the aragonite single crystal at nucleation and physically limits its size and shape, though this is not yet well understood. There is evidence that the aragonite platelets are bound by a layer of amorphous CaCO$_3$ (Nassif *et al.*, 2005). Histochemical analysis shows that sites of aragonite nucleation are rich in both carboxyl and sulfate groups of the soluble proteins, suggesting that the two functionalities could work cooperatively to direct nucleation (Nudelman *et al.*, 2006).

One working model (Fig. 10.11) for how molluscs nucleate and grow aragonite crystals of the nacreous layer proposes that mineralization occurs within a compartment whose boundaries are defined by intertabular/ interlammelar matrix molecules. The composition of this material is discussed later. Recent investigations have demonstrated that the volume to be mineralized is

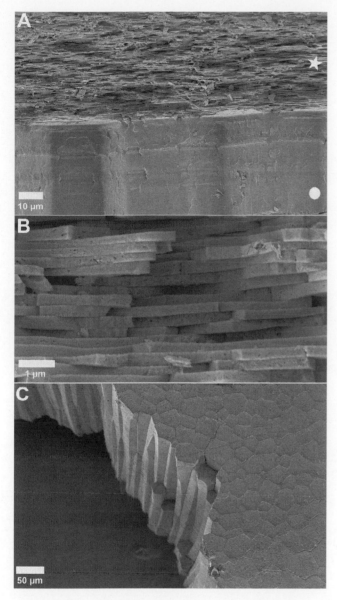

Figure 10.10 (a) Cross-sectional SEM image of a shell from the mollusc *Atrina rigida*, showing the nacreous (white star) and prismatic (white circle) layers. (b) SEM image showing a cross-section of the aragonite tablets in the nacreous layer. (c) An SEM image showing the prismatic layer, comprised of calcite prisms (used with permission, Nudelman *et al.*, 2007).

prefilled by a framework of silk proteins that likely exist in a gel-like state (Addadi *et al.*, 2006; Nudelman *et al.*, 2008; Levi-Kalisman *et al.*, 2001). That $CaCO_3$ is mineralizing from a gel phase in the absence of a bulk solution suggests that basic concepts in biomineral formation currently founded upon solution-based crystal growth studies may need to be revisited.

Prismatic layer- The calcite crystals that develop prismatic morphologies are each delimited by a relatively thick layer of organic matrix (Nakahara and Bevelander,

1971; Grégoire, 1961). β-chitin and Gly-rich proteins are present as framework macromolecules. Although these proteins are similar to those found in nacre, the structures and locations within the mineral differ significantly – prism chitin is highly disordered and gel-like phases have not been observed. The soluble matrix macromolecules are highly acidic (even more so than those in nacre) and associated with both the chitin and the mineral phase; several of these unusually acidic proteins have been isolated and characterized in last five years (Tsukamoto *et al.*, 2004; Marin *et al.*, 2005; Nudelman *et al.*, 2007). The processes involved in prismatic layer formation have received less attention than the nacreous layer. There are significant challenges ahead for understanding the conditions that reliably direct specific carbonate polymorphs to form during nucleation and how the organic matrix promotes development of these columnar morphologies.

10.3.3 Soluble and insoluble matrix components

In a very simplified model, the organic matrix may be viewed as a two-component system. The primary constituents are typically highly cross-linked, insoluble, and slightly hydrophobic macromolecules; these entities form the structural, three-dimensional framework for mineralization and are ultimately less intimately associated with the resultant biomineral than their soluble counterparts (Meldrum, 2003). Besides providing the physical arena for mineralization, these 'framework macromolecules' may also act as a substrate on which other proteins can interact with the mineral phase.

The second set of matrix elements is made up of soluble proteins whose functions are only beginning to be understood. They may be unbound, adsorbed onto a matrix substrate, or occluded within a growing mineral. Soluble matrix macromolecules fall roughly into three categories based on their primary structures and occurrences (Weiner and Addadi, 1997): (1) Aspartic acid/asparagine rich proteins and glycoproteins – associated with crystalline $CaCO_3$ phases; (2) Glutamic acid/glutamine and serine-rich glycoproteins – associated with amorphous calcium carbonate (ACC); (3) Polysaccharide-rich macromolecules – not unusually acidic, found in the biogenic $CaCO_3$ of many organisms. The first two categories are comprised of unusually acidic macromolecules due to the abundance of carboxylate functional groups. The high negative charge associated with these proteins and glycoproteins makes them good candidates for facilitating interactions at charged crystal faces as well as for accumulating calcium ions at nucleation sites. For these reasons, the acidic macromolecules are presumed to be most influential in $CaCO_3$ nucleation

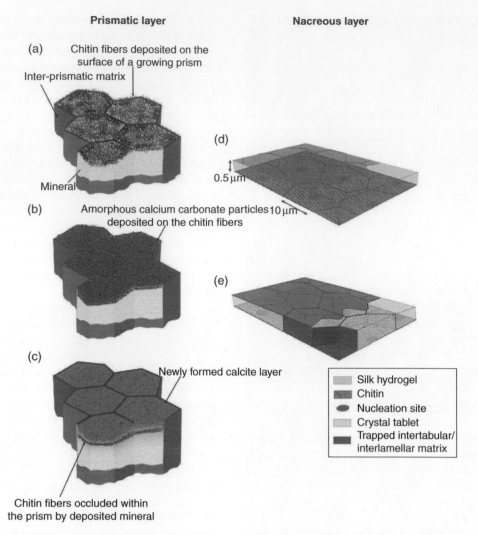

Prismatic layer

Nacreous layer

(a) Chitin fibers deposited on the surface of a growing prism
Inter-prismatic matrix
Mineral

(b) Amorphous calcium carbonate particles deposited on the chitin fibers

(c) Newly formed calcite layer
Chitin fibers occluded within the prism by deposited mineral

(d) 0.5 µm 10 µm

(e)

Silk hydrogel
Chitin
Nucleation site
Crystal tablet
Trapped intertabular/ interlamellar matrix

Figure 10.11 Schematic showing the interplay of organic and inorganic elements and the proposed mechanisms of prism (a–c) and nacre (d–e) formation in the mollusc *Atrina rigida*. (used with permission from Nudelman *et al.*, 2007).

and growth. The soluble matrix macromolecules have been shown to exert remarkable control over crystal nucleation, growth, polymorph, and morphology, even when separated from the rest of the matrix (Belcher *et al.*, 1996; Thompson *et al.*, 2000; Raz *et al.*, 2003). When studying biomineralization processes, however, it is important to consider the matrix in its entirety; several studies have demonstrated the importance of the interplay of framework macromolecules with acidic proteins and mineral phases (Albeck *et al.*, 1993; Falini *et al.*, 2003; Addadi *et al.*, 2006; Tong *et al.*, 2002; Nudelman *et al.*, 2007).

10.3.3.1   Characterization of matrix macromolecules

A significant challenge for elucidating the individual roles of matrix macromolecules lies in the task of protein characterization (Marin, 2004). Matrix proteins are typically separated by gel electrophoresis under denaturing conditions or a combination of HPLC and ion exchange after extraction from the demineralized organism (Falini *et al.*, 1996; Cariolou, 1988). Proteins that are not highly acidic are readily stained with Coomassie blue or silver; sequences for many proteins of this type have been reported (Miyamoto *et al.*, 1996). The more interesting acidic proteins, however, have proved difficult to extract and purify while keeping their structure and function intact due to their high charge and close association with the mineral phase (Gotliv *et al.*, 2003).

Recently, new electrophoretic strategies and the application of cDNA sequencing have allowed the characterization of some new acidic matrix proteins (Tsukamoto *et al.*, 2004; Gotliv *et al.*, 2005). Further borrowing of molecular genetic techniques may be useful in identifying more matrix elements, an important step toward understanding the roles of individual proteins and chemical functionalities.

10.3.3.2 Modes of regulation

A wide variety of experimental techniques have been applied to examine the functions of biomolecules in $CaCO_3$ biomineralization from the angstrom scale to the macroscale. To date, the functions of organic matrices are often inferred from observations obtained with scanning and transmission electron microscopy. Because calcium carbonate is easily grown from solution or deposited on ordered substrates that mimic the framework macromolecules seen in nature (Mann *et al.*, 1990; Falini *et al.*, 2000), a number of chemical additives have been studied in the laboratory as simple models for how components of the organic matrix could influence mineralization. Compounds that range from synthetic polymers (Wang *et al.*, 2005; Jada and Jradi, 2006; Naka, 2007) to soluble macromolecules extracted from organisms (Wheeler *et al.*, 1981; Walters *et al.*, 1997; Kim *et al.*, 2006) have been tested. *In situ* experiments based on atomic force microscopy (De Yoreo and Dove, 2004; Elhadj *et al.*, 2006b; Fu *et al.*, 2005) as well as X-ray and neutron scattering methods (Pontoni *et al.*, 2003; DiMasi *et al.*, 2006; Lee *et al.*, 2007; Pipich *et al.*, 2008) have been key in uncovering the dynamics of macromolecular controls on $CaCO_3$ nucleation and growth. These findings have lead to two modes for how matrix molecules influence biomineral formation, as presented below.

Electrostatic accumulation- by biomolecules An obvious function of negatively charged biomolecules in $CaCO_3$ mineralization is the electrostatic attraction of calcium ions to the site of crystallization. These interactions act to concentrate ions at specific nucleation sites on the matrix, potentially inducing crystal nucleation by increasing the local supersaturation (Greenfield *et al.*, 1984; Mann, 1988). Matrix macromolecules are known to bind Ca^{2+} at levels that exceed the number of potential anionic binding sites on the matrix; this phenomenon can be explained by local anion binding and secondary calcium binding triggered by the initial binding of calcium to the protein, an effect that has been demonstrated in biomineral systems (Lee *et al.*, 1983).

The specific chemical interaction of calcium with protein functional groups is an important consideration when thinking about the role of any specific matrix protein (Aizenberg *et al.*, 1999). To induce crystal nucleation, macromolecule surfaces must exhibit a high charge density sufficient enough to attract ions from solution. In general, calcium binding to carboxyl groups is cooperative and involves at least two or three ligands

(Kretsinger, 1976). It has been suggested that amino acid sequences resembling Asx-X-Asx, where X is a neutral residue, may provide an optimal binding configuration for calcium in order to induce nucleation (Weiner and Hood, 1975). These acidic sequences are common in molluscan matrix proteins.

Stereochemical matching of crystal faces- Biomineral morphology is constrained by both the physical aspects of the bounding macromolecules and the chemical interactions of soluble proteins with the mineral phase. The stereochemical recognition model provided a first explanation of chemical control over crystal shape by proteins. By this approach, specific morphologies were stabilized by binding proteins to otherwise unstable crystal faces due to stereochemical matching to the crystal lattice (Addadi and Weiner, 1985). This match is not necessarily achieved by crystallization onto a rigid substrate, but rather through cooperative interactions between a somewhat flexible matrix and the developing mineral in its embryonic state (Lee *et al.*, 2007). However, recent work has refocused this discussion away from faces, emphasizing that morphological changes are rooted in step-specific interactions. That is, proteins and other macromolecules bind to surface sites to alter the local free energy landscape (De Yoreo and Dove, 2004; Orme *et al.*, 2001).

10.3.4 The sea urchins: model organisms for the study of non-classical crystallization processes

Sea urchins, members of the phylum Echinodermata, hold remarkable promise for understanding cellular environments and the detailed chemistry and roles of their constituent macromolecules during mineralization. Studies of sea urchins are motivated in part by their prevalence in the fossil record and the modern ocean, but again, efforts to interpret embedded compositional signatures for climate reconstruction are complicated by vital effects (see discussion of forams). From a biological point of view, spicule structure influences the shape, orientation and motility of larvae (Pennington and Strathmann, 1990). Of particular interest in biomineralization studies are the factors that control the onset of spicule calcification and morphological development. Compared to most multicellular organisms, sea urchin larvae offer unique advantages to biomineralization research that include small physical size, short time scale for spicule development, ease for probing cellular processes by confocal microscopy, and relative ease of extracting spicule-associated macromolecules (Wilt, 1999, 2002; Wilt *et al.*, 2003; Cheers and Ettensohn, 2005; Wilt and Ettensohn, 2007). Here we consider how the organism develops in tandem with spicule mineralization.

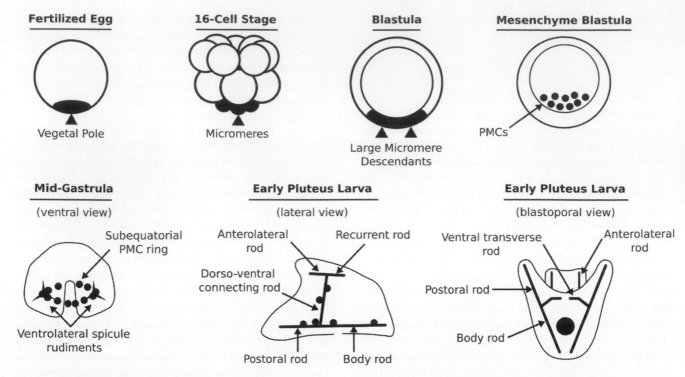

Figure 10.12 Developmental stages of sea urchin larva. The micromeres, which are ultimately responsible for mineralization arise in the 16-cell stage, ingress during formation of the blastula and become primary mesenchyme cells (PMCs). During the mid-gastrula stage the PMCs organize and fuse to form a syncytium cable where triradiate spicules form. The spicules continue to grow through the pluteus phase where mature calcite skeletal rods are observed (used with permission from Wilt and Ettensohn, 2007).

10.3.4.1 Embryo development

The process of embryo development involves the synchronous formation of the organism and its skeleton (Fig. 10.12). Upon fertilization, the cell quickly divides; at the 16-cell stage an unequal equatorial division results in small cells at the vegetal pole, called large micromeres. These specialized cells are ultimately responsible for controlling mineralization within the larvae (Summers *et al.*, 1993). Cell cleavage ends at the 128-cell stage when the cells reorganize to form the blastula, a hollow sphere of cells with a central cavity. During blastula formation, 16 or 32 descendents of the large micromeres are incorporated into the epithelial wall and undergo a round of cell division resulting in groups of 32 or 64 cells (Takahashi and Okazaki, 1979). In the mid-blastula phase the large micromeres ingress into the vegetal pole of the central cavity, and are subsequently known as primary mesenchyme cells (PMCs). While the PMCs are reorienting, the simultaneous invagination of the embryo results in gut formation. During this stage of development the PMCs organize into two clusters and form protrusions that fuse to form a network known as the syncytium (Fig. 10.13), which is where mineral formation occurs (Wolpert and Gustafson, 1961). It is within the two syncytia that the initial calcite deposits appear as single rhombic crystals. As the larvae mature into the pluteus stage, these crystalline seeds grow through the addition of amorphous carbonate material along the symmetrically equivalent directions parallel to [100] (Wolpert and Gustafson, 1961; Okazaki and Inoue, 1976). The amorphous regions in the tri-radiate spicule rudiments later convert to calcite (Politi *et al.*, 2008).

10.3.4.2 Primary mesenchyme cells: directors of spiculegenesis

PMCs receive ectoderm-derived signals that regulate the synthesis of proteins directly related to spiculogenesis and also allow for the cell-to-cell fusion that forms the syncytial networks (Armstrong *et al.*, 1993; Guss and Ettensohn, 1997; Hodor and Ettensohn, 1998). These signals also direct PMCs to reorient during spiculogenesis, which influences the morphology of growing spicules (Wolpert and Gustafson, 1961; Malinda and Ettensohn, 1994). Through protein labeling methods, Wilt *et al.* (2008) demonstrated that spicule-associated proteins produced and excreted by PMCs do not travel more than 5–10 μm within the syncytium, indicating that the local crystallization environment is strongly influenced by protein regulation

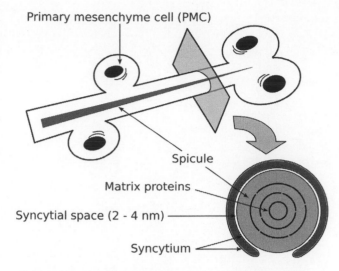

Primary mesenchyme cell (PMC)

Spicule

Matrix proteins

Syncytial space (2 - 4 nm)

Syncytium

Figure 10.13 A schematic representation of the calcium carbonate depositional environment in a sea urchin larva. The spicule develops from calcium and carbonate ions and proteins that are transported to the growing spicule from nearby PMCs. Matrix proteins like SM50 and SM30 are incorporated into the spicule as it forms (after Wilt, 1999).

within nearby PMCs. PMCs carry out all tasks related to spiculogenesis in the urchin embryo by producing an extracellular site for mineral deposition, supplying the growing spicule with calcium carbonate, and producing and transporting proteins necessary for controlling mineral growth.

In the initial phase of mineral deposition, the syncytial network forms a microenvironment that is conducive to the nucleation and growth of crystalline calcium carbonate. Ultrastructural studies indicate that the syncytium and the central cavity are interconnected; introduction of calcium chelators and low pH conditions results in spicule dissolution (Decker *et al.*, 1987). Beniash *et al.* (1999) carefully characterized this environment and found conclusive evidence that at most there could be 2–4 nm of solution between the sheath and the crystalline component. Incidentally, this thickness correlates with the amount of water that would condense on a calcite surface in a humid environment (Chiarello *et al.*, 1993).

10.3.4.3 Integral spicule matrix proteins

To develop a rigorous model for biomineralization in sea urchin embryos, it is necessary to understand the role of PMC synthesized proteins in the syncytium, since they are likely responsible for stabilizing amorphous precursors to crystalline calcium carbonate (ACC), and initiating calcite nucleation. The first step towards this end is to differentiate those proteins involved in biomineralization from those with other functions in the cell. The traditional method has been to study those proteins directly associated with the spicule

(Killian and Wilt, 1996; Ameye *et al.*, 1999, 2001; Kitajima and Urakami, 2000; Urry *et al.*, 2000; Seto *et al.*, 2004;). However, it is known that there are membrane proteins that may serve as substrates for crystal formation that are not contained within spicules (Cheers and Ettensohn, 2005). The function(s) of specific proteins within a cell or organism are generally determined by employing strategies to knockout or suppress the expression of the gene(s) that code for the protein(s) of interest. For eukaryotes, development of this technology can be a challenging process. To meet this need, a new loss of function tool has been utilized, Morpholino Oligos (MO), which can block mRNA translation in specific genes and effectively inhibit protein synthesis when introduced to cells at the proper stage (Heasman, 2002). When applied to sea urchin embyos these tools clarify the function of certain proteins in spiculegenesis (Cheers and Ettensohn, 2005; Peled-Kamar *et al.*, 2002).

The main goals of studies on spicule-associated proteins have been to characterize their structures and functions, their localization in both PMCs and the spicule, and their effects on the growth of the spicule *in vivo*. Killian and Wilt (1996) used 2D-gel electrophoresis to identify 40–45 proteins associated with urchin spicules. The majority of these proteins (35 out of 45) are acidic (with isolelectric points below 6.0), and contain large quantities of Asp/Asn and Glu/Gln amino acid residues. The remaining ten proteins are basic in nature. Within the basic group, five proteins SM50, SM37, SM32, SM29, and PM27, have been characterized, of which SM50 and PM27 have been the focus of most investigations (Zhu *et al.*, 2001; Illies *et al.*, 2002; Livingston *et al.*, 2006; Harkey *et al.*, 1995; Lee *et al.*, 1999). These proteins are abundant in the extra-cellular matrix and on the surface of the spicule (Seto *et al.*, 2004; Kitajima and Urakami, 2000; Wilt *et al.*, 2008). MO knock-out experiments show that spicule elongation is not observed in the absence of SM50 (Peled-Kamar *et al.*, 2002) which strongly indicates that SM50 plays an integral role in directing spicule growth along the <1 0 0> directions (Kitajima and Urakami, 2000). Wustman *et al.* (2002) suggested that a glycine loop motif found in PM27 may play a role in altering the mechanical properties of the carbonate spicule, as suggested by Berman *et al.* (1990 and 1993).

From the acidic group, the glycosylated SM30 protein (and its derivatives) has been characterized, revealing the presence of a highly conserved C-type lectin calcium binding motif (Livingston *et al.*, 2006). Investigations with immunogold protein labeling and green fluorescent protein tagged SM30 mutants show that these proteins are a major component of the spicule-associated protein assemblage (Kitajima and Urakami, 2000; Seto *et al.*, 2004; Wilt *et al.*, 2008), suggesting that SM30 promotes elongation in the *c* crystallographic direction.

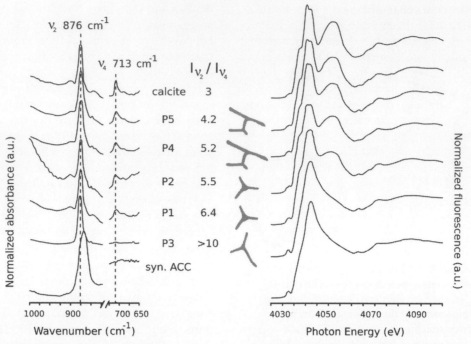

Figure 10.14 Infrared spectra and X-ray absorbtion near edge spectroscopy (XANES) spectra for geologic calcite, sea urchin larval spicules, and synthetic ACC. Samples P1 and P2 represent triradiate spicules isolated at the mid-gastrula phase, P3 is from late-gastrula phase, P4 and P5 represent fully developed body rods from pluteus larva. The ratio of the v_2 (out of plane carbonate bending mode) and v_4 (in-plane carbonate bending mode) represents a measure of the relative abundances of ACC and calcite in the material. The XANES spectra show P5, P4, P2 and P1 are similar in character to calcite and P3 is very similar to ACC (used with permission from Politi *et al.*, 2006).

However, some have suggested that its main function is to make the spicule more resistant to cleavage or fracture (Wilt and Ettensohn, 2007).

The discovery of P16 represents a new mode of discovering biomineralization related proteins, because it was found with a complementary DNA (cDNA) library which was used to screen proteins that are up-regulated within the PMCs during spiculegenesis rather than by traditional protein purification methods (Illies *et al.*, 2002). Subsequent MO investigations show that PMC function remains normal when P16 expression is suppressed, but spicule elongation does not occur. Its precise chemical role in the mineralization process is, however, not known (Cheers and Ettensohn, 2005). P16 is a small protein (16 kDa, 172 residues) that includes a glycine, aspartic acid-rich domain, a transmembrane domain and short C-terminal domain and has a predicted isoelectric point of ~3.6 (Illies *et al.*, 2002).

10.3.4.4 Development of the spicule

There is convincing evidence that spicule formation proceeds through a metastable disordered calcium carbonate phase (ACC) en route to the final crystalline product, rather than by direct nucleation of calcite from solution (Politi *et al.*, 2008). The resulting calcite spicule contains about 5 mole percent Mg and 0.1% organic material (Okazaki and Inoue, 1976). Under optical crossed nichols, spicules behave as a single crystals, however studies using synchrotron-based XRD suggest a reduction in the coherence length of the signal, which implies the presence of oriented micro crystals that may be held in place by organic material localized at grain boundaries (Berman *et al.*, 1988, 1990, 1993). Beniash *et al.* (1997) reports that the tri-radiate spicule rudiments isolated from developing larva contain significant amounts of ACC. Furthermore, Politi *et al.* (2006) conclusively demonstrated with X-ray absorption spectroscopy that the initially deposited tri-radiate spicules were up to 90% ACC, and that the ACC to calcite ratio decreased at more advanced stages of larval development (Fig. 10.14). There is evidence demonstrating that this strategy is generalized in the sea urchin, as researchers have made similar observations in both adult spines (Politi *et al.*, 2004) and teeth (Nudelman *et al.*, 2007; Killian *et al.*, 2009; Ma *et al.*, 2009).

To understand the source of the ACC precursor phase, researchers have looked for physiological structures in PMCs that are responsible for calcium and carbonate storage. Decker *et al.* (1987) found electron dense granules around the nucleus and suggested that these granules are necessary to support the mineralization process. Beniash *et al.* (1997) later confirmed this finding by identifying the granules in TEM sections of PMCs. In a more recent study, Wilt *et al.* (2008) stained these granules with a fluorescent calcium binding molecule, calcein and demonstrated that the PMCs actively concentrate ACC in vacuoles and most likely transport them to mineralization sites in the synctium.

These kinds of studies are leading to a more detailed picture of calcium carbonate biomineralization in which non-classical crystal growth mechanisms are central. The nature of the ACC to calcite transformation remains controversial because the physical and chemical conditions *in vivo* are substantially different from those that have been employed in classical studies of calcite nucleation and growth (Politi *et al.*, 2008). However, much of the community's efforts are currently focused on understanding the mechanics of crystallization through disordered precursor phases *ex vivo*. Recent studies of the early stages of calcium carbonate using cryo-TEM (Pouget *et al.*, 2009; Dey *et al.*, 2010), titration / ultracentrifugation (Gebauer *et al.*, 2008), and electrospray mass spectrometry (Wolf *et al.*, 2011) suggest that a significant population of calcium carbonate and phosphate prenucleation clusters may be present in solution, perhaps even at equilibrium. The structure and energetics of these clusters and bulk amorphous calcium carbonate (both synthetic and biogenic) are also being explored with calorimetry (Radha *et al.*, 2010), X-ray scattering/ spectroscopy and NMR (Michel *et al.*, 2008; Gebauer *et al.*, 2010; Pancera *et al.*, 2010), Reverse Monte Carlo structural modeling (Goodwin *et al.*, 2010) and molecular simulation (Raiteri and Gale, 2010). It is as yet unclear the extent to which findings obtained from this new generation of inorganic precipitation studies will enlighten our understanding of biologically-mediated mineralization, although the potential is great. Additional experimental approaches need to be developed to explore the effects proteins, low water activity, and spatial confinement have on crystal growth (Stephens *et al.*, 2010; Lotse *et al.*, 2010). Moreover, additional work is required to more accurately determine the roles matrix proteins play *in vitro* (Fu *et al.*, 2005; Elhadj *et al.*, 2006a; Gayathri *et al.*, 2007; Piana *et al.*, 2007).

10.4   A brief history of skeletons

Eukaryotic biomineralization encompasses a diverse suite of geobiological processes, and it also provides key perspectives in geobiological history. The reasons for this are obvious. As noted in previous sections, skeleton formation reflects the physiological interplay between organism and their ambient environments, making mineralized skeletons potential recorders of paleoenvironmental conditions. More fundamentally, the Phanerozoic fossil record of animals and many protists *is*, to a first order, a record of skeletons. Eukaryotes have been precipitating mineralized skeletal structures for nearly 800

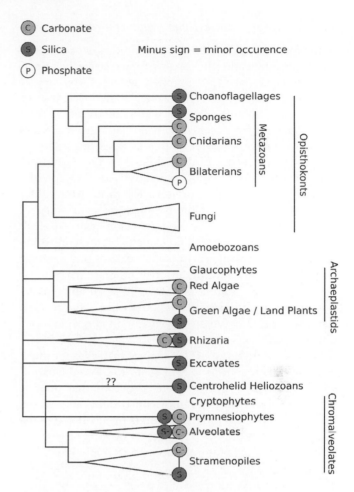

Figure 10.15 The phylogenetic distribution of mineralized skeletons within the eukaryotic domain. Occurrences of silica, calcium carbonate, and phosphatic skeletons are shown on an unrooted phylogenetic tree (after Fehling *et al.*, 2007). In some cases, biomineralization originated multiple times within a clade; e.g. carbonate skeletons in cnidarian and bilaterian animals, silica skeletons within the Rhizaria. Minor occurrences are indicated using a minus sign; the question marks denote uncertainty in the phylogenetic placement of centrohelid heliozoans.

million years. Perhaps, then, advances in our understanding of biomineralization processes can illuminate evolutionary history. Conversely, perhaps greater geobiological insight into skeletal evolution can help us to understand our evolutionary present and foreseeable future.

Figure 10.15 shows the phylogenetic distribution of mineralized scales, tests, shells and bones in the Eukarya (molecular phylogeny from Fehling *et al.*, 2007). This distribution focuses discussion on key contrasts explored in previous sections: those between silica and calcium biominerals, and between unicellular and multicellular organisms.

10.4.1   *SiO₂ and intracellular biomineralization*

In the macroscopic world of our everyday existence, calcium carbonate and phosphates appear to be the minerals of choice for skeletal biomineralization. Even cursory examination of Fig. 10.15, however, tells a different story. Skeletons of opaline silica are widely distributed across the eukaryotic tree, occurring in all major groups except the amoebozoans. Much of this escapes our attention because it is a record of unicellular eukaryotes. Indeed, save for siliceous sponges, which nucleate SiO_2 spicules within cells and then complete growth in extracellular compartments, silica biomineralization is a story of intracellular processes. Indeed, in a statistical sense, silica is the *preferred* material for skeletal biomineralization in unicellular protists. In some cases, phylogenetic uncertainties make it difficult to enumerate the number of independent origins of intracellular biomineralization using opaline silica, but ten would seem to be a minimum number (versus two major and one minor origins for $CaCO_3$).

Why should this be? The advantage of silica for intracellular skeletogenesis would seem to lie in the exquisite control over micron-scale structure that opaline silica affords. Silica is a minor constituent of seawater, but the cost of skeleton formation is proportional to the degree of saturation with respect to the mineral in question, and not absolute abundance (Raven, 1983). For most of Earth's history, the oceans have probably been at or near saturation with respect to opaline silica (Siever, 1992), so intracellular SiO_2 biomineralization made good environmental sense. Because of the remarkable evolutionary success of diatoms, this is no longer true, with important consequences for evolution in radiolaria, siliceous sponges and diatoms, themselves (e.g. Maliva *et al.*, 1989; see below). The paucity of macroscopic silica skeletons may find sufficient explanation in the low absolute abundance of monosilicic acid in seawater; a macroscopic animal would have to pump a lot of seawater to make an external shell of silica.

10.4.2   *Calcium phosphate skeletons*

Skeleton formation using carbonated hydroxylapatite (dahllite) is not our focus in this chapter, but it is worth noting the restricted phylogenetic distribution of eukaryotes that precipitate skeletons of this material. The major biological use of dahllite is skeletogenesis by us – vertebrate animals. Lingulid brachiopods also precipitate dahllite in their shells and several now extinct invertebrate clades appear to have done so in Cambrian oceans, as well (Bengtson and Conway Morris, 1992). Why such limited occurrence? First, algae are unlikely candidates for dahllite biomineralization, as phosphate

is a potentially limiting nutrient for growth. Eukaryotes cannot fix nitrogen, and so algae are nutritionally dependent on ambient seawater for P and N. Primary producers depress phosphate contents of the surface ocean, further reducing the attractiveness of phosphate as a skeletal material. Some animals clearly do use phosphate, and this must be viewed in stoichiometric perspective as well. Given the near-Redfield composition of the foods that animals consume, the sequestration of large amounts of phosphate in skeletons requires an efficient means of excreting excess nitrogen. Might the success of phosphate biomineralization in vertebrates depend, at least in part, on the kidney? In any event, the biomechanical advantage of dahllite for bone biomineralization may lie in the ease with which it is remolded during growth (e.g. Knoll, 2003).

10.4.3   *Calcium carbonate skeletons*

The overwhelming predominance of calcium carbonate minerals in the skeletons of marine animals and multicellular algae probably lies in both the availability of carbonate (the modern surface ocean is about sixfold oversaturated with respect to calcite and aragonite) and the structural properties of organic-carbonate composites.

As noted above, some foraminiferans and coccolithophorid algae precipitate calcite intracellularly, but multicellular clades form mineralized skeletons in tissue-delimited compartments. Carbonate skeletons evolved several times in both green and red seaweeds, and have a minor occurrence among the brown algae. Carbonate skeletons are widespread among marine invertebrates; uncertainties introduced by unresolved phylogenetic relationships and extinct taxa make it hard to know how many times carbonate skeletons evolved in the Metazoa, but 20 would seem a conservative estimate (see Knoll, 2003, for discussion). (Note that this estimate is for independent origins of carbonate structures; the number of independently evolved physiological pathways for carbonate biomineralization in animals is almost undoubtedly much lower.)

10.4.4   *The timeline of skeletal evoluton*

The oldest known evidence for skeletal biomineralization occurs in 750–800 million-year-old rocks. Not surprisingly, this consists of minute scales secreted interacellularly by unicellular protists. Vase-shaped microfossils from the Grand Canyon include tests with geometrically regular wall patterns closely comparable to those made by rhizarian testate amoebae that precipitate silica scales (Porter *et al.*, 2003), while coeval rocks from Alaska contain phosphatic scales of uncertain original composition (Allison and Hilgert, 1986; Cohen *et al.*, 2011) much like those made by some extant centrohelid heliozoans (Preisig, 1994).

Silica biomineralization gained biogeochemical importance with the Ediacaran-Cambrian radiations of siliceous sponges and Radiolaria, but it was the Mesozoic origin and, especially, the Cenozoic expansion of diatoms that changed the evolutionary trajectories of all silica-users (Maliva *et al.*, 1989). Diatoms govern the low abundance of silica in modern surface oceans, and the fossil record shows that sponges (Maldonado *et al.*, 1999), Radiolaria (Harper and Knoll, 1975; Lazarus *et al.*, 2009), and diatoms themselves (Finkel *et al.*, 2005) all responded by using less silica in skeletons or by retreating to deeper waters where silica concentrations are higher.

Carbonate biomineralization is first recorded in eukaryotic multicellular organisms; sponge- or cnidarian grade animal problematica occur in microbial reefs of Ediacaran age (548–542 Ma; Grant, 1990; Grotzinger *et al.*, 2000; Wood *et al.*, 2002). Carbonate skeletogenesis increased markedly as part of the broader Cambrian diversification of animals, but curiously, after the extinction of archaeocyathids and other major skeleton formers near the end of the Early Cambrian, skeletons contributed little to accumulating limestones for more than 40 million years, until renewed animal radiation began late in the Early Ordovician (Pruss *et al.*, 2010, and references therein). Extracellular formation of calcite skeletons by foraminifera and coccolithophorids began in the Devonian and Late Triassic, respectively; the latter, and the broadly coeval radiation of planktonic Foraminifera, exerted a strong influence on environmental and geographic patterns of carbonate deposition in the oceans.

As illustrated by the later Cambrian decline of skeletal abundances, the course of skeletonization never did run smoothly. Several mass extinctions are documented in Phanerozoic rocks and these, by definition, record major extinctions of skeleton-forming organisms. In particular, patterns of extinction and survival during the largest known mass extinction, at the end of the Permian Period 252 Ma suggest the decisive importance of skeletal physiology. Organisms that produced massive carbonate skeletons but had limited physiological control over the fluids from which minerals precipitate disappeared almost entirely at the P–Tr boundary – more than 90% of latest Permian genera have no Triassic record (Knoll *et al.*, 2007). In contrast, latest Permian groups that fashioned skeletons from materials other than $CaCO_3$ lost only 10% of their genera. In some detail, observed patterns of extinction and survival match predictions made on the basis of physiological research on present day ocean acidification (Knoll *et al.*, 2007, and references cited therein). Therefore, skeletal biomineralization appears to have played a major role in determining organismic responses to end-Permian environmental change. More broadly, selective extinction and not sea-

water Mg/Ca may govern the proportional abundances of calcite and aragonite-precipitating taxa through time (Kiessling *et al*, 2008).

Indeed, organisms that produce massive carbonate skeletons have waxed and waned over the past 500 million years; metazoan reefs, for example, have appeared and disappeared half a dozen times during the Phanerozoic Eon, including the Cambrian decline noted earlier (Copper, 1994). As at the P–Tr boundary, long-term reef dynamics may reflect the physiological responses of calcifying organisms to physical changes in ocean chemistry (in many cases, redox-mediated; Higgins *et al.*, 2009).

The overall conclusion, then, is that both silica- and carbonate-precipitating organisms have clear trajectories that reflect interactions between skeletal physiology and dynamic environments. In silica-precipitating organisms, the principal environmental pressures appear to have been biological – especially the radiation of the diatoms and its consequences for silica availability in surface water masses. No carbonate skeleton-former has had a comparable influence on saturation states in the world's oceans, but physical changes in marine chemistry through time have episodically altered the cost/benefit ratio of skeleton formation and, therefore, the biological composition of ocean life (e.g. Knoll, 2003; Raven and Giordano, 2009).

10.5 Summary

In this chapter we highlighted areas where our understanding of skeletal formation is rapidly advancing while also identifying frontier questions and new research areas. We showed that biologically mineralized tissues are associated with, and often intimately linked to organic components that are concentrated within specialized compartments during mineral formation. While the presence of these mineral-associated organics implies a role in the mineralization process, assigning a function to a particular molecule within the organic matrix is not a straight-forward procedure. Some mineralizing organic matrices harbor a diverse population of molecules, and the time required to screen each for a particular function is often prohibitively daunting. Also, as protein–protein interactions may be important *in vivo*, many biomolecules are unlikely to retain their function when isolated. This can be especially problematic when novel proteins are identified, which bear little or no similarity to members of protein structure databases. To bridge this gap, future research efforts in this fast-moving area must be focused towards developing techniques for inferring protein function *in vivo*. For those molecules that appear to retain their function upon isolation (e.g. silicatein, silicase, silaffins, etc.), new insights into mineralization processes may be obtained by using *in situ* techniques, often in conjunction with model systems, to monitor organic–mineral interactions. At the same time however, new efforts are also needed to determine whether the behaviour of biomolecular species in *in vitro* experiments actually reflects their *in vivo* activity. To meet these objectives, complementary investigations targeted at understanding communication and signaling schemes utilized by cells to coordinate the synthesis and transport of these macromolecular species to designated locations at specific times will be required. At first glance, this level of biochemical detail may seem excessive; however, information garnered from these investigations will ultimately shed light on some of the most critical questions currently challenging the geoscience community.

Future investigations will undoubtedly focus on the chemical, physical, and genetic controls that regulate the nucleation, growth, and maintenance/remodeling of permineralized biomaterials. Of particular interest are the roles of amorphous and transient metastable phases during biomineralization processes, and the nature of their interactions with macromolecules in the biologically-mediated deposition environment. Related to these issues is the question of how macromolecular species may inhibit or promote the amorphous to crystalline transition while also controlling the expression of certain polymorphs and crystal faces. Indeed, with findings that many organisms use amorphous precursors to mineralization (Gotliv *et al.*, 2003; Nassif *et al.*, 2005; Politi *et al.*, 2006, 2008), many of the earlier assumptions about biomineral nucleation and growth will likely be fully revisited in the coming years. As our understanding of the non-classical crystallization pathways utilized during skeletogenesis increases, so too will our ability to understand which chemical and isotopic signatures within biominerals are faithful indicators of ambient environmental conditions. And, skeletal fossils will reveal newly recognized influences on long-term evolution in the oceans. Such information will help inform environmental issues likely to be important in the 21st century, such as how climate and acidification will affect marine organisms in a higher CO_2 world (e.g. Iglesias-Rodriguez *et al.*, 2008). This offers the promise of directly benefiting current and future efforts to interpret how the Earth–climate system responds to natural and anthropogenic perturbations. With the means to decouple signatures arising from the extraorganism environment from those imposed *in vivo*, an increasingly accurate record of past climate conditions will also emerge.

Acknowledgements

This material is based upon work supported by the National Science Foundation (EAR-0545166; OCE-052667) and the Department of Energy (FG02–00ER15112). A.H.K.'s contributions were supported by the NASA Astrobiology Institute. A.F.W. thanks Jim Schiffbauer for the SEM image of pennate diatoms in Fig. 10.4d.

References

Addadi L, Joester D, Nudelman F, Weiner S (2006) Mollusk shell formation: A source of new concepts for understanding biomineralization processes. *Chemistry – A European Journal*, **12**, 981–987.

Addadi L, Weiner S (1985) Interactions between acidic proteins and crystals: stereochemical requirements in biomineralization. *Proceedings of the National Academy of Sciences of the USA*, **82**, 4110–4114.

Addadi LW, Weiner S, Geva, M (2001) On how proteins interact with crystals and their effect on crystal formation. *Zeitschrift für Kardiologie*, **90**, 92–98.

Aizenberg J, Black AJ, Whitesides GM (1999) Control of crystal nucleation by patterned self-assembled monolayers. *Nature*, **398**, 495–498.

Aizenberg J, Weaver JC, Thanawala MS, Sundar VC, Morse DE, Fratzl P (2005) Skeleton of *Euplectella* sp.: Structural hierarchy from the nanoscale to the macroscale. *Science*, **309**, 275–278.

Al-Horani FA, Al-Moghrabi SM, De Beer D (2003) The mechanism of calcification and its relation to photosynthesis and respiration in the scleractinian coral *Galaxea fascicularis*. *Marine Biology*, **142**, 419–426.

Albeck S, Aizenberg J, Addadi L, Weiner S (1993) Interactions of various skeletal intracrystalline components with calcite crystals. *Journal of the American Chemical Society*, **115**, 11691–11697.

Allison CW, Hilgert JW (1986) Scale microfossils from the early Cambrian of northwest Canada. *Journal of Paleontology*, **60**, 973–1015.

Ameye L, De Becker G, Killian C, *et al.* (2001) Proteins and saccharides of the sea urchin organic matrix of mineralization: Characterization and localization in the spine skeleton. *Journal of Structural Biology*, **134**, 56–66.

Ameye L, Hermann R, Killian C, Wilt F, Dubois P (1999) Ultrastructural localization of proteins involved in sea urchin biomineralization. *Journal of Histochemistry and Cytochemistry*, **47**, 1189–1200.

Anand P, Elderfield H, Conte MH (2003) Calibration of Mg/Ca thermometry in planktonic foraminifera from a sediment trap time series. *Paleoceanography*, **18**, 1050.

Anderson OR (1976) A cytoplasmic fine-structure study of two spumellarian Radiolaria and their symbionts. *Marine Micropaleontology*, **1**, 81–89.

Anderson OR (1980) Radiolaria. In: *Biochemistry and Physiology of Protozoa* (eds Levandowski M, Hunter S). Academic Press, New York, pp. 1–40.

Anderson OR (1981) Radiolarian fine structure and silica deposition. In: *Silicon and Siliceous Structures in Biological Systems* (eds Simpson TL, Volcani BE). Springer-Verlag, New York, pp. 347–380.

Anderson OR (1983) *Radiolaria*, Springer-Verlag, New York.

Anderson OR, Nigirini C, Boltovskoy D, Takahashi K, Swanberg NR (2002) Class Polycystina. In: *An Illustrated*

Guide to the Protozoa (eds Lee JJ, Leedale GF, Bradbury P). Allen Press, Lawrence, Kansas, USA, pp. 994–1022.

Armstrong HA, Brasier MD (2006) *Microfossils,* Blackwell, Malden.

Armstrong N, Hardin J, Mcclay DR (1993) Cell-cell interactions regulate skeleton formation in the sea urchin embryo. *Development,* 119, 833–840.

Azam F, Hemmingsen BB, Volcani BE (1974) Role of silicon in diatom metabolism. 5. Silicic acid transport and metabolism in the heterotrophic diatom *Nitzschia alba. Archives of Microbiology,* 97, 103–114.

Basak C, Rathburn AE, Perez ME, *et al.* (2009) Carbon and oxygen isotope geochemistry of live (stained) benthic foraminifera from the Aleutian Margin and the Southern Australian Margin. *Marine Micropaleontology,* 70, 89–101.

Beck JW, Edwards RL, Ito E, *et al.* (1992) Sea-surface temperature from coral skeletal strontium calcium ratios. *Science,* 257, 644–647.

Becker ML, Cole JM, Rasbury ET, Pedone VA, Montanez IP, Hanson GN (2001) Cyclic variations of uranium concentrations and oxygen isotopes in tufa from the middle Miocene Barstow Formation, Mojave Desert, California. *Geology,* 29, 139–142.

Behrens P, Jahns M, Menzel H (2007) The polyamine silica system: A biomimetic model for the biomineralization of silica. In: *Handbook of Biomineralization* (eds Behrens P, Bauerlein E). Wiley-VCH, Weinheim, pp. 3–18.

Belcher AM, Wu XH, Christensen RJ, Hansma PK, Stucky GD, Morse DE (1996) Control of crystal phase switching and orientation by soluble mollusc-shell proteins. *Nature,* 381, 56–58.

Bellwood DR, Hughes TP, Folke C, Nystrom M (2004) Confronting the coral reef crisis. *Nature,* 429, 827–833.

Beltran C, de Rafelis M, Minoletti F, Renard M, Sicre MA, Ezat U (2007) Coccolith delta O-18 and alkenone records in middle Pliocene orbitally controlled deposits: High-frequency temperature and salinity variations of sea-surface water. *Geochemistry Geophysics Geosystems,* 8, Q05003.

Bemis BE, Spero HJ, Lea DW, Bijma J (2000) Temperature influence on the carbon isotopic composition of *Globigerina bulloides* and *Orbulina universa* (planktonic foraminifera). *Marine Micropaleontology,* 38, 213–228.

Bengtson S, Conway Morris S (1992) Early radiation of biomineralizing phyla. In: *Origin and Early Evolution of the Metazoa* (eds Lipps JH, Signor PW). Plenum, New York, p 447–481.

Beniash E, Aizenberg J, Addadi L, Weiner S (1997) Amorphous calcium carbonate transforms into calcite during sea urchin larval spicule growth. *Proceedings of the Royal Society of London Series B-Biological Sciences,* 264, 461–465.

Beniash E, Addadi L, Weiner S (1999) Cellular control over spicule formation in sea urchin embryos: A structural approach. *Journal of Structural Biology,* 125, 50–62.

Benner K, Klufers P, Vogt M (2003) Hydrogen-bonded sugar-alcohol trimers as hexadentate silicon chelators in aqueous solution. *Angewante Chemie.-International Edition,* 42, 1058–1062.

Bentov S, Erez J (2005) Novel observations on biomineralization processes in foramifera and implications for Mg/Ca ratio in the shells. *Geology,* 33, 841–844.

Bentov S, Erez J (2006) Impact of biomineralization processes on the Mg content of foraminiferal shells: A biological perspective. *Geochemistry, Geophysics, Geosystems,* 7, Q01P08.

Berman A, Addadi L, Weiner S (1988) Interactions of sea-urchin skeleton macromolecules with growing calcite crystals – a study of intracrystalline proteins. *Nature,* 331, 546–548.

Berman A, Addadi L, Kvick A, Leiserowitz L, Nelson M, Weiner S (1990) Intercalation of sea urchin proteins in calcite: study of a crystalline composite material. *Science,* 250, 664–667.

Berman A, Hanson J, Leiserowitz L, Koetzle TF, Weiner S, Addadi L (1993) Biological control of crystal texture: A widespread strategy for adapting crystal properties to function. *Science,* 259, 776–779.

Bhattacharyya P, Volcani BE (1980) Sodium-dependent silicate transport in the apochlorotic marine diatom *Nitzschia alba. Proceedings of the National Academy of Sciences of the USA,* 77, 6386–6390.

Bhattacharyya P, Volcani BE (1983) Isolation of silicate ionophore(s) from the apochlorotic diatom *Nitzschia alba. Biochemical and Biophysical Resesarch Communications,* 114, 365–372.

Billups K, Schrag DP (2002) Paleotemperatures and ice volume of the past 27 Myr revisited with paired Mg/Ca and $^{18}O/^{16}O$ measurements on benthic foraminifera. *Paleoceanography,* 17, 1003.

Binder BJ, Chisholm SW (1980) Changes in the soluble silicon pool size in the marine diatom *Thalassiosira weisflogii. Marine Biology Letters,* 1, 205–212.

Birchall JD (1995) The essentiality of silicon in biology. *Chemistry Society Review,* 24, 351–357.

Borremans C, Hermans J, Baillon S, Andre L, Dubois P (2009) Salinity effects on the Mg/Ca and Sr/Ca in starfish skeletons and the echinoderm relevance for paleoenvironmental reconstructions. *Geology,* 37, 351–354.

Boyle EA, Keigwin LD (1985) Comparison of atlantic and pacific paleochemical records for the last 215,000 years – Changes in deep ocean circulation and chemical inventories. *Earth and Planetary Science Letters,* 76, 135–150.

Boyle EA, Labeyrie L, Duplessy JC (1995) Calcitic foraminiferal data confirmed by cadmium in aragonitic *Hoeglundina* - application to the last glacial maximum in the northern Indian-ocean. *Paleoceanography,* 10, 881–900.

Brasier M, Green O, Shields G (1997) Ediacarian sponge spicule clusters from southwest Mongolia and the origins of the Cambrian fauna. *Geology,* 25.

Brunner E, Lutz K, Sumper M (2004) Biomimetic synthesis of silica nanospheres depends on the aggregation and phase separation of polyamines in aqueous solution. *Physical Chemistry Chemical Physics,* 6, 854–857.

Bryan SP, Marchitto TM (2008) Mg/Ca-temperature proxy in benthic foraminifera: New calibrations from the Florida Straits and a hypothesis regarding Mg/Li. *Paleoceanography,* 23, PA2220.

Cariolou M (1988) Purification and characterization of calcium-binding conchiolin shell peptides from the mollusk, *Haliotis rufescens,* as a function of development. *Journal of Comparative Physiology,* 157, 717–729.

Cha JN, Shimizu K, Zhou Y *et al.* (1999) Silicatein filaments and subunits from a marine sponge direct the polymerization of silica and silicones *in vitro. Proceedings of the National Academy of Sciences of the USA,* 96, 361–365.

Chang VTC, Williams RJP, Makishima A, Belshawl NS, O'Nions RK (2004) Mg and Ca isotope fractionation during CaCO₃ biomineralisation. *Biochemical and Biophysical Research Communications,* **323,** 79–85.

Chase Z, Anderson RF, Fleisher MQ (2001) Evidence from authigenic uranium for increased productivity of the glacial Subantarctic Ocean. *Paleoceanography,* **16,** 468–478.

Cheers MS, Ettensohn CA (2005) P16 is an essential regulator of skeletogenesis in the sea urchin embryo. *Developmental Biology,* **283,** 384–396.

Cheng XG, Varona PL, Olszta MJ, Gower LB (2007) Biomimetic synthesis of calcite films by a polymer-induced liquid-precursor (PILP) process 1. Influence and incorporation of magnesium. *Journal of Crystal Growth,* **307,** 395–404.

Chiarello RP, Wogelius RA, Sturchio NC (1993) In-situ synchrotron X-ray reflectivity measurements at the calcite-water interface. *Geochimica et Cosmochimica Acta,* **57,** 4103–4110.

Chisholm SW, Azam F, Eppley RW (1978) Silicic acid incorporation in marine diatoms on light:dark cycles: Use as an assay for phased cell division. *Limnology and Oceanography,* **23,** 518–529.

Chivas AR, de Dekker P, Shelley JMG (1986) Magnesium and Strontium in nonmarine ostracod shells as indicators of paleosalinity and paleotemperature. *Hydrobiologia,* **143,** 135–142.

Claquin P, Martin-Jézéquel V (2005) Regulation of the Si and C uptake and of the soluble free-silicon pool in a synchronised culture of *Cylindrotheca fusiformis* (*Bacillariophyceae*): Effects on the Si/C ratio. *Marine Biology,* **146,** 877–886.

Cohen AL, Mcconnaughey TA (2003) Geochemical perspectives on coral mineralization. In: *Biomineralization* (eds Dove PM, De Yoreo JJ, Weiner S). Mineralogical Society of America, Washington, DC, pp. 151–187.

Cohen PA, Schopf JW, Butterfield NJ, Kudryavtsev, AB, Macdonald FA (2011) Phosphate biomineralization in mid-Neoproterozoic protists. *Geology,* **39,** 539–542.

Conway HL, Harrison PJ (1977) Marine diatoms grown in chemostats under silicate or ammonium limitation .4. Transient-response of *Chaetoceros debilis, Skeletonema costatum,* and *Thalassiosira gravida* to a single addition of limiting nutrient. *Marine Biology,* **43,** 33–43.

Conway HL, Harrison PJ, Davis CO (1976) Marine diatoms grown in chemostats under silicate or ammonium limitation .2. Transient-response of *Skeletonema costatum* to a single addition of limiting nutrient. *Marine Biology,* **35,** 187–199.

Coombs J, Volcani BE (1968) Studies on the biochemistry and fine structure of silica-shell formation in diatoms. Chemical changes in the wall of *Navicula pelliculosa* during its formation. *Planta (Berlin),* **82,** 280–292.

Copper, P (1994) Ancient reef ecosystem expansion and collapse. *Coral Reefs* 13: 3–11.

Crawford RM (1981) The siliceous components of the diatom cell wall and their morphological variation. In: *Silicon and Siliceous Structures in Biological Systems* (eds Simpson TL, Volcani BE). Springer-Verlag, New York, pp. 129–156.

Crawford SA, Higgins MJ, Mulvaney P, Wetherbee R (2001) Nanostructure of the diatom frustule as revealed by atomic force and scanning electron microscopy. *Journal Phycol.,* **37,** 543–554.

Crick RE (1989) *Origin, Evolution, and Modern Aspects of Biomineralization in Plants and Animals.* Plenum Press, New York, pp. 536.

Cronin TM, Dowsett HJ, Dwyer GS, Baker PA, Chandler MA (2005) Mid-Pliocene deep-sea bottom-water temperatures based on ostracod Mg/Ca ratios. *Marine Micropaleontology,* **54,** 249–261.

Danelian T, Moreira D (2004) Palaeontological and molecular arguments for the origin of silica-secreting marine organisms. *Comptes Rendus Palevol,* **3,** 229–236.

Davis KJ, Dove PM, Wasylenki LE, De Yoreo JJ (2004) Morphological consequences of differential Mg²⁺ incorporation at structurally distinct steps on calcite. *American Mineralogist,* **89,** 714–720.

Dekens PS, Lea DW, Pak DK, Spero HJ (2002) Core top calibration of Mg/Ca in tropical foraminifera: Refining paleotemperature estimation. *Geochemistry Geophysics Geosystems,* **3,** 1022.

De La Rocha CL, Hutchins DA, Brzezinski MA, Zhang YH (2000) Effects of iron and zinc deficiency on elemental composition and silica production by diatoms. *Marine Ecology-Progress Series,* **195,** 71–79.

De Wever P, Dumitrica P, Caulet JP, Nigrini C, Caridroit M (2001) *Radiolarians in the Sedimentary Record.* Gordon and Breach Science Publishers, Amsterdam.

Dey A, Bomans PHH, Müller FA, *et al.* (2010) The role of prenucleation clusters in surface-induced calcium phosphate crystallization. *Nature Materials,* **9,** 1010 – 1014.

De Yoreo JJ, Dove PM (2004) Materials science. Shaping crystals with biomolecules. *Science,* **306,** 1301–1302.

Decker GL, Morrill JB, Lennarz WJ (1987) Characterization of sea urchin primary mesenchyme cells and spicules during biomineralization in vitro. *Development,* **101,** 297–312.

Del Amo Y, Brzezinski MA (1999) The chemical form of dissolved Si taken up by marine diatoms. *Journal of Phycology,* **35,** 1162–1170.

Delaney M, Bé AWH, Boyle EA (1985) Li, Sr, Mg and Na in foraminiferal calcite shells from laboratory culture, sediment traps, and sediment cores. *Geochimica et Cosmochimica Acta,* **49,** 1327–1341.

Dickson JAD (2002) Fossil echinoderms as monitor of the Mg/Ca ratio Phanerozoic oceans. *Science,* **298,** 1222–1224.

DiMasi E, Kwak SY, Amos FF, Olszta MJ, Lush D, Gower LB (2006) Complementary control by additives of the kinetics of amorphous CaCO₃ mineralization at an organic interface: *In-situ* synchrotron x-ray observations. *Physics Review Letters,* **97,** 045503.

Dove PM, De Yoreo JJ, Weiner S (2003) Biomineralization. *Reviews in Mineralogy and Geochemistry 54* (ed Rosso JJ). Mineralogical Society of America, Washington, D.C., pp. 381.

Dunn CW, Hejnol A, Matus DQ, *et al.* (2008) Broad phylogenomic sampling improves resolution of the animal tree of life. *Nature,* **452,** 745–U745.

Eggins S, De Deckker P, Marshall J (2003) Mg/Ca variation in planktonic foraminifera tests: Implications for reconstructing palaeo-seawater temperature and habitat migration. *Earth and Planetary Science Letters,* **212,** 291–306.

Ehrlich H, Heinemann S, Heinemann C, *et al.* (2008a) Nanostructural organization of naturally occurring composites – Part I: Silica-collagen-based biocomposites. *Journal of Nanomaterials,* 1–8, 10.1155/2008/623838.

Ehrlich H, Janussen D, Simon P, *et al.* (2008b) Nanostructural organization of naturally occurring composites – Part II: Silica-chitin-based biocomposites. *Journal of Nanomaterials*, 1–8, 10.1155/2008/670235.

Elderfield H, Ganssen G. (2000) Past temperature and delta ^{18}O of surface ocean waters inferred from foraminiferal Mg/Ca ratios. *Nature*, **405**, 442–445.

Elhadj S, De Yoreo JJ, Hoyer JR, Dove PM (2006a) Role of molecular charge and hydrophilicity in regulating the kinetics of crystal growth. *Proceedings of the National Academy of Sciences of the USA*, **103**, 19237–19242.

Elhadj S, Salter EA, Wierzbicki A, De Yoreo JJ, Han N, Dove PM (2006b) Peptide controls on calcite mineralization: Polyaspartate chain length affects growth kinetics and acts as a stereochemical switch on morphology. *Crystal Growth and Design*, **6**, 197–201.

Epstein S, Lowenstam HA (1954) Temperature-shell-growth relations of recent and interglacial Pleistocene shoal-water biota from Bermuda. *Journal of Geology*, **61**, 424–438.

Erez J (2003) The source of ions for biomineralization in foraminifera and their implications for paleoceanographic proxies. In: *Biomineralization* (eds Dove PM, De Yoreo JJ, Weiner S). Mineralogical Society of America, Washington, D.C., pp.115–149.

Fairhead M, Johnson KA, Kowatz T, *et al.* (2008) Crystal structure and silica condensing activities of silicatein alpha-cathepsin L chimeras. *Chemical Communications*, 1765–1767.

Falini G, Albeck S, Weiner S, Addadi L (1996) Control of aragonite or calcite polymorphism by mollusk shell macromolecules. *Science*, **271**, 67–69.

Falini G, Fermani S, Gazzano M, Ripamonti A (2000) Polymorphism and architectural crystal assembly of calcium carbonate in biologically inspired polymeric matrices. *Journal of the Chemical Society, Dalton Transactions*, 3983–3987.

Falini G, Weiner S, Addadi L (2003) Chitin-silk fibroin interactions: Relevance to calcium carbonate formation in invertebrates. *Calcified Tissue International*, **72**, 548–554.

Fehling J, Stoecker D, Baldauf SL (2007) Photosynthesis and the Eukaryotic tree of life. In: *Evolution of Aquatic Photoautotrophs*. (eds Falkowski P, Knoll AH). London, Academic Press, pp. 75–107.

Finkel ZV, Katz ME, Wright JD, Schofield OME, Falkowski PF (2005) Climatically driven macroevolutionary patterns in the size of marine diatoms over the Cenozoic. *Proceedings of the National Academy of Sciences of the USA*, **102**, 8927–8932.

Fischer WW, Higgins JA, Pruss SB, 2007, Delayed biotic recovery from the Permian-Triassic extinction may have been influenced by a redox-driven reorganization of the marine carbonate system. *Geological Society of America, Abstracts with Programs*, **39**(6), 420.

Fu G, Qiu SR, Orme CA, Morse DE, De Yoreo JJ (2005) Acceleration of calcite kinetics by abalone nacre proteins. *Advanced Materials*, 17, 2678–2683.

Garrone R (1978) *Phylogenesis of Connective Tissue. Morphological aspects and biosynthesis of sponge intercellular matrix*, Karger, Basel.

Garrone R, Simpson TL, Pottu-Boumendil J (1981) Ultrastructure and deposition of Silica in sponges. In: *Silicon and Siliceous Structures in Biological Systems* (eds Simpson TL, Volcani BE). Springer-Verlag, New York, pp. 495–526.

Gattuso JP, Frankignoulle M, Wollast R (1998) Carbon and carbonate metabolism in coastal aquatic ecosystems. *Annual Review of Ecology and Systematics*, **29**, 405–434.

Gayathri S, Lakshminarayanan R, Weaver JC, Morse DE, Kini RM, Valiyaveettil S (2007) In vitro study of magnesium-calcite biomineralization in the skeletal materials of the seastar *Pisaster giganteus*. *Chemistry – A European Journal*, **13**, 3262–3268.

Gehling JG, Rigby JK (1996) Long expected sponges from the Neoproterozoic Ediacara fauna of South Australia. *Journal of Paleontology*, **70**, 185–195.

Gebauer D, Gunawidjaja PN, Ko JYP, *et al.* (2010) Proto-calcite and proto-vaterite in amorphous calcium carbonates. *Angewante Chemie International Edition*, **49**, 8889–8891.

Gebauer D, Völkel A, Cölfen H (2008) Stable prenucleation calcium carbonate clusters. *Science*, **322**, 1819–1822.

Goodwin AL, Michel FM, Phillips BL, Keen DA, Dove MT, Reeder RJ (2010) Nanoporous structure and medium-range order in synthetic amorphous calcium carbonate. *Chemistry of Materials*, **22**, 3197–3205.

Gotliv BA, Addadi L, Weiner S (2003) Mollusk shell acidic proteins: In search of individual functions. *Chembiochem*, **4**, 522–529.

Gotliv BA, Kessler N, Sumerel JL, *et al.* (2005) Asprich: A novel aspartic acid-rich protein family from the prismatic shell matrix of the bivalve *Atrina rigida*. *Chembiochem*, **6**, 304–314.

Grachev M, Sherbakova T, Masyukova Y, Likhoshway Y (2005) A potential zinc-binding motif in silicic acid transport proteins of diatoms. *Diatom Research*, **20**, 409–411.

Grant SWF (1990) Shell structure and distribution of *Cloudina*, a potential index fossil for the terminal Proterozoic. *American Journal of Science*, **290A**, 261–294.

Greenfield EM, Wilson DC, Crenshaw MA (1984) Ionotropic nucleation of calcium carbonate by molluscan matrix. *American Zoologist*, **24**, 925–932.

Grégoire C (1961) Structure of the conchiolin cases of the prisms in *Mytilus edulis* Linne. *Journal of Biophysical and Biochemical Cytology*, **9**, 395–400.

Griffith EM, Paytan A, Kozdon R, Eisenhauer A, Ravelo AC (2008) Influences on the fractionation of calcium isotopes in planktonic foraminifera. *Earth and Planetary Science Letters*, **268**, 124–136.

Grotzinger JP, Watters W, Knoll AH (2000) Calcareous metazoans in thrombolitic bioherms of the terminal Proterozoic Nama Group, Namibia. *Paleobiology*, **26**, 334–359.

Gueta R, Natan A, Addadi L, Weiner S, Refson K, Kronik L (2007) Local atomic order and infrared spectra of biogenic calcite. *Angewandte Chemie International Edition*, **46**, 291–294.

Guss K, Ettensohn C (1997) Skeletal morphogenesis in the sea urchin embryo: Regulation of primary mesenchyme gene expression and skeletal rod growth by ectoderm-derived cues. *Development*, **124**, 1899–1908.

Haeckel E (1887) Report on radiolaria by H.M.S. challenger during the years 1873–1876. In: *The Voyage of the H.M.S. Challenger* (eds Thompson CW, Murray J). Her Majesty's Stationery Office, London, pp. 1–1760.

Hardie LA (1996) Secular variation in seawater chemistry: An explanation for the coupled secular variation in the mineralogies of marine limestones and potash evaporates over the past 600 m.y. *Geology*, **24**, 279–283.

Harper HE, Knoll AH (1975) Silica, diatoms, and Cenozoic radiolarian evolution. *Geology*, 3, 175–177.

Hall JM, Chan LH (2004) Ba/Ca in *Neogloboquadrina pachyderma* as an indicator of deglacial meltwater discharge into the western Arctic Ocean. *Paleoceanography*, 19, PA1017.

Harkey MA, Klueg K, Sheppard P, Raff RA (1995) Structure, expression, and extracellular targeting of PM27, a skeletal protein associated specifically with growth of the sea urchin larval spicule. *Developmental Biology*, 168, 549–566.

Harrison CC, Loton N (1995) Novel routes to designer silicas – Studies of the decomposition of $(M^+)_2[Si(C_6H_4O_2)_3]$ •XH_2O – Importance of M^+ identity of the kinetics of oligomerization and the structural characteristics of the silicas produced. *Journal of the Chemical Society-Faraday Transactions*, 91, 4287–4297.

Hartman WD, Goreau TF (1970) Jamaican coralline sponges: their morphology, ecology and fossil relatives. In: *The Biology of Porifera. Symposium, Zoological Society of London* (ed Fry WG). Academic Press, London, pp. 205–243.

Hastings DW, Emerson SR, Mix AC (1996) Vanadium in foraminiferal calcite as a tracer for changes in the areal extent of reducing sediments. *Paleoceanography*, 11, 665–678.

Healey SL, Thunell RC, Corliss BH (2008) The Mg/Ca-temperature relationship of benthic foraminiferal calcite: New core-top calibrations in the < 4 degrees C temperature range. *Earth and Planetary Science Letters*, 272, 523–530.

Heasman J (2002) Morpholino Oligos: Making Sense of Antisense? *Developmental Biology*, 243, 209–214.

Hecky RE, Mopper K, Kilham P, Degens ET (1973) The amino acid and sugar composition of diatom cell-walls. *Mar. Biol.*, 19, 323–331.

Henderson GM (2002) New oceanic proxies for paleoclimate. *Earth and Planetary Science Letters*, 203, 1–13.

Higgins JA, Fischer WW, Schrag DP (2009) Oxygenation of the ocean and sediments: Consequences for the seafloor carbonate factory. *Earth and Planetary Science Letters*, 284, 25–33.

Hildebrand M (2000) Silicic acid transport and its control during cell wall silicification in diatoms. In: *Biomineralization: Progress in Biology, Molecular Biology and Application* (ed Bauerlein E). Wiley-VCH, Weinheim, pp. 159–176.

Hildebrand M, Wetherbee R (2003) Components and control of silicification in diatoms. In: *Silicon Biomineralization: Biology-Biochemistry-Molecular Biology-Biotechnology* (ed Müller WEG). Springer-Verlag, Berlin, pp. 11–57.

Hildebrand M, Volcani BE, Gassmann W, Schroeder JI (1997) A gene family of silicon transporters. *Nature*, 385, 688–689.

Hildebrand M, Dahlin K, Volcani BE (1998) Characterization of a silicon transporter gene family in *Cylindrotheca fusiformis*: Sequences, expression analysis, and identification of homologs in other diatoms. *Molecular and General Genetics*, 260, 480–486.

Hintz CJ, Shaw TJ, Chandler GT, Bernhard JM, Mccorkle DC, Blanks JK (2006) Trace/minor element: calcium ratios in cultured benthic foraminifera. Part 1. Inter-species and inter-individual variability. *Geochimica et Cosmochimica Acta*, 70, 1952–1963.

Hodor PG, Ettensohn CA (1998) The dynamics and regulation of mesenchymal cell fusion in the sea urchin embryo. *Developmental Biology*, 199, 111–124.

Iglesias-Rodriguez MD, Halloran PR, Rickaby REM, *et al.* (2008) Phytoplankton calcification in a high-CO_2 world. *Science*, 320, 336–340.

Iler RK (1979) *The Chemistry of Silica: Solubility, Polymerization, Colloid and Surface Properties, and Biochemistry,* John Wiley & Sons, New York.

Illies MR, Peeler MT, Dechtiaruk AM, Ettensohn CA (2002) Identification and developmental expression of new biomineralization proteins in the sea urchin *Strongylocentrotus purpuratus*. *Development Genes and Evolution*, 212, 419–431.

Jada A, Jradi K (2006) Role of polyelectrolytes in crystallogenesis of calcium carbonate. *Macromolecular Symposia*, 233, 147–151.

Jones DS and Quitmyer IR (1989) Marking time with bivalve shells: Oxygen isotopes and season of annual increment formation. *Palaios*, 11, 340–346.

Jones WC (1970) The composition, development, form and orientation of calcareous sponge spicules. In: *The Biology of the Porifera Symposium, Zoological Society of London* (ed Fry WG). Adademic Press, London, pp. 91–123.

Kiessling W, Aberhan M, Villier L (2008) Phanerozoic trends in skeletal diversity driven by mass extinctions. *Nature Geoscience*, 1, 527–530.

Killian CE, Wilt FH (1996) Characterization of the proteins comprising the integral matrix of *Strongylocentrotus purpuratus* embryonic spicules. *Journal of Biological Chemistry*, 271, 9150–9159.

Killian CE, Metzler RA, Gong YUT, *et al.* (2009) Mechanism of calcite co-orientation in the sea urchin tooth. *Journal of the American Chemical Society*, 131, 18404 – 18409.

Kim IW, Darragh, MR, Orme C, Evans JS (2006) Molecular 'tuning' of crystal growth by nacre-associated polypeptides. *Crystal Growth and Design*, 6, 5–10.

King K (1974) Preserved amino acids from silicified protein in fossil Radiolaria. *Nature,* 252, 690–692.

King K (1975) Amino acids composition of the silicified organic matrix in fossil polycystine Radiolaria. *Micropaleontology*, 21, 215–226.

King K (1977) Amino acid survey of recent calcareous and siliceous deep-sea microfossils. *Micropaleontology*, 23.

Kinrade SD, Del Nin JW, Schach AS, Sloan TA, Wilson KL, Knight CTG (1999) Stable five- and six-coordinated silicate anions in aqueous solution. *Science*, 285, 1542–1545.

Kinrade SD, Hamilton RJ, Schach AS, Knight CTG (2001a) Aqueous hypervalent silicon complexes with aliphatic sugar acids. *Journal of the Chemical Society-Dalton Transactions*, 961–963.

Kinrade SD, Schach AS, Hamilton RJ, Knight CTG (2001b) NMR evidence of pentaoxo organosilicon complexes in dilute neutral aqueous silicate solutions. *Chemical Communications*, 1564–1565.

Kinrade SD, Gillson AME, Knight CTG (2002) Silicon-29 NMR evidence of a transient hexavalent silicon complex in the diatom Navicula pelliculosa. *Journal of the Chemical Society-Dalton Transactions*, 307–309.

Kitajima T, Urakami H (2001) Differential distribution of spicule matrix proteins in the sea urchin embryo skeleton. *Development Growth and Differentiation*, 42, 295–306.

Klein RT, Lohmann KC, Thayer CW (1996) Sr/Ca and C-13/C-12 ratios in skeletal calcite of *Mytilus trossulus*: Covariation with metabolic rate, salinity, and carbon isotopic composition of seawater. *Geochimica et Cosmochimica Acta*, **60**, 4207–4221.

Knoll AH (2003) Biomineralization and evolutionary history. In: *Biomineralization* (eds Dove PM, De Yoreo JJ, Weiner S). Mineralogical Soc. America, Washington, D.C., pp. 329–356.

Knoll AH, Bambach RK, Payne J, Pruss S, Fischer W (2007) A paleophysiological perspective on the end-Permian mass extinction and its aftermath. *Earth and Planetary Science Letters*, **256**, 295–313.

Kooistra WHCF, Gersonde R, Medlin LK, Mann DG (2007) The origin and evolution of the diatoms: their adaptation to a planktonic existence. In: *Evolution of Aquatic Photoautotrophs* (eds Falkowski P, Knoll AH). London, Academic Press, pp. 207–249.

Kotachi A, Miura T, Imai H (2006) Polymorph control of calcium carbonate films in a poly(acrylic acid)-chitosan system. *Crystal Growth and Design*, **6**, 1636–1641.

Kouduka M, Matuoka A, Nishigaki K (2006) Acquisition of genome information from single-celled unculturable organisms (radiolaria) by exploiting genome profiling (GP). *BMC Genomics*, **7**, 10.

Kozdon R, Eisenhauer A, Weinelt M, Meland MY, Nurnberg D (2009) Reassessing Mg/Ca temperature calibrations of *Neogloboquadrina pachyderma* (sinistral) using paired delta Ca-44/40 and Mg/Ca measurements. *Geochemistry, Geophysics, Geosystems*, **10**, Q03005.

Krautter M, Conway KW, Barrie JV, Neuweiler M (2001) Discovery of a 'living dinosaur': Globally unique modern hexactinellid sponge reefs off British Columbia, Canada. *Facies*, **44**, 265–282.

Krautter M, Conway KW, Barrie JV (2006) Recent hexactinosidan sponge reefs (silicate mounds) off British Columbia, Canada: Frame-building processes. *Journal of Paleontology*, **80**, 38–48.

Kretsinger RH (1976) Calcium-binding proteins. *Annual Reviews of Biochemistry*, **45**, 239–266.

Kröger N, Sumper M (1998) Diatom cell wall proteins and the cell biology of silica biomineralization. *Protist*, **149**, 213–219.

Kröger N, Sumper M (2000) The molecular basis of diatom biosilica formation. In: *Biomineralization: Progress in Biology, Molecular Biology and Application* (ed Bauerlein E). Wiley-VCH, Weinheim, pp. 138–158.

Kröger N, Bergsdorf C, Sumper M (1994) A new calcium-binding glycoprotein family constitutes a major diatom cell-wall component. *EMBO Journal*, **13**, 4676–4683.

Kröger N, Lehmann G, Rachel R, Sumper M (1997) Characterization of a 200-kDa diatom protein that is specifically associated with a silica-based substructure of the cell wall. *European Journal of Biochemistry*, **250**, 99–105.

Kröger N, Deutzmann R, Sumper M (1999) Polycationic peptides from diatom biosilica that direct silica nanosphere formation. *Science*, **286**, 1129–1132.

Kröger N, Deutzmann R, Bergsdorf C, Sumper M (2000) Species-specific polyamines from diatoms control silica morphology. *Proceedings of the National Academy of Sciences of the USA*, **97**, 14133–14138.

Kröger N, Deutzmann R, Sumper M (2001) Silica-precipitating peptides from diatoms - The chemical structure of silaffin-1A from Cylindrotheca fusiformis. *Journal of Biological Chemistry*, **276**, 26066–26070.

Kröger N, Lorenz S, Brunner E, Sumper M (2002) Self-assembly of highly phosphorylated silaffins and their function in biosilica morphogenesis. *Science*, **298**, 584–586.

Kunitomo Y, Sarashina I, Iijima M, Endo K, Sashida K (2006) Molecular phylogeny of acantharian and polycystine radiolarians based on ribosomal DNA sequences, and some comparisons with data from the fossil record. *European Journal of Protistology*, **42**, 143–153.

Lambert JB, Lu G, Singer SR, Kolb VM (2004) Silicate complexes of sugars in aqueous solution. *Journal of the American Chemical Society*, **126**, 9611–9625.

Lazarus DB, Kotrc B, Wulf G, Schmidt DN (2009) Radiolarians decreased silicification as an evolutionary response to reduced Cenozoic ocean silica availability. *Proceedings of the National Academy of Sciences, USA*, **106**, 9333–9338.

Lea DW (2006) The oceans and marine geochemistry. In: *Treatise on Geochemistry* (ed Elderfield H). Elsevier, Oxford, pp. 365–390.

Lea DW, Spero HJ (1994) Assessing the reliability of paleochemical tracers–barium uptake in the shells of planktonic foraminifera. *Paleoceanography*, **9**, 445–452.

Lea DW, Mashiotta TA, Spero HJ (1999) Controls on magnesium and strontium uptake in planktonic foraminifera determined by live culturing. *Geochimica et Cosmochimica Acta*, **63**, 2369–2379.

Lea DW, Pak JK, Paradis G (2005) Influence of volcanic shards on foraminferal Mg/Ca in a core from the Galapagos region. *Geochemistry, Geophysics, Geosystems*, **6**, Q11P04.

Leadbeater BSC, Riding R (1986) *Biomineralization in Lower Plants and Animals*. The Systematics Association Special Volume 30. Clarendon Press, Oxford, pp. 401.

Lear CH, Elderfield H, Wilson PA (2000) Cenozoic deep-sea temperatures and global ice volumes from Mg/Ca in benthic foramaniferal calcite. *Science*, **287**, 269–272.

Lear CH, Rosenthal Y, Slowey N (2002) Benthic foraminifera Mg/Ca-paleothermometry: A revised core-top calibration. *Geochimica et Cosmochimica Acta*, **66**, 3375–3387.

Lee JRI, Han YJ, Willey TM, Wang D, *et al.* (2007) Structural development of mercaptophenol self-assembled monolayers and the overlying mineral phase during templated $CaCO_3$ crystallization from a transient amorphous film. *Journal of the American Chemical Society*, **129**, 10370–10381.

Lee SL, Glonek T, Glimcher MJ (1983) ^{31}P nuclear magnetic resonance spectroscopic evidence for ternary complex formation of fetal dentin phosphoprotein with calcium and inorganic orthophosphate ions. *Calcified Tissue International*, **35**, 815–818.

Lee YH, Britten RJ Davidson EH (1999) *SM37*, a skeletogenic gene of the sea urchin embryo linked to the *SM50* gene. *Development Growth and Differentiation*, **41**, 303–312.

Levi Y, Albeck S, Brack A, Weiner S, Addadi L (1998) Control over aragonite crystal nucleation and growth: An in vitro study of biomineralization. *Chemistry – A European Journal*, **4**, 389–396.

Levi-Kalisman Y, Falini G, Addadi L, Weiner S (2001) Structure of the nacreous organic matrix of a bivalve mollusk shell examined in the hydrated state using cryo-TEM. *Journal of Structural Biology*, **135**, 8–17.

Liang YC, Si J, Romheld V (2005) Silicon uptake and transport is an active process in Cucumis sativus. *New Phytologist*, **167**, 797–804.

Livingston BT, Killian CE, Wilt F, *et al.* (2006) A genome-wide analysis of biomineralization-related proteins in the sea urchin *Strongylocentrotus purpuratus*. *Developmental Biology*, **300**, 335–348.

Lotse E, Park RJ, Warren J, Meldrum C (2010) Precipitation of calcium carbonate in confinement. *Advanced Functional Materials*, **14**, 1211 – 1220.

Love GD, Grosjean E, Stalvies C, *et al.* (2008) Fossil steroids record the appearance of Demospongiae during the Cryogenian period. *Nature*, **457**, 718–721.

Lowenstam HA, Weiner S (1989) *On Biomineralization*, Oxford University Press, New York.

Lutz K, Gröger C, Sumper M, Brunner E (2005) Biomimetic silica formation: Analysis of the phosphate-induced self-assembly of polyamines. *Physical Chemistry Chemical Physics*, **7**, 2812–2815.

Ma JF, Tamai K, Yamaji N, *et al.* (2006) A silicon transporter in rice. *Nature*, **440**, 688–691.

Ma JF, Yamaji N, Mitani N, *et al.* (2007) An efflux transporter of silicon in rice. *Nature*, **448**, 209–U212.

Ma YR, Aichmayer B, Paris O, *et al.* (2009) The grinding tip of the sea urchin tooth exhibits exquisite control over calcite crystal orientation and Mg distribution. *Proceedings of the National Academy of Sciences of the USA.*, **106**, 6048–6053.

Mackinder L, Wheeler G, Schroeder D, Riebesell U, Brownlee C (2010) Molecular mechanisms underlying calcification in Coccolithophores. *Geomicrobiology Journal*, **27**, 585–595.

Maldonado M, Carmona MG, Uriz MJ, Cruzado A (1999) Decline in Mesozoic reef-building sponges explained by silicon limitation. *Nature*, **401**, 785–788.

Malinda KM, Ettensohn CA (1994) Primary mesenchyme cell migration in the sea urchin embryo: Distribution of directional cues. *Developmental Biology*, **164**, 562–578.

Maliva R, Knoll AH, Siever R (1989) Secular change in chert distribution: a reflection of evolving biological participation in the silica cycle. *Palaios*, **4**, 519–532.

Mann S (1988) Molecular recognition in biomineralization. *Nature*, **332**, 119–124.

Mann S (2001) *Biomineralization: Principles and Concepts in Bioinorganic Materials Chemistry*, Oxford University Press, New York.

Mann S, Webb J, Williams RJP (1989) *Biomineralization: Chemical and Biochemical Perspectives*. VCH, New York, pp. 541.

Mann S, Didymus JM, Sanderson, NP, Heywood BR, Aso Samper EJ (1990) Morphological influence of functionalized and non-functionalized-alpha,omega-dicarboxylates on calcite crystallization. *Journal of the Chemical Society, Faraday Transactions*, **86**, 1873–1880.

Marchitto TM, Curry WB, Oppo DW (2000) Zinc concentrations in benthic foraminifera reflect seawater chemistry. *Paleoceanography*, **15**, 299–306.

Marin F, Amons R, Guichard N, *et al.* (2005) Caspartin and calprismin, two proteins of the shell calcitic prisms of the Mediterranean fan mussel *Pinna nobilis. Journal of Biological Chemistry*, **280**, 33895–33908.

Marin F (2004) Molluscan shell proteins. *Comptes Rendus Palevol*, **3**, 469–492.

Martin-Jézéquel V, Hildebrand M, Brzezinski MA (2000) Silicon metabolism in diatoms: Implications for growth. *Journal of Phycology*, **36**, 821–840.

Matsunaga S, Sakai R, Jimbo M, Kamiya H (2007) Long-chain polyamines (LCPAs) from marine sponge: Possible implication in spicule formation. *ChemBioChem*, **8**, 1729–1735.

McCrea JM (1950) On the isotopic chemistry of carbonates and a paleotemperature scale. *Journal of Physical Chemistry*, **18**, 849–857.

Medina-Elizalde M, Lea DW (2005) The mid-Pleistocene transition in the tropical pacific. *Science*, **310**, 1009–1012.

Medina-Elizalde M, Lea DW, Fantle MS (2008) Implications of seawater Mg/Ca variability for Plio-Pleistocene tropical climate reconstruction. *Earth and Planetary Science Letters*, **269**, 585–595.

Meibom A, Yurimoto H, Cuif J-P, *et al.* (2006) Vital effect in coral skeletal composition display strict three-dimensional control. *Geophysical Research Letters*, **33**, L11608, doi:10.1029/2006GL025968.

Meibom A, Cuif J-P, Houlbreque F, *et al.* (2008) Compositional variations at ultra-structure length scales in coral skeleton. *Geochemica et Cosmochimic Acta*, **72**, 1555–1569.

Meldrum FC (2003) Calcium carbonate in biomineralisation and biomimetic chemistry. *International Materials Reviews*, **48**, 187–224.

Michel FM, MacDonald J, Feng J, *et al.* (2008) Structural characteristics of synthetic amorphous calcium carbonate. *Chemistry of Materials*, **20**, 4720–4728.

Milliman JD (1993) Production and accumulation of calcium-carbonate in the ocean – budget of a nonsteady state. *Global Biogeochemical Cycles*, **7**, 927–957.

Milliman JD, Droxler AW (1996) Neritic and pelagic carbonate sedimentation in the marine environment: Ignorance is not bliss. *Geologische Rundschau*, **85**, 496–504.

Mitani N, Ma JF (2005) Uptake system of silicon in different plant species. *Journal of Experimental Botany*, **56**, 1255–1261.

Miyamoto H, Miyashita T, Okushima M, Nakano S, Morita T, Matsushiro A (1996) A carbonic anhydrase from the nacreous layer in oyster pearls. *Proceedings of the National Academy of Science of the USA*, **93**, 9657–9660.

Müller WEG, Belikov SI, Tremel W, *et al.* (2006) Siliceous spicules in marine demosponges (example *Suberites domuncula*). *Micron*, **37**, 107–120.

Müller WEG, Boreiko A, Schlossmacher U, *et al.* (2008a) Identification of a silicatein(-related) protease in the giant spicules of the deep-sea hexactinellid *Monorhaphis chuni. Journal of Experimental Biology*, **211**, 300–309.

Müller WEG, Schlossacher U, Wang X, *et al.* (2008b) Poly(silicate)-metabolizing silicatein in siliceous spicules and silicasomes of demosponges comprises dual enzymatic activities (silica polymerase and silica esterase). *FEBS Journal*, **275**, 362–370.

Müller WEG, Wang XH, Kropf K, *et al.* (2008c) Bioorganic/inorganic hybrid composition of sponge spicules: Matrix of the giant spicules and of the comitalia of the deep sea hexactinellid *Monorhaphis. Journal of Structural Biology*, **161**, 188–203.

Naka K (2007) Delayed action of synthetic polymers for controlled mineralization of calcium carbonate. *Topics in Current Chemistry*, **271**, 119–154.

Nakahara H, Bevelander G (1971) The formation and growth of the prismatic layer of *Pinctada radiata. Calcification Tissue Research*, **7**, 31–45.

Nakajima T, Volcani BE (1969) 3,4 Dihydroxyproline: A new amino acid in diatom cell walls. *Science*, **164**, 1400–1406.

Nakajima T, Volcani BE (1970) Σ-N-Trimethyl-L-™-hydroxylysine phosphate and its nonphosphorylated compound in diatom cell walls. *Biochemical and Biophysical Research Communications*, **39**, 28–33.

Nassif N, Pinna N, Gehrke N, Antonietti M, Jäger C, Cölfen H (2005) Amorphous layer around aragonite platelets in nacre.

Proceedings of the National Academy of Science of the USA, **102,** 12653–12655.

Nelson DM Tréguer P, Brzezinski MA, Leynaert A, Queguiner B (1995) Production and dissolution of biogenic silica in the ocean – Revised global estimates, comparison with regional data and relationship to biogenic sedimentation. *Global Biogeochemical Cycles,* **9,** 359–372.

Nudelman F, Gotliv BA, Addadi L, Weiner S (2006) Mollusk shell formation: mapping the distribution of organic matrix components underlying a single aragonitic tablet in nacre. *Journal of Structural Biology,* **153,** 176–187.

Nudelman F, Chen HH, Goldberg HA, Weiner S, Addadi L (2007) Spiers Memorial Lecture. Lessons from biomineralization: comparing the growth strategies of mollusc shell prismatic and nacreous layers in *Atrina rigida. Faraday Discusions,* **136,** 9–25.

Nudelman F, Shimoni E, Klein E, *et al.* (2008) Forming nacreous layer of the shells of the bivalves *Atrina rigida* and *Pinctada margaritifera:* An environmental- and cryo-scanning electron microscopy study. *Journal of Structural Biology,* **162,** 290–300.

Nurnberg D, Bijma J, Hemleben C (1996) Assessing the reliability of magnesium in foraminiferal calcite as a proxy for water mass temperatures. *Geochimica et Cosmochimica Acta,* **63,** 2369–2379.

Ogane K, Tuji A, Suzuki N, Matsuoka A, Kurihara T, Hori RS (2010) Direct observation of the skeletal growth patterns of polycystine radiolarians using a fluorescent marker. *Marine Micropaleontology,* **77,** 137–144.

Okazaki K, Inoue S (1976) Crystal property of the larval sea urchin spicule. *Development Growth and Differentiation,* **18,** 413–434.

Oliver S, Kuperman A, Coombs N, Lough A, Ozin Ga (1995) Lamellar aluminophosphates with surface patterns that mimic diatom and radiolarian microskeletons. *Nature,* **378,** 47–50.

Olstza MJ, Cheng X, Jee SS, Kumar R, Kim YY, Kaufman MJ, Douglas EP, Gower LB (2007) Bone structure and formation: A new perspective. *Materials Science and Engineering* **58,** 77–116.

Orme CA, Noy A, Wierzbicki A, Mcbride MT, Grantham M, Teng HH, Dove PM, De Yoreo JJ (2001) Formation of chiral morphologies through selective binding of amino acids to calcite surface steps. *Nature,* **411,** 775–779.

Ortiz JD, Mix AC, Rugh W, Watkins JM, Collier RW (1996) Deep-dwelling planktonic foraminifera of the northeastern Pacific Ocean reveal environmental control of oxygen and carbon isotopic disequilibria. *Geochimica et Cosmochimica Acta,* **60,** 4509–4523.

Paasche E (1973a) Silicon and the ecology of marine plankton diatoms. 1. *Thalassiosira pseudonana (cyclotella nana)* growth in a chemostat with silicate as limiting nutrient. *Marine Biology,* **19,** 117–126.

Paasche E (1973b) Silicon and the ecology of marine plankton diatoms. 2. Silicate uptake kinetics in five diatom species. *Marine Biology,* **19,** 262–269.

Pancera JL, Boyko V, Shukla A, Narayanan T, Huber K (2010) Evaluation of particle growth of amorphous calcium carbonate in water by means of the parod invariant from SAXS. *Langmuir,* **26,** 17405–17412.

Pawlowski J, Holzmann M, Berney C, Gooday AJ, Cedhagen T, Habura A, Bowser SS (2003) The evolution of early foraminif-

era. *Proceedings of the National Academy of Science of the USA,* **100,** 11494–11498.

Pearson PN, Ditchfield PW, Singano J, Harcourt-Brown KG, Nicholas CJ, Olsson RK, Shackleton NJ, Hall MA (2001) Warm tropical sea surface temperatures in the Late Cretaceous and Eocene epochs. *Nature,* **413,** 481–487.

Peled-Kamar M, Hamilton P, Wilt FH (2002) Spicule matrix protein LSM34 is essential for biomineralization of the sea urchin spicule. *Experimental Cell Research,* **272,** 56–61.

Pennington JT, Strathmann RR (1990) Consequences of the calcite skeletons of planktonic echinoderm larvae for orientation, swimming, and shape. *Biological Bulletin,* **179,** 121–133.

Perry CC (2003) Silicification: The processes by which organisms capture and mineralize silica. In: *Biomineralization* (eds Dove PM, De Yoreo JJ, Weiner S). Mineralogical Soc. America, Washington, D.C., pp. 291–327.

Piana S, Jones F, Gale JD (2007) Aspartic acid as a crystal growth catalyst. *Crystal Engineering Commications,* **9,** 1187–1191.

Pickett-Heaps J (1991) Cell-division in diatoms. *International Review of Cytology-a Survey of Cell Biology,* **128,** 63–108.

Pipich V, Balz M, Wolf SE, Tremel W, Schwahn D (2008) Nucleation and growth of $CaCO_3$ mediated by the egg-white protein ovalbumin: A time-resolved *in situ* study using small-angle neutron scattering. *Journal of the American Chemial Society,* **130,** 6879–6892.

Politi Y, Arad T, Klein E, Weiner S, Addadi L (2004) Sea urchin spine calcite forms via a transient amorphous calcium carbonate phase. *Science,* **306,** 1161–1164.

Politi Y, Levi-Kalisman Y, Raz S, *et al.* (2006) Structural characterization of the transient amorphous calcium carbonate precursor phase in sea urchin embryos. *Advanced Functional Materials,* **16,** 1289–1298.

Politi Y, Metzler RA, Abrecht M, *et al.* (2008) Transformation mechanism of amorphous calcium carbonate into calcite in the sea urchin larval spicule. *Proceedings of the National Academy of Sciences of the USA* **105,** 17362–17366.

Pontoni DP, Bolze J, Dingenouts N, Narayanan T, Ballauff M (2003) Crystallization of calcium carbonate observed in-situ by combined small- and wide-angle x-ray scattering. *The Journal of Physical Chemistry B,* **107,** 5123–5125.

Porter SM, Meisterfeld R, Knoll AH (2003) Vase-shaped microfossils from the Neoproterozoic Chuar Group, Grand Canyon: a classification guided by modern testate amoebae. *Journal of Paleontology,* **77,** 205–225.

Pouget EM, Bomans PHH, Goos JAC, Frederik PM, de With G, Sommerkijk JM (2009) *Science,* **323,** 1455–1458.

Preisig HR (1994) Siliceous structures and silicification in flagellated protists. *Protoplasma,* **181,** 29–42.

Preston RR (1998) Transmembrane Mg^{2+} currents and intracellular free Mg^{2+} concentration in *Parmecium tetraurmelia. Journal of Membrane Biology,* **164,** 11–24.

Pruss S, Finnegan S, Fischer WW, Knoll AH (2010) Carbonates in skeleton-poor seas: New insights from Cambrian and Ordovician strata of Laurentia. *Palaios,* **25,** 73–84.

Puceat E, Lecuyer C, Donnadieu Y, *et al.* (2007) Fish tooth delta [18]O revising Late Cretaceous meridional upper ocean water temperature gradients. *Geology,* **35,** 107–110.

Racki G, Cordey F (2000) Radiolarian palaeoecology and radiolarites: is the present the key to the past? *Earth-Science Reviews,* **52,** 83–120.

Radha AV, Forbes TZ, Killian CE, Gilbert PUPA, Navrotsky A (2010) Transformation and crystallization energetics of synthetic and biogenic amorphous calcium carbonate. *Proceedings of the National Academy of Sciences, USA,*107, 16438–16443.

Ragueneau O, Tréguer P, Leynaert A, *et al.* (2000) A review of the Si cycle in the modern ocean: Recent progress and missing gaps in the application of biogenic opal as a paleoproductivity proxy. *Global and Planetary Change,* 26, 317–365.

Rains DW, Epstein E, Zasoski RJ, Aslam M (2006) Active silicon uptake by wheat. *Plant Soil,* 280, 223–228.

Raiteri P, Gale JD (2010) Water is the key to nonclassical nucleation of amorphous calcium carbonate. *Journal of the American Chemical Society,* 132, 17623–17634.

Rathmann S, Kuhnert H (2007) Carbonate ion effect on Mg/Ca, Sr/Ca, and stable isotopes on the benthic foraminifera *Oridorsalis umbonatus* off Namibia. *Marine Micropaleontology,* 66, 120–133.

Raven JA (1983) The transport and function of silicon in plants. *Biological Reviews of the Cambridge Philosophical Society,* 58, 179–207.

Raven JA, Giordano M (2009) Biomineralisation by photosynthetic organisms: Evidence of co-evolution of the organisms and their environment? *Geobiology,* 7, 140–154.

Raz S, Weiner S, Addadi L (2000) Formation of high magnesian calcites via an amorphous precursor phase: Possible biological implications. *Advanced Materials,* 12, 38–42.

Raz S, Hamilton PC, Wilt FH, Weiner S, Addadi L (2003) The transient phase of amorphous calcium carbonate in sea urchin larval spicules: The involvement of proteins and magnesium ions in its formation and stabilization. *Advanced Functional Materials,* 13, 480–486.

Raz-Bahat M, Erez J, Rinkevich B (2006) In vivo light-microscopic documentation for primary calcification processes in the hermatypic coral *Stylophora pistillata*. *Cell Tissue Research,* 325, 361–368.

Riedel GF, Nelson DM (1985) Silicon uptake by algae with no known Si requirement 2. Strong pH-dependence of uptake kinetic-parameters in Phaeodactylum-Tricornutum (Bacillariophyceae). *Journal of Phycology,* 21, 168–171.

Ries, JB (2004) Effect of ambient Mg/Ca ratio on Mg fractionation in calcareous marine invertebrates: A record of the oceanic Mg/Ca ratio over the Phanerozoic. *Geology,* 32, 981–984.

Ripperger S, Schiebel R, Rehkamper M, Halliday AN (2008) Cd/Ca ratios of in situ collected planktonic foraminiferal tests. *Paleoceanography,* 23, 16.

Rollion-Bard C, Erez J, Zilberman T (2008) Intra-shell oxygen isotope ratios in the benthic foraminifera genus *Amphistegina* and the influence of seawater carbonate chemistry and temperature on this ratio. *Geochimica et Cosmochimica Acta,* 72, 6006–6014.

Romani A, Maguire ME (2002) Hormonal regulation of Mg^{2+} transport and homeostasis in eukaryotic cells. *Biometals,* 15, 271–283.

Rosenthal Y, Lohmann GP (2002) Accurate estimation of sea surface temperatures using dissolution-corrected calibrations for Mg/Ca paleothermometry. *Paleoceanography,* 17, 1044.

Rosenthal Y, Boyle EA, Slowey N (1997) Temperature control on the incorporation of mangesium, strontium, fluorine, and cadmium into benthic foraminiferal shells from Little Bahama Bank: Prospects for thermocline paleoceanography. *Geochimica et Cosmochimica Acta,* 61, 3633–3643.

Round FE (1981) Morphology and phyletic relationships of the silicified algae and the archetypal diatom-monophyly or polyphyly In: *Silicon and Siliceous Structures in Biological Systems* (eds Simpson TL, Volcani BE). Springer-Verlag, New York, pp. 97–128.

Ruggeberg A, Fietzke J, Liebetrau V, Eisenhauer A, Dullo WC, Freiwald A (2008) Stable strontium isotopes (delta Sr-88/86) in cold-water corals – A new proxy for reconstruction of intermediate ocean water temperatures. *Earth and Planetary Science Letters,* 269, 569–574.

Russell AD, Hönisch B, Spero HJ, Lea DW (2004) Effects of seawater carbonate ion concentration and temperature on shell U, Mg, and Sr in cultured planktonic foraminifera. *Geochimica et Cosmochimica Acta,* 68, 4347–4361.

Scala S, Bowler C (2001) Molecular insights into the novel aspects of diatom biology. *Cellular and Molecular Life Sciences,* 58, 1666–1673.

Schmid AMM, Schulz D (1979) Wall morphogenesis in diatoms – Deposition of silica by cytoplasmic vesicles. *Protoplasma,* 100, 267–288.

Schneider K, Erez J (2006) The effect of carbonate chemistry on calcification and photosynthesis in the herematypic coral *Acropora eurystoma*. *Limnology Oceanography,* 51, 1284–1293.

Schröder HC, Perovic-Ottstadt S, Rothenberger M, *et al.* (2004) Silica transport in the demosponge *Suberites domuncula*: Fluorescence emission analysis using the PDMPO probe and cloning of a potential transporter. *Biochemistry Journal,* 381, 665–673.

Schröder HC, Boreiko A, Korzhev M, *et al.* (2006) Co-expression and functional interaction of silicatein with galectin – matrix-guided formation of siliceous spicules in the marine demosponge *Suberites domuncula*. *Journal Biological Chemistry,* 281, 12001–12009.

Schröder HC, Brandt D, Schlossmacher U, *et al.* (2007a) Enzymatic production of biosilica glass using enzymes from sponges: Basic aspects and application in nanobiotechnology (material sciences and medicine). *Naturwissenschaften,* 94, 339–359.

Schröder HC, Natalio F, Shukoor I, *et al.* (2007b) Apposition of silica lamellae during growth of spicules in the demosponge *Suberites domuncula*: Biological/biochemical studies and chemical/biomimetical confirmation. *Journal of Structural Biology,* 159, 325–334.

Schröder HC, Wang XH, Tremel W, Ushijima H, Müller WEG (2008) Biofabrication of biosilica-glass by living organisms. *Natural Products Reports,* 25, 455–474.

Segev E, Erez J (2006) Effect of Mg/Ca ratio in seawater on shell composition in shallow benthic foraminifera. *Geochemistry, Geophysics, Geosystems,* 7, doi:10.1029/2005GC000969.

Sen Gupta BK (1999) *Modern Foraminifera*. Kluwer Academic Publishers, Dordrecht.

Seto J, Zhang Y, Hamilton P, Wilt F (2004) The localization of occluded matrix proteins in calcareous spicules of sea urchin larvae. *Journal of Structural Biology,* 148, 123–130.

Shen GT, Dunbar RB (1995) Environmental controls on uranium in reef corals. *Geochimica et Cosmochimica Acta,* 59, 2009–2024.

Sherbakova TA, Masyukova YA, Safonova TA, *et al.* (2005) Conserved motif CMLD in silicic acid transport proteins of diatoms. *Molecular Biology*, **39**, 269–280.

Shi L, Sun XD, Li HD, Weng D (2006) Hydrothermal growth of novel radiolarian-like porous ZnO microspheres on compact TiO_2 substrate. *Materials Letters*, **60**, 210–213.

Shimizu K, Cha J, Stucky GD, Morse DE (1998) Silicatein alpha: Cathepsin L-like protein in sponge biosilica. *Proceedings of the National Academy of Sciences of the USA*, **95**, 6234–6238.

Shimizu K, Del Amo Y, Brzezinski MA, Stucky GD, Morse DE (2001) A novel fluorescent silica tracer for biological silicification studies. *Chemistry and Biology*, **8**, 1051 – 1060.

Siever R (1992) The silica cycle in the Precambrian. *Geochimica et Cosmochimica Acta*, 56, 3265–3272.

Silverman J, Lazar B, Erez J (2007a) Community metabolism of a coral reef exposed to naturally varying dissolved inorganic nutrient loads. *Biogeochemistry*, **84**, 67–82.

Silverman J, Lazar B, Erez J (2007b) Effect of aragonite saturation, temperature, and nutrients on the community calcification rate of a coral reef. *Journal of Geophysical Research*, **112**, C05004.

Sime NG, De La Rocha CL, Tipper ET, Tripati A, Galy A, Bickle MJ (2007) Interpreting the Ca isotope record of marine biogenic carbonates. *Geochimica et Cosmochimica Acta*, **71**, 3979–3989.

Simkiss K, Wilbur K (1989) *Biomineralization. Cell Biology and Mineral Deposition*. Academic Press, Inc., San Diego pp. 337.

Simpson TL, Volcani BE (1981) *Silicon and Siliceous Structures in Biological Systems*. Springer-Verlag, New York, pp. 587.

Smetacek V (1999) Diatoms and the ocean carbon cycle. *Protist*, **150**, 25–32.

Spero HJ, Lea DW (1996) Experimental determination of stable isotope variability in Globigerina bulloides: Implications for paleoceanographic reconstructions. *Marine Micropaleontology*, **28**, 231–246.

Stephens CJ, Ladden SF, Meldrum FC, Christenson K (2010) Amorphous calcium carbonate is stabilized in confinement. *Advanced Functional Materials*, **20**, 2108– 2115.

Stephenson AE, De Yoreo JJ, Wu L, Wu KJ, Hoyer J, Dove PM (2008) Peptides enhance magnesium signature in calcite: Insights into origins of vital effects. *Science*, **322**, 724–727.

Stoll HM, Rosenthal Y, Falkowski P. (2002) Climate proxies from Sr/Ca of coccolith calcite: Calibrations from continuous culture of *Emiliania huxleyi*. *Geochimica et Cosmochimica Acta*, **66**, 927–936.

Stott L, Tachikawa K, Tappa E, *et al.* (2004) Interlaboratory comparison study of Mg/Ca and Sr/Ca measurements in planktonic foraminifera for paleoceanographic research. *Geochemistry Geophysics Geosystems*, **5**, Q04D09.

Sullivan CW (1976) Diatom mineralization of silicic-acid .1. $Si(OH)_4$ transport characteristics in *Navicula pelliculosa*. *Journal of Phycology*, **12**, 390–396.

Sullivan CW (1979) Diatom mineralization of silicic acid .4. Kinetics of soluble Si pool formation in exponentially growing and synchronized *Navicula pelliculosa*. *Journal of Phycology*, **15**, 210–216.

Sullivan CW (1986) Silicifiction by diatoms. In: *Silicon Biochemistry* (ed O'Connor EDM). John Wiley & Sons, Ltd, Chichester.

Summers RG, Morrill JB, Leith A, Marko M, Piston DW, Stonebraker AT (1993) A stereometric analysis of karyokinesis, cytokinesis and cell arrangements during and following 4th cleavage period in the sea-urchin, *Lytechninus variegatus*. *Development Growth and Differentiation*, **35**, 41–57.

Sumper M (2002) A phase separation model for the nanopatterning of diatom biosilica. *Science*, **295**, 2430–2433.

Sumper M (2004) Biomimetic patterning of silica by long-chain polyamines. *Angewandte Chemie – International Edition*, **43**, 2251–2254.

Sumper M, Kröger N (2004) Silica formation in diatoms: The function of long-chain polyamines and silaffins. *Journal of Materials Chemistry*, **14**, 2059–2065.

Sumper M, Lehmann G (2006) Silica pattern formation in diatoms: Species-specific polyamine biosynthesis. *ChemBioChem*, **7**, 1419–1427.

Sumper M, Lorenz S, Brunner E (2003) Biomimetic control of size in the polyamine-directed formation of silica nanospheres. *Angewandte Chemie-International Edit.*, **42**, 5192–5195.

Swift DM, Wheeler AP (1990) Evidence of a biosilica organic matrix in a fresh-water diatom and its role in regulation of silica polymerization. *Abstracts of Papers of the American Chemical Society*, **200**, 181-COLL.

Swift DM, Wheeler AP (1992) Evidence of an organic matrix from diatom biosilica. *Journal of Phycology*, **28**, 202–209.

Takahashi MM, Okazaki K (1979) Total cell number and number of the primary mesenchyme cells in whole, 1/2 and 1/4 larve of *Clypeaster japonicus*. *Development Growth and Differentiation*, **21**, 553–566.

Taylor NJ (1985) Silica incorporation in the diatom Coscinodiscus-Granii as affected by light-intensity. *British Phycological Journal*, **20**, 365–374.

Thamatrakoln K, Alverson AJ, Hildebrand M (2006) Comparative sequence analysis of diatom silicon transporters: Toward a mechanistic model of silicon transport. *Journal of Phycology*, **42**, 822–834.

Thamatrakoln K, Hildebrand M (2007) Analysis of *Thalassiosira pseudonana* silicon transporters indicates distinct regulatory levels and transport activity through the cell cycle. *Eukaryotic Cell*, **6**, 271–279.

Thompson DW (1942) *On Growth and Form*, Macmillan, New York.

Thompson JB, Paloczi GT, Kindt JH, *et al.* (2000) Direct observation of the transition from calcite to aragonite growth as induced by abalone shell proteins. *Biophysical Journal*, **79**, 3307–3312.

Tong H, Hu J, Ma W, Zhong G, Yao S, Cao N (2002) In situ analysis of the organic framework in the prismatic layer of mollusc shell. *Biomaterials*, **23**, 2593–2598.

Towe KM, Rutzler K (1968) Lepidocrite iron mineralization in keratose sponge granules. *Science*, **162**, 268–269.

Toyofuku T, Kitazato H, Kawahata H, Tsuchiya M, Nohara M (2000) Evaluation of Mg/Ca thermometry in foraminifera: Comparison of experimental results and measurements in nature. *Paleoceanography*, **15**, 456–464.

Tréguer P, Nelson DM, Vanbennekom AJ, Demaster DJ, Leynaert A, Queguiner B (1995) The silica balance in the world ocean – a reestimate. *Science*, **268**, 375–379.

Tribovillard N, Algeo TJ, Lyons T, Riboulleau A (2006) Trace metals as paleoredox and paleoproductivity proxies: An update. *Chemical Geology*, **232**, 12–32.

Tsukamoto D, Sarashina I, Endo K (2004) Structure and expression of an unusually acidic matrix protein of pearl oyster shells. *Biochemical and Biophysical Research Communications,* **320,** 1175–1180.

Urey HC, Lowenstam HA, Epstein S, Mckinney CR (1951) Measurement of paleotemperatures and temperatures of the Upper Cretaceous of England, Denmark, and the southeastern United States. *Bulletin of the Geological Society of America,* **62,** 399–416.

Uriz MJ (2006) Mineral skeletogenesis in sponges. *Canadian Journal of Zoology – Reviews of Canadian Zoology,* 84, 322–356.

Urry LA, Hamilton PC, Killian CE, Wilt FH (2000) Expression of spicule matrix proteins in the sea urchin embryo during normal and experimentally altered spiculogenesis. *Developmental Biology,* **225,** 201–213.

Van Cappellen P (2003) Biomineralization and global biogeochemical cycles. In: *Biomineralization* (eds Dove PM, De Yoreo JJ, Weiner S). Mineralogical Soc. America, Washington, D.C., pp.357–381.

Veizer J (1989) Strontium isotopes in seawater through time. *Annual Reviews of Earth and Planetary Science,* **17,** 141–167.

Volcani BE (1981) Cell wall formation in diatoms: morphogenesis and biochemistry In: *Silicon and Siliceous Structures in Biological Systems* (eds Simpson TL, Volcani BE). Springer-Verlag, New York, pp. 157–200.

Volkmer D, Tugulu S, Fricke M, Nielsen T (2003) Morphosynthesis of star-shaped titania-silica shells. *Angewandte Chemie-International Edition,* **42,** 58–61.

von Langen PJ, Pak JK, Spero HJ, Lea DW (2005) Effects of temperature on Mg/Ca in neogloboquadrinid shells determined by live culturing. *Geochemistry, Geophysics, Geosystems,* **6,** Q10P03.

Vrieling EG, Gieskes WWC, Beelen TPM (1999) Silica deposition in diatoms: control by the pH inside the silica deposition vesicle. *Jounal of Phycology,* **35,** 548–559.

Wada N, Yamashita K, Umegaki T (1999). Effects of carboxylic acids on calcite formation in the presence of Mg2+ ions. *Journal of Colloid Interface Chemistry,* **212,** 357–364.

Waldeab S, Lea DW, Schneider R, Anderson N (2007) 155,000 years of West African monsoon and ocean thermal evolution. *Science,* **316,** 1303–1307.

Wallace AF, De Yoreo JJ, Dove PM (2009) Kinetics of silica nucleation on carboxyl- and amine-terminated surfaces: Insights for biomineralization. *Journal of the American Chemical Society,* **131,** 5244–5250.

Wallace AF, Gibbs GV, Dove PM (2010) Influence of ion-associated water on the hydrolysis of Si–O bonded interactions. *Journal of Physical Chemistry A,* **114,** 2534 – 3542.

Walters DA, Smith BL, Belcher AM, *et al.* (1997) Modification of calcite crystal growth by abalone shell proteins: An atomic force microscope study. *Biophysical Journal,* **72,** 1425–1433.

Wanamaker AD, Kreutz KJ, Wilson T, Borns HW, Introne DS, Feindel S (2008) Experimentally determined Mg/Ca and Sr/Ca ratios in juvenile bivalve calcite for *Mytilus edulis*: implications for paleotemperature reconstructions. *Geo-Marine Letters,* **28,** 359–368.

Wang D, Wallace AF, De Yoreo JJ, Dove PM (2009) Carboxylated molecules regulate magnesium content of amorphous calcium carbonates during calcification. *Proceedings of the National Academy of Sciences, U.S.A.,* **106,** 21511–21516.

Wang JU, Xiao QA, Zhou HJ, *et al.* (2007) Radiolaria-like silica with radial spines fabricated by a dynamic self-organization. *Journal of Physical Chemistry C,* **111,** 16544–16548.

Wang T, Colfen H, Antonietti M (2005) Nonclassical crystallization: Mesocrystals and morphology change of $CaCO_3$ crystals in the presence of a polyelectrolyte additive. *Journal of the American Chemical Society,* **127,** 3246–3247.

Wasylenki LE, Dove PM, De Yoreo JJ (2005) Effects of temperature and transport conditions on calcite growth in the presence of Mg^{2+}: Implications for paleothermometry. *Geochimica et Cosmochimica Acta,* **69,** 4227–4236.

Weaver JC, Morse DE (2003) Molecular biology of demosponge axial filaments and their roles in biosilicification. *Microscopy Research and Technique,* **62,** 356–367.

Weaver JC, Aizenberg J, Fantner GE, Kisailus D, Woesz A, Allen P, Fields K, Porter MJ, Zok FW, Hansma PK, Fratzl P, Morse DE (2007) Hierarchical assembly of the siliceous skeletal lattice of the hexactinellid sponge *Euplectella aspergillum. Journal of Structural Biology,* **158,** 93–106.

Weber JN, Raup DN (1966) Fractionation of the stable isotopes of carbon and oxygen in marine calcareous organisms—the Echinoidea. Part II. Environmental and genetic factors. *Geochimica et Cosmochimica Acta,* **30,** 705–736.

Weiner S (2006) Transient precursor strategy in mineral formation of bone. *Bone,* **39,** 431–433.

Weiner S, Addadi L (1997) Design strategies in mineralized biological materials. *Journal of Materials Chemistry,* **7,** 689–702.

Weiner S, Dove PM (2003) An overview of biomineralization processes and the problem of the vital effect. In: *Biomineralization* (eds Dove PM, De Yoreo JJ, Weiner S). Mineralogical Soc. America, Washington, D.C., pp. 1–29.

Weiner S, Hood L (1975) Soluble protein of the organic matrix of mollusk shells: a potential template for shell formation. *Science,* **190,** 987–989.

Weiner S, Traub W, Lowenstam, HA (1983) Organic matrix in calcified exoskeletons. In *Biomineralization and Biological Metal Accumulation, Biological and Geological Perspective* (eds Westbroek P, De Jong EW). Reidel, Dordrecht, pp. 205–224.

Wendt J, Wu XC, Reinhardt JW (1989) Deep-water hexactinellid sponge mounds from the upper Triassic of northern Sichuan (China). *Palaeogeography Paleoclimatology Palaeoenvironments,* **76,** 17–29.

Wenzl S, Hett R, Richthammer P, Sumper M (2008) Silacidins: Highly acidic phosphopeptides from diatom shells assist in silica precipitation in vitro. *Angewandte Chemie – International Edit.,* **47,** 1729–1732.

Werner D (1966) Silicic acid in he meabolism of *Cyclotella cryptica. Archives of Microbiology,* **55,** 278–308.

Werner D (1977) *The Biology of Diatoms.* Blackwell Scientific Publications, Oxford, pp. 498.

Wheeler AP, George JW, Evans CA (1981) Control of calcium carbonate nucleation and crystal growth by soluble matrix of oyster shell. *Science,* **212,** 1397–1398.

Wilt F, Ettensohn CA (2007) The morphogensis and biomineralization of the sea urchin larval skeleton. In: *Handbook of*

Biomineralization (eds Baeuerlein E, Behrens P). Wiley-VCH, pp. 183–210.

Wilt FH (1999) Matrix and mineral in the sea urchin larval skeleton. *Journal of Structural Biology,* **126,** 216–226.

Wilt FH (2002) Biomineralization of the spicules of sea urchin embryos. *Zoological Science,* **19,** 253–261.

Wilt FH, Killian CE, Hamilton P, Croker L (2008) The dynamics of secretion during sea urchin embryonic skeleton formation. *Experimental Cell Research,* **314,** 1744–1752.

Wilt FH, Killian CE, Livingston BT (2003) Development of calcareous skeletal elements in invertebrates. *Differentiation,* **71,** 237–250.

Wischmeyer AG, Del Amo Y, Brzezinski M, Wolf-Gladrow DA (2003) Theoretical constraints on the uptake of silicic acid species by marine diatoms. *Marine Chemistry,* **82,** 13–29.

Wolf SE, Müller L, Barrea R, *et al.* (2011) Carbonate-coordinated metal complexes precede the formation of liquid amorphous mineral emulsions of divalent metal carbonates. *Nanoscale,* DOI: 10.1039/c0nr00761g.

Wolpert L, Gustafson T (1961) Studies on the cellular basis of morphogenesis of the sea urchin embryo: Development of the skeletal pattern. *Experimental Cell Research,* **25,** 311–325.

Won MZ, Below R (1999) Cambrian radiolaria from the Georgina Basin, Queensland, Australia. *Micropaleontology,* **45,** 325–363.

Wood RA, Grotzinger JP, Dickson JAD (2002) Proterozoic modular biomineralized metazoan from the Nama Group, Namibia. *Science,* **296,** 2383–2386.

Wustman BA, Santos R, Zhang B, Evans JS (2002) Identification of a 'glycine-loop' like coiled structure in the 34 AA Pro, Gly, Met repeat domain of the biomineral-associated protein, PM27. *Biopolymers,* **65,** 362–372.

Xiao S, Hu J, Yuan X, Parsley RL, Cao R (2005) Articulated sponges from the Lower Cambrian Hetang Formation in southern Anhui, South China: Their age and implications for the early evolution of sponges. *Palaeogeography, Palaeoclimatology and Palaeoecology,* **220,** 89–117.

Xu AW, Dong WF, Antonietti M, Colfen H (2008) Polymorph switching of calcium carbonate crystals by polymer-controlled crystallization. *Advanced Functional Materials,* **18,** 1307–1313.

Yamaji N, Ma JF (2007) Spatial distribution and temporal variation of the rice silicon transporter Lsi1. *Plant Physiology,* **143,** 1306–1313.

Yu JM, Elderfield H, Honisch B (2007) B/Ca in planktonic foraminifera as a proxy for surface seawater pH. *Paleoceanography,* **22,** PA2202.

Zaremba CM, Belcher AM, Fritz M, Li YL, Mann S, Hansma PK, Morse DE, Speck JS, Stucky GD (1996) Critical transitions in the biofabrication of abalone shells and flat pearls. *Chemical Materials,* **8,** 679–690.

Zeebe RE (2005) Stable boron isotope fractionation between dissolved $B(OH)_3$ and $B(OH)_4^-$. *Geochimica et Cosmochimica Acta,* **69,** 2753–2766.

Zhu XD, Mahairas G, Illies M, Cameron RA, Davidson EH, Ettensohn CA (2001) A large-scale analysis of mRNAs expressed by primary mesenchyme cells of the sea urchin embryo. *Development,* **128,** 2615–2627.

Zhu YQ, Hsu WK, Terrones M, *et al.* (1998) 3D silicon oxide nanostructures: from nanoflowers to radiolaria. *Journal of Materials Chemistry,* **8,** 1859–1864.

Zurzolo C, Bowler C (2001) Exploring bioinorganic pattern formation in diatoms. A story of polarized trafficking. *Plant Physiology,* **127,** 1339–1345.

11

PLANTS AND ANIMALS AS GEOBIOLOGICAL AGENTS

David J. Beerling[1] and Nicholas J. Butterfield[2]

[1]Department of Animal and Plant Sciences, University of Sheffield, Sheffield S10 2TN, UK
[2]Department of Earth Sciences, University of Cambridge, Cambridge, CB2 2EQ, UK

11.1 Introduction

Unlike the world of prokaryotes and protists, plants and animals have a relatively limited metabolic repertoire – essentially oxygenic photoautotrophs or aerobic heterotrophs – but they make up for this shortfall through their enormous, seemingly inexhaustible, capacity to exploit morphology. Plants and animals are unique in having independently acquired a tissue and organ-grade level of multicellularity, thereby taking intra-organismal divisions of labour to fundamentally new levels, not least the invention of large size, complex life histories and complex behaviour. Morphological, metabolic and tissue chemistry developments, therefore, collectively open doors to entirely novel approaches for exploiting ecospace, which in turn exert major influences on global biogeochemical cycles and the environment.

One of the most direct and effective means of assessing the role of plants and animals as geobiological agents is to examine evidence for the expression of these effects on geological and evolutionary timescales (i.e. over millions of years). The geological record offers a unique view of ancient conditions and alternative worlds, including those partially or entirely devoid of the modern complement of plants and animals. In this chapter, we review the influence of terrestrial plants and animals on the physical and biological processes affecting the Earth system. Insofar as the record of land plants appears to extend no further than the Ordovician (Wellman *et al.*, 2003), and eumetazoans no further than the Ediacaran (Peterson and Butterfield, 2005), this review is limited to the relatively recent past (ca. 635 million years), though comparison with pre-embryophyte- and pre-eumetazoan-worlds offers a powerful insight into the progressive geobiological impact of these two remarkable clades.

11.2 Land plants as geobiological agents

The geobiological activities of plants and vegetation operate on short (minutes to year) and very long (millions of years) timescales. On short timescales, they alter the energy balance of the landscape and the chemistry of atmospheric greenhouse gases and their precursors, as well as affecting the atmospheric loading of organic aerosols (e.g. Tunved *et al.*, 2006; Claeys *et al.*, 2004; Spracklen *et al.*, 2008). On longer timescales, plants' geobiological actions are expressed by their accumulated influence on the operation of the organic and inorganic carbon cycles. Here, the net effects are on global atmospheric O_2 and CO_2 levels respectively, with an influence on the evolutionary trajectory of Earth's climate and terrestrial biota (Berner, 2004; Beerling, 2007).

11.2.1 Short-term feedback processes (10^0 to 10^3 years)

11.2.1.1 Land surface energy balance

Terrestrial vegetation, especially forests and grasslands, control the exchange of water, energy and momentum between the land surface and the atmosphere (Bonan, 2008). Ecosystems can therefore exert geobiophysical forcings and feedbacks that dampen or amplify regional and global climate change. In turn these effects also play a role in determining the structure, function and distribution of vegetation. Albedo and evapotranspiration

Fundamentals of Geobiology, First Edition. Edited by Andrew H. Knoll, Donald E. Canfield and Kurt O. Konhauser.
© 2012 Blackwell Publishing Ltd. Published 2012 by Blackwell Publishing Ltd.

(canopy transpiration + soil evaporation) are two key ecosystem processes influencing land surface energy budgets. The global energy budget is given by the simple energy balance equation (Crowley and North, 1991):

$$\frac{S_c}{4(1-\alpha)} = \varepsilon\sigma T^4 \tag{11.1}$$

where S_c is the solar constant (1367 W m^{-2} at 0 Ma), σ is the planterary albedo (viewed from the top of the atmosphere), ε is the planetary emissivity, i.e. capacity to emit long-wave radiation, and σ is the Stefan–Boltzman constant (5.67 × 10^{-8} W m^{-2} K^{-4}), and T is planetary temperature (K). The division of S_c by 4 accounts for the fact that the Earth absorbs radiation like a two dimensional disc, but in fact this is spread over the surface area of a sphere (because of the Earth's rotation) which has four surface area of times the area of a disk. Equation 11.1 can be modified to be more relevant to terrestrial ecology and predict mean terrestrial surface temperature from incoming solar energy, albedo and atmospheric CO_2, as discussed by Beerling and Woodward (2001).

Terrestrial albedo depends primarily on vegetation type and the leaf area index (LAI), the number of leaf layers in the canopy, a trait which can vary seasonally depending on climate and the dominant functional type of vegetation. Incoming solar energy absorbed by the canopy is dissipated by evaporation (latent heat) or convective radiation (sensible heat). Evaporation of water from plant canopies and soils therefore increases the flux of latent heat from the land surface. In tropical rainforests of Amazonia, for example, about 80% of the intercepted net radiation is dissipated as latent heat and 20% lost by convection (Grace *et al.*, 1995). Vegetation height also influences local climate by changing the degree of aerodynamic coupling between the land surface and the atmosphere; taller trees are aerodynamically rougher, enhancing the transfer of mass and energy through increased turbulence.

Soil water content plays a critical role in regulating ecosystem evapotranspiration. As the soils dries out, and the moisture zone around roots is depleted, stomatal pores on the surface of leaves close, reducing transpirational cooling by latent heat loss (Buckley, 2005). This, in turn, leads to elevated canopy temperatures with greater upwards transport of sensible heat to the atmosphere, to an extent governed by the surface-to-air temperature gradient.

The maximum influence of these vegetation properties on global climate has been assessed by comparing climate model simulations between two hypothetical extremes, 'desert world' and a 'green world' (Kleidon *et al.*, 2000). Desert world is characterized by a smooth land surface with a high albedo and poor soil water storage capacity, whereas a green world with forests everywhere has the opposite characteristics. In this extreme comparison, vegetation tripled land surface evapotranspiration, enhanced atmospheric moisture and cloud cover leading to a doubling of land-surface precipitation. These effects combined to lower near-surface seasonal mean temperatures by 8 K in the 'green world' simulation compared that of 'desert world'. In the context of these idealized calculations, the net cooling effect of vegetation indicates its effects on the hydrological cycle overwhelm any increased net absorption of radiation at the surface due to a lower albedo compared to deserts (Kleidon *et al.*, 2000).

In real deserts, the absence of vegetation feedbacks is proposed to promote their own existence (Charney, 1975; Charney *et al.*, 1975). Deserts are stable entities because their sandy non-vegetated surfaces reflect solar radiation back to space, effectively allowing them to act as a net heat sink relative to surrounding areas, which in turn cools the air above. This cooling draws airflow inwards over the Sahara, which warms and dries as it descends, greatly reducing the chances of rainfall and the establishment of vegetation.

Palaeoclimate modelling studies quantifying the importance of vegetation-climate interactions are generally restricted to the Cretaceous, with more comprehensive studies focusing on vegetation feedbacks in a future 'greenhouse' world (e.g. Sellers *et al.*, 1996; Betts *et al.*, 1997). In the warm Late Cretaceous environment (80–65 Ma ago), forest distribution was extensively modified compared to the present-day situation, notably by their extending throughout the high northern and southern latitude continental land masses (Upchurch *et al.*, 1998; Beerling and Woodward, 2001). Climate modelling indicates that the presence of high-northern latitude deciduous forests warmed January and July land-surface temperatures by 2–4 °C and up to 8 °C, respectively, compared to simulations with bare-ground or tundra (Otto-Bliesner and Upchurch, 1997; Upchurch *et al.*, 1998; DeConto *et al.*, 2000). The wintertime warming occurs because the trees mask snow cover, decreasing the albedo of the land surface. Some of the warming is advected to the nearby high-latitude oceans and initiates sea-ice loss (Otto-Bliesner and Upchurch, 1997; Upchurch *et al.*, 1998). Nevertheless, even after accounting for vegetation-land surface feedbacks, reproducing the warm wintertime continental climates of past greenhouse eras in the current generation of climate models, and even next generation 'Earth systems' models, continues to remain a major challenge and implies key processes are missing or poorly represented.

Investigations of the direction and magnitude of vegetation-climate feedbacks in a future high CO_2 'greenhouse' world reveal a more complex picture with competing physiological and structural feedbacks (Sellers *et al.*, 1996; Betts *et al.*, 1997). Doubling the atmospheric CO_2 concentration could decrease in stomatal conductance of vegetation (Ainsworth and Rogers, 2007), with attendant reductions in transpiration and possible warming through a decreased latent heat flux. Global simulations characterising this 'physiological feedback' effect suggest widespread warming over the northern hemisphere land surface by 1 °C (Betts *et al.*, 1997). However, a CO_2-rich atmosphere could also alter vegetation structural properties, like LAI, by stimulating photosynthetic and water-use efficiency of growth (Ainsworth and Long, 2005; Ainsworth and Rogers, 2007). This structural response of vegetation to CO_2 could largely offset the warming effects of physiology by increasing evapotranspiration with a greater LAI and total canopy conductance to water vapour (Betts *et al.*, 1997). The net sign of the feedback of predicted changes in vegetation physiology and structure in a near-term future (i.e. decades), without significant migration of biomes under high CO_2 atmosphere is likely to be a cooling. However, climate-carbon cycle simulations indicate that significant afforestation of the high latitudes, often proposed as a measure to counteract global warming, may be counter-productive leading to warming through decreases in land surface albedo (Bala *et al.* 2007). Analyses of this sort raise questions concerning the efficacy of such efforts to mitigate climate change.

Recent evidence from the high arctic suggests vegetation is already exerting an effect on climate through changing the land surface albedo. Accelerated warming over the high-latitude northern continents has reduced seasonal snow cover shifting the albedo of the landscape from very reflective snow to the darker underlying vegetation and soils (Chapin *et al.*, 2005). The darker vegetation absorbs more solar radiation, warming the land surface and heating the atmosphere. At the same time, rapid vegetation change is also evident, with tree and shrub expansion into more northerly regions replacing low-lying tundra ecosystems (Chapin *et al.*, 2005; Sturm *et al.*, 2005), with a profound significance for regional climatic warming by increased heat transfer to the atmosphere (Strack *et al.*, 2007).

We emphasize that our understanding of 'geobiological' feedbacks of vegetation for past or future climates and atmospheres is largely derived from theory, applied and embedded in large complex computer models. Evaluating models against palaeobotanical, geochemical and sedimentary evidence is therefore a critical endeavour.

11.2.1.2 Atmospheric composition and aerosols

Greenhouse gases in the atmosphere absorb long-wave radiant energy emitted from the Earth's surface rather than letting it escape into space. On geological timescales, terrestrial ecosystems influence the atmospheric CO_2 concentration by accelerating weathering of Ca-Mg

rocks and to lesser extent through organic carbon burial (see next section). On human timescales vegetation plays an important role in modulating anthropogenic CO_2 emissions from fossil fuel burning and deforestation. Net primary productivity (NPP) is defined as the uptake of CO_2 by photosynthesis less that released by autotrophic (plant) respiration. This primary production provides the energy sources and substrates for virtually all major ecosystems on Earth. Net ecosystem productivity is NPP less carbon lost by decomposition via heterotrophic (microbial and fungal) respiration. In addition, organic carbon is also lost through disturbance by, for example, fire. At the global scale, terrestrial NPP today is about 60 Pg C yr^{-1}, comparable to that of the ocean biosphere (65 Pg C yr^{-1}) (Falkowski *et al.*, 1998). Accounting for heterotrophic respiration reduces NPP considerably to about 5 Pg C yr^{-1}, and allowing for carbon losses by disturbances such as fire and soil erosion lowers it further to about 2–3 Pg C yr^{-1}. This carbon sink is small but of a similar magnitude to annual emissions due to humanity's burning of fossil fuels and deforestation (9 Pg yr^{-1}) (Woodward, 2007). Vegetation therefore represents an important sink for about 20–30% of human emissions to an extent that varies from year to year due to climatic variability. Because atmospheric CO_2 is well-mixed globally, vegetation–carbon cycle feedbacks are usually manifested globally whereas the land surface biophysical feedbacks discussion occur most strongly at regional scales.

Terrestrial ecosystems play an important role in determining atmospheric concentrations of the three important trace greenhouse gases, methane (CH_4), ozone (O_3) and nitrous oxide (N_2O) in the lower atmosphere (troposphere) (Beerling *et al.*, 2007). Collectively these trace greenhouse gases are fundamentally important components of the global climate system. On a per molecule basis, for example, methane and ozone are approximately 25 and 1000 times, respectively, more effective at planetary warming than CO_2 over a 100-year timeframe but are present in far lower concentrations. Biogenic sources of methane also lead to indirect climatic warming because atmospheric methane oxidation supplies the relatively dry lower stratosphere with water vapour, where it acts as a very effective greenhouse gas.

Terrestrial ecosystems are the largest natural sources of methane and nitrous oxide and therefore strongly influence the concentrations of both gases in the lower atmosphere. The activities of methane-producing Archaea (methanogens) in anaerobic natural wetland soils supply about 200 Tg CH_4 annually, while microbial activities in the soils of tropical and temperate forests annually release about 10 Tg N_2O. Soil microbial activity, particularly in the tropics and sub-tropics, also release nitrogen oxides (NO_x = NO and NO_2) (5–8 Tg yr^{-1}), with further emissions from biomass/biofuel burning (ca. 6–12 Tg yr^{-1}) and lightning (ca. 3–7 Tg yr^{-1}) (Denman *et al.*, 2007). NO$_x$ is an important atmospheric constituent for many key chemical reactions and its production by terrestrial ecosystems, either directly or indirectly, in a preindustrial atmosphere prior to anthropogenic sources, highlights another important indirect geobiologial influence of vegetation.

Polar ice core records of Earth's past atmospheric composition indicate that the concentration of both methane and nitrous oxide underwent significant variations on glacial-interglacial timescales, as well as on millennial timescales since the last glacial epoch, some 20 kyr ago (Flückiger *et al.*, 2004). In contrast with atmospheric CO$_2$, these variations are in phase with temperature, indicating the high sensitivity of the sources of methane and nitrous oxide to climate change. Further back in Earth's history, during much of the Mesozoic and early Palaeogene, climates were considerably warmer than even the warmest interglacial climate (Huber *et al.*, 2000; Beerling and Woodward, 2001). It follows, therefore, that past greenhouse climates, characterized by a more vigorous hydrological cycle due to largely ice-free polar regions, and warmer temperatures, offer conditions expected to enhance methane and nitrous oxide emissions from terrestrial ecosystems. It is likely, therefore, that the concentration of methane and nitrous oxide in the atmosphere at these times far exceeded levels for which we have direct measurements or experience, with significant climate change potential.

Forests also alter the chemistry of the atmosphere by releasing biogenic volatile organic compounds (BVOC) (Fehsenfeld *et al.*, 1992). Isoprene is the most reactive BVOC, with a chemical lifetime ranging from a few minutes to hours. At the global scale, the terrestrial biosphere annually emits 500–750 Tg yr^{-1} of isoprene (Guenther *et al.*, 2006), over double the mass of methane produced by natural wetlands. Isoprene is an important natural precursor for the formation of tropospheric ozone, when oxidized by hydroxyl radicals in the presence of sufficient NO$_x$. Because the principal sink for methane is oxidation by hydroxyl radicals, BVOC emissions from forests can alter the lifetime and concentration of methane and its potential for planetary warming. BVOCs constitute only 1% of the total carbon flux into the atmosphere, yet act as significant drivers of climate change by influencing the photochemistry of methane and ozone, and the formation of secondary organic aerosols (SOAs) in the atmosphere.

Secondary organic aerosols arise when the oxidation products of BVOCs, especially monoterpene emitted from vegetation, condense on existing aerosol particles, and can be an important component of the aerosol loading of the atmosphere in many regions of the globe. Newly formed particles can grow rapidly to gain a diameter >70 nm and act as effective cloud condensation nuclei (CCN) (Claeys *et al.*, 2004). Aerosols influence climate both directly by scattering and absorbing incoming solar radiation and indirectly by forming CCN which influence a range of cloud properties, including albedo, lifetime, and precipitation efficiency (Spracklen *et al.*, 2008). Above northern European boreal forests, BVOCs emitted in late spring and early autumn contribute to 12–50% of the aerosol mass and sustain CCN of 200 cm^{-3}, double that of maritime air masses (Tunved *et al.*, 2006), whilst over the Amazonian rainforest photo-oxidation of isoprene produces considerable quantities of SOA, with important implications for the radiation budget (Claeys *et al.*, 2004). The magnitude and sign of terrestrial, and marine (Kump and Pollard, 2008), aerosol-cloud feedbacks during warm, high CO$_2$ pre-Quaternary climates are at very early stages of investigation.

11.2.1.3 Evolving short-term geobiological feedbacks

The capacity of vegetation land surfaces to alter the physical and chemical environment of the Earth system through this suite of processes is likely to have been most strongly exerted during the early evolutionary diversification of terrestrial plants between the Ordovician and Devonian (450–360 Ma ago). Initial stages of land colonization were conducted by primitive shallow-rooting non-vascular plants during the Ordovician (470–450 Ma ago), and were followed by small rhizomatous vascular plants in the late Silurian and early Devonian. The shallow-rooting, coarse, hairless root structures, of these plants, together with their dependency on homospory for reproduction, confined them to heterogeneous patches in lowland habitats. By the Late Devonian (385–360 Ma ago), the appearance of arborescence vegetation reaching heights of 30 m, and the development of seed habit, allowed colonization of a broader range of environments including uplands and primary successional habitats (Algeo and Scheckler, 1998). Reproduction by seed was a key factor in the rapid spread and diversification of ancient gymnosperm forests during the Late Devonian and Early Carboniferous. Rapid increases in vegetation height were associated with the development of deeper and more complex rooting systems penetrating up to 1 m into the regolith (Algeo and Scheckler, 1998; Raven and Edwards, 2001).

The spread and diversification of the above-ground terrestrial biomass in the form of deep-rooting trees and forests, likely strengthened several feedbacks with opposing signs. The two primary warming effects would be exerted through decreased land surface albedo and the evolution of methane-emitting wetland ecosystems. Counteracting such effect might be the competing with cooling effects from increasing rates of evapotranspiration and cloud cover. The sign of biological aerosol-cloud feedbacks remains to be assessed. Compared to the algal and cyanobacterial mats that existed prior to the Ordovician, the lowering of continental albedo as forests expanded their geographical distributions which could have enhanced atmospheric heating, particularly in the high latitudes where the potential for masking seasonal snow cover is greatest. The evolution and spread of persistent floras in a diversity of wetland habitats (Greb *et al.*, 2006) would have increased methane emissions raising tropospheric concentrations to those approaching the contemporary atmosphere, with exceptionally high levels predicted during the Carboniferous when peat-forming plant communities were geographically very extensive (Bartdorff *et al.*, 2008; Beerling *et al.*, 2009). The taxonomic distribution of extant isoprene emitting groups of plants is broad, and includes mosses and gymnosperms (Sharkey *et al.*, 2008), implying that an expanding biomass increased emissions of isoprene and other BVOCs, with possible increases in methane lifetime and concentration by reduced hydroxyl radical concentrations. If wildfires produced sufficient NO_x in the Palaeozoic (Scott and Glasspool, 2006), increased isoprene emissions may have also increased tropospheric ozone concentrations, causing further warming.

Countering any warming effects from these processes, an expanding cover, density and height of vegetation, and deepening of the soil profile, throughout the Palaeozoic would have permitted a more vigorous hydrological cycle, greater rates of evapotranspiration and cloud formation. In addition, a greater atmospheric loading of biological VOCs may also have increased SOA formation over these ancient forested ecosystems, inducing further seasonal cooling both by reflecting incoming short-wave radiation back to space and by seeding cloud formation. All of these geobiological feedbacks remain to be investigated and quantified.

11.2.2   Long-term feedback processes (>10⁶ years)

Operating in parallel with the processes described above are the geobiological activities of plants that, when summed over millions of years, influence atmospheric O_2 and CO_2 levels (Berner, 2004; Beerling, 2007). Oxygen is a physiologically important gas, and therefore substantial fluctuations in the past are implicated in altering biotic evolutionary trajectories.

11.2.2.1   Organic carbon cycle, atmospheric oxygen and biotic evolution

The organic carbon cycle involves the conversion of inorganic (atmospheric) CO_2 to plant biomass via photosynthesis, and the burial of that organic matter in sediments preventing its decomposition by consumption of oxygen. Carbon burial leads to the net addition of O_2 to the atmosphere and is reversed when sediments and rocks are uplifted or exposed by falling sea-levels to allow oxidation of the organic matter. Only when organic matter is buried and prevented from decomposing can O_2 accumulate in the atmosphere. Rates of organic matter burial therefore significantly determine the Earth's O_2 content over millions of years (Berner and Canfield, 1989).

On an annual time-step, O_2 production by photosynthesis and O_2 consumption by heterotrophic organisms (i.e. those that acquire carbon from other organic matter rather than synthesizing it themselves) are finely balanced; only about 1% of photosynthesized organic carbon is actually buried. The overall process is represented by the following global equation:

$$CO_2 + H_2O \leftrightarrow CH_2O + O_2 \qquad (11.2)$$

Moving from left to right, Equation 11.2 shows inorganic carbon (CO_2) fixed by net photosynthesis being converted to organic matter (CH_2O) that becomes buried in sediments, releasing O_2. The reverse reaction, in Equation 11.2 represents the two processes of oxidation of old sedimentary organic matter subjected to chemical weathering on continents, and the thermal breakdown of organic matter. Both processes release reduced carbon compounds and lead to its oxidation, consuming O_2.

Organic carbon burial occurs on land most obviously in tropical or temperate mires, or peatland ecosystems, where the anoxic sedimentary environment retards decomposition by aerobic heterotrophic microbes. The evolutionary appearance, spread and diversification of land plants during the mid- to late-Palaeozoic (440–255 Ma ago) massively increased organic carbon production and burial worldwide both on land and at sea (Berner and Canfield, 1989; Robinson, 1990a,b; Berner *et al.*, 2003a). Enhanced carbon burial during this interval in Earth's history was probably linked to three factors: (i) increased terrestrial biomass, (ii) the evolutionary appearance of lignin, a microbially resistant compound plants adopted for structural integrity, and (iii) physical aspects of the depositional environment favouring the appearance of coal swamp habitats including tectonism, expansive epicontinental seas and wet climates (Berner and Canfield, 1989; Robinson, 1990a,b).

Marine burial of terrestrially derived organic carbon in anoxic muds of continental margins occurs when rainfall on continental surfaces flushes soil minerals loaded with organic matter out to sea. Burial efficiency of the organically-coated particles on continental margins is dependent on sedimentation rate. The Bengal fan system, for example, has a high organic carbon burial efficiency because the high erosion rate in the Himalayas generates high sedimentation rates and low oxygen

Figure 11.1 Schematic representation of the long-term carbon cycle.

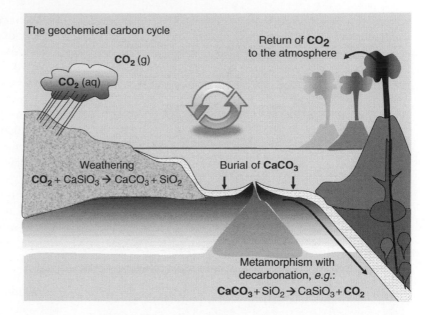

The geochemical carbon cycle

CO_2 (g)

CO_2 (aq)

Return of CO_2 to the atmosphere

Weathering
$CO_2 + CaSiO_3 \rightarrow CaCO_3 + SiO_2$

Burial of $CaCO_3$

Metamorphism with decarbonation, *e.g.*:
$CaCO_3 + SiO_2 \rightarrow CaSiO_3 + CO_2$

availability (Galy *et al.*, 2007). In contrast, the Amazon basin experiences far lower sedimentation rates and 70% of the riverine organic carbon is returned to the atmosphere before burial in marine sediments (Hedges *et al.*, 1997).

Geochemical models of the Earth's oxygen cycle simulating changes in the organic carbon and sulphur cycles over the Phanerozoic predict the consequences for Earth's atmospheric O_2 concentration (Berner and Canfield, 1989; Berner, 2004). The burial and weathering of sulphur is included because it influences atmospheric O_2 over geological time, though quantitatively it is less significant than the role of organic carbon. Although aspects of the calculations are uncertain, geochemical models consistently predict that O_2 levels rose to a peak value of ~27–35% during the Permo-Carboniferous (centred at around 300 Ma ago), driven chiefly by the enhanced burial of terrestrial and marine organic debris worldwide. Atmospheric O_2 concentrations then fell over the following 50 Ma to an unprecedented low of ~13–15% during the late Permian and early Triassic (250 Ma ago), when continental uplift and climatic drying reduced the geographical extent of lowland forests and swamps, reducing burial of organic carbon and pyrite (Berner, 2005).

To the extent to which the models are validated by a range of different proxies (e.g. Berner *et al.*, 2003a; Scott and Glasspool, 2006), plant geophysical effects on atmospheric O_2 appear substantial and are controversially linked with a range of biological evolutionary radiations and events (Ward *et al.*, 2006; Falkowski *et al.*, 2005; Berner *et al.*, 2007; Labandeira, 2007). In this latter regard, plants may act as geobiological agents of terrestrial biotic evolution. Most notably, the Permo-Carboniferous rise in O_2 levels is coincident with the

appearance in the fossil record of a spectacular episode of gigantism in insects, as well as other arthropods and terrestrial vertebrates (Graham *et al.*, 1995). At this time, Carboniferous dragonfly wing-spans reached 71 cm and thorax widths 2.8 cm, and millipedes attained lengths of over 1.5 m (Graham *et al.*, 1995). The subsequent late Permian–early Triassic fall in O_2 (13–15%) saw a decrease in the level of animal gigantism, though it is worth noting that dragonflies twice as large as the largest extant forms thrived during the late Triassic and early Jurassic oxygen minimum (Okajima, 2008), an interval that also saw the appearance of high oxygen-demand groups: flying reptiles and mammals (Butterfield, 2009).

11.2.2.2 Plants, mycorrhizae, weathering and atmospheric CO_2

The geochemical carbon cycle involves the uptake of CO_2 from the atmosphere and its transformation during the weathering of continental Ca and Mg silicate minerals to dissolved bicarbonate ions in rivers that are flushed out to the oceans and precipitated as $CaCO_3$ and $MgCO_3$ minerals (Fig. 11.1) (Berner, 2004). Over tens-to-hundreds of millions of years, thermal breakdown at depth of carbonate minerals via metamorphism, diagenesis and volcanism, transfers CO_2 back to the atmosphere (Berner, 2004). Because the atmospheric carbon reservoir is several orders of magnitude smaller than that of the rock reservoir, inputs by volcanic degassing and outputs by silicate rock weathering are thought to be finely balanced, otherwise atmospheric CO_2 levels would rise or fall dramatically (Berner and Caldeira, 1997; Zeebe and Caldeira, 2008). Rates of continental silicate rock weathering are dependent upon climate (temperature and hydrology) to create a negative

Earth's thermodynamic thermostat

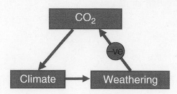

Figure 11.2 Simplified systems diagram for the thermostatic control of Earth's climate via the geochemical carbon cycle. Plain arrows are positive feedbacks, those with a –ve symbol are negative feedbacks. The loop between the CO_2, climate and weathering boxes represents a negative feedback, i.e. only one –ve label. See text for further details.

feedback loop (Fig. 11.2), – that is a thermodynamic feedback control on Earth's temperature – whereby rising CO_2 causes warming and moister climates that together enhance weathering and promote CO_2 removal from the atmosphere (Walker *et al.*, 1981; Berner *et al.*, 1983c). Co-evolutionary partnerships of plants and symbiotic fungi likely accelerated mineral weathering processes removing CO_2 from the atmosphere (Berner, 1998; Berner *et al.*, 2003b; Taylor *et al.*, 2009).

The weathering of Ca silicate minerals can be simply represented by the overall reaction (Berner, 2004):

$$CO_2 + CaSiO_3 \rightarrow CaCO_3 + SiO_2 \qquad (11.3)$$

This type of reaction liberates Ca^{2+} ions that are carried to the oceans by rivers, and with dissolved bicarbonate ions, leads to the precipitation of calcium carbonate minerals. With magnesium silicates, Mg^{2+} is liberated, but exchanged with calcium in marine basalts, leading to the deposition of $CaCO_3$ rather than $MgCO_3$ (Berner, 2004). Ultimately this flux represents the net transfer of carbon from the surficial system to the rock reservoir until it is released millions of years later during metamorphism or weathering (i.e. the reverse of reaction represented by Equation 11.3). Photosynthesis by land plants, algae, and phytoplankton also transfers carbon from the ocean–atmosphere system into the lithosphere, but the masses of this organic carbon sub-cycle are smaller than those involved in the inorganic subcycle and are less important from a climatic perspective.

The advent of land plants some 470 Ma ago, and the subsequent spread of rooted trees and forests likely introduced, for the first time in Earth's history, significant biotic regulation of the global atmospheric CO_2 concentration through their collective effects on mineral weathering (e.g. Berner, 1997, 1998). The current paradigm recognizes that vascular plant activities enhance silicate mineral weathering by dissolution of bedrocks in five major ways. See Berner *et al.* (2003b) for an extensive review of the field and laboratory evidence, and Taylor *et al.* (2009) for a critical outline of the geochemical arguments:

Mechanism (i). Acidification due to hydrogen ion and organic exudates.

Mechanism (ii). Acidification due to respiration and elevated $CO_2(g)$.

Mechanism (iii). Litter decomposition and carbon transfer to heterotrophs.

Mechanisms (iv). Recycling of evapotranspiration and repeat flushing of soils.

Mechanisms (v) Bank stabilization and retardation of soil erosion.

Mechanisms (i) to (v) above were probably strengthened by: (1) the evolution, diversification and spread of deeply rooting trees throughout upland areas from the Silurian through the Devonian (416–359 Ma ago), and (2) more contentiously, by the replacement of gymnosperms with the more advanced angiosperms from the early Cretaceous onwards (145–65 Ma ago) (Berner, 1997, 1998, 2004). Empirical representation of the effect of these two plant evolutionary axes on mineral weathering processes in long-term carbon cycle models (e.g. Berner and Kothavala, 2001, Bergman *et al.*, 2004; Berner, 1997, 1998; Volk, 1989) indicates the potential capacity of each to promote CO_2 removal from the atmosphere and alter the trajectory of Earth's CO_2 history and climate.

Current thinking about the role of plants in altering chemical weathering processes has been dominated by the prevailing view outlined above. However, this view gives only passing recognition to the possible role of fungi in weathering processes (Hoffland *et al.*, 2004; Leake *et al.*, 2008; Bonneville *et al.*, 2009; Taylor *et al.*, 2009). Yet approximately 80–90 % of plant species form symbiotic associations with mycorrhizal fungi (Read, 1991), which act as the sink for a large organic carbon flux received from plants (Högberg and Read, 2006).

A revised framework incorporating fungal and soil physiological ecology proposes that rates of biotic weathering can be more usefully conceptualized as being 'driven by, and proportional to, autotrophic carbon fixation, especially the fraction allocated belowground' (Taylor *et al.*, 2009). This suggestion is underpinned by the realization that rates of biotic weathering are 'fundamentally controlled by the energy supply to photosynthetic organisms which, in turn, controls their biomass, surface area of contact, and their capacity to interact physically and chemically with the minerals' (Taylor *et al.*, 2009). In this revised view, it is important to emphasize that it is the combined effects of vegetation, especially root activities and associated mycorrhizal fungi, that are implicated as key biotic components of weathering processes, all of which are fuelled by photosynthate fixed by the above-ground biomass.

The two major functional groups of mycorrhizal fungi have different origination dates and differ in their modes of nutrient acquisition (Taylor *et al.*, 2009). Such differences may translate into differences in their effectiveness in driving biotic weathering processes. Arbuscular mycorrhizal fungi (AMF), exclusively of the order Glomales (Glomeromycota), are the ancestral group. Molecular clock evidence indicates they originated between 462 and 353 Ma ago (Simon *et al.*, 1993), and there is evidence in the fossil record for their association with plants by the Early Devonian (400 Ma ago) (Remy *et al.*, 1994). Ectomycorrhizal fungi (EMF) arose considerably later, probably sometime in the Jurassic, about 180 Ma ago (Berbee and Taylor, 2001). AMF penetrate cells of plant roots by forming invaginations in the cell membrane, and provide nutrients, especially phosphorus, to plants that would otherwise be inaccessible to the root systems. During growth, AMF reduce the pH of fluids around the hyphal tip both through proton release during the uptake of inorganic cations (e.g. NH_4^+) to maintain electrostatic neutrality, and by supporting AMF-associated bacteria (Villegas and Fortin, 2001, 2002). This is a crucial factor in increasing mineral dissolution and P uptake. AMF mycelial networks also secrete glycoproteinaceous compounds into the soil that increase soil particle aggregation, which could alter water cycling and cation flushing.

The plant–fungus interface of EMF is fundamentally different from that of AMF, consisting of mycelia sheaths around the absorptive root-tips of EMF-forming angiosperm and gymnosperm trees. Typically over 90% of root-tips are sheathed in mycelium so that virtually all of the labile carbon released into soils from roots is channelled through these fungi (Leake *et al.*, 2004). EMF mycelia networks extend outwards from the root-fungal mantle into the soil to mobilize and absorb nutrients by actively secreting organic acids and protons at their hyphal tips altering localized pH values and accelerating mineral dissolution (Landeweert *et al.*, 2001). AMF are not known to secrete organic acids or chelators for active mineral dissolution (Taylor *et al.*, 2009).

The proposed involvement of mycorrhizae in driving biotic weathering calls for a re-evaluation of how the rise and spread of deep-rooting trees, and the rise of angiosperms, are represented in geochemical carbon cycle models. In particular, it questions the relevance of contemporary field studies on areas of vegetation dominated by EMF-forming trees for developing empirical modelling functions describing early forest development that likely involved AMF associations (Strullu-Derrien and Strullu, 2007). Further, it also implies that falling atmospheric CO_2 levels during the Mesozoic and Cenozoic may actually be linked to the rise and spread of EMF-forming trees rather than simply being linked to the rise of angiosperms at the expense of gymnosperms. A critical appraisal of the frequently cited evidence for angiosperms being more effective at weathering than gymnosperms indicates it to be equivocal (Taylor *et al.*, 2009).

11.3 Animals as geobiological agents

Given their enormous standing biomass and domination of most terrestrial environments, it is hardly surprising that land plants serve as powerful geobiological agents. Often less appreciated is the correspondingly large impact made by animals. The key to metazoan influence lies in their underlying physiology, which in most instances combines heterotrophy and motility with organ-grade multicellularity. By tapping into an effectively inexhaustible source of novel morphology and behaviour, motile multicellular heterotrophs have revolutionized the exchange between biosphere and geosphere over the past 600 million years (Butterfield, 2007, 2011).

The essence of organ-grade animals (= eumetazoans) is the gut – a specialized multicellular chemical reactor adapted for the digestion of food, usually in collaboration with symbiotic gut microbes (Penry and Jumars, 1986). At one level, there is nothing particularly novel about this apparatus, neither the chemical nor microbial processes differing substantially from those in the external environment. However, the active maintenance of 'optimal' digestive conditions by the gut sets these phenomena in a fundamentally more efficient and purposeful context, particularly when they are integrated with other organ systems optimized for sensing, pursuing, capturing, filtering and comminuting food.

By diverting productivity away from default microbial processing to that of a roving chemical reactor, animals might be expected to increase rates of biogeochemical cycling – which in many ways they certainly do. Even so, it is worth appreciating that metazoans direct a significant proportion of digested food – generally considered to be around 10% – into a loop with fundamentally lower levels of return. Due to the three-quarter scaling relationship between body mass and metabolic rate, respiration in larger organisms is exponentially slower than in smaller ones, making animals the repositories of relatively long-lived, non-cycling biomass. In aquatic food webs, where predators tend be much larger than their prey, this size-specific metabolism extends up the trophic structure with stepwise increases in predator size balanced by correspondingly lower levels of respiration – to the extent that marine food chains exhibit broadly equivalent biomasses at each of ca. five trophic levels, from phytoplankton to sea monsters (Sheldon *et al.*, 1972; Kerr and Dickie, 2001; Brown *et al.*, 2004). With four of these five levels occupied exclusively by metazoans, it is clear the majority of marine biomass is represented by animals – the equivalent of trees on land.

Like all organisms, animals package biomass; but the stoichiometry of these packages differs substantially from that of primary producers, varying with environment, bodysize and taxon-specific physiology (Sterner and Elser, 2002). In terrestrial ecosystems, where plants are embellished with C-rich structural macromolecules, the conversion to herbivore biomass entails a pronounced reduction in C:N/P ratios, though the effect is generally swamped by the predominance of plant biomass. In aquatic ecosystems, however, the fundamentally higher proportion of consumer biomass – an inverted trophic pyramid – means that animal stoichiometry can impinge directly on ecosystem nutrient flux (Polis, 1999).

In their review of ecological stoichiometry in lakes, Elser *et al.* (1996) compared the effects of differing types of herbivorous mesozooplankton on overall nutrient status. For zooplankton with relatively high N:P ratios such as calanoid copepods (N:P ~ 30:1), associated phytoplankton growth is typically N-limited due to the preferential retention of N in consumer biomass. By contrast, the much lower N:P of cladocerans (e.g. *Daphnia* N:P ~ 15:1) tend to induce P-limitation. The reason for such pronounced differences reflects the differing life histories of these two groups, with the very rapid growth rates and parthenogenic reproduction of *Daphnia* requiring substantially greater rates of protein synthesis and accompanying ribosomal RNA (ca. 10% of the dry weight of *Daphnia* is RNA, vs. just 2% in copepods).

The more fundamental reason for cladoceran vs. copepod domination in lacustrine systems relates to ecological tradeoffs. Thus, under severe P-limitation, Elser and colleagues found that slower growing, high N:P copepods always outcompeted cladocerans. Under more equable conditions, however, the principal control on secondary production switches from the 'bottom-up' availability of nutrients to the 'top-down' effects of predation. Unlike copepods, which are capable of inertial escape jumps, *Daphnia* are mostly constrained to viscous flow and are consequently much more susceptible to visual predation (Naganuma 1996). Thus the introduction of zooplankton-feeding minnows to *Daphnia*-dominated lakes can lead to a replacement by high N:P calenoid copepods, which in turn converts primary productivity from P-limited to N-limited conditions. Minnows, of course, are also susceptible to predation, and the introduction of higher level predators (piscivores) can induce a return to *Daphnia*-dominated zooplankton and P-limited primary production. This is a classic account of a 'trophic cascade' in which the activities of predators cascade down food chains to control both primary production and overall ecosystem function (Pace *et al.*, 1999).

Trophic cascades are most strongly expressed in simple systems with strong trophic links, such as in lakes, and often fail to propagate faithfully through more complex foodwebs due to various compensatory effects (Pace *et al.*, 1999). Even so, there is an increasing recognition that the top-down activities of consumers control the structure of most modern ecosystems. Recent over-fishing in the Black Sea, for example, is thought to be responsible for its increasingly frequent jellyfish blooms and rising turbidity as the removal of apex predators (bonito, mackerel, bluefish, dolphins) cascades down through zooplanktivorous fish ($\uparrow$), crustacean mesozooplankton ($\downarrow$) and phytoplankton ($\uparrow$) (Daskalov 2002). Instances of trophic cascades in open marine conditions are less easily recognized, in part because of the larger scale and more generally oligotrophic conditions, although the catastrophic and seemingly permanent shifts in population structure associated with commercial fisheries demonstrates that it does happen (e.g. Frank *et al.* 2005, Casini *et al.* 2009). A happier cascade saw the return of Pacific kelp forest communities following the 20th century recovery of sea otter populations and their top-down control on sea urchin grazing (Estes and Duggins 1995).

The fundamentally greater levels of plant biomass in terrestrial ecosystems tend to attenuate the top-down effects of consumers, though these become more apparent in simplified microcosms. In low-diversity arctic tundra, for example, the grazing activities of muskox and caribou have been shown to suppress the spread of woody shrubs (Post and Pederson 2008), and in tropical forest communities that have lost their top predators, plant biomass becomes decimated due to order-of-magnitude increases in herbivorous howler monkeys, iguanas and leaf-cutter ants (Terborgh *et al.*, 2001). Such observations support the so-called 'green world' hypothesis, which holds that the terrestrial biosphere maintains its vegetated cover through the top-down control of herbivory by predators (Hairston *et al.*, 1960).

There are, of course, other constraints on the activity of herbivores, not least the pronounced stoichiometric disparity between land plants and herbivores (Polis, 1999). But in terms of geobiological impact, terrestrial predators still punch well above their weight. In a temperate grassland ecosystem, for example, Schmitz (2008) demonstrated a 14% reduction in plant diversity, a 33% increase in nitrogen mineralization and a remarkable 163% increase in above ground primary productivity when spider predation on grasshoppers switched from 'sit and wait' ambush-type hunting to a more actively roaming strategy. By presenting a persistent and detectible danger, ambush hunters appear to induce a behavioural response in their prey which cascades down to affect the structure and function of the entire system.

Such pronounced, seemingly idiosyncratic amplification of individual behaviour greatly complicates the geobiological accountability of trophic interactions. One might ask, for example, what controls the presence or absence of 'sit and wait' type spiders in temperate grasslands. Both higher-level predation and within-level competition are possibilities, but the more useful response might be to recognize the overarching effect of diversity itself (Hooper *et al.*, 2005; Duffy *et al.*, 2007). Species richness provides multiple responses to particular circumstances which can enhance productivity within a particular trophic level (Tilman *et al.*, 2001), while the presence of multiple links between trophic levels strongly attenuates the transmission of trophic cascades (McCann, 2000). But diversity is also about ecological novelty, which, under certain circumstances, can be accompanied by dramatic non-linear responses – witness any number of (un)natural experiments involving invasive species (Mooney and Cleland, 2001), from cats and cane toads to rabbits and zebra mussels.

11.3.1   Ecosystem engineering

The geobiological impact of animals extends well beyond immediate trophic effects. Like the vascular land plants discussed earlier, animals are capable of altering physical environments and thereby defining the larger-scale ecological context. The classic case of such ecosystem 'engineering' (Jones *et al.*, 1994) is the dam-building behaviour of beavers, which not only creates vast expanses of temperate wetland, but also increases habitat heterogeneity, sediment and nutrient retention, and associated biodiversity (Naiman *et al.*, 1994). A comparable degree of engineering occurs in marine settings, where coral reefs and kelp forests dissipate wave energy, stabilize sediments, and provide the framework for unique, multidimensional ecosystems (Estes and Duggins 1995, Idjadi and Edmunds, 2006; Alvarez-Filip *et al.* 2009).

Smaller animals have less individual engineering potential, but can more than make up for this through their collective activities. The addition of minnows to clear-water lakes, for example, can switch the principal zooplankton from cladocerans to slower-growing copepods, leading to increased phytoplankton densities, increased water-column turbidity, eutrophication, and bottom water anoxia. In the Lower Great Lakes (Ontario, Erie, Michigan), it was the invasion of suspension-feeding dreissenid (zebra) mussels – in concert with a multibillion dollar investment in sewage treatment – that finally restored their historical clear-water conditions (Higgins *et al.*, 2008). Ironically, the lakeshores have recently become more fouled than ever by *Cladophora glomerata* blooms, despite the fundamentally lower levels of nutrient loading. By clearing the water column of suspended phytoplankton the mussels greatly expanded the photic-zone habitat for benthic *Cladophora*, at the same time as their shell-beds created abundant new hard-substrate on which these algae can become established.

All ecosystems are engineered at some level by the organisms they support, which in turn are dependent on the nature of the ecosystem. Such interdependency gives rise to positive feedback effects and hysteresis, with ecosystems sometimes shifting dramatically between alternative stable states (Scheffer and Carpenter, 2003). Thus the cladocerans and zebra mussels responsible for clear-water conditions in lakes tend to reinforce the clear-water conditions in which they thrive, thereby increasing the system's resilience to turbidity-inducing nutrient loading. Even so, at some level of nutrient input (and/or reduced grazing) the system can be overloaded, tipping it into a stratified, turbid-water state, which is further reinforced by the exclusion of benthic suspension feeders. Under such positive feedback conditions, the original clear-water state is not recovered by simply returning to the original conditions; rather, some destabilizing overshoot is required – such as the introduction of voracious suspension-feeding mussels. But there are always other complexifying engineers: benthic feeding fish, for example, can induce significant levels of non-algal turbidity by resuspending sediment and removing macrophytes, thereby excluding benthic suspension feeders. It is this complex feedback between biologically and physically defined environment that makes the business of restoration ecology so unpredictable (Byers *et al.*, 2006).

11.3.2   Alimentation, bioturbation and biomineralization

All heterotrophic organisms are in the business of feeding which, when conducted by relatively large organisms such as animals, tends to move, mix or package aspects of the environment that would not otherwise be moved, mixed or packaged. Migratory fish, for example, transport significant amounts of marine-derived nutrients to freshwater lakes and streams (e.g. Schindler *et al.*, 2005), and the guano of sea-birds often has profound effects on the functioning of island ecosystems (Anderson and Polis, 1999).

In a similar fashion, zooplankton fertilize much of the marine shelf benthos by repackaging phytoplankton into rapidly sinking, nutrient-rich faecal pellets. Although faecal pellets are not the only means of exporting phytoplankton out of the water-column (Turner, 2002), their contribution to the 'biological pump' is both quantitatively and qualitatively significant (Wassmann, 1998, Hernández-León *et al.*, 2008). The faecal pellets and pseudo-faeces of benthic suspension-feeders also serve as an important link between the plankton and benthos, not only with respect to nutrients but by reconstituting enormous volumes of suspended clay into dense sand-sized particles (Pryor, 1978). Moreover, the osmoregulatory requirement of marine fish to continuously drink results in enormous quantities of physiologically excreted carbonate – representing as much as 15%

of total carbonate production in the modern oceans (Wilson *et al.*, 2009).

Metazoans larger than about a millimetre also introduce substantial amounts of turbulence to the environment while pursuing their individual ecologies. Although trivial in subaerial settings, it has been argued that metazoans contribute as much mechanical energy to the world's oceans as winds and tides (Huntley and Zhou, 2004, Dewar *et al.*, 2006). Whether or not such calculations are quantitatively realistic (Visser, 2007), the activities of zooplankton, fish and aquatic mammals represent a qualitatively distinct style and distribution of marine turbulence, with potentially enormous knock-on effects for marine stratification, nutrient regeneration and phytoplankton ecology.

By far the most substantial environmental mixing by animals is their bioturbation of sediments and soils. Such activity is dominated by relatively large organisms capable of moving sedimentary grains and ventilating otherwise anoxic environments (Jumars *et al.*, 1990). Like the subsurface components of vascular land plants, infaunal animals have a profound effect on substrate permeability, erodibility, carbon burial, nutrient recycling and habitat facilitation.

The geobiological consequences of animal bioturbation depend on particular infaunal behaviours. Deposit feeding, for example, results in the mixing (and/or unmixing) of constantly ingested sediment, whereas the construction of open burrow systems is likely to have greater impact on advective ventilation (Aller, 1982). Bioturbation in soils is dominated by earthworms, ants, termites and burrowing mammals, which are responsible for a wide range of engineering effects – from increased nutrient regeneration, primary productivity and carbon burial, to textural modification and aeration (e.g. Folgarait, 1998, Reichman and Seabloom, 2002, Wilkinson *et al.* 2009). By volume, it is the highly motile, non-selective deposit feeders that do the most work, while habitat heterogeneity and its compounding effects on biodiversity tend to derive from more localized 'intentional' engineers (Joquet *et al.*, 2006). Geophagous earthworms, for example, homogenize soil environments, whereas the nest-building behaviour of ants and termites gives rise to islands of concentrated influence, often in collaboration with other behaviourally complex species. Terrestrial bioturbators also play an important role in controlling landscape topography by altering soil coherence and mobility (Deitrich and Perron, 2006, Wilkinson *et al.* 2009).

Subaqueous soft sediments are fundamentally more homogeneous than soils, and are generally supplied with more digestible, nutrient-rich organic material. As such, they are intensively and (almost) ubiquitously processed by infaunal deposit and detritus feeders. Shallow water infaunal echinoids, for example,

completely rework surface sediments over the course of a few days (Lohrer *et al.*, 2004), and virtually all marine sediments are thought to pass through the guts of deposit feeders at least once before final burial (Jumars *et al.*, 1990). Bioturbating marine infauna – primarily polychaetes, bivalves, echinoids and crustaceans – are also responsible for marked increases in oxygen demand, sulfide oxidation and ammonium generation, while organic carbon burial may be enhanced by sediment mixing. At a structural level, bioturbation disrupts both the physical and microbial binding of sediments resulting in relatively 'soupy' substrates and increased bottom-water turbidity. With submarine soft-sediments covering well over half the planet – and subaerial soft-sediments covering much of the rest – bioturbating animals represent one of the planet's most influential engineering guilds.

Although burrowing and necto-benthic animals tend to disrupt sub-aqueous sediments, those that live epifaunally often contribute substantially to sediment stabilization (Reise 2002). The most obvious stabilizers are colonial/modular invertebrates with a capacity to construct, baffle and bind various types of wave-resistant reefs (e.g. corals, sponges, bryozoans, oysters). As with plants on land, animal-induced stabilization in the marine realm is usually associated with relatively large body-sizes and sessile habits, though it also occurs in certain motile forms such as sediment-climbing mussels (van Leeuwen *et al.* 2010). By engineering unique high-energy environments, reef-forming metazoans also bear directly on water turbulence, wave dissipation, and shoreline dynamics (Idjadi and Edmunds, 2006, Alverez-Filip *et al.*, 2009).

One of the principal factors contributing to modern marine environments is the enormous capacity of marine invertebrates to precipitate biominerals from seawater. In addition to their structural function in reefs, biomineralized skeletons play a defining role in many level-bottom settings. The unique hydraulic properties and accompanying taphonomic feedback of metazoan bioclasts create a range of structurally and ecologically unique sedimentary environments, including spicule mats, shell beds and cheniers (Kidwell and Jablonski, 1983, Liu and Walker, 1989, Bett and Rice, 1992, Gutierrez *et al.*, 2003).

Metazoan biomineralization also contributes significantly to ocean chemistry through the enzymatically enhanced precipitation of carbonate and silica, though it is unicellular protists – most notably coccolithophores and diatoms – that currently determine the larger-scale, biogeochemical impact of biomineralization. Even so, there is a strong case for linking the evolution of biomineralizing phytoplankton to the escalatory effects of grazing metazoan zooplankton (Smetacek, 2001; Hamm *et al.*, 2003), which would make animals responsible – either directly or indirectly – for the majority of biomineral production in the modern ocean.

11.3.3 Evolutionary engineering and the Ediacaran–Cambrian radiations

The overarching effect of animal ecology on environment identifies the modern biosphere as fundamentally and pervasively metazoan. This was not always the case, however, and it is clear that the style and degree of influence has varied over evolutionary time (Erwin, 2008). At some point in the past beavers did not build dams, earthworms did not mix soil, and there were no coccolithophores or coral reefs. Bioturbation has been a feature of most marine sediments for the past 550 million years, but the intense 'biological bulldozing' typical of the modern oceans is a relatively recent, Mesozoic, invention (Thayer, 1983). Indeed, it was not so long ago that animals had yet to evolve, and ecosystems were engineered exclusively by microbes. Consideration of this pre-metazoan 'microbial world,' and its transition into the Phanerozoic, offers a unique view of how animals came to dominate the structure and function of the modern biosphere (Butterfield, 2007, 2009, 2011).

Whether animals – specifically organ-grade eumetazoans – have deep Proterozoic roots is a matter of long-standing debate, but it is clear that they were of no geobiological significance until the Ediacaran. Not only is there no fossil record of pre-Ediacaran eumetazoans, but the profound evolutionary stasis expressed by contemporaneous microfossils reflects the absence of any significant co-evolutionary drivers, at least with respect to organism size or morphology (Peterson and Butterfield, 2005; Butterfield, 2007). The break comes in the earliest Ediacaran, which sees an unprecedented radiation of large, morphologically complex microfossils and an order-of-magnitude increase in rates of evolutionary turnover – indirect but compelling evidence for the appearance of organ-grade eumetazoans. A further early Cambrian radiation of microfossils – this time representing phytoplankton – is most reasonably interpreted as a co-evolutionary response to the appearance of metazoan zooplankton, with myriad feedback effects driving the Cambrian explosion of large animals (Butterfield, 1997, 2007, 2011).

Interestingly, the Ediacaran and early Cambrian also represent an interval of pronounced change in oceanic chemistry and structure, including a switch from cyanobacteria- to algae-dominated export production (Knoll et al., 2007), increased rates of vertical transport (Logan et al., 1995), and a progressive ventilation/oxygenation of the deep ocean (Canfield et al., 2007). The conventional interpretation of these data is that they reflect 'bottom-up' shifts in biogeochemical cycling, giving rise to an oxygenated atmosphere and the 'permissive environments' that allowed both eukaryotic algae and animals to pursue their full evolutionary potential. There is, however, an alternative 'top down' interpretation, which recognizes animals as powerful, context-altering, geobiological agents – the cause rather than the consequence of geochemical perturbation (e.g. Logan et al., 1995; Butterfield, 1997, 2007, 2009, 2011). Qualitative shifts in the Neoproterozoic-Cambrian sulfur record, for example, have recently been ascribed to the onset of bioturbation-induced sulphide oxidation (Canfield and Farhquar 2009).

Insofar as animals play a fundamental role in structuring aquatic ecosystems, it is clear that the pre-Cambrian and pre-Ediacaran oceans must have operated in a fundamentally different manner than their modern counterparts. Certainly the absence of Phanerozoic-style biomineralization and bioturbation would have imparted a distinctly non-uniformitarian quality to early benthic environments, with important implications for sediment stability, benthic ecology and biogeochemistry (Aller, 1982, Seilacher, 1999, Droser et al., 2005, Canfield and Farquhar, 2009). In the pelagic realm, the absence of metazoans would have precluded the faecal export of surface productivity (Logan et al., 1995), as well as the calcium carbonate currently generated by fish (Wilson et al., 2009).

But animals do more than engineer physical environments. They are also powerful agents of co-evolutionary change, directly responsible for the Phanerozoic radiations of organism size, morphology and biomineralization (Vermeij, 1994, Butterfield, 2007, 2011). Body size is of particular significance in aquatic ecology, and in the case of phytoplankton there is a clear advantage – in terms of buoyancy, nutrient scavenging and self-shading – for maintaining small cell size. The paradox in the modern oceans is that the vast majority of export production is represented by relatively large-celled, morphologically diverse, and often substantially biomineralized eukaryotic phytoplankton (Hutchinson, 1961) – for which the best explanation is a top-down ecological trade-off imposed by suspension-feeding zooplankton (Smetacek, 2001; Jiang et al., 2005). In the (pre-metazoan) absence of such selective pressures, phytoplankton are expected to be both small and non-biomineralizing, leading to fundamentally lower rates of sinking, increased surface-water turbidity, and a compounding positive feedback in favour of stratified, cyanobacteria-dominated oceans – all of which are features of the Proterozoic record (Knoll et al., 2007; Canfield et al., 2007; Butterfield, 2009). Given the sudden, often profound regime shifts associated with ecological perturbation in modern ecosystems (Scheffer and Carpenter, 2003), the Ediacaran introduction of organ-grade metazoans can be usefully viewed as the tip-off point between two alternative stable states in planetary function

(Butterfield, 2007, 2009, 2011): on the one hand a microbial world dominated by prokaryotic ecologies and evolutionary stasis, and on the other the extraordinarily diverse and dynamic world of the Phanerozoic.

11.4 Conclusions

Multicellular plants and animals clearly have an overarching impact on ecosystem function in the Phanerozoic, and there is little doubt that the biosphere would have worked in fundamentally different way prior to their appearance. All organisms act as geobiological agents, but the influence of any one component is highly dependent on the overall context, with a potential for powerful, highly idiosyncratic and non-linear feedbacks. Certainly the evolution of land plants gave rise to conditions that encouraged the proliferation of land plants and exerted profound impacts on the Earth system (Beerling, 2007), while the rise of animals inevitably constructed the opportunities for ever more animals.

Both animals and plants engineer novel environments, but they also engineer one another over evolutionary time scales. Land plant diversity is pervasively linked to the co-evolutionary effects of animals – most obviously in the form of pollination mutualisms and top-down responses to herbivory (Fenster _et al._, 2004; Holdo _et al._, 2009) – but so, too has terrestrial animal diversity been facilitated and driven by plants. Thus the great radiations of Tertiary mammals were as much a product of evolving grasses and expanding grasslands as the expansion of grasslands was driven by the co-evolutionary radiation of mammals, disease and fire (Retallack, 2001; Holdo _et al._, 2009; Gill _et al._, 2009). More recent evolutionary and technological innovations have of course expanded the geobiological influence of plants and animals out of all proportion, reengineering planetary environments, climate and biodiversity as they pursue their own evolutionary advantage.

Acknowledgements

We thank Lyla Taylor for kindly drafting Fig. 11.1. DJB gratefully acknowledges funding of geobiological research in his lab, through the NERC, UK, and a Royal Society-Wolfson Research Merit Award.

References

Ainsworth EA, Long SP (2005) What have we learned from 15 years of free-air CO_2 enrichment (FACE)? A meta-analytic review of the responses of photosynthesis, canopy properties and plant production to rising CO_2. _New Phytologist_ **165**, 351–372.

Ainsworth EA, Rogers A (2007) The response of photosynthesis and stomatal conductance to rising [CO_2]: mechanisms and environmental interactions. _Plant, Cell and Environment_ **30**, 258–270.

Algeo TJ, Scheckler SE (1998) Terrestrial-marine teleconnections in the Devonian: links between the evolution of land plants, weathering processes, and marine anoxic events. _Philosophical Transactions of the Royal Society_ **B353**, 113–130.

Aller RC (1982) The effects of macrobenthos on chemical properties of marine sediment and overlying water. In: _Animal-Sediment Relations_ (eds McCall PL, Tevesz MSS). Plenum Press, New York, pp., 53–102.

Alvarez-Filip L, Dulvy NK, Gill JA, Côté IM, Watkinson AR (2009) Flattening of Caribbean coral reefs: region-wide declines in architectural complexity. _Proceedings of the Royal Society B_ **276**, 3019–3025.

Anderson WB, Polis GA (1999) Nutrient fluxes from water to land: seabirds affect plant nutrient status on Gulf of California islands. _Oecologia_ **118**, 324–332.

Bala G, Caldeira K, Wickett M., et al. (2007) Combined climate and carbon-cycle effects of large-scale deforestation. _Proceedings of the National Academy of Sciences_ **104**, 6550–6555.

Bartdorff O, Wallmann K, Latif M, Semenov V (2008) Phanerozoic history of atmospheric methane. _Global Biogeochemical Cycles_ **22**, doi:10.1029/2007GB002985.

Beerling DJ (2007) _The Emerald Planet. How Plants Changed Earth's History_. Oxford University Press, Oxford.

Beerling DJ, Berner RA (2005) Feedbacks and the coevolution of plants and atmospheric CO_2. _Proceedings of the National Academy of Sciences_ **102**, 1302–1305.

Beerling DJ, Woodward FI (2001) _Vegetation and the Terrestrial Carbon Cycle. Modelling the First 400 Million Years_. Cambridge University Press, Cambridge.

Beerling DJ, Hewitt CN, Pyle JA, Raven JA (2007) Critical issues in trace gas biogeochemistry and global change. _Philosophical Transactions of the Royal Society_ **A365**, 1629–1642.

Beerling DJ, Berner RA, MacKenzie FT, Harfoot MB, Pyle JA (2009) Methane and the CH_4-related greenhouse over the past 400 million years. _American Journal of Science_ **309**, 97–113.

Berbee ML, Taylor JW (1993) Dating the evolutionary radiations of the true fungi. _Canadian Journal of Botany_ **71**, 1114–1127.

Bergman NM, Lenton TM, Watson AJ (2004) COPSE: a new model of biogeochemical cycling over Phanerozoic time. _American Journal of Science_ **304**, 397–437.

Berner RA (1997) The rise of plants and their effect on weathering and atmospheric CO_2. _Science_ **276**, 544–546.

Berner RA (1998) The carbon cycle and CO_2 over Phanerozoic time: the role of land plants. _Philosophical Transactions of the Royal Society_ **B353**, 75–82.

Berner RA (2005) The carbon and sulphur cycles and atmospheric oxygen from middle Permian to middle Triassic. _Geochimica et Cosmochimica Acta_ **69**, 3211–3217.

Berner RA (2004) _The Phanerozoic Carbon Cycle: CO_2 and O_2_. Oxford University Press, Oxford.

Berner RA, Caldeira K (1997) The need for mass balance and feedback in the geochemical carbon cycle. _Geology_ **25**, 955–956.

Berner RA, Canfield DE (1989) A new model of atmospheric oxygen over Phanerozoic time. _American Journal of Science_ **289**, 333–361.

Berner RA, Kothavala Z (2001) GEOCARB III: a revised model of atmospheric CO_2 over Phanerozoic time. _American Journal of Science_ **301**, 182–204.

Berner RA, Lasaga AC, Garrels RM (1983c) The carbonate-silicate geochemical cycle and its effect on atmospheric carbon dioxide over the past 100 million years. *American Journal of Science* **283**, 641–683.

Berner RA, Beerling DJ, Dudley R, Robinson JM, Wildman RA (2003a) Phanerozoic atmospheric oxygen. *Annual Reviews of Earth and Planetary Sciences* **31**, 105–134.

Berner EK, Berner RA, Moulton KL (2003b) Plants and mineral weathering: present and past. *Treatise on Geochemistry* **5**, 169–188.

Berner RA, VandenBrooks JM, Ward PD (2007) Oxygen and evolution. *Science* **316**, 557–558.

Bett BJ, Rice AL (1992) The influence of hexactinellid sponge (*Pheronema carpenteri*) spicules on the patchy distribution of macrobenthos in the Porcupine Seabight (bathyal NE Atlantic). *Ophelia* **36**, 217–226.

Betts RA, Cox PM, Lee SE, Woodward FI (1997) Contrasting physiological and structural vegetation feedbacks in climate change simulations. *Nature* **387**, 796–799.

Bonan GB (2008) Forests and climate change: forcings, feedbacks and the climate benefits of forests. *Science* **320**, 1444–1449.

Bonneville S, Smits MM, Brown A, *et al.* (2009) Plant-driven fungal weathering: early stages of mineral alteration at the nanometre scale. *Geology* **37**, 615–618.

Brown JH, Gillooly JF, Allen AP, Savage VM, West GB (2004) Toward a metabolic theory of ecology. *Ecology* **85**, 1771–1789.

Buckley TN (2005) The control of stomatal water balance. *New Phytologist* **168**, 275–291.

Burd AB, Jackson GA (2009) Particle aggregation. *Annual Review of Marine Science* **1**, 65–90.

Butterfield NJ (1997) Plankton ecology and the Proterozoic–Phanerozoic transition. *Paleobiology* **23**, 247–262.

Butterfield NJ (2007) Macroecovolution and macroecology through deep time. *Palaeontology* **50**, 41–55.

Butterfield NJ (2009) Oxygen, animals and oceanic ventilation – and alternative view. *Geobiology* **7**, 1–7.

Butterfield NJ (2011) Animals and the invention of the Phanerozoic Earth system. Trends in *Ecology & Evolution*, **26**, 81–87.

Byers JE, Cuddington K, Jones C.G, *et al.* (2006) Using ecosystem engineers to restore ecological systems. *Trends in Ecology and Evolution* **21**, 493–500.

Canfield DE, Farquhar JE (2009) Animal evolution, bioturbation, and the sulfate concentration of the oceans. *Proceedings of the National Academy of Sciences, USA* **106**, 8123–8217.

Canfield DE, Poulton SW, Narbonne GM (2007) Late-Neoproterozoic deep ocean oxygenation and the rise of animal life. *Science* **315**, 92–95.

Casini M, Hjelm J, Molinero J.-C, *et al.* (2009) Trophic cascades promote threshold-like shifts in pelagic marine ecosystems. *Proceedings of the National Academy of Sciences, USA* **106**, 197–202.

Chapin FS, Sturm M, Serreze MC (2005) Role of land-surface changes in Arctic summer warming. *Science* **310**, 627–628.

Charney JG (1975) Dynamics of deserts and droughts in the Sahel. *Quarterly Journal of the Meteorological Society* **101**, 193–202.

Charney JG, Stone PH, Quirk WJ (1975) Drought in the Sahara: a biogeophysical feedback mechanism. *Science* **187**, 434–435.

Claeys M, Graham B, Vas G, *et al.* (2004) Formation of secondary organic aerosols through photooxidation of isoprene. *Science* **303**, 1173–1176.

Crowley TJ, North GR (1991) *Paleoclimatology*. Oxford University Press, Oxford.

Daskalov, G. M. (2002) Overfishing drives a trophic cascade in the Black Sea. *Marine Ecology Progress Series* **225**, 53–63.

DeConto RM, Brady EC, Bergengren J, Hay WW (2000) Late-Cretaceous climate, vegetation and ocean interactions. In: *Warm climates in Earth history* (eds Huber BT, MacLeod KG, Wing SL). Cambridge University Press, Cambridge, pp. 275–296.

Denman KL, Brasseur G, Chidthaisong A, *et al.* (2007) Couplings between changes in the climate system and biogeochemistry. In: *Climate Change 2007: The Physical Science Basis. Contribution of Working Group I to the Fourth Assessment Report of the Intergovernmental Panel on Climate Change* (eds Solomon S, *et al.*). Cambridge University Press, Cambridge, pp.499–587.

Dewar WK, Bingham RJ, Iverson RL, Nowacek DP, St. Laurent LC, Wiebe PH (2006) Does the marine biosphere mix the ocean? *Journal of Marine Research* **64**, 541–561.

deYoung B, Harris R, Alheit J, Beaugrand G, Mantua N, Shannon L, (2004) Detecting regime shifts in the ocean: data considerations. *Progress in Oceanography* **60**, 143–164.

Dietrich WE, Perron JT (2006) The search for a topographic signature of life. *Nature* **439**, 411–418.

Droser ML, Jensen S, Gehling JG (2002) Trace fossils and substrates of the terminal Proterozoic–Cambrian transition: implications for the record of early bilaterians and sediment mixing. *Proceedings of the National Academy of Sciences, USA* **99**, 12572–12576.

Duffy JE, Cardinale BJ, France KE, McIntyre PB, Thébault E, Loreau M (2007) The functional role of biodiversity in ecosystems: incorporating trophic complexity. *Ecology Letters* **10**, 522–538.

Elser JJ, Dobberfuhl D, Mackay NA, Schampel JH (1996) Organism size, life history, and N:P stoichiometry: towards a unified view of cellular and ecosystem processes. *BioScience* **46**, 674–684.

Erwin DH (2008) Macroevolution of ecosystem engineering, niche construction and diversity. *Trends in Ecology and Evolution* **23**, 304–310.

Estes JA, Duggins DO (1995) Sea otters and kelp forests in Alaska: generality and variation in a community ecological paradigm. *Ecological Monographs* **65**, 75–100.

Falkowski PG, Barber RT, Smetacek V (1998) Biogeochemical controls and feedbacks on ocean primary productivity. *Science* **281**, 200–206.

Falkowksi PG, Katz ME, Milligan AJ, *et al.* (2005) The rise of oxygen over the past 205 million years and the evolution of large placental mammals. *Science* **309**, 2202–2204.

Fehsenfeld F, Calvert J, Fall R, *et al.* (1992) Emissions of volatile organic compounds from vegetation and their implications for atmospheric chemistry. *Global Biogeochemical Cycles* **6**, 389–430.

Fenster CB, Armbruster WS, Wilson P, Dudash MR, Thomson JD (2004) Pollination syndromes and floral specialization. *Annual Review of Ecology, Evolution and Systematics* **35**, 375–403.

Flückiger J, Blunier T, Stauffer B, *et al.* (2004) N_2O and CH_4 variations during the last glacial episode: insight into global processes. *Global Biogeochemical Cycles* **18**, doi:10.1029/2003GB 002122.

Folgarait PJ (1998) Ant biodiversity and its relationship to ecosystem functioning: a review. *Biodiversity and Conservation* **7**, 1221–1244.

Frank KT, Petrie B, Choi JS, Leggett WC (2005) Trophic cascades in a formerly cod-dominated ecosystem. *Science* **308**, 1621–1623.

Galy V, France-Lanord C, Beyssac O, Faure P, Kudrass H, Palhol F (2007) Efficient carbon burial in the Bengal fan sustained by the Himalyan erosional system. *Nature* **450**, 407–410.

Gill JL, Williams JW, Jackson ST, Lininger KB, Robinson GS (2009) Plant communities, and enhanced fire regimes in North America. *Science* **326**, 1100–1103.

Grace JC, Lloyd J, McIntyre J, *et al.* (1995) Fluxes of carbon dioxide and water vapour over an undisturbed tropical forest in south-west Amazonia. *Global Change Biology* **1**, 1–12.

Graham JB, Dudley R, Aguilar NM, Gans C (1995) Implications of the late Palaeozoic oxygen pulse for physiology and evolution. *Nature* **375**, 117–120.

Greb, SF, DiMichele WA, Gastaldo, RA (2006) Evolution and importance of wetlands in Earth history. *Geological Society of America Special Paper* **399**, 1–40.

Gragnani A, Scheffer M, Rinaldi S (1999) Top-down control of cyanobacteria: a theoretical analysis. *American Naturalist* **153**, 59–72.

Guenther A, Karl T, Harley P, Wiedinmyer C, Palmer PI, Geron C (2006) Estimates of global terrestrial isoprene emissions using MEGAN (Model of Emissions of Gases and Aerosols from Nature). *Atmospheric Chemistry and Physics* **6**, 107–173.

Gutiérrez JL, Jones CG, Strayer DL Iribarne OO (2003) Mollusks as ecosystem engineers: the role of shell production in aquatic habitats. Oikos, **101**, 79–90.

Hairston NG, Smith FE, Slobodkin LB (1960) Community structure, population control, and competition. *American Naturalist* **94**, 421–425.

Hamm CE, Merkel R, Springer O, Jurkojc P, Maier C, Prechtel K, Smetacek V (2003) Architecture and material properties of diatom shells provide effective mechanical protection. *Nature* **421**, 841–843.

Hedges JI, Keil RG, Benner R (1997) What happens to terrestrial organic matter in the ocean? *Organic Geochemistry* **27**, 195–212.

Hernádez-León S, Fraga C, Ikeda T (2008) A global estimation of mesozooplankton ammonium excretion in the open ocean. *Journal of Plankton Research* **30**, 577–585.

Higgins SN, Malkin SY, Howell ET, *et al.* (2008) An ecological review of *Cladophora glomerata* (Chlorophyta) in the Laurentian Great Lakes. *Journal of Phycology* **44**, 839–854.

Hoffland E, Kuyper TW, Wallander H, *et al.* (2004) The role of fungi in weathering. *Frontiers in Ecology and the Environment* **2**, 258–264.

Högberg P, Read DJ (2006) Towards a more plant physiological perspective on soil ecology. *Trends in Ecology and Evolution* **21**, 548–554.

Holdo RM, Sinclair ARE, Dobson AP, *et al.* (2009) A disease-mediated trophic cascade in the Serengeti and its implications for ecosystem C. *PLoS Biology* **7**, e1000210, doi:10.1371/journal.pbio.1000210.

Hooper DU, Chapin FS. III, Ewel, JJ, *et al.* (2005) Effects of biodiversity on ecosystem functioning: a consensus of current knowledge. *Ecological Monographs* **75**, 3–35.

Huber BT, MacLeod KG, Wing SL (2000) *Warm Climates in Earth History*. Cambridge University Press, Cambridge.

Huntley ME, Zhou M (2004) Influence of animals on turbulence in the sea. *Marine Ecology Progress Series* **273**, 65–79.

Hutchinson GE (1961) The paradox of the plankton. *American Naturalist* **95**, 137–145.

Idjadi JA, Edmunds PJ (2006) Scleractinian corals as facilitators for other invertebrates on a Caribbean reef. *Marine Ecology Progress Series* **319**, 117–127.

Jiang L, Schofield OME, Falkowski PG (2005) Adaptive evolution of phytoplankton cell size. *American Naturalist* **166**, 496–505.

Jones CG, Lawton JH, Shachak M (1994) Organisms as ecosystem engineers. *Oikos* **69**, 373–386.

Jouquet P, Dauber J, Lagerlöf J, Lavelle P, Lepage M (2006) Soil invertebrates as ecosystem engineers: intended and accidental effects on soil and feedback loops. *Applied Soil Ecology* **32**, 153–164.

Jumars PA, Mayer LM, Deming JW, Baross JA, Wheatcroft RA (1990) Deep-sea deposit-feeding strategies suggested by environmental and feeding constraints. *Philosophical Transactions of the Royal Society of London* **A331**, 85–101.

Kerr SR, Dickie L (2001) *The Biomass Spectrum. A Predator-Prey Theory of Aquatic Production*. Columbia University Press, New York, 320pp.

Kidwell SM, Jablonski D (1983) Taphonomic feedback: Ecological consequences of shell accumulation. *In* McCall PL, Tevesz MSS (eds) Animal-sediment relations. The biogenic alteration of sediments. *Topics in Geobiology*, **2**, p. 195–248.

Kleidon A, Fraedrich K, Heimann M (2000) A green planet versus a desert world: estimating the maximum effect of vegetation on the land surface climate. *Climatic Change* **44**, 471–493.

Knoll AH, Summons RE, Waldbauer JR, Zumberge JE (2007) The geological succession of primary producers in the oceans. In: *Evolution of Primary Producers in the Sea* (eds Falkowski PG, Knoll AH). Elsevier Academic Press, Oxford, pp. 134–163.

Kump LR, Pollard D (2008) Amplification of Cretaceous warmth by biological cloud feedbacks. *Science* **320**, 195.

Labandeira C (2007) The origin of herbivory on land: initial patterns of plant tissue composition by arthropods. *Insect Science* **14**, 259–275.

Landeweert R, Hoffland E, Finlay RD, Kuyper TW, van Breemen N (2001) Linking plants to rocks: ectomycorrhizal fungi mobilize nutrients from minerals. *Trends in Ecology and Evolution* **16**, 248–254.

Leake J, Johnson D, Donnelly D, Muckle G, Boddy L, Read DJ (2004) Networks of power and influence: the role of mycorrhizal mycelium in controlling plant communities and agroecosystem functioning. *Canadian Journal of Botany* **82**, 1016–1045.

Leake JR, Duran AL, Hardy KE, *et al.* (2008) Biological weathering in soil: the role of symbiotic root-associated fungi biosensing minerals and directing photosynthate-energy into grain-scale mineral weathering. *Minerological Magazine* **72**, 85–89.

Liu C, Walker HJ (1989) Sedimentary characteristics of cheniers and the formation of the chenier plains of east China. *Journal of Coastal Research* **5**, 353–368.

Logan GA, Hayes JM, Hieshima GB, Summons RE (1995) Terminal Proterozoic reorganization of biogeochemical cycles. *Nature* **376**, 53–56.

Lohrer AM, Thrush SF, Gibbs MM (2004) Bioturbators enhance ecosystem function through complex biogeochemical interactions. *Nature* **431**, 1092–1095.

McCann K (2000) The diversity-stability debate. *Nature* **405**, 228–233.

Mooney HA, Cleland EE (2001) The evolutionary impact of invasive species. *Proceedings of the National Academy of Sciences, USA* **98**, 5446–5451.

Naganuma T (1996) Calanoid copepods: linking lower-higher trophic levels by linking lower-higher Reynolds numbers. *Marine Ecology Progress Series* **136**, 311–313.

Naiman RJ, Pinay G, Johnston CA, Pastor J (1994) Beaver influences on the long-term biogeochemical characteristics of boreal forest drainage networks. *Ecology* **75**, 905–921.

Okajima R (2008) The controlling factors limiting maximum body size of insects. *Lethaia* **41**, 423–430.

Otto-Bliesner BL, Upchurch GR (1997) Vegetation-induced warming of the high latitude regions during the Late Cretaceous period. *Nature* **385**, 804–807.

Pace ML, Cole JJ, Carpenter SR, Kitchell JF (1999) Trophic cascades revealed in diverse ecosystems. *Trends in Ecology and Evolution* **14**, 483–488.

Penry DL, Jumars PA (1986) Chemical reactor analysis and optimal digestion. *BioScience* **36**, 310–315.

Peterson KJ, Butterfield NJ (2005) Origin of the Eumetazoa: testing ecological predictions of molecular clocks against the Proterozoic fossil record. *Proceedings of the National Academy of Sciences, USA* **102**, 9547–9552.

Polis GA (1999) Why are parts of the world green? Multiple factors control productivity and the distribution of biomass. *Oikos* **86**, 3–15.

Post E, Pedersen C (2008) Opposing plant community responses to warming with and without herbivores. *Proceedings of the National Academy of Sciences, USA* **105**, 12353–12358.

Pryor WA (1975) Biogenic sedimentation and alteration of argillaceous sediments in shallow marine environments. *Geological Society of America Bulletin* **86**, 1244–1254.

Raven JA, Edwards D (1998) Roots: evolutionary origins and biogeochemical significance. *Journal of Experimental Botany* **52**, 381–401.

Read DJ (1991) Mycorrhizas in ecosystems. *Experimentia* **47**, 376–391.

Reichman OJ, Seabloom EW (2002) The role of pocket gophers as subterranean ecosystem engineers. *Trends in Ecology and Evolution* **17**, 44–49.

Reise K (2002) Sediment mediated species interactions in coastal waters. *Journal of Sea Research*, **48**, 127–141.

Remy W, Taylor TN, Hass H, Kerp H (1994) Four hundred-million-year-old vesicular arbuscular mycorrhizae. *Proceedings of the National Academy of Sciences, USA* **91**, 11841–11843.

Retallack GJ (2001) Cenozoic expansion of grasslands and climatic cooling. *Journal of Geology*, **109**, 407–426.

Robinson JM (1990a) Lignin, land plants, and fungi: biological evolution affecting Phanerozoic oxygen balance. *Geology* **15**, 607–610.

Robinson JM (1990b) The burial of organic carbon as affected by the evolution of land plants. *Historical Biology* **5**, 189–201.

Sellers PJ, Dickinson R, Randall D, *et al.* (1996) Comparison of radiative and physiological effects of doubled atmospheric CO_2 on climate. *Science*, **271**, 1402–1406.

Scheffer M, Carpenter SR (2003) Catastrophic regime shifts in ecosystems: linking theory to observation. *Trends in Ecology and Evolution* **18**, 648–656.

Schindler DE, Leavitt PR, Brock CS, Johnson SP, Quay PD (2005) Marine-derived nutrients, commercial fisheries, and production of salmon and lake algae in Alaska *Ecology* **86**, 3225–3231.

Schmitz OJ (2008) Effects of predator hunting mode on grassland ecosystem function. *Science* **319**, 952–954.

Scott AC, Glasspool IJ (2006) The diversification of Paleozoic fire systems and fluctuations in atmospheric oxygen concentration. *Proceedings of the National Academy of Sciences, USA* **109**, 10861–10865.

Seilacher A (1999) Biomat-related lifestyles in the Precambrian. *Palaios*, **14**, 86–93.

Sharkey TD, Wiberley AE, Donohue AR (2008) Isoprene emissions from plants: why and how. *Annals of Botany* **101**, 5–18.

Sheldon RW, Prakash A, Sutcliffe WH (1972) The size distribution of particles in the ocean. *Limnology and Oceanography* **17**, 327–340.

Simon L, Bousquet J, Levesque RC, Lalonde M (1993) Origin and diversification of endomycorrhizal fungi and coincidence with vascular land plants. *Nature* **363**, 67–69.

Smetacek V (2001) A watery arms race. *Nature* **411**, 745.

Spracklen DV, Bonn B, Carslaw KS (2008) Boreal forests, aerosols, and the impact on clouds and climate. *Philosophical Transactions of the Royal Society* **A366**, 4613–4626.

Sterner RW, Elser JJ (2002) *Ecological Stoichiometry, the Biology of Elements from Molecules to the Biosphere.* Princeton University Press, Princeton, NJ.

Strack JE, Pielke RA, Liston GE (2007) Arctic tundra shrub invasion and soot deposition: consequences for spring snow melt and near-surface air temperatures. *Journal of Geophysical Research* **112**, doi: G04S44.

Strullu-Derrien C, Strullu DG (2007) Mycorrhization of fossil and living plants. *Comptes Rendus Palevol.* **6**, 483–494.

Sturm M, Douglas T, Racine C, Liston BE (2005) Changing snow and shrub conditions affect albedo with global implications. *Journal of Geophysical Research* **110**, doi: 10.1029/2005JG000013.

Taylor LL, Leake JR, Quirk J, Hardy K, Banwart SA, Beerling DJ (2009) Biological weathering and the long-term carbon cycle: integrating mycorrhizal evolution into the current paradigm. *Geobiology* **7**, 171–191.

Terborgh J, Lopez L, Nunez V, *et al.* (2001) Ecological meltdown in predator-free forest fragments. *Science* **294**, 1923–1926.

Thayer CW (1983) Sediment-mediated biological disturbances and the evolution of the marine benthos. In: *Biotic Interactions in Recent and Fossil Benthic Communities* (eds Tevesz MJS, McCall PL). Plenum Press, New York, pp. 480–625.

Tilman D, Reich PB, Knops J, Wedin D, Mielke T, Lehman C (2001) Diversity and productivity in a long-term grassland experiment. *Science* **294**, 843–845.

Tunved P, Hansson HC, Kerminen VM, *et al.* (2006) High natural aerosol loading over boreal forests. *Science* **312**, 261–263.

Turner JT (2002) Zooplankton fecal pellets, marine snow and sinking phytoplankton blooms. *Aquatic Microbial Ecology* **27**, 57–102.

Upchurch GR, Otto-Bliesner BL, Scotese C (1998) Vegetation-atmosphere interactions and their role in global warming during the latest Cretaceous. *Philosophical Transactions of the Royal Society* **B353**, 97–112.

van Leeuwen B, Augustijn DCM, van Wesenbeeck, BK, Hulscher SJMH de Vries MB (2010) Modeling the influence of a young mussel bed on fine sediment dynamics on an intertidal flat in the Wadden Sea. *Ecological Engineering*, 36, 145–153.

Vermeij GJ (1994) The evolutionary interaction among species: selection, escalation, and coevolution. *Annual Review of Ecology and Systematics* **25**, 219–236.

Villegas J, Fortin JA (2001) Phosphorus solubilization and pH changes as a result of the interactions between soil bacteria and arbuscular mycorrhizal fungi on a medium containing NH_4 as nitrogen source. *Canadian Journal of Botany* **79**, 865–870.

Villegas J, Fortin JA (2002) Phosphorus solubilization and pH changes as a result of the interactions between soil bacteria and arbuscular mycorrhizal fungi on a medium containing NH_3^- as nitrogen source. *Canadian Journal of Botany* **80**, 571–576.

Visser AW (2007) Biomixing of the oceans? *Science* **316**, 838–839.

Volk T (1989) Rise of angiosperms as a factor in long-term climate cooling. *Geology* **17**, 107–110.

Walker JCG, Hays PB, Kasting JF (1981) A negative feedback mechanism for the long-term stabilization of earth's surface temperature. *Journal of Geophysical Research* **86**, 9776–9782.

Ward P, Labandeira C, Laurin M, Berner RA (2006) Romer's gap as a low oxygen interval constraining the timing of initial arthropod and vertebrate terrestrialization. *Proceedings of the National Academy of Sciences, USA* **103**, 16818–16822.

Wassmann P (1998) Retention versus export food chains: processes controlling sinking loss from marine pelagic systems. *Hydrobiologia* **363**, 29–57.

Wellman CH, Osterloff PL, Mohiuddin U (2003) Fragments of the earliest land plants. *Nature* **425**, 282–285.

Wilkinson MT, Richards PJ, Humphreys GS (2009) Breaking ground: pedological, geological, and ecological implications of soil bioturbation. *Earth-Science Reviews* **97**, 254–269.

Wilson RW, Millero FJ, Taylor JR, *et al.* (2009) Contribution of fish to the marine inorganic carbon cycle. *Science* **323**, 359–362.

Woodward FI (2007) Global primary production. *Current Biology* **17**, R269–R273.

Zeebe RE, Caldeira K (2008) Close mass balance of long-term carbon fluxes from ice-core CO_2 and ocean chemistry records. *Nature Geoscience* **1**, 312–315.

12

A GEOBIOLOGICAL VIEW OF WEATHERING AND EROSION

Susan L. Brantley[1], Marina Lebedeva[1] and Elisabeth M. Hausrath[2]

[1]Center for Environmental Kinetics Analysis, Earth and Environmental Systems Institute, Pennsylvania State University, University Park PA 16802, USA
[2]Department of Geosciences, University of Nevada, Las Vegas, 4505 s. Maryland Parkway, Las Vegas, NV 89154, United States

12.1 Introduction

Weathering of crustal rocks can be conceptualized as the equilibration of relatively basic, often electron-rich rocks with more acid, electron-poor atmospheric gases and fluids. This chemical process is accompanied by physical phenomena that disaggregate and fracture the rock into particles, increasing the surface area. Many of these processes are influenced by the biota: weathering has thus long been recognized to include chemical, physical, and biological processes (Dokuchaev, 1883; Jenny, 1980; Buol *et al.*, 1989; Huggett, 1995; Amundson, 2004). Researchers have built models that can reproduce some of the important features of weathering and soil formation. However, no model adequately quantifies the inter-relationships and coupling between all of the important processes (Godderis *et al.*, 2006; Minasny *et al.*, 2008; Steefel, 2008). In this paper, we discuss a few laboratory-, field- and model-derived observations that have been used to explore chemical, physical, and biological interactions in the weathering environment over geologic time.

One indicator of weathering over geologic time is the concentration of CO_2 in the atmosphere (Fig. 12.1). Atmospheric CO_2 represents the balance between rates of weathering and volcanic degassing over 10^5–10^6 y timescales. The chemical reaction describing how silicate weathering fixes CO_2 from the atmosphere into bicarbonate ions that are in turn transported to the oceans and precipitated as carbonates can be written as follows (for example, Berner and Kothavala 2001):

$$CO_2 + (Ca_x, Mg_{1-x})SiO_3 \rightarrow SiO_2 + (Ca_x, Mg_{1-x})CO_3$$

$$(12.1)$$

This reaction summarizes how the weathering of Ca–Mg-silicate is accompanied by precipitation of carbonate minerals at the sea floor, sequestering CO_2 over geological time. Burial, subduction, metamorphism and volcanism can then return that CO_2 to the atmosphere.

According to these arguments, changes in concentration of CO_2 (Fig. 12.1) document changes in the weathering fluxes on the continents relative to magmatic degassing of carbon. Such processes have been modeled with various approaches. For example, in Fig. 12.1, model predictions of atmospheric CO_2 are plotted over the last 600 million years in comparison to paleo-reconstructions of atmospheric CO_2 over the Phanerozoic (542 Ma) (Berner and Kothavala, 2001; Royer *et al.*, 2004). Importantly, the modelling is in general agreement with estimates of CO_2 in the atmosphere based on proxies in the rock and fossil record as shown. This model (GEOCARB III) incorporates an important negative feedback hypothesized to regulate climate over geological time periods (Walker *et al.*, 1981; Berner *et al.*, 1983; Kump *et al.*, 2000): as concentrations of CO_2 increase and warm the atmosphere due to the greenhouse effect, global temperatures and chemical weathering rates increase and draw down atmospheric CO_2. Alternately, if rates of weathering draw down CO_2 faster than it can be maintained in the atmosphere, cooling occurs and weathering rates slow.

Also included in such models of atmospheric CO_2 are the effects of tectonic cycles on weathering (Berner *et al.*, 1983; Sundquist, 1991). For example, several lines of evidence suggest that tectonic uplift and the creation of new rock surfaces accelerate rates of weathering (Stallard and Edmond, 1983; Raymo *et al.*, 1988; Raymo

Fundamentals of Geobiology, First Edition. Edited by Andrew H. Knoll, Donald E. Canfield and Kurt O. Konhauser.

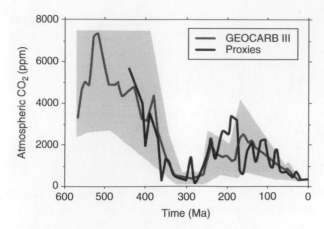

Figure 12.1 A comparison of model predictions based on the GEOCARB III model (Berner and Kothavala, 2001) plotted with paleo-concentrations of CO_2 in the atmosphere inferred from proxy reconstructions. The shaded area represents the range of error for model predictions. See original publication for a description of the model and the proxies (the latter are based on isotopic composition of pedogenic minerals and phytoplankton, the stomatal distribution in leaves, and the boron isotopic composition of foraminifera (Royer *et al.* 2004). Figure adapted from Royer *et al.* (2004).

and Ruddiman, 1992; Riebe *et al.*, 2001; Millot *et al.*, 2002; Waldbauer and Chamberlain, 2005; Hilley and Porder, 2008). In fact, the relative importance of the kinetics of chemical reaction rates versus the tectonically-induced physical production and erosive transport of mineral particles has been vigorously debated. For many watersheds, denudation – the rate of removal of mass from the watershed by all chemical and physical processes – is transport-limited, that is, the kinetics of mineral–water weathering reactions do not influence the rate of denudation. However, denudation rates for some systems correlate with temperature, a finding that some argue is more consistent with weathering control (i.e. a dependence upon the kinetics of mineral–water reactions). Globally, denudation varies from transport limitation in regimes of low erosion to weathering limitation in regimes of high erosion (Stallard, 1995; West *et al.*, 2005; Hren *et al.*, 2007).

All of these regimes of erosion and weathering are also heavily affected by the presence of biota. Even long before colonization of land surfaces by plants, microbiota could have affected weathering (Schwartzman and Volk, 1989, 1991; Raven, 1995; Berner *et al.*, 2004; Neaman *et al.*, 2005a, b). The effect of the biota must have grown with time: for example, both the development of land plants (470–450 Ma ago) and the growth of forests (395–360 Ma ago) probably had large chemical and physical effects on weathering. Indeed, large drops in atmospheric CO_2 have been attributed to colonization of land by deeply rooted land plants and later to enhanced burial of organic matter derived from plants (Fig. 12.1)

(Berner and Kothavala, 2001; Berner *et al.*, 2004). Furthermore, the development of deep-rooting vascular plants in the Cretaceous may have considerably enhanced biotic weathering effects (Moulton *et al.*, 2000).

Effects of the biota on silicate mineral weathering include phenomena such as the secretion of protons and organic ligands, uptake and recycling of porefluids, regional effects on fluxes of meteoric water, uptake and recycling of nutrient species, effects on soil temperature and redox potential, rock disaggregation, soil stabilization, aggregation of soil particles, retention of water by organic matter, accumulation of atmospheric dust, bioturbation, and production and uptake of CO_2 in the soil atmosphere (e.g. Drever, 1994; Kump *et al.*, 2000; Berner *et al.*, 2004; Schwartzman, 2008). Authors who have tried to quantify the biotic effect on weathering have argued that biota accelerate weathering-driven rates of drawdown of CO_2 by factors of ~2 (Berner *et al.*, 2003) to 100 (Schwartzman, 2008). However, vegetation can *increase* or *decrease* net chemical weathering rates for various elements when considered over different scales of space and time. Furthermore, vegetation tends to *decrease* erosion rates. No model has emerged to deconvolve the effects on denudation of these opposing tendencies over various scales and to relate these to the regimes of weathering- and transport-limited denudation.

In this chapter, we explore some of these effects in terms of the complexity of chemical versus physical versus biological weathering. Although we do not provide the exhaustive description of earlier reviews (Drever, 1994; Kump *et al.*, 2000; Berner *et al.*, 2004; Schwartzman, 2008), in Section 12.2 we discuss a few of the effects of biota on weathering and erosion in order to emphasize that biotic impacts can differ markedly when considered over short or long timeframes. In Section 12.3, we focus on how organic molecules affect element mobility and mineral dissolution. These observations lead to the idea of an organomarker – an element whose mobility is strongly affected by the presence of organic ligands. Intriguingly, the chemistry of such ligands has changed over geologic time (Neaman *et al.*, 2005b). To test the utility of organomarkers, we first look for evidence of such elements in rivers in Section 12.4. Then, in Section 12.5, we discuss what soil profiles of organomarkers look like. In contrast to rivers, we find that soil profiles can be powerful records of the presence of biota and organic ligands during pedogenesis, and we discuss the use of the organomarker idea in ancient soils. In Sections 12.6 and 12.7 we use regolith evolution models to explore interactions of chemical, physical, and biological weathering on regolith over time.

The models are broadly consistent with the conclusion that biota catalyse the evolution of surficial Earth systems toward transport control. However, such transport control results eventually in extremely thick, nutrient-poor regolith where biota find it difficult to find

Table 12.1 Some effects of biota on chemical weathering fluxes from soil

Factor	How biota (B) affects factor	Effect on chemical efflux (F) from soil	Length of time soil barren	Example Reference
Physical erosion (W)	B↓ W↑			Vanacker et al., 2007
Temperature (T)	B↓ T↑	T↑ F↑	<3 years	Keller et al., 2006
Water throughput (Q)	B↓ Q↑	Q↑ F↑	<3 years	Keller et al., 2006
Water throughput (Q)	B↓ Q↓	Q↓ F↓	>60, >10 000 years	Moulton et al., 2000
Porewater pH	B↓ pH↔	pH↔ F↔	<3 years	Keller et al., 2006
Porewater pH	B↓ pH↑	pH↑ F↓	>60, >10 000 years	Moulton et al., 2000
Dust accumulation (D)	B↓ D↓	D↓ F↓		Pye, 1987

↓ factor or rate decreases; ↑ factor or rate increases; ↔ factor or rate not observed to increase or decrease consistently.

rock-derived nutrients. Tectonic activity can push such systems back toward weathering control by exposing new mineral surfaces. The balance of transport-limited versus weathering-limited land surfaces within a given catchment or over the globe controls the overall rate of chemical weathering and the global average thickness of soil. In large measure, this balance is related to the counterbalancing of tectonism versus biology.

12.2 Effects of biota on weathering

Table 12.1 gives a qualitative summary of a small number of parameters or phenomena influencing weathering and how they are affected by biota. We use Table 12.1 to introduce some of the inter-relationships and timescales of importance. Of the factors in Table 12.1 (erosion rate, temperature, water throughput, porewater pH, dust accumulation), the one that may be most unequivocally important is the observation made by many researchers for many systems that growth of vegetation lowers the physical erosion rate of a catchment or hillslope by stabilizing soil. This effect is related to both the 'holding' action of roots as well as the binding and formation of aggregates by organic matter in the soil. Such organic matter retains not only soil particles but also moisture, which then further enhances weathering and soil formation. For example, dense vegetation has been observed to decrease erosion in tropical mountain

areas impacted by changing human land use (Vanacker et al., 2007). Those workers suggest that erosion rates estimated as sediment yields in units of mass per unit area per unit time (M L^{-2} T^{-1}), W, decrease exponentially with an increase in the fraction of land surface covered by vegetation, C:

$$W = ae^{-bC} \tag{12.2}$$

Here, a and b are constants. In that research, modern sediment yields from catchments in the southern Ecuadorian Andes were accelerated 100-fold when vegetation cover in the upstream area dropped from ~80 to <20%, regardless of whether the cover was native forest or grassland/disturbed forest. In fact, on a nonvegetated hillslope, downslope transport of soil may occur quickly due to mostly physical processes whereas for a vegetated hillslope the downslope transport may occur more slowly. In fact, erosion on hillslopes may often be dictated by biologically controlled processes (Roering et al., 2002). Many other studies have quantified the effect of vegetation and cropping practices on erosion through empirical soil loss equations (Wischmeier and Smith, 1978).

Biota also affect the temperature of the soil (Table 12.1) because shading of soils by vegetation causes cooling. Thus, harvesting of forests may lead to increases in the temperature of weathering. However, removal of aboveground vegetation can cause either faster or slower microbial activity which also impacts the temperature of the soil. In some cases over some timescales, these effects work together. For example, for three years after forest harvest in the White Mountains of New Hampshire (USA), warming of the soil was observed and was attributed especially to increases in the rates of microbial respiration driven by the acceleration of decomposition reactions (Keller et al., 2006). In that study, the temperature effect was inferred to contribute to increased chemical weathering after harvest.

As indicated in Table 12.1, however, such short timescale biotic effects can be countered by effects that become important over longer timescales. For example, the effect of plants on the flux of water through soil (e.g. Q in Table 12.1) varies with the timescale of observation. Over short timescales, throughput of water through soils can increase after harvesting since water uptake into trees stops (Keller et al., 2006). Later, during the interval when trees grow back, uptake of water may decrease the water flux through the soil. Nonetheless, over decadal timescales, plants increase recirculation of water within ecosystems through local evapotranspiration-related effects. Thus, chemical weathering, which generally increases with water throughput, can be increased or decreased by a change in biota in the system depending upon the timescale of observation (Table 12.1).

Finally, as discussed in the next section, biota also increase the rate of solublization of many rock-forming

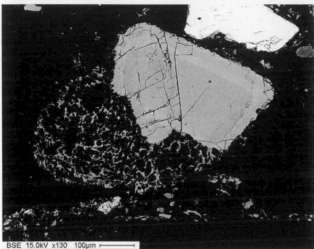

Figure 12.2 Back-scattered electron micrographs (BSEM) of a conifer root and associated soil imaged from a sample from a soil core collected from the Cascade Mountains of Washington State (images courtesy of M. Andrews). Minerals comprised of elements of higher atomic number appear brighter under BSEM. Both samples were epoxied to preserve root-mineral associations. The parent material of the soil is granodiorite. Upper image: the dominant minerals observed are plagioclase and feldspar. The image demonstrates the physical proximity of the root (centre) to the weathered plagioclase feldspar. Lower image: Back-scattered electron image of plagioclase: the weathered half is contiguous to a conifer root, which is outlined in small mineral pieces at the bottom of the image. Further information is summarized elsewhere (Andrews, 2008).

and trace elements through biological effects on soil water chemistry (Berner *et al.*, 2004; Neaman *et al.*, 2005b). In basaltic watersheds in Iceland for example, Moulton and others have observed that chemical fluxes of Ca and Mg are 2–5 times larger in areas vegetated by vascular plants than in barren areas when factors such as microclimate, slope, and lithology are held constant

(Moulton and Berner, 1998; Moulton *et al.*, 2000). These same authors have reported that the rate of plagioclase and pyroxene weathering can be increased by factors of 2 and 10 respectively by the presence of trees. Other authors (Oliva *et al.*, 1999) have argued based on riverine flux data that organic complexation can enhance chemical weathering by factors of about 20% (see also Section 12.4).

Such enhancements of chemical weathering are at least partially caused by changing pH in soil pore waters to values as low as 3 near fine roots (Arthur and Fahey, 1993). Values of pH can drop by up to a unit compared to bulk pH in microbial biofilms on minerals (Liermann *et al.*, 2000). Furthermore, high CO_2 concentrations in the soil atmosphere due to respiration and decay leads to pore fluids that are rich in dissolved carbonic acid. Organic acids also drive dissolution through chelation of metal ions. The effect of these acids and ligands is to etch minerals (April and Keller, 1990; Cochran and Berner, 1996; Berner and Cochran, 1998) in close proximity to rootlets or fungal hyphae (Fig. 12.2; Andrews, 2008). Of especial importance are mycorrhizas and the symbiotic association of fungi with plants: plants provide a reduced carbon source for the fungi while fungi secrete chelating ligands that mobilize rock-derived nutrients (Sterflinger, 2000; Landeweert *et al.*, 2001; Van Scholl *et al.*, 2008).

Of course, elements are not only solubilized by biota but are also taken up or recycled by biota (Bormann and Likens, 1967). In this regard, a distinction must be made between rates of chemical weathering measured as the rate of loss of minerals in the regolith (Fig. 12.2) versus weathering measured as the rate of loss of solutes. Of course, for systems where the composition and mass of biota are constant, these rates must be the same over some interval of time long enough to average out diurnal or seasonal variations. In other words, for systems where biota are at steady state, the chemical loss of elements from regolith can be related to the chemical loss as solutes. However, systems can be far from steady state if ecosystems are aggrading or degrading. Furthermore, the relative 'leakiness' of biota (i.e. how much nutrient is lost as outflux to groundwater as a function of total nutrient mobilized from minerals in the regolith) varies from one species to another. Moulton and others have argued for example that angiosperms leak more nutrients than gymnosperms (Moulton and Berner, 1998; Moulton *et al.*, 2000).

Generally, chemical weathering losses recorded in regolith differ from losses measured in solute outflux even for well-constrained systems. This observation can either mean that the system has not been operating in steady state (i.e. aggrading or degrading biota) or that the comparison is affected by issues related to timescales (Stonestrom *et al.*, 1998; White, 2002). As an example, immediately after a disturbance such as fire or infestation, chemical fluxes out of a watershed may be relatively large, but as the forest re-establishes and grows, total loss of solutes from the watershed may become relatively low due to uptake by biota (Bormann *et al.*, 1998; Berner *et al.*, 2004; Balogh-Brunstad *et al.*, 2008). While such solute losses may change rapidly over years to decades, the regolith chemistry changes at much smaller rates and thus records time-integrated fluxes that can differ significantly from the solute data at any point in time (White and Brantley, 2003).

To assess these time-integrated rates of weathering, several researchers have used chronosequences – soils developed on a single lithology but representing different exposure ages within the same climate. For example, in a comparison of several chronosequences, it was observed that depletion of regolith caused ecosystem degradation over 1000 to 10,000 y as ecosystems evolved from N- to P-limited (Wardle *et al.*, 2004). The change in nutrient limitation was attributed to increasingly lower supplies of P due to P loss into groundwater over time. P was lost even though the efficiency of recycling of P increased as the food webs transitioned from bacteria- to fungi-based. Such complex changes can take place over tens to hundreds of thousands of years and have been explored in chronosequences using molecular techniques to assess temporal changes in microbiota in soils (Moore *et al.*, 2009).

Although some roots can penetrate tens of meters into regolith (Jackson *et al.*, 1999), in very old soils, regolith can become so deep and nutrient-poor that it eventually isolates the vegetation from bedrock, limiting the rate of extraction of nutrients. In such cases, dust inputs become important sources of nutrients. For example, over time in the Hawaiian islands, the source of nutrients for vegetation changes from rock- to atmospherically derived (Chadwick *et al.*, 1999; Kurtz *et al.*, 2001). Once regolith has thickened to the extent that vegetation is largely isolated from bedrock-derived nutrients, above-ground vegetation no longer impacts rates of advance of the saprolite–bedrock interface except through indirect processes. In fact, for such thick regolith, biota may decelerate rather than accelerate rates of chemical weathering. For example, biota serve to accumulate dust (Pye, 1987). In systems with mineral-depleted regolith, dust-derived minerals dissolve at the surface, making weathering solutions even less corrosive and less likely to weather the underlying bedrock.

In such older chronosequences, tectonic uplift can enhance erosion and regenerate the ecosystem by providing new access to nutrients. The inter-relationships among tectonics, erosion, and P availability have been explored through modelling (Porder *et al.*, 2007). Where uplift is very slow, such as in the Amazon lowlands, Porder *et al.* concluded that P depletion may become

important. In contrast, in areas such as Central America and Southeast Asia where uplift rates are moderate and the rate of delivery of rock-derived P is adequate, P limitation is unlikely.

In contrast to older soils, on exposed bedrock with little to no soil, the accumulation of dust in crevices or depressions is accelerated by plants and fungi. Thus, on bedrock, biota can accelerate soil formation and weathering by holding regolith in place. The rate of weathering increases from small rates on exposed bedrock surfaces to a maximum rate for some optimal soil thickness. Above that optimum, the rate of weathering then decreases again as regolith thickens and shields underlying bedrock from corrosive fluids. Such a humped function describing the rate of weathering advance as a function of regolith thickness has not been experimentally confirmed but has been suggested by many workers (e.g. Cox, 1980). Perhaps soil depths less than the optimal value are unstable (Dietrich *et al.*, 1995). Some of the implications of such biotic effects on regolith formation rates are considered throughout this chapter.

12.3 Effects of organic molecules on weathering

Many authors have discussed the nature or concentration of low molecular weight organic molecules in soil and aquatic environments (Stevenson, 1967, 1991; Powell *et al.*, 1980; Stone, 1997; Jones, 1998; Ganor *et al.*, 2009). In this section, we summarize some of the chemical controls on the solubilization of elements in the presence of organic compounds. As discussed by Neaman *et al.* (2005b), before the ultraviolet shield developed on early Earth, concentrations of organic molecules in prebiotic soil solutions must have been generally low since the abiotic synthesis of these molecules was limited and the lifetimes of these molecules must have been short. As the biomass of organisms increased over geological time, however, the concentrations of organic acids in soil solutions may have increased to modern values in soil solutions. Today, concentrations of aliphatic and aromatic organic ligands range from ~1 to 4×10^{-3} M and from ~8 to 30×10^{-5} M, respectively (Stevenson, 1991).

Although it has been suggested that by about a billion or billion and a half years ago land-colonizing organisms probably produced the full suite of organic compounds that are secreted today (Raven, 1995), the relative abundance of different organic molecules in soil solutions must have changed over geologic time. The first organisms to colonize land were most likely Bacteria or Archaea, and these organisms today often secrete aliphatic carboxylic acids (Neaman *et al.*, 2006). Less commonly, they secrete aromatic compounds such as siderophores, molecules that complex Fe strongly and that allow the organism to extract Fe (Hersman, 2000) or other metals such as molybdenum from soil minerals (Liermann *et al.*, 2005).

In contrast to bacteria, secretion of aromatic acids is considerable for fungi, lichens, and vascular plants. Terrestrial fungi and lichens are thought to have evolved some 565 million years ago (see Neaman *et al.* (2005b) for a summary of lines of evidence) and vascular plants about 400 million years ago. Once the production of lignin and tannin became common during the evolution of vascular plants, the abundance of aromatic organic molecules is likely to have increased due to degradation of these high molecular-weight compounds. As argued by Neaman *et al.*, therefore, the proportion of aromatic carboxylic acids in soil solutions must have increased with the evolution of fungi, lichens, and vascular plants. These aromatic acids have therefore had an increasingly important effect on soil chemistry over time.

The effect of low molecular weight organic acids on silicate dissolution has been discussed and reviewed repeatedly (e.g. Furrer and Stumm, 1986; Chin and Mills, 1991; Wogelius and Walther, 1991; Drever and Vance, 1994; Welch and Ullman, 1996; Drever and Stillings, 1997; Brantley, 2004; Brantley, 2008; Ganor *et al.*, 2009). Although the importance of organic anions present at mM concentrations are pH- and ligand-dependent, their effects on dissolution rates are generally less than a factor of ~15 for most silicates (Wogelius and Walther, 1991; Brantley and Chen, 1995; Drever and Stillings, 1997; Ganor *et al.*, 2009). However, at the lower concentrations observed in natural soils, the effect of organic ligands on dissolution rates of silicates is probably even less than a factor of ~5 (van Hees *et al.*, 2002). Ganor *et al.* (2009) point out that often the effect of organic acids in accelerating silicate dissolution is small enough that it is within inter-laboratory reproducibility of rate measurements.

Despite the relatively small effects of ligands on dissolution rates, some minor elements become particularly mobile in the presence of organic matter. It has been hypothesized that such leaching may produce signatures in the regolith diagnostic of the presence of biota (Neaman *et al.*, 2005a, b). For example, Y and the rare earth elements (REE), like Hf and Th, are often considered 'particle-reactive' in that they show little solubility and are controlled largely by the availability of surfaces for complexation or precipitation. (Y and the REE generally show similar behaviour due to their occurrence as trivalent ions and the systematic decrease in ionic radii with increasing mass number from light REE (LREE: La to Sm) to heavy REE (HREE: Gd to Lu)). Although these elements are generally particle-reactive, Wood (2000) has summarized published complexation data consistent with the importance of REE– and Th–humic complexes in natural systems. Therefore, in some cases,

solubilization of such elements may be influenced by the presence of organic molecules.

Many of the effects of organic molecules on element leaching can be predicted from knowledge of solution complexation as observed in experiments. For example, the tendency of metal ions to complex with negatively charged oxygen atoms in organic ligands correlates with the tendency of metals to hydrolyse (e.g. Morel and Hering, 1993; Hernlem *et al.*, 1996). More specifically, the tendency for divalent metals to complex with a wide variety of organic ligands is well known to increase according to the Irving–Williams series in the order, $Mn^{2+} < Fe^{2+} < Co^{2+} < Ni^{2+} < Cu^{2+}$. The Zn^{2+} ion is also considered as part of the Irving–Williams series, but Zn is observed to have a relatively low tendency to complex with organic ligands.

These trends from solution complexation reactions are similar to the trends for complexation of ligands with metal cations on mineral surfaces (Furrer and Stumm, 1986; Zinder *et al.*, 1986; Amrhein and Suarez, 1988; Welch and Ullman, 1993, 1996; Stillings *et al.*, 1998). Although the exact mechanism is controversial, several authors have suggested that when organic ligands complex cations at a mineral surface, they have a direct effect on mineral dissolution due to weakening of bonds between the cation and the underlying mineral lattice (see recent summary by Ganor *et al.*, 2009).

Consistent with these arguments, the fraction of an element that is released from a rock interacting with water often increases following the trends in cation-ligand stability constants for aqueous systems (Fig. 12.3). For example, Eick *et al.* performed dissolution studies on basalt with and without citric and oxalic acids at pH 7 (Eick *et al.*, 1996a, b). They observed that in ligand-free conditions, elemental release of Ca was much larger than for Fe and Al. In the presence of citric acid, however, elemental release for both Fe and Al increased relative to Ca, as predicted by the strong affinity of aqueous organic molecules for Fe and Al. In other batch dissolution experiments, organic ligands were shown to also increase the percent of element released from basalt following general trends in metal complexation (Fig. 12.3) (Neaman *et al.*, 2005b; Hausrath *et al.*, 2009). As an example, Cu, which is high in the Irving–Williams series, was observed to leach significantly from basalt in experiments with organic acids under oxic conditions (Neaman *et al.*, 2005b). The work of Neaman and others has also highlighted the effect of organic acids on leaching of P from basalt and granite. Due to organic-ligand promoted dissolution of the P-hosting mineral apatite, P mobility was observed to be enhanced in experiments with both rock types in the presence of organic ligands.

The effects of low molecular-weight aromatic ligands on element release have been observed to be smaller than that of aliphatic ligands under some conditions

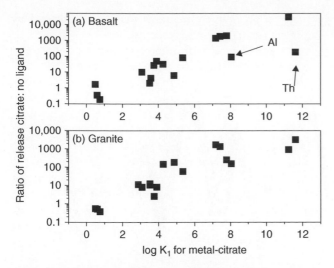

Figure 12.3 The ratio of the cumulative percentage of a given element released from a column of packed basalt powder leached with solutions containing 0.01 M citrate or no citrate plotted versus the stability constant for the element with citrate. Here, K_1 = [ML]/[M][L] where [ML], [M] and [L] are concentrations of the element–ligand complex, free ion, and free ligand respectively. For Fe, the stability constant for Fe^{3+} was used as the columns were not isolated from the atmosphere. Elements include Na, K, Rb, Mg, Ca, Sr, Ba, Mn, Fe, Ni, Cu, Zn, Al, Y, La, Ce, Pb, Th. In these experiments, citrate was assumed to be deprotonated. The column leaching experiments were maintained for 45 weeks. In the last five weeks of the experiments, pH averaged 7.33 and 7.04 for citrate-containing and – free outlet solutions respectively. Figure reproduced with permission from Hausrath *et al.* (2009).

(Furrer and Stumm, 1986; Chin and Mills, 1991; Neaman *et al.*, 2005b, 2006). Neaman *et al.* (2005a, b) attributed the lower efficiency of extraction of metals from basalts by such aromatic ligands under acidic and neutral pH to the differences in values of pK_a for the acids. For example, for the aliphatic compound oxalic acid, shown below on the left, deprotonation of both carboxylic groups occurs above pH 4. At that pH and above, the ligand can form strong bidentate complexes and can therefore affect dissolution. In contrast, for the aromatic compound gallic acid, shown on the right, deprotonation of the first two groups only occurs above pH 8 and the effect of such a molecule on leaching is thus minimized at low pH.

In the experiments by Neaman *et al.*, the most effective of the low molecular-weight aromatic ligands tested was gallate (Neaman *et al.*, 2005b, 2006).

Of course, the nature of the organic ligand is not the sole determinant of the extent of elemental leaching from a given rock. For example, Cu release from basalt was enhanced by the presence of organic acids under oxic but not anoxic conditions. In contrast, in similar leach experiments with granite, Cu was leached under both oxic and anoxic conditions. The disparity was attributed to differences in the Fe content of the basalt and granite and the effects of Fe mobilization on solubility of Cu/Fe sulfides. Such differences in leaching of Cu from two rock types highlight the importance of mineralogy, elemental composition and partial pressure of oxygen as well as the nature of the organic ligand.

Discrepancies between granite and basalt in organic-promoted leaching of elements have also been documented with respect to Y and Zr. Neaman *et al.* (2005b) showed that release of Y from basalt in dissolution experiments was more important in organic-containing as compared to organic-free solutions. In contrast, release of Y from granite by leaching with and without citrate was at the detection limit. This disparity was attributed to the effect of organics on dissolution of the Y-hosting minerals apatite (basalt) versus sphene (granite). Similarly, although Ti and Zr are often considered immobile elements, they can both be mobilized in the presence of organic ligands. In experimental leaching of a granite where zircons hosted Zr, this element was observed to be more immobile than Ti, whereas in similar leaching of a basalt where Fe/Ti oxides hosted the Zr, Ti was observed to be more immobile (Neaman *et al.*, 2006).

12.4 Organomarkers in weathering solutions

Based on leaching experiments such as those just described, Neaman *et al.* (2005a, b, 2006) pointed out that the mobility of some elements during weathering might be useful in documenting the presence of organic ligands in weathering solutions. They termed such elements 'organomarkers' – elements that dissolve preferentially to provide signatures of the influence of organics. For example, for both granite and basalt, release of Fe, P, and Y was observed to be enhanced by organic ligands. As mentioned in the previous section, Cu release was similarly enhanced for granite and for oxic conditions for basalt. On the basis of such observations, it was suggested that P, Fe, Cu and Y might be useful as organomarkers. However, the identification of organomarkers in natural systems is complicated because the effects of organic–metal complexation are superimposed over natural variations in elemental abundance, mineralogy of the dissolving and precipitating phases, and chemical conditions such as pH and redox state in surficial geochemical systems.

One way to investigate whether such organomarker elements identified in the laboratory might be useful for natural systems is to consider the chemistry of

Figure 12.4 Rivers draining basalt lithology normalized to the concentrations reported in basalt standard BCR-1. River concentrations are from Tosiani *et al.* (2004) and Pokrovsky *et al.* (2006) as compiled by Hausrath *et al.* (2009). The different concentrations of organic carbon in the rivers are represented by the lines: the dotted line is plotted for rivers with <10 mg l^{-1} DOC, 10–25 mg l^{-1} = dashed line, and >25 mg l^{-1} = solid line, with corresponding increases in line thickness with DOC concentration. Where elements are not plotted, concentrations were not reported for the rivers. Elements are organized after the periodic table, with first *s* block elements, from smallest to largest atomic radius, then *d* block elements, then *p* block and *f* block elements. Figure adapted from Hausrath *et al.* (2009).

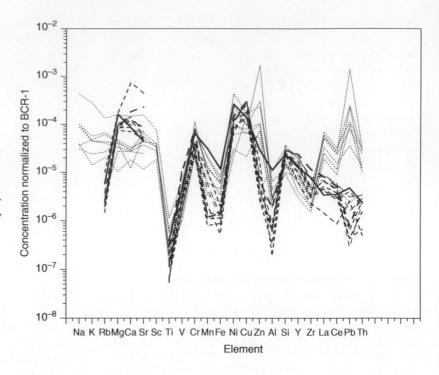

organomarkers in rivers. Consistent with some of the laboratory experiments, the concentrations of elements such as Al, Fe, Th, Zr, Y and REE are often higher in rivers containing significant organic ligands (Dai and Martin, 1995; Rousseau *et al.*, 1996; Viers *et al.*, 1997, 2000; Braun *et al.*, 1998, 2005; Oliva *et al.*, 1999; Dupre *et al.*, 2003; Tosiani *et al.*, 2004; Pokrovsky *et al.*, 2006). For example, in a study in South America (Tosiani *et al.*, 2004), the partitioning of elements between the dissolved and suspended loads in rivers varied from 30–90%, depending upon the element and the concentration of dissolved organic carbon (DOC). The partitioning of elements between dissolved and suspended loads was described by a coefficient that correlated with the hydrolysis constant of each element. This latter observation may be consistent with trends in organic–metal complexation, as discussed previously.

Given these effects, researchers have suggested that in catchments in areas such as Cameroon, weathering may dominantly be occurring in organic-rich swamps rather than in the organic-poor fluids filtering through hillslopes (Oliva *et al.*, 1999). The major and trace elements mobilized by DOC in swamps are later mobilized into rivers during high rainfall (Viers *et al.*, 2000). Swamp organic matter especially may contribute to the mobilization of Fe, Al, Zr, Ti, and Th in the top meter of regolith underlying swamps (Braun *et al.*, 2005).

However, not all the enhanced mobility in rivers is due to increased solubility. For example, in studies of small catchments in Cameroon, river waters were sampled and analysed for cations and DOC (Viers *et al.*, 1997). Samples with low DOC contained low

concentrations of many trace and major elements. In contrast, in organic-rich waters, Al, Ga, Fe, Ti, Th, Zr, Y and REE were present in higher concentrations. This observation was not strictly due to aqueous organic complexation, however, because when samples were treated by successive filtration to submicron pore sizes it was discovered that only in the low-organic waters were many of the elements present in dissolved form. In contrast, in the high-organic waters, elements were largely present as organic colloids – particles 1 nm to 1 μm in size – that were removed by filtration. The high specific surface area and high densities of reactive surface sites on colloids interact with the 'particle reactive' elements REE, Y, and Th to increase mobility even when solubilities are low (Douglas *et al.*, 1999). In some cases, the presence of organics may stabilize colloids (e.g. Buffle, 1988).

To eliminate the effect of the chemistry of the underlying lithology, the data for elemental concentrations in organic-poor and -rich rivers draining basalt lithologies were compiled and plotted together in Fig. 12.4 after normalization to a basalt standard, BCR-1 (Hausrath *et al.*, 2009). In this figure, elements such as Ca, Mg, Fe, and Cu, along with Zr and Y to lesser extents, are generally (but not always) higher in concentration in rivers with higher organic content, possibly due to organic complexation. Regardless of organic content, Ti concentrations are relatively low – presumably related to the lack of solubility of Ti-containing phases in basalts even in the presence of organics (Neaman *et al.*, 2005b). Based on these observations, Ti acts as a relatively reliable immobile element in basalt watersheds. La, Ce, Pb, and Th – all particle reactive elements – show different behaviours in two distinct families of rivers. Their mobility may actually be lower when DOC is high, an effect perhaps related to colloid aggregation and precipitation. Similarly, a few rivers show anomalously high Zn in the presence of low organic carbon.

Thus, like the laboratory experiments which document relatively modest rate enhancements for mineral dissolution in the presence of organics (Ganor *et al.*, 2009), the enhancement by organic molecules of riverine fluxes of organomarker elements appears to be relatively small. Of course, the small effect of organic complexation when integrated over the timescales of formation of regolith should be larger and easier to discern, as discussed in the next section.

12.5 Elemental profiles in regolith

Elemental depth profiles in regolith have been studied by soil scientists and geochemists for as long as geochemical measurements have been made. Indeed, these profiles can often be fruitfully investigated using simple models (White, 1995, 2008; Brantley *et al.*, 2008a, b).

The type of chemical profile that can be modeled most simply is a profile developed on one parent lithology situated at a ridgetop, terrace, or similar landscape that is not experiencing significant input of sediment. Such a profile can be conceptualized as a column characterized by one-dimensional flows of water and earth material with insignificant inputs from upslope sources. Here, we also concentrate on relatively wet climates where profiles are characterized predominantly by downward net fluid flow, that is evapotranspiration is less than precipitation. When such profiles are considered over thousands to millions of years, the flow of water is downward but the flow of solid material is upward as erosion removes material at the land surface and weathering advances downward (Fig. 12.5).

For one-dimensional, fluid-downward, solid-upward systems, several endmember types of element profiles (Fig. 12.6) can be discussed. However, to understand these element-depth profiles first requires normalization of the raw elemental concentration data to take into account the effects of changes in multiple elements happening concurrently in the soil (Brantley and White, 2009). Normalization for the concentration of an element j in weathered regolith ($C_{j,w}$) is accomplished using the concentration of j in the unweathered parent material ($C_{j,p}$) as well as the concentration of an immobile element in weathered and parent material ($C_{i,w}$ versus $C_{i,p}$). These concentrations are used to calculate the coefficient $\tau_{i,j}$ (defined in Fig. 12.6 (Brimhall and Dietrich, 1987; Anderson *et al.*, 2002)). $\tau_{i,j} < 0$ indicates depletion of element j as compared to element i in the parent material, and $\tau_{i,j} > 0$ indicates addition. For a depletion profile, the

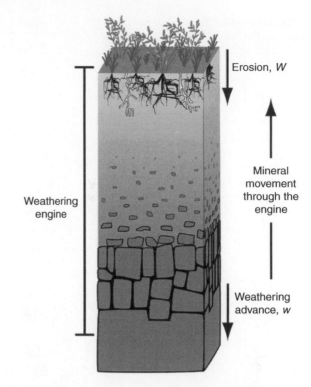

Figure 12.5 A schematic picture of rock weathering in the 'weathering engine' at a ridgetop site at the Earth's surface. The rate of downward movement of the saprolite-bedrock surface is the weathering advance rate, *w*. The rate of removal of material at the surface is the erosion rate, *W*. For a steady state profile, *w* = *W*. From the perspective of the land surface, mineral fragments flow upward. Where precipitation is larger than potential evapotranspiration, fluid flows downward. Figure adapted from Brantley *et al.* (2008).

fraction of element j removed from the profile at the surface is equal to $|\tau_{i,j}|$ at the surface.

Element profiles such that $\tau_{i,j} = 0$ at all depths document *immobile element profiles*. As discussed in the previous section on organomarkers, an immobile element is a relatively unreactive element, such as Zr (often found in zircon), Ti (rutile, anatase), or Nb (columbite, tantalite). These elements commonly become enriched in regolith as more soluble constituents are depleted (e.g. White, 2008). Importantly, however, the complexation of both Ti and Zr by organic ligands sometimes makes these elements problematic as immobile elements and every regolith profile must be individually evaluated with respect to which element is most immobile (see previous section).

To identify an immobile element, it is best to use measured bulk density to calculate strain due to expansion or contraction of the soil (Brimhall and Dietrich, 1987; White, 2008). However, even without bulk density measurements, it is possible to determine the most immobile element by trial and error. For example, $\tau_{i,j}$ can be calculated for different elements i that might be immobile as a function of depth and then these values

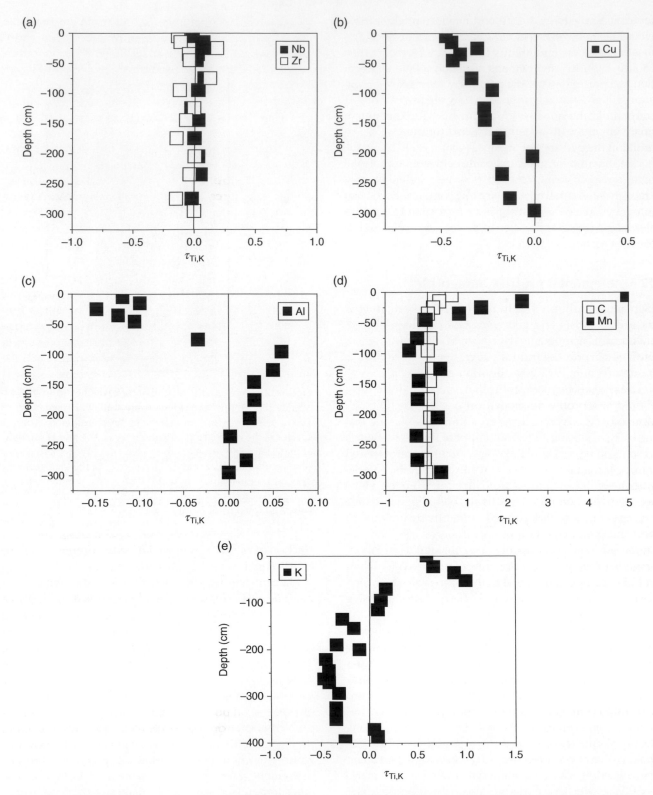

Figure 12.6 $\tau_{i,j}$ plotted versus depth below the mineral soil – organic horizon interface for (a) immobile, (b) depletion, (c) depletion-enrichment, (d) addition, and (e) biogenic profiles. All data derive from regolith in central Pennsylvania USA developed on Rose Hill shale in the Susquehanna Shale Hills Critical Zone Observatory. These data were calculated for different soil cores within Shale Hills Observatory and thus show different depths to bedrock. The definition of $\tau_{i,j}$ is

$$\tau_{i,j} = \frac{C_{j,w}}{C_{j,p}} \frac{C_{i,p}}{C_{i,w}} - 1 .$$ See text for discussion of this coefficient.

Figure reproduced with permission from Brantley and White (2009).

for different elements can be analysed to determine which element choice is consistent with depletion for most of the other highly soluble elements. Even with the choice of the most immobile element, some elements will not be depleted in the profile but will appear to be added to the soil (see below): such elements that are commonly added to profiles through atmospheric inputs should not be considered as candidates for the immobile element (e.g. C, N, Zn, Pb).

Once element concentrations are normalized, four other types of soil profiles are observed, of which two are most basic: (i) *depletion profiles* exhibiting depletion of the element of interest with respect to the immobile element; and (ii) *addition profiles* exhibiting enrichment of the element with respect to the immobile element. Element profiles may also comprise a mixture of these categories: indeed, some mixtures of depletion and addition profiles are so common that we refer to them as follows: (iii) *depletion–enrichment profiles* exhibit depletion toward the top and enrichment at depth; and (iv) *biogenic profiles* exhibit enrichment at the top overlying depletion at depth. Each of these four soil profile types is discussed more fully below, with an emphasis on how these profiles may show evidence for biotic influences.

In contrast to immobile profiles, *depletion profiles* (Fig. 12.6b) document loss of a mobile element with respect to the immobile element in the parent material, often due to mineral dissolution followed by solute or colloidal transport out of the system. A depletion profile is generally characterized by $\tau_{i,j} < 0$ at all depths down to parent, i.e., – net depletion.

Organomarker elements discussed previously may be depleted in a regolith profile where organic ligands are important. For example, Fig. 12.6b shows a profile depicting leaching of Cu from the Rose Hill shale in central Pennsylvania in the Shale Hills watershed (the Susquehanna-Shale Hills Critical Zone Observatory). The high affinity of Cu for organic ligands (consistent with the Irving-Williams series discussed earlier) may explain the leaching of Cu from this soil developed on shale parent. Na also shows a depletion profile in the Rose Hill shale soil (data not shown): in fact, geochemists often observe depletion profiles for Na in regolith, and these are generally due to dissolution of the Na-containing mineral plagioclase which is accelerated by the presence of organic ligands (White, 1995; Brantley *et al.*, 2008a). Depletion profiles can be characterized as 'partially depleted' or 'incompletely developed' *profiles* ($-1 < \tau_{i,j} < 0$ at the surface) or 'completely depleted' or 'completely developed' profiles ($\tau_{i,j} = -1$ at the surface) (Brantley and White, 2009). Figure 12.6b depicts an incompletely developed Cu profile.

Inputs to soils due to the processes of dust deposition or fixation from the atmosphere create *addition profiles* where $\tau_{i,j} > 0$ toward the surface. Such profiles are characterized by net addition when integrated over the whole profile. Figure 12.6d shows the addition profiles of Mn and C developed on Rose Hill shale weathering in central Pennsylvania. Such profiles commonly document inputs of wet or dry deposition (e.g. Mn in Fig. 12.6; see Herndon *et al.*, 2011), or inputs of elements due to biological fixation (e.g. carbon in Fig. 12.6d). Biological addition is also important for N (Amundson, 2004). Addition profiles demonstrate both addition at the surface and re-distribution downward by leaching and reprecipitation or mixing due to bioturbation. These profiles are therefore characterized by concentrations that are highest at the surface, but that grade back to parent concentration at depth.

When an element that is leached at the surface reprecipitates at depth in the regolith, the profile can be described as a depletion–enrichment profile (Fig. 12.6c). Depletion–enrichment profiles can document the distribution of organic complexation in the soil. An example of a depletion–enrichment profile is the E-Bs, E-Bt sequence commonly observed in forest soils where dissolution of Al occurs near the top but precipitation of Al-containing secondary minerals occurs deeper in the profile. Some of this redistribution of Al is due to the contribution of organic ligands: such ligands can be present at high concentrations in surficial layers where production of organic molecules is high compared to decomposition, but can be low at depth where decomposition of organic molecules proceeds faster than production. Fe often exhibits a depletion–enrichment profile, as do other elements that are affected by the chemistries of Fe and Al (for example, P). Depletion–enrichment profiles may be such that the element is conservative within the whole profile (no net loss or gain of the element when integrated over the whole profile), or may demonstrate a net loss if organic complexes are transported out of the soil column.

The processes of mobilization and redeposition of metals due to the presence of organic ligands have long been recognized as important during supergene ore deposition (Wood, 2000). Specifically, when dissolution, mobilization, and precipitation occur over relatively long time periods that include changes in the depth of the water table, economic minerals can be enriched in regolith. Wood points out that organic interactions in the supergene environment are important for metals such as Al, Au, Cr, Cu, Fe, Hg, Pb, Th, U, and V due to complexation and redox effects. In contrast, the mobilities of economic metals such as Co, Mn, and Zn are less impacted by the presence of organic molecules largely due to their electronic structure (compare the Irving–Williams series).

Biogenic profiles characterize elements such as K or Ca that, although released from parent material, are redistributed because they are tightly retained in the soil and they are used as nutrients (Fig. 12.6e). Profiles characterized as biogenic show depletion at depth but enrichment at the surface. Biogenic profiles can result due to the secretion of organic acids by plants or fungi through roots to dissolve primary minerals near the rooting depth (Fig. 12.2). If the nutrients are taken up by fungi and plants, used in biomass, and recycled in the upper layers of the soil, then a biogenic profile results due to enrichment at the surface. Global soil compilations document that topsoil profiles are often characterized by the effects of biocycling (Jobbagy and Jackson, 2001). Jobbagy and Jackson investigated published data for the upper meter of 10 000 soils worldwide and observed that nutrients such as P and K that are strongly cycled by plants were more concentrated in the upper 20 cm of soil than nutrients whose concentration did not generally limit growth. The more scarce the nutrient, the higher the topsoil concentration. The authors attributed such observations to the upward transport of nutrients by plants: they concluded that cycling by plants is a dominant control on the vertical distribution of the elements that are most required as nutrients and least available in soils. Some elements such as Fe may be depleted at depth and concentrated at the surface due to multiple factors including redox stratification, water saturation related to clay layers, and biological depletion of oxygen. Finally, it is intriguing to note also that nutrient profiles are also affected by human impact. For example, 200 years of farming by the Romans affected nutrient gradients in soils that can still be seen 2000 years later (Dupouey *et al.*, 2002). Given that humans often apply fertilizer to soils, element profiles in agricultural soils often have characteristics of addition as well as biocycling.

Higher concentrations of nutrient elements at the surface can result in lower rates of chemical weathering of minerals containing those elements and can even result in precipitation of minerals that might not form without biocycling. It has been hypothesized for example that biocycling of elements may be partially responsible in the tropics for precipitation of Si, Al, and Fe as clays and oxides in upper soil layers (Lucas *et al.*, 1993). Furthermore, just like in depletion–enrichment profiles, elements of interest in a biogenic profile may be conservative (i.e. no net loss or gain) or net depleted when the concentrations are integrated over the whole profile.

Importantly, a given element does not always show the same type of profile in every soil. Furthermore, many profiles do not fall easily into one of the simple categories. Perhaps the best example of this is one element described earlier as a possible organomarker – P. In many profiles, P is ultimately derived from the primary mineral apatite. This mineral is often dissolved by organic acids secreted by organisms (Taunton *et al.*, 2000; Goyne *et al.*, 2006). But P is taken up into biomass, driving the P profile to often appear like a biogenic profile. When P is released upon decomposition of biomass, it is often sorbed or occluded into Fe oxyhydroxides in the soil. These processes can imprint a depletion-enrichment character to the P profile, overprinting the biogenic character and often yielding a very complex P profile.

It is important to reiterate that the concepts discussed with respect to Fig. 12.6 are best applied to systems where lateral inputs to the system from sediments are minimal and where net fluid flow is dominantly downward. These types of sites can be thought of as so-called one-dimensional weathering profiles (Fig. 12.5). However, profiles for two-dimensional (non-convergent hillslope) or three-dimensional (convergent hillslope) weathering systems may also demonstrate characteristics similar to those in Fig. 12.6. For example, peat bogs, one of the best examples of vegetation-dominated environments, are complex in terms of their hydrology and sediment history. Peat bogs represent environments where plants such as sphagnum build up over time in a valley or other setting. It has long been known that in such bogs, natural organic matter (NOM) may increase the dissolved concentration of trace and major elements through formation of organic complexes (Steinmann and Shotyk, 1997a, b). As mentioned previously, NOM also stabilizes colloids that adsorb solutes and enhance mobility of particle-reactive elements (Buffle, 1988; Vaughan *et al.*, 1993; Wells *et al.*, 1998; Douglas *et al.*, 1999).

Although they are generally not simple systems like Fig. 12.5 with respect to their hydrology, peat bogs can demonstrate depletion profiles (Shotyk *et al.*, 1990). For example, for Fe and Na, low pH (Na, Fe) and organic complexation (Fe) are important in solubilization and mobility. In contrast, K and Mn have been observed to be enriched in the upper layers of bogs but depleted at depth –biogenic profiles (Shotyk *et al.*, 1990). Zn and Pb profiles in bog materials are attributed largely to atmospheric inputs, and thus can comprise addition profiles. Peat bogs are clearly an important case example of the strong effects of biota on elemental composition.

Given the strong effects of organic matter on element mobilization, paleosol chemistry has also been investigated to search for evidence of organomarkers over geologic time. For example, the chemistry of hypothetical organomarker elements was studied in the weathered surface of two of the oldest-known basalt-derived paleosols – the Mount Roe (2.76 Ga) and the Hekpoort (2.25 Ga) (Neaman *et al.*, 2005a). In those paleosols, P was depleted relative to the parent material, consistent with the possible presence of organic ligands during early weathering. The depletion of P in the soil cannot be definitive in documenting the presence of biota in a paleosol (Neaman *et al.*, 2005a) given that abiotic dissolution could also have occurred. However, Fe was also depleted but Al was not, and these observations are also diagnostic of organic ligand-promoted dissolution. Retention of Cu in the Mount Roe paleosol but not in the Hekpoort was attributed to soil formation under anoxic and oxic atmospheres, respectively. Others came to similar conclusions on the basis of Fe concentrations in the two paleosols (Yanai, 1992; MacFarlane *et al.*, 1994a, b; Yang and Holland, 2003). The immobility of Al in both paleosols was attributed to formation under relatively low rates of precipitation. On the basis of concentrations of organomarker elements, the authors concluded that organic acids – and possibly microbiota – may have been present in soils as early as 2.76 Ga.

In many cases, then, organic complexation creates a depletion profile for an organomarker element in regolith. In some cases, as discussed above, organomarker elements re-sorb or precipitate at some depth in the profile. In such cases, the solute chemistry in the rivers may not document the impact of the organic ligands on the weathering. However, the pattern of redistribution that is documented in the regolith profile (Fig. 12.6) may still document that organic molecules were present even if little to no net loss of the organomarker element occurred as a solute flux. Furthermore, some organomarkers as well as nutrient elements are taken up into plants and recycled within biomass and topsoil. For these elements, biocycling may actually decrease the rate of solute loss from regolith as compared to a hypothetical biota-free control. Nonetheless, the presence of a biogenic profile may be a good indicator that biota have affected the soil.

12.6 Time evolution of profile development

The discussion in the previous section generally ignored the time evolution of concentration-depth profiles. As discussed in Section 12.2, however, over geologically long time periods, regolith profiles generally show evidence of net depletion unless uplift or erosive processes rejuvenate the profile. Even for biogenic profiles where biocycling may retard the loss of organomarker elements and nutrients, over the long term, some net loss occurs, typically leading to nutrient depletion that can only be rejuvenated through erosion, uplift, or dust input (Kennedy *et al.*, 1998; Chadwick *et al.*, 1999; Wardle *et al.*, 2004; Porder *et al.*, 2005).

The easiest type of profile to investigate with respect to temporal evolution is the depletion profile. Where erosion is insignificant, a depletion profile such as that shown in Fig. 12.6b will deepen with time at the *weathering advance rate* or *regolith production rate*. The geometry of the elemental depth profile will change with time as more and more of the element is depleted at the surface: the profile will vary from an incompletely developed profile ($-1 < \tau_{i,j} < 0$ at the land surface) to a completely developed profile ($\tau_{i,j} = -1$ at surface) (Fig. 12.7a and c).

Once the value of the uppermost concentration stops changing in a completely developed profile, the profile is said to be quasi-stationary (Lichtner, 1988) such that it retains its geometry (i.e., geometry of the concentration-depth curve) but moves downward with respect to the land surface at the weathering advance rate.

In contrast, where erosion is significant, a steady-state profile can develop wherein the concentration-depth curve geometry remains constant without moving downward (Fig. 12.7b and d). For such a system, the rate of erosion equals the rate of weathering advance and the solid material can be conceptualized as moving upward through the weathering engine (Fig. 12.5). As mentioned earlier, such steady-state profiles have implications for ecosystems: movement of material through the weathering zone is ongoing and can support an ecosystem with constant nutrient delivery at a rate equal to the weathering advance rate. Erosion, which can be episodic or relatively constant in time, rejuvenates the weathering-derived nutrient supply (Porder *et al.*, 2005).

Conceivably, even a biogenic profile could attain a steady state. For example, Fig. 12.7e shows a model treatment of a steady state biogenic profile developed within regolith with advective–diffusive solute transport while undergoing erosion. For this model, it was assumed that the element of interest was depleted at the regolith–bedrock boundary and was re-released in the upper layers at a rate $p_0 \exp(-p_1 h)$. This is a hypothetical function used to describe mineral decomposition that decreased downward with depth, where p_0 and p_1 are constants, and h is the depth of regolith. In contrast to the steady-state biogenic profile, in a non-eroding quasistationary profile that deepens with time, regolith at the surface becomes nutrient-poor, and ecosystems become limited by the rate of nutrient extraction from parent material (Wardle *et al.*, 2004). In such nutrient-limited regimes, atmospheric additions (e.g. Fig. 12.6d) may ultimately support biota (Chadwick *et al.*, 1999).

The rate of movement of the weathering front, defined as the weathering advance rate or regolith production rate, w (White and Brantley, 1995), equals the erosion rate, W, in a steady state system (Figs 12.5 and 12.7). The thickness of any layer in that steady-state weathering system (partially weathered bedrock, saprolite, soil) and the rate of transformation of each layer into the overlying layer must be constant in time over the timescale of interest. Furthermore, over that same timescale, the chemistry of the biomass and organic matter reservoirs must also be constant in time. Generally, over geological timeframes, it is assumed that biota maintains such a steady-state reservoir. However, given that biota change as regolith changes (Wardle *et al.*, 2004), the concept of steady-state biomass

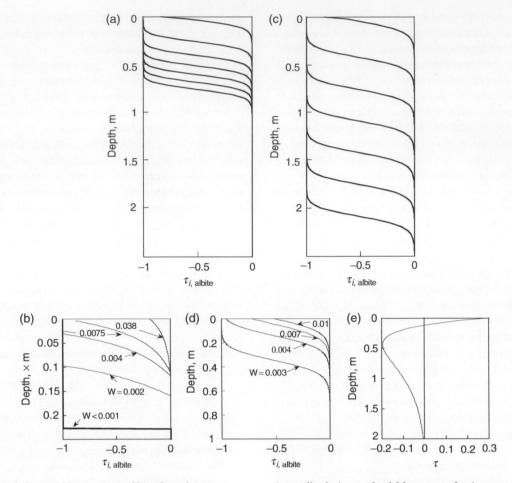

Figure 12.7 Depletion and biogenic profiles plotted versus depth for albite weathering to kaolinite as calculated within isovolumetric regolith containing only feldspar + quartz (Lebedeva *et al.*, 2010). The model is described by one-dimensional reactive transport equations. Two cases are analysed: transport by diffusion only and transport by diffusion and vertical advection through regolith and bedrock, assuming constant Darcy velocity. (a) (diffusive transport only) and (c) (advective-diffusive transport) show the time evolution of such a profile where increasing duration of weathering without significant erosion results in profiles that deepen as shown. If erosion is significant such that the rate of downward movement of the profile is balanced by the rate of erosion, then steady-state profiles can result as shown in (b), (d), and (e). For (b), the specific surface area for reacting albite was assumed to be constant and equal to the geometric surface area (i.e., all curves represent 'rock' with the same grain size albite). The curves document depletion of albite as a function of depth, contoured for various rates of erosion, W ($L\ T^{-1}$), at steady state, as indicated. When the erosion rate W

is small relative to the feldspar weathering rate constant, k, a very sharp reaction front – the depth zone over which feldspar dissolves – is observed; in contrast, when W is fast compared to k, a steep reaction front is observed. A sharp (thin) reaction front within a completely developed profile is characteristic of the local equilibrium regime (e.g. $W < 0.001$ m per ky) while a partially developed profile ($W > 0.0075$ and 0.004 m per ky, b and d, respectively) is characteristic of the kinetic regime. (e) A biogenic profile for advective–diffusive transport with erosion. The re-release of an element is modeled as $p_0 \exp(-p_1 x)$ where x is the regolith depth, p_0 and p_1 are constants. This term is added to the equation describing evolution of the mineral concentration (see text). These regimes observed for simulated profiles for model 'rocks' containing albite + quartz + kaolinite are similar to the transport-limited regime (local equilibrium) and the weathering-limited regimes (kinetic regime) hypothesized for catchments. (a and b) After Lebedeva *et al.* (2010); Brantley and White (2009).

may be problematic. It is in fact unknown whether surface systems can be maintained at steady state – or, if a steady state is possible, how long this steady state can be maintained. Regardless, the steady-state concept is a useful model for comparison to real systems as shown in the next section.

12.7 Investigating chemical, physical, and biological weathering with simple models

The erosional and weathering regimes of soil profiles, hillslopes, and catchments have been classified by many workers (Carson and Kirkby, 1972; Stallard, 1992, 1995;

Kump *et al.*, 2000; Riebe *et al.*, 2004; West *et al.*, 2005). Here, we use the terminology of Stallard (1995). Stallard points out that chemical processes largely determine the rate that weathered material is produced (i.e. the weathering advance rate). For some systems, the capacity for transport of weathered materials out of the system is large compared to the rate of weathering advance: these are weathering-limited systems. In contrast, systems that are described as transport-limited are characterized by rates of weathering advance that are fast compared to the capacity to transport material away by physical processes. In general, transport-limited systems are characterized by thick, primary mineral-depleted regolith and weathering-limited systems are characterized by thin, primary mineral-containing regolith (Stallard, 1995; Kump *et al.*, 2000).

Recently, we have explored similar concepts by modeling (Lebedeva *et al.*, 2007, 2010) simplified 'rock' systems (Fig. 12.7). Lebedeva *et al.* (2010) analysed 'rock' models containing albite + quartz weathering chemically to kaolinite + quartz while undergoing erosive loss of material at the top. When the erosion rate is fast ($W = 0.038$ and 0.01 m per ky in Fig. 12.7b and d, respectively), an *incompletely developed* profile develops for the albite ($-1 < \tau_{i,\, albite} < 0$ at the surface), and the chemical weathering solute flux out of the model system is calculated to be a function of the albite dissolution rate constant (Lebedeva *et al.*, 2010). Note that here we are calculating $\tau_{i,j}$ values on a mineralogical rather than elemental basis, by using the concentration of albite (j) and quartz (i). Lebedeva *et al.* describe dissolution of albite in such a system as kinetically controlled.

In contrast, when $\tau_{i,albite}$ for a system $= -1$ at the surface (a completely developed profile) and a relatively thin reaction front develops for albite transformation to kaolinite (e.g. $W < 0.001$ m per ky in Fig. 12.7b), the chemical weathering solute flux out of the system does not vary with the rate constant for dissolution of albite. Lebedeva *et al.* describe albite dissolving in such a system as experiencing local equilibrium control. Under local equilibrium control, increasing the rate constant for dissolution does not increase the chemical weathering flux. A mixed-control regime where both rates of erosional transport and chemical reaction contribute and neither is rate-limiting was termed by Lebedeva *et al.* (2010) the transition regime.

The kinetic and local equilibrium regimes described for these models of 'rock' weathering are similar in many ways to weathering- and transport-limited regimes identified by Stallard (1995) for hillslopes or catchments. However, the one-dimensional soil profiles in the model are significantly less complex than hillslopes and catchments in that they do not include laterally derived inputs or outputs of sediments or solutes. Furthermore, in real systems, many minerals react

sequentially or simultaneously and one mineral could be under local equilibrium control while another mineral is under kinetic control.

Nonetheless, Brantley and White (2009) previously hypothesized that for profiles developed on bedrock, the profile of the mineral that weathers deepest may be "profile-controlling". They argued that if this deepest-weathering profile-controlling mineral displays an incompletely developed profile then the system is weathering limited; in contrast, if the profile-controlling mineral is experiencing local equilibrium control, then this profile may document that the system is transport-limited. For example, in quartz diorite weathering in Puerto Rico, the deepest weathering mineral is biotite. Oxidation of the ferrous component in the biotite occurs deep within the pristine bedrock (Fletcher *et al.*, 2006; Buss *et al.*, 2008). Complete oxidation occurs across a relatively thin zone (tens of cms) of bedrock weathering that is delineated by spheroidal fracturing. Lebedeva *et al.* (2010) characterized this thin front for oxidation of biotite as developing within the local equilibrium regime, consistent with weathering occurring in a largely transport-limited system.

Intriguingly, even though the interface between bedrock and saprolite in the Puerto Rico system lies at 5–8 m depth, the interface is inhabited by a microbial ecosystem that may be driven by bacteria deriving energy from Fe(II) oxidation (Buss *et al.*, 2005). It is thus possible that even though this reaction occurs deep in the subsurface, weathering is accelerated by bacteria. No biotic effects were included in the models of Lebedeva *et al.* (2007, 2010).

Here we take a first step toward exploring biological processes in models of weathering and erosion in steady-state systems. We investigate biotic influences by again using the simple system of albite + quartz weathering to kaolinite + quartz modelled in Fig. 12.7a (Lebedeva *et al.*, 2007, 2010). Calculated chemical weathering solute fluxes are plotted for this model system versus physical erosion in Fig. 12.8. The diagram shows the regime of transport limitation (local equilibrium control), on the left-hand side of the diagram. For example, this is where we infer that the Puerto Rico quartz diorite example discussed previously would plot. In contrast, the regime of weathering control (kinetic limitation), appears on the right-hand side of the diagram. The diagram was calculated for a system with both advective and diffusive transport. The slanted line on the left-hand side of the diagram indicates where the chemical weathering flux varies directly with the rate of physical erosion. The horizontal lines show chemical weathering fluxes in the weathering-limited regime for four different values of the albite dissolution rate constant.

For these models, the chemical weathering flux, Ω, is defined as the rate of release of soluble ions during

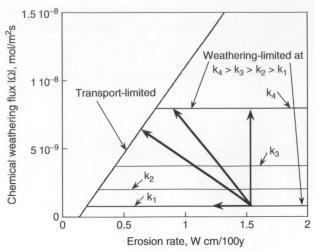

Figure 12.8 Model calculations of the regimes of local equilibrium control (left), versus kinetic limitation (right). This diagram shows the absolute value of the weathering flux, $|\Omega|$, plotted versus erosion rate (W), as calculated for a model albite + quartz 'rock' weathering to kaolinite + quartz (Lebedeva *et al.*, 2010). The calculation assumes combined diffusive and advective transport. The slanted line on the diagram indicates the points where the chemical weathering flux is not dependent upon the rate constant for mineral dissolution but only varies with rate of erosion. In contrast, the horizontal lines show how chemical weathering fluxes in the kinetic-limited regime are independent of increasing erosion rate. Four different chemical weathering fluxes are shown as described by rate constants $k_4 > k_3 > k_2 > k_1$. The diagram shows schematically how vegetation might affect either chemical weathering fluxes or erosion rate or both. For example, a weathering system operating in the kinetic regime with a chemical weathering flux of about 1×10^{-9} mol m^{-2} s^{-1} and erosion rate of about 1.5 cm per 100 years might experience either faster chemical weathering due to production of organic acids without a change in erosion rate (vertical arrow) or, more likely, faster chemical weathering accompanied by lower erosion rate (oblique arrows). Alternately, erosion rate might be lowered while chemical weathering flux might remain constant (horizontal arrow). In every case, vegetation would drive a weathering system from kinetic control toward local equilibrium control. (Note that the intersection of the horizontal kinetic-limited lines with the slanted line indicates systems experiencing mixed control where chemical weathering fluxes depend on both the rate constant for dissolution and the erosion rate; however, this diagram was made based on analytical solutions (Lebedeva *et al.*, 2010) that did not include this mixed regime so the intersections are strictly drawn in by inspection.)

isovolumetric weathering of bedrock and is expressed as a flux out of the system in units of mass per unit area per unit time (M L^{-2} T^{-1}). Chemical weathering fluxes were calculated by summing the fluxes of solutes estimated by the model. We reproduce here, in simplified form, some of the equations derived by Lebedeva *et al.* (2010) to exemplify the controls on weathering fluxes. For example, in the local equilibrium regime, the flux,

Ω_{eq}, depends upon the erosion rate W but does not depend upon the mineral dissolution rate constant k:

$$\Omega_{eq} = p - WQ^0 \qquad (12.3)$$

The erosion rate, W, (L T^{-1}), is defined as the rate of lowering of the Earth's surface due to removal of material by physical processes (Fig. 12.5). In contrast, in the kinetic regime, the solute flux (Ω_{kin}) does not depend on the erosion rate W but is dependent upon the mineral dissolution rate constant k:

$$\Omega_{kin} = q + r\left(1 - \sqrt{1 + \lambda k}\right)\Delta C \qquad (12.4)$$

Here p, q, r and λ are constants that depend on the mass transport in the pore fluid, Q^0 (mass per unit volume, M L^{-3}) is the concentration of albite in the bedrock, k (T^{-1}) is an effective rate constant, and ΔC(M L^{-3}) is the difference between concentrations of a species in the pore fluid at the base and at the top of the regolith. The effective rate constant is calculated as $k = k_{ab}sQ^0\Psi$ where k_{ab} is the dissolution rate constant for albite in mol m^{-2} s^{-1} measured in the laboratory (Bandstra *et al.*, 2008), and s is the specific surface area of albite (m^2 m^{-3}) based on a geometric model for particles. The rate constant is corrected with a factor, Ψ(m^3 mol^{-1}), included to account for the fact that in a real system, other minerals and species in solution affect the dissolution rate. That factor was calculated from the numerical solution of a model that contains more minerals (Lebedeva *et al.*, 2007).

In Fig. 12.8, weathering fluxes are plotted versus the erosion rate for four different rate constants such that $k_4 > k_3 > k_2 > k_1$. Note that in equations 12.3 and 12.4, chemical weathering fluxes are expressed as negative values (i.e. larger fluxes are more negative); in contrast, for simplicity we have plotted these fluxes as positive values in Fig. 12.8. The intersection of the horizontal kinetic-limited line for each of these rate constants with the slanted line indicates a system experiencing mixed control (i.e. the transition regime). Figure 12.8 was calculated from analytical solutions for the model system (Lebedeva *et al.*, 2010). Since no analytical solutions were calculated for the mixed regime, the figure simply shows the analytical solutions for local equilibrium and for kinetic limited regimes merged together.

In natural systems characterized by very fast erosion compared to chemical weathering, bare bedrock becomes exposed. Exposed bedrock is often observed in high relief systems where erosion rates are high. The exposed bedrock of such an endmember weathering-limited system typically only supports minor biomass. As vegetation grows on the exposed bedrock, however, regolith thickness can increase due to (i) increased rates of chemical weathering, (ii) decreased physical erosion, and (iii) dust accumulation. As regolith grows and thickens, eventually a local equilibrium regime can develop very deep regolith that insulates vegetation from bedrock-derived nutrients.

In such an endmember regime, nutrient availability begins to again limit biomass. Clearly, an optimal thickness of regolith provides enough regolith and organic matter to retain water and anchor biota without isolating that biota from nutrients in bedrock.

These ideas are shown in Fig. 12.8. Arrows on the diagram show schematically how growth of vegetation might affect chemical weathering flux or erosion rate. For example, a weathering system without biota operating in the kinetic regime with a chemical weathering flux of about 1×10^{-9} mol m^{-2} s^{-1} and erosion rate of about 1.5 cm per 100 years would be expected to be perturbed if vegetation began to grow: vegetation would most like increase the chemical weathering rate due to production of organic acids without a change in erosion rate (vertical arrow) or faster chemical weathering flux might be accompanied by a lower erosion rate (oblique arrows). Growth of vegetation might even lower the erosion rate without affecting the chemical weathering flux (horizontal arrow). Based upon these arrows, the vegetation would drive the weathering system from kinetic (weathering-limited) toward local equilibrium (transport-limited) control.

Similarly, if a steady-state weathering system without biota started on the transport-limited side of the diagram and then began to support growth of vegetation, the result would be a decrease in erosion rate. This perturbation would push the system away from the steady state and would result in an increase in regolith thickness. Presumably, since the system started in the regime of transport limitation, even if the vegetation caused an increase in chemical dissolution rate due to organic acids or other factors, the chemical weathering flux would not increase at steady state (because Equation 12.3 does not include k, the dissolution rate constant). As regolith thickened over time, a new steady state and new steady-state thickness of regolith would be attained in the transport-controlled regime. Thus for systems starting in either the weathering- or transport-limited parts of the diagram, vegetation growth would generally move the system toward transport control of weathering and most likely, deeper regolith.

Apparently, while tectonic activity and volcanism can create kinetically limited systems on the right-side of Fig. 12.8, weathering reactions drive these systems toward transport-limitation. Biota catalyse this movement toward transport-limitation. On the vegetation-covered landscapes on Earth today, we thus expect to see a combination of systems experiencing both weathering- and transport-limitation depending upon where they sit on a diagram such as Fig. 12.8.

Indeed, many researchers have looked for evidence of weathering- or transport-limitation on mobilization of individual elements in world rivers by assessing solute and sediment fluxes at different scales. It has been argued that, if the chemical weathering fluxes are weathering-limited, then factors such as temperature, pH, mineralogy, and ligand concentration should affect mineral dissolution and CO_2 consumption rates. In contrast, if the weathering fluxes are transport-limited, then weathering may be more affected by rates of uplift and soil erosion than by chemical factors.

We have already shown how the concentrations of a few elements in some small rivers sometimes document the effects of organic acids (Fig. 12.4). The importance of organic acids in some rivers is consistent with, although not necessarily unique to, weathering-limited systems. In larger watersheds, low fluxes of solutes derived from silicate weathering are observed in systems such as the Congo with highly weathered sediments. In contrast, large watersheds characterized by high chemical fluxes of solutes derived from silicate weathering generally have poorly weathered sediments that have not lost all of their primary minerals (Gaillardet *et al.*, 1999a). These observations could be consistent with chemical weathering occurring in the local equilibrium regime in catchments with low chemical fluxes and weathering occurring in the kinetic-limited regime in catchments with high chemical fluxes. Such a conclusion is broadly consistent with Figs 12.8 and 12.9.

Regardless of the size of the river, however, it has been pointed out that the effect of temperature on chemical weathering flux is weak, an observation some have attributed to transport limitation (Fig. 12.9a, c). In addition, several authors have pointed out that the chemical weathering fluxes from world rivers increase in direct proportion to their physical erosion fluxes (Fig. 12.9b, d). This has again been attributed to transport limitation. Indeed, by comparing Figs 12.8 and 12.9, it is apparent that chemical weathering fluxes for silicates in the world's largest 60 rivers are not consistent with denudation occurring strictly in a weathering-limited regime (i.e. weathering fluxes are not independent of erosion rates).

However, given the scatter in data such as that compiled in Fig. 12.9, denudation may be both transport- and weathering-controlled as discussed by West *et al.* (2005). In a figure from their work (reproduced here as Fig. 12.10), chemical weathering fluxes (plotted on the *y* axis) lie on the 1:1 line with respect to total denudation for various systems situated on continental cratons. West *et al.* attribute this observation to fast chemical weathering and soil production compared to erosion in those regions. In contrast, for the submontane, montane, and alpine catchments, the chemical weathering fluxes do not lie on the 1:1 line. West *et al.* argue that Fig. 12.10 is consistent with young, weathering-limited landscapes evolving toward transport-limited regimes on older, less tectonically active terrains. Similar arguments have been advanced based on other data (e.g. Stallard, 1995; Hren *et al.*, 2007). We argue here that biota catalyse this evolution from weathering- to transport-limitation by increasing chemical weathering rates and decreasing erosion rates, driving systems from the right part of Fig. 12.8 toward the left-hand side.

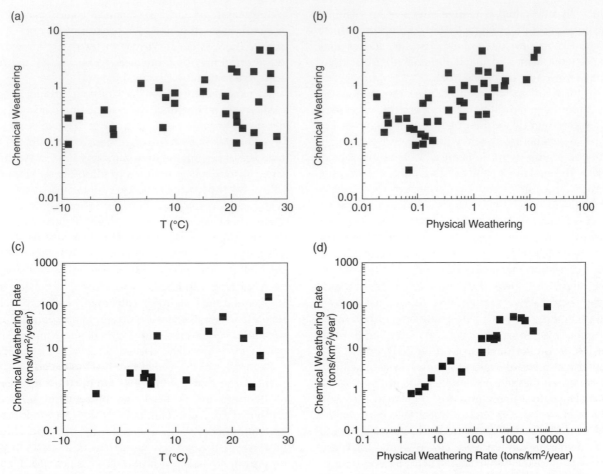

Figure 12.9 Correlation between silicate chemical weathering fluxes and temperature (a) and physical erosion fluxes (b) as compiled for large watersheds. Here, silicate fluxes are riverine fluxes of major solute cations + Si for the 60 largest world rivers, corrected to reflect only silicate weathering. The values are normalised to the Amazon watershed (Gaillardet *et al.*, 1999b). The Huanghe river was removed by the original authors due to the strong effect of deforestation on the mechanical weathering in that watershed. Similar relationships for small monolithological watersheds are shown in (c) and (d). Figure reproduced with permission (Dupre *et al.*, 2003).

12.8 Conclusions

Weathering is the process whereby exhumed crustal rocks move toward chemical equilibration with surface fluids. This process produces both solutes and sediments and partially controls the concentrations of electrons and protons in surficial Earth systems while playing a major role in controlling the chemistry of Earth's atmosphere over geological timescales.

The biota affect chemical and physical weathering. Ongoing rates of weathering are documented in the riverine fluxes we measure today while time-integrated rates are documented in element-depth profiles in regolith. Such profiles in both modern regolith and in ancient paleosols provide records of weathering and erosion over time. The influence of the biota on weathering rates can be seen in some riverine fluxes and, especially, in regolith-depth profiles for nutrients and 'organomarker'

elements. Organomarker elements – elements whose mobility is especially affected by the presence of organic molecules – and nutrients can show depletion, depletion–enrichment, and biogenic profiles in regolith depending upon the interplay between leaching, precipitation, and biotic uptake. The effect of the biota on organomarker elements often produces a larger signal in regolith than in riverine chemistry. This is because small rate differences manifest as small signatures in riverine chemistry but manifest as large signatures when integrated over the geologically long time periods that are recorded in chemical depth profiles of regolith.

Interpreting regolith profiles requires modeling approaches to interpret the regimes of weathering and the influence of biota. For example, regolith varies from weathering- to transport-limitation as evidenced by areas of bedrock exposure and very deep regolith,

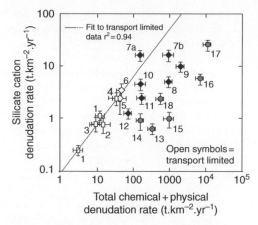

Continental cratons
1. Canadian shield
2. Siberian shield
3. African shield
4. Guyana shield

Alpine catchments
13. Colorado Rockies
14. Sierra Nevada
15. Svalbard
16. High Himalaya
17. West Southern Alps
18. Swiss Alps

Submontane catchments
5. British Columbia
6. Sabah Malaysia
7. Puerto Rico
   a. Long term erosion
   b. Modern day erosion
8. East Southern Alps
9. Lesser Himalaya
10. Cote d'Ivoire
11. Idaho Batholith
12. Appalachians

Figure 12.10 Silicate cation solute fluxes plotted versus the total denudation rate as published by West *et al.* (2005). In transport-limited catchments (open symbols), chemical weathering fluxes are limited by the transport of weathered material away through physical erosion. In this regime, a thick layer of regolith may become depleted in cations and primary minerals. At higher erosion rates, the transport of weathered material is fast compare to the rate of production of weathered material. Instead, weathering may be limited by the kinetics of mineral–water reactions (closed symbols). In the kinetic regime, the weathering fluxes may depend on mineral–water reaction kinetics. Error bars represent ±1σ. The regression line indicates a least-squares fit to transport-limited catchments. Figure reproduced with permission from the original (West *et al.*, 2005).

respectively. Weathering limitation is best observed in tectonically active or fast-eroding systems. Mountain-building activity exposes new, nutrient-containing rock surfaces. Biotic colonization of such newly exposed rock surface catalyses the transformation of the system from weathering-limited (the kinetic regime) toward transport limitation (the local equilibrium regime).

Changes in climatic conditions in the future will cause shifts in the relative areas of the crust that experience weathering-limited versus transport-limited regimes (West *et al.*, 2005). Ongoing human activities are also impacting both weathering and erosion. In contrast to most biotic effects, humans increase chemical weathering *and* physical erosion (Johnson *et al.*, 1981; Raymond and Cole, 2003; Wilkinson, 2005; Wilkinson and McElroy, 2007). Significant challenges remain for us in terms of predicting the effects of such changes on the Earth's surface.

Acknowledgements

S.L.B.'s interest in the effects of biota on regolith chemistry was fostered through funding from the NASA Astrobiology Institute Cooperative Agreement NCC2–1057 to H. Ohmoto and C. House at Penn State. The research on organomarkers was funded by NASA Exobiology grant NAG5–12330 that represented a collaborative effort with K. Goyne and J. Chorover. E.M.H. was partially supported while at Penn State by a NASA Mars Fundamental Research grant NNG05GN72G. The modeling efforts were all funded by a grant to S.L.B. from the Department of Energy's Office of Basic Energy Science, DE-FG02–05ER15675: this grant supported M. L. The authors benefitted from discussions with P. Lichtner and C. Steefel as part of the Penn State Center for Environmental Kinetics, an Environmental Molecular Sciences Institute supported by the National Science Foundation under Grant No. CHE-0431328 and the US Department of Energy, Biological and Environmental Research (BER). Financial support for work at the Susquehanna Shale Hills Critical Zone Observatory summarized in Fig. 12.6 was provided by National Science Foundation under Grant No. EAR-0725019. S.L.B. acknowledges many discussions with A. White, H. Buss, O. Chadwick, L. Jin, J. Moore, and A. Navarre-Sitchler and M.L. acknowledges V. Balashov. Reviews from K. Konhauser and L. Kump helped improve the manuscript.

References

Amrhein C, Suarez DL (1988) The use of a surface complexation model to describe the kinetics of ligand-promoted dissolution of anorthite. *Geochimica et Cosmochimica Acta* **52**, 2785–2793.

Amundson R (2004) Soil formation. In: *Treatise on Geochemistry* (ed Drever JI). Elsevier Pergamon, Oxford. **5**, 1–35

Anderson SP, Dietrich WE, Brimhall GH (2002) Weathering profiles, mass balance analysis, and rates of solute loss: Linkages between weathering and erosion in a small, steep catchment. *Geological Society of America Bulletin* **114**(9), 1143–1158.

Andrews MY (2008) Quantification of the effects of angiosperms and gymnosperms on silicate weathering and related soil nutrient cycling: implications for Phanerozoic atmospheric CO_2 and modern soil trace metal pollution. PhD Dissertation, Yale University, New Haven, CT.

April R, Keller D (1990) Mineralogy of the rhizosphere in forest soils of the eastern United States. *Biogeochemistry* **9**, 1–18.

Arthur MA, Fahey TJ (1993) Controls on soil solution chemistry in a subalpine forest in north-central Colorado. *Soil Science Society of America Journal* **57**, 1123–1130.

Balogh-Brunstad Z, Keller CK, Bormann BT, O'Brien R, Wang D, Hawley G (2008) Chemical weathering and chemical denudation dynamics through ecosystem development and disturbance. *Global Biogeochemical Cycles* **22**, doi:10.1029/2007GB002957.

Bandstra JZ, Buss HL, Campen RK, *et al.* (2008) Compilation of mineral dissolution rates. In: *Kinetics of Water-Rock Interaction* (eds Brantley SL, Kubicki JD, White AF). Springer, New York. 737–823

Berner RA, Cochran MF (1998) Plant-induced weathering of Hawaiian basalts. *Journal of Sedimentary Research* **68**, 723–726.

Berner RA, Kothavala Z (2001) GEOCARB III: a revised model of atmospheric CO_2 over Phanerozoic time. *American Journal of Science* **301**, 182–204.

Berner RA, Lasaga AC, Garrels RM (1983) The carbonate-silicate geochemical cycle and its effect on the atmospheric carbon dioxide over the past 100 million years. *American Journal of Science* **283**, 641–683.

Berner EK, Berner RA, Moulton KL (2004) Plants and mineral weathering: past and present. In: *Treatise in Geochemistry*. ElsevierPergamon ed.Drever JI, Amsterdam. **5**, 169–188

Bormann FH, Likens GE (1967) Nutrient cycling. *Science* **155**, 424–429.

Bormann BT, Wang D, Bormann FH, Benoit G, April R, Snyder MC (1998) Rapid, plant-induced weathering in an aggrading experimental ecosystem. *Biogeochemistry* **43**, 129–155.

Brantley SL (2004) Reaction kinetics of primary rock-forming minerals under ambient conditions. In: *Treatise on Geochemistry* (eds Drever JI, Turekian KK, Holland HD). Elsevier Pergamon, Oxford. 73–118

Brantley SL (2008) Kinetics of mineral dissolution. In: *Kinetics of Water–Rock Interaction* (eds Brantley SL, Kubicki JD, White AF). Springer-Kluwer, New York.

Brantley SL, Chen Y (1995) Chemical weathering rates of pyroxenes and amphiboles. In: *Chemical Weathering Rates of Silicate Minerals* (eds White AF, Brantley SL). Mineralogical Society of America, Washington, DC. 119–172

Brantley SL White AF (2009) Approaches toward modelling weathering regolith. In: *Water–Rock Interaction* (ed Oelkers E). Mineralogical Society of America, Washington DC. 435–484

Brantley SL, Bandstra J, Moore J, White AF (2008a) Modelling chemical depletion profiles in regolith. *Geoderma* **145**, 494–504.

Brantley SL, Goldhaber MB, Ragnarsdottir V (2008b) Crossing disciplines and scales to understand the Critical Zone. *Elements* **3**, 307–314.

Braun JJ, Viers J, Dupre B., Polve M, Ndam J, Muller J-P (1998) Solid/liquid REE fractionation in the lateritic system of Goyoum, East Cameroon: the implication for the present dynamics of the soil covers of the humid tropical regions. *Geochimica et Cosmochimica Acta* **62**, 273–300.

Braun J-J, Ngoupayou JRN, Viers J, *et al.* (2005) Present weathering rates in a humid tropical watershed: Nsimi, South Cameroon. *Geochimica et Cosmochimica Acta* **69**, 357–387.

Brimhall G, Dietrich WE (1987) Constitutive mass balance relations between chemical composition, volume, density, porosity, and strain in metasomatic hydrochemical systems: results on weathering and pedogenisis. *Geochimica et Cosmochimica Acta* **51**, 567–587.

Buffle J (1988) *Complexation Reactions in Aquatic Systems. An Analytical Approach*. Ellis Horwood, Chichester.

Buol SW, Hole FD, McCracken RJ (1989) *Soil Genesis and Classification*. Iowa State University Press, Ames, IA.

Buss HL, Bruns MA, Schultz MJ, Moore J, Mathur CF, Brantley SL (2005) The coupling of biological iron cycling and mineral weathering during saprolite formation, Luquillo Mountains, Puerto Rico. *Geobiology* **3**, 247–260.

Buss HL, Sak PB, Webb SM, Brantley SL (2008) Weathering of the Rio Blanco quartz diorite, Luquillo Mountains, Puerto Rico: coupling oxidation, dissolution and fracturing. *Geochimica Cosmochimica Acta* **72**, 4488–4507.

Carson MA Kirkby MJ (1972) *Hillslope Form and Process*. Cambridge University Press, Cambridge.

Chadwick OA, Derry LA, Vitousek PM, Huebert BJ, Hedin LO (1999) Changing sources of nutrients during four million years of ecosystem development. *Nature* **397**, 491–497.

Chin PKF, Mills GL (1991) Kinetics and mechanism of kaolinite dissolution – effects of organic ligands. *Chemical Geology* **90**, 307–317.

Cochran MF, Berner RA (1996) Promotion of chemical weathering by higher plants: field observations on Hawaiian islands. *Chemical Geology* **132**, 71–77.

Cox NJ (1980) On the relationship between bedrock lowering and regolith thickness. *Earth Surface Processes* **5**, 271–274.

Dai M-H, Martin J-M (1995) First data on trace metal level and behaviour in two major Arctic river-estuarine systems (Ob and Yenisey) and in the adjacent Kara Sea, Russia. *Earth and Planetary Science Letters* **131**, 127–141.

Dietrich WE, Reiss R, Hsu M-L, Montgomery DR (1995) A process-based model for colluvial soil depth and shallow landsliding using digital elevation data. *Hydrological Processes* **9**, 383–400.

Dokuchaev VV (1883) Russian Chernozem. In: *Selected Works of V.V. Dokuchaev*, Vol. 1. S. Monson, Jerusalem, pp. 14–419.

Douglas GB, Hart BT, Beckett R, Gray CM, Oliver RL (1999) Geochemistry of suspended particulate matter (SPM) in the Murray-Darling river system: a conceptual isotopic/geochemical model for the fractionation of major, trace and rare earth elements. *Aquatic Geochemistry* **5**, 167–194.

Drever JI (1994) The effect of land plants on weathering rates of silicate minerals. *Geochimica et Cosmochimica Acta* **58**, 2325–2332.

Drever JI, Stillings L (1997) The role of organic acids in mineral weathering. *Colloids and Surfaces* **120**, 167–181.

Drever JI, Vance GF (1994) Role of soil organic acids in mineral weathering processes. In: *Organic Acids in Geological Processes* (eds Pittman ED, Lewan MD). Springer-Verlag, New York.138–161

Dupouey JL, Dambrine E, Laffite JD, Moares C (2002) Irreversible impact of past land use on forest soils and biodiversity. *Ecology* **83**, 2978–2984.

Dupre B, Dessert C, Oliva P, *et al.* (2003) Rivers, chemical weathering and Earth climate. *C.R. Geoscience* **335**, 1141–1160.

Eick MJ, Grossl PR, Golden DC, Sparks DL, Ming DW (1996a) Dissolution kinetics of a lunar glass simulant at 25°C: the effect of pH and organic acids. *Geochimica et Cosmochimica Acta* **60**, 157–170.

Eick MJ, Grossl PR, Golden DC, Sparks DL, Ming DW (1996b) Dissolution of a lunar basalt simulant as affected by pH and organic anions. *Geoderma* **74**, 139–160.

Fletcher RC, Buss HL, Brantley SL (2006) A spheroidal weathering model coupling porewater chemistry to soil thicknesses during steady-state denudation. *Earth and Planetary Science Letters* **244**, 444–457.

Furrer G, Stumm W (1986) The coordination chemistry of weathering; I. Dissolution kinetics of δ-Al_2O_3 and BeO. *Geochimica et Cosmochimica Acta* **50**, 1847–1860.

Gaillardet J, Dupre B, Allegre CJ (1999a) Geochemistry of large river suspended sediments: silicate weathering or recycling tracer? *Geochimica Cosmochimica Acta* **63**, 4037–4051.

Gaillardet J, Dupre B, Louvat P, Allegre CJ (1999b) Global silicate weathering and CO_2 consumption rates deduced from the chemistry of large rivers. *Chemical Geology* **159**, 3–30.

Ganor J, Reznik IJ, Rosenberg YO (2009) Organics in water–rock interactions. In: *Thermodynamics and Kinetics of Water–Rock Interaction* (eds Oelkers EH, Schott J), Chapter 7. Mineralogical Society of America, Geochemical Society, Washington, DC.259–369

Godderis Y, Francois LM, Probst A, *et al.* (2006) Modelling weathering processes at the catchment scale. *Geochimica Cosmochimica Acta* **70**, 1128–1147.

Goyne KW, Brantley SL, Chorover J (2006) Effects of organic acids and dissolved oxygen on apatite and chalcopyrite dissolution: implications for using elements as organomarkers and oxymarkers. *Chemical Geology* **234**, 28–45.

Hausrath EM, Neaman A, Brantley SL (2009) Elemental release rates from dissolving basalt and granite with and without organic ligands. *American Journal of Science* **309**(8), 633–660.

Herndon E, Jin L, Brantley SL (2011). Soils reveal widespread manganese enrichment from industrial inputs. *Environmental Science & Technology* **45**(1), 241–247.

Hernlem BJ, Vane LM, Sayles GD (1996) Stability constants for complexes of the siderophore desferrioxamine B with selected heavy metal cations. *Inorganica Chimica Acta* **244**, 179–184.

Hersman LE (2000) The role of siderophores in iron oxide dissolution. In: *Environmental Microbe–Metal Interactions* (ed. Lovley D). ASM Press, Washington, DC.145–157

Hilley GE, Porder S (2008) A framework for predicting global silicate weathering and CO_2 drawdown rates over geologic time-scales. *Proceedings of the National Academy of Sciences of the USA* **105**, 16855–16859.

Hren MT, Hilley GE, Chamberlain CP (2007) The relationship between tectonic uplift and chemical weathering rates in the Washington Cascades: field measurements and model predictions. *American Journal of Science* **307**, 1041–1063.

Huggett RJ (1995) *Geoecology, An Evolutionary Approach.* Routledge, London.

Jackson RB, Moore LA, Hoffmann WA, Pockman WT, Linder CR (1999) Ecosystem rooting depth determined with caves and DNA. *Ecology* **96**, 11387–11392.

Jenny H (1980) *The Soil Resource: Origin and Behavior.* Springer-Verlag, New York.

Jobbagy EG, Jackson RB (2001) The distribution of soil nutrients and depth: Global patterns and the imprint of plants. *Biogeochemistry* **53**, 51–77.

Johnson NM, Driscoll CT, Eaton JS, Likens GE, McDowell WH (1981) 'Acid rain', dissolved aluminum and chemical weathering in the Hubbard Brook Experimental Forest, New Hampshire. *Geochimica Cosmochimica Acta* **45**, 1421–1437.

Jones DL (1998) Organic acids in the rhizosphere: A critical review. *Plant and Soil* **205**, 25–44.

Keller CK, White TM, O'Brien R, Smith JL (2006) Soil CO_2 dynamics and fluxes as affected by tree harvest in an experimental sand ecosystem. *Journal of Geophysical Research* **111**, doi:10.1029/2005JG000157.

Kennedy BM, Chadwick OA, Vitousek PM, Derry LA, Hendricks D (1998) Changing sources of base cations during ecosystem development, Hawaiian Islands. *Geology* **26**, 1015–1018.

Kump LR, Brantley SL, Arthur MA (2000) Chemical weathering, atmospheric CO_2 and climate. *Annual Review of Earth and Planetary Sciences* **28**, 611–667.

Kurtz AC, Derry LA, Chadwick OA (2001) Accretion of Asian dust to Hawaiian soils. *Geochimica Cosmochemica Acta* **65**, 1971–1983.

Landeweert R, Hoffland E, Finlay RD, Kuyper TW, van Breemen N (2001) Linking plants to rocks: ectomycorrhizal fungi mobilize nutrients from minerals. *Trends in Ecology and Evolution* **16**, 248–254.

Lebedeva MI, Fletcher RC, Balashov VN, Brantley SL (2007) A reactive diffusion model describing transformation of bedrock to saprolite. *Chemical Geology* **244**, 624–645.

Lebedeva MI, Fletcher RC, Brantley SL (2010) A mathematical model for steady-state regolith production at constant erosion rate. *Earth Surface Processes and Landforms*, doi: 10.1002/esp. 1954.**35**,508–524

Lichtner PC (1988) The quasi stationary state approximation to coupled mass transport and fluid-rock interaction in a porous medium. *Geochimica Cosmochimica Acta* **52**, 143–165.

Liermann LJ, Barnes AS, Kalinowski BE, Zhou X, Brantley SL (2000) Microenvironments of pH in biofilms grown on dissolving silicate surfaces. *Chemical Geology* **171**, 1–6.

Liermann LJ, Guynn RL, Anbar A, Brantley SL (2005) Production of a molybdophore during metal-targeted dissolution of silicates by soil bacteria. *Chemical Geology* **220**, 285–302.

Lucas Y, Luizao FJ, Chauvel A, Rouiller J, Nahon D (1993) The relation between biological activity of the rain forest and mineral composition of soils. *Science* **260**, 521–523.

MacFarlane AW, Danielson A, Holland HD (1994a) Geology and major and trace element chemistry of late Archean weathering profiles in the Fortescue Group, Western Australia: Implications for atmospheric P_{O2}. *Precambrian Research* **65**, 297–317.

MacFarlane AW, Danielson A, Holland HD, Jacobsen SB (1994b) REE chemistry and Sm-Nd systematics of the late Archaean weathering profiles in the Fortescue Group, western Australia. *Geochimica et Cosmochimica Acta* **58**, 1777–1794.

Millot R, Gaillardet J, Dupre B, Allegre CJ (2002) The global control of silicate weathering rates and the coupling with physical erosion: new insights from rivers of the Canadian Shield. *Earth and Planetary Science Letters* **196**, 83–93.

Minasny B, McBratney AB, Salvador-Blanes S (2008) Quantitative models for pedogenesis – a review. *Geoderma* **144**, 140–157.

Moore J, Macalady JL, Schulz MS, White AF, Brantley SL (2009) Evolution of microbial community structure across a marine terrace grassland chronosequence, Santa Cruz, California. *Soil Biology and Biochemistry* **42**, 21–31.

Morel FMM Hering JG (1993) *Principles and Applications of Aquatic Chemistry*. John Wiley & Sons, Inc., New York.

Moulton KL Berner RA (1998) Quantification of the effect of plants on weathering: studies in Iceland. *Geology* **26**, 895–898.

Moulton KK, West J, Berner RA (2000) Solute flux and mineral mass balance approaches to the quantification of plant effects on silicate weathering. *American Journal of Science* **300**, 539–570.

Neaman A, Chorover J, Brantley SL (2005a) Element mobility patterns record organic ligands in soils on early Earth. *Geology* **33**, 117–120.

Neaman A, Chorover J, Brantley SL (2005b) Implications of the evolution of organic acid moieties for basalt weathering over geological time. *American Journal of Science* **305**, 147–185.

Neaman A, Chorover J, Brantley SL (2006) Effects of organic ligands on granite dissolution in batch experiments at pH 6. *American Journal of Science* **306**, 451–473.

Oliva P, Viers J, Dupre B, *et al.* (1999) The effect of organic matter on chemical weathering: study of a small tropical watershed: Nsimi-Zoetele site, Cameroon. *Geochimica et Cosmochimica Acta* **63**, 4013–4035.

Pokrovsky OS, Schott J, Dupre B (2006) Trace element fractionation and transport in boreal rivers and soil porewaters of permafrost-dominated basaltic terrain in Central Siberia. *Geochimica et Cosmochimica Acta* **70**, 3239–3260.

Porder S, Paytan A, Vitousek PM (2005) Erosion and landscape development affect plant nutrient status in the Hawaiian islands. *Oecologia* **142**, 440–449.

Porder S, Vitousek PM, Chadwick OA, Chamberlain CP, Hilley GE (2007) Uplift, erosion, and phosphorus limitation in terrestrial ecosystems. *Ecosystems* **10**, 158–170.

Powell PE, Cline GE, Reid CPP, Szaniszlo PJ (1980) Occurrence of hydroxamate siderophore iron chelators in soils. *Nature* **287**, 833–834.

Pye K (1987) *Aeolian Dust and Dust Deposits*. Academic Press, London.

Raven JA (1995) The early evolution of land plants: aquatic ancestors and atmospheric interactions. *Botanical Journal of Scotland* **47**, 151–175.

Raymo ME Ruddiman WF (1992) Tectonic forcing of late Cenozoic climate. *Nature* **359**, 117–122.

Raymo ME, Ruddiman WF, Froelich PN (1988) Influence of late Cenozoic mountain building on ocean geochemical cycles. *Geology* **16**, 649–653.

Raymond PA Cole JJ (2003) Increase in the export of alkalinity from North America's largest river. *Science* **301**, 88–91.

Riebe CS, Kirchner JW, Granger DE, Finkel RC (2001) Strong tectonic and weak climatic control of long-term chemical weathering rates. *Geology* **29**, 511–514.

Riebe CS, Kirchner JW, Finkel RC (2004) Erosional and climatic effects on long-term chemical weathering rates in granitic landscapes spanning diverse climate regimes. *Earth and Planetary Science Letters* **224**, 547–562.

Roering JJ, Almond P, Tonkin P, McKean J (2002) Soil transport driven by biological processes over millennial time scales. *Geology* **30**, 1115–1118.

Rousseau D, Dupre B, Gaillardet J, Allegre CJ (1996) Major and trace elements of river-borne material: the Congo Basin. *Geochimica et Cosmochimica Acta* **60**, 1301–1321.

Royer DL, Berner RA, Montanez IP, Tabor NJ, Beerling DJ, (2004) CO_2 as a primary driver of Phanerozoic climate. *GSA Today* **14**, doi: 10.1130/1052–5173(2004)014<4:CAAPDO> 2.0.CO;2.4–9

Schwartzman DW (2008) Coevolution of the Biosphere and Climate. In: *Global Ecology* (eds Jorgensen SE, Fath BD). Elsevier, Oxford.648–658

Schwartzman DW Volk T (1989) Biotic enhancement of weathering and the habitability of Earth. *Nature* **340**, 457–460.

Schwartzman DW Volk T (1991) Biotic enhancement of weathering and surface temperatures on Earth since the origin of life. *Global and Planetary Change* **90**, 357–371.

Shotyk W, Nesbitt HW, Fyfe WS (1990) The behavior of major and trace elements in complete vertical peat profiles from three *Sphagnum* bogs. *International Journal of Coal Geology* **15**, 163–190.

Stallard RF (1992) Tectonic processes, continental freeboard, and the rate-controlling step for continental denudation. In: *Global Biogeochemical Cycles* (Butcher SS, Charlson RJ, Orians GH, Wolfe GV). Academic Press, London. 93–121

Stallard R (1995) Relating chemical and physical erosion. In: *Chemical Weathering Rates of Silicate Minerals* (eds White AF, Brantley SL). Mineralogical Society of America, Washington, DC. 534–564

Stallard RF, Edmond JM (1983) Geochemistry of the Amazon 2. The influence of geology and weathering environment on the dissolved load. *Journal of Geophysical Research* **88**, 9671–9688.

Steefel C (2008) Geochemical kinetics and transport. In: *Kinetics of Water–Rock Interaction* (Brantley SL, Kubicki JD, White AF). Springer, New York. 554–590

Steinmann P, Shotyk W (1997a) Chemical composition, pH and redox state of sulfur and iron in complete vertical porewater profiles from two sphagnum peat bogs, Jura Mountains, Switzerland. *Geochimica et Cosmochimica Acta* **61**, 1143–1163.

Steinmann P, Shotyk W (1997b) Geochemistry, mineralogy, and geochemical mass balance on major elements in two peat bog profiles (Jura Mountains, Switzerland). *Chemical Geology* **138**, 25–53.

Sterflinger K (2000) Fungi as geological agents. *Geomicrobiology Journal* **17**, 97–124.

Stevenson FJ (1967) Organic acid in soil. In: *Soil Biochemistry* (eds McLaren AD, Peterson GH). John Wiley & Sons, Inc., New York. 119–146

Stevenson FJ (1991) Organic matter-micronutrient reactions in soil. In: *Micronutrients in Agriculture* (ed. Mortvedt JJ). Soil Science Society of America, Madison, WI.

Stillings LL, Drever JI, Brantley S, Sun Y, Oxburgh R (1996) Rates of feldspar dissolution at pH 3–7 with 0–8 M oxalic acid. *Chemical Geology* **132**, 79–89.

Stillings L, Drever JI, Poulson SR (1998) Oxalate adsorption at a plagioclase (An(47)) surface and models for ligand-promoted dissolution. *Environmental Science & Technology* **32**, 2856–2864.

Stone AT (1997) Reactions of extracellular organic ligands with dissolved metal ions and mineral surfaces. In: *Geomicrobiology: Interactions Between Microbes and Minerals* (eds Banfield JF, Nealson KH). Mineralogical Society of America, Washington, DC. 309–344

Stonestrom DA, White AF, Akstin KC (1998) Determining rates of chemical weathering in soils-solute transport versus profile evolution. *Journal of Hydrology* **209**, 331–345.

Sundquist ET (1991) Steady- and non-steady-state carbonate-silicate controls on atmospheric CO_2. *Quaternary Science Reviews* **10**, 283–296.

Taunton AE, Welch SA, Banfield JF (2000) Microbial controls on phosphate and lanthanide distributions during granite weathering and soil formation. *Chemical Geology* **169**, 371–382.

Tosiani T, Loubet M, Viers J, *et al.* (2004) Major and trace elements in river-borne materials from the Cuyuni basin (southern Venezuela): evidence for organo-colloidal control on the dissolved load and element redistribution between the suspended and dissolved load. *Chemical Geology* **211**, 305–334.

van Hees PAW, Lundstrom US, Morth C-M (2002) Dissolution of microcline and labradorite in a forest O horizon extract: the effect of naturally occurring organic acids. *Chemical Geology* **189**, 199–211.

Van Scholl L, Kuyper TW, Smits MM, Landeweert R, Hoffland E, Van Breemen N (2008) Rock-eating mycorrhizas: their role in plant nutrition and biogeochemical cycles. *Plant Soil* **35**, 35–47.

Vanacker V, Von Blanckenburg F, Govers G, Molina A, Poesen J, Deckers J (2007) Restoring dense vegetation can slow mountain erosion to near natural benchmark levels. *Geology* **35**, 303–306.

Vaughan D, Lumsdon DG, Linehan DJ (1993) Influence of dissolved organic matter on the bioavailability and toxicity of metals in soils and aquatic systems. *Chemistry and Ecology* **8**, 185–201.

Viers J, Dupre B, Polve M, Schott J, Dandurand J-L, Braun J-J (1997) Chemical weathering in the drainage basin of a tropical watershed (Nsimi-Zoetele site, Cameroon): comparison between organic-poor and organic-rich waters. *Chemical Geology* **140**, 181–206.

Viers J, Dupre B, Braun J-J, *et al.* (2000). Major and trace element abundances, and strontium isotopes in the Nyong basin rivers (Cameroon): constraints on chemical weathering processes and elements transport mechanisms in humid tropical environments. *Chemical Geology* **169**, 211–241.

Waldbauer JR, Chamberlain CP (2005) Influence of uplift, weathering, and base cation supply on past and future CO_2 levels. In: *A History of Atmospheric CO_2 and its Effects on Plants, Animals, and Ecosystems* (eds Ehleringer JR, Cerling TE, Dearing MD). Springer Verlag, Berlin. 166–184

Walker JCG, Hays PB, Kasting JF (1981) A negative feedback mechanism for the long-term stabilization of Earth's surface temperature. *Journal of Geophysical Research* **86**, 9776–9782.

Wardle DA, Walker LR, Bardgett RD (2004) Ecosystem properties and forest decline in contrasting long-term chronosequences. *Science* **305**, 509–513.

Welch SA, Ullman WJ (1993) The effect of organic acids on plagioclase dissolution rates and stoichiometry. *Geochimica et Cosmochimica Acta* **57**, 2725–2736.

Welch SA, Ullman WJ (1996) Feldspar dissolution in acidic and organic solutions: compositional and pH dependence of dissolution rate. *Geochimica Cosmochimica Acta* **60**, 2939–2948.

Wells ML, Kozelka PB, Bruland KW (1998) The complexation of 'dissolved' Cu, Zn, Cd and Pb by soluble and colloidal organic matter in Naragansett Bay, RI. *Marine Chemistry* **62**, 203–217.

West AJ, Galy A, Bickle M (2005) Tectonic and climatic controls on silicate weathering. *Earth and Planetary Science Letters* **235**, 211–228.

White AF (1995) Chemical weathering rates of silicate minerals in soils. In: *Chemical Weathering Rates of Silicate Minerals* (eds White AF, Brantley SL). Mineralogical Society of America, Washington, DC. 407–461

White AF (2002) Determining mineral weathering rates based on solid and solute weathering gradients and velocities: application to biotite weathering in saprolites. *Chemical Geology* **190**, 69–89.

White AF (2008) Quantitative approaches to characterizing natural chemical weathering rates. In: *Kinetics of Water–Rock Interaction* (eds Brantley SL, Kubicki JD, White AF). Springer, New York. 469–544

White AF, Brantley SL (1995) *Chemical Weathering Rates of Silicate Minerals*. Mineralogical Society of America, Washington, DC.

White AF, Brantley SL (2003) The effect of time on the weathering of silicate minerals: why do weathering rates differ in the laboratory and field? *Chemical Geology* **202**, 479–506.

Wilkinson BH (2005) Humans as geologic agents: A deeptime perspective. *Geology* **33**, 161–164.

Wilkinson BH, McElroy BJ (2007) The impact of humans on continental erosion and sedimentation. *Geological Society of American Bulletin* **119**, 140–156.

Wischmeier WH, Smith DD (1978) *Predicting Rainfall Erosion Losses–A Guide to Conservation Planning*. Agriculture Handbook no. 537. U.S. Department of Agriculture, Washington, DC.

Wogelius RA, Walther JV (1991) Olivine dissolution at 25°C: effects of pH, CO_2, and organic acids. *Geochimica et Cosmochimica Acta* **55**, 943–954.

Wood SA (2000) Organic matter: supergene enrichment and dispersion. In: *Ore Genesis and Exploration: The Roles of Organic Matter* (eds Giordano TH, Kettler RM, Wood SA). Society of Economic Geology, Littleton, CO.157–192

Yanai RD (1992) Phosphorus budget of a 70-year old northern hardwood forest. *Biochemistry* **17**, 1–22.

Yang WB, Holland HD (2003) The Hekpoort paleosol profile in Strata 1 at Gaborone, Botswana: soil formation during the great oxidation even. *American Journal of Science* **303**, 187–220.

Zinder B, Furrer G, Stumm W (1986) The coordination chemistry of weathering: II. Dissolution of Fe(III) oxides. *Geochimica et Cosmochimica Acta* **50**, 1861–1869.

13

MOLECULAR BIOLOGY'S CONTRIBUTIONS TO GEOBIOLOGY

Dianne K. Newman[1], Victoria J. Orphan[2] and Anna-Louise Reysenbach[3]

[1]Howard Hughes Medical Institute, California Institute of Technology, 1200 E. California Blvd.,
Pasadena, CA 91125, USA
[2]California Institute of Technology, Pasadena, CA 91125 USA
[3]Portland State University, Portland, OR 97207, USA

13.1 Introduction

On August 7, 1996, US President Bill Clinton held a press conference to announce the possibility that the Allan Hills 84001 meteorite might provide insight into ancient life on Mars. With soaring rhetoric, he declared:

'Today, rock 84001 speaks to us across all those billions of years and millions of miles. It speaks of the possibility of life. If this discovery is confirmed, it will surely be one of the most stunning insights into our universe that science has ever uncovered. Its implications are as far-reaching and awe-inspiring as can be imagined. Even as it promises answers to some of our oldest questions, it poses still others even more fundamental.'

Shortly thereafter, NASA expanded its support for astro-and geobiological research, which marked the beginning of a renaissance in geobiology. Seemingly overnight, geobiology was transformed from a somewhat arcane discipline to a glamorous field that promised to reveal the secrets of life. While today, most geobiologists would agree that the evidence for past life in AH84001 is inconclusive at best, and find the hype surrounding its discovery to be comical, nonetheless, the excitement it engendered has had a long-lasting and positive impact on our science. The enduring consequence of Clinton's press conference was that it called attention to the fact that life has been leaving signatures in its environment (be it earthly or extraterrestrial) for billions of years. In the years following the meteorite's discovery, it has become clear that to understand life's traces and—more importantly—effects on its environment, it is necessary to understand *how* life leaves its imprint and whether this can be distinguished from similar imprints left by abiotic processes. This is a central challenge in geobiology.

Whether we speak about traces of Martian life in ancient rocks or the roles of microorganisms in (paleo) geochemical cycles, inexorably we confront the necessity of understanding biological mechanisms in order to draw conclusions about how life has shaped its environment. Molecules more often than shapes are what remain in ancient rocks that provide us with a glimpse into ancient ecosystems. These molecules can be organic or inorganic, but if the latter, their utility as biosignatures rests upon an appreciation of how living organisms transform them in recognizable ways (see chapter 14 in this volume). In modern environments, many geochemical cycles are shaped by enzymatic activities of living cells, and an understanding of what these enzymes are, what controls their expression, how they function and turn over, and how many organisms possess them is essential if we seek to identify and quantify pathways where life makes a significant impact on its environment. In addition, an understanding of these enzymes (both sequence-based and structural) can provide us with information about their history and inform our appreciation of metabolic evolution. Not surprisingly, molecular biology provides a variety of methods that can be informative in these contexts.

In this chapter, we will overview the development and application of molecular biological techniques to geobiological problems. In the past several decades, the application of these techniques in microbial ecological studies has transformed our understanding of the diversity of natural microbial communities and their role in

Fundamentals of Geobiology, First Edition. Edited by Andrew H. Knoll, Donald E. Canfield and Kurt O. Konhauser.
© 2012 Blackwell Publishing Ltd. Published 2012 by Blackwell Publishing Ltd.

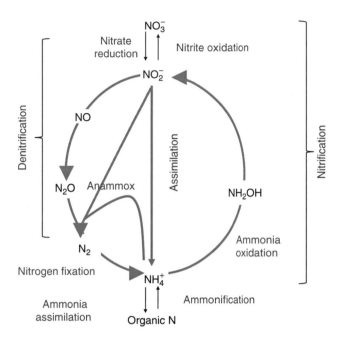

Plate 1 Diagram of the biological nitrogen cycle showing the main inorganic forms in which nitrogen occurs in natural and anthropogenically influenced environments.

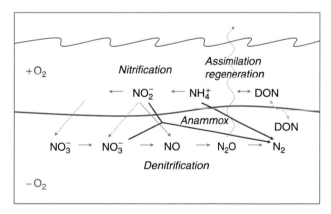

Plate 2 Transformations in the nitrogen cycle emphasizing the linkages between processes that occur in aquatic environments. Denitrification and anammox both occur in anoxic conditions, but are linked ultimately to nitrification (favoured under oxic conditions) for supply of substrates.

Ammonium, the initial substrate for nitrification, is regenerated by mineralization of organic matter, but is also assimilated as the preferred N source by many organisms. Solid arrows signify microbial transformations while dotted arrows imply diffusion.

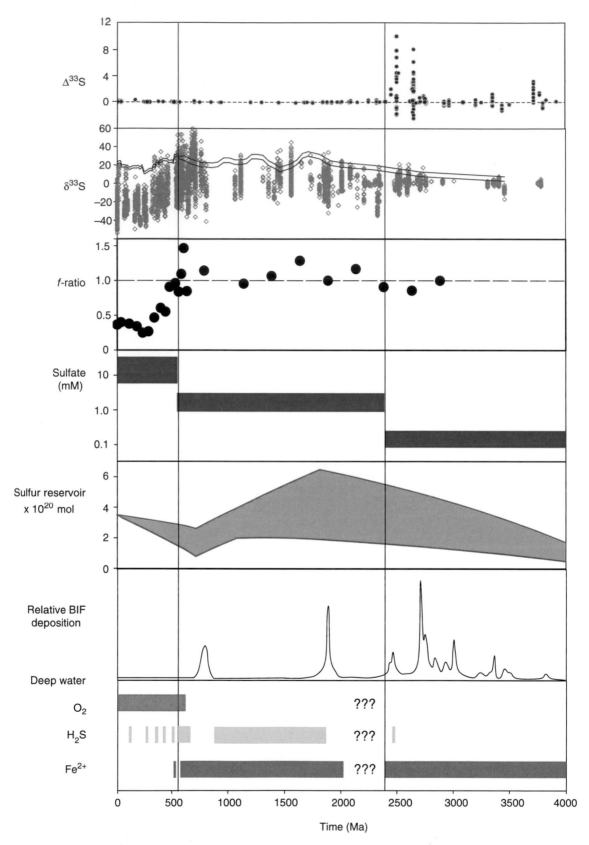

Plate 3 A compilation of data relevant to the evolution of the sulfur cycle through time, as well as the evolution of marine chemistry. Vertical lines represent the Cambrian-Ediacaran boundary at 542 Ma and the 'Great Oxidation Event' at 2400 Ma. The term 'deep water' chemistry refers to chemistry below the mixed upper layer and into the deep sea including oxygen-minimum zone settings. Question marks indicate suspected chemistry, but lack of data either direct or indirect. Two or more deep water chemistries at the same time could indicate either spatial or vertical structure in the chemistry of the water column. See text for further details.

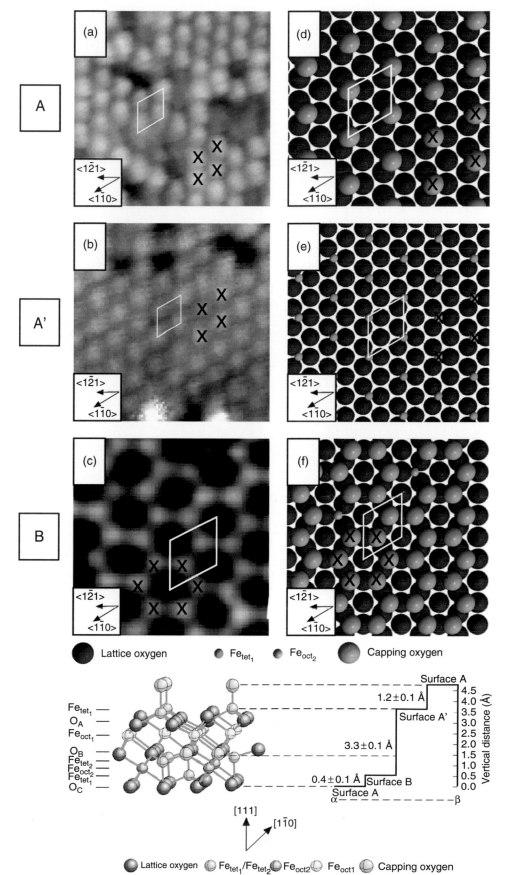

Plate 7 Scanning tunneling microscope (STM) images of the (111) surface of magnetite obtained under UHV conditions: (a) $10 \times 10\,\text{nm}^2$ ($100 \times 100\,\text{Å}^2$) image of the A surface; (b) $12 \times 12\,\text{nm}^2$ ($120 \times 120\,\text{Å}^2$) image of the A' surface; (c) $1.6 \times 1.6\,\text{nm}^2$ ($16 \times 16\,\text{Å}^2$) image of the B surface, with corresponding models of the surfaces (d,e and f, respectively) which are interpreted as arising from different level slices through the bulk structure of magnetite, as illustrated on the right hand side of the figure (redrawn after Cutting *et al.*, 2006).

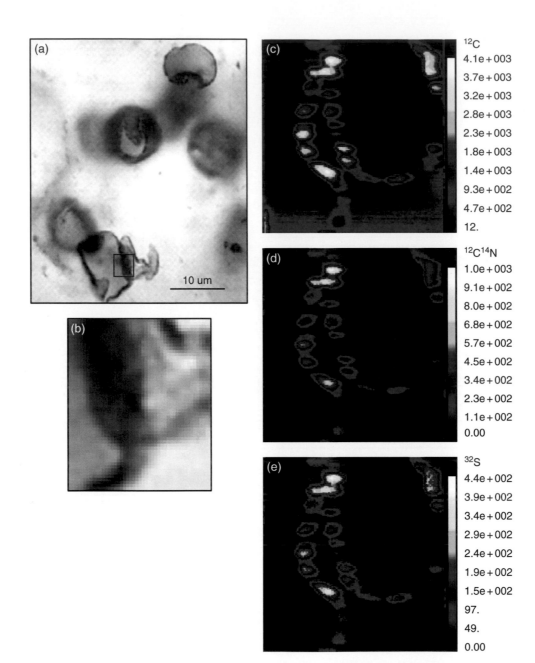

Plate 8 NanoSIMS of Late Precambrian microfossils from the Bitter Springs Formation of Australia: (a, b) are transmitted light optical photomicrographs of a polished thin section of chert; (b) is the area within the rectangle marked on (a) where two cells are in contact; (c, d, e) are NanoSIMS elements maps of ^{12}C, $^{12}C^{14}N$, and ^{32}S, respectively, of the cell contact imaged in (b). The maps show a one-to-one correspondence between these isotopes. This correspondence, along with the globular aligned character, are strong indicators of biogenicity (after Oehler *et al.*, 2006, 2009).

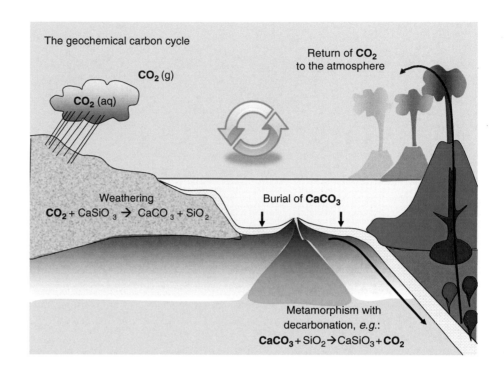

Plate 11 Schematic representation of the long-term carbon cycle.

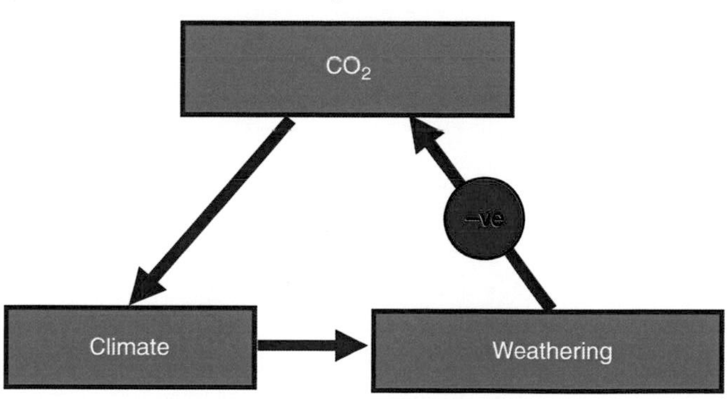

Earth's thermodynamic thermostat

CO₂

Climate

Weathering

–ve

Plate 12 Simplified systems diagram for the thermostatic control of Earth's climate via the geochemical carbon cycle. Plain arrows are positive feedbacks, those with a –ve symbol are negative feedbacks. The loop between the CO_2, climate and weathering boxes represents a negative feedback, i.e. only one –ve label. See text for further details.

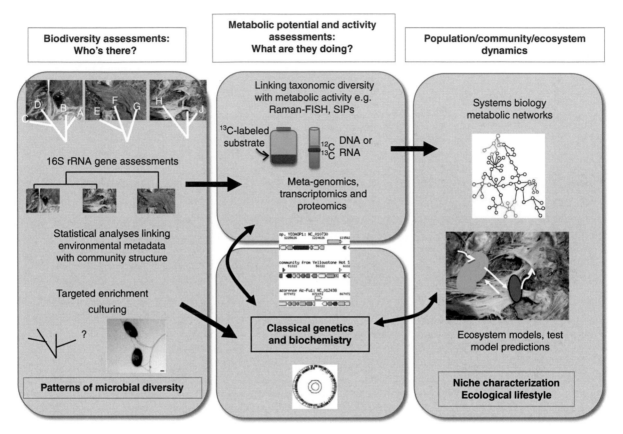

Biodiversity assessments: Who's there?

Metabolic potential and activity assessments: What are they doing?

Population/community/ecosystem dynamics

16S rRNA gene assessments

Statistical analyses linking environmental metadata with community structure

Targeted enrichment culturing

Patterns of microbial diversity

Linking taxonomic diversity with metabolic activity e.g. Raman-FISH, SIPs

^{13}C-labeled substrate

^{12}C
^{13}C

DNA or RNA

Meta-genomics, transcriptomics and proteomics

Classical genetics and biochemistry

Systems biology metabolic networks

Ecosystem models, test model predictions

Niche characterization Ecological lifestyle

Plate 13 Overview of molecular approaches used in geobiology. Rapid phylogenetic assessments can be accomplished through high throughput sequencing, providing sufficient sequence depth and breadth for testing the statistical significance of observed patterns. Additionally, linking these patterns to environmental metadata assists in targeting enrichment culturing of organisms of interest. In order to obtain insights in the functional roles populations have in a community, numerous single gene, in situ approaches or metagenomic approaches can be used. Complementary classical genetic and biochemical approaches enable specific hypotheses about function to be tested.

Genomes of environmentally important isolates can serve as reference genomes for metagenomics. The accumulated environmental, activity and sequence data can be incorporated into dynamic models that explore the interactions within the communities and their biogeochemical outputs under specific environmental conditions. These in turn provide new hypotheses that can be tested both in the environment and in the laboratory. Arrows link complementary methods that can be applied together to enhance the ability to discover and characterize novel microbial symbioses within complex microbial communities in the environment.

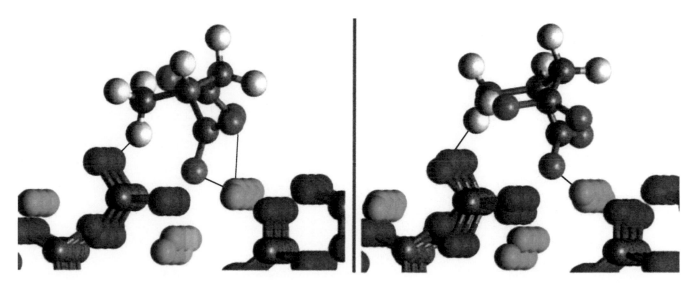

Plate 18 The most stable configurations for L- and D-aspartate on the calcite (21–34) surface (left and right, respectively). The D enantiomer, which requires significantly less calcite surface relaxation and aspartate distortion, is favoured by 8 kcal mol^{-1} – the largest known enantiospecific effect (after Hazen 2006).

Plate 19 The polymerase chain reaction (PCR) copies a sequence of DNA. (a) A strand of DNA is mixed in solution with DNA nucleotides (precursors), a primer that targets a specific piece of DNA, and an enzyme (polymerase) that helps to assemble DNA. The mix is heated to about 90 °C to separate DNA strands. (b) When cooled to about 60 °C, primers attach to the DNA strands. (c) At 70 °C, nucleotides begin to attach to the DNA strands. (d) At the end you have two copies of the desired DNA (from Trefil and Hazen 2009).

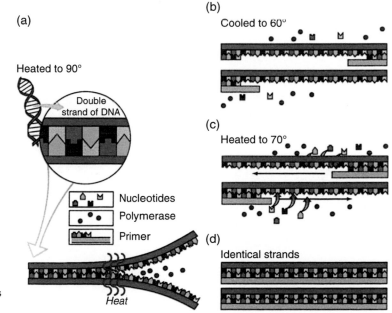

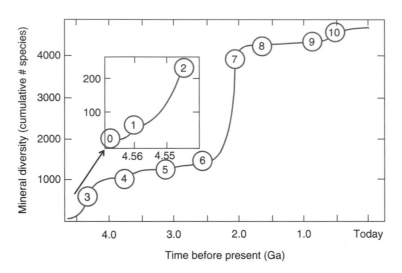

Plate 20 Estimated cumulative number of different mineral species versus time, with key events in Earth history. Numbers correspond to stages, as outlined in Table 18.1.

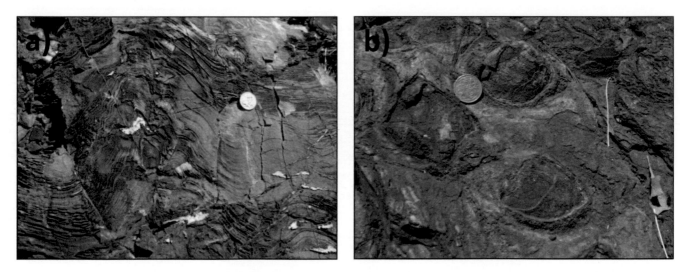

Plate 21 (a) The oldest known stromatolitic dolomite from the Trendall locality of the ~3.35 Ga Strelley Pool Formation, Western Australia. (b) Three-dimensional view of stomatolitic chert-barite from the 3.49 Ga Dresser Formation, Western Australia (photos by D. Papineau).

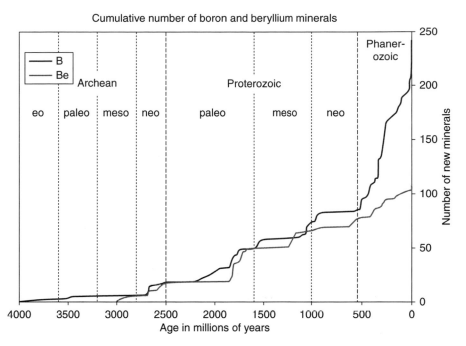

Cumulative number of boron and beryllium minerals

Plate 24 Plot of the reported oldest occurrences of 106 Be minerals and 261 B minerals based on literature search (Grew and Hazen, 2009, 2010a, 2010b). The plot is cumulative because each reported new appearance is added to the number of minerals having been reported prior the age of the appearance. The plot is not meant to indicate the totality of minerals forming in the Earth's near surface at any given time, including the present; i.e. some minerals formed once or over a limited time interval, and have not formed since. Note the approximate doubling of Be and B mineral species during the time interval from 2.0 to 1.7 Ga. (courtesy of Edward Grew).

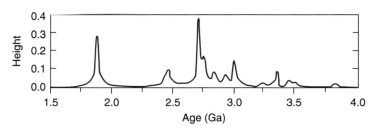

Plate 25 Temporal distribution of iron formations. 'Height' is an approximation of the relative abundance of IFs, taking into account the frequency of occurrence and uncertainties in the age estimates. Ages are reported in billions of years before present (Ga). After Isley and Abbott (1999).

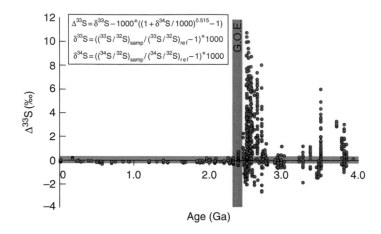

Plate 26 Compilation of Δ^{33}S data from many sources (available on request). Clearly delineated is the disappearance of non-mass-dependent (NMD) fractionation at the Great Oxidation Event (GOE), which fingerprints the first persistent and appreciable accumulation of oxygen in the atmosphere.

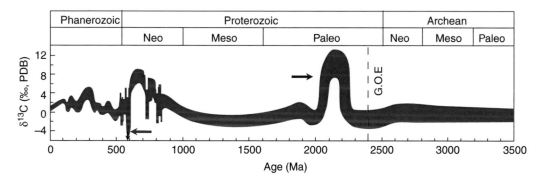

Plate 27 δ^{13}C of marine carbonates through time as a proxy for the isotopic composition of dissolved inorganic carbon in the ocean. After Karhu (1999). Recent work on the late Neoproterozoic suggests an extended period of strongly negative δ^{13}C values (even lower than those shown) – comprising the 'Shuram–Wonoka'. anomaly (left arrow).

Also apparent is the long-lived positive Paleoproterozoic anomaly of the Lomagundi Event (right arrow) and the extended period of mid-Proterozoic δ^{13}C stability. The approximate position of the Great Oxidation Event (GOE) is indicated. Ages are reported in millions of years before present (Ma).

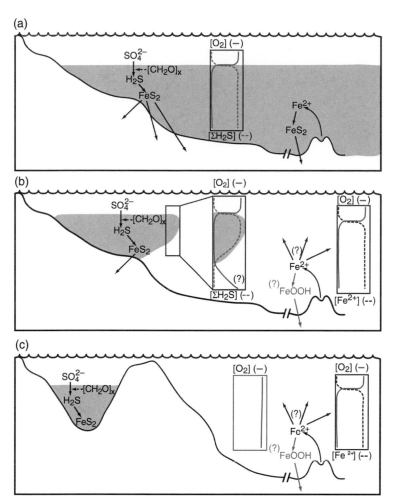

Plate 32 Conceptual models for the redox structure of Proterozoic ocean ranging from (a) pervasively and persistent 'whole-ocean' deep euxinia to (b) recent arguments for more localized and perhaps transient euxinia, possibly as mid-water 'wedges' (similar to modern oxygen minimum zones) that rimmed the global ocean as controlled, for example, by upwelling regions of high biological productivity. In the latter case, the sulfidic waters may have given way to dominantly ferruginous deeper waters throughout the Proterozoic. Alternatively, euxinia may have been deep but limited to restricted, marginal marine basins analogous to the modern Black Sea (c). In each case, surface waters were likely to have been well oxygenated. We can predict that all three possibilities existed at different times and likely at the same time in different places of the ocean. Modified from Lyons *et al.* (2009b). Additional details are available in the text and Lyons *et al.* (2009a, b).

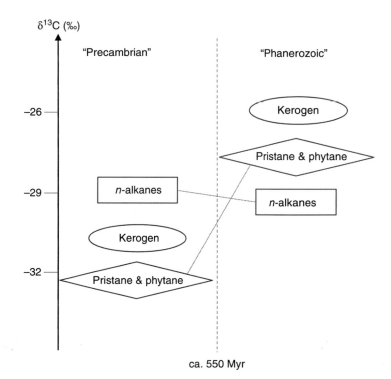

Plate 33 A fundamental switch in stable carbon isotopic patterns ($\delta^{13}C$) is found for hydrocarbon constituents of extractable rock bitumen (free *n*-alkanes and the acyclic isoprenoids, pristine and phytanc) vcrsus kerogen (from Logan *et al.*, 1995) for ancient sedimentary rocks. The switch in isotopic ordering appears to occur around 550 million years ago (Kelly *et al.*, 2008). Absolute $\delta^{13}C$ (‰) values vary from sample to sample.

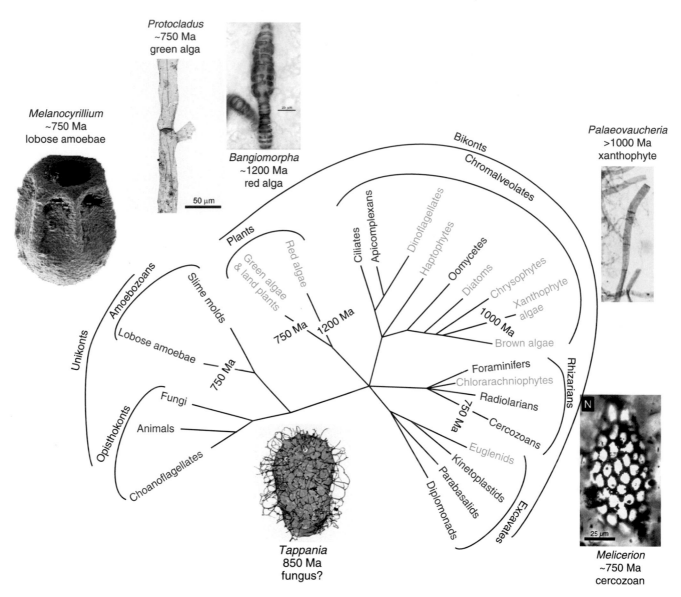

Melanocyrillium
~750 Ma
lobose amoebae

Protocladus
~750 Ma
green alga

Bangiomorpha
~1200 Ma
red alga

50 μm

Palaeovaucheria
>1000 Ma
xanthophyte

Bikonts

Chromalveolates

Plants

Amoebozoans

Slime molds

Red algae

Green algae
& land plants

Ciliates

Apicomplexans

Dinoflagellates

Haptophytes

Oomycetes

Diatoms

Chrysophytes

Xanthophyte
algae

Lobose amoebae

750 Ma

1200 Ma

1000 Ma

Brown algae

Unikonts

Fungi

Foraminifers

Chlorarachniophytes

Radiolarians

Rhizarians

N

750 Ma

Animals

750 Ma

Cercozoans

Opisthokonts

Euglenids

Choanoflagellates

Kinetoplastids

Parabasalids

Diplomonads

Excavates

Tappania
850 Ma
fungus?

Melicerion
~750 Ma
cercozoan

25 μm

Plate 34 Eukaryote phylogeny based on molecular and cytological data. The following lineages are photosynthetic: green algae and land plants, red algae, dinoflagellates, haptophytes, diatoms, chrysophytes, xanthophyte algae, brown algae, chlorarachniophytes, and euglenids (some ciliates and apicomplexans may have been photosynthetic but subsequently lost their photosynthetic capabilities). Six mid-Proterozoic fossils are used to date the divergence time of major eukaryote lineages. Modified from Baldauf (2000) and Porter (2006). Fossil images courtesy of N.J. Butterfield, A.H. Knoll, and S.M. Porter.

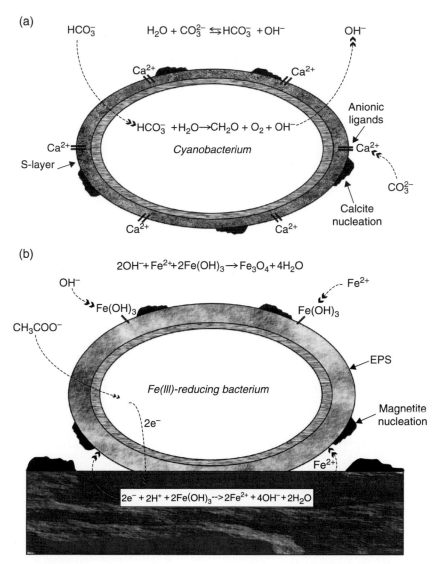

(a)

$$HCO_3^-$$

$$H_2O + CO_3^{2-} \leftrightarrows HCO_3^- + OH^-$$

$$OH^-$$

$$Ca^{2+} \qquad Ca^{2+}$$

$$HCO_3^- + H_2O \rightarrow CH_2O + O_2 + OH^-$$

Cyanobacterium

Anionic
ligands

$$Ca^{2+}$$

$$Ca^{2+}$$

S-layer

$$Ca^{2+}$$

$$CO_3^{2-}$$

Calcite
nucleation

$$Ca^{2+} \qquad Ca^{2+}$$

(b)

$$2OH^- + Fe^{2+} + 2Fe(OH)_3 \rightarrow Fe_3O_4 + 4H_2O$$

$$OH^-$$

$$Fe(OH)_3$$

$$Fe^{2+}$$

$$Fe(OH)_3$$

$$CH_3COO^-$$

EPS

Fe(III)-reducing bacterium

$$2e^-$$

Magnetite
nucleation

$$Fe^{2+}$$

$$2e^- + 2H^+ + 2Fe(OH)_3 \dashrightarrow 2Fe^{2+} + 4OH^- + 2H_2O$$

Plate 4 Schematic of metabolically-induced biomineralization in (a) cyanobacteria and (b) Fe(III)-reducing heterotrophs. In the cyanobacteria, uptake of the bicarbonate anion leads to excretion of OH^-, which in turn changes the alkalinity and inorganic carbon speciation proximal to the cell surface. The generation of carbonate anions and the pre-adsorption of calcium cations to the cell's sheath can then induce calcification. In the Fe(III)-reducing bacterium, release of Fe(II) from the ferric hydroxide substratum can promote magnetite formation both on the pre-existing mineral surfaces, but also on the cell surface if deprotonated ligands had adsorbed ferric iron. In both examples, the secondary minerals form as by-products of microbial metabolism – the cell itself did not control the mineralization process.

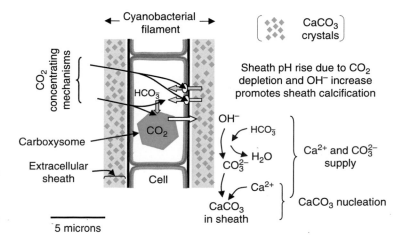

Plate 5 Schematic of *in vivo* cyanobacterial sheath calcification driven by CCM-enhanced photosynthesis (from Riding 2006, based on information from Miller and Colman, 1980; Thompson and Ferris, 1990; Merz, 1992; Price *et al.*, 1998; Kaplan and Reinhold, 1999; Badger and Price, 2003). Cyanobacterial CCMs include active uptake of HCO_3^- and its conversion to CO_2 for fixation by Rubisco in the carboxysome. This releases OH^-, which elevates the pH of fluids within the sheath pH, and ultimately promotes calcification.

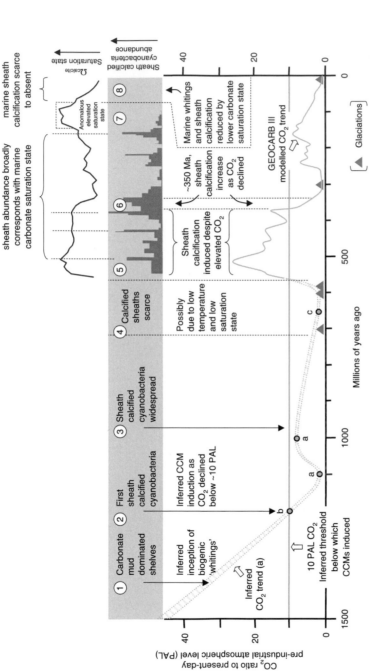

Plate 6 Conjectural history of cyanobacterial marine sheath calcification and picoplanktic 'whiting' precipitation. The Proterozoic inferred CO_2 trend is based on (a) Sheldon (2006), (b) Kah and Riding (2007), (c) Hyde *et al.* (2000) and (d) Ridgwell *et al.* (2003); the continuous trend line is from Berner & Kothavala (2001, fig. 13); the Neoproterozoic glaciations is from Walter *et al.* (2000); the occurrences of marine sheath calcified cyanobacteria are from Riding and Liang (2005b, fig. 5); and the threshold below which CCMs are induced is based on Badger *et al.* (2002). Several key developments can be inferred from the figure. (1) Photosynthetic 'whitings', as reflected by widespread carbonate mud sedimentation, may have been triggered as CO_2 reduced pH buffering (see Arp *et al.*, 2001; Riding, 2006). (2) A further decline below ~10 PAL CO_2 induced CCM development and sheath calcification at ~1200 Ma (Kah and Riding, 2007). (3) Calcified sheaths were widespread in the early Neoproterozoic (see references in Knoll and Semikhatov 1998), but (4) became scarce during 'Snowball' glaciations, possibly due to reduction in CCM development as low temperatures favoured diffusive entry of CO_2 into cells, and due to lower seawater saturation state reflecting reduction in both temperature and p_{CO_2}. (5) Sheath calcification was common in marine environments during the early-mid Palaeozoic despite elevated CO_2, suggesting that once CCMs had developed they were readily induced where carbon limitation developed, such as microbial mats. Throughout much of the Palaeozoic and early Mesozoic, calcified sheath abundance appears to vary with carbonate saturation state. (6) As CO_2 declined in the Late Devonian-Early Mississippian, calcified sheath abundance temporarily increased, possibly reflecting enhanced CCM induction, but then declined as the saturation state dropped in the Mississippian-Pennsylvanian (Riding, 2009). (7) Despite a high calculated saturation state in the Late Cretaceous-Palaeogene, planktic calcifiers probably reduced the actual saturation state sufficiently to inhibit cyanobacterial calcification. Thus, calcified sheaths were scarce in marine environments. (8) Since the Eocene, low carbonate saturation – due to low levels of both Ca ions and pCO_2^- is reflected in extreme scarcity in sheath calcification in marine environments.

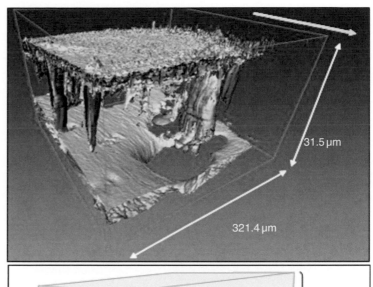

Plate 9 Confocal scanning laser microscope (CSLM) image of a biofilm grown between two quartz glass plates by introducing a nutrient solution and inoculating with *Pseudomonas aeruginosa*. Below the image is a schematic diagram showing the various components of the biofilm (after Brydie *et al.*, 2004, 2009).

31.5 μm

321.4 μm

Bulk solution

Carapace

Carapace support structure

Basal biofilm

Mineral surface

25.3 μm

321.4 μm

Uncolonised mineral surface

Intra-biofilm fluid flow

Primary colonising bacteria

Intra-biofilm bacterial colonies

Carapace bacterial colonies

Carapace polysaccharides/EPS

Basal polysaccharides/EPS

Fluid flow

(a)

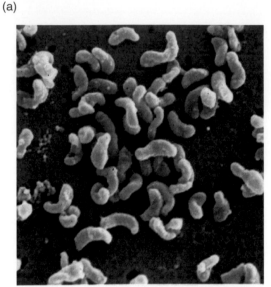

000004 10KV X15.0K 2.00um

(b)

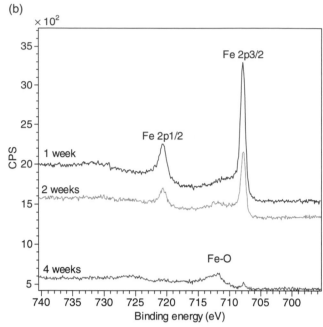

(c)

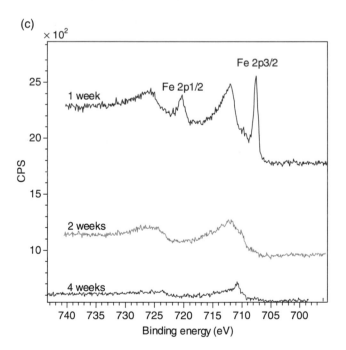

(d)

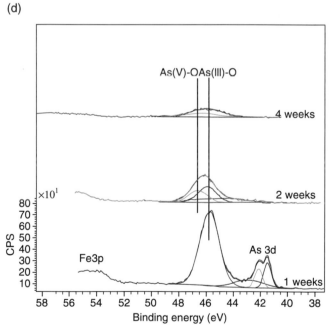

Plate 10 Oxidative breakdown of arsenopyrite (FeAsS) involving the organism *Leptospirillum ferrooxidans*: (a) ESEM image of *Leptospirillum*; (b) Fe2p XPS data for the arsenopyrite surface abiotically oxidized for periods of 1, 2 and 4 weeks; (c) the same experiment in the presence of *Leptospirillum*; (d) As 3D XPS data for the arsenopyrite surface oxidized in the presence of *Leptospirillum* for 1, 2 and 4 weeks. The Fe2p spectra show development of Fe^{3+} and Fe oxide peaks over time with much more rapid oxidation in the biotic system. The As3d spectra show evidence for the later onset of As^{3+} to As^{5+} oxidation (after Corkhill *et al.*, 2008).

(a)

(b)

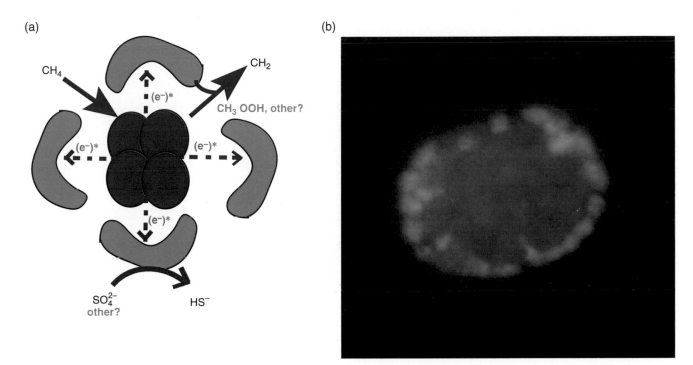

Plate 14 Model and data illustrating the AOM consortia. (a) Cartoon of the AOM consortia and hypothesized metabolic interactions, where red coloured cells are the methane-oxidizing archaea 'ANME' and green cells represent the sulfate-reducing bacterial partner. Currently undetermined components of the pathway are in gray, e-represents an as yet unknown electron transfer intermediate produced by the methane-oxidizing ANME archaea and consumed by the sulfate-reducing bacterial partner. (b) Fluorescence *in situ* hybridization (FISH) micrograph of an ANME/SRB aggregate. ANME archaea are stained red and the sulfate-reducing bacteria are green. Diameter of aggregate ~6 μm. Modified from Dekas *et al.* (2009).

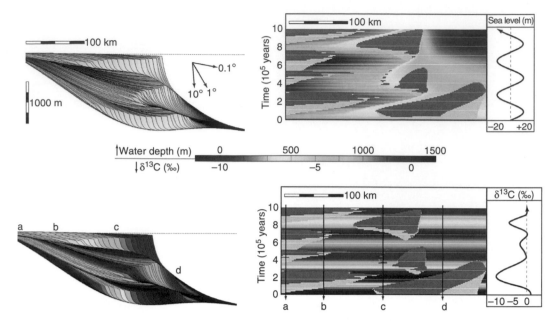

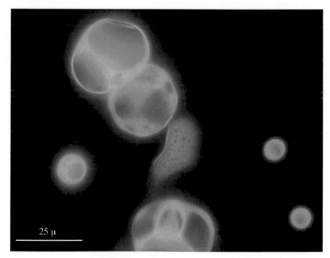

Plate 15 A forward model simulating a carbonate-rich succession on the scale of a passive margin. Upper left – a cross-section orthogonal to the strike of the margin. Note the strong vertical exaggreation. Timelines are shown in black, and coloured according to water depth at the time of deposition. Upper right – a Wheeler plot showing the stratigraphy across the basin as a function of time rather than thickness. In this plot light grey corresponds to areas of non-deposition whereas dark grey marks areas that underwent erosion. The sea level curve input used to create the stratigraphy is plotted alongside. The bottom two panels show the same simulation as above, this time coloured with contours corresponding to carbon isotope ratio. The secular trend in seawater DIC $\delta^{13}C$ used as model input is plotted alongside. Note that any one stratigraphic section is incomplete. To sample the entire history recorded in this sedimentary basin, one must measure and sample multiple stratigraphic sections arrayed across the basin, for example, at locations a, b, c, and d. See text for details.

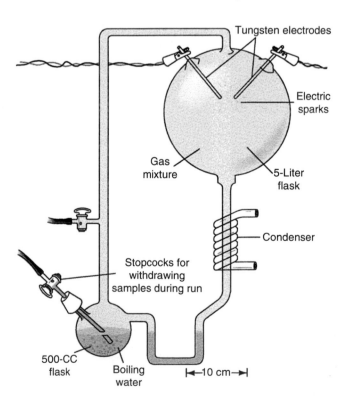

Plate 16 The Miller–Urey Experiment, which attempted to simulate an early Earth environment with water, methane, hydrogen and ammonia, incorporated temperature gradients, fluid fluxes, interfaces, and periodic electric spark discharges (Miller 1953; Miller and Urey 1959; Wills and Bada 2000).

Plate 17 Lipid molecules can self-organize in aqueous solutions to form cell-like enclosures called vesicles (from Hazen and Deamer 2007).

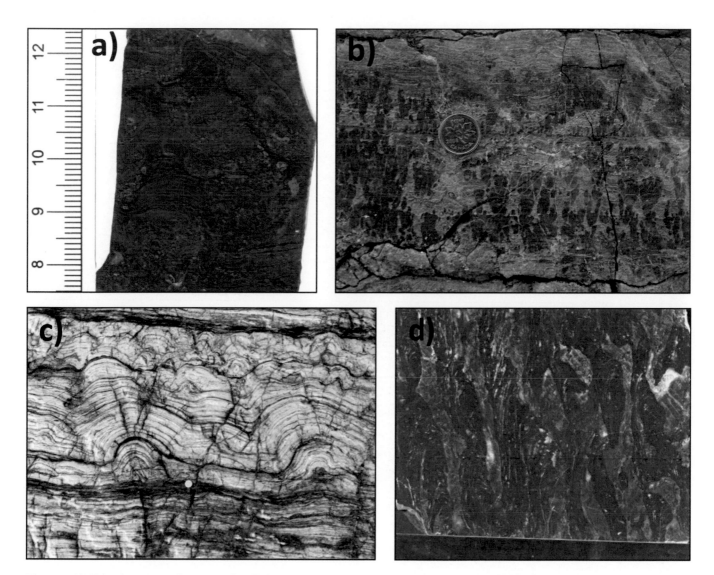

Plate 22 (a) Slab of a stromatolitic jasper banded iron formation from the ~1.88 Ga Biwabik Formation, Minnesota, USA, (b) stromatolitic chert-dolomite from the ~1.9 Ga McLeary Formation, Belcher Islands, Canada, (c) stromatolitic dolomite from the 2.06–2.09 Ga Rantamaa Formation, Finland, (d) stromatolitic phosphorite (carbonate-fluorapatite and dolomite) from the 2.0 Ga Jhamarkotra Formation, Rajasthan, India (photos by D. Papineau).

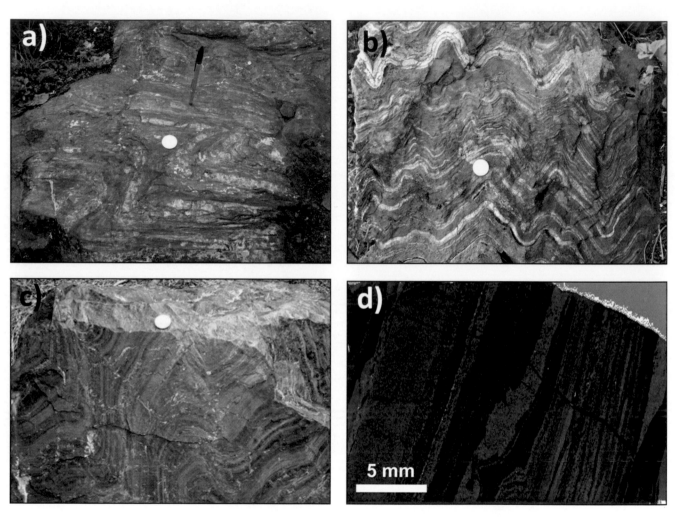

Plate 23 (a) Chert-magnetite-siderite banded iron formation from the Neoarchean Hunter Mine Group, Canada, (b) chert-magnetite banded iron formation from the Neoarchean Bababudan Supracrustal Belt, India, (c) jasper-chert banded iron formation from the Paleoproterozoic Negaunee Formation, Michigan, United States, (d) sulfide facies banded iron formation from the ~2.0 Ga Pathavaara Formation, Finland. All coins are about 2.5 cm in diameter (photos by D. Papineau).

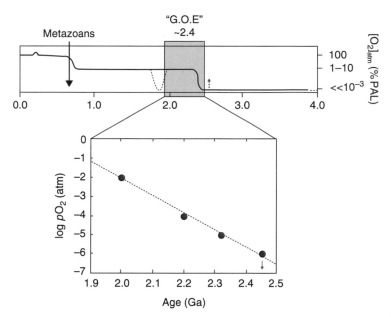

Plate 28 Estimates of atmospheric O_2 content (pO_2) through time (Ga) reported as % of the present atmospheric level (PAL). The arrow at 2.5 Ga indicates the pre-GOE 'whiff' of oxygen described in the text. The suggestion of a decrease at ~2.0 to 1.8 Ga reflects the recurrence of IFs and the associated Cr isotope arguments of Frei *et al.* (2009). Modified from Canfield (2005), Kump (2008), and Lyons and Reinhard (2009b). Proterozoic pO_2 values are not well constrained and may have been substantially lower than those shown. The inset is a summary of literature constraints, perhaps better described as best guesses given the inherent uncertainties, on pO_2 as a function of age – at and immediately following the GOE. The estimates are as follows: $pO_2 = 10^{-6}$ atm at 2.45 Ga (the arrow indicates that pre-GOE pO_2 may have been lower than indicated; Pavlov and Kasting, 2002; Bekker *et al.*, 2004), $pO_2 = 10^{-5}$ atm at 2.32 Ga (Bekker *et al.*, 2004), $pO_2 = 10^{-4}$ atm at 2.2 Ga (Rye and Holland, 1998), and $pO_2 = 10^{-2}$ atm at 2.0 Ga (Rye and Holland, 1998).

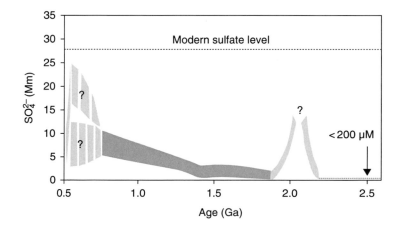

Plate 29 Highly schematic summary of seawater sulfate concentrations for the Proterozoic. Apparently conflicting estimates for sulfate in the Neoproterozoic range from high to very low, and the model emerging for the Paleoproterozoic predicts temporal swings between high and low values. Modified from Lyons and Gill (2010).

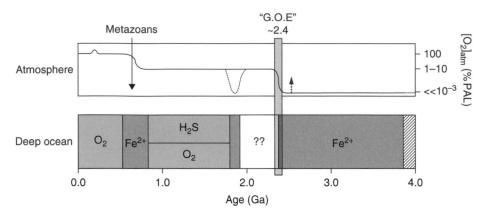

Plate 30 Classic views of evolving deep-ocean chemistry arguing that either oxic (Holland, 2006) or euxinic conditions (the so-called 'Canfield ocean,' Canfield, 1998) prevailed over much of the Proterozoic. Note the return to ferruginous conditions assumed for the latter part of the Neoproterozoic (Canfield *et al.*, 2008). These conceptual models are displayed relative to the backdrop of best estimates for the evolving O_2 content of the atmosphere (described in Fig. 20.4). Recent work, discussed at length in the text, is pointing to a more textured deep-ocean redox across time and space, including pervasive if not dominant ferruginous conditions in the deep mid-Proterozoic ocean. Absolute atmospheric oxygen levels are only approximations included to highlight the relative trends through time.

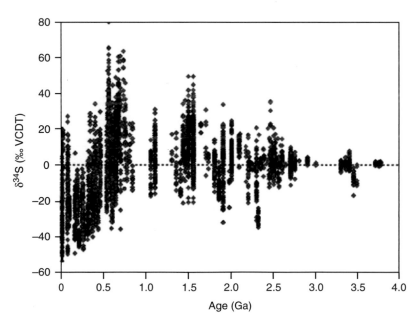

Plate 31 Summary of sedimentary pyrite sulfur isotope data. The data are presented using the standard $\delta^{34}S$ notation, and VCDT refers to the Vienna Canyon Diablo troilite standard. Modified from Canfield (2005). See the text and Lyons and Gill (2010) for additional details. Data are from many sources (available on request).

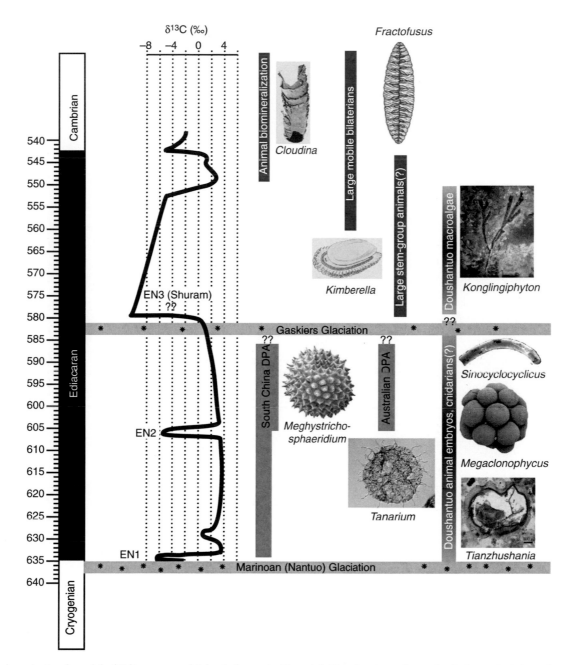

Plate 35 A conjectural model of Ediacaran geobiological events. Uncertainties about geochronological constraints and phylogenetic interpretations are indicated by question marks. Modified from Xiao (2008).

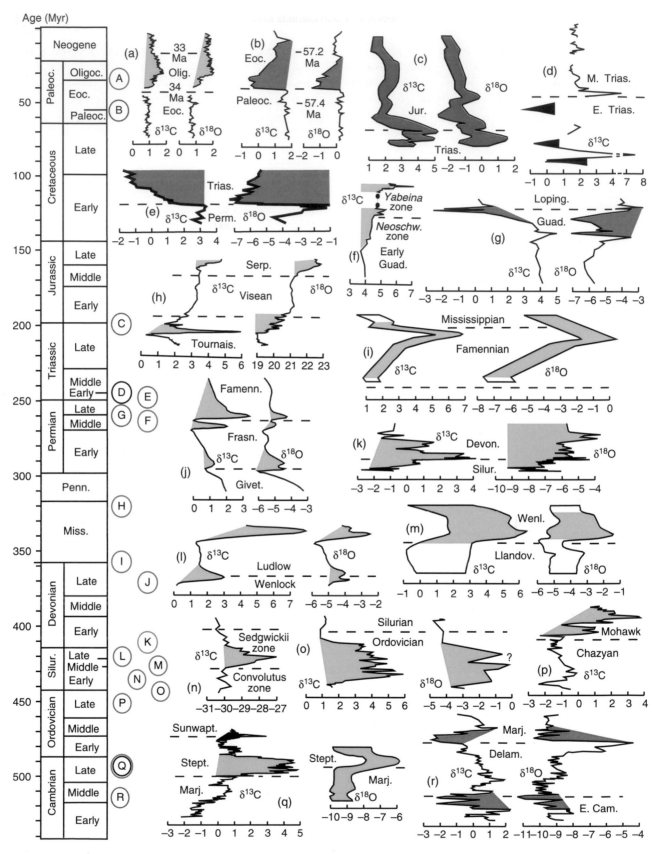

Plate 36 Stable isotope excursions that have been documented in shallow marine strata in association with mass extinctions. Eighteen intervals (A–R) contain a total of 26 such δ¹³C excursions. Corresponding to these, and trending in the same direction, are 19 published δ¹⁸O excursions, which are displayed in the plots to the right of those depicting δ¹³C. Encircled letters on the left indicate temporal positions of excursions. Blue indicates association with global cooling and

red, with global warming; black indicates absence of published evidence of associated climate change. Horizontal scales represent magnitudes of δ¹³C and δ¹⁸O excursions in ‰. Light δ¹³C in N is for organic carbon rather than carbonates, and heavy δ¹⁸O in H is for conodonts rather than bulk or skeletal carbonate. Ordinates represent stratigraphic positions of samples and are neither precisely linear with respect to time nor scaled the same for all graphs (after Stanley, 2010).

biogeochemical cycles. Although we will provide a broad overview of how this has been done, by way of illustration, we will focus in depth on how various molecular approaches have been used to provide insights into a novel biogeochemical process, anaerobic methane oxidation.

13.2 Molecular approaches used in geobiology

While molecular biological approaches are now commonplace in geobiology, it has only been ~30 years since they began to be applied to our discipline. In this time, there has been an exponential progression in our ability to rapidly and inexpensively sequence DNA that has led to an explosion of opportunities in collecting molecular data. Indeed, we now find ourselves in a historical moment where the rate of sequencing has outpaced our ability to analyse the data it generates. A significant opportunity for future geobiological work resides in creating and applying bioinformatic algorithms to analyse these data, as we will discuss at the conclusion of this chapter. In this section, we will chronicle the milestones in the application of molecular tools to geobiology. Our intent is to provide the beginning student with a historical framework for understanding how we arrived at where we are today, a clear picture of the types of questions that molecular tools can help us answer, and an appreciation for the limitations of these tools and where important gaps in our understanding remain. We also hope to illustrate how gaining insight into geobiological systems requires complementary, iterative approaches (Fig. 13.1).

13.2.1 The culture-dependence of 'culture independent' work

The single most important advance in modern evolutionary systematics came in 1977 when Carl Woese and George Fox reported that ribosomal RNA (rRNA) sequences could be used to classify life (Woese and Fox, 1977). Using rRNAs that had been isolated from diverse organisms by themselves and many other colleagues, Woese and Fox made the profound observation that the sequences of these rRNAs grouped into three major categories (what we now refer to as the domains of the Bacteria, the Archaea and the Eucarya). This discovery lay in stark contrast to the previous classification scheme by Whittaker that posited five kingdoms of life (Plantae, Animalia, Protista, Fungi and Monera) with the Monera (i.e. bacteria) at the base of the evolutionary tree. With Woese's new method for inferring evolutionary relationships, it became clear that the phylogenetic distance between man and fungi was miniscule relative to the diversity of life contained within the microbial world

(Fig. 13.2). The epiphany that rRNA was a better molecular chronometer than proteins for inferring evolutionary relationships came from the facts that: (i) it is a component of all self-replicating systems, (ii) its sequence changes slowly with time, and (iii) it is readily isolated. This latter aspect is often forgotten by modern molecular ecologists and geobiologists who utilize 16S rRNA as a tool for 'culture independent' work. While indeed today we can sequence DNA, RNA and proteins straight from the environment (see sections that follow), it is important to keep in mind that our ability to know which molecules to sequence in the first place rests upon work done with cultured organisms where the biochemical role of these molecules was established. So there truly is no such thing as 'culture independent' work. Moreover, our ability to understand the meaning of novel sequences that we find today using metagenomic and metaproteomic approaches ultimately requires classical genetic and biochemical studies to illuminate their function, as we describe below.

13.2.2 Applying 16S rRNA sequencing to microbial ecology

One of the first questions a geobiologist wants to answer about a given environment is 'which organisms are present'? The challenge has been that conventional enrichment culturing and isolation approaches to address this question often do not capture the most geobiologically significant organisms (rather, they select for those that grow best in the enrichment medium). Thus, a more accurate appreciation for environmental microbial diversity has been gained from approaches based on environmental 16S rRNA gene inventories (Pace, 1997). Pace and colleagues' recognition that one could relatively easily document much of the >99.9% uncultivated microbial diversity by extracting the environmental 16S rRNA genes/amplicons using the polymerase chain reaction, PCR (Olsen *et al.*, 1986; Pace, 1997), represented a breakthrough in our understanding of the natural world. Because of this, our appreciation for the diversity of microbial life has grown exponentially in the past two decades. In addition, these 16S rRNA gene inventories have produced many surprises from a previously hidden biosphere (Sogin *et al.*, 2006), resulting in the detection of novel phyla and new lineages, and have helped guide new approaches to isolate and cultivate ecologically important microorganisms. In fact, in the bacterial domain alone, over 36 bacterial divisions were clearly identifiable from these initial studies (Hugenholtz *et al.*, 1998).

Perhaps the first realization of how little we knew about the microbial world came from studies of hot spring archaeal diversity. In one such study, Barns *et al.* (1996), identified a new putative phylum in the Archaea,

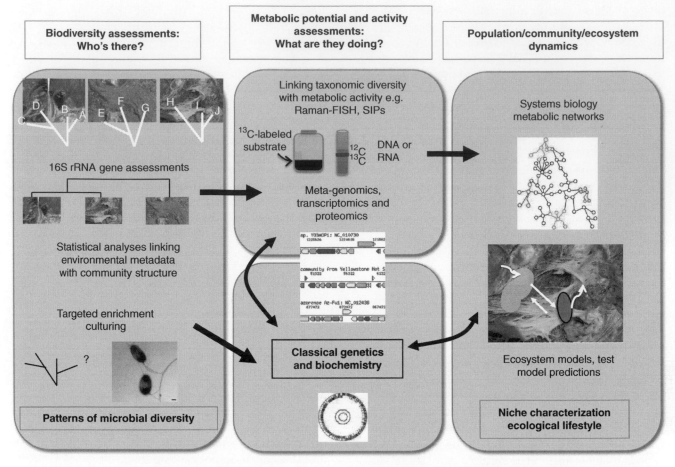

Figure 13.1 Overview of molecular approaches used in geobiology. Rapid phylogenetic assessments can be accomplished through high throughput sequencing, providing sufficient sequence depth and breadth for testing the statistical significance of observed patterns. Additionally, linking these patterns to environmental metadata assists in targeting enrichment culturing of organisms of interest. In order to obtain insights in the functional roles populations have in a community, numerous single gene, in situ approaches or metagenomic approaches can be used. Complementary classical genetic and biochemical approaches enable specific hypotheses about function to be tested.

Genomes of environmentally important isolates can serve as reference genomes for metagenomics. The accumulated environmental, activity and sequence data can be incorporated into dynamic models that explore the interactions within the communities and their biogeochemical outputs under specific environmental conditions. These in turn provide new hypotheses that can be tested both in the environment and in the laboratory. Arrows link complementary methods that can be applied together to enhance the ability to discover and characterize novel microbial symbioses within complex microbial communities in the environment.

the Korarchaeota. Subsequently, they have been detected in most global terrestrial hot springs, shallow and deep-sea marine vents (e.g. Hjorleifsdottir *et al.*, 1997; Reysenbach *et al.*, 2000; Baker *et al.*, 2003; Auchtung *et al.*, 2006). In addition to detecting Korarchaeota from deep-sea vents, numerous novel lineages have been detected with no known isolates in culture. One of these lineages (DHVE2) was detected on vent deposits at almost every vent environment studied (Reysenbach *et al.*, 2006 and references therein) and appears to be endemic to deep-sea vents, although a deeply diverging sequence was recently obtained from hot springs in Nevada, USA (Costa *et al.*, 2009). Likewise, in studies of marine,

aquatic, soil and sediment environments, many novel Crenarchaeota and Euryarchaeota, not related to thermoacidophiles or other extremophiles have been detected (Hershberger *et al.*, 1996; DeLong, 1998; Béjà *et al.*, 2002; Nicol and Schleper, 2006).

16S rRNA assessments have also transformed our understanding of the geobiology of the open ocean. One of the biggest surprises was the discovery of abundant free living Archaea in this habitat. In addition, these diversity studies also revealed an abundant alpha-proteobacterium designated SAR11, which often accounts for over 35% of the total Bacteria and Archaea in the marine surface waters. It took many years to

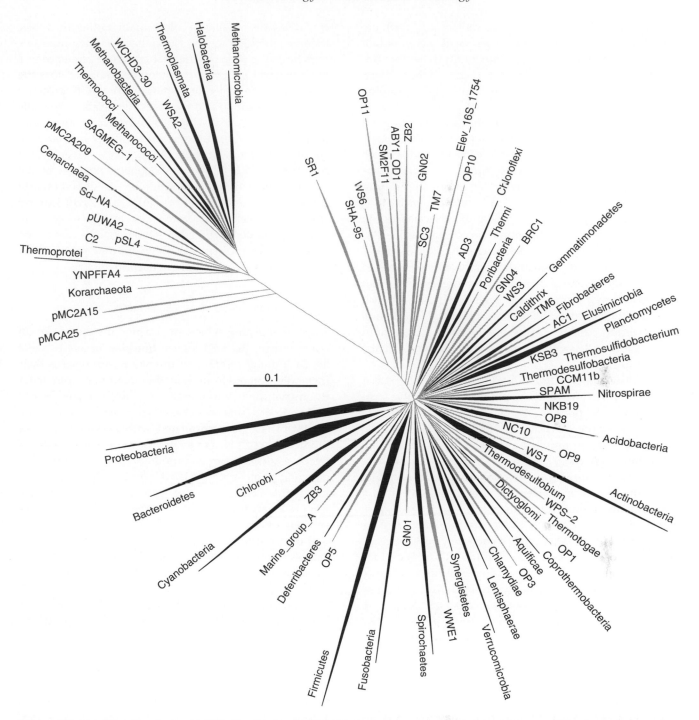

Figure 13.2 The diversity of microbial life. Small subunit (16S) rRNA phylogenetic tree of major lineages within the Archaea and Bacteria. Gray lineages have no known representatives in culture yet and were identified based on environmental 16S rRNA gene sequences. Courtesy Phil Hugenholtz.

determine the role that these important members of the ocean ecosystem play (Morris *et al.*, 2002), and was enabled by careful molecular tracking of the 16S rRNA gene and then eventually its isolation in very dilute culturing medium. Once the organism was isolated in the laboratory, its genome was sequenced and the role of its unusual oligotrophic lifestyle started to unfold (Giovannoni *et al.*,

2005). One other interesting geobiological system that yielded some diversity surprises is an acid mine drainage system in Northern California, called Iron Mountain. Prior to molecular ecological studies of this environment, thiobacilli were thought to be key players in acid mine drainage and mobilizers of metals in these systems. As it turned out, *Leptospirillum* (Schrenk *et al.*, 1998) and an archaeon _Ferroplasma_ (Edwards _et al._, 2000) are actually far more important members of the mine tailings community of Iron Mountain, and possibly worldwide in these systems. Prior to these 'who's there' surveys, Archaea had not been implicated in mining operations.

Although they represent a significant leap forward into understanding the diversity of the microbial world, these initial approaches were still restricted by the depth at which we could sequence (sampling of sequence space). Recent developments in high-throughput sequencing techniques (called 'pyrosequencing') – where thousands to tens of thousands of sequences can be obtained from a single sample (in contrast to 50–1000 with more traditional molecular cloning approaches) – have alleviated this constraint and have revealed an even greater diversity than previously recognized from numerous environments (e.g. Sogin _et al._, 2006; Roesch _et al._, 2007; Fierer _et al._, 2008; Huse _et al._, 2008; Turnbaugh and Gordon, 2009). Undoubtedly, high throughput sequencing will become routine in geobiology in the near future.

A cautionary tale did emerge from this 16S rRNA genetic diversity revolution though. Because the technique uses the polymerase chain reaction, conserved sequences (primers) within the 16S RNA gene have been designed that can amplify most 16S rRNA genes from Bacteria and Archaea. However, the accuracy of these primers is only as good as the ribosomal sequence databases that are used to develop them. In some elegant experiments, Stetter and colleagues had noticed very small cocci associated with a new archaeal isolate, _Ignicoccus_ (Huber _et al._, 2002). When they separated out the small cocci and sequenced the 16S rRNA genes, they noticed that the primers routinely used to amplify Archaea and Bacteria would not have detected this new small coccoid organism, now named _Nanoarcheum equitans_. One can only wonder how many other organisms might have fallen through this selective 'PCR net' because of the bias created through primer design, not to mention other sources of bias and error inherent in PCR (Polz and Cavanaugh, 1998; Acinas _et al._, 2005).

Until recently, none of the novel lineages described above had known isolates in culture, which limited our ability to understand their role in the environment. Yet these microbial diversity inventories provide targets of important geomicrobial organisms for culturing and genomic analyses. For example, after a decade of trying to get the ubiquitous and enigmatic Korarchaeota to grow in the lab, Elkins _et al._ (2008), managed to get an enrichment culture with high abundance of Korarchaeota. The researchers stumbled on something interesting in their attempts to grow this organism. They noticed that it was very resistant to lysis by sodium dodecyl sulfate (SDS). Using this information, Elkins _et al._ (2008) were able to select against other microbes, thereby enriching for the Korarchaeota. They used this enrichment to then get the first complete sequence of a korarchaeotal genome. The genome indicated that the organism is an anaerobic heterotroph able to grow on peptides and amino acids, yet lacking the ability to make a variety of cofactors, vitamins and purines, and therefore reliant on the environment (or other microbes) to supply these for it. The genome sequence also enhanced our appreciation of the phylogenetic position of the Korarchaeota. Based on some of its evolutionarily conserved genes such as the large and small subunit rRNA sequences, the Korarchaeota appear to be more closely affiliated with the Crenarchaeota, however, based on such molecules as DNA-binding proteins, the Korarchaeota are more euryarchaeal-like.

13.2.3 Going beyond the 'who is there?' question

After establishing which organisms are present in an environment, the next logical question to ask is 'what are they doing'? Often, by this we mean 'what is their metabolic potential?' or 'how does their metabolic activity impact the environment?' 16S rRNA sequences can only help with this when the phylogenetic identity of an organism is tightly correlated with its metabolic program. For example, if one identifies a cyanobacterial 16S rRNA sequence in an environment, it is a good bet that oxygenic photosynthesis is taking place there. Similarly, guilds that catalyse various steps in the nitrogen cycle are phylogenetically tight (e.g. the autotrophic aerobic nitrifying bacteria, which oxidize ammonium to nitrite and then nitrite to nitrate), and the appearance of these 16S rRNA sequences in a sample suggests that these metabolisms may be active in that environment (Ward, 2005). However, the majority of geobiologically significant organisms are either metabolically diverse (i.e, capable of multiple metabolisms, whose expression/activity depends on what substrates are available in the environment) or poorly characterized (i.e. not yet available in culture), making 16S rRNA a poor predictor of metabolism. In the latter case, when genomic sequences have been inferred for the organism of interest (either through sequencing enrichment cultures, as described above for a korarcheaon, or through modern day metagenomic reconstructions or single-cell sequencing, see below) this can help us guess what metabolisms the organism might be capable of performing, but falls short of proving they are doing so.

To circumvent these limitations, geobiologists have found creative ways to link phylogenetic identity (16S rRNA) to metabolic function. Combining biogeochemical measurements with dynamics of 16S rRNA gene counts can help predict key players associated with specific observed geochemical processes. If certain known processes are of interest, which have conserved diagnostic genes for that function (such as methanogenesis, sulfate reduction, nitrogen fixation, ammonia oxidation, and various carbon fixation pathways), the potential of these processes (gene presence) and the expression of the genes (mRNA) or their gene products (e.g. protein) can be monitored. However, more often than not, the inferred process (from geochemical measurements) is an undescribed biological pathway and other methods are required to illuminate them.

Once genetic systems are established in organisms that perform this process and the genes encoding them are identified (see below), then, if the genes are well conserved, their presence and/or expression can be monitored in the environment. An example of where this was done is for microbial arsenate respiration: first a genetic system was established in a model organism (Saltikov *et al.*, 2003), the genes encoding the enzyme that catalyses respiratory arsenate reduction were identified (Saltikov and Newman 2003), and because it is highly conserved (i.e. most bacteria that respire arsenate use the same enzyme), mRNA transcripts were employed to detect its expression in an anaerobic field-site where arsenic is abundant (Malasarn *et al.*, 2004).

Beyond making gross geochemical measurements and correlating them with 16S rRNA or metabolic gene inventories, there are several innovative methods that integrate molecular biological tools with more precise chemical measurements to better describe the ecophysiology of microbial communities. For example, techniques such as lipid or RNA-based stable-isotope probing (SIP; Radajewski *et al.*, 2003; Friedrich, 2007; Kreutzer-Martin, 2007), isotope arrays (Adamcyk *et al.*, 2003), fluorescence *in situ* hybridization (FISH)–microautoradiography (FISH-MAR; Rogers *et al.*, 2007), Raman-FISH (reviewed in Wagner, 2009), FISH–secondary-ion mass spectrometry (FISH-SIMS; reviewed in Orphan and House, 2009) and variations thereof, link specific processes and substrate utilization to a species or 16S rRNA lineage (or a lipid biosignature). For example, the key organism that was involved in naphthalene degradation in a contaminated sediment was initially identified by SIP and the organism was later cultivated (Jeon *et al.*, 2003).

FISH is a whole-cell hybridization method that uses a 16S rRNA targeted fluorescent oligonucleotide probe and epifluorescence microscopy to specifically identify and quantify microorganisms within natural samples based on their ribosomal RNA signature (DeLong *et al.*, 1989; Amann, 1995). Combining FISH with radioisotopes (e.g. ^{14}C, ^{3}H) or stable isotope (e.g. ^{13}C, ^{15}N) incubations, one can specifically couple a cell's phylogenetic identity with a specific uptake of a labeled substrate of interest. For example, Herndl *et al.* (2005) showed that uptake of ^{14}C-labeled inorganic carbon by Crenarchaeota (recently reclassified as Thaumarchaeota; Brochier-Armanet *et al.*, 2008) increases with depth, which suggested these organisms may be growing autotrophically in the environment. In addition to radioactively labeled substrates, a number of recently developed techniques such as FISH-SIMS and Raman-FISH also enable researchers to track the incorporation and flow of specific compounds through phylogenetically-stained microorganisms within natural communities using stable isotope labeled substrates, with ^{13}C and ^{15}N being the most commonly applied. Raman-FISH is a microanalytical method that relies on the detection of a spectral shift, known as the 'red shift', between ^{13}C or ^{15}N labeled amino acids (e.g. phenylalanine) and their unlabeled counterparts as an indicator of the degree of isotopic enrichment within individual, FISH-stained cells (Huang *et al.*, 2007a; Wagner, 2009). This method has been used to track naphthalene degradation by *Pseudomonas* in groundwater (Huang *et al.*, 2007b). These studies showed that different individual cells had different ^{13}C contents, indicating metabolic heterogeneity between different *Pseudomonas* cells (see Wagner, 2009 and references therein). Like SIP, Raman-FISH requires a high degree of isotope labeling for successful detection. Secondary ion mass spectrometry (SIMS, nanoSIMS, and ToF-SIMS) combined with FISH are also powerful approaches for linking phylogeny with metabolic function, offering a versatile platform for measuring a wide spectrum of elements and isotope ratios in biological materials at the micron or submicron scale (see Orphan and House, 2009 and references therein). We will discuss these approaches in more detail in the case study that follows. One caveat to these community-based substrate-labeling approaches is that they are unable to separate the effects of consumption of the original substrate vs. consumption of a down-stream metabolic product (often termed 'cross-feeding' or 'trophic transfer'). This is not necessarily a negative (see Murase and Frenzel, 2007), but merely something that can complicate their interpretation.

Despite the power of these methods, to date they have been applied exclusively in combination with phylogenetic gene markers and fall short of providing mechanistic insight into the genes/gene products that are responsible for the metabolisms of interest. Recent efforts to link metabolic genes with specific taxonomic groups have involved visualization techniques for detection of metabolic genes or gene products within single cells (e.g. immunostaining (Lanoil *et al.*, 2001), mRNA FISH (Pernthaler and Thiel, 2004) and gene FISH (Moraru *et al.*, 2010)) and targeted single cell genome sequencing. A recent example of where single cell analysis was used to link specific phylotypes to specific metabolic genes comes from the work of Ottesen *et al.* (2006) who used microfluidics to separate individual cells from the complex termite gut microbial community. Digital PCR was used to then amplify multiple genes from single cells. Alternatively, flow cytometry has been used to sort single cells from the environment and sequence their genomes using whole genome amplification techniques (Stepnauskas and Sieracki, 2007) thereby 'reconstructing' the metagenome of the environment one cell at a time (Woyke *et al.*, 2009). Yet single-cell genome sequencing is still in its infancy, and significant technological challenges remain to be overcome (Marcy *et al.*, 2007; Binga *et al.*, 2008; Ishoey *et al.*, 2008). No doubt as single-cell sequencing technologies improve, new methods will continue to build on these pioneering efforts to develop high throughput genomic analyses of the spatial and temporal dynamics of complex microbial communities.

13.2.4 Metagenomics, metatranscriptomics and metaproteomics

The approaches discussed in the preceding subsections depend on knowing what to look for – be it a specific gene, geochemical profile, or the consumption of a specific metabolic substrate. In the absence of a hypothesis about what might be relevant in a particular environment, one can perform an unbiased 'fishing expedition' to let the organisms do the talking. 'Omics' approaches (metagenomics, metatranscriptomics and metaproteomics) have been used in geobiological studies to accomplish this, and they can greatly enable hypothesis generation.

Environmental 'shotgun sequencing' allows for the study of sequence data directly from the environment; this is commonly referred to as 'metagenomics' or 'environmental genomics' (see *Nature Microbial Reviews* Volume 3, 2005 for reviews). This random sequencing of all the members of a microbial community provides a glimpse into its functional potential and ecological structure. For example, in a metagenomic library from pelagic samples obtained from the Sargasso Sea in 2004, Venter and colleagues (2004) found an archaeal-associated scaffold containing distant homologues of the bacterial ammonia monooxygenase genes. These archaeal *amo* genes were 70% similar to those also found in a crenarchaeal soil metagenomic fragment (Venter *et al.*, 2004; Schleper *et al.*, 2005), suggesting that these Archaea were nitrifiers. In 2005, the first example of these mesophilic Crenarchaeota was cultivated from an ammonia-enriched aquarium. These and other studies have pointed to the global importance of crenarchaeal ammonia oxidation, a process previously thought to be restricted to Bacteria (Konneke *et al.*, 2005). Another interesting example of the deployment of metagenomics is its application to the Iron Mountain mine tailings microbial community. In reconstruction of the

community metagenome, it appeared that one of the minor *Leptospirillum* members of the community (*Leptospirillum* sp. Group III) had the genes to fix nitrogen, which gave rise to the hypothesis that this organism could be a key player in the community because of this capacity (Tyson *et al.*, 2004). While it remains to be demonstrated, such partitioning of functions in natural microbial communities is likely to increase the robustness of microbial ecosystems in dynamic environments; support for this notion comes from simplified studies with experimental systems (Venturi *et al.*, 2010). Beyond insights into community metabolic potential, metagenomic sequencing can reveal novel mechanisms of microbial evolution. In taking a closer look at the Iron Mountain metagenomic sequences, Banfield and colleagues noticed that the archaeal members of this community, multiple strains of *Ferroplasma acidamanus*, had mosaic genomes due to extensive homologous recombination of three sequence types (Tyson and Banfield, 2008). They observed that some of the high population-level diversity occurred at c̲lustered r̲egularly i̲nterspaced s̲hort p̲alindromic r̲epeats (CRISPR), which are a family of repetitive DNA sequences that are present in Archaea and Bacteria but not in Eucarya. (Jansen *et al.*, 2002). Intriguingly, these bacterial and archaeal repeats are use to defend the cell against viruses (Barrangou *et al.*, 2007) and are analogous to RNA interference (RNAi) systems in eukaryotes (Marraffini and Sontheimer, 2010). How they function in shaping microbial genomes in the environment is a fascinating and open question (Denef *et al.*, 2010).

While metagenomic data sets have revealed many important new insights into the structure, function and evolution of microbial communities, the metagenomic approach is limited by several factors. First among them is how deeply and broadly one can sequence. As in general in complex systems, the metagenome may only provide a partial genomic picture and some of the complexity of the community can be lost from a single sample analysis. For example, Thompson *et al.* (2004) explored the sequence and genome variability of *Vibrio* isolates taken from one marine location over time, and showed that even though the 16S rRNA gene sequences varied less that 1%, there was extensive variation within the genomes, which could explain why genome reconstructions from complex metagenomic communities is still extremely difficult. Nonetheless, having known genomes of isolates obtained from these sorts of environments, greatly assist in gene assignments (Giovannoni *et al.*, 2005). For example, the metagenomic study of the communities of three Aquificales-dominated hot spring environments in Yellowstone National Park was greatly enhanced due to the availability of three draft and closed genome sequences from isolates obtained from Yellowstone (Inskeep *et al.*, 2010). These genomes had been carefully annotated (Reysenbach *et al.*, 2009) and thus could serve as the reference (anchor) for gene assignment and assemblies of the metagenomes. Likewise, Wilhelm *et al.* (2007), were able to use the genome of the ubiquitous marine oligotrophic heterotroph, '*Candidatus pelagibacter*' (or SAR11) as the reference genome to analyse the SAR11 populations associated with the Sargasso Sea metagenome. They were able to show that although there are core genome features shared by this group across oceanic scales, significant variation within the genomes leads to expansion of SAR11 diversity.

As sequencing approaches are increasing in throughput, and viable single-cell genomics is on the horizon (see above), we can now envision studies that will address community level gene dynamics across various spatial and temporal scales. Along these lines (although working with cultured isolates rather than single cells), Rocap and coworkers (2003) sequenced the genomes of three strains of *Prochlorococcus*, important photosynthetic members of the marine phytoplankton. One strain is high-light adapted (strain MED4, high chlorophyll *b/a* ratios; a second strain (strain SS120, low chlorophyll *b/a* ratios) grows best under very low light conditions and is found at depths of greater than about 50 m. The third strain (MIT9313) is best adapted between the MED4 and SS120 niches in the water column. Comparing the genomes of these three strains provided many interesting observations regarding their ecological partitioning in the ocean. In particular, the SS120 has the smallest genome and can only use ammonium and amino acids for nitrogen sources, whereas MIT9313 has the largest genome, which suggests that it has greater genome versatility, enabling it to occupy the transition zone between the two different preferred light regimes.

Perhaps the greatest caveat to metagenomics work, however, is that gene presence (DNA) does not tell us which proteins present in a system are 'doing the work' to change the geochemistry of the environment. Two crucial steps beyond DNA need to be considered: transcription (converting DNA to mRNA) and translation (converting mRNA to protein). Let us consider transcription first. In microbial systems, the RNA extracted from the environment typically includes both rRNA and mRNA, and major difficulty with bacterial and archaeal metatranscriptome analyses is the high concentrations of co-extracted rRNA with the mRNA (>90% of the extracted RNA can be rRNA). Nevertheless, methods are being developed to overcome this problem (e.g. Frias-Lopez *et al.*, 2008; DeLong, 2009; Stewart *et al.*, 2010). Other difficulties are that many of the sequences may not have matches in databases (Frias-Lopez *et al.*, 2008) and in some cases automated assignments can sometimes be misleading. For example, the gene encoding for the enzyme involved in demethylating an important organic

sulfur compound, dimethylsulfoniopropionate (DMSP) was originally placed with a family of genes that are though to be involved in the degradation of glycine (Howard *et al.*, 2006). The emerging picture from several metatranscriptomic studies from marine (Frias-Lopez *et al.*, 2008) and terrestrial environments (Urich *et al.*, 2008) is that, as expected, genes associated with maintenance of basic cell function and metabolism are highly expressed as are genes required for energy transduction. However, although comparisons between expression at geographically distinct marine areas and in samples collected during day/night showed similar patterns, some differences could be explained by differences in the community composition and responses to light or dark conditions (Hewson *et al.*, 2010). Likewise, Poretsky *et al.* (2009) demonstrated that transcripts for photosynthesis, C_1 metabolism and oxidative phosporylation were highest in the day at the Hawaiian Ocean Time Series (HOTS) surface water. Metatranscriptomes can also provide insights into organism-specific expression and function in the community. For example, Ulrich *et al.* (2008), detected expression of genes for ammonia oxidation and CO_2 fixation and attributed this activity to the less abundant soil Crenarchaeota. Additionally, Hewson *et al.* (2009), detected members of the marine cyanobacterium, *Crocosphaera watsonii*, expressing nitrogen fixation genes in the southwest Pacific, which pointed to their importance as keystone species in the global nitrogen cycle.

It is also important to recognize that mRNA content does not always accurately predict protein abundance. An elegant illustration of this comes from recent work in yeast using genome-wide measurements of translation, which showed that variations in the efficiency of translation can profoundly effect the dynamic range of gene expression for different genes (Ingolia *et al.*, 2009). Thus, if we ultimately seek to describe the presence (and stability) of enzymes that catalyse geochemically significant reactions, the most direct measure of this is to study the ability of proteins being expressed in the environment to catalyse the reactions we care about. As a step in this direction, we can capture mRNAs being translated to protein (Ingolia *et al.*, 2009) or we can attempt to quantify the proteins themselves. The latter approach defines metaproteomics, and is increasingly being applied to geobiological studies (Wilmes and Bond, 2006; Maron *et al.*, 2007). One of the first metaproteomic studies was done on samples collected from the Iron Mountain acid mine drainage system described above (Denef *et al.*, 2010). Because of its low diversity, this natural community provided an excellent opportunity to troubleshoot the approach for geobiological applications, but since then, metaproteomic methods have improved even further (VerBerkmoes *et al.*, 2009). In this initial study, only 5% of the proteins from the least abundant community members were identified and only about 50% of the predicted proteins were identified for the dominant organism; the abundant proteins of unknown function served as interesting targets for further investigation (Ram *et al.*, 2005). This led to the characterization of two novel cytochromes involved in a new iron oxidation pathway (Jeans *et al.*, 2008; Singer *et al.*, 2008). Another example of mechanistic insights enabled by metaproteomics comes from a study of Sargasso Sea samples (Sowell *et al.*, 2008). In particular, periplasmic substrate-binding proteins from SAR11 were highly abundant, which, given the ultra-small cell size (<500 nm) of this organism and its large periplasmic space, suggests that this may be a way SAR11 maximizes nutrient uptake in a low nutrient marine environment. However, in all these examples, the metaproteome data are only as good as the corresponding metagenomes or reference genomes available to identify peptide fragments. In addition, we must not confuse the abundance of proteins identified by mass spectra with their activity. At present, we understand so little about the kinetics of mRNA and protein synthesis and degradation in complex environments, that metatranscriptomic and metaproteomic data should only be seen as rich hypotheses-generating tools. Experimental transcriptomics and proteomics, performed under standardized conditions in the lab, are necessary to interpret what environmental metatranscriptomics and proteomics actually mean.

As high throughput DNA sequencing and proteomic technologies continue to be developed, we predict that 'omics' approaches will become routine tools in most geobiology laboratories, much as 16S rRNA gene sequencing is today. These different 'omics' tools help address synergistic questions, for example, transcriptomic and proteomic profiles capture a different kinetic expression of intracellular information transfer. However, for all of these community-based 'omics' approaches, controlled genomic, transcriptional and translational studies with single-cells or environmental isolates will remain necessary to enable their interpretation.

13.2.5   The need for classical genetics and biochemistry

As previously mentioned, there are far more genes in genomes that are 'genes of unknown function' than there are genes whose function is known. Even in *Escherichia coli*, arguably the best understood microorganism on the planet, only 76% of its genes have been assigned functions, and only 66% of these have had their functions determined experimentally (Karp *et al.*, 2007). Imagine the challenge, then, in interpreting metagenomic or metaproteomic data that contain sequences from organisms that have never before been cultured. How do we give these orphan genes a home? Happily, classical genetic and biochemical approaches that have been honed over the years in model organisms such as

E. coli and *Salmonella* can be applied to geobiological problems. There are two ways this can be done: (1) to establish genetic systems in geobiologically significant organisms or (2) to express genes from the environment in genetically-tractable foreign hosts.

13.2.5.1   Creating genetic systems in geobiologically relevant organisms

'Ain't nothing like the real thing, baby' goes the refrain to Marvin Gaye's classic soul hit. These words are fitting when it comes to creating experimental systems to elucidate the biological function of novel molecules. Whenever possible, if one seeks to understand their function deeply, it is best to develop a genetic system in the organism(s) one cares about. This, in turn, enables:

1 Genes to be identified through mutational analyses that catalyse or are somehow otherwise involved in enabling reactions of interest.
2 Determination of the environmental cues that trigger their expression and the signal transduction/regulatory machinery that is required for this to happen.
3 Protein over-expression, which can be very helpful in producing enough material to allow the structure, trace-metal content, and kinetic parameters of interesting biomolecules to be solved, and
4 Cell-biological studies, which can provide insight into where these biomolecules localize within their hosts and with what they interact, which can enhance our understanding of their function.

While genetically tractable 'model organisms' cannot be expected to faithfully represent every organism in nature that performs a geobiological function of interest, more often than not, the insights one can gain from them are broadly applicable. That said, it is important to remember that a model is only a model, and that what one discovers in a model system cannot be assumed to be relevant for other systems until it is directly demonstrated.

In the case study on anaerobic oxidation of methane that follows, we will briefly mention recent progress in the development of genetic systems for methanogens, close cousins to the methane-oxidizing Archaea, as well as some of the important discoveries that these systems have enabled. In addition to the genetic and biochemical work that has performed with these Archaea, we note that many other geobiologically significant organisms today have well-established genetic systems (whose sophistication improves every year), including halophiles (Bjornsdottir *et al.*, 2006; Allers *et al.*, 2010), sulfur-oxidizing thermophilic Archaea (e.g. *Sulfolobobus*, *Pyrococcus*, *Thermococcus*) (Rother and Metcalf, 2005), cyanobacteria (Cohen *et al.*, 1998; Bhaya *et al.*, 2000, 2001; Holten *et al.*, 2005), algae (Davies and Grossman, 1998), bacteria that oxidize and/or reduce metals and metalloids (e.g. *Geobacter* (Coppi *et al.*, 2001; Rollefson *et al.*, 2009), *Shewanella* (Saffarini *et al.*, 1994), *(Acidi) thiobacillus* (Rawlings and Kusano, 1994; Liu *et al.*, 2000), *Rhodopseudomonas and Rhodobacter* species (Donohue and Kaplan, 1991; Jiao *et al.*, 2005), *Alkalilimnicola ehrlichii* (Zargar *et al.*, 2010)), reduce sulfate (e.g. *Desulfovibrio* (Wall *et al.*, 2008)), fix nitrogen (e.g. *Azotobacter* (Bishop *et al.*, 1990)), and make magnetite (e.g. *Magnetospirillum* (Schuler, 2008)).

The type of approaches that have been used in these organisms for introducing DNA, generating mutants (using both targeted and random gene disruption), restoring wild type phenotypes using plasmid or chromosomal-based complementation, and performing expression analyses using reporter gene fusions, are generic and can be implemented in diverse organisms. A *sine qua non* of making a practical genetic system is having the ability to segregate individual mutants. This is typically achieved by being able to plate low-density dilutions on agar plates and select for the growth of single colonies from individual mutant cells (see Newman and Gralnick, 2005) for a description of practical aspects of performing genetics). While getting cells to form colonies on plates can be non-trivial, creative recent efforts to achieve this for *Prochlorococcus* (Morris *et al.*, 2008) demonstrate that with sufficient resourcefulness, investigators can surmount this obstacle. This is not to say that it is all downhill from getting cells to form colonies on plates to establishing a workable genetic system. If one's organism of choice grows slowly, or for whatever reasons is unable to be easily transformed, establishing a genetic system may be just too impractical to be worth the effort. If so, it is important to reflect on whether the questions one is asking need to be answered with that particular organism. If not, selecting a genetically tractable strain for future work would be wise. If, however, the questions demand that particular strain be used, then other genetic approaches (such as those described below) would be the next best option. The reason genetic analysis is so important is because genetics uniquely provides the ability to prove that a gene is required for a process. Genetics provides the springboard from which to address other important aspects of how a process works, including biochemical and cell biological aspects of a problem. Yet for processes that are novel, genetic approaches are the only ones that can directly demonstrate the *in vivo* relevance of a particular gene/gene product. All other molecular methods ultimately can only show correlations, not causation.

13.2.5.2 Expressing genes from the environment in genetically-tractable foreign hosts

Ideally, every geobiologist would be able to study organisms that are both environmentally significant and genetically tractable, yet this is not always the case. Some organisms, while culturable, thrive under such extreme conditions that classical genetics is impractical, such as in the case of acidophilic Fe(II)-oxidizing bacteria, where antibiotics are typically used as markers that enable the selection of mutant strains degrade at low pH (Woods *et al.*, 1986). Given this experimental limitation, molecular methods such as quantitative PCR, and a variety of 'omics' techniques (transcriptomics, proteomics, metabolomics) can be applied to gain insight into what genes/gene products are associated with the process of interest. These approaches, together with traditional biochemical and physiological studies, have enabled a depth of understanding of how acidophilic bacteria oxidize Fe(II) (Brasseur *et al.*, 2002; Yarzabal *et al.*, 2002, 2004; Nouailler *et al.*, 2006; Quatrini *et al.*, 2006, 2009). It should be noted, however, that the menu of selectable markers for genetic studies is expanding (Allers *et al.*, 2004; Barrett *et al.*, 2008), and in the future, geobiologists may be able to make progress in establishing genetic systems in organisms where it was previously thought impossible. Inspiration for this can be found in work done in developing genetic systems for *Thermoanaerobacterium saccharolyticum* and *Clostridium thermocellum*, organisms involved in the degradation of lignocellulose (Shaw *et al.*, 2008, 2010).

Even when culturing conditions do not constrain the choice of selectable markers for use in genetic studies, other realities can preclude creating a genetic system in a geomicrobe of interest. For example, slow growth rate. In a typical PhD lifetime, if a student wants to do genetics, (s)he would be wise to pick an organism that can make colonies on plates overnight or within a few days (note: that bacterial genetics was developed in *E. coli*, which doubles every 20 minutes, is not random). If a geomicrobial system operates on a more sedate time frame, such as in the case of the consortia involved in the anaerobic oxidation of methane which double every three months (Nauhaus *et al.*, 2007; Orphan *et al.*, 2009)), setting up a classical genetic system is impractical. Finally, if an organism is only known through environmental genomic reconstruction but does not yet exist in culture, obviously, classical genetics is out of the question. So where does this leave the geobiologist who seeks certainty about gene function?

One approach to dealing with this challenge is to express environmental DNA in a genetically-tractable foreign host. The story of the discovery of proteorhodopsin – a protein that contributes to energy generation by many bacteria in the world's oceans (Béjà *et al.*, 2000) – provides an excellent example of how this can be done. In sequencing libraries of environmental DNA containing sequences from uncultivated marine gamma Proteobacteria (known as the 'SAR86' group), Beja *et al.* noticed sequences that encoded a homolog to bacteriorhodopsin genes from extremely halophilic Archaea (Mukohata *et al.*, 1999). In these organisms, bacteriorhodopsin had been shown to function as a light-driven proton pump (Hoff *et al.*, 1997). Hypothesizing that the environmental rhodopsin-like sequence, which they dubbed 'proteorhodopsin', might play a similar function in the SAR86 bacteria, they cloned this gene into a vector for expression in *E. coli*. Biophysical and functional characterization of the gene product confirmed its ability to function as a light-driven proton pump (Béjà *et al.*, 2000). This is an example of a 'gain of function' assay, where the function of a gene can be determined by expressing it in a foreign host, in tandem with follow-up biochemical/ biophysical/physiological studies. Similar 'gain of function' approaches have been utilized to identify genes responsible for catalyzing phototrophic Fe(II) oxidation; in this case, genes were cloned from an organism that is not genetically tractable (*Rhodobacter ferroxidans* SW2) and expressed in a close relative that is (*R. capsulatus* SB1003) (Croal *et al.*, 2007). Another good example of using complementation assays to reveal gene function is that of Martinez *et al.* (2009), who were able to assign previously uncharacterized ORFs from metagenomic libraries' roles in phosphonate metabolism by this approach. Whenever possible, it is desirable to express genes in a host that is as similar as possible to the genetically-intractable organism of interest, to increase the chances that the gene(s) one seeks to characterize will be able to be expressed and/or their gene product processed. For example, if one has a gene that appears to encode a multi-heme *c*-type cytochrome, it is necessary to express this gene in a strain that also contains the machinery for heme insertion into the protein backbone (also known as the 'apo-protein').

An important caveat to any heterologous expression experiment is that while genes catalyzing an activity may be selected, this does not establish that these genes are responsible for this activity in the organism from which they derived. For example: it may be that an organism has various ways of catalyzing a particular enzymatic reaction, and that the one that happened to be selected in the heterologous complementation experiment is not the primary one that is used in the native host. That said, gain of function assays are as close as we can get to establishing that a particular gene has a specific function in the absence of being able to make mutants in the original host. Follow-up studies to determine whether that particular gene is expressed under the conditions one would expect if it played the inferred function in the original host can be performed to test the hypothesis.

13.3 Case study: anaerobic oxidation of methane

A classic example of where molecular methods have contributed to solving a global geochemical enigma is the story of the anaerobic oxidation of methane (AOM). In this penultimate section, we summarize the background to the AOM problem and explain how a set of interesting geochemical observations set the stage for geobiological discoveries that were enabled by advances in molecular microbial ecology and classical biochemistry and genetics.

From a geobiological perspective, the production and oxidation of CH_4 is believed to have played an important role in the co-evolution of the atmosphere and biota during the Archean and early Proterozoic eons. Methanogens, and by inference, the more recently discovered methanotrophic Archaea (known as ANME), evolved relatively early, prior to the rise of oxygen (Fox *et al.*, 1980; House *et al.*, 2003; Battistuzi *et al.*, 2004; Konhauser *et al.*, 2009). The distinctively light carbon isotopic signature of methane-oxidation (e.g. Summons *et al.*, 1994; Whiticar, 1999; Pancost and Sinninghe Damsté, 2003; Peckman and Thiel, 2004; Thomazo *et al.*, 2009) is one of the more recognizable forms of microbial respiration in the rock record, with the potential for generating organics (kerogen) and minerals (e.g. carbonates) that are extremely depleted in ^{13}C. Kerogens as light as −60‰ have been identified in samples dating as far back as 2.7 billion years (Schoell and Wellmer, 1981; Hayes, 1983, 1994; Rye and Holland, 2000; Hinrichs, 2002; Peckman and Thiel, 2004), while in comparatively younger settings (Mesozoic/Cenozoic aged deposits), evidence of methane-based ecosystems are recorded in ^{13}C-depleted lipid biomarkers (Elvert *et al.*, 1999; Peckman and Thiel, 2004; Birgel *et al.*, 2006, 2008). These lipids are structurally and isotopically similar to those produced by methane-consuming Archaea and sulfate-reducing bacteria observed in modern methane seep environments, again suggesting that the process of AOM and relatives of the extant methane-oxidizing microorganisms have persevered over millions, perhaps billions of years (Elvert *et al.*, 1999; Peckman and Thiel, 2004; Birgel *et al.*, 2006, 2008).

Today, the process of anaerobic oxidation of methane coupled to sulfate reduction is estimated to oxidize up to 80% of CH_4 that would otherwise be released from marine sediments, keeping oceanic methane contributions to the atmosphere in check. Substantial progress in identifying the microbial players involved in AOM and biochemical mechanisms underpinning this process have been made in the last decade, largely thanks to the application of molecular tools to this problem. The evolutionary path of anaerobic microorganisms capable of oxidizing methane and the breadth of possible electron acceptors coupled to this process are areas of active research (Beal *et al.*, 2009; Ettwig *et al.*, 2009, 2010; Crowe *et al.*, 2011; Sivan *et al.*, 2011). Here we highlight how multidisciplinary investigations that incorporate molecular methods to the study of modern analogue ecosystems and living anaerobic methanotrophs can help constrain and identify biosignatures of archaeal methanotrophy and provide insight into the dynamics and biological coupling of carbon, sulfur, nitrogen, and metal cycles on early Earth.

13.3.1   Hints from geochemistry

While today we have a good explanation for what drives AOM, this was not always the case. Indeed, that methane was oxidized anaerobically posed something of a geochemical conundrum. The potential for microbially-driven AOM was predicted by geochemists decades ago (Barnes and Goldberg, 1976; Reeburgh, 1976; Martens and Berner, 1977; Alperin and Reeburgh, 1985; Iversen and Jorgensen *et al.*, 1985), however the specific identity of these anaerobic methane-consuming microorganisms until about 10 years ago remained a mystery. Geochemical modeling and thermodynamic predictions had indicated that methanogens (methane-producing Archaea) might also be responsible for its oxidation. Indeed, physiological studies of methanogens in pure culture suggested the potential for methane-oxidation, however this oxidation was reported to be a minor component of the total methane production and thus could not account for the net AOM documented in marine sediments (Zehnder and Brock, 1979). Hoehler *et al.* (1994) proposed that under certain environmental conditions, methanogens could form syntrophic associations with sulfate-reducing bacteria, promoting thermodynamically favourable conditions for running the methanogenic biochemical pathway in reverse. In this model, the methanogenic archaeon oxidizes methane through an obligate metabolic partnership with a syntrophic sulfate-reducing bacterial partner, serving as a sink for the methane-sourced metabolic intermediates (e.g. hydrogen or other electron shuttle) (Fig. 13.3a). While geochemical and thermodynamic data pointed to the involvement of specific microbial 'guilds' (defined as a metabolically coherent microbial group), cultivation of anaerobic methanotrophic microorganisms that are capable of net methane-oxidation coupled to sulfate-reduction have been unsuccessful.

13.3.2   Insights provided from 16SrRNA, stable isotopic signatures, and FISH

Only recently through the application of molecular methods was the identity of the microorganisms mediating this process revealed from their DNA 'fingerprint'

(16S rRNA). The initial multi-disciplinary studies characterizing the phylogenetic and chemotaxonomic identity of these anaerobic methane-oxidizing microorganisms were conducted in 1999, using comparative analysis of the diversity of archaeal 16S rRNA genes and the diversity and $\delta^{13}C$ values of archaeal lipid biomarkers believed to be indigenous to methane seep sediments where AOM was actively occurring (Hinrichs *et al.*, 1999). In the Hinrichs *et al.* (1999) study and follow-up work by Orphan *et al.* (2001a), sediment from active methane seeps was shown to contain abundant archaeal lipids similar to those recovered from ancient seep carbonates (Peckman and Thiel, 2004) and cultured methanogens (e.g. sn-2-hydroxyarchaeol and PMI; Koga *et al.*, 1993), that also had an unusually light carbon-13 signature ($\delta^{13}C$ value of −105‰). Such light ^{13}C values suggested that these biomarkers were sourced from an extant archaeon (or multiple Archaea) that had consumed ^{13}C-depleted methane within the reduced anoxic seep sediment (in this case, environmental CH_4 had a low $\delta^{13}C$ value of −50‰). Parallel cloning and sequencing of 16S rRNA genes from Archaea and Bacteria in these samples revealed an abundance of phylotypes that clustered among methanogenic lineages, but were distinct from known cultured representatives and sequences previously reported from other environments.

Based on the co-occurrence of ^{13}C-depleted lipid biomarkers and these newly discovered archaeal 16S rRNA clades (called ANME, or ANaerobic MEthane-oxidizing Archaea), the ANME groups were hypothesized to be involved in AOM (Hinrichs *et al.*, 1999; Orphan *et al.*, 2001a). In addition to archaeal 16S rRNA genes, the analysis of bacterial ribosomal genes in these samples revealed a phylogenetic clade of deltaproteobacteria related to the sulfate-reducing *Desulfosarcina/Desulfococcus* (DSS) that was common to all of the methane seep samples analysed (Orphan *et al.*, 2001a). This widespread DSS lineage was proposed to also play a role in AOM, in this case functioning as the syntrophic sulfate-reducing partner hypothesized by Hoehler *et al* (1994). Using the 16S rRNA gene sequences from the ANME Archaea and sulfate-reducing bacteria recovered from seep sediments, specific fluorescently labeled oligonucleotide probes were designed for FISH experiments to determine the abundance and distribution of these putative methanotrophic microorganisms by epifluorescence microscopy (Boetius *et al.*, 2000; Orphan *et al.*, 2001b; Knittel *et al.*, 2005). FISH results from a study by Boetius *et al.* (2000) revealed that these uncultured ANME phylotypes comprised up to 80% of the biomass within the AOM active seep sediment, and, interestingly, were found in well-structured, layered cell aggregations with the DSS sulfate-reducing deltaproteobacteria (Fig. 13.3b).

At present there are at least three known uncultured Euryarchaeotal lineages capable of anaerobic methane

(a)

(b)

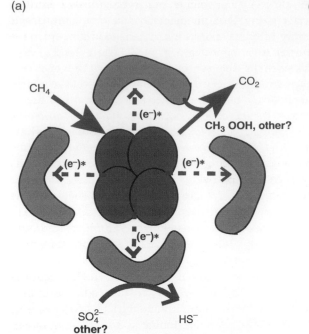

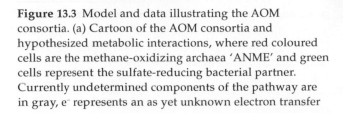

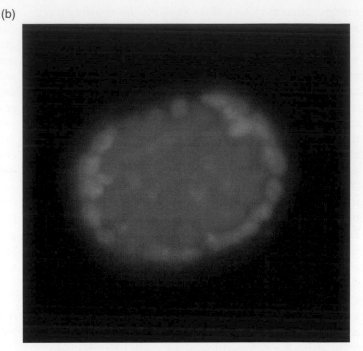

Figure 13.3 Model and data illustrating the AOM consortia. (a) Cartoon of the AOM consortia and hypothesized metabolic interactions, where red coloured cells are the methane-oxidizing archaea 'ANME' and green cells represent the sulfate-reducing bacterial partner. Currently undetermined components of the pathway are in gray, e⁻ represents an as yet unknown electron transfer intermediate produced by the methane-oxidizing ANME archaea and consumed by the sulfate-reducing bacterial partner. (b) Fluorescence *in situ* hybridization (FISH) micrograph of an ANME/SRB aggregate. ANME archaea are stained red and the sulfate-reducing bacteria are green. Diameter of aggregate ~6 μm. Modified from Dekas *et al.* (2009).

oxidation in marine environments, called ANME-1, ANME-2 and ANME-3 (Hinrichs *et al.*, 1999; Orphan *et al.*, 2001a; Knittel *et al.*, 2005; Niemann *et al.*, 2006). Surveys of archaeal 16S rRNA gene diversity and transcribed ribosomal RNA (Inagaki *et al.*, 2004; Mills *et al.*, 2004; Martinez *et al.*, 2006) from methane-influenced environments from around the globe suggest ANME groups are common in near seafloor methane vents and seeps as well as other methane-rich environments with detectable sulfate levels including select deep subseafloor sediment horizons (Reed *et al.*, 2002; Inagaki *et al.* 2003; Parkes *et al.*, 2007), hydrates (Lanoil *et al.*, 2001; Orcutt *et al.*, 2005), petroleum reservoirs (Head *et al.*, 2003), hypersaline habitats (Baker *et al.*, 2003; Orcutt *et al.*, 2005; Lloyd *et al.*, 2006) euxinic basins (Vetriani *et al.*, 2003), terrestrial mud volcanoes (Alain *et al.*, 2006), and sedimentary (Teske *et al.*, 2002) and serpentinite-hosted hydrothermal vent systems (Brazelton *et al.*, 2006).

13.3.3 *Applying biochemical knowledge to gene discovery in the ANME groups*

Further evidence of the similarities between the ANME groups and methanogens was revealed through studies of metabolic genes associated with the methanogenic pathway. For example, the methyl coenzyme M reductase (MCR), the enzyme responsible for the terminal methane production step in the methanogenic pathway (Thauer, 1998), is a commonly used metabolic gene marker for methanogen diversity in environmental samples, producing a phylogeny which, in most cases, is congruent with 16S rRNA (Springer *et al.*, 1995). PCR amplification of the *mcr*A subunit from methane seep sediments harboring different 16S rRNA ANME lineages (ANME-1 or ANME-2), resulted in two important discoveries: (1) the uncultured ANME's contained *mcr*A, similar to methanogens (2) different ANME clades have distinct *mcr*A genes that appear to be diagnostic for each group (Hallam *et al.*, 2003; Inagaki *et al.*, 2004; Friedrich, 2005). The recovery of *mcr*A genes in environments containing ANME Archaea provided independent evidence that these putative methanotrophic Archaea are evolutionary related to methanogens. Targeted PCR-based metabolic gene analysis has also been extended to the analysis of other genes and gene transcripts (e.g. mRNA) believed to be associated with the ANME/sulfate-reducing bacteria consortium, including *dsr*AB, required for dissimilatory sulfate reduction (Thompson *et al.*, 2001; Dhillon *et al.*, 2003; Teske *et al.*, 2003; Harrison *et al.*, 2009; Lloyd *et al.*, 2010) and *nif*H and *nif*D, involved in nitrogen fixation (Pernthaler *et al.*, 2008; Dang *et al.*, 2009; Dekas *et al.*, 2009; Miyazaki *et al.*, 2009).

The analysis of expressed gene products associated with methane-oxidation, sulfate-reduction and nitrogen-fixation in methane seeps in tandem with geochemical analyses and/or radiotracer rate measurements have expanded our understanding of the potential physico-chemical variables influencing the activity of specific microbial guilds in the environment. For example, a recent study by Lloyd *et al* (2010) nicely demonstrates the strong correlation between the active component of the microbial community (via mRNA analysis and rate measurements) and steep geochemical gradients established through sulfate-dependent methane cycling. Here, the visible boundaries of advective fluid seepage, demarcated by a sulfide-oxidizing microbial mat, also set the boundaries of community composition and activity near the seabed, illustrating that the expression of *mcr*A and *dsr*AB varies on the scale of a few centimeters both laterally and with increasing sediment depth, depending on position within or just outside of the seep. In another RNA based study, Miyazaki *et al.* (2009) examined the expression of nitrogen fixation genes (*nif*H and *nif*D) along a depth profile within methane seep sediments, revealing unexpected patterns between expressed *nif* transcripts and pore water ammonium, methane, and sulfate concentrations.

13.3.4 Contributions from metagenomics and new questions raised

Further progress in understanding the underlying biology and potential mechanism of syntrophically mediated AOM has been accomplished through environmental metagenomics. Published metagenomic studies of the methanotrophic ANME's have applied different cloning and sequencing strategies, each with method accompanied by unique benefits and challenges. In the first metagenome study by Hallam *et al.* (2004), density gradients to enrich methane-oxidizing ANME-1 microorganisms from methane seep sediment followed by fosmid (approximately 40 000 base pairs long) and whole genome shot gun library (3000–4000 base pair inserts) construction, resulting in 111.3 million base pairs (Mbp) of shot gun sequence and 7.4 Mbp from screened archaeal fosmids (Hallam *et al.*, 2004). Fosmid screening and sequencing provided the first glimpse of the genomic make-up of the methanotrophic ANME lineages. While this work examined only a small fraction of the ANME-1 and ANME-2 genomes, a number of significant findings were gained through this approach. The most notable discovery was the occurrence of a nearly complete suite of genes in the archaeal ANME-1 that are highly similar to the canonical seven-step methanogenesis pathway previously documented in sequenced methanogens.

The homology between enzymes previously described from conventional methanogens and the anaerobic methane-oxidizing ANME Archaea lends support to the reverse methanogenesis hypothesis (Hoehler *et al.*, 1994), and, suggests that substantial evolutionary innovation was not necessarily required prior to the development of methanotrophy in the ANME Archaea, but rather appears to be derived, at least in part, from the methanogenic pathway. The apparent absence of the *mer* gene, encoding for the F_{420} dependent N^5, N^{10}-methenyltetrahydromethanopterin (methylene-H4MPT) reductase, and presence of F_{420} reducing hydrogenases and F_{420}-dependent quinone oxidoreductase (*fqo*) in the ANME-1 metagenome library, led Hallam *et al.* (2004) to propose an alternative mechanism of energy conservation for these organisms linked to an F_{420}-dependent respiratory chain.

The details of how the inert methane molecule is initially activated during AOM is still somewhat of a mystery. The genes encoding the methyl coenzyme M reductase (*mcr*) were shown to be present in the ANME genome, however the conditions enabling net methane oxidation over the conventional CH_4 production by the MCR enzyme, were not understood, leaving open the potential for an as-yet-unknown pathway for methane oxidation. Follow-up protein-based analyses, however, clearly demonstrated that the methyl coenzyme M reductase (MCR) and the associated nickel porphinoid cofactor F_{430}, were not only translated in the methanotrophic Archaea, but also comprised a significant percentage of the total proteins extracted from ANME-1-dominated methanotrophic microbial mats in the Black Sea, implying functional significance (Kruger *et al.*, 2003). Using the translated sequences from the ANME-1 metagenome (e.g. Meyerdierks *et al.*, 2005), Kruger and colleagues were able to link the distinct MCR to the ANME-1 lineage. Notably, the Ni-F_{430} cofactor, serving as the MCR active site, contained an unusual modification, yielding a larger protein with a molecular mass of 951 Da, distinct from the smaller (905 Da) F_{430} cofactors previously described from cultured methanogenic strains (Shima and Thauer, 2005). This prominent modification associated with the MCR active site spurred speculation that this alteration may in some way increase the catalytic efficiency of MCR for methane oxidation in some, but not all, of the ANME Archaeal lineages (Mayr *et al.*, 2008; Thauer and Shima, 2008).

Comparisons of orthologous genes recovered from the metagenomes of the major ANME lineages by Meyerdierks *et al.* (2005) provided further insight into the similarities and inherent diversity between members of the methanotrophic Archaea. Similar to reports by Hallam, analysis of metagenomic fosmid libraries associated with ANME-2, ANME-3, and ANME-1 from geographically distinct methane seeps confirmed the genetic relationship with methanogens including genes involved in energy metabolism (e.g. *mch* and *hdr*) and carbon fixation (CODH; Meyerdierks *et al.*, 2005). Additionally, analysis of 16 rRNA containing fosmids revealed notable variations in the ribosomal operon structure between members of the ANME-1b, ANME-2a, ANME-2c, and ANME-3, with members of the ANME-2a and ANME-3 possessing the canonical operon structure observed in most Bacteria and Archaea (i.e. 16S rRNA, 23S rRNA and 5S rRNA; Jinks-Robertson and Nomura, (1987)), while ANME-2c and ANME-1b microorganisms possess an unlinked 16S rRNA and 5S rRNA, respectively (Meyerdierks *et al.*, 2005).

Further sequencing efforts using a method known as 'genome walking' from fosmids constructed from an ANME-1b dominated microbial ecosystem in the Black Sea resulted in a near complete (82–90%) composite genome for this uncultured methanotrophic archaeon (Meyerdierks *et al.*, 2009). Comparative sequence analysis of relatives of the ANME-1 indicated differences in the electron transfer components of these methanotrophs, harboring genes homologous with FeFe-hydrogenases, while apparently lacking the more traditional NiFe-hydrogenases common in methanogenic relatives. Additional insights into the phylogenetic potential and possible mode of electron transfer by this organism included acetate, formate, and, interestingly, a putative secreted multiheme cytochrome C oxidase.

Expanding these metagenomic studies to include the sulfate-reducing symbiont in addition to the methanotrophic Archaea, Pernthaler *et al.* (2008) used a modified whole-cell immunofluorescence capture method called Magneto-FISH to selectively enrich for ANME-2c Archaea and their physically-associated bacterial partners directly from methane seep sediment, effectively reducing the overall complexity (i.e. diversity) of the sample prior to metagenome sequencing. In this study, over 40 000 ANME-2c cell aggregates (each containing on average 200–300 cells) were captured from a small volume of paraformaldehyde fixed methane seep sediment. Through the application of new technologies in whole genome amplification (i.e. multiple displacement amplification, based on the activity of phi 29 polymerase) and high throughput pyrosequencing, metagenomic sequence data from the ANME-2c and associated bacteria was obtained. This magneto-FISH enabled study expanded the known diversity of bacteria capable of forming associations with the ANME-2c Archaea, including two distinct sulfate-reducing deltaproteobacteria and a betaproteobacterium, and suggests that these different microbial associations may occupy distinct niches and possess unique physiological traits. For example, the detection of nitrogenase-associated gene fragments and genes encoding the capability to respire and assimilate nitrate inspire new avenues of research

regarding the involvement of ANME Archaea and their bacterial partners in nitrogen cycling within marine methane seeps (Pernthaler *et al.*, 2008).

Although no genetic system exists for the methanogens involved in AOM (or is likely to exist in the near future, until they can be isolated and cultured), it is important to point out that significant progress has been made in the past decade in the development of genetic systems for methanogens (for details, see reviews by Metcalf: Metcalf, 1999; Rother and Metcalf, 2005). Not only do these efforts illustrate that sophisticated genetics can be performed even in organisms that are challenging to cultivate, but they have led to discoveries about methanogenic pathways that would not have been possible otherwise. For example, genetic analysis in metabolically versatile *Methanosarcina* species lead to the identification of reduced ferredoxin as the electron donor in the first step of methanogenesis from H_2/CO_2 (Meuer *et al.*, 2002), the recognition that acetogenesis can also be used for energy conservation by some of these organisms under certain conditions (Rother and Metcalf, 2004), and that hydrogen-cycling is a preferred means of energy conservation for many *Methanosarcina* species (Kulkarni *et al.*, 2009). Finally, of direct relevance to the AOM story, a recent *in vitro* investigation of purified methyl coenzyme M reductase from a cultured methanogen has now provided direct evidence for methane activation and conversion to methyl coenzyme M by this enzyme (Scheller *et al.*, 2010).

13.4   Challenges and opportunities for the next generation

As we hope this chapter has illustrated, molecular biology has profoundly impacted geobiology over the past few decades. Many of the most important findings in our field have been enabled by molecular methods, and there is every reason to believe that these methods will continue to drive future discoveries. We sit in an opportune historical moment, where our ability to couple precise geochemical measurements with sequence information (be it DNA, mRNA or protein) in natural environments is more achievable than ever before. As automated processes for data collection are designed and implemented, we predict these data sets will begin to be collected over vast arrays of time and space.

In the midst of this data-acquisition revolution, it is increasingly important to step back and take stock of the problems we wish to solve. What are the most important questions? Just because we can measure something, should we? What are the best measurements to make for the questions we are asking? There is no doubt that environmental science – whether it goes by the name of 'geobiology', 'biogeochemistry', or something else – is a complex field that depends on thoughtful collaborations between investigators with diverse backgrounds. The importance of physical, chemical and biological forces in shaping the Earth and influencing each other through complex feedback loops is widely, albeit vaguely, appreciated, and there is much enthusiasm for integration of these disciplines.

Going forward, an important challenge for geobiologists will be to define tractable questions where individuals with complementary backgrounds can work together towards a rigorous, holistic understanding of particular biogeochemical cycles. Because many problems could be tackled in theory, we recommend the following criteria be used to select the best ones: (1) the simplest systems should be studied first, or, if complex systems are chosen, simpler analogs should be available to study in parallel so that the driving variables may be elucidated; (2) the dominant environmental perturbations to these systems should be known (or knowable), and they should be practical to measure over time; (3) the most important actors should be identified (or identifiable), and a testable model should be in place (or developed) to explain how they interact; (4) gaps in our understanding/ability to make predictive models should be ones that can be addressed by making the appropriate measurements; and (5) robust analytical tools should already be in place to make these measurements, and environmental measurements should be calibrated by controlled laboratory experiments.

We leave it to future students of geobiology to define these questions for themselves, but we close with a few friendly words of advice. First, chose problems that you and others can investigate deeply over many years. Do not be seduced by problems that are overly complex; in the end, it may be better to start with something simple where you can learn something unambiguously, with the hope that what you discover will translate to more complex systems. Second, understand in detail how the methods you use work, and always keep in mind the limitations of what they can tell you. Try to exhaust all other possible explanations for your favoured hypothesis before you believe it, and even then, be skeptical. Third, become an expert in something. Do not fall into the trap of being a jack-of-all-trades and a master of none. While geobiology demands a liberal scientific education so that you can appreciate the many components that contribute to our world, if you are to make lasting contributions to understanding it, you will need depth in a particular sub-specialty. Fourth, remember that even as molecular biology gets increasingly sophisticated, classical culture-based approaches (physiological, genetic, and biochemical) will always be necessary to gain mechanistic understanding into geobiological systems. Fortunately, developing genetic systems in geobiologically important organisms is getting easier to achieve as the tools for performing genetics in unconventional

organisms are improving, which will facilitate such efforts in the future. And finally, participate in driving the development of rigorous yet user-friendly bioinformatics platforms to keep up with the massive amounts of sequence data that are being generated as sequencing capacity enlarges and costs plummet. This includes software to help users annotate (meta)genomic information, as well as manage (meta)transcriptomic and (meta)proteomic data sets and identify interesting patterns/correlations between them. Opportunities abound for students interested in computational biology to contribute to geobiology, and it will be exciting to see 21st century molecular geobiology develop into an increasingly rigorous and mature discipline.

Acknowledgements

The authors gratefully acknowledge the Howard Hughes Medical Institute (D.K.N.), Gordon and Betty Moore Foundation and DOE Early Career Grant (V.J.O.), and NSF OCE-0937404 (A.L.R) for support. We also thank our students, postdocs, and colleagues for shaping our thinking on these topics over the years.

References

Acinas SG, Sarma-Rupavtarm R, Klepac-Ceraj V, Polz MF (2005) PCR-induced sequence artifacts and bias: insights from comparison of two 16S rRNA clone libraries constructed from the same sample. *Applied and Environmental Microbiology* **71**, 8966–8969.

Adamczyk J, Hesselsoe M, Iversen N, Horn M, Lehner A, Nielsen PH, Schloter M, Roslev P, Wagner M (2003) The isotope array, a new tool that employs substrate-mediated labeling of rRNA for determination of microbial community structure and function. *Applied and Environmental Microbiology* **69**, 6875–6887.

Alain K, Holler T, Musat F, Elvert M, Treude T, Kruger M (2006) Microbiological investigation of methane- and hydrocarbon-discharging mud volcanoes in the Carpathian Mountains, Romania. *Environmental Microbiology* **8**, 574–590.

Allers T, Ngo HP, Mevarech M, Lloyd RG (2004) Development of additional selectable markers for the halophilic Archaeon *Haloferax volcanii* based on the leuB and trpA genes. *Applied and Environmental Microbiology* **70**, 943–953.

Allers T, Barak S, Liddell S, Wardell K, Mevarech M (2010) Improved strains and plasmid vectors for conditional overexpression of His-tagged proteins in *Haloferax volcanii*. *Applied and Environmental Microbiology* **76**, 1759–1769.

Alperin MJ, Reeburgh WS (1985) Inhibition experiments on anaerobic methane oxidation. *Applied and Environmental Microbiology* **50**, 940–945.

Amann RI (1995) Fluorescently labeled, ribosomal-rRNA-targeted oligonucleotide probes in the study of microbial ecology. *Molecular Ecology* **4**, 543–553.

Auchtung TA, Takacs-Vesbach CD, Cavanaugh CM (2006) 16S rRNA phylogenetic investigation of the candidate division

'Korarchaeota'. *Applied and Environmental Microbiology* **72**, 5077–5082.

Baker GC, Smith JJ, Cowan DA (2003) Review and re-analysis of domain-specific 16S rRNA primers. *Journal of Microbiological Methods* **55**, 541–555.

Barnes RO, Goldberg ED (1976) Methane production and consumption in anaerobic marine sediments. *Geology* **4**, 297–300.

Barns, SM, Delwiche CF, Palmer JD, Pace NR (1996) Perspectives on archaeal diversity, thermophily and monophyly from environmental rRNA sequences. *Proceedings of the National Academy of Sciences, USA* **93**, 9188–9193.

Barrangou R, Fremaux C, Deveau H, Richards M, Boyaval P, Moineau S, Romero DA, Horvath P (2007) CRISPR provides acquired resistance against viruses in prokaryotes. *Science* **315**, 1709–1712.

Barrett AR, Kang Y, Inamasu KS, Son MS, Vukovich JM, Hoang TT (2008) Genetic tools for allelic replacement in Burkholderia species. *Applied and Environmental Microbiology* **74**, 4498–4508.

Battistuzzi FU, Feijao A, Hedges SB (2004) A genomic timescale of prokaryote evolution: insights into the origin of methanogenesis, phototrophy, and the colonization of land. *BMC Evolutionary Biology* **4**, 44.

Beal EJ, House CH, Orphan VJ (2009) Manganese and iron dependent marine methane oxidation: novel microbial sinks for methane. *Science* **325**, 184–187.

Béjà O, Aravind L, Koonin EV, *et al.* (2000) Bacterial rhodopsin: evidence for a new type of phototrophy in the sea. *Science* **289**, 1902–1906.

Béjà O, Koonin EV, Aravind L, *et al.* (2002) Comparative genomic analysis of archaeal genotypic variants in a single population and in two different oceanic provinces. *Applied and Environmental Microbiology* **68**, 335–345.

Bhaya D, Bianco NR, Bryant D, Grossman A (2000) Type IV pilus biogenesis and motility in the cyanobacterium Synechocystis sp PCC6803. *Molecular Microbiology* **37**, 941–951.

Bhaya D, Takahashi A, Shahi P, Grossman AR (2001) Novel motility mutants of Synechocystis strain PCC 6803 generated by in vitro transposon mutagenesis. *Journal of Bacteriology* **183**, 6140–6143.

Binga EK, Lasken RS, Neufeld JD (2008) Something from (almost) nothing: the impact of multiple displacement amplification on microbial ecology. *ISME Journal* **2**, 233–241.

Birgel D, Thiel V, Hinrichs KU, *et al.* (2006) Lipid biomarker patterns of methane-seep microbialites from the Mesozoic convergent margin of California. *Organic Geochemistry* **37**, 1289–1302.

Birgel D, Himmler T, Freiwald A, Peckmann J (2008) A new constraint on the antiquity of anaerobic oxidation of methane: late Pennsylvanian seep limestones from southern Namibia. *Geology* **36**, 543–546.

Bishop PE, Macdougal SI, Wolfinger ED, Shermer CL (1990) *Genetics of Alternative Nitrogen-Fixation Systems in Azotobacter vinelandii. In Nitrogen Fixation: Achievements and Objectives (ed Gresshof M). Chapman and Hall, New York. 789–795.* Routledge, Chapman & Hall Inc., New York.

Bjornsdottir SH, Blondal T, Hreggvidsson GO, *et al.* (2006) *Rhodothermus marinus*: physiology and molecular biology. *Extremophiles* **10**, 1–16.

Boetius A, Ravenschlag K, Schubert CJ, *et al.* (2000) A marine microbial consortium apparently mediating anaerobic oxidation of methane. *Nature* **407**, 623–626.

Brasseur G, Bruscella P, Bonnefoy V, Lemesle-Meunier D (2002) The bc(1) complex of the iron-grown acidophilic chemolithotrophic bacterium *Acidithiobacillus ferrooxidans* functions in the reverse but not in the forward direction. Is there a second bc(1) complex? *Biochimica et Biophysica Acta-Bioenergetics* **1555**, 37–43.

Brazelton WJ, Schrenk MO, Kelley DS, Baross JA (2006) Methane- and sulfur-metabolizing microbial communities dominate the lost city hydrothermal field ecosystem. *Applied and Environmental Microbiology* **72**, 6257–6270.

Brochier-Armanet C, Boussau B, Gribaldo S, Forterre P (2008) Mesophilic crenarchaeota: proposal for a third archaeal phylum, the Thaumarchaeota. *Nature Reviews Microbiology* **6**, 245–252.

Cohen MF, Meeks JC, Cai YA, Wolk CP (1998) Transposon mutagenesis of heterocyst-forming filamentous cyanobacteria. In: *Photosynthesis: Molecular Biology of Energy Capture*, Vol. 297 (ed McIntosh L). Academic Press Inc., San Diego, CA, pp. 3–17.

Coppi MV, Leang C, Sandler SJ, Lovley DR (2001) Development of a genetic system for *Geobacter sulfurreducens*. *Applied and Environmental Microbiology* **67**, 3180–3187.

Costa KC, Navarro JB, Shock EL, Zhang CL, Soukup D, Hedlund BP (2009) Microbiology and geochemistry of great boiling and mud hot springs in the United States Great Basin. *Extremophiles* **13**, 447–459.

Croal LR, Jiao YQ, Newman DK (2007) The fox operon from Rhodobacter strain SW2 promotes phototrophic fe(II) oxidation in *Rhodobacter capsulatus* SB1003. *Journal of Bacteriology* **189**, 1774–1782.

Crowe SA, Katsev S, Leslie K, *et al.* (2011) The methane cycle in ferruginous Lake Matano. *Geobiology* **9**, 61–78.

Dang H, Luan X, Zhao J, Li J (2009) Diverse and novel nifH and nifH-like gene sequences in the deep-sea methane seep sediments of the Okhotsk Sea. *Applied and Environmental Microbiology* **75**, 2238–2245.

Davies JP, Grossman AR (1998) The use of Chlamydomonas (Chlorophyta : Volvocales) as a model algal system for genome studies and the elucidation of photosynthetic processes. *Journal of Phycology* **34**, 907–917.

Dekas AD, Poretsky RS, Orphan VJ (2009) Deep-sea archaea fix and share nitrogen in methane-consuming microbial consortia. *Science* **326**, 422–426.

DeLong EF (1998) Everything in moderation: archaea as 'non-extremophiles'. *Current Opinion in Genetics & Development* **8**, 649–654.

DeLong EF (2009) The microbial ocean from genomes to biomes. *Nature* **259**, 200–206.

Delong EF, Wickham GS, Pace NR (1989) Phylogenetic stains – ribosomal RNA-based probes for the identification of single cells. *Science* **243**, 1360–1363.

Denef VJ, Kalnejais LH, Mueller RS, *et al.* (2010) Proteogenomic basis for ecological divergence of closely related bacteria in natural acidophilic microbial communities. *Proceedings of the National Academy of Sciences, USA* **107**, 2383–2390.

Denef VJ, Mueller RS, Banfield JF (2010) AMD biofilms: using model communities to study microbial evolution and ecological complexity in nature. *ISME Journal* **4**, 599–610.

Dhillon A, Teske A, Dillon J, Stahl DA, Sogin ML (2003) Molecular characterization of sulfate-reducing bacteria in the Guaymas Basin. *Applied and Environmental Microbiology* **69**, 2765–2772.

Donohue TJ, Kaplan S (1991) Genetic techniques in Rhodospirillaceae. *Methods in Enzymology* **204**, 459–485.

Edwards KJ, Bond PL, Gihring TM, Banfield JF (2000) An archaeal iron-oxidizing extreme acidophile important in acid mine drainage. *Science* **287**, 1796–1799.

Elkins JG, Podar M, Graham DE, *et al.* (2008) A korarchaeal genome reveals insights into the evolution of the Archaea. *Proceedings of the National Academy of Sciences, USA* **105**, 8102–8107.

Elvert M, Suess E, Whiticar MJ (1999) Anaerobic methane oxidation associated with marine gas hydrates: superlight C-isotopes from saturated and unsaturated C20 and C25 irregular isoprenoids. *Naturwissenschaften* **86**, 295–300.

Ettwig KF, van Alen T, van de Pas-Schoonen KT, Jetten MSM, Strous M (2009) Enrichment and molecular detection of denitrifying methanotrophic bacteria of the NC10 Phylum. *Applied and Environmental Microbiology* **75**, 3656–3662.

Ettwig KF, Butler MK, Le Paslier D, *et al.* (2010) Nitrite-driven anaerobic methane oxidation by oxygenic bacteria. *Nature* **464**, 543–548.

Fierer N, Hamady M, Lauber CL, Knight R (2008) The influence of sex, handedness, and washing on the diversity of hand surface bacteria. *Proceedings of the National Academy of Sciences, USA* **105**, 17994–17999.

Fox GE, Stackebrandt E, Hespell RB, *et al.* (1980) The phylogeny of prokaryotes. *Science* **209**, 457–463.

Frias-Lopez J, Shi Y, Tyson GW, *et al.* (2008) Microbial community gene expression in ocean surface waters. *Proceedings of the National Academy of Sciences, USA* **105**, 3805–3810.

Friedrich MW (2005) Methyl-coenzyme M reductase genes: unique functional markers for methanogenic and anaerobic methane-oxidizing Archaea. *Environmental Microbiology* **397**, 428–442.

Friedrich MW (2006) Stable-isotope probing of DNA: insights into the function of uncultivated microorganisms from isotopically labeled metagenomes. *Current Opinion in Biotechnology* **17**, 59–66.

Giovannoni SJ, Tripp HJ, Givan S, *et al.* (2005) Genome streamlining in a cosmopolitan oceanic bacterium. *Science* **309**, 1242–1245.

Hallam SJ, Girguis PR, Preston CM, Richardson PM, DeLong EF (2003) Identification of methyl coenzyme M reductase A (mcrA) genes associated with methane-oxidizing archaea. *Applied and Environmental Microbiology* **69**, 5483–5491.

Hallam SJ, Putnam N, Preston CM, *et al.* (2004) Reverse methanogenesis: testing the hypothesis with environmental genomics. *Science* **305**, 1457–1462.

Harrison BK, Zhang H, Berelson W, Orphan VJ (2009) Variations in archaeal and bacterial diversity associated with the sulfate-methane transition zone in continental margin sediments (Santa Barbara Basin, California). *Applied and Environmental Microbiology* **75**, 1487–1499.

Hayes JM (1983) Geochemical evidence bearing on the origin of aerobiosis, a speculative interpretation. In: *The Earth's Earliest Biosphere: Its Origin and Evolution* (ed. Schopf JW). Princeton University Press, Princeton, NJ, 93–134.

Hayes, JM (1994) Global methanotrophy at the Archaean-Proterozoic transition. In: *Early Life on Earth*, Nobel Symposium, Vol. 84 (ed Bengtson S). Columbia University Press, New York, 220–236.

Head IM, Jones DM, Larter SR (2003) Biological activity in the deep subsurface and the origin of heavy oil. *Nature* **426**, 344–352.

Herndl GJ, Reinthaler T, Teira E, *et al.* (2005) Contribution of Archaea to total prokaryotic production in the Deep Atlantic Ocean. *Applied and Environmental Microbiology* **71**, 2303–2309.

Hershberger KL, Barns SM, Reysenbach AL, Dawson SC, Pace NR (1996) Wide diversity of Crenarchaeota. *Nature* **384**, 420.

Hewson I, Poretsky RS, Beinart RA, *et al.* (2009) In situ transcriptomic analysis of the globally important keystone N2-fixing taxon *Crocosphaera watsonii. ISME Journal* **3**, 618–631.

Hewson I, Poretsky RS, Tripp HJ, Montoya JP, Zehr JP (2010) Spatial patterns and light-driven variation of microbial population gene expression in surface waters of the oligotrophic open ocean. *Environmental Microbiology* **12**, 1940–1956.

Hinrichs KU (2002) Microbial fixation of methane carbon at 2.7 Ga: was an anaerobic mechanism possible? **3**, 1042 *DOI: 10.1029/2001GC000286.*

Hinrichs KU, Hayes JM, Sylva SP, Brewer PG, DeLong EF (1999) Methane-consuming archaebacteria in marine sediments. *Nature* **398**, 802–805.

Hjorleifsdottir S, Skirnisdottir S, Hreggvidsson GO, Holst O, Kristjansson JK (2001) Species composition of cultivated and noncultivated bacteria from short filaments in an Icelandic hot spring at 88 degrees C. *Microbiology Ecology* **42**, 117–125.

Hoehler T, Alperin M, Albert D, Martens C (1994) Field and laboratory studies of methane oxidation in an anoxic marine sediment: evidence for a methanogen-sulfate reducer consortium. *Global Biogeochemical Cycles* **8**, 451–463.

Hoff WD, Jung KH, Spudich JL (1997) Molecular mechanism of photosignaling by archaeal sensory rhodopsins. *Annual Review of Biophysics and Biomolecular Structure* **26**, 223–258.

Holtman CK, Chen Y, Sandoval P, *et al.* (2005) High-throughput functional analysis of the *Synechococcus elongatus* PCC 7942 genome. *DNA Research* **12**, 103–115.

House CH, Runnegar B, Fitz-Gibbon ST (2003) Geobiological analysis using whole genome-based tree building applied to the Bacteria, Archaea, and Eukarya. *Geobiology* **1**, 15–26.

Howard EC, Henriksen JR, Buchan A, *et al.* (2006) Bacterial taxa that limit sulfur flux from the ocean. *Science* **314**, 649–652.

Huang WE, Stoecker K, Griffiths R, *et al.* (2007a) Raman-FISH: combining stable-isotope Raman spectroscopy and fluorescence in situ hybridization for the single cell analysis of identity and function. *Environmental Microbiology* **9**, 1878–1889.

Huang WE, Stoecker K, Griffiths R, *et al.* (2007b) Raman-FISH: combining stable-isotope Raman spectroscopy and fluorescence in situ hybridization for the single cell analysis of identity and function. *Environmental Microbiology* **9**, 1878–1889.

Huber H, Hohn MJ, Rachel R, Fuchs T, Wimmer VC, Stetter KO (2002) A new phylum of Archaea represented by a nanosized hyperthermophilic symbiont. *Nature* **417**, 63–67.

Hugenholtz P, Pitulle C, Hershberger KL, Pace NR (1998) Novel division level bacterial diversity in a Yellowstone hot spring. *Journal of Bacteriology* **180**, 366–376.

Huse SM, Dethlefsen L, Huber JA, Welch DM, Relman DA, Sogin ML (2008) Exploring microbial diversity and taxonomy using SSU rRNA hypervariable tag sequencing. *PLoS Genetics* **4**, e1000255.

Inagaki F, Suzuki M, Takai K, *et al.* (2003) Microbial communities associated with geological horizons in coastal subseafloor sediments from the Sea of Okhotsk. *Applied and Environmental Microbiology* **69**, 7224–7235.

Inagaki F, Tsunogai U, Suzuki M, *et al.* (2004) Characterization of C-1-metabolizing prokaryotic communities in methane seep habitats at the Kuroshima Knoll, southern Ryukyu arc, by analyzing pmoA, mmoX, mxaF, mcrA, and 16S rRNA genes. *Applied and Environmental Microbiology* **70**, 7445–7455.

Ingolia NT, Ghaemmaghami S, Newman JRS, Weissman JS (2009) Genome-wide analysis in vivo of translation with nucleotide resolution using ribosome profiling. *Science* **324**, 218–223.

Inskeep WP, Rusch DB, Jay ZJ, *et al.* (2010) Metagenomes from high-temperature chemotrophic systems reveal geochemical controls on microbial community structure and function. *PLoS ONE* **5**, e9773.

Ishoey T, Woyke T, Stepanauskas R, Novotny M, Lasken RS (2008) Genomic sequencing of single microbial cells from environmental samples. *Current Opinion in Microbiology* **11**, 198–204.

Iversen N, Jorgensen BB (1985) Anaerobic methane oxidation rates at the sulfate-methane transition in marine sediments from Kattegat and Skagerrak (Denmark). *Limnology and Oceanography* **30**, 944–955.

Jansen R, van Embden JDA, Gaastra W, Schouls LM (2002) Identification of genes that are associated with DNA repeats in prokaryotes. *Molecular Microbiology* **43**, 1565–1575.

Jeans C, Singer SW, Chan CS, *et al.* (2008) Cytochrome 572 is a conspicuous membrane protein with iron oxidation activity purified directly from a natural acidophilic microbial community. *ISME Journal* **2**, 542–550.

Jeon CO, Park W, Padmanabhan P, DeRito C, Snape JR, Madsen EL (2003) Discovery of a bacterium, with distinctive dioxygenase, that is responsible for in situ biodegradation in contaminated sediment. *Proceedings of the National Academy of Sciences, USA* **100**, 13591–13596.

Jiao YYQ, Kappler A, Croal LR, Newman DK (2005) Isolation and characterization of a genetically tractable photo autotrophic Fe(II)-oxidizing bacterium, *Rhodopseudomonas palustris* strain TIE-1. *Applied and Environmental Microbiology* **71**, 4487–4496.

Jinks-Robertson S, Nomura M (1987) *Escherichia coli and Salmonella typhimurium: Cellular and Molecular Biology*. American Society for Microbiology, Washington, DC.

Karp PD, Keseler IM, Shearer A, *et al.* (2007) Multidimensional annotation of the Escherichia coli K-12 genome. *Nucleic Acids Research* **35**, 7577–7590.

Knittel K, Losekann T, Boetius A, Kort R, Amann R (2005) Diversity and distribution of methanotrophic archaea at cold seeps. *Applied and Environmental Microbiology* **71**, 467–479.

Koga Y, Nishihara M, Morii H, Akagawa-Matsushita M (1993) Ether polar lipids of methanogenic bacteria: structures, comparative aspects, and biosyntheses. *Microbiology and Molecular Biology Reviews* **57**, 164–182.

Konhauser KO, Pecoits E, Lalonde SV, *et al.* (2009) Oceanic nickel depletion and a methanogen famine before the Great Oxidation Event. *Nature* **458**, 750-U85.

Konneke M, Bernhard AE, de la Torre JR, Walker CB, Waterbury JB, Stahl DA (2005) Isolation of an autotrophic ammonia-oxidizing marine archaeon. *Nature* **437**, 543–546.

Kreuzer-Martin HW (2007) Stable isotope probing: linking functional activity to specific members of microbial communities. *Soil Science Society of America Journal* **71**, 611–619.

Kruger M, Meyerdierks A, Glockner FO, *et al.* (2003) A conspicuous nickel protein in microbial mats that oxidize methane anaerobically. *Nature* **426**, 878–881.

Kulkarni G, Kridelbaugh DM, Guss AM, Metcalf WW (2009) Hydrogen is a preferred intermediate in the energy-conserving electron transport chain of *Methanosarcina barkeri*. *Proceedings of the National Academy of Sciences of the United States of America* **106**, 15915–15920.

Lanoil BD, Sassen R, La Duc MT, Sweet ST, Nealson KH (2001) Bacteria and archaea physically associated with Gulf of Mexico gas hydrates. *Applied and Environmental Microbiology* **67**, 5143–5153.

Lin S, Henze S, Lundgren P, Bergman B, Carpenter EJ (1998) Whole-cell immunolocalization of nitrogenase in marine diazotrophic cyanobacteria, trichodesmium spp. *Applied and Environmental Microbiology* **64**, 3052–3058.

Liu Z, Guiliani N, Appia-Ayme C, Borne F, Ratouchniak J, Bonnefoy V (2000) Construction and characterization of a recA mutant of *Thiobacillus ferrooxidans* by marker exchange mutagenesis. *Journal of Bacteriology* **182**, 2269–2276.

Lloyd KG, Lapham L, Teske A (2006) An anaerobic methane-oxidizing community of ANME-1b archaea in hypersaline Gulf of Mexico sediments. *Applied and Environmental Microbiology* **72**, 7218–7230.

Lloyd KG, Albert DB, Biddle JF, Chanton JP, Pizarro O, Teske A (2010) Spatial structure and activity of sedimentary microbial communities underlying a Beggiatoa spp. mat in a Gulf of Mexico hydrocarbon seep. *PLoS ONE* **5**, e8738.

Malasarn D, Saltikov W, Campbell KM, Santini JM, Hering JG, Newman DK (2004) arrA is a reliable marker for As(V) respiration. *Science* **306**, 455–455.

Marcy Y, Ishoey T, Lasken RS, *et al.* (2007) Nanoliter reactors improve multiple displacement amplification of genomes from single cells. *PLoS Genetics* **3**, 1702–1708.

Maron PA, Ranjard L, Mougel C, Lemanceau P (2007) Metaproteomics: a new approach for studying functional microbial ecology. *Microbial Ecology* **53**, 486–493.

Marraffini LA, Sontheimer EJ (2010) CRISPR interference: RNA-directed adaptive immunity in bacteria and archaea. *Nature Reviews Genetics* **11**, 181–190.

Martens CS, Berner RA (1977) Interstitial water chemistry of Long Island Sound sediments, I, dissolved gases. *Limnology and Oceanography* **22**, 10–25.

Martinez RJ, Mills HJ, Story S, Sobecky PA (2006) Prokaryotic diversity and metabolically active microbial populations in sediments from an active mud volcano in the Gulf of Mexico. *Environmental Microbiology* **8**, 1783–1796.

Martinez A, Tyson GW, DeLong EF (2009) Widespread known and novel phosphonate utilization pathways in marine bacteria revealed by functional screening and metagenomic analyses. *Environmental Microbiology* **12**, 222–238.

Mayr S, Latkoczy C, Kruger M, *et al.* (2008) Structure of an F430 variant from archaea associated with anaerobic oxidation of methane. *Journal of the American Chemical Society* **130**, 10758–10767.

Metcalf WW (1999) Genetic analysis in the domain Archaea. In: *Methods in Microbiology*, Vol. 29 (eds Smith M, Sockett L). Academic Press Inc., San Diego, CA, 277–326.

Meuer J, Kuettner HC, Zhang JK, Hedderich R, Metcalf WW (2002) Genetic analysis of the archaeon *Methanosarcina barkeri* Fusaro reveals a central role for Ech hydrogenase and ferredoxin in methanogenesis and carbon fixation. *Proceedings of the National Academy of, USA* **99**, 5632–5637.

Meyerdierks A, Kube M, Lombardot T, *et al.* (2005) Insights into the genomes of archaea mediating the anaerobic oxidation of methane. *Environmental Microbiology* **7**, 1937–1951.

Meyerdierks A, Kube M, Kostadinov I, *et al.* (2009) Metagenome and mRNA expression analyses of anaerobic methanotrophic archaea of the ANME-1 group. *Environmental Microbiology* **12**, 422–439.

Mills HJ, Martinez RJ, Story S, Sobecky PA (2004) Identification of members of the metabolically active microbial populations associated with Beggiatoa species mat communities from Gulf of Mexico cold-seep sediments. *Applied and Environmental Microbiology* **70**, 5447–5458.

Miyazaki J, Higa R, Toki T, *et al.* (2009) Molecular characterization of potential nitrogen fixation by anaerobic methane-oxidizing archaea in the methane seep sediments at the number 8 Kumano Knoll in the Kumano Basin, offshore of Japan. *Applied and Environmental Microbiology* **75**, 7153–7162.

Moraru C, Lam P, Fuchs BM, Kuypers MM, Amann R (2010) GeneFISH – an in situ technique for linking gene presence and cell identity in environmental microorganisms. *Environmental Microbiology* **12**, 3057–3073.

Morris RM, Rappe MS, Connon SA, *et al.* (2002) SAR11 clade dominates ocean surface bacterioplankton communities. *Nature* **420**, 806–810.

Morris JJ, Kirkegaard R, Szul MJ, Johnson ZI, Zinser ER (2008) Facilitation of robust growth of Prochlorococcus colonies and dilute liquid cultures by 'Helper' heterotrophic bacteria. *Applied and Environmental Microbiology* **74**, 4530–4534.

Mukohata Y, Ihara K, Tamura T, Sugiyama Y (1999) Halobacterial rhodopsins. *Journal of Biochemistry* **125**, 649–657.

Murase J, Frenzel P (2007) A methane-driven microbial food web in a wetland rice soil. *Environmental Microbiology* **9**, 3025–3034.

Nauhaus K, Albrecht M, Elvert M, Boetius A, Widdel F (2007) In vitro cell growth of marine archaeal-bacterial consortia during anaerobic oxidation of methane with sulfate. *Environmental Microbiology* **9**, 187–196.

Newman DK, Gralnick JA (2005) What genetics offers geobiology. *Reviews in Molecular Geomicrobiology*, Vol. 59 (eds Banfield JF, Nealson KH). Mineralogical Society of America, Chantilly, VA, pp. 9–26.

Nicol GW, Schleper C (2006) Ammonia-oxidising Crenarchaeota: important players in the nitrogen cycle? *Trends in Microbiology* **14**, 207–212.

Niemann H, Losekann T, de Beer D, *et al.* (2006) Novel microbial communities of the Haakon Mosby mud volcano and their role as a methane sink. *Nature* **443**, 854–858.

Nouailler M, Bruscella P, Lojou E, Lebrun R, Bonnefoy V, Guerlesquin F (2006) Structural analysis of the HiPIP from

the acidophilic bacteria: *Acidithiobacillus ferrooxidans*. *Extremophiles* **10**, 191–198.

Olsen GJ, Lane DJ, Giovannoni SJ, Pace NR, Stahl DA (1986) Microbial ecology and evolution: a ribosomal RNA approach. *Annual Review of Microbiology* **40**, 337–365.

Orcutt B, Boetius A, Elvert M, Samarkin V, Joye SB (2005) Molecular biogeochemistry of sulfate reduction, methanogenesis and the anaerobic oxidation of methane at Gulf of Mexico cold seeps. *Geochimica et Cosmochimica Acta* **69**, 4267–4281.

Orphan VJ, House CH (2009) Geobiological investigations using secondary ion mass spectrometry: microanalysis of extant and paleo-microbial processes. *Geobiology* **7**, 360–372.

Orphan VJ, Hinrichs KU, Ussler W, *et al.* (2001a) Comparative analysis of methane-oxidizing archaea and sulfate-reducing bacteria in anoxic marine sediments. *Applied and Environmental Microbiology* **67**, 1922–1934.

Orphan VJ, House CH, Hinrichs KU, McKeegan KD, DeLong EF (2001b) Methane-consuming archaea revealed by directly coupled isotopic and phylogenetic analysis. *Science* **293**, 484–487.

Orphan VJ, Turk KA, Green AM, House CH (2009) Patterns of N-15 assimilation and growth of methanotrophic ANME-2 archaea and sulfate-reducing bacteria within structured syntrophic consortia revealed by FISH-SIMS. *Environmental Microbiology* **11**, 1777–1791.

Ottesen EA, Hong JW, Quake SR, Leadbetter JR (2006) Microfluidic digital PCR enables multigene analysis of individual environmental bacteria. *Science* **314**, 1464–1467.

Pace NR (1997) A molecular view of microbial diversity and the biosphere. *Science* **276**, 734–740.

Pancost RD, Sinninghe Damsté JS (2003) Carbon isotopic compositions of prokaryotic lipids as tracers of carbon cycling in diverse settings. *Chemical Geology* **195**, 29–58.

Parkes RJ, Cragg BA, Banning N, *et al.* (2007) Biogeochemistry and biodiversity of methane cycling in subsurface marine sediments (Skagerrak, Denmark). *Environmental Microbiology* **9**, 1146–1161.

Peckmann J, Thiel V (2004) Carbon cycling at ancient methane–seeps. *Chemical Geology* **205**, 443–467.

Pernthaler A, Amann R (2004) Simultaneous fluorescence in situ hybridization of mRNA and rRNA in environmental bacteria. *Applied and Environmental Microbiology* **70**, 5426–5433.

Pernthaler A, Dekas AE, Brown CT, Goffredi SK, Embaye T, Orphan VJ (2008) Diverse syntrophic partnerships from deep-sea methane vents revealed by direct cell capture and metagenomics. *Proceedings of the National Academy of Sciences, USA* **105**, 7052–7057.

Polz MF, Cavanaugh CM (1998) Bias in template-to-product ratios in multitemplate PCR. *Applied and Environmental Microbiology* **64**, 3724–3730.

Poretsky RS, Hewson I, Sun S, Allen AE, Zehr JP, Moran MA (2009) Comparative day/night metatranscriptomic analysis of microbial communities in the North Pacific subtropical gyre. *Environmental Microbiology* **11**, 1358–1375.

Quatrini R, Appia-Ayme C, Denis Y, *et al.* (2006) Insights into the iron and sulfur energetic metabolism of *Acidithiobacillus ferrooxidans* by microarray transcriptome profiling. *Hydrometallurgy* **83**, 263–272.

Quatrini R, Appia-Ayme C, Denis Y, Jedlicki E, Holmes DS, Bonnefoy V (2009) Extending the models for iron and sulfur oxidation in the extreme acidophile Acidithiobacillus ferrooxidans. *BMC Genomics* **10**, 394.

Radajewski S, McDonald IR, Murrell JC (2003) Stable-isotope probing of nucleic acids: a window to the function of uncultured microorganisms. *Current Opinion in Biotechnology* **14**, 296–302.

Ram RJ, VerBerkmoes NC, Thelen MP, *et al.* (2005) Community proteomics of a natural microbial biofilm. *Science* **308**, 1915–1920.

Rawlings DE, Kusano T (1994) Molecular genetics of *Thiobacillus ferroxidans*. *Microbiological Reviews* **58**, 39–55.

Reeburgh WS (1976) Methane consumption in Cariaco Trench waters and sediments. *Earth and Planetary Sciences Letters* **28**, 337–344.

Reed DW, Fujita Y, Delwiche ME, *et al.* (2002) Microbial communities from methane hydrate-bearing deep marine sediments in a forearc basin. *Applied and Environmental Microbiology* **68**, 3759–3770.

Reysenbach AL, Ehringer M, Hershberger K (2000) Microbial diversity at 83 degrees C in Calcite Springs, Yellowstone National Park: another environment where the Aquificales and 'Korarchaeota' coexist. *Extremophiles* **4**, 61–67.

Reysenbach AL, Liu Y, Banta AB, *et al.* (2006) A ubiquitous thermoacidophilic archaeon from deep-sea hydrothermal vents. *Nature* **442**, 444–447.

Reysenbach AL, Hamamura N, Podar M, *et al.* (2009) Complete and draft genome sequences of six members of the Aquificales. *Journal of Bacteriology* **191**, 1992–1993.

Rocap G, Larimer FW, Lamerdin J, *et al.* (2003) Genome divergence in two Prochlorococcus ecotypes reflects oceanic niche differentiation. *Nature* **424**, 1042–1047.

Roesch LF, Fulthorpe RR, Riva A, *et al.* (2007) Pyrosequencing enumerates and contrasts soil microbial diversity. *ISME Journal* **1**, 283–290.

Rogers SW, Moorman TB, Ong SK (2007) Fluorescent in situ hybridization and micro-autoradiography applied to ecophysiology in soil. *Soil Science Society of America Journal* **71**, 620–631.

Rollefson JB, Levar CE, Bond DR (2009) Identification of genes involved in biofilm formation and respiration via Mini-Himar transposon mutagenesis of *Geobacter sulfurreducens*. *Journal of Bacteriology* **191**, 4207–4217.

Rother M, Metcalf WW (2004) Anaerobic growth of *Methanosarcina acetivorans* C2A on carbon monoxide: an unusual way of life for a methanogenic archaeon. *Proceedings of the National Academy of Sciences, USA* **101**, 16929–16934.

Rother M, Metcalf WW (2005) Genetic technologies for Archaea. *Current Opinion in Microbiology* **8**, 745–751.

Rye R, Holland HD (2000) Life associated with a 2.76 Ga ephemeral pond?: evidence from Mount Roe #2 paleosol. *Geology* **28**, 483–486.

Saffarini DA, Dichristina TJ, Bermudes D, Nealson KH (1994) Anaaerobic respiration of *Shewanella putrefaciens* requires both chromosomal and plamid-borne genes. *FEMS Microbiology Letters* **119**, 271–277.

Saltikov CW, Newman DK (2003) Genetic identification of a respiratory arsenate reductase. *Proceedings of the National Academy of Sciences, USA* **100**, 10983–10988.

Saltikov CW, Cifuentes A, Venkateswaran K, Newman DK (2003) The ars detoxification system is advantageous but not required for As(V) respiration by the genetically tractable Shewanella species strain ANA-3. *Applied and Environmental Microbiology* **69**, 2800–2809.

Schleper C, Jurgens G, Jonuscheit M (2005) Genomic studies of uncultivated archaea. *Nature Reviews Microbiology* **3**, 479–488.

Scheller S, Goenrich M, Boecher R, Thauer RK, Jaun B (2010) The key nickel enzyme of methanogenesis catalyses the anaerobic oxidation of methane. *Nature* **465**, 606–608.

Schoell M, Wellmer FW (1981) Anomalous C-13 depletion in early Precambrian graphites from Superior-Province, Canada. *Nature* **290**, 696–699.

Schrenk MO, Edwards KJ, Goodman RM, Hamers RJ, Banfield JF (1998) Distribution of *Thiobacillus ferrooxidans* and *Leptospirillum ferrooxidans*: implications for generation of acid mine drainage. *Science* **279**, 1519–1522.

Schuler D (2008) Genetics and cell biology of magnetosome formation in magnetotactic bacteria. *FEMS Microbiology Reviews* **32**, 654–672.

Shaw AJ, Podkaminer KK, Desai SG, *et al.* (2008) Metabolic engineering of a thermophilic bacterium to produce ethanol at high yield. *Proceedings of the National Academy of Sciences, USA* **105**, 13769–13774.

Shaw AJ, Hogestt DA, Lynd LR (2010) Natural competence in Thermoanaerobacter and Thermoanaerobacterium species. *Applied and Environmental Microbiology* **76**, 4713–4719, doi:10.1128/AEM.00402-10.

Shima S, Thauer RK (2005) Methyl-coenzyme M reductase and the anaerobic oxidation of methane in methanotrophic Archaea. *Current Opinion in Microbiology* **8**, 643–648.

Singer SW, Chan CS, Zemla A, *et al.* (2008) Characterization of cytochrome 579, an unusual cytochrome isolated from an iron-oxidizing microbial community. *Applied and Environmental Microbiology* **74**, 4454–4462.

Sivan O, Adler M, Pearson A, Gelman F, Bar-Or I, John SG, Eckert W (2011) Geochemical evidence for iron-mediated anaerobic oxidation of methane. *Limnology and Oceanography* **56**, 1536–1544.

Sogin ML, Morrison HG, Huber JA, *et al.* (2006) Microbial diversity in the deep sea and the underexplored 'rare biosphere'. *Proceedings of the National Academy of Sciences, USA* **103**, 12115–12120.

Sowell SM, Wilhelm LJ, Norbeck AD, *et al.* (2008) Transport functions dominate the SAR11 metaproteome at low-nutrient extremes in the Sargasso Sea. *ISME Journal* **3**, 93–105.

Springer E, Sachs MS, Woese CR, Boone DR (1995) Partial gene sequences for the A subunit of methyl-coenzyme M reductase (mcrI) as a phylogenetic tool for the family Methanosarcinaceae. *International Journal of Systematic Bacteriology* **45**, 554–559.

Stepanauskas R, Sieracki ME (2007) Matching phylogeny and metabolism in the uncultured marine bacteria, one cell at a time. *Proceedings of the National Academy of Sciences of the United States of America* **104**, 9052–9057.

Stewart FJ, Ottesen EA, DeLong EF (2010) Development and quantitative analyses of a universal rRNA-subtraction protocol for microbial metatranscriptomics. *ISME Journal* **4**, 896–907.

Summons RE, Jahnke LJ, Rsksandic Z (1994) Carbon isotopic fractionation in lipids from methanotrophic bacteria:

relevance for interpretation of the geochemical record of bio-markers. *Geochem et Cosmochem Acta* **58**, 2853–2863.

Teske A, Hinrichs KU, Edgcomb V, *et al.* (2002) Microbial diversity of hydrothermal sediments in the Guaymas Basin: evidence for anaerobic methanotrophic communities. *Applied and Environmental Microbiology* **68**, 1994–2007.

Teske A, Dhillon A, Sogin ML (2003) Genomic markers of ancient anaerobic microbial pathways: sulfate reduction, methanogenesis, and methane oxidation. *The Biological Bulletin* **204**, 186–191.

Thauer RK (1998) Biochemistry of methanogenesis: a tribute to Marjory Stephenson. 1998 Marjory Stephenson Prize Lecture. *Microbiology* **144**(Pt 9), 2377–2406.

Thauer RK, Shima S (2008) Methane as fuel for anaerobic microorganisms. *Annals of the New York Academy of Sciences* **1125**, 158–170.

Thomazo C, Ader M, Farquhar J, Philippot P (2009) Methanotrophs regulated atmospheric sulfur isotope anomalies during the Mesoarchean (Tumbiana Formation, Western Australia), *Earth and Planetary Science Letters* **279**(1–2), 65–75.

Thompson JR, Randa MA, Marcelino LA, Tomita-Mitchell A, Lim E, Polz MF (2004) Diversity and dynamics of a north atlantic coastal Vibrio community. *Applied and Environmental Microbiology* **70**, 4103–4110.

Thomsen TR, Finster K, Ramsing NB (2001) Biogeochemical and molecular signatures of anaerobic methane oxidation in a marine sediment. *Applied and Environmental Microbiology* **67**, 1646–1656.

Turnbaugh PJ, Gordon JI (2009) The core gut microbiome, energy balance and obesity. *Journal of Physiology* **587**, 4153–4158.

Tyson GW, Banfield JF (2008) Rapidly evolving CRISPRs implicated in acquired resistance of microorganisms to viruses. *Environmental Microbiology* **10**, 200–207.

Tyson GW, Chapman J, Hugenholtz P, *et al.* (2004) Community structure and metabolism through reconstruction of microbial genomes from the environment. *Nature* **428**, 37–43.

Urich T, Lanzen A, Qi J, Huson DH, Schleper C, Schuster SC (2008) Simultaneous assessment of soil microbial community structure and function through analysis of the meta-transcriptome. *PLoS ONE* **3**, e2527.

Venter JC, Remington K, Heidelberg JF, *et al.* (2004) Environmental genome shotgun sequencing of the Sargasso Sea. *Science* **304**, 66–74.

Venturi V, Bertani I, Kerenyi A, Netotea S, Pongor S (2010) Co-swarming and local collapse: quorum sensing conveys resilience to bacterial communities by localizing cheater mutants in Pseudomonas aeruginosa. *PLoS ONE* **5**, e9998 DOI: 10.1371/journal.pone.0009998.

VerBerkmoes NC, Denef VJ, Hettich RL, Banfield JF (2009) Systems biology: functional analysis of natural microbial consortia using community proteomics. *Nature Reviews Microbiology* **7**, 196–205.

Vetriani C, Tran HV, Kerkhof LJ (2003) Fingerprinting microbial assemblages from the oxic/anoxic chemocline of the Black Sea. *Applied and Environmental Microbiology* **69**, 6481–6488.

Wagner M (2009) Single-cell ecophysiology of microbes as revealed by Raman microspectroscopy or secondary ion mass spectrometry imaging. *Annual Review of Microbiology* **63**, 411–429.

Wall JD, Arkin AP, Balci NC, Rapp-Giles B (2008) *Genetics and Genomics of Sulfate Respiration in Desulfovibrio.* Springer-Verlag, Berlin.

Ward BB (2005) Molecular approaches to marine microbial cology and the marine nitrogen cycle. *Annual Review of Earth and Planetary Sciences* **33**, 301–333.

Whiticar MJ (1999) Carbon and hydrogen isotope systematics of bacterial formation and oxidation of methane. *Chemical Geology* **161**, 291–314.

Wilhelm L, Tripp HJ, Givan S, Smith D, Giovannoni S (2007) Natural variation in SAR11 marine bacterioplankton genomes inferred from metagenomic data. *Biology Direct* **2**, 27.

Wilmes P, Bond PL (2006) Metaproteomics: studying functional gene expression in microbial ecosystems. *Trends in Microbiology* **14**, 92–97.

Woese CR, Fox GE (1977) Phylogenetic structure of the prokaryotic domain: the primary kingdoms. *Proceedings of the National Academy of Sciences,USA* **74**, 5088–5090.

Woods DR, Rawlings DE, Barros ME, Pretorius I, Ramesar R (1986) Molecular genetic studies of *Thiobacillus ferrooxidans*: the development of genetic systems and the expression of cloned genes. *Biotechnology and Applied Biochemistry* **8**, 231–241.

Woyke T, Xie G, Copeland A, *et al.* (2009) Assembling the marine metagenome, one cell at a time. *PLoS ONE* **4**, e5299.

Yarzabal A, Brasseur G, Ratouchniak J, *et al.* (2002) The high-molecular-weight cytochrome c Cyc2 of Acidithiobacillus ferrooxidans is an outer membrane protein. *Journal of Bacteriology* **184**, 313–317.

Yarzabal A, Appia-Ayme C, Ratouchniak J, Bonnefoy V (2004) Regulation of the expression of the *Acidithiobacillus ferrooxidans* rus operon encoding two cytochromes c, a cytochrome oxidase and rusticyanin. *Microbiology-SGM* **150**, 2113–2123.

Zargar K, Hoeft S, Oremland RS, Saltikov CW (2010) Identification of a novel arsenite oxidase gene, arxA, in the haloalkaliphilic, arsenite-oxidizing bacterium *Alkalilimnicola ehrlichii* strain MLHE-1. *Journal of Bacteriology* **192**, 3755–3762.

Zehnder AJB, Brock TD (1979) Methane formation and methane oxidation by methanogenic bacteria. *Journal of Bacteriology* **137**, 420–432.

14

STABLE ISOTOPE GEOBIOLOGY

D.T. Johnston[1] and W.W. Fischer[2]

[1]Department of Earth and Planetary Sciences, Harvard University, Cambridge, MA, USA
[2]Division of Geological and Planetary Sciences, MC 100-23 California Institute of Technology, Pasadena, CA 91125, USA

14.1 Introduction

Stable isotopes are tools that geobiologists use to investigate natural and experimental systems, with questions ranging from those aimed at modern environments and extant microorganisms, to deep time and the ancient Earth. In addition to a wide range of timescales, stable isotopes can be applied over a vast range of spatial scales. Isotope studies inform our understanding of bulk planetary compositions, global element cycles, down to the metabolism of single cells. With roots extending back to the first half of the twentieth century (Urey, 1947), stable isotope geobiology recently emerged as a scientific discipline with questions rooted in geology and Earth history, with methods adopted from nuclear chemistry.

Stable isotope studies have strongly shaped our understanding of the environmental history of Earth's surface and its interplay with an evolving biology. Studies ranging from the analysis of carbon isotopes in both carbonate and organic carbon (e.g. Knoll *et al.*, 1986; Shields and Veizer, 2002), sulfur isotope studies of sulfate and sulfide (e.g. Canfield and Teske, 1996: Fike *et al.*, 2006), to a litany of metals and complementary isotope systems (Arnold *et al.*, 2004; Rouxel *et al.*, 2005) provide information about seawater chemistry, atmospheric oxygen, and evolution of the biosphere. Driven in part by the advent of new technologies (e.g. multi-collector ICP-MS, or inductively coupled plasma mass spectrometry; see also Box 14.1), and the relative ease with which most light stable biogeochemical elements can routinely be measured, stable isotope geobiology will only continue to gain momentum.

The primary impetus behind using stable isotope ratios (see Box 14.2) to study geobiological problems is that isotope ratios become important tracers of mass flux and process in systems where absolute measurements are extremely challenging (either because the scale is too small and eludes measurement, or because the geologic record offers imperfect preservation). That said, there are a number of 'big questions' in geobiology, and isotopic studies will likely be involved in many of their solutions; but how? We see three specific features driving the evolution of isotope studies. The first is technological innovation. As mass-spectrometers become capable of making better measurements faster (or even at all), our understanding of the natural world stands to benefit. The direct beneficiaries of increased precision are the metal isotopes (e.g. Fe, Mo) and multiple isotope systems of biogeochemically active elements (namely O and S). Second, more accurate stories can be written by linking stable isotope systems from different element cycles, which allows multiple processes that operate on wide-ranging time-scales to be quantitatively connected. For instance, numerous studies have recently interpreted the co-variation of isotope data for multiple element systems, namely carbon and sulfur (Fike *et al.*, 2006; McFadden *et al.*, 2008). This approach also includes measuring the isotopic composition of different elements in the same compound or mineral (see FeS_2: Archer and Vance, 2006; $BaSO_4$: Turchyn and Schrag, 2004; Bao *et al.*, 2007). The final direction is more rigorous stratigraphic study of isotope records with insight gained from the patterns and processes captured in the geologic record. The methods of sedimentology and

Fundamentals of Geobiology, First Edition. Edited by Andrew H. Knoll, Donald E. Canfield and Kurt O. Konhauser.

Box 14.1

Isotope ratio mass-spectrometry

Regardless of the element of interest, there are a number of similarities in the instrumentation used to make isotope ratio measurements. For those unfamiliar with these analytical methods, here we provide a primer on gas source mass spectrometry, the technique behind most light stable isotope measurements. A mass-spectrometer (MS) has three fundamental components: a source, a magnet, and a collector (described below: see Fig. B14.1)[1].

- [The source]: The purpose of the source is to ionize the sample (which is a gas) with a filament. The sample, which now carries a charge, is then accelerated through a series of focusing slits into the flight tube of the MS. This acceleration is driven by the difference in pressure between the source and flight tube and the draw of the magnet on the ion.
- [The magnet]: Once in the flight tube, the flight of the charged sample is controlled by the interaction between its own charge and the magnetic field produced by the magnet. At this stage, the degree to which the magnet deflects the charged ion will be a function of the mass of the molecule (and the charge). For example, singly charged $^{12}CO_2$ will be deflected more strongly than singly charged $^{13}CO_2$.
- [The collectors]: The relative deflection separates the sample by isotopes, directing $^{12}CO_2$ and $^{13}CO_2$ along slightly different flight paths. An array of collectors, which are simply cups that record the impact of charged molecules, then measure the intensity of each incoming species (i.e. $^{12}CO_2$ versus $^{13}CO_2$). Comparing the relative intensities of $^{12}CO_2$ to $^{13}CO_2$ yields ^{13}r (see Equation 14.1).

Two primary configurations

A GSMS can be run in either dual inlet mode (DI) or in continuous flow mode (CF). In the case of DI, sample gas is loaded into a bellows (compressible and calibrated volume; think of an accordion) where the pressure is balanced by another bellows containing reference gas. During DI, the instrument alternates between drawing gas from the sample and standard bellows. The relative intensity of the sample and standard signals (height of the peaks) determines R (see Equation 14.1). Alternatively, CF streams a carrier gas (often He or Ar) continuously into the MS, and then injects an aliquot of known standard gas followed by an aliquot of sample gas. Rather than measuring the intensity (as is done with DI), CF integrates under the entire sample or standard peak, allowing for accurate measurement of smaller gas volumes and thus smaller sample sizes. Though they can be made on very small amounts of sample, CF measurements often make a slight tradeoff in measurement precision, compared to DI methods.

Front-end (sample prep)

All of the action at the front end of a MS (or sometimes done off line) goes to convert a sample from its natural state (be it gas, liquid, or solid) into a simple measureable gas molecule containing a given number of isotopologs. This is often the least uniform step in the process. Cryogenic or chromatographic treatments are commonly used to concentrate and/or separate particular compounds of interest and remove possible mass interferences (e.g. CO has several isotopologs with similar masses to N_2). In the case of certain measurements like carbonate carbon, this conversion is simple. Calcium carbonate ($CaCO_3$: a solid) is reacted with acid to produce CO_2 gas; this process faithfully converts carbonate to CO_2 and retains the isotope composition of the original carbonate. Other compounds/elements are less straightforward. A common system used for measuring many geobiological elements is an elemental analyser, or EA, where samples can be quickly converted via high temperature combustion (and additional *in situ* chemistry) to easily measurable gases. The reader is best served by surveying the primary literature for front-end techniques and sample preparation, as it ranges widely and is important for obtaining rapid, accurate, and precise results.

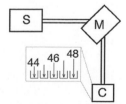

Figure B14.1 S = source, M = magnet, C = collectors. The 5-collector assembly captured in the inset is for CO_2, which is measured at masses 44–48 (more specifically mass/charge), where isotopologues of CO_2 comprise the mass range. This allows for measurements of $^{13}C/^{12}C$ as well as additional rare isotopologues, for instance Δ^{47} (see Eiler and Schauble, 2004).

---

[1] For the sake of this discussion, we focus on gas source mass spectrometry (GSMS) since this instrumentation is commonly used for the geobiological elements (H, N, C, O, S). Alternative instruments (thermal ionization mass-spectrometers [TIMS] and multi-collector inductively coupled plasma mass-spectrometers [MC-ICPMS]) are used for other stable and radiogenic isotopes of interest in Fig. 14.1.

Box 14.2

A survey of published stable isotopic data (Coplen *et al.*, 2002) for a range of elements (see Fig. B14.2) reveals an interesting relationship, whereby the elements central to biology commonly display larger relative fractionations in nature (see B.A. Wing, pers. comm.). Plotted as 'drive' for fractionation, we note that S and N are the most information-rich isotope systems, with C, O, and H falling close behind. Further, S and O have multiple stable isotopes and thus carry more potential information than elements with only two isotopes. This observation is important because in order to use isotopes to interpret a geochemical cycle, there has to be an interpretable level of variability (with favorable signal/noise). The values here are not set in stone, though there is some fundamental chemistry at work. As metal isotope systems receive more attention, the observed range of fractionation will likely increase. Also note that these relationships may change as more isotopic measurements are made at yet smaller length scales (e.g. with SIMS and nanoSIMS technology).

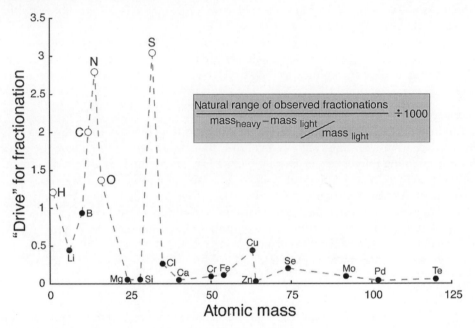

$$\frac{\text{Natural range of observed fractionations}}{\frac{\text{mass}_{heavy} - \text{mass}_{light}}{\text{mass}_{light}}} \div 1000$$

Figure B14.2 'Drive' for fractionation plotted against atomic mass. The 'drive' term is calculated from the inset equation and represents the fractionation observed in nature to that predicted by simple mass difference.

stratigraphy offer the promise of understanding and sampling the ancient world along both vectors of time and environment. Re-animating ancient ecosystems in this context (e.g. paleo-depth gradients, shelf and slope environments) is an approach that will continue to best inform geobiological processes, interactions, and long-term evolution. Before we pursue any of these directions, however, it is important to cover the rules of the stable isotope game.

14.1.1 Overview

Stable isotope geobiology is, often, an exercise in studying ancient isotope records in the context of an understanding of extant processes, microorganisms, and their associated metabolisms. Thus, it requires an understanding of chemistry (the elements and isotopes), biology (the organisms and metabolism), and geology

(the history and context). Other chapters will discuss specific processes, cycles, and records; here we will focus on describing isotope *systematics* and how stable isotope ratio data can be studied in ancient rocks. We begin by describing the geobiological elements (and their isotopes), reviewing relevant notations (and conventional shorthand), and outlining how to quantitatively assess isotopic fractionation in a system (introduction to model formulation). Following on this discussion, we will move to describe a series of treatments that range from cutting-edge to classical. Broadly, this discussion will cover how isotopes can be used to:

- identify microbial processes in the rock record, including experimental calibration and geological application,
- track elements or molecules as they flow through a system (including labelling techniques), be that system experimental or natural, and

- ask an appropriate geobiological question using stable isotope ratio data.

It is important that, although the following examples are given with regard to specific isotope systems, the general principles outlined below apply to all other stable isotope systems[2]. Understand these and you can apply the same approach to any system of interest.

14.2 Isotopic notation and the biogeochemical elements

14.2.1 Fundamentals of fractionation

Many of the bioessential elements have more than one stable isotope (see Fig. 14.1). The chemical properties of a given element are largely controlled by the number of protons that element contains and its configuration of electrons. In addition to protons, the nucleus of an atom can also contain neutrons (a chargeless quantity with a mass similar to a proton). Isotopes are thus defined by their overall mass, or the sum of protons and neutrons. Since they all contain the same number of protons, all isotopes of a particular element display similar chemistry. So if neutrons do not carry a charge, or greatly influence the reactivity, what are the implications of having them in variable numbers?

The answer rests with the additional mass. With the addition or subtraction of a neutron, the mass of the nucleus of an atom can change significantly. In the case of carbon, for example, moving from ^{12}C to ^{13}C results in ~8% change in the mass of the nucleus. Now, although ^{12}C and ^{13}C have nearly the same chemical properties, the strength of the bonds they form, and the rates at which they react, are related to the mass of the isotope. Differences in bond energies, caused by isotopic substitution (i.e. ^{13}C taking the place of ^{12}C), can be expressed with the concept of zero-point energies (ZPE). Shown in Fig. 14.2, the energy associated with a given bond (and isotopes of the elements involved) can be approximated as a function of the distance between the two nuclei ('nuclear separation' in Fig. 14.2). (There are several simplifications associated with this approach, and readers are encouraged to look at Urey, 1947, and Bigeleisen and Mayer, 1947, for a more thorough discussion.)

With isotopic substitution, the energy required to break a bond, or dissociate, changes such that heavier isotopes form stronger bonds. This often results in the preferential breaking of bonds with the lighter isotopes. The influence of mass on bond energy is the central principle driving isotope effects. An example of

mass-dependent variability is presented for carbon and oxygen isotopic substitution in a C–O molecule (inset in Fig. 14.2). Using ^{12}C–^{16}O as the reference isotopologue, the isotopic enrichment is systematic and increases proportionally as the difference between ^{12}C–^{16}O and the mass of the other isotopologs (for instance ^{12}C–^{18}O) increases. For example, the difference between the preferential breaking of bonds between $^{17}O/^{16}O$ would be roughly half that of $^{18}O/^{16}O$. This is explained by the fact that the mass difference between ^{17}O and ^{16}O is half that of ^{18}O and ^{16}O. Note that this approach only *approximates* the C–O isotope effect, as the exact calculation includes a series of additional terms (for further discussion, see Chacko *et al.*, 2001).

14.2.2 Notations and short-hand

Because the natural abundances of some isotopes are low, and the variability in isotope ratios displayed by many elements is on the order of parts per thousand (or per mill, ‰), geobiologists commonly report and discuss stable isotope ratio data using *delta notation* (δ, see Equations 14.1 and 14.2). Using carbon (^{12}C, ^{13}C) as an example, the ratio of ^{12}C and ^{13}C in a sample ($^{13}r_{sample}$) can be related to that same ratio of an internal reference ($^{13}r_{ref.}$) of known composition. The resulting ratio provides the composition of the sample on the scale determined by the reference ($^{13}R_{sample}$):

$$^{13}R_{sample} = {^{13}r_{sample}} \big/ {^{13}r_{ref.}} \tag{14.1}$$

International standards allow data from different laboratories to be directly compared, resulting in $^{13}R_{standard}$ ($^{13}r_{standard}/^{13}r_{ref.}$) and allowing $\delta^{13}C$ to be calculated:

$$\delta^{13}C_{sample} = 1000 \times \left[\frac{^{13}R_{sample}}{^{13}R_{standard}} - 1 \right] \tag{14.2}$$

Thus, $\delta^{13}C$ is the composition of a sample on a ‰ scale established by multiple standards.

The fractionation difference between two materials can also be calculated as a fractionation factor (α), which is the quotient of two isotope ratios:

$$^{13}\alpha_{A/B} = \left[\frac{^{13}R_{sample\_A}}{^{13}R_{sample\_B}} \right] = \left[\frac{\frac{\delta^{13}C_{sample\_A}}{1000} + 1}{\frac{\delta^{13}C_{sample\_B}}{1000} + 1} \right]$$

$$\tag{14.3}$$

The term α carries a precise meaning, however, and approximations often simplify mathematical treatments of isotope ratio data. The term epsilon (ε)[3] is one such

---

[2] For an understanding of radiogenic, as opposed to stable, isotope systems, see Dicken (2004).

[3] In trace element geochemistry, ε is used to denote parts per ten thousand.

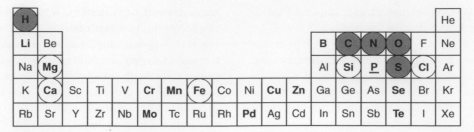

Figure 14.1 The periodic table of geobiological elements (inspired by Anbar and Rouxel, 2007). Within the grey circles are the major elements found in biological materials. In white circles are additional important elements common to sedimentary successions of all ages. The remaining elements, written in black text, have stable isotopes and are redox sensitive, pH sensitive, or are important micronutrients. Note that, although P is certainly a geobiologically-interesting element, it is also mono-isotopic. The isotope systems commonly studied in geobiology are carbon (^{12}C, ^{13}C), nitrogen (^{14}N, ^{15}N), oxygen (^{16}O, ^{17}O, ^{18}O), sulfur (^{32}S, ^{33}S, ^{34}S, and ^{36}S), iron (^{54}Fe, ^{56}Fe, ^{57}Fe, and ^{58}Fe), copper (^{63}Cu and ^{65}Cu), zinc (^{64}Zn, ^{66}Zn, ^{67}Zn, ^{68}Zn, and ^{70}Zn), and molybdenum (^{92}Mo, ^{94}Mo, ^{95}Mo, ^{96}Mo, ^{97}Mo, ^{98}Mo, and ^{100}Mo).

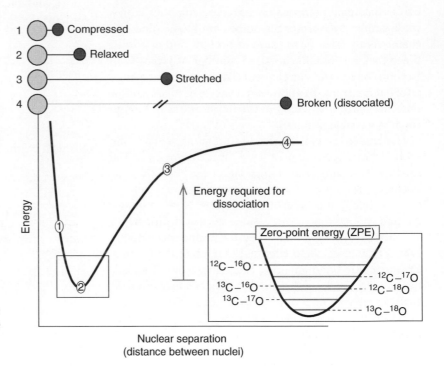

Figure 14.2 A schematic diagram showing the concept of zero point energies for isotopic substitution, with energy on the *y*-axis and distance between bound nuclei on the *x*-axis. As heavier isotopes are substituted into a molecule, the 'energy required for dissociation' increases, and as a result, the bonds are more difficult to break. The cartoon shown above the ZPE well illustrates how a diatomic molecule behaves, moving from compressed (point 1), into the bottom of the ZPE well where the bond is relaxed (point 2), through a stretching (point 3) and finally to dissociation (point 4). Included in the inset are approximate differences in ZPE for each of the isotopologues, approximated via changes in vibrational energies.

approximation and is calculated as the linear difference (in ‰) between δ values:

$$\varepsilon_{A-B} = \delta^{13}C_{sample\_A} - \delta^{13}C_{sample\_B} \qquad (14.4)$$

The approximation is apparent if we consider the isotopic difference between two distinct compositions: $\delta^{13}C_a = +5‰$ and $\delta^{13}C_b = -35‰$. Our approximation yields $\varepsilon_{a-b} = -40‰$, but if we instead substitute these values into Equation 14.3 (and with $\delta = 1000 \times (\alpha - 1)$, we see that the exact isotopic offset is $\delta^{13}C_{a-b} = 41.45‰$. The difference between ε_{a-b} and $\delta^{13}C_{a-b}$ is small, but important. This is especially true in multiple isotope systems, where slight (but resolvable) variability between multiple isotope ratios is being interpreted.

An additional isotopic notation, Δ (capital delta), is commonly used in the literature to represent one of two possible features. First, the Δ notation is used as an equivalent to ε_{a-b}, and is commonly applied to the fractionation observed between paired measurements of isotope ratios in organic carbon and carbonate carbon (e.g. Knoll *et al.*, 1986; Rothman *et al.*, 2003) or pyrite sulfur and sulfate sulfur in ancient rock samples (e.g. Fike *et al.*, 2006, McFadden *et al.*, 2008). This approach is discussed briefly below. The other use for Δ appears in treatments of isotope systems with more than two stable isotopes (namely O and S, with a slightly difference use for CO_2). Here the notation represents the deviation of observed isotope ratios from an expected fractionation relationship (see Luz *et al.* (1999) for an example with oxygen isotopes and Farquhar and Wing, (2003) or Johnston *et al.* (2008) for an example using sulfur isotopes). For the remainder of the chapter, we use either ε_{a-b} or δ_{a-b} to denote the fractionation between two species, and reserve Δ for describing multiple isotope systems (also see Chapter 5).

Finally, when reporting, discussing, and interpreting isotopic data, we ask that the reader note the difference between the terms 'isotope effects' and 'isotope fractionation'. Conceptually, isotope effects derive from a physical process via differences in chemical reaction rate constants for the isotopes of interest. We do not measure these effects directly. Rather in stable isotope geobiology, we measure isotope fractionation, that is the differences in isotope ratios between different pools. We then infer something about the underlying process in the context (often implicitly) of mathematical models. As our interpretations are only ever as good as the methods used to generate them, below we discuss some of the models commonly employed to interpret measurements of stable isotope ratios in geobiological problems.

14.2.3 *Flavors of isotopic fractionations*

Models constructed to evaluate isotopic data come in several common flavors. For isotope geobiology, we limit the discussion to two generalized categories of fractionation processes: *equilibrium* and *kinetic* isotope effects. With equilibrium isotope effects, the preferential breaking of bonds is related to the equilibrium constant (K_{eq}) of the system[4]. For instance, the following represents the chemical and isotopic equilibrium for carbon between bicarbonate and aqueous CO_2:

$$^{13}CO_{2(aq)} + H^{12}CO_3^- \Leftrightarrow {}^{12}CO_{2(aq)} + H^{13}CO_3^-$$

In a manner similar to how bicarbonate and aqueous CO_2 undergo exchange until equilibrium is reached, the

carbon isotopes (^{12}C and ^{13}C from above) exchange until isotopic equilibrium is reached. For tropical surface seawater at equilibrium, bicarbonate ends up approximately 9‰ heavier than the aqueous CO_2 (Zhang *et al.* 1995). Isotopic equilibrium can also exist between the same chemical species in different phases (O'Neil and Epstein, 1966):

$$C^{18}O_{2(aq)} + C^{16}O_{2(gas)} \Leftrightarrow C^{16}O_{2(aq)} + C^{18}O_{2(gas)}$$

This equilibrium would describe atmosphere-ocean exchange of carbon. Considering that equilibrium isotope effects are closely related to K_{eq}, they are often subject to all of the things that affect the equilibrium reaction, particularly temperature, and can be used in paleothermometry. In the case of equilibrium isotope effects, fractionation factors (α) are often used to represent the isotopic difference between species.

A majority of the fractionations associated with biological processes are considered kinetic (see Kaplan 1975). Rather than being related to K_{eq}, kinetic isotope effects are related to the ratio of the isotope-specific rate constants (k) for a unidirectional reaction; rate constants quantify the speed of a reaction in time. For example, one of the steps involved in dissimilatory sulfate (SO_4^{2-}) reduction is the intracellular reduction of sulfite (SO_3^{2-}) to hydrogen sulfide (H_2S) by the enzyme dissimilatory sulfite reductase, or *Dsr*. Since reaction rates are isotope-specific, and lighter isotopes react more quickly, the products of a reaction preferentially accumulate the light isotopes. Using the example above, the ratio of the rate constants will be less than 1, or $^{34}k_{dsr}/_{32}k_{dsr} < 1$, where

$^{34}k_{dsr}$ is the rate constant for *Dsr* involving ^{34}S. Kinetic isotope effects associated with enzymatic processes are often difficult to define exactly, since local (intracellular) conditions influence both substrate concentrations and the activity of the enzyme and subsequently alter the fractionation. Furthermore, most biological processes are composed of numerous enzymatic steps, meaning that the net fractionation for a given process is the sum of a series of intermediate steps (Hayes, 2001). The overall process of sulfate reduction, for example, of which *Dsr* catalysis comprises one step, is composed of numerous enzymatic, active, and passive transport processes (all of which have an associated rate). In fact, even the *Dsr* example above is likely composed of a set of intermediate reactions.

14.3 Tracking fractionation in a system

The use of stable isotopes as a tool requires more than an understanding of the magnitude of an effect, or determining a fractionation factor. Their use requires the ability to constrain how an effect, or a series of effects, propagate through a natural system and are

---

[4] A proper treatment of the relationship between K_{eq} and isotopic fractionation involves a parameter used to describe a partition function (Q), which incorporates all possible energies associated with a molecule. Different sorts of relevant energies include translational, rotational, and vibrational, the latter of which is central to the inset in Fig. 14.2. For worked examples (^{18}O exchange between $CO-CO_2$), see Chacko *et al.* (2001)).

preserved in the geologic record. Geobiologists use mass-balance (both elemental and isotopic) to place quantitative constraints on a system or geochemical cycle. This is an accounting exercise. For example, the marine carbon cycle can be tracked on an elemental scale by placing estimates on the amount of carbon that enters the oceans (via weathering, atmosphere–ocean exchange, and submarine volcanism), the amount that resides within the ocean (the sum of the aqueous species) and the amount that exits the ocean (via burial and subduction). If the flux of carbon into the ocean exceeds carbon removal, then the size of the marine reservoir will increase. The opposite is also true. Thus, the size of the marine reservoir will remain constant when the inputs equal the outputs (see residence time discussion below). Another important consideration in trying to understand a geochemical/isotopic record is 'time,' as certain time-scales lend to approximations that greatly simplify mathematical treatments. This approach is common in studies of Precambrian chemostratigraphy, where 10s to 100s of millions of years are considered. In all cases, similar principles hold for isotopic mass-balance, whereby rather than just C, we track ^{12}C and ^{13}C.

14.3.1 Non-steady state approaches

There are numerous methodological approaches that allow quantitative constraints to be placed on isotopic fractionations in a system. The description given below is in the spirit of Mariotti *et al.* (1981) and Hayes (2001). If we adopt a generic system, represented by a single box (Fig. 14.3), overall mass-balance on the system can be described as:

$$\frac{dM}{dt} = \sum J_{in} - \sum J_{out} \tag{14.5}$$

where dM is the change in mass per change in unit time (dt), and J represents the molar flux of material in (J_{in}) and out (J_{out}). In Fig. 14.3, $J_{in} = J_1$ and $J_{out} = J_2 + J_3$. Equation 14.5 can be extended to include isotope ratios:

$$\frac{d\delta_m}{dt} = \frac{\sum [(J_{in}(\delta_{in} - \delta_m)] - \sum [J_{out}(\delta_{out} - \delta_m)]}{M} \tag{14.6}$$

In Equation 14.6, the change in mass and isotopic composition is tracked per unit time. As stated above, using δ notation introduces an approximation, but for the sake of simplicity, we proceed with this expression and acknowledge that using R (see Equation 14.1), rather then δ, results in an exact solution. Overall, these definitions (Equations 14.5 and 14.6) represent *non steady-state* solutions; this simply means that the size of the reservoir (M: Fig. 14.3) and isotopic composition (δ_M: Fig. 14.3) of the system is not held

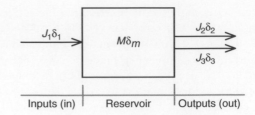

Figure 14.3 A simple box model representation of the equations presented in the text (see Equations 14.5 and 14.6). Here, material influx is noted as J_1 with an isotopic composition of δ_1. The mass and composition of the reservoir is noted as M and δ_m, with two sinks (J_2 and J_3) with compositions δ_2 and δ_3. Such a simply topology is commonly invoked by isotope studies of the carbon and sulfur cycles over geological timescales.

constant and can change as a function of the specific fluxes of inputs and outputs.

14.3.1.1   Rayleigh fractionation

One common non-steady state scenario is distillation, where material is fractionally removed from a reservoir. Quantifying the isotopic evolution of this style of system is often done using a Rayleigh fractionation equation, where the composition of the residual reactant (δ_r) as material is removed can be calculated by:

$$\delta_r = 1000 \times \left\{ \left[\left(1 + \frac{\delta_0}{1000} \right) \times f^{\alpha - 1} \right] - 1 \right\} \tag{14.7}$$

Here, δ_0 represents the starting composition of the reactant pool, f is the fraction of reactant remaining, and α is the fractionation factor associated with the transformation. Given this formulation, there are two products of interest. The first represents the instantaneous product (δ_{ip}), or the composition of the material being generated at any given point (along) during the reaction:

$$\delta_{ip} = \delta_r + (\alpha - 1) \times 1000 \tag{14.8}$$

The other statistic of interest is δ_{pp}, the cumulative product reservoir, which complements the evolving reactant pool to give the overall mass-balance constraint on the system:

$$\delta_{pp} = 1000 \times \left[\left(1 + \frac{\delta_0}{1000} \right) \times \left(\frac{1 - f^\alpha}{1 - f} \right) \right] \tag{14.9}$$

An example is given in Fig. 14.4(a), where $\alpha = 0.98$ and $\delta_0 = 0‰$ and the relevant compositions change as a function of the fractional removal (f). This type of system behaviour can be very useful for laboratory experiments wherein f can be carefully controlled and α can be calculated using Equation 14.9 from measurements of δ_{pp} and δ_r (e.g. Johnston *et al.*, 2007).

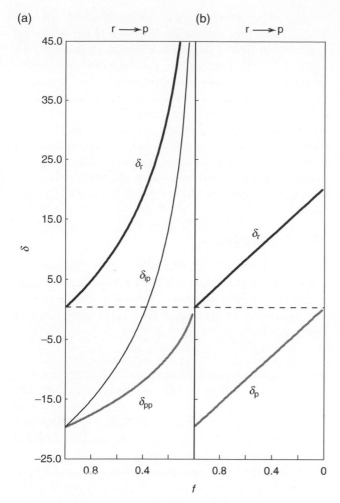

Figure 14.4 Isotopic mass-balance calculations for non steady-state (a, Equations 14.7–14.9) and steady-state (b, Equation 14.12) systems. Reactions proceed from reactant (r) to product (p) as values of f increase. Isotopic compositions of the residual reactant pool are noted by δ_r. The instantaneous product and pooled product in 'a' are further described in the text and noted here as δ_{ip} and δ_{pp}, respectively. In 'b', the product pool is noted as δ_p. Note how mass balance is satisfied in both cases (the product composition at $f = 0$ returns to the original value of the reactant pool at $f = 1$).

14.2.1 Steady-state approaches

In examining questions of geobiological interest over sufficiently long timescales, steady-state can often be assumed. This means that the overall size of the reservoir of interest can be fixed, which leads to a simplification of Equations 14.5 and 14.6:

$$\frac{dM}{dt} = 0 \text{ and } \frac{d\delta_m}{dt} = 0 \tag{14.10}$$

By setting these expressions to zero, the overall mass and isotopic composition of the system does not change. If we again examine the system of a product (p) and evolved reactant (r), we find:

$$(J_0 \times \delta_0) = (J_r \times \delta_r) + (J_p \times \delta_p) \tag{14.11}$$

where J_0 and δ_0 represent the original starting mass and isotopic composition. This can be further simplified by considering that $J_0 = J_r + J_p$ and by setting $J_0 = 1$. With the mass of the system normalized to 1, a relative flux term (f_p) can be introduced and substituted for J_p, leaving J_r equal to $(1 - f_p)$. These terms are substituted into Equation 14.11 to produce:

$$\delta_0 = f_p \delta_p + (1 - f_p)\delta_r \tag{14.12}$$

This represents a typical example of a *steady-state* solution of a branching pathway with one reservoir, one input, and two outputs (Fig. 14.4b).

Real world examples using this type of approach are common. For instance, models of the behaviour of the carbon cycle over long (geological) timescales begin with such an expression, where Equation 14.12 is modified:

$$\delta^{13}C_{out.+weath.} = f_{org.}\delta^{13}C_{org.} + (1 - f_{org.})\delta^{13}C_{carb.} \tag{14.13}$$

and further rearranged,

$$f_{org.} = \frac{\delta^{13}C_{out.+weath.} - \delta^{13}C_{carb.}}{\delta^{13}C_{org.} - \delta^{13}C_{carb.}}. \tag{14.14}$$

Here, $\delta^{13}C$ values of volcanic outgassing and weathering ($\delta^{13}C_{out.+weath.}$; thought to be -6‰), marine carbonate ($\delta^{13}C_{carb.}$; a measurable quantity) and organic carbon in sediments ($\delta^{13}C_{org.}$; a measurable quantity) can be used to estimate the relative burial flux of carbon as organic matter ($f_{org.}$) from the oceans. Such a statistic is interesting because organic matter burial is linked, via primary production, in a stoichiometric fashion to O_2 production (or more generally, oxidant production, see Canfield, 2005). Thus, when the temporal isotope records of organic carbon and carbonate are related in Equation 14.14, we find that the proportional burial of organic carbon has remained relatively constant throughout much of Earth's history (Schidlowski *et al.*, 1975; Strauss *et al.*, 1992; Shields and Veizer, 2002, Kump *et al.* 2001, Holland 2002, Fischer *et al.* 2009), and suggests that oxidant production was always an important byproduct of the carbon cycle (Hayes and Waldbauer, 2006).

The same set of equations can also be expanded (or nested within one another) to solve for more complex networks or reaction schemes (see Hayes, 2001; Johnston *et al.*, 2005; 2007; Farquhar *et al.*, 2007). For instance, if given a set of reactions A → B → C → D, by 'nesting' you can solve C → D in terms of D, and substitute in the solution, yielding A → B → D. The same exercise can be performed repeatedly until finally solving the last reaction of A → D, which would be the solution for A → B → C → D. Using this approach, one can demonstrate that it is principally the *rate-limiting reaction* that controls the isotopic fractionation in a given system (Hayes, 2001).

14.3.2.1 Residence time considerations

When considering stable isotopes of natural systems, the residence time of a particular species (how long material resides in reservoir *M*) is of central importance. Residence time is a measure of the average amount of time that a particular chemical species spends in a specified reservoir. Calculations of residence time make the assumption that a reservoir's inputs equal the outputs. This fixes the mass of the reservoir, resulting in a steady state system (discussed above). The mathematical definition of residence time (τ),

$$\tau = \frac{C \times V}{f}, \tag{14.15}$$

involves the concentration (*C* in mol l^{-1}) of a species or element, the volume of the reservoir (*V* in l), and an input or output flux (*f* in mol t^{-1}). Solving this equation results in a solution in units of time (*t*), hence residence time. The flux term, often constrained by measurements of riverine inputs for marine systems, can also be estimated by placing constraints on the outputs from a reservoir. For practical purposes in isotope geobiology, this expression quantifies how resistant to change a given reservoir is, with longer residence times indicating stable reservoirs that are robust to perturbations.

As an example, consider sulfate in modern seawater. We begin by setting the volume of the oceans at 1.37×10^{21} l (Broecker and Peng, 1982) and assuming a modern concentration of seawater sulfate of ~28 mM (Vairavamurthy *et al.*, 1995). Riverine fluxes of sulfate into the ocean have been estimated to be between 2–3.5×10^{12} mol S per year (Walker, 1986), resulting in a residence time of ~10 million years (Ma). The steady state assumption is powerful because it allows us to make the same calculation using fluxes of sulfate leaving the oceans rather than riverine input. Sulfate has two major sinks, one in calcium sulfate evaporite minerals and a second (following dissimilatory sulfate reduction) in pyrite. Exact values for these fluxes are challenging to pin down (e.g. Alt and Shanks, 2003; Strauss, 1999; Holser *et al.*, 1979; Walker, 1986), but are generally of the same order as the inputs and yield a similar estimate of residence time (Berner 1982, McDuff and Edmond 1982, Paytan *et al.* 1998).

The concept of residence time is valuable because it predicts the timescale over which reservoir size and isotopic composition will respond to changes in either the input or output fluxes. In the example of seawater sulfate given above we can note that the estimated residence time is much longer than the characteristic physical mixing time of the oceans (on the order of 1000 years). From this exercise we learn not to expect strong gradients in $[SO_4^{2-}]$ in seawater and consequently vanishingly little sulfate isotopic variability throughout the world's oceans. Such a condition need not be the case in times past. Indeed it is possible, using stratigraphic data of stable isotope ratios, to assess the time scale over which the isotopic composition of seawater sulfate changes. This is one approach to indirectly measuring the concentration of sulfate in ancient seawater (a quantity not readily offered by other geological proxies), by rearranging Equation 14.15 (e.g. Kah *et al.* 2004).

14.4 Applications

Three broad divisions define stable isotope geobiology: experimental geobiology, modern environmental studies, and historical geobiology. Different reference frames of observation in space and time accompany each of these divisions.

- Experimental geobiology is often aimed at targeting the biological mechanisms that drive isotopic fractionation; consequently, investigations focus on quantifying signatures produced by mixed or pure cultures of extant microorganisms and derivatives thereof (i.e. enzyme-specific studies). Most of these processes occur on the order of seconds to days or weeks.
- Modern environmental studies provide a larger-scale survey of geobiological processes in a variety of aqueous environments. Put differently, modern studies capture the operation of seasonal and annual cycles through experimental work at the community or ecosystem scale. Two commonly studied modern analog environments, as they pertain to Precambrian and geobiological research (and capturing several orders of magnitude in length scale), are microbial mats (Archean biosphere analog: Jørgensen and Cohen, 1977; Teske et al., 1998; Fike et al., 2008) and the Black Sea (Proterozoic studies: Lyons and Berner, 1992; Lyons and Severmann, 2006; Severmann et al., 2008).
- The third division of isotope geobiology involves reconstruction of time-series behaviour of geobiological processes from ancient sedimentary environments (often referred to as historical geobiology). Isotope records allow historical geobiologists to judge the relatedness of isotope signatures between ancient and modern environments, to investigate similarities between the two settings, and ask questions about how geobiological processes have varied non-uniformly with the coupled evolution of Earth surface environments.

The diversity of spatial scales that accompanies these categories is represented schematically in Fig. 14.5. Here, we demonstrate how an elemental cycle and

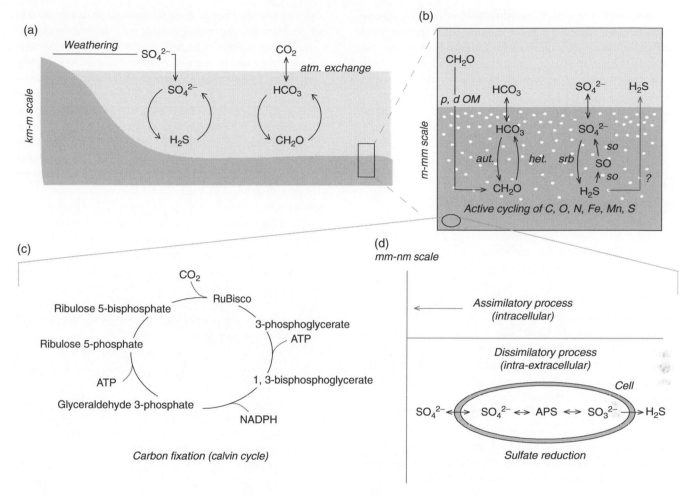

Figure 14.5 A series of figures outlining the variety of scales of interest to a geobiologist, all of which can be studied profitably using isotopes. In (a) we examine large-scale geochemical cycles using carbon and sulfur as the examples. For sulfur, oxidative and physical weathering of the continents delivers sulfate to the oceans via rivers. For carbon, CO_2 emitted from the solid Earth exchanges between the atmosphere and ocean, as well as being involved in crustal weathering reactions. Both cycles contain strong biological fractionation mechanisms and have oxidized and reduced sinks in marine sediments. In (b), we zoom in to look at a vertical cross-section of a sediment package, where communication between the water-column (frame a) and sedimentary microbial processes control element and isotopic cycling. Autotrophy and heterotrophy are noted by 'aut.' and 'het.', respectively. Sulfate reducers and sulfide oxidizers are listed as 'srb' and 'so.' Together, these processes account for the cycling and indirect removal of carbon and sulfur from the active marine environment. Continuing down to the cellular scale, assimilatory and dissimilatory processes also contribute significantly to the isotopic compositions present. In (c) we illustrate carbon fixation by the Calvin cycle, a carbon fixation pathway identified deep in the geologic record (see text for discussion). In (d) we show dissimilatory sulfate reduction, one of the most important processes for respiring organic carbon in anoxic marine environments.

isotope system can be used to ask questions at a variety of scales. Global or basin scale geochemical cycles can be assessed through investigating the inputs and outputs to a marine system (Fig. 14.5a). Within this cycle is a sedimentary component (Fig. 14.5b) with its own inputs, outputs, and associated fractionations. Furthermore, within the sediments, aerobic and anaerobic bacteria catalyse a variety of the chemical transformations. The array of active microbial processes within these environments is vast, but can generally be subdivided as either assimilatory (anabolic), where material is accumulated as biomass (Fig. 14.5c), or dissimilatory (catabolic), where the material is used for direct energy gain (Fig. 14.5d). Interpreting isotope ratio data from a marine environment thus requires a quantitative understanding of how each of these processes (and the numerous other processes operating on each of those scales) influences the overall composition as measured in a single sample or mineral grain.

Given the spectrum of possible applications and scales, there are two ways that geobiological hypotheses can be tested. The first tracks the natural abundances of isotopes and relies on variability within a given system to generate an interpretable signature. The second relies on spiking the experiment or system with an excess of a certain isotope and following the spike as it is distributed though out the system. Both have benefits and disadvantages, but with careful thought and experimental design, each provides powerful (and often complementary) information. Below we detail a series of case studies that employ each of these approaches.

14.4.1 Natural abundance case studies

14.4.1.1 Carbon fixation

Take the four common carbon fixation pathways as an example ($CO_2 \rightarrow$ organic carbon: OC). The Calvin cycle (Fig. 14.5C) commonly produces OC that is depleted relative to the reactant CO_2 by ~20–35‰ in $\delta^{13}C$ (Schidlowski *et al.*, 1983). The reductive acetyl coenzyme-A pathway often produces a similar fractionation, but carries the capability to produce even more depleted OC (Schidlowski *et al.*, 1983, Hayes *et al.* 2001). Conversely, the reductive tricarboxylic acid and the 3-hydroxypropionate cycles produce much smaller fractionations (<20‰). By measuring the $\delta^{13}C$ of OC in sedimentary samples, hypotheses about the dominant mode of C-fixation can be tested and constrained. For example, the isotope ratios of OC preserved in sedimentary rocks are broadly similar through much of Earth history (Strauss *et al.*, 1992), supporting the notion that the Calvin cycle has a long history of use, despite evolving biology and environmental conditions. Studies of rocks 2.7 to 2.5 billion years old, however, have revealed that OC is, on average, much more ^{13}C-depleted, often showing $\delta^{13}C$ values of −40 to −60‰ (Hayes 1994; Fischer *et al.*, 2009). These low values generally fall outside of the range expected for the metabolisms listed above (though some of this data is consistent with carbon fixation via the reductive acetyl CoA pathway; Fischer *et al.*, 2009). The pattern led to the interesting hypothesis that methanotrophy (methane commonly displays very low $\delta^{13}C$ values) was a fundamental process for assimilating carbon into biomass at that time (Schoell and Wellmer, 1981; Hayes, 1994; Hinrichs, 2002).

14.4.1.2 The Precambrian sulfur cycle

Data accumulated over decades of research indicate that the range of $\delta^{34}S$ in sedimentary sulfides increased near the Archean–Proterozoic (2500 million years ago) boundary (for instance, see Canfield *et al.*, 2000). Habicht *et al.* (2002), using a suite of laboratory and environmental experimental data, suggested that this transition represented a biological response of sulfate reducing bacteria (Fig. 14.5d) to increasing seawater sulfate

concentrations. They argued that at low seawater sulfate concentrations, the rate-limiting step in the biological reduction of sulfate to sulfide shifts from *Dsr* to sulfate uptake processes in the cell (the first step in Fig. 14.5d). This process is facilitated by active transport by membrane-bound proteins and is thought to carry a small (~3‰) fractionation. When seawater sulfate concentrations increase beyond the critical threshold, the rate-limiting step changes to an internal conversion (possibly *Dsr*) that results in much larger fractionations (see Chapter 5). Through careful experimental work, Habicht and coauthors (2002) pinpointed the sulfate concentration at which this switch occurred and 'calibrated' the concentration of seawater sulfate to be <200 µM (modern sulfate is ~100 × greater) in seawater older than ~2500 Ma. These researchers used a modern experiment, coupled to a geochemical observation, to convincingly argue that the change in the range of isotopic compositions preserved in the rock record near the Archean–Paleoproterozoic boundary is the result of a biochemical response at the cellular level.

14.4.2 Isotope spike case studies

Next we target two case studies where synthetic isotope enrichments allow for the quantification of small effects or processes that are otherwise difficult to assay. These applications utilize a different approach and employ isotope spikes.

14.4.2.1 Stable isotope probing (SIP)

In studying natural microorganisms and/or consortia, molecular biological techniques directed at amplification and/or sequencing of genetic material (see Chapters 12–13) answer the question of 'who is there?' SIP, on the other hand, helps to answer the question of 'what are they doing and how?' SIP works as follows: an isotopically labelled substrate is introduced to a microorganism or community (through injection into growth media or local environment) and is subsequently tracked as the cell processes it. The fate of the spike informs our understanding of how that original substance is processed throughout the cell and what biochemical (or physiological) processes are active. Further, the mass difference between labelled and unlabelled compounds allows for their separation (e.g. by centrifugation) so that each can be analysed individually, via molecular or geochemical techniques.

This approach can be used in a variety of fashions. Boschker and colleagues (1998) targeted the accumulation of an isotope spike in fatty acids, linking *Desulfotomaculum acetoxidans* (a sulfate-reducing firmicute) to ^{13}C-labelled acetate oxidation. Most commonly, SIP has been applied to study methane cycling (see Chapter 24: Lin *et al.*, 2004; Hutchens *et al.*, 2004). Here, for instance, Morris *et al.* (2002) incubated soils with ^{13}C-labelled methane ($^{13}CH_4$) and identified the transfer of the ^{13}C enrichment into DNA. Further, this ^{13}C-labelled DNA was extracted, purified, and used as a template for molecular biological techniques that were later able to identify the affinity of the specific methanotrophs incorporating the spike to known organisms. The incorporation of SIP into DNA also allows for new insights into key enzymes, such as CH_4 oxidation genes associated with methane monooxygenase (Lin *et al.*, 2004; Hutchens *et al.*, 2004). Field-based SIP techniques even extend to investigating microbial activity in contaminated (or extreme) environments (Jeon *et al.*, 2003). Though a majority of the SIP studies have focused on carbon isotopes, potential avenues of research are large (especially those involving nanoSIMS technology), and thus this methodology is in a state of rapid development limited only by the creativity of the researcher.

14.4.2.2 Anaerobic ammonium oxidation

One elegant application of isotope spikes was their role in the discovery of the anaerobic oxidation of ammonium (termed anammox). There was a long established imbalance in the marine N cycle; anaerobic microbial respiration processes predicted a substantial NH_4^+ reservoir that was not observed (Richards, 1965). This led to the prediction that a missing N_2 flux out of the ocean could account for the lack of a NH_4^+ pool (Codispoti, 1995). From thermodynamic considerations, it was later proposed that nitrite (the electron acceptor) and ammonium (the electron donor) could be metabolized to produce this N_2; a flux that could account for the observed imbalance. In 2002, Thamdrup and Daalsgard (2002) used a ^{15}N spike in nitrite to test this hypothesis. If the stochiometry of the reaction is described by $NO_2^- + NH_4^+ \rightarrow N_2$ and both species are dominantly ^{14}N (the most abundant isotope of nitrogen), then the N_2 in the environment should have a mass of ~28 ($^{14}NO_2^- + ^{14}NH_4^+ \rightarrow ^{14}N^{14}N$). Conversely, by labelling one of the aqueous N species, they could track this potential reaction by the accumulation of mass 29 N_2 (for instance, $^{15}NO_2^- + ^{14}NH_4^+ \rightarrow ^{14}N^{15}N$). This is exactly what was observed. In years since, this approach has been applied widely and demonstrated the predominant role of anammox in anoxic basins (Dalsgaard *et al.*, 2003; Jensen *et al.*, 2008) and coastal upwelling zones off western S. America (Thamdrup *et al.*, 2006; Hamersley *et al.*, 2007) and western Africa (Kuypers *et al.*, 2003, 2005; Lam *et al.*, 2009).

14.5 Using isotopes to ask a geobiological question in deep time

In this section, we examine the methods and logic used to probe the operation of geobiological processes in deep time, with data generated from the sedimentary rock record. We are fortunate that the Earth has captured its own environmental and biological history in sedimentary successions, but this record is far from perfect. The most challenging aspects of these approaches lie in the fragmentary nature and unequal preservation inherent to the geologic record. Studies rooted in data from ancient rocks require solving an inverse problem, often implicitly. With inverse problems, it is not possible to run the experiment again to observe what occurs (i.e. the Earth has but one history); here, model parameters and conceptual understanding flow from observations of historical data. We will introduce several important aspects of working with sedimentary rock record isotope ratio data: selecting appropriate lithologies for analysis, generating high-resolution data with respect to time and paleoenvironment from an incomplete geologic record, and heeding the effects of post-depositional alteration (diagenesis and metamorphism).

In sedimentary rocks, elements of interest to stable isotope geobiology fall into major, minor, or trace element categories depending on their abundance. The mass of sample required to make a precise isotopic measurement on a particular lithology will vary accordingly. *In general, you should strive to work with as little sample mass as is required to obtain precise data.* Processing more rock to obtain enough mass will always make a given measurement possible, but it integrates many mineral phases and textures present in a rock (many of which are not primary), and additional problems concerning external contamination can arise. This means that only a subset of available lithologies are appropriate for a particular isotopic analysis. Take, for example, marine limestone [$CaCO_3$] or dolomite [$CaMg(CO_3)_2$] (often referred to collectively as carbonates). Carbonates are ideal targets for carbonate carbon isotopes because C is a major component in the rock and they are, quantitatively, the most important sink in the carbon cycle (Knoll *et al.* 1986; Hoffman *et al.* 1998; Halverson *et al.* 2005; Fischer *et al.* 2009). Carbonates can also carry sufficient OC to make meaningful isotopic measurements (requiring generally >0.1 weight % OC; e.g. Strauss *et al.* 1992). Siliciclastic shale, specifically OC-rich shale (often referred to as black shale) can contain high concentrations of elements that are insoluble in their reduced form or reactive with other reducing species, such as OC. For instance, molybdenum isotope studies (Arnold *et al.*, 2004; Anbar *et al.*, 2007) often focus on black shale, since this lithology concentrates Mo[5]. Along with Fe, which binds with sulfur to form pyrite (FeS_2), chalcophile (or sulfur loving) elements like copper and zinc will also concentrate in

---

[5] Readers are referred to Anbar and Rouxel (2007) for a thorough review of stable transition metal isotopes.

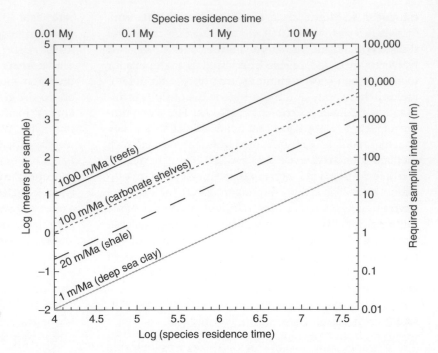

Figure 14.6 A calculation illustrating the minimum allowable density for stratigraphic sampling, given a particular species residence time and rock type (with a characteristic sedimentation rate). Values are from modern systems (Einsele, 1992) and are similar to those estimated for Precambrian environments (Altermann and Nelson, 1998; Trendall *et al.*, 2004).

this lithology (black shales are often S-rich; Canfield, 1998; Shen *et al.*, 2002, 2003). Organic-rich sediments are also good targets for OC isotopic analyses and lipid biomarker compound-specific isotope studies, such as nitrogen (^{14}N and ^{15}N) in porphyrin compounds. Thus, after choosing an element and isotope system of interest, researchers can identify the rock-type that best suits a specific geobiological question. In general, there are several important keys to lithology selection: (1) a researcher needs to be able to easily and accurately connect the element in a particular lithology to a geo-biological value or process operating in the sedimentary environment (e.g. seawater SO_4^{2-} or CO_3^{2-}, pore fluid S^{2-} from sulfate reduction, OC from photic zone primary production); (2) the lithology must have sufficient concentrations of a particular element to make a precise measurement without homogenizing and integrating too much material; and (3) the isotopic composition of major elements are more difficult to alter after deposition during burial diagenesis and metamorphism than minor or trace elements.

With an idea of what lithologies are appropriate for certain isotopic studies, the next task is to obtain a suite of samples from a stratigraphic succession arrayed along vectors of time and paleoenvironment. The sedimentary processes responsible for creating stratigraphy are strongly non-linear; because of this, thoughtful sampling strategies are required. Rather than use examples from the literature, our pedagogy here is to illustrate how sedimentary rocks record geobiological data using forward model simulations that capture these dynamics.

From the principle of superposition, we recognize that sedimentary packages offer time series data from samples collected from the bottom (oldest) to the top (youngest) of a stratigraphic section. In general, we want to collect enough samples from a section to produce smooth (small changes between data points) time-series data. In order to determine what sampling density (samples per metre) is required, recall the concept of residence time (Section 15.3.2.1). A good rule of thumb is that the timescale required for a reservoir to change composition and reach a new steady state following a perturbation is on the order of ~4× the estimated residence time. As stated previously, in practice (particularly in Precambrian-age rocks), a good estimate of the residence time is not always known, and so it is often better to sample at a higher density than you think is required. Presented in Fig. 14.6 is a set of calculations designed to determine a *minimum* sample density requirement, given a rock type and isotopic species of interest. Axes at the right and top relate to geobiological 'questions,' whereas axes at the left and bottom were used in the construction of the plot. We read this figure as follows: if, for instance, we are interested in the isotopic composition of modern seawater sulfate with a residence time of 10 Myr as recorded by sedimentary sulfides in shale deposits, we need less than 1 sample per 100 m of shale to capture each residence time once. In actuality, a sample density much greater (say 4×, or 1 sample per 25 m) is ideal, which is why we outline the relationships in Fig. 14.6 as placing strict minimum estimates. If we continue with this example, but now take an estimate of Archean seawater sulfate, with a loosely estimated residence time of 0.1 Myr[6], the minimum sample density is 1 sample per metre. High-resolution studies of Precambrian carbonate carbon (for example, Halverson *et al.*, 2005; Fischer *et al.*, 2009) often exceed this minimum requirement, as have recent studies of multiple S isotopes (Kaufman *et al.*, 2007).

In reality, the sedimentary record is not a simple uniform layer cake, but rather dynamic in space and time. To illustrate this point, we will look at two forward simulations of carbonate bearing sedimentary successions and their carbonate carbon isotope ratio data. The synthetic stratigraphy was generated using a modified version of STRATA (Flemings and Grotzinger, 1996), a basin modelling tool that approximates process sedimentology using diffusion behaviour.

Figure 14.7 depicts the results of a simulation of metre-scale shallowing-upward cycles in peritidal carbonates; this sedimentary dynamic is ubiquitous on carbonate platforms of all age (Sadler, 1994). The model forcing is a periodic sea level curve composed of two waves of similar amplitude (1 m) but with different periods. As sea level rises, accommodation is produced and a sequence of carbonate rock (a, b, c, d, and e), from subtidal facies to supratidal facies, is deposited. As sea level drops and the available accommodation goes to zero, the system enters a period of local non-deposition, characterized by a terminal hiatus (a′, b′, c′, d′, and e′). The episodic deposition of carbonate is clear from the stair step-shaped plot of accumulation verses time. It is noteworthy within this simulation that there is more time missing (apportioned along the surfaces of hiatuses) than preserved by the carbonate sequences. This is a general feature of the geologic record at all length and time scales, regardless of environment (Sadler, 1981). Furthermore, between blocks c and d note the missing 'beat' or sea level cycle wherein insufficient accommodation was created to leave behind a sequence of rock, effectively doubling the time of the hiatus. Non-linearities like these are common in the geologic record, even in 'cyclic' sedimentary rocks. In this simulation, we can also examine the effect that non-steady sedimentation and paleoenvironment can have on stratigraphic isotope ratio data. We consider three possible isotopic scenarios: X, Y, and Z. In X, there is a linear trend in seawater dissolved inorganic carbon (DIC) $\delta^{13}C$ from 0 to 6‰ over the length of the simulation. In Y, we apply no secular trend, but add differences in the $\delta^{13}C$ of specific facies, wherein supratidal facies are 3‰ ^{13}C-enriched compared to subtidal facies (this could be due to differences in primary geobiological processes producing

carbonate in these different environments, or due to secondary diagenetic processes that are correlated with features like grain size and permeability). In Z, we examine the combined effects X and Y. We see that unsteady sedimentation produces isotope records that can have sharp breaks corresponding to sequence stratigraphic surfaces. These breaks are the strongest in scenario Y, where the superposition of different facies highlights the discordances. In addition, it is important to be able to deconvolve the effects of paleoenvironment and diagenesis from those of true secular trends. If we were using the isotopic data from scenario Z to estimate f_{org} (again, the fraction of total C buried as OC) and the flux of oxidant production during the deposition of this stratigraphy, our estimates would be off by nearly 20%.

Figure 14.8 shows the results of a different simulation, working this time on the length scale of a large sedimentary basin; these results illustrate how unsteady accumulation leads the development of complex sedimentary sequences in a carbonate-rich passive margin succession. The primary model forcing is periodic sea level change with 20 m amplitude and several million-year period, and we take into account isostatic compensation of the crust due to the sedimentary load. The upper left panel in this figure shows a cross-section orthogonal to the basin margin (with sediment moving from left to right). Note that the thickness of this succession attenuates toward the basin (where the sediment flux is too low to fill the available accommodation) and toward the craton (where the rate of accommodation was insufficient and so much of the sediment bypassed). The upper right panel shows a complementary relationship, but rather than plot thickness on the y-axis, here time is shown (this if often referred to as a Wheeler plot or chronostratigraphic plot). By comparing these two panels, we can see that sediment was not deposited everywhere at the same time, and erosion has helped shape the resulting sedimentary sequences. The bottom panels show the same simulation but with $\delta^{13}C$ data. Here we have imposed a complex secular trend with three negative $\delta^{13}C$ excursions of varying magnitude on the isotopic composition of seawater DIC (this record is not unlike that reported for Ediacaran seawater; e.g. Fike *et al.*, 2006).

It is critical to recognize that all possible stratigraphic sections throughout the basin contain hiatuses and unconformities. There is no single section where one can recover the complete isotopic history. We can combat this challenge by measuring and sampling multiple sections throughout an outcrop belt or in subsurface drill core (preferably oriented perpendicular to the strike of the basin) and constructing a composite isotope history, or curve. To construct a composite curve and accurately capture the $\delta^{13}C$ history recorded in this succession, we require measurements from at least

---

[6] This estimate is reached by relating modern seawater concentrations to estimates on Archean seawater (e.g. Habicht *et al.*, 2002) (28 mM/0.2 mM) and with the assumption that inputs and outputs will scale similarly.

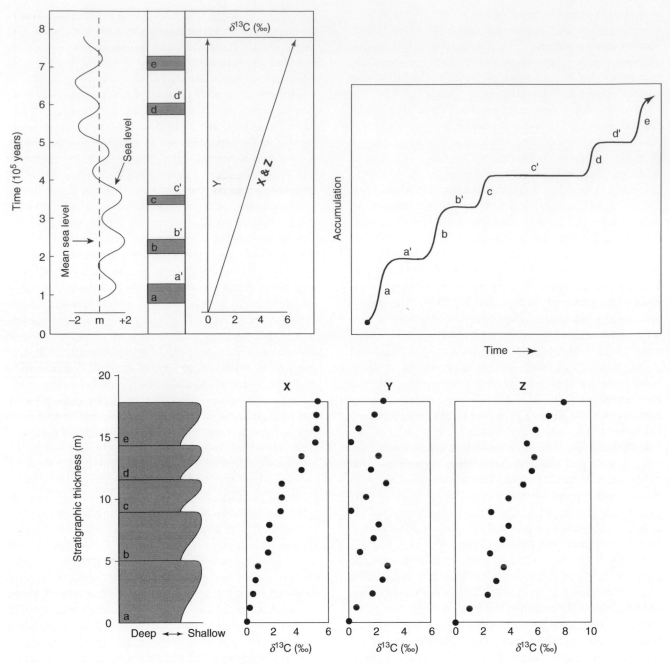

Figure 14.7 A forward model simulating the deposition of metre-scale shallowing-upward cycles in peritidal carbonates. The upper left panel shows the sea level forcing used to generate the stratigraphy. The grey blocks a, b, c, d, and e correspond to time recorded by carbonate cycles, and the intervening white blocks (a', b', c', d', and e') denote time not recorded by rock, but rather apportioned to bounding hiatuses, or times of non-deposition, shown in a plot of accumulation against time in the upper right panel. Note the missing 'beat' or sea level cycle between blocks c and d. Three possible isotopic scenarios are simulated. X – secular seawater DIC δ^{13}C trend from 0 to 6‰. Y – no secular seawater trend, but shallow water facies are 3‰ heavier than deep water facies. Z – combines both the secular trend and facies trend. See text for details.

four sections, for example at locations a, b, c, and d. More would be ideal for redundancy and confidence. We also expect that the ideal sampling density between these sections will be different. Section d, for example, is condensed and will require a more intense sampling effort (see Myrow and Grotzinger, 2000). By connecting your sampling effort to the inferences gained from sequence stratigraphy, it is possible to mitigate the tendency of sedimentary processes to offer an incomplete historical record.

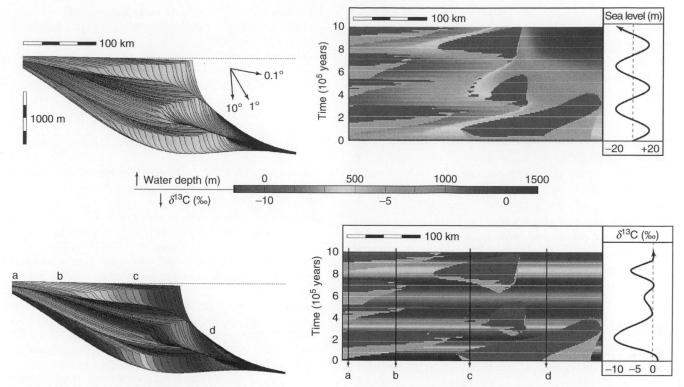

Figure 14.8 A forward model simulating a carbonate-rich succession on the scale of a passive margin. Upper left – a cross-section orthogonal to the strike of the margin. Note the strong vertical exaggeration. Timelines are shown in black, and coloured according to water depth at the time of deposition. Upper right – a Wheeler plot showing the stratigraphy across the basin as a function of time rather than thickness. In this plot light grey corresponds to areas of non-deposition whereas dark grey marks areas that underwent erosion. The sea level curve input used to create the stratigraphy is plotted alongside. The bottom two panels show the same simulation as above, this time coloured with contours corresponding to carbon isotope ratio. The secular trend in seawater DIC δ^{13}C used as model input is plotted alongside. Note that any one stratigraphic section is incomplete. To sample the entire history recorded in this sedimentary basin, one must measure and sample multiple stratigraphic sections arrayed across the basin, for example, at locations a, b, c, and d. See text for details.

14.6 Conclusions

Let us step back for a moment and review the tools that geobiological researchers can use to understand: (1) the underlying biochemical roots of isotopic fractionations, (2) how precise mathematical formulations underlie the study of all isotope systems, and (3) the considerations necessary in asking a geobiological question with isotope ratio data available from the sedimentary record.

In any experimental or natural system, there will likely be a number of components that contribute to the cumulative isotope fractionation. These components will include both biological (enzymes) and abiological (physical transport and mixing) mechanisms. As a rule of thumb, however, it is the rate-limiting step within a system that controls the net isotopic fractionation, and about which the most information can be gained.

Detailed isotopic studies of microbial metabolism can provide a great deal of information about the operation of the organism itself. For instance, we outlined above how, when grown under low sulfate concentrations, sulfate-reducing bacteria produce much smaller δ^{34}S fractionations between the product sulfide and residual sulfate than under sulfate replete conditions (Rees, 1973). This information was used to both calibrate levels of Archean seawater sulfate (Habicht *et al.*, 2002), and inform the relative rate of enzymatic and diffusive processes within the cell (Canfield *et al.*, 2006; Johnston *et al.*, 2007). Other cases, such as those employing stable isotope probes, are also directed at understanding material flow in a system, but seek to quantify and understand a modern geobiological process. Future work placing constraints on the magnitude of the fractionation associated with key microbial or biochemical processes is necessary for the richness of the natural records (be they modern or ancient) to be understood. These constraints call for dedicated laboratory work, motivated by questions from patterns in the geological record.

Whereas work to date has often focused on a particular measurement from a single isotope system, the future of this field lies in working within numerous isotope systems simultaneously. The biggest challenge facing isotope studies is that competing hypotheses do not always have unique predictions for the behaviour of isotopic data. By combining several isotopic systems, however, this problem of non-uniqueness can be overcome. Whereas paired measurements of carbon and sulfur isotope ratios (at present, the core geobiological isotopes) are beginning to be allied more commonly, companion records from complementary element systems (such as the relationship between Mo and nitrogenase, or Zn and carbonic anhydrase) will add an extra layer of sophistication to these hypotheses. In the end, the primary goal of isotope geobiology is to understand the interactions between life and environments today as well as in times past. Stable isotopes remain a powerful tool to be levied in this pursuit.

Acknowledgements

We appreciate the wonderful education provided by our mentors (J. Farquhar, D. Canfield, and A. Knoll). Financial support was provided by NASA (NNX07AV51G: DTJ), Microbial Science Institute at Harvard (DTJ) and The Agouron Institute (WWF). Comments from T. Mauck, N. Tosca, P. Cohen and S. D. Wankel greatly improved this chapter.

References

Alt JC, Shanks WC (2003) Serpentinization of abyssal peridotites from the MARK area, Mid-Atlantic Ridge: Sulfur geochemistry and reaction modeling. *Geochimica et Cosmochimica Acta* **67**, 641–653.

Altermann W, Nelson DR (1998) Sedimentation rates, basin analysis and regional correlations of three Neoarchaean and Palaeoproterozoic sub-basins of the Kaapvaal craton as inferred from precise U-Pb zircon ages from volcaniclastic sediments. *Sedimentary Geology* **120**, 225–256.

Anbar AD, Rouxel O (2007) Metal stable isotopes in paleoceanography. *Annual Review of Earth and Planetary Sciences* **35**, 717–746.

Anbar AD, Duan Y, Lyons TW, *et al.* (2007) A whiff of oxygen before the Great Oxidation Event? *Science* **317**, 1903–1906.

Archer C, Vance D (2006) Coupled Fe and S isotope evidence for Archean microbial Fe(III) and sulfate reduction. *Geology* **34**, 153–156.

Arnold GL, Anbar AD, Barling J, Lyons TW (2004) Molybdenum isotope evidence for widespread anoxia in mid-proterozoic oceans. *Science* **304**, 87–90.

Bao HM, Rumble D, Lowe DR (2007) The five stable isotope compositions of Fig Tree barites: Implications on sulfur cycle in ca. 3.2 Ga oceans. *Geochimica et Cosmochimica Acta* **71**, 4868–4879.

Berner RA (1982) Burial of organic carbon and pyrite sulfur in modern ocean – its geochemical and environmental significance. *American Journal of Science*, **282**, 451–473.

Bigeleisen J, Mayer MG (1947) Calculation of equilibrium constants for isotopic exchange reactions. *Journal of Chemical Physics* **15**, 261–267.

Boschker HTS, Nold SC, Wellsbury P, *et al.* (1998) Direct linking of microbial populations to specific biogeochemical processes by C-13-labelling of biomarkers. *Nature* **392**, 801–805.

Broecker WS, Peng TH (1982) *Tracers in the Sea*. Lamont-Doherty Earth Observatory, Columbia University, NY, pp. 690.

Canfield DE (1998) A new model for Proterozoic ocean chemistry. *Nature* **396**, 450–453.

Canfield DE (2005) The early history of atmospheric oxygen: Homage to Robert A. Garrels. *Annual Review of Earth and Planetary Sciences* **33**, 1–36.

Canfield DE, Teske A (1996) Late Proterozoic rise in atmospheric oxygen concentration inferred from phylogenetic and sulphur-isotope studies. *Nature* **382**, 127–132.

Canfield DE, Habicht KS, Thamdrup B (2000) The Archean sulfur cycle and the early history of atmospheric oxygen. *Science* **288**, 658–661.

Canfield DE, Olesen CA, Cox RP (2006) Temperature and its control of isotope fractionation by a sulfate-reducing bacterium. *Geochimica et Cosmochimica Acta* **70**, 548–561.

Chacko T, Cole DR, Horita J (2001) Equilibrium oxygen, hydrogen and carbon isotope fractionation factors applicable to geologic systems. In: Stable Isotope Geochemistry, pp. 1–81.

Codispoti LA (1995) Biogeochemical cycles – is the ocean losing nitrate. *Nature* **376**, 724–724.

Coplen TB, Bohlke JK, De Bievre P, *et al.* (2002) Isotope-abundance variations of selected elements – (IUPAC Technical Report). *Pure and Applied Chemistry* **74**, 1987–2017.

Dalsgaard T, Canfield DE, Petersen J, Thamdrup B, Acuna-Gonzalez J (2003) N-2 production by the anammox reaction in the anoxic water column of Golfo Dulce, Costa Rica. *Nature* **422**, 606–608.

Dicken AP (2004) *Radiogenic Isotope Geology*. Cambridge University Press, Cambridge, pp. 425.

Eiler JM, Schauble E (2004) (OCO)-O-18-C-13-O-16 in Earth's atmosphere. *Geochimica et Cosmochimica Acta* **68**, 4767–4777.

Einsele G (1992) *Sedimentary Basins: Evolution, Facies, and Sediment Budget*. Springer-Verlag, Berlin.

Farquhar J, Wing BA (2003) Multiple sulfur isotopes and the evolution of the atmosphere. *Earth and Planetary Science Letters*, **213**, 1–13.

Farquhar J, Johnston DT, Wing BA (2007) Implications of conservation of mass effects on mass-dependent isotope fractionations: Influence of network structure on sulfur isotope phase space of dissimilatory sulfate reduction. *Geochimica et Cosmochimica Acta* **71**, 5862–5875.

Fike DA, Grotzinger JP, Pratt LM, Summons RE (2006) Oxidation of the Ediacaran Ocean. *Nature* **444**, 744–747.

Fike DA, Gammon CL, Ziebis W, Orphan VJ (2008) Micron-scale mapping of sulfur cycling across the oxycline of a cyanobacterial mat: a paired nanoSIMS and CARD-FISH approach. *ISME Journal* **2**, 749–759.

Fischer WW, Schroeder S, Lacassie JP, *et al.* (2009) Isotopic constraints on the late Archean carbon cycle from the Transvaal Supergroup along the western margin of the Kaapvaal Craton, South Africa. *Precambrian Research*, **169**, 15–27.

Flemings PB, Grotzinger JP (1996) STRATA: Freeware for Solving Classic Stratigraphic Problems. *GSA Today*, 6, 1–7.

Habicht KS, Gade M, Thamdrup B, Berg P, Canfield DE (2002) Calibration of sulfate levels in the Archean Ocean. *Science* 298, 2372–2374.

Halverson GP, Hoffman PF, Schrag DP, Maloof AC, Rice AHN (2005) Toward a Neoproterozoic composite carbon-isotope record. *Geological Society of America Bulletin* 117, 1181–1207.

Hamersley MR, Lavik G, Woebken D, et al. (2007) Anaerobic ammonium oxidation in the Peruvian oxygen minimum zone. *Limnology and Oceanography*, 52, 923–933.

Hayes JM (1994) Global methanotrophy at the Archean-Proterozoic transition. In: *Early Life on Earth* (ed Bengston S), vol. 84. Columbia University Press, New York, pp. 200–236 (Nobel Symposium).

Hayes JM (2001) Fractionation of carbon and hydrogen isotopes in biosynthetic processes. In: *Stable Isotope Geochemistry* (eds Valley J, Cole DR). Mineralogical Society of America, Washington DC, pp. 225–277.

Hayes JM, Waldbauer JR (2006) The carbon cycle and associated redox processes through time. *Philosophical Transactions of the Royal Society B-Biological Sciences* 361, 931–950.

Hinrichs KU (2002) Microbial fixation of methane carbon at 2.7Ga: was an anaerobic mechanism possible? *Geochemistry Geophysics Geosystems G³*, 2001GC000286.

Hoffman PF, Kaufman AJ, Halverson GP, Schrag DP (1998) A Neoproterozoic snowball earth. *Science* 281, 1342–1346.

Holland HD (2002) Volcanic gases, black smokers, and the great oxidation event. *Geochimica et Cosmochimica Acta* 66, 3811–3826.

Holser WT, Kaplan IR, Sakai H, Zak I (1979) Isotope geochemistry of oxygen in the sedimentary sulfate cycle. *Chemical Geology* 25, 1–17.

Hutchens E, Radajewski S, Dumont MG, Mcdonald IR, Murrell JC (2004) Analysis of methanotrophic bacteria in Movile Cave by stable isotope probing. *Environmental Microbiology* 6, 111–120.

Jensen MM, Kuypers MMM, Lavik G, Thamdrup B (2008) Rates and regulation of anaerobic ammonium oxidation and denitrification in the Black Sea. *Limnology and Oceanography*, 53, 23 36.

Jeon CO, Park W, Padmanabhan P, Derito C, Snape JR, Madsen EL (2003) Discovery of a bacterium, with distinctive dioxygenase, that is responsible for in situ biodegradation in contaminated sediment. *Proceedings of the National Academy of Sciences of the USA* 100, 13591–13596.

Johnston DT, Farquhar J, Canfield DE (2007) Sulfur isotope insights into microbial sulfate reduction: When microbes meet models. *Geochimica et Cosmochimica Acta* 71, 3929–3947.

Johnston DT, Farquhar J, Habicht KS, Canfield DE (2008) Sulphur isotopes and the search for life: strategies for identifying sulphur metabolisms in the rock record and beyond. *Geobiology* 6, 425–435.

Johnston DT, Wing BA, Farquhar J, et al. (2005) Active microbial sulfur disproportionation in the Mesoproterozoic. *Science* 310, 1477–1479.

Jorgensen BB, Cohen Y (1977) Solar Lake (Sinai) .5. Sulfur cycling of benthic cyanobacterial mats. *Limnology and Oceanography* 22, 657–666.

Kah LC, Lyons TW, Frank TD (2004) Low marine sulphate and protracted oxygenation of the proterozoic biosphere. *Nature* 431, 834–838.

Kaplan IR (1975) Stable isotopes as a guide to biogeochemical. *Proceedings of the Royal Society of London Series B-Biological Sciences* 189, 183–211.

Kaufman AJ, Johnston DT, Farquhar J, et al. (2007) Late Archean biospheric oxygenation and atmospheric evolution. *Science* 317, 1900–1903.

Knoll AH, Hayes JM, Kaufman AJ, Swett K, Lambert IB (1986) Secular variation in carbon isotope ratios from upper Proterozoic successions of Svalbard and east Greenland. *Nature* 321, 832–838.

Kump LR, Kasting, JF, Barley ME (2001) Rise of atmospheric oxygen and the 'upside-down' Archean mantle. *Geochemistry Geophysics Geosystems G³*, 2000GC000114.

Kuypers MMM, Sliekers AO, Lavik G, Schmid M, Jorgensen BB, Kuenen JG, Damste JSS, Strous M, Jetten MSM (2003) Anaerobic ammonium oxidation by anammox bacteria in the Black Sea. *Nature* 422, 608–611.

Kuypers MMM, Lavik G, Woebken D, et al. (2005) Massive nitrogen loss from the Benguela upwelling system through anaerobic ammonium oxidation. *Proceedings of the National Academy of Sciences of the USA* 102, 6478–6483.

Lam P, Lavik G, Jensen MM, et al. (2009) *Proceedings of the National Academy of Sciences of the USA* 106, 4752–4757.

Lin JL, Radajewski S, Eshinimaev BT, Trotsenko YA, Mcdonald IR, Murrell JC (2004) Molecular diversity of methanotrophs in Transbaikal soda lake sediments and identification of potentially active populations by stable isotope probing. *Environmental Microbiology* 6, 1049–1060.

Luz B, Barkan E, Bender ML, Thiemens MH, Boering KA (1999) Triple isotope composition of atmospheric oxygen as a tracer of biosphere productivity. *Nature*, 400, 547–551.

Lyons TW, Berner RA (1992) Carbon sulfur iron systematics of the uppermost deep-water sediments of the Black Sea. *Chemical Geology* 99, 1–27.

Lyons TW, Severmann S (2006) A critical look at iron paleoredox proxies: New insights from modern euxinic marine basins. *Geochimica et Cosmochimica Acta* 70, 5698–5722.

Mariotti A, Germon JC, Hubert P, et al. (1981) Experimental determination of nitrogen kinetic isotope fractionation – some principles – illustration for the denitrification and nitrification processes. *Plant and Soil* 62, 413–430.

McDuff RE, Edmond JM (1982) On the fate of sulfate during hydrothermal circulation at mid-ocean ridges. *Earth and Planetary Science Letters* 57, 117–132.

McFadden KA, Huang J, Chu XL, et al. (2008) Pulsed oxidation and biological evolution in the Ediacaran Doushantuo Formation. *Proceedings of the National Academy of Sciences of the USA* 105, 3197–3202.

Morris SA, Radajewski S, Willison TW, Murrell JC (2002) Identification of the functionally active methanotroph population in a peat soil microcosm by stable-isotope probing. *Applied and Environmental Microbiology* 68, 1446–1453.

Myrow P, Grotzinger JP (2000) Chemostratigraphic proxy records: Forward modeling the effects of unconformities, variable sediment accumulation rates, and sampling-interval bias. In: SEPM Special Publication – Carbonate Sedimentation and Diagenesis in the Evolving Precambrian World (eds Grotzinger J, James N), pp. 43–58.

Oneil JR, Epstein S (1966) Oxygen isotope fractionation in system dolomite-calcite-carbon dioxide. *Science*, **152**, 198–201.

Paytan A, Kastner M, Campbell D, Thiemens MH (1998) Sulfur isotopic composition of Cenozoic seawater sulfate. *Science* **282**, 1459–1462.

Rees CE (1973) Steady-state model for sulfur isotope fractionation in bacterial reduction processes. *Geochimica et Cosmochimica Acta* **37**, 1141–1162.

Richards, FA (1965) Anoxic basins and fjords. In Riley JP, Skirrow G (eds) *Chemical Oceanography* Academic Press, New York, pp. 611–645.

Rothman DH, Hayes JM, Summons RE (2003) Dynamics of the Neoproterozoic carbon cycle. *Proceedings of the National Academy of Sciences of the USA* **100**, 8124–8129.

Rouxel OJ, Bekker A, Edwards KJ (2005) Iron isotope constraints on the Archean and Paleoproterozoic ocean redox state. *Science* **307**, 1088–1091.

Sadler PM (1981) Sediment accumulation rates and the completeness of stratigraphic sections. *Journal of Geology* **89**, 569–584.

Sadler PM (1994) The expected duration of upward-shallowing peritidal carbonate cycles and their terminal hiatuses. *Geological Society of America Bulletin*, **106**, 791–802.

Schidlowski M (1983) Evolution of photoautotrophy and early atmospheric oxygen levels. *Precambrian Research*, **20**, 319–335.

Schidlowski M, Eichmann R, Junge CE (1975) Precambrian sedimentary carbonates – carbon and oxygen isotope geochemistry and implications for terrestrial oxygen budget. *Precambrian Research* **2**, 1–69.

Schidlowski M, Hayes JM, Kaplan IR (1983) Isotopic inferences of ancient biochemistries: Carbon, sulfur, hydrogen, and nitrogen. In: *The Earth's Earliest Biosphere* (ed Schopf JW). Princeton University Press, Princeton, NJ, pp. 149–185.

Schoell M, Wellmer FW (1981) Anomalous ^{13}C depletion in early Precambrian graphite's from Superior Province, Canada. *Nature* **290**, 696–699.

Severmann S, Lyons TW, Anbar A, Mcmanus J, Gordon G (2008) Modern iron isotope perspective on the benthic iron shuttle and the redox evolution of ancient oceans. *Geology*, **36**, 487–490.

Shen Y, Knoll AH, Walter MR (2003) Evidence for low sulphate and anoxia in a mid-Proterozoic marine basin. *Nature*, **423**, 632–635.

Shen YN, Canfield DE, Knoll AH (2002) Middle proterozoic ocean chemistry: Evidence from the McArthur Basin, northern Australia. *American Journal of Science*, **302**, 81–109.

Shields G, Veizer J (2002) Precambrian marine carbonate isotope database: Version 1.1. *Geochemistry Geophysics Geosystems*, 3.

Strauss H, DesMarais DJ, Hayes JM, Summons RE (1992) The carbon isotopic record. In: *The Proterozoic Biosphere: A Multidisciplinary Study* (eds Schopf JW, Klein C). Cambridge University Press, Cambridge, pp. 117–127.

Strauss H (1999) Geological evolution from isotope proxy signals – sulfur. *Chemical Geology* **161**, 89–101.

Teske A, Ramsing NB, Habicht KS, *et al.* (1998) Sulfate-reducing bacteria and their activities in cyanobacterial mats of Solar Lake (Sinai, Egypt). *Applied and Environmental Microbiology* **64**, 2943–2951.

Thamdrup B, Dalsgaard T (2002) Production of N-2 through anaerobic ammonium oxidation coupled to nitrate reduction in marine sediments. *Applied and Environmental Microbiology*, **68**, 1312–1318.

Thamdrup B, Dalsgaard T, Jensen MM, Ulloa O, Farias L, Escribano R (2006) Anaerobic ammonium oxidation in the oxygen-deficient waters off northern Chile. *Limnology and Oceanography*, **51**, 2145–2156.

Trendall AF, Compston W, Nelson DR, De Laeter JR, Bennett VC (2004) SHRIMP zircon ages constraining the depositional chronology of the Hamersley Group, Western Australia. *Australian Journal of Earth Sciences* **51**, 621–644.

Turchyn AV, Schrag DP (2004) Oxygen isotope constraints on the sulfur cycle over the past 10 million years. *Science* **303**, 2004–2007.

Urey HC (1947) The thermodynamic properties of isotopic substances. *Journal of the Chemical Society* 562–581.

Vairavamurthy MA, Orr Wl, Manowitz B (1995) Geochemical transformations of sedimentary sulfur: An introduction. In: Geochemical Transformations of Sedimentary Sulfur (eds Vairavamurthy MA, Schoonen Maa), pp. 1–14.

Walker JCG (1986) Global geochemical cycles of carbon, sulfur and oxygen. *Marine Geology* **70**, 159–174.

Zhang J, Quay PD, Wilbur DO (1995) Carbon isotope fractionation during gas-water exchange and dissolution of CO_2. *Geochimica et Cosmochimica Acta*, **59**, 107–114.

15

BIOMARKERS: INFORMATIVE MOLECULES FOR STUDIES IN GEOBIOLOGY

Roger E. Summons and Sara A. Lincoln

Massachusetts Institute of Technology, Department of Earth and Planetary Sciences, 77 Massachusetts Ave., Cambridge MA 02139-4307, USA

Where have all life's molecules gone?
Recycled to carbon dioxide every one!
Ah! But look – a hardy few remain
As biomarkers imprisoned in the fossil domain

Geoffrey Eglinton (2004)

15.1 Introduction

Biomarker molecules are natural products that have distinctive structural and/or isotopic attributes. These attributes convey information about the origins of the molecule with various degrees of specificity. For environmental and geological studies the most effective biomarkers have a limited number of well-defined sources, are recalcitrant, are readily measured in environmental samples and can be tracked through diagenesis, that is, the chemical changes that ensue after death. Accordingly, biomarkers can be proxies for different biogeochemical processes taking place in modern environments as well as chemical fossils that afford a paleobiological record of the past activities of living communities (Brocks and Summons, 2003; Gaines et al., 2009). This chapter briefly describes the chemistry and biological origins of some commonly encountered biomarkers, illustrated with examples of their use in environmental geomicrobiology and in paleoreconstruction.

15.2 Origins of biomarkers

In geobiology and astrobiology, the most commonly studied molecular biomarkers are lipids. These include pigments as well as cell wall and interior membrane constituents essential to the integrity and physiology of cells. Other biochemicals, such as carbohydrates, proteins and nucleic acids, are accessible food sources for bacteria and fungi and generally do not survive for long after the death of a cell. In contrast, lipids have recalcitrant, hydrophobic hydrocarbon moieties that are much more difficult to degrade, especially in the absence of oxygen. Thus, even fatty acids that were linked to glycerol by hydrolysable ester linkages have some preservation and diagnostic potential. Polycyclic or highly branched structures, such as are found in polyisoprenoid lipids, are even more resistant to biodegradation and have the greatest preservation potential. In this respect, cyclic triterpenoids, such as steroids (Fig. 15.1) and hopanoids (Fig. 15.2), are well established as excellent biomarkers for Eukarya and Bacteria, respectively (Brassell et al., 1983). Further, their hydrocarbon cores can be preserved in sediments and petroleum for hundreds of millions to billions of years (Dutkiewicz et al., 1998, 2006; Brocks et al., 2003a; Eigenbrode et al., 2008; George et al., 2008; Grosjean et al., 2009; Waldbauer et al., 2009). Hydrocarbon biomarkers can be diagnostic, with varying degrees of confidence, for the physiological activities of organisms, paleoenvironmental conditions and the thermal history of sediments. Biomarkers have long been used in exploration for fossil fuels (Peters et al., 2005).

15.3 Diagenesis

Biomarkers undergo transformations once an organism dies and its cellular contents are released into the

Fundamentals of Geobiology, First Edition. Edited by Andrew H. Knoll, Donald E. Canfield and Kurt O. Konhauser.
© 2012 Blackwell Publishing Ltd. Published 2012 by Blackwell Publishing Ltd.

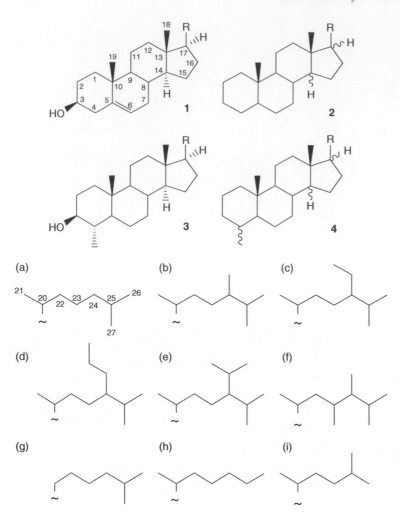

Figure 15.1 Generic structures of a Δ5-sterol (**1**) and a 4-methyl stanol (**3**) along with a selection of side-chains (**a–i**) that commonly are attached to C17. These steroids, including dinostanol (**3f**), sitosterol (**1c**) and cholesterol (**1a**), are typical of those found in the membranes of protists, plants and animals. The 4-methylsterane dinosterane (**4f**), stigmastane (sitostane) (**2c**) and cholestane (**2a**) are formed from their precursor sterols during burial and diagenesis and are representative of the diversity of eukaryotic molecular fossils found in sediments and petroleum.

environment. These alteration processes are termed diagenesis. Only the most recalcitrant structures survive. In the case of intact polar lipids (IPLs) the stable entities are the hydrocarbon cores, that is, the alkyl chains that are bound to glycerol by ester or ether bonds. Preservation of carbon compounds is strongly aided by the exclusion of electron acceptors such as oxygen and sulfate (Canfield, 1989) and a tight association with minerals (Hedges and Keil, 1995), and is enhanced by the presence of sulfide which is a reducing agent and can contribute to the removal of unstable features such as double bonds (Adam *et al.*, 2000; Hebting *et al.*, 2006; Kohnen *et al.*, 1989). Complex terpenoid lipids such as sterols, bacteriohopanepolyols (BHPs) and carotenoids can be reduced to their hydrocarbon skeletons without compromising the basic structural features that confer diagnostic utility. One important and potentially very useful aspect of diagenesis is the propensity of some lipids to undergo predictable stereochemical modifications as well as structural ones. These changes tend to be progressive and a sample in the earliest stage of diagenesis can already have a dizzying array of compounds derived

from a single original precursor (Fig. 15.3). Biomarker stereochemistry can be used to assess the maturity of the sample and to investigate the possibility of contamination by younger organic matter.

15.4 Isotopic compositions

Accurate elucidation of the chemical structures of lipids from cultured organisms provided the original basis for assignments of the biological sources of chemical fossils (Eglinton, 1970). However, given the conservative aspects of lipid biochemistry, and the paucity of lipids that are exclusive to a single organism, such assignments are often ambiguous. Early on it was also realized that metabolites inherit C, H, N and O isotopic compositions that are set when the organism acquires its carbon and the precursor molecule is biosynthesized (Abelson and Hoering, 1961). Thus, isotopic data is another means of refining the assignments of biomarkers to their sources provided that we have sufficient knowledge to reliably interpret isotopic variations (Hayes, 2001, 2004).

The first-order controls on the isotopic compositions of an entire organism lie in how it acquires its carbon,

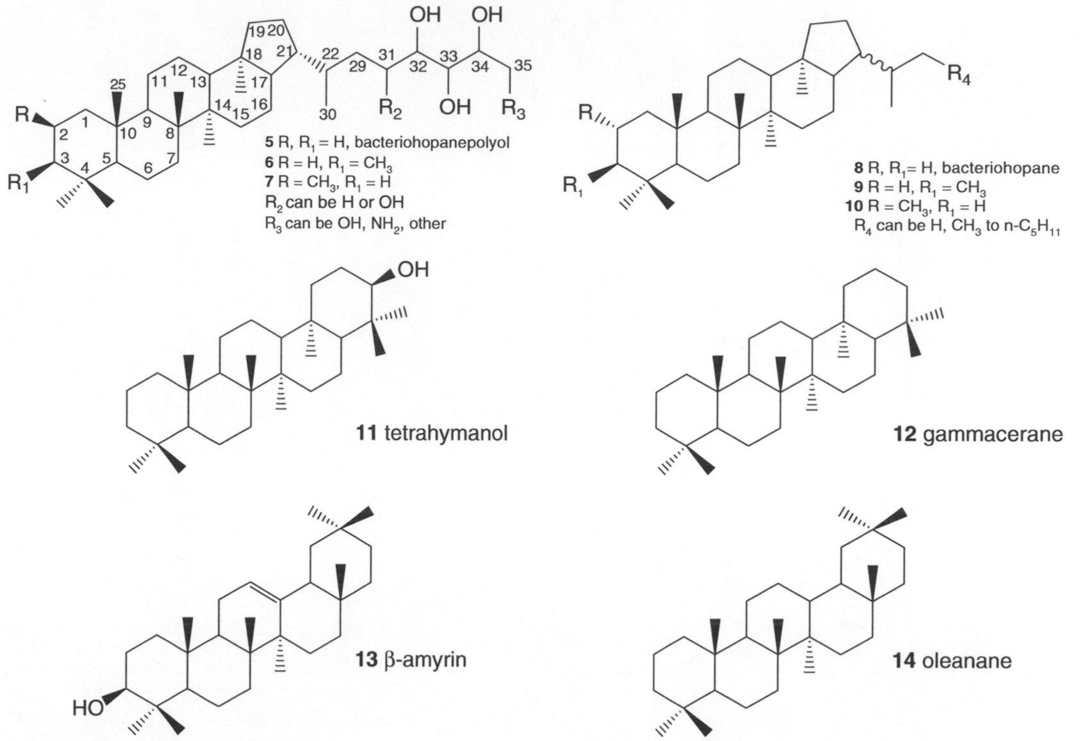

5 R, R$_1$ = H, bacteriohopanepolyol
6 R = H, R$_1$ = CH$_3$
7 R = CH$_3$, R$_1$ = H
R$_2$ can be H or OH
R$_3$ can be OH, NH$_2$, other

8 R, R$_1$ = H, bacteriohopane
9 R = H, R$_1$ = CH$_3$
10 R = CH$_3$, R$_1$ = H
R$_4$ can be H, CH$_3$ to n-C$_5$H$_{11}$

11 tetrahymanol

12 gammacerane

13 β-amyrin

14 oleanane

Figure 15.2 Structural diversity in bacteriohopanepolyols (BHP; **5–7**) and the derived hopanoid hydrocarbons (**8–10**) that are found in sediments and petroleum. 3β-Methylbacteriohopanepolyols (**6**) and 2β-methylbacteriohopanepolyols (**7**), and the derived hydrocarbons (**9** and **10**) have been have been proposed as biomarkers for methanotrophic bacteria and cyanobacteria respectively. Tetrahymanol (**11**) and β-amyrin (**13**) are represented in the fossil record by the hydrocarbons gammacerane (**12**) and oleanane (**14**) which are considered to be good biomarkers for bacteriovorous ciliates and flowering plants, respectively.

hydrogen, nitrogen and other elements, as well as the isotopic compositions of the substrates for these elements. In the case of carbon, this is determined by physiology, namely whether the organism is autotrophic or heterotrophic and, in the case of the former, the assimilation pathway (House *et al.*, 2003). Once assimilated, carbon is further fractionated during respiration and biosynthesis. The isotopic fractionations that take place as carbon flows through biosynthetic pathways are of particular interest (Hayes, 2001) and are discussed further below. Other variables known to affect the C-isotopic composition of biomass and individual components include the size of an organism, the rate of growth, the stage of the growth cycle and how the carbon is compartmentalized.

The burgeoning technical capability to conduct isotope analysis on individual compounds (i.e. compound-specific isotope analysis, CSIA) on trace amounts of lipid has greatly enhanced the accuracy of lipids as proxy for both the organisms that produced them and/or their physiologies (Bradley *et al.*, 2009; Ohkouchi *et al.*, 2008; Zhang *et al.*, 2009). Isotopic data also increase the depth of historical information that can be derived from fossil molecules and builds confidence in our interpretations of this record (Hayes, 2004). C-isotopic compositions of molecules are likely highly conserved over long periods of geological time because, without destroying a molecule, there are no known processes that allow exchange of its carbon atoms. Hydrogen isotopic compositions are also set during biosynthesis (Schimmelmann *et al.*, 1999; Sessions *et al.*, 1999) and, in the case of bacteria, reflect the physiology of NADPH production in central metabolism (Zhang *et al.*, 2009). However, δD values are also prone to progressive corruption through exchange reactions over long timescales; some H-atoms are more easily exchanged than others (Sessions *et al.*, 2004). The N-isotopic compositions of chlorophyll-derived biomarkers (Chikaraishi *et al.*, 2005, 2008 Ohkouchi *et al.*, 2006) provide important data on nitrogen sources for biosynthesis and can serve as a proxy for reconstructing ancient nitrogen cycling. Compound-specific oxygen isotopic measurements on lipids are more difficult because few lipids have high O/C ratios; for this reason, the O-isotopic composition of lipids has not been exploited to any extent in contrast to measurements on oxygen-rich molecules such as cellulose (Sternberg *et al.*, 1984). Most recently, a continuous flow technique utilizing inductively coupled plasma mass spectrometry (ICPMS) has been developed to measure the δ^{34}S values of individual compounds (Amrani *et al.*, 2009) thereby completing the array of light, bioactive

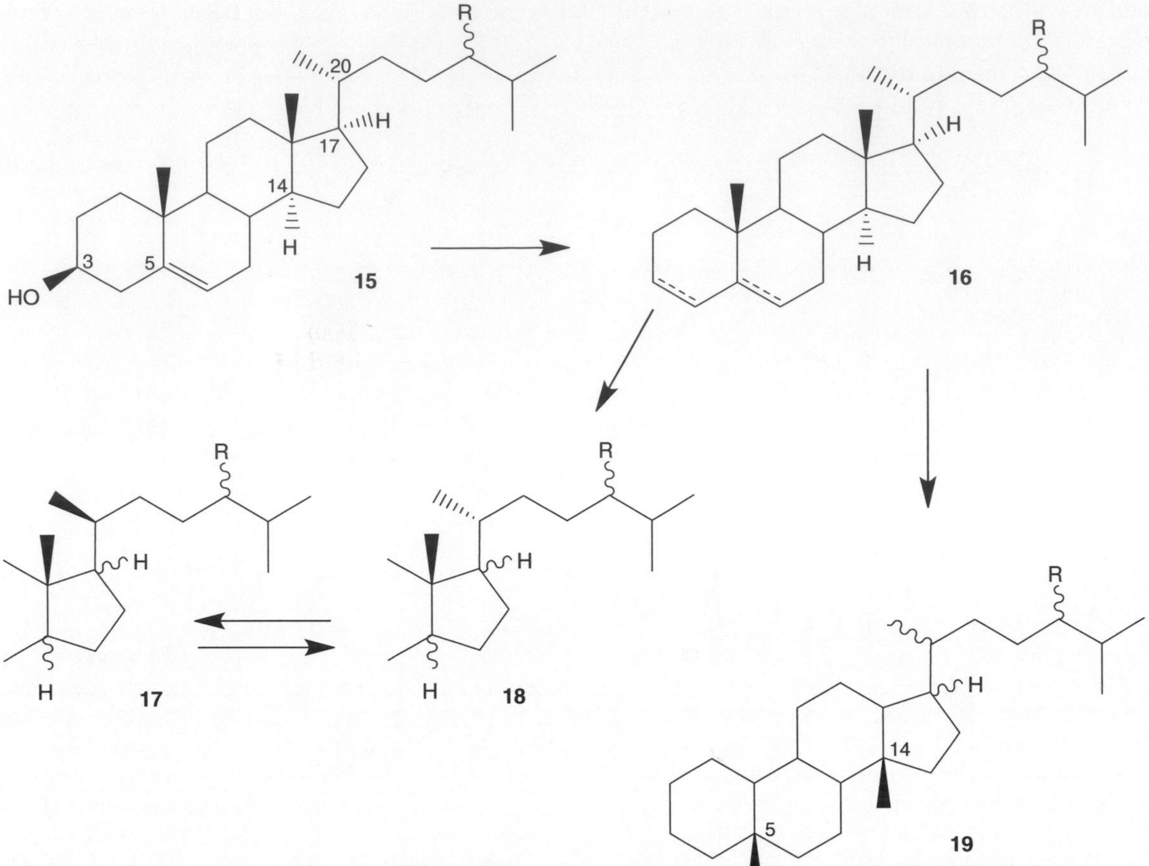

Figure 15.3 Some of the steps in the complex diagenetic pathway from sterols to steranes (Mackenzie *et al.*, 1982) and illustrating the major stereochemical changes. Sterenes and diasterenes (**16**) are initially formed from sterols and steryl esters by dehydration or elimination reactions and these are then reduced to the sterane isomers **17** and **18**. Alternatively, they may undergo a backbone rearrangement and reduction to form diasteranes (rearranged steranes; **19**).

elements amenable to isotopic analysis. Clearly, simultaneous measurement of C-, H-, O-, N- and S-isotopic compositions of an individual biomarker compound is challenging but would provide the ultimate characterization (Hinrichs *et al.*, 2001).

15.5 Stereochemical considerations

The vast majority of biomarkers exhibit some kind of stereospecificity. In some ways this is analogous to the homochirality that is characteristic of amino acids with the important difference being that biomarkers generally have multiple asymmetric centres. The common sterol cholesterol (Fig. 15.1, 1a) has eight. The plant sterol sitosterol (Fig. 15.1, 1c) has nine as does the dinoflagellate-specific dinosterol. Bacteriohopanepolyols (BHPs) are even more complex with bacteriohopanetetrol (BHtetrol) having at least eight in the C_{30} pentacyclic ring system and four in the side chain. This could result in a bewildering collection of potential stereoisomers except for the fact that biosynthesis generally produces

one isomer exclusively (Figs 15.1–15.4). This fact immediately leads to a method for distinguishing the biomarkers of extant and recent organisms from those in the geologic past because the latter are invariably mixtures, often quite complex, of stereoisomers.

The post mortem stereochemical changes that biomarkers undergo encode valuable information about diagenetic conditions and thermal histories. Acyclic isoprenoids readily isomerize very early in their burial so that the fossil forms comprise mixtures of diastereoisomers (aka diastereomers) (Patience *et al.*, 1978). Steroids may form pairs of diastereoisomers at former sites of unsaturation (double bonds). Some hydrogen atoms at ring junctions are prone to isomerize while the whole steroid backbone may also rearrange especially if acidic clays are abundant in the enclosing sediment (Mackenzie *et al.*, 1982; van Kaam-Peters *et al.*, 1998). Diagenesis in carbonate sediments proceeds differently and such re-arrangements are often suppressed (Rullkötter *et al.*, 1984). As mentioned above, the changes are progressive so that in geological samples, which have experienced long periods of burial, all biomarkers will come to comprise mixtures dominated by the most thermodynamically stable stereoisomers. This is illustrated for steroids (structures **1–4**) and triterpenoids (**5–14** and **15–19**) in Figs 15.1–15.3.

15.6 Lipid biosynthetic pathways

The hydrocarbon skeletons of lipids fall into two broad categories. First there are the acetogenic lipids which are those with straight chains (normal) or simple branching patterns (e.g. iso-, anteiso-, etc.). As the name implies, acetogenic lipids are constructed from acetate units originally donated from acetyl coenzyme-A (De Niro and Epstein, 1977; Hayes, 1993). Additional carbons can be acquired from other short-chain fatty acid precursors and from methyl groups donated by *S*-adenosyl methionine (SAM). The second major class of lipids comprises the polyisoprenoids. These are constructed from a five-carbon precursor, dimethylallyl diphosphate (DMAPP; **20**) or isopentyl diphosphate (IPP; **21**) leading to a diverse array of structures determined by number of 'isoprene' building blocks and the manner in which they are assembled (Fig. 14.4). For example, geranyldiphosphate (GPP; **22**) is formed when two C5 units are connected head to tail; phytol (**23**) is derived from four C5 units linked head to tail. A notable feature of acetogenic and polyisoprenoid lipids, and a point of contrast to other biopolymers such as nucleic acids and proteins, is that only a few specific classes of lipids contain hydrolysable linkages.

The acetyl building block of acetogenic lipids can also have multiple origins with consequences for the isotopic composition of its products. In photosynthetic organisms and heterotrophs acetate is formed primarily through the decarboxylation of pyruvate, itself a product of the synthesis or degradation of carbohydrates. In one of the first observations of intermolecular 'isotopic order,' DeNiro and Epstein (1977) made site-specific measurements of the isotopic compositions of each carbon atom in pyruvate and acetate in cultures of the bacterium *Escherichia coli*. They determined that a major fractionation accompanies the decarboxylation of pyruvate leading to a depletion of the carboxyl carbon atom depleted of approximately 15‰ relative to the methyl carbon atom. Overall, the acetate produced by *E. coli*, and the lipids that were formed from it, were depleted by 7–8‰ relative to glucose and the pyruvate precursors. Upon further investigation, and working with both yeast and *E. coli*, Monson and Hayes confirmed that individual fatty acids were depleted in ^{13}C by 7‰ relative to glucose carbon source and that the carbon atoms along the chain comprised two isotopically-distinct groups (Monson and Hayes, 1980, 1982). Odd-numbered carbon atoms, derived from the carboxyl carbon of acetate, were depleted by around 6‰ relative to the glucose while the even-numbered carbons

were enriched by about 0.5‰. Research conducted since these seminal observations has further verified the phenomenon of intramolecular isotopic order and confirmed that acetogenic lipids are almost invariably depleted relative to total biomass and other biochemicals (Hayes, 2001; Schouten *et al.*, 2008c). However, additional factors can affect the isotopic composition of acetyl-CoA in organisms grown under different conditions and in environmental samples with consequences for correctly interpreting the results from analyses of geologic samples (Heuer *et al.*, 2006).

Polyisoprenoid lipids are constructed from C5 (isoprene) building blocks. The IPP and DMAPP precursors can be made from acetyl-CoA following the mevalonic acid (MVA) pathway. The elucidation of the MVA pathway was based on studies of yeast, and other easily manipulated organisms and tissues, and was long assumed to be the exclusive route (Bloch, 1992). However, in studying the biosynthesis of hopanoid triterpenoids in bacteria, Rohmer and colleagues (Rohmer, 2003) discovered evidence for a second pathway. They observed anomalous labeling patterns when $[1\text{-}^{13}C]$ and $[2\text{-}^{13}C]$ acetate were fed to hopanoid-producing bacteria such as *Rhodopseudomonas palustris* (Flesch and Rohmer, 1988). Further experiments with ^{13}C isotopomers of glucose (Rohmer *et al.*, 1993) enabled the unraveling of the mevalonate-independent methylerythritol phosphate (MEP) pathway (Rohmer, 2003), also known as the DOXP pathway. In this pathway, IPP and DMAPP are the products of condensation of a C_2 subunit formed by pyruvate decarboxylation and a C_3 moiety in the form of a triose phosphate. Methylerythritol 4-phosphate is the first committed intermediate, hence the naming of the pathway as MEP. MVA is the pathway common to Archaea and non-plastid bearing Eukarya. Bacteria can have either MEP or MVA while plastid-bearing Eukarya can have either or both (Hayes, 2001). However, MEP is widely present in the chloroplasts of plants (Eisenreich *et al.*, 1998) and their green algal cousins presumably reflecting inheritance from the cyanobacterial ancestor (Lichtenthaler *et al.*, 1997). A consequence of having two pathways for isoprenoid biosynthesis is that complexities abound in the interpretation of isotopic data for isoprenoids. At the same time, the MEP versus MVA dichotomy can be informative and useful in interpretations of the natural variability in the isotopic compositions of polyisoprenoids (Schouten *et al.*, 2008c).

15.7 Classification of lipids

Biological lipids are diverse cellular chemicals that are largely soluble in organic solvents and insoluble in water. Some of the important compound classes are discussed below.

15.7.1   Hydrocarbons

Hydrocarbons are a diverse compound class. They encompass some of the simplest and least specific biomarker compounds such as *n*-alkanes as well as complex isoprenoid hydrocarbons of the botryococcene type which, so far, appear to be exclusive to a single species of chlorophyte alga, *Botryococcus braunii*. Interestingly, methane, the simplest hydrocarbon of all, can be a biomarker in certain circumstances because its biosynthesis is exclusive to the methanogenic Archaea. However, caution is warranted because methane can have other origins, for example, the thermal cracking of complex organic matter. Isotopic and other data are crucial for understanding the origins of methane (Schoell, 1988).

Important biogenic hydrocarbons include the *n*-alkanes found in the leaf wax of plants (Eglinton, 1970), short-chain normal (*n*-) and branched alkanes produced by cyanobacteria (Han and Calvin, 1969; Jahnke *et al.*, 2004) and some algae, long-chain branched alkanes found in the waxy cuticles of insects and an array of distinctive isoprenoid hydrocarbons produced by algae of various types.

Hydrocarbons are also prominent components of petroleum and sedimentary bitumen (Peters *et al.*, 2005). Here, they are the fossilized remains of former functionalized lipids. Hydrocarbons are the main chemical form in which biomarkers are preserved in the geologic record and can be extracted from sediments with organic solvents (yielding bitumen), or released from macromolecular sedimentary organic matter (kerogen) by heating in a vacuum, in an inert gas such as helium or in reactive fluids such as water or hydrogen (Lewan *et al.*, 1979, 1985; Love *et al.*, 1995).

15.7.2   Fatty acids, alcohols and ketones

These are essentially hydrocarbons bearing one or more oxygen-containing functional groups. Fatty acids are most commonly found as the acyl chains in the IPLs of Bacteria and Eukarya. Their carbon numbers and sites of branching and/or unsaturation confer diagnostic utility (Guckert *et al.*, 1985). Besides the steroids and terpenoids discussed below, biomarker alcohols include components of plant leaf wax (Eglinton and Hamilton, 1967) and the esterifying moieties of chlorophyll. Phytol (**23**) is one of the most prominent alcohol biomarkers found in nature. It originates mainly from the side chain of chlorophylls a and b common to plants, algae and cyanobacteria. The hydrocarbons phytane and pristane, both of which can be derived from phytol, are amongst the most ubiquitous and abundant fossil biomarkers (Volkman and Maxwell, 1984).

The best studied ketones are the long-chain alkenones produced by haptophyte algae (Volkman *et al.*, 1980). The patterns of unsaturation in the C_{37} alkenones, presently known to be derived from haptophyte algae of the class Prymnesiophyceae, specifically *Emiliania huxleyi*, *Gephyrocapsa oceanica*, and members of the genera *Isochrysis* and *Chrysotila* (Rontani *et al.*, 2004), have been shown to vary systematically with the temperature of the waters in which they grew (Prahl and Wakeham, 1987; Sikes and Volkman, 1993). The C_{37} alkenones, therefore, encode a record of sea surface temperature (Brassell *et al.*, 1986) and this has resulted in the development, testing and widespread application of the $U_{37}^{k\prime}$ sea surface temperature proxy (Rosell-Melé *et al.*, 1995; Sikes *et al.*, 1997).

15.7.3   Chlorophylls and carotenoids

The landmark 1936 paper by Alfred Treibs (Treibs, 1936) reporting discovery of fossilized organic nitrogen compounds laid foundations for the entire field of biomarker research. Tetrapyrroles preserved in rocks were so complex that they could only have been sourced from a few biological precursors, either hemes or chlorophylls. Consideration of their structures, natural abundances and distributions all pointed to an origin from the photosynthetic pigments such as the chlorophyll a (Chl a) that is ubiquitous in plants and algae. Although among the least stable of lipids, chlorophylls and bacteriochlorophylls (Bchls) can undergo a sequence of diagenetic transformations to ultimately form vanadyl and nickel porphyrins. Their structures, in combination with specific C- and N-isotopic compositions, can sometimes be exceptionally diagnostic (Boreham *et al.*, 1990; Chikaraishi *et al.*, 2005; Ohkouchi *et al.*, 2008). Even after oxidation leads to fragmentation of the tetrapyrrole ring-system some of the resulting products, maleimides (1*H*-pyrrole-2,5-diones), can be preserved. In most instances, these have simple methyl and ethyl substitution at positions 3 and 4, thereby rendering their precursor indistinct. However, maleimides having Me *n*-propyl and Me *i*-butyl substitutions are highly specific since these are only observed in Bchl *c*, *d* and *e*, common to the filamentous bacteria (Chloroflexaceae) the green sulfur bacteria (Chlorobiaceae) and the recently described Yellowstone isolate *Chloracidobacterium thermophilum* (Bryant *et al.*, 2007). In an elegant application of C-isotopic measurements, Grice and colleagues (Grice *et al.*, 1996) were able to measure the C-isotopic compositions of Me *n*-propyl and Me *i*-butyl maleimides in the Permian Kupferschiefer to show that, in part, they can derived from the bacteriochlorophylls of the green sulfur bacteria.

15.7.4   Steroids and triterpenoids

Steroids (tetracyclic triterpenoids; Fig. 15.1) and pentacyclic triterpenoids (Fig. 15.2) are of great interest in geobiology because they are demonstrably the most recalcitrant classes of biomarkers and also because they are related in having a common biosynthetic precursor, the ubiquitous C_{30} acyclic hydrocarbon squalene (**24**). The enzymes responsible for cyclization of squalene are also hypothesized to have a shared evolution (Ourisson *et al.*, 1979, 1987; Rohmer *et al.*, 1979). This has become known as the sterol–triterpenoid biosynthetic bifurcation (Nes, 1974). The bacterial enzyme (squalene–hopene cyclase: SHC) and the eukaryotic enzyme (oxidosqualene cyclase or OSC) have sequence similarities that betray a shared origin but this alone does not allow one to unambiguously identify which is the more primitive (Fischer and Pearson, 2007). However, a connection to

the oxygenation of Earth's surface seems clear. Oxygen is required for biosynthesis of sterols but not hopanoids. In sterol biosynthesis (Fig. 15.4), both the conversion of squalene to oxidosqualene via the enzyme squalene monooxygenase (SQMO) and the downstream removal of the angular methyl groups at C4 and C14 of the protosterols, lanosterol (**27**) in the case of fungi and metazoa or cycloartenol (**28**) in plants, en-route to the functional

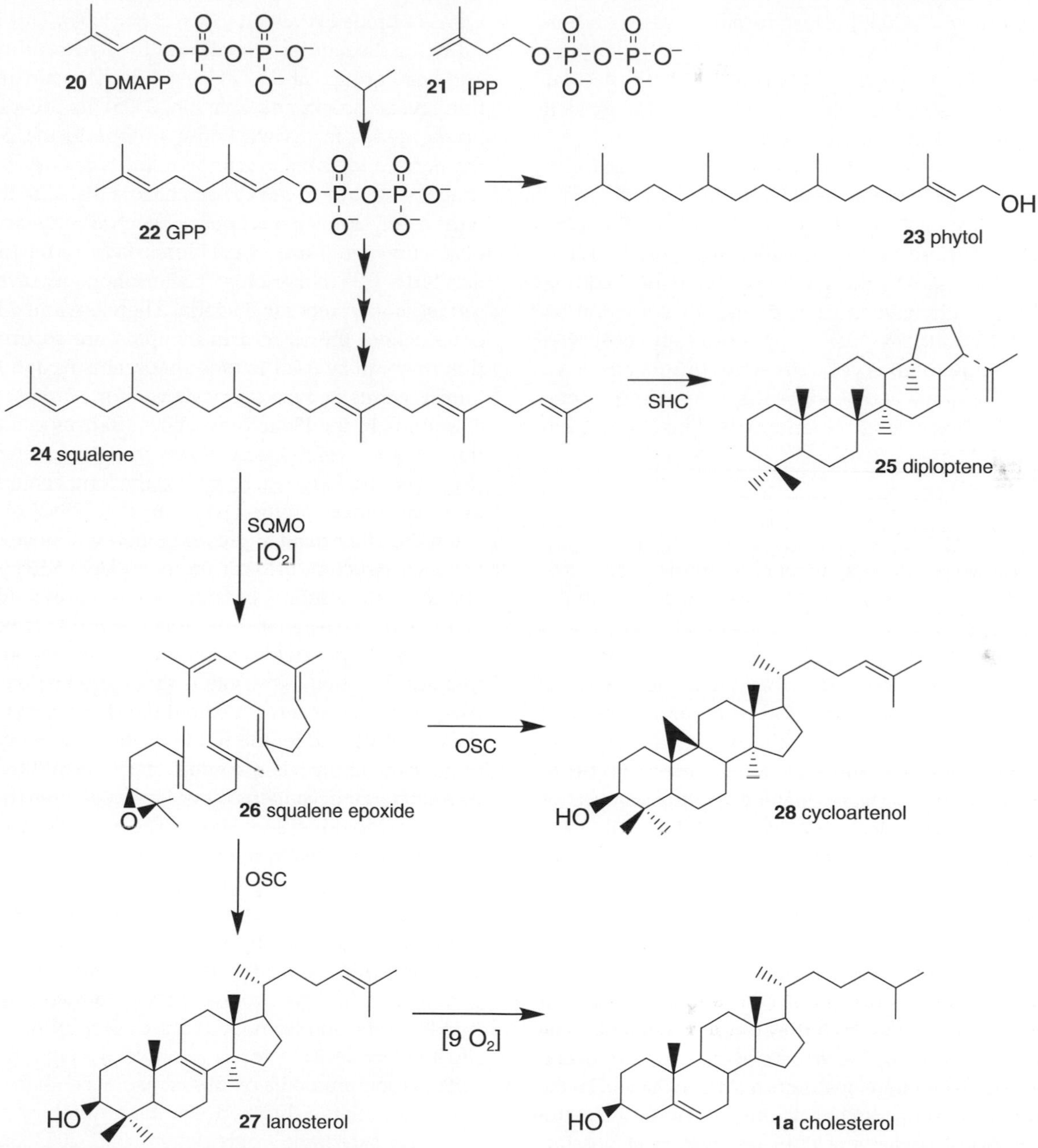

Figure 15.4 Key steps in the biosynthetic pathway to diterpenoids and triterpenoids such as phytol (**23**) and squalene (**24**) respectively. Squalene is the biosynthetic precursor of hopanoids such as diplopterol and diploptene (**25**) which are formed without involvement of molecular oxygen. One molecule of oxygen is required for the transformation of squalene to squalene epoxide (aka oxidosqualene; **26**) which is,

in turn, the precursor of the either one of two protosterols lanosterol (**27**) or cycloartenol (**28**). An additional nine molecules of oxygen is required to complete the steroid biosynthetic pathway to cholesterol (**1a**). Oxidosqualene is also the precursor of β-amyrin (**13**) in the flowering plants. In contrast, oxygen is not required to form tetrahymanol (**11**); the oxygen in this sterol surrogate is derived from water.

steroids including cholesterol (**1a**), ergosterol and sitosterol (**1c**) require oxygen. In all, up to eleven moles of oxygen are necessary for the biosynthesis of a mole of cholesterol (Summons *et al.*, 2006). Bloch and colleagues hypothesized that significant functional improvement in sterols accompanied the evolution of the pathway from lanosterol to cholesterol (Bloch, 1979, 1987) while others have noted that sterol structure profoundly influences curvature and lipid phase separation in artificial membranes (Bacia *et al.*, 2005). More recent research on biochemical networks is consistent with this and identifies other areas where an increasing availability of molecular oxygen over geological time has resulted in elaboration of new pathways (Raymond and Segrè, 2006).

Other intriguing aspects of the evolutionary histories of SHCs and OSCs are encoded into the overall amino acid sequences, and especially the degree to which the active sites of these protein families are conserved. For example, alignment of the sequences of genes encoding for OSCs reveals a very high degree of conservation with the active site residues being absolutely conserved across the known eukaryotic diversity (Summons *et al.*, 2006), and the same is true for SHCs as applies to bacterial diversity (excluding the enigmatic Planctomycetes) (Fischer and Pearson, 2007).

15.7.5 Intact polar lipids

For environmental studies, intact polar lipids (IPLs) carry the most taxonomic information as they include both the core lipid (Figs 15.5 and 15.6), with its preservable carbon skeleton, along with potentially diagnostic polar head groups (Fig. 15.7). Recent advances in technical capabilities and instrumentation for liquid chromatography (LC) and mass spectrometry (MS) with soft ionization such as electrospray ionization and atmospheric pressure chemical ionization (APCI) have enabled routine analysis of IPLs including BHPs (Rutters *et al.*, 2002; Sturt *et al.*, 2004; Talbot *et al.*, 2003, 2007). This has stimulated development of a growing database for the taxonomic distributions of eukaryotic (McDonald *et al.*, 2007), bacterial (Van Mooy *et al.*, 2006) and archaeal IPLs (Hopmans *et al.*, 2000; Jahn *et al.*, 2004). LC-MS technology development has also increased the range of lipids that can be routinely studied and the precision with which they can be measured while considerably reducing the workload. In fact, the revelation of the diverse range of structures that exist within the membrane-spanning archaeal and bacterial glycerol dialkyl glycerol tetraethers (GDGTs) was only possible with the advent of LC-MS technologies (Hopmans *et al.*, 2000). IPL analyses are particularly powerful when combined with genomic information (e.g, ribosomal DNA clone libraries and metagenomes) that inform us about the taxonomic diversity of microbial communities. IPL analyses are also potentially quantitative thereby giving

information on the absolute abundances of microbes in a sample (Lipp *et al.*, 2008).

15.7.6 Biomarkers for physiology or phylogeny?

Inspection of the phylogenetic distributions of lipid classes confirms that, for organisms known to date, there are significant correspondences between taxonomic positions of organisms and the predominant types of lipids expressed (Sturt *et al.*, 2004). This is especially the case for terpenoids and the hydrocarbon cores and head groups of IPL, at least at the domain and phylum levels (Brocks and Pearson, 2005). Isoprenoid ether lipids are, as far as we know, unique to the Archaea. Similarly, and as far as is known, the core GDGT crenarchaeol is unique to the Crenarchaeota. Because there are only a few known exceptions, such as Myxobacteria, desmethylsterols are robust biomarkers for the Eukarya. Similarly, C35 hopanoids (bacteriohopanepolyols) are robust biomarkers for Bacteria. There are a few notable cases where the structures of lipids are so distinctive that they can be used to identify organisms at the order, family, genus or even species level, for example ladderane lipids in the Planctomycetales (Rattray *et al.*, 2008), the botryococcene-related lipids in *Botryococcus braunii* (Metzger and Largeau, 2005) and the long-chain ketones of the prymnesiophytes (Volkman *et al.*, 1995).

On the other hand, a preoccupation with relationships between structure of biomarkers and phylogeny draws attention away from a potentially more profound aspect of biomarker chemistry: the relationship between lipid structure and physiology. Lipids always serve some physiological function. Nowhere is this more evident that in the process of anaerobic oxidation of ammonia (anammox) and the ladderane lipids of the Planctomycetales. Organisms living via anammox catabolism have anammoxosomes that contains the critical hydrazine/hydroxylamine oxidoreductase enzyme. The anammoxosome has to be impermeable to hydrazine and oxygen and this is presumed to be accomplished with the aid of membranes constructed from phospholipids known as ladderane lipids. The ladderane hydrocarbon cores, with their distinctive linearly concatenated cyclobutane moieties, are connected to the glycerol moiety via both ester and ether bonds. Protection from O_2 and provision of the essential nitrite electron acceptor is afforded through syntrophy with aerobic ammonia oxidizers. As far as is known, there is a direct relationship between the physiology of anammox and ladderane lipids (Jetten *et al.*, 2003; Sinninghe Damsté *et al.*, 2005). The recent paleorecord of ladderanes has recently been exploited as a tool for studying variations in the intensity of the oxygen minimum zone in the Arabian Sea over the past 1000 years (Jaeschke *et al.*, 2009). It is therefore somewhat ironic that such a direct connection cannot be explored in deep time because, despite the robustness of anammoxosome membranes in living cells, the lipids themselves are thermally unstable on geological timescales (Jaeschke *et al.*, 2008).

Rather than asking questions like 'What organisms makes what lipids?' biomarker researchers are turning their attention to learning more about the physiological role of a specific compound, its cellular localization, the environmental conditions that require or allow its production and the phylogenetic distribution of its biosynthetic pathway(s) (Doughty *et al.*, 2009, 2011; Sessions *et al.*, 2009; Welander *et al.*, 2009, 2010).

15.8 Lipids diagnostic of Archaea

Three characteristics distinguish archaeal membrane lipids from those of bacteria: (1) hydrophobic cores are composed of isoprenoidal hydrocarbon rather than fatty acyl chains; (2) cores are linked to glycerol moieties by

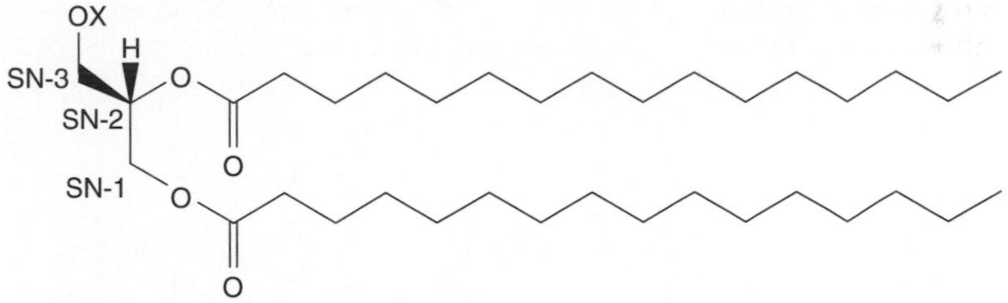

29 X = H, *sn*-1, 2-diacylglycerol

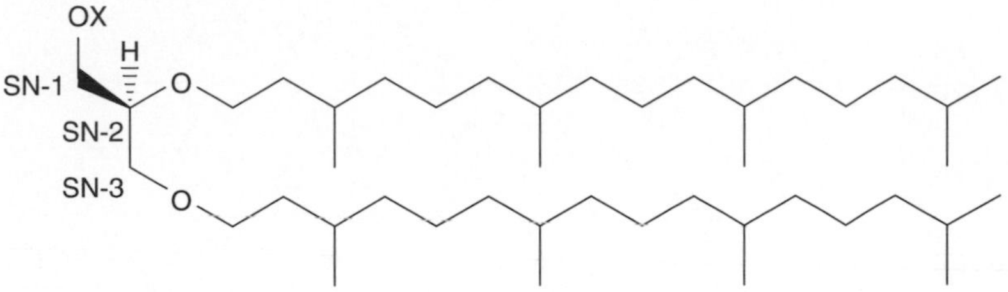

30 X = H, archaeol
sn-2,3-di-O-phytanylglycerol

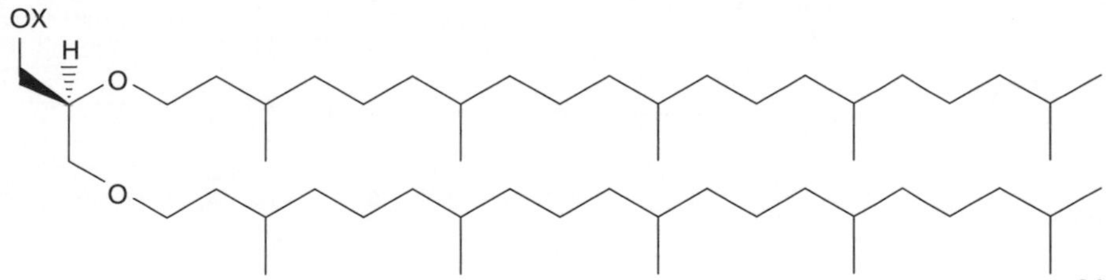

31 X = H, C_{25}, C_{25}-archaeol
sn-2,3-di-O-sesterterpanylglycerol

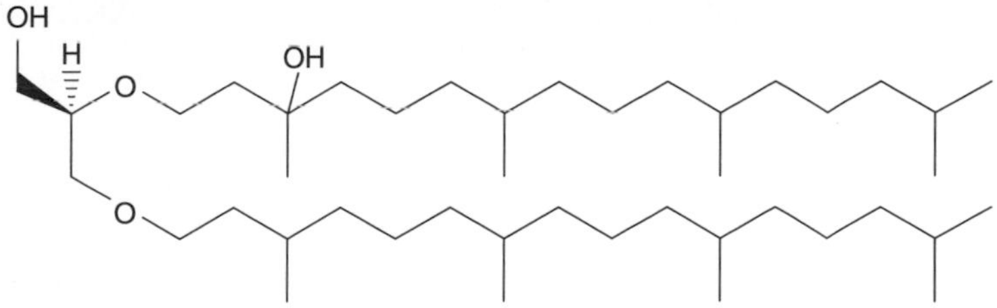

32 X = H,*sn*-2-hydroxyarchaeol

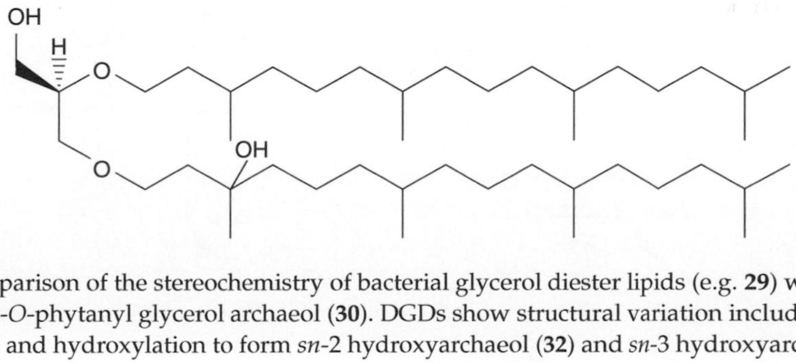

33 X = H, *sn*-3-hydroxyarchaeol

Figure 15.5 Comparison of the stereochemistry of bacterial glycerol diester lipids (e.g. **29**) with the dialkyl ether lipids of Archaea typified by the di-*O*-phytanyl glycerol archaeol (**30**). DGDs show structural variation including replacement of the C_{20} phytanyl chains by C_{25} (**31**) and hydroxylation to form *sn*-2 hydroxyarchaeol (**32**) and *sn*-3 hydroxyarchaeol (**33**).

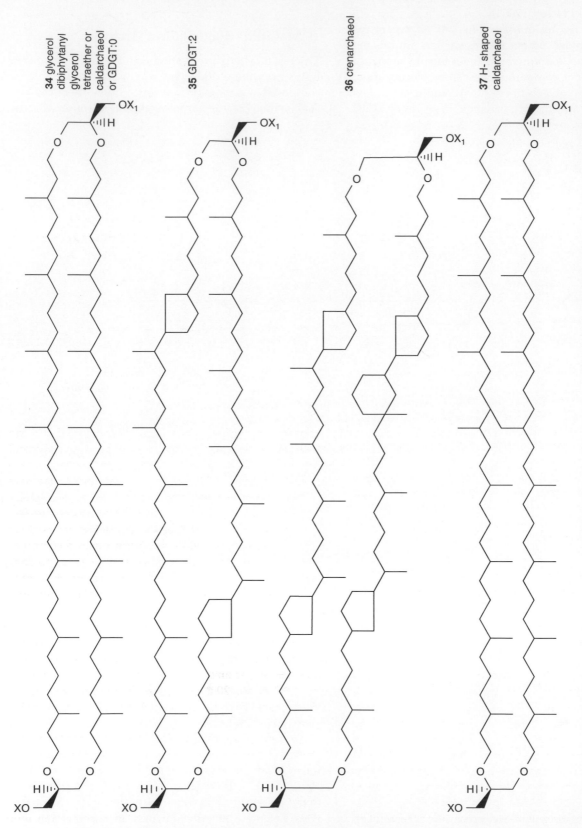

Figure 15.6 Structural variation in the glycerol dialkyl glycerol tetraethers (GDGTs) of Archaea. Caldarchaeols (**34–37**) can have multiple pentacyclic rings as illustrated by a GDGT-2 (**35**). Many of the polycyclic GDGTs are found in both the Euryarchaeota and the Crenarchaeota. However, crenarchaol (**36**) is thought to be confined to the Crenarchaeota while the 'H-shaped' caldarchaeol (**37**) has been found in thermophilic Euryarchaeota. X and X1 denote polar head groups which, in the case of the Archaea, are typically phosphate, phosphoglycosyl or glycosyl groups (Fig. 15.7 **i**, **q** or **p**).

34 glycerol dibiphytanyl glycerol tetraether or caldarchaeol or GDGT:0

35 GDGT:2

36 crenarchaeol

37 H- shaped caldarchaeol

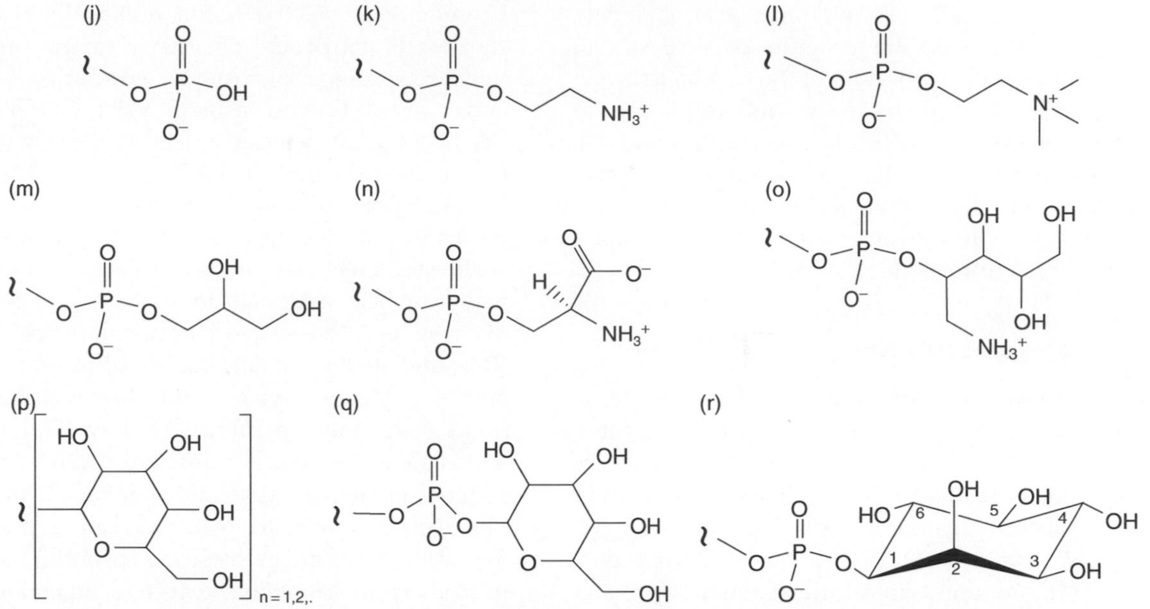

Figure 15.7 A selection of polar head groups that occur in the intact polar lipids of Archaea, Bacteria and Eukarya.

ether rather than ester bonds; and (3) glycerol stereochemistry is 2,3-*sn*-glycerol stereochemistry instead of the bacterial 1,2-*sn*-glycerol (Figs 15.5 and 15.6). However, it is important to recognize that ether lipids are not exclusive to Archaea and have been isolated from bacteria including Aquificales (Jahnke *et al.*, 2001), Mycoplasma (Wagner *et al.*, 2000) and Thermotogales (Sinninghe Damsté *et al.*, 2007). Some Archaea are also known to synthesize fatty acids (Gattinger *et al.*, 2002), but 2,3-*sn*-glycerol stereochemistry is an exclusive trait of Archaea (Peretó *et al.*, 2004) and the synthesis of lipids with both ether bonds and isoprenoid cores also appears confined to this domain (Koga and Morii, 2005).

15.8.1 Intact polar lipids of Archaea

Systematic investigations of the specific patterns of the cores and polar headgroups in archaeal lipids are relatively recent (Koga and Morii, 2005; Nishihara and Koga, 1995; Sturt *et al.*, 2004), and many gaps exist in our knowledge of them. It is not yet clear to what extent head group composition follows taxonomic lines. Complicating this picture is the apparent ability of single clades to synthesize GDGTs with a variety of headgroups. For example, the crenarchaeaon *Nitrosopumilus maritimus* (Schouten *et al.*, 2008b) synthesizes GDGTs with hexose, dihexose and phosphohexose head groups. Until a broader selection of cultures has been analysed it will be difficult to determine whether biosynthetic overlap between physiological groups is extensive; accordingly, the biomarker potential of archaeal IPLS cannot be fully realized at this time. However, isotopic analysis of IPLs is likely to provide additional information about archaeal metabolism in natural populations and will enable more extensive tracer experiments.

15.8.2 Core lipids

Broadly speaking, archaeal membrane core lipids can be divided into two classes: C43 diphytanyl glycerol diethers (DGDs; Fig. 15.5) and C_{86} membrane spanning glycerol dibiphytanyl glycerol tetraethers (GDGTs; Fig. 15.6), also called caldarchaeols. GDGTs have varying degrees of cyclization, containing 0–6 cyclopentane rings and 0–1 cyclohexane rings. One GDGT, crenarchaeol (**36**), has been given a common, non-IUPAC name; it contains one hexacyclic and four pentacyclic rings. The observation that temperature is one control on the degree of cyclization of caldarchaeol GDGTs (Gliozzi *et al.*, 1983) led to the development of the TEX-86 paleotemperature proxy based on their distribution in marine sediments (Schouten *et al.*, 2002). Crenarchaeol may alternatively be a biomarker specifically for ammonia-oxidizing Crenarchaeota (de la Torre *et al.*, 2008) or more generally for Group I Crenarchaeota, including heterotrophic species (Biddle *et al.*, 2006).

An unusual GDGT in which the biphytanyl moieties are covalently linked was first detected in the hyperthermophilic methanogen *Methanothermus fervidus* (Morii *et al.*, 1998). This 'H-shaped' core (**37**) has since been found in several euryarchaeal hydrothermal vent isolates and was recently detected in both marine and lacustrine sediments (Schouten *et al.*, 2008a), where its sources are, as yet, unknown.

DGDs, like GDGTs, show structural diversity (Fig. 15.5). Variations on the structure of archaeol (**30**) exist. Replacement of one or both of the C_{20} chains by C_{25} is common (e.g, **31**) while hydroxylation and unsaturation of the isoprenoid moieties is also seen (**32** and **33**). Macrocyclic archaeol, a diether containing 0–2 internal cyclopentane rings, has been identified in *Methanococcus janaschii* and in mud volcano carbonates (Comita and Gagosian, 1983; Stadnitskaia *et al.*, 2005).

15.8.3 *Taxonomic distribution of archaeal core lipids*

In Archaea certain lipid structural motifs appear to be conserved along phylogenetic lines. Within the cultured Euryarchaeota, halophiles synthesize DGDs such as archaeol (often as an unsaturated archaeol analogue, and sometimes substituting one or two C_{25} chains for C_{20}; **31**) (Kates, 1978). Archaeol (**30**) itself is broadly distributed, occurring in trace amounts in cultured Crenarchaeota and co-occurring with GDGTs in some methanogens. It has been suggested that DGDs may be intermediates in the biosynthesis of GDGTs, but the biosynthetic pathways leading to tetraethers are unresolved. Functionalized archaeols, such as *sn*-2 hydroxyarchaeol (**32**) and *sn*-3 hydroxyarchaeol (**33**) together with a dihydroxyarchaeol are prevalent in methanogenic and methanotrophic archaea (Sprott *et al.*, 1990; Nishihara and Koga, 1995; Bradley *et al.*, 2009). Pentamethylicosane (PMI) is also known from cultivated methanogens and methanotrophic microbial mats (Tornabene *et al.*, 1979; Schouten *et al.*, 1997; Thiel *et al.*, 1999). These particular lipids appear to be exclusively synthesized by Euryarchaeota.

Less phylogenetic specificity is apparent among the GDGTs. Some methanogens synthesize caldarchaeol GDGTs containing 1–3 cyclopentane rings (Gattinger *et al.*, 2002) and thermoacidophiles synthesize GDGTs with as many as 3 cyclopentane rings (Macalady *et al.*, 2004). Cultured Crenarchaeota, e.g. *Sulfolobus solfataricus*, previously known as *Caldariella acidophila*; (Gliozzi *et al.*, 1983) synthesize GDGTs with 0–6 cyclopentane rings, with up to four rings per biphytane moiety. Crenarchaeol, containing four cyclopentane and one cyclohexane ring, is ubiquitous in marine sediments and thought to be biosynthesized by marine Crenarchaeota. However, it has also been detected in soils and terrestrial hydrothermal environments (Pearson *et al.*, 2004) and in a cultured thermophile (de la Torre *et al.*, 2008).

15.8.4 *Preservation potential, geologic record*

Archaeal lipids, like all biomolecules, become defunctionalized during diagenesis. However, under favourable conditions archaeal core lipids can be preserved in the geologic record for hundreds of millions of years.

Kuypers *et al.* (2001) found archaeal lipids to be a significant component of organic matter preserved in mid-Cretaceous shales deposited during an oceanic anoxic event. Caldarchaeol, GDGT:1, GDGT:2 and crenarchaeol were detected intact in these samples, and remain the oldest intact GDGTs known to date.

GDGT-derived isoprenoid hydrocarbons provide a means to trace the geologic history of Archaea in older sediments. The ether bonds of GDGTs and DGDs were broken when subjected to extreme temperatures in hydrous pyrolysis experiments simulating diagenesis (Rowland, 1990), yielding the more resistant hydrocarbons biphytane, phytane, and shorter-chained isoprenoids. Biphytane is specific to GDGT-synthesizing archaea and is therefore a useful biomarker. Biphytane and other archaeal biomarkers crocetane, pentamethylicosane and phytane occur in Late Pennsylvanian seep limestones, extending the record of the process of anaerobic oxidation of methane to 300 million years ago, an antiquity nearly double previous estimates based on macrofossil, isotopic and biomarker evidence (Birgel *et al.*, 2008).

15.9 Lipids diagnostic of Bacteria

The lipids of Bacteria have many features in common with those of Eukarya. Most bacterial membranes comprise alkyl chains linked to the *sn*-1 and *sn*-2 carbons of glycerol through ester bonds (Fig. 15.5; **29**). The polar head groups are also similar with phospholipids, glycolipids and phosphoglycolipids being common. Important differences occur in the length and functionalization of the acyl chains with 2-, 3- and 10-methyl branching, hydroxylation, distinctive unsaturation, cyclopropyl groups and ω-cyclohexyl groups common. In addition to lipids with hydrolysable acyl linkages, some bacteria biosynthesize glycerol ether lipids. These can be glycerol diethers, glycerols with one ether-linked alkyl and one acyl chain and membrane-spanning non-isoprenoidal GDGTs.

Lipids that appear to be diagnostic for bacterial taxa include the ladderanes specific to Planctomycetales (Sinninghe Damsté *et al.*, 2005), bacteriochlorophylls and the aromatic carotenoids that, although found in cyanobacteria (Graham *et al.*, 2008; Maresca *et al.*, 2008) and actinobacteria, are widely considered characteristic of the green and purple sulfur bacteria (Brocks and Summons, 2003; Brocks *et al.*, 2005). The stable hydrocarbon cores of the aromatic carotenoids have been found in rocks as old as 1.64 Ga (Brocks *et al.*, 2005). They occur widely in Phanerozoic sediments and are especially prevalent during oceanic anoxic events, attesting to the presence of anoxic and euxinic water columns, the conditions in which their modern counterparts thrive today (Schouten *et al.*, 2001; Smittenberg *et al.*, 2004; Wakeham *et al.*, 2007).

Box 15.1   Example study: Marine archaea

Biomarkers have proven to be useful tools for probing the physiology of the marine archaea, abundant but poorly understood players in marine microbial communities.

Since the discovery that members of the domain Archaea are widespread in the oceans (DeLong, 1992; Fuhrman *et al.*, 1992; DeLong *et al.*, 1998), numerous studies have confirmed their abundance and ubiquity. Once thought to be obligate extremophiles requiring high temperatures, acidic environments, or hypersaline conditions, archaea are now known to be cosmopolitan. Cultivation-independent genomic surveys have revealed that representatives of the Archaea are common in non-hydrothermal terrestrial environments and inhabit all marine environments, spanning the euphotic zone (Frigaard *et al.*, 2006), the mesopelagic (Karner *et al.*, 2001; Massana *et al.*, 1997), and the benthos (Biddle *et al.*, 2006; Lipp *et al.*, 2008) while their range extends from tropical to polar waters. Extrapolating from data obtained through fluorescence *in situ* hybridization at the North Pacific Subtropical Gyre and coastal sites, Karner *et al.* (2001) estimated that archaea comprise nearly a quarter of planktonic marine microbes. Of that number, the majority appear to be Marine Group I Crenarchaeota. Three marine euryarchaeal groups (II–IV) have been identified, but the Marine Group II Euryarchaeota are typically the most abundant of these (DeLong, 2006). Together the Marine Groups I and II account for the bulk of extractable archaeal rRNA in the ocean.

Although their importance to marine biogeochemistry is inevitable if only because of their sheer numbers, little is known about the carbon and energy metabolism of marine archaea. Few representatives exist in culture, so direct study is difficult. Studies of the isolates *Nitrosopumilus maritimus* and *N. yellowstonii* have shown that both marine and terrestrial Crenarchaeota are capable of oxidizing ammonium to nitrite (de la Torre *et al.*, 2008; Konneke *et al.*, 2005), while the presence of the light-driven proton pumping pigment proteorhodopsin in many photic zone Euryarchaeota points to a light-dependent lifestyle. Despite clues such as these, fundamental questions about marine archaea remain unanswered; importantly, the balance of autotrophy and heterotrophy is unknown. The stable carbon isotopic composition of GDGTs found in marine sediment did not provide conclusive evidence for either; their $\delta^{13}C$ value of −20 to −23‰ could indicate consumption of an isotopically heavy carbon source potentially derived from algae, or autotrophic uptake of bicarbonate that is generally enriched relative to CO_2 (Schouten *et al.*, 2008b). Benthic GDGTs also contain levels of radiocarbon that point in favour of autotrophic dominance (Pearson *et al.*, 2001), but which remain ambiguous due to potential confounding effects of the mixing of pre-aged material (Shah *et al.*, 2008).

Incubation studies have found evidence for both autotrophy and heterotrophy. Uptake of tritium-labelled amino acids has been documented in Archaea through a hybridized microautoradiography-FISH approach (Ouverney and Fuhrman, 2000; Teira *et al.*, 2006) while the uptake of isotopically labelled bicarbonate in mesocosms has also been seen (Herndl *et al.*, 2005; Wuchter *et al.*, 2003).

To investigate the importance of chemoautotrophy among marine archaea in the mesopelagic, Ingalls *et al.* (2006) conducted an *in situ* study in the North Pacific Gyre using natural radiocarbon as a tracer. Measuring the carbon isotopic composition of dissolved inorganic carbon (DIC) and that of individual GDGTs in >30,000, they found that surface Archaea incorporate modern carbon into GDGTs, but it was not possible to determine heterotrophy/autotrophy because the ^{14}C content of organic matter and DIC are similar in surface waters. At depth, however, the ^{14}C content becomes more informative. In the 670 m sample GDGTs contained more ^{14}C than expected for a purely autotrophic community. Using an isotopic mass balance model with end members of DIC (151‰) and surface-derived organic carbon (71‰), Ingalls *et al.* (2006) determined the mesopelagic archaeal community was ~80% autotrophic. Whether this figure represents different archaeal groups with different metabolisms or a mixotrophic community remains uncertain. In any case, the predominance of autotrophy suggests that the impact of marine Archaea on marine carbon and nitrogen cycling is significant.

The most ubiquitous of lipids diagnostic for bacteria, in both modern and ancient environments are, however, the hopanoids. These are C_{30}–C_{35} pentacyclic triterpenoids and, because of similarities in their structures, physical dimensions and biosynthetic origins, they have been hypothesized to fulfill a comparable function to the sterols that are obligatory membrane components of Eukarya (Ourisson *et al.*, 1987). However, it is important to stress that the role of hopanoids as direct sterol surrogates is hypothetical as there have been few successful studies that unambiguously link hopanoids to a specific physiological function. Furthermore, while all eukaryotes require sterols (or a sterol-like molecule such as tetrahymanol; **11**) for correct membrane function, hopanoids are not universally required by bacteria. Even within particular bacterial phyla, for example, the cyanobacteria, one

---

Box 15.2 Example Study: Bacterial and archaeal lipids of the Lost City hydrothermal system

Actively venting carbonate chimneys at the Lost City hydrothermal field (Kelley *et al.*, 2001, 2005) contain significant abundances of organic carbon (~0.6%) that include intact polar lipids diagnostic for a metabolically viable microbial community. Ribosomal DNA analyses show that the living microbial community includes Methanosarcinales and various groups of bacteria (Brazelton *et al.*, 2006). The $\delta^{13}C$ values of the total organic carbon in these chimneys can be as high as −2.8‰ VPDB and they contain high abundances of isoprenoidal and non-isoprenoidal ether lipids. These lipids also display extraordinary ^{13}C enrichments (Bradley *et al.*, 2009). Isoprenoidal ether lipid biomarkers with structures typical of methanogenic and methanotrophic Archaea, including *sn*-2 and *sn*-3 hydroxyarchaeols (**31** and **32**), have $\delta^{13}C$ values that range from −2.9 to +6.7‰ VPDB. In the same samples, non-isoprenoidal ether lipids, with structures similar to those produced by sulfate reducing bacteria in culture have $\delta^{13}C$ values between −11.8‰ and +3.6‰.

Assemblages of hydroxyarchaeols and nonisoprenoidal glycerol ether lipids have been reported in natural environments where methane is being oxidized anaerobically (Hinrichs *et al.*, 1999, 2000; Pancost *et al.*, 2000; Blumenberg *et al.*, 2004) and a feature of these lipids is their characteristic depletion in ^{13}C. Yet these same lipids in the Lost City carbonate towers are manifestly enriched in ^{13}C.

Biogeochemical cycles at Lost City environment are a reflection of the chemical environment created when ultramafic rocks interact with sea water in a process known as serpentinization. This leads to very high hydrogen concentrations in the vent fluids. Values of $[H_2]$ in excess of $14\,mmol\,kg^{-1}$ have been measured in some cases. These high hydrogen concentrations make it thermodynamically feasible for an unusual co-existence of methanogenesis and sulfate reduction, while, at the same time, making it unfavourable for anaerobic methane oxidation. Further, the lipids specific to methane-cycling Archaea are enriched in ^{13}C relative to the methane in vent fluids, thereby making methane an unlikely carbon source. These observations lead one to surmise that the Methanosarcinales inhabiting the Lost City towers are methanogens and not methane-oxidizing methanotrophs (Bradley *et al.*, 2009). Moreover, microorganisms living in the chemical environment of Lost City, unlike the sulfate-requiring organisms of AOM consortia, are operating totally independently of surface photosynthesis and its by-products of oxygen, nitrate and sulfate. As such, they are a useful model for a biosphere that existed on Earth prior to the advent of oxygenic photosynthesis (Martin *et al.*, 2008; Martin and Russell, 2007).

---

finds closely related species with and without the capacity to make hopanoids. In a recent study of the purple non-sulfur bacterium *Rhodopseudomonas palustris* strain TIE-1, an absence of hopanoids induced by deletion of the *shc* gene encoding for SHC, the enzyme responsible for cyclization of squalene to the C_{30} hopanoids diploptene and diplopterol, results in a mutant that no longer produced any polycyclic triterpenoids (Welander *et al.*, 2009). Although this mutant is able to grow both heterotrophically and phototrophically, it also has a severe growth defect in that it incurs significant membrane damage when cells are grown at high or low pH. It is not yet known if compromising pH homeostasis is a direct or indirect effect of the inability to produce hopanoids (Welander *et al.*, 2009). Beyond this, the location of hopanoids in some specific kinds of bacterial membranes suggests a physiological connection to membrane behaviour (Doughty *et al.*, 2009). In the case of the nitrogen-fixing bacterium *Frankia*, the concentration of hopanoids in the vesicles housing the nitrogenase suggests an oxygen-protection function, at least in this organism (Berry *et al.*, 1993).

The biosynthesis of hopanoids is widely considered to have an evolutionary connection to the biosynthesis of sterols and the oxygen-carrying triterpenoids of plants. Amino acid sequences suggest that bacterial SHCs and the eukaryotic OSCs are homologous (Fischer and Pearson, 2007) despite their different substrates. Further, the domains that correspond to active sites required for cyclization are highly conserved. Phylogenetic trees constructed for the hopanoid and steroid cyclases suggest that they diverged from a common ancestor although, at the present time, there is no way to distinguish which appeared first (Fischer and Pearson, 2007). Studies of *sqhC* gene sequences drawn from environmental metagenomes now provides an important new window from which we might observe which bacteria produce hopanoids and where they are produced. A recent comparison of SHC amino acid sequences for cultured bacteria with the sequences for *sqhC* fragments in publicly available metagenomic libraries revealed the presence of many novel *sqhC* sequences putatively encoding for cyclase proteins, suggesting that the full diversity of hopanoid-producers remains to be discovered. The same data also suggested that hopanoid biosynthesis may be relatively uncommon, with <10% of the bacteria detected in these environmental metagenomes capable of producing hopanoids (Pearson *et al.*, 2007).

The precursors of sedimentary hopanoids are the BHPs, which have a primary carbon skeleton composed of 35 carbon atoms. BHPs are constructed from one of the C_{30} precursors, diplopterol or diploptene, linked to a moiety derived from a C_5 sugar, likely ribose. Adenosylhopane is a possible intermediate (Neunlist *et al.*, 1988). In fact, correlation of the known stereochemistry of adenosylhopane with a derivative of bacteriohopanetetrol from *Methylobacterium organophilum* enabled the determination of the configuration of all asymmetric centres of the bacteriohopanetetrol side-chain as 22R, 32R, 33R and 34S. While tetrasubstituted BHPs such as bacteriohopanetetrol are probably the most ubiquitous hopanoids in the organisms that have been studied (Rohmer *et al.*, 1984; Talbot *et al.*, 2008), the advent of LC-MS methods has revealed considerable structural diversity in the BHP side-chains (Talbot *et al.*, 2007, 2008). When viewed together with modifications to the basic ring system, where one finds unsaturation and additional methyl groups at either C2 or C3 (Rohmer *et al.*, 1984; Zundel and Rohmer, 1985a, b, c), there is tremendous richness in the chemistry of BHPs.

Although the vast majority of known hopanoid-producing bacteria are aerobes, oxygen is not required for any step in the biosynthesis of these triterpenoids (Rohmer *et al.*, 1984). Anaerobic hopanoid producers are now known through studies of cultured organisms' analyses of their genomes for biosynthetic capability (Sinninghe Damsté *et al.*, 2004a; Fischer *et al.*, 2005; Blumenberg *et al.*, 2006; Rashby *et al.*, 2007). Hopanoids bearing an extra methyl substituent on the A-ring are of particular interest in respect to O_2 in the environment. The 2β-methylbacteriohopanepolyols (2-MeBHPs; **7**) have been proposed as biomarkers for the cyanobacteria (Summons *et al.*, 1999) which are, in the main, oxygen-producing photoautotrophs. The 3β-methylbacteriohopanepolyols (3-MeBHPs; **6**) have been found in cultured acetic acid bacteria (*Acetobacter* sp.) and the methanotrophic bacteria (Type 1 methanotrophs; Zundel and Rohmer, 1985b) both of which are aerobes. Isotopic analyses allow one to distinguish between these possibilities in natural environmental samples and in chemical fossils because, when their isotopic compositions have been measured, 3-methylhopanoids (**9**) invariably have light $\delta^{13}C$ values consistent with an origin from the microaerophilic methanotrophs (Blumenberg *et al.*, 2007; Collister *et al.*, 1992).

2-Methylhopanoids have been found in the Rhizobiales order of α-Proteobacteria (specifically families Methylobacteriaceae and Rhizobiaceae) and in the Cyanobacteria where they are widely distributed across the taxonomy. A study of genomic data suggests that an *Acidobacterium* sp. is also able to make 2-methylhopanoids (Welander *et al.*, 2010). Within the Rhizobiales, the 2-methylhopanoids to be identified first were the 2β-methyldiplopterols and 2β-methyldiploptenes

(Vilcheze *et al.*, 1994; Zundel and Rohmer, 1985c). These compounds, together with the extended side-chain version, 2β-methylbacteriohopanetetrol (**7**) were subsequently identified in a strain, TIE-1, of *Rhodopseudomonas palustris*, a member of the Rhizobiaceae (Rashby *et al.*, 2007). *R. palustris*, a metabolically versatile purple non-sulfur bacterium, is also distinctive in that, in addition to hopanoids, it also produces triterpenoids of the gammacerane type, including methylated analogs, (Bravo *et al.*, 2001). *R. palustris* can grow as an anoxygenic photoautotroph, a photoheterotroph and as an aerobic heterotroph. Based on its ability to produce 2-methylhopanoids under all these conditions, it is now clear that 2-MeBHPs are not directly required for oxygen-producing photosynthesis and are not exclusive to cyanobacteria. On the other hand, the presence of 2-MeBHPs in marine cyanobacterial mat samples (Allen *et al.*, 2010), and the distributions of 2-methylhopanoid hydrocarbons in sediments and petroleum, and especially their high abundances in marine carbonate rocks and sediments deposited during oceanic anoxic events suggests that cyanobacteria are the predominant source of these compounds in the marine sedimentary record (Farrimond *et al.*, 2004; Kuypers *et al.*, 2004; Knoll *et al.*, 2007; Talbot *et al.*, 2008; Cao *et al.*, 2009).

15.10 Lipids of Eukarya

Biosynthesis of sterols, with very few exceptions, is a characteristic of eukaryotic organisms (Rohmer *et al.*, 1979). Although sterols (or surrogates such as tetrahymanol) are required by all eukaryotes, animal, plant and fungal membranes have distinct complements of sterols that strongly suggest a connection between their structures and physiological roles (Volkman, 2005). Usually the sterols vary in their patterns of unsaturation and side-chain alkylation (Fig. 15.1; 1a–i). A variety of different sterol esters and steryl glycosides are found in plants and yeast. All eukaryotic organisms require sterols, or a sterol surrogate, for survival.

15.11 Preservable cores

Steroidal hydrocarbons (**2a–i**), most commonly those with 27, 28 and 29 carbon atoms are ubiquitous in sediments and oils. Less abundant are the C_{26} and C_{30} steranes which have more restricted distributions. Of these, C_{26} steranes with three distinct motifs are common (Moldowan *et al.*, 1991). Isomers of 21-norsterane (**2g**) are common in rocks of all ages but are particularly abundant in bitumens from hypersaline depositional environments (Bao and Li, 2001; Grosjean *et al.*, 2009). 27-Norsteranes (**2h**) are also ubiquitous and, along with the 21-norsteranes, likely reflect specific but unknown biological inputs (Moldowan *et al.*, 1991; Peters *et al.*, 2005). 24-Norsteranes

(2i) are the least common but also the most biologically diagnostic since culture studies have demonstrated that one of their sources are diatoms (Rampen *et al.*, 2007) consistent with their increased abundances in Cenozoic rocks and oils relative to older samples (Holba *et al.*, 1998a, b). Dinosteranes (**4f**) are known to be robust biomarkers for dinoflagellates (Moldowan and Talyzina, 1998; Summons *et al.*, 1987) while 24-*n*-propyl cholestanes (**2d**) are derived from the sterols of marine pelagophyte algae and useful for recognizing petroleums of marine origin (Moldowan, 1984). The isomeric 24-*i*-propylcholestanes (**2e**) are the only fossil steroids so far known to be derived from animals. Their sterol precursors (e.g, **1e**) are characteristic of sponges while the steranes are notably abundant in Terminal Neoproterozoic and Cambrian sediments and oils from (McCaffrey *et al.*, 1994; Love *et al.*, 2009).

15.11.1 Forms of bias in biomarker records

It is important to pay close attention to the numerous forms of bias when using biomarker records for paleoreconstruction. A prime example stems from operational considerations that lead to the heavy focus on the tractable and volatile saturated and aromatic hydrocarbon fractions of the solvent extractable component (bitumen) of sedimentary organic matter. It is rare to see as much attention paid to the dominant and demonstrably *in situ* insoluble component (i.e. the kerogen). Even microscopic analyses of kerogen can reveal much about the sources, heterogeneity and maturity of preserved organic matter.

To be detectable through molecular, isotopic or even microscopic analyses, source organisms have to be prolific and comprise a significant fraction of the environmental biomass, and/or their biomarkers of interest must be highly resistant to degradation. This leads to a preponderance of preserved organic matter derived from the dominant photoautotrophic components of an ecosystem such as plants, algae and cyanobacteria as well as robust records of the more stable isoprenoid lipids relative to more vulnerable acetogenic lipids. There are very few examples of known biomarkers representative of organisms at the top of trophic structures (e.g, most animals) as is the case for microbes that occupy highly specialized environmental niches (e.g, many symbionts). Ballasting, and rapid transport and burial in anoxic sediments, is an important pathway to organic matter preservation. Organics entrained in rapidly settling fecal pellets or adhering to biogenic minerals and clays (Hedges and Hare, 1987; Hedges and Keil, 1995) will be preserved in preference to organics dissolved in water.

The environment in which organic matter is formed has to be conducive to preservation. Organic matter produced, transported and sedimented subaerially is less well represented in the record compared to organics formed and sedimented subaqueously. Formation and transport through oxygen-rich waters is very destructive with roughly 95–99% of neoformed organic matter in the ocean being recycled through aerobic respiration. Bioturbated sedimentary environments, which are irrigated with waters containing oxygen, also tend to be poor sites for preservation. Respiration with nitrate and sulfate acting as electron acceptors consumes much of the organic matter remaining after sediments become anoxic. In contrast, sulfidic (euxinic) conditions, which can be associated with stratified water bodies and evaporitic environments, lead to exceptional preservation because the sulfide is toxic to most grazers. Sulfide is also a chemical reductant that assists in removing unstable functional groups and replaces them with C-S cross-linkages (Adam *et al.*, 2000; Kohnen *et al.*, 1989). Incorporation of sulfide is one of the most important diagenetic pathways through which organic matter is preserved but it introduces another source of bias by way of the preservation of especially susceptible entities. An understanding of the issues surrounding selective preservation is critical to accurate paleoenvironmental reconstructions.

The sedimentary rock record is, itself, biased. Sediments deposited on the passive margins of continents and in intracratonic basins are well represented in the record. Sediments deposited in the middle of ocean basins tend to be subducted and destroyed as are those continental settings that are uplifted and more readily eroded. Accordingly, the biota inhabiting passive continental margins, marine and continental evaporitic environments large rift lakes, deltas and estuarine systems are well-represented in the biomarker record.

15.11.2 Biomarkers in ancient sediments

Abundantly preserved organic matter is, in itself, a biomarker. There are no known examples of high TOC sediments where the organics are demonstrably abiogenic. The isotopic composition of biogenic organic matter overlaps with that of organic matter formed by abiotic processes (McCollom and Seewald, 2006, 2007). Although we know of abiogenic catalytic processes that can naturally form small hydrocarbons (Sherwood Lollar *et al.*, 2006; McCollom and Seewald, 2007; Proskurowski *et al.*, 2008), evidence for significant geological accumulations of carbonaceous matter formed by abiotic processes is lacking. Because of their potential as archives of early life, analysis and interpretations of molecular fossils from ancient sediments has been a prime focus of organic geochemists. This is especially so in the microbial world where the vast majority of the protagonists lack any preservable and recognizable entities apart from their lipids. The prime enemies that conspire to prevent our having a representative sedimentary record of past microbial life are uplift and weathering, thermal metamorphism and damage from ionizing radiation. This leads to a rock record that is heavily skewed toward more recent geological periods and events. The Quaternary, Cenozoic and Mesozoic eras are all well documented in respect to the molecular fossil records of microbes and environments. Paleozoic sediments are also well represented, as are the major transitions in Earth's history including the Neoproterozoic–Cambrian interval that saw the rise of complex, multicellular life and the major mass extinction events of the Phanerozoic Eon.

The Proterozoic Eon, during which Earth's ocean, atmosphere and surface environments became oxygenated, is recorded on a few stable cratons with sedimentary rock sequences that contain abundant and well-preserved organic matter that has proved to be amenable to molecular and isotopic analysis. However, this record is temporally and/or geographically discontinuous and there are large gaps when momentous occurrences, including the 2.5–2.2 Ga 'Great Oxidation Event' and the 'Snowball Earth' intervals, have not been adequately evaluated using the tools of molecular organic geochemistry.

The Archean Eon is recorded in just a handful of sedimentary rock sequences that are amenable to molecular investigation. All the known ones, on the Pilbara, Kaapvaal and North American cratons have been heated well beyond the 'oil window', that is, they have time/temperature histories where hydrocarbon generation would have been complete. Only small contents of extractable hydrocarbons remain and their study is at the very edge of feasibility. Consequently, investigations of these rocks have attracted controversy because of the potential for contamination by younger hydrocarbons that have migrated through the rocks or which have been introduced inadvertently during sampling, handling storage and laboratory analysis (Grosjean and Logan, 2007; Sherman *et al.*, 2007; Brocks *et al.*, 2008). The rocks of the Hamersley Basin overlying the Pilbara Craton have received most attention and yielded suites of diagnostic, microbially derived hydrocarbons that were initially assessed as 'probably syngenetic' (Brocks *et al.*, 1999, 2003a, b). Interpretations of isotopic data by Rasmussen and colleagues (Rasmussen *et al.*, 2008) have been proposed to falsify the findings of Brocks *et al.*, 1999–2003 and others (Summons *et al.*, 1999). Unfortunately, the stratigraphic sampling intervals of the rocks analysed by Rasmussen *et al.* (2008) were not reported. Furthermore, conventional approaches to biomarker analysis in ancient rocks destroy them in the process so it is not possible to reproduce the exactly the same experiments on exactly the same samples.

Therefore, we must turn to different cores and/or rock units and improve techniques and methodological approaches in order to test and retest this most challenging and contentious aspect of the biogeochemical record.

Different approaches to testing the provenance of the steroids and triterpenoids detected in Archean and Paleoproterozoic rocks have appeared recently. In one that circumvents the potential contamination problems inherent in studies of shale-hosted hydrocarbons Dutkiewicz, George and coworkers have extracted and analysed biomarkers from Palaeoproterozoic oil-bearing fluid inclusions (Dutkiewicz *et al.*, 2006; George *et al.*, 2008). They studied ca. 2.45 Ga fluvial sediments of the Matinenda Fm., Elliot Lake, Canada and, although the origin of oil is not completely constrained, the timing of its emplacement is. Most likely its source was the conformably overlying deltaic McKim Fm. Hydrocarbons trapped in quartz and feldspar crystals during early metamorphism of the host rock, probably before 2.2 Ga, have been entombed in a closed system, and protected from thermal cracking, evaporative loss and contamination ever since. The methodology is founded on identification and isolation of the inclusion-bearing crystals, fastidious cleaning protocols and achievement of system blank levels near to zero (George *et al.*, 1997).

In further studies of the Hamersley Basin sedimentary sequence that had earlier been investigated by Brocks, Eigenbrode employed statistical methods to investigate relationships between different classes of biomarkers extracted from 15 Neoarchean sediments and bulk geochemical properties of the host rocks (Eigenbrode, 2007). Of particular note were the observation of significant correlations between dolomite abundance and the 2-methylhopane index $(100 \times [\mathbf{10}/(\mathbf{10} + \mathbf{8})]\%)$, a proxy for cyanobacteria, on one case and the $\delta^{13}C$ of kerogen and the 3-methylhopane index $(100 \times [\mathbf{9}/(\mathbf{9} + \mathbf{8})]\%)$, a proxy for methane oxidizing bacteria, on the other. The relative abundances of 2α-methylhopanes in both shale and carbonate from shallow-water sediments showed a strong correlation to carbonate abundance as would be expected if these compounds were largely derived from organisms, such as cyanobacteria, that inhabit shallow water lagoonal environments where carbonate tends to be a predominant lithology (Eigenbrode, 2007; Eigenbrode *et al.*, 2008). Analyses of Phanerozoic petroleum yields a similar result; 2α-methylhopanes tend to be relatively more abundant in oils sourced from carbonates and marls deposited in low latitude paleoenvironments (Knoll *et al.*, 2007). In the same Hamersley Basin rocks, the relative abundances of 3β-methylhopanes were strongly anticorrelated to the $\delta^{13}C$ of coeval kerogen; more 3β-methylhopane was found in rocks having insoluble organic matter with higher $\delta^{13}C$ values (Eigenbrode *et al.*, 2008). This result, although unexpected, might be explained if the kerogens most enriched in ^{13}C are derived from organic matter formed in the shallow water settings and, assuming that oxygenic photosynthesis was extant, this would also have been where oxygen was produced and available as an electron acceptor. In the modern world, at least, 3β-methylhopanoids are known to be biosynthesized by microaerophilic bacteria. Observations such as these will need to be further tested in other late Archean terrains as well as in the younger rock record to determine if they are reproducible.

In a third approach, the spatial and stratigraphic distributions of hydrocarbons were measured in two cores drilled 24 km apart on the Campbellrand Platform of the Neoarchean Kapvaal Craton, South Africa. Sediments of diverse lithologies, including iron formations, stromatolitic carbonates and shallow to deepwater shales were recovered under clean drilling, sampling, handling protocols (Knoll and Beukes, 2009; Sherman *et al.*, 2007). This was one of the prime goals of the Agouron Griqualand Drilling Project where over 2500 m of thermally mature, yet well-preserved Transvaal Supergroup sediments, dating from ca. 2.67 to 2.46 Ga were fully cored. The same stratigraphic sequence, representing different water depths, was intersected in the two wells and the various units could be correlated lithologically and from the presence of multiple impact spherule beds (Simonson *et al.*, 2009). Biomarker patterns from the correlated intervals showed similarities and differences which provide support for their syngenetic nature. Notably, these sediments were devoid of hydrocarbons known to be of anthropogenic origin (Brocks *et al.*, 2008) and there was a sharp discontinuity in molecular maturity parameters across the ~2 Gyr unconformity between the Neoarchean units and the overlying Permian Dwyka Formation. The suite of molecular fossils identified in these Neoarchean bitumens included hopanes attributable to bacteria, potentially including cyanobacteria and methanotrophic proteobacteria, and steranes of eukaryotic origin. The results, should they be reproducible in other, similarly ancient rocks, speak to the existence of cellular life comprising the three domains and, at that time, the existence of oxygenic photosynthesis and the anabolic use of O_2 (Waldbauer *et al.*, 2009).

15.11.3 Biomarkers in petroleum and bitumen

Petroleum is ubiquitous in the geological record of the Phanerozoic and there are traces of oil in rocks that are much older (Dutkiewicz *et al.*, 1998., 2006; Jackson *et al.*, 1986). Its status as a fossil fuel derives from the fact that petroleum constituents are the molecular remains of past organic life. Many of the molecules that occur abundantly in petroleum (acyclic isoprenoids, steroids and triterpenoids) are so structurally complex that they could not have been formed through non-biological pro-

cesses. There are no robust data to support alternative theories (Gold, 1992) that specify a primordial nature for subsurface hydrocarbons.

The patterns of hydrocarbons in petroleum hold a detailed account of the history of photosynthetic organisms in the oceans, lakes and rivers of the past (Knoll *et al.*, 2007). However, it is a biased history since oil-prone sedimentary rocks reflect the preferential preservation of coastal margin and intracratonic basin environments. Sediments deposited at low paleolatitude are more disposed to preserve organic matter. Petroleum is also unevenly distributed through geological time with certain periods being biologically, climatically and tectonically favourable for the deposition of petroleum source rocks (Klemme and Ulmishek, 1991). Logically, it is also comprised of the remains of the most quantitatively significant organisms in sedimentary environments, being cyanobacteria and algae in the ocean and in lakes and vascular plants on land and coastal marine settings. Seemingly contrary to this, if one is interested in long-term evolutionary trends, analyses of petroleum samples avoids many of the biases inherent on looking at small pieces of rock from core or outcrop. Petroleum accumulates as fluids that are expelled over millions of years from large volumes of deeply buried sediment that were themselves deposited across aerially extensive basins and over multi-million year timescales (Tissot and Welte, 1978). Therefore a reservoir rock contains hydrocarbon that is spatially and temporally integrated. Volumetrically large samples of petroleum, provided that they have been carefully collected and curated, should also be less prone to contamination.

From analyses of petroleum samples one can discern secular successions in the ocean plankton (Grantham and Wakefield, 1988; Schwark and Empt, 2006). For example, the Triassic radiation and rise to prominence of dinoflagellates can be observed in the increasing quantitative importance of dinosteroids in bitumens and oils (Summons *et al.*, 1987, 1992; Moldowan and Talyzina, 1998). The Cenozoic rise of some diatom taxa is evident in the abundances of HBI and 24-norcholestanes (**2i**) in oils (Holba *et al.*, 1998a, b; Sinninghe Damsté *et al.*, 2004b). Aromatic hydrocarbons that are derived from land-plant debris have made an indelible imprint on the isotopic compositions of petroleum hydrocarbons (Sofer, 1984; Murray *et al.*, 1998). An evolving biochemistry of vascular plants, together with the Cenozoic rise of angiosperms and their paleogeographic distributions is also preserved in the patterns of petroleum hydrocarbons (Moldowan *et al.*, 1994). Lastly, the hydrocarbons found in bitumen and petroleum attest to the antiquity of specific taxa such as the green and purple sulfur bacteria (Brocks *et al.*, 2005) as well as the biosynthetic pathways they employed to build their lipids (Ourisson and Albrecht, 1992; Ourisson *et al.*, 1979).

Box 15.3   Example study: Mass extinctions

Biomarkers offer considerable promise for elucidating biogeochemical processes associated with mass extinction events. This is because much of the knowledge about these events stems from the study of the rich and diverse record of macroscopic and microscopic fossils; the entire Phanerozoic timescale is fundamentally underpinned by descriptions of profound changes in abundance and diversity of body organisms. In contrast, the record of molecular and isotopic fossils speaks to the successions of organisms that do not leave morphologically recognizable remains, the redox structure of water columns and, possibly, to atmospheric perturbations.

William Holser (Holser, 1977) was one of the first to link the overt disruption of ocean chemistry to biological mass extinction. Subsequent research has provided many examples of shifts in the isotopic compositions of carbon, oxygen and sulfur in marine sediments (Holser *et al.*, 1989) and their links to extinction-radiation of marine plankton, animals and plants. The term 'Strangelove Ocean' has been used to dramatize rapid trends near to the Proterozoic-Phanerozoic transition (Hsu *et al.*, 1985; Kump, 1991) that were interpreted as signifying a great reduction in biological activity. Various geological, paleoceanographic and biological processes have been explored in an effort to test the linkages between these chemical phenomena and the mass extinction and ensuing recovery in biodiversity. At the present time, there is no real consensus on the precise causal links between the chemical 'events' and biotic extinction and re-radiation. However, there is consensus about the existence of C-cycle isotopic anomalies associated with the Permian–Triassic Boundary (PTB) (Magaritz *et al.*, 1988, 1992; Cao *et al.*, 2002; Payne *et al.*, 2004), Cretaceous-Paleogene (K-P) (Arthur *et al.*, 1987; Zachos *et al.*, 1989; D'Hondt *et al.*, 1998; Arens and Jahren, 2000) and Paleocene-Eocene Thermal Maximum (PETM) (Kennett and Stott, 1991; Dickens, 2003; Zachos *et al.*, 2005) events.

One tool that has been applied only relatively recently is the study of secular trends in biomarker lipids. Time series of changes in biomarkers and their carbon isotopic compositions opens a window onto plankton successions and geomicrobiological processes which appear to be driving the C- and S-isotopic excursions that characterize many mass extinction events, and especially the PTB (Sephton *et al.*, 2002, 2005; Xie *et al.*, 2005; Grice *et al.*, 2006; Hays *et al.*, 2007; Wang, 2007; Wang and Visscher, 2007).

Precise correlations of geochronologic, biostratigraphic and chemostratigraphic data for the PTB have become available relatively recently and particularly relate to the Global Stratotype, Section and Point (GSSP) at Meishan, China. Biomarker data collected from sediment samples through a core recently drilled through the Meishan GSSP shows there was a protracted and global euxinic ocean prior to and throughout the Permian to Triassic transition. Of particular note are the trends in the abundances of biomarkers derived from the green sulfur bacteria (Chlorobiaceae) throughout sediments deposited throughout the last few million years of the Permian and into the Early Triassic implying shallow water euxinic conditions were pervasive and protracted (Cao *et al.*, 2009). Independent modelling shows that toxic hydrogen sulfide would have been upwelling from the deep ocean onto continental shelves and entering the atmosphere (Kump *et al.*, 2005; Riccardi *et al.*, 2006; Meyer *et al.*, 2008). Together, these results suggest sulfide toxicity may have contributed to the biological extinction in the marine realm as well as on land. Well-documented sections of PTB Event distributed across the Tethys, Panthalassic and Boreal oceans all show molecular evidence for euxinic conditions in the upper water column (Grice *et al.*, 2005; Hays *et al.*, 2006, 2007; Cao *et al.*, 2009). Isotopic and biomarker data from the Meishan section suggest there was a long-term disruption to the N-cycle and C-cycles. The relative abundances of hopanoids and steroids suggest that bacteria, and not algae, may have been the dominant primary producers in planktonic communities for significant periods of time presaging and following the extinction in the vicinity of Meishan (Cao *et al.*, 2009). Biomarker data, therefore, may record the existence of a 'bacterial ocean' at the end of the Permian Period with strong ocean redox stratification.

15.12   Outlook

Established biomarkers can be exploited to gain insight into nearly any question in geobiology. Increased exploration of unusual environments, as well as those as comparatively 'mundane' as the soil in our cities or the waters at our shores, gives us the opportunity to make progress toward understanding the transformations of matter and energy that occur there and to connect them to processes of the past. The geologic record is also largely unexplored, containing numerous paleoenvironmental transitions that have yet to be investigated using biomarkers. It is also important to expand the temporal breadth of our studies from the known perturbations and transitions to sediments deposited during the more stable times between in order to better position such extraordinary events in a larger geological context.

Box 15.4   Example study: Neoproterozoic radiation of sponges

Very few biomarkers qualify as proxies for animal life. One exception is a class of sedimentary steranes, the 24-iso-propylcholestanes (McCaffrey *et al.*, 1994). Using sensitive GC-MS methodologies, these can be distinguished from the 24-*n*-propylcholestanes (**2d**) that are ubiquitous in, and diagnostic for, marine oils and marine petroleum source rocks through geological time (Moldowan, 1984). The 24-*n*-propylcholestanes are derived from 24-*n*-propyl sterols (**1d**) that are characteristic of marine pelagophyte algae. On the other hand, 24-isopropylcholestanes originate from sterols with 24-isopropyl substituents that are biosynthesized *de novo* (Silva *et al.*, 1991) by some demosponges (Bergquist *et al.*, 1991). Notably, the isopropylcholestanes were absent from a choanoflagellate, a member of the unicellular sister group of animals, that was investigated for its sterol content (Kodner *et al.*, 2008). Nor have isopropyl sterols been found in the calcisponges or hexactinellids (Love *et al.*, 2009).

A ratio of 24-isopropylcholestanes/24-*n*-propylcholestanes (*i/n*; **2e/2d**) ≥1 was first observed as an anomalous feature Neoproterozoic to Early Cambrian oils and bitumens from Oman, Australia, Siberia, the Urals and India (McCaffrey *et al.*, 1994) where it was attributed to the rise of sponges and/or their extinct reef-forming cousins, the Archaeocyathids. An *i/n* ratio ≤0.4, and usually much lower, is normal for oils sourced from rocks younger than Cambrian. Further, in an exhaustive study of bitumens and oils of the Huqf Supergroup, South Oman Salt Basin, an *i/n* ratio >0.4 was observed throughout the entire stratigraphy including in sediments laid down between the Sturtian (ca. 713 Ma) and Marinoan (ca. 635 Ma) 'Snowball Earth' glacial events (Love *et al.*, 2009) was. The appearance of these anomalously high levels of isopropylcholestanes during the Cryogenian period is consistent with divergence age estimates for Demospongiae based on molecular clocks calibrated to the first appearances of Paleozoic fauna (Peterson and Butterfield, 2005; Peterson *et al.*, 2007).

The demosponge steranes identified in Oman are a quantitatively significant component of the steroids in all samples analysed and may document the first significant accumulation of animal biomass in the sedimentary record. As molecular fossils, they predate the oldest Ediacaran megascopic animals of ca. 575 Ma (Narbonne and Gehling, 2003). They are also older than the first trace fossils of ca. 555 Ma (Droser *et al.*, 2002) and the likely animal embryos of c. 632 Ma (Xiao *et al.*, 1998; Yin *et al.*, 2007). If these observations are correct, dissolved oxygen concentrations (Fike *et al.*, 2006; Canfield *et al.*, 2007) of the shallow waters of the Cryogenian period were sufficient to support metazoan life 100 Ma, or more, in advance of the rapid diversification of the extant animal phyla during the Cambrian Explosion (Knoll and Carroll, 1999; Marshall, 2006).

Perhaps a greater challenge and opportunity in biomarker research lies in developing our understanding of the genetics responsible for biomarker biosynthesis. Even the genes responsible for key steps in biomarker synthesis in many cultured and genetically sequenced organisms remain unknown; moreover, the environment is replete with microbes of which we have far less knowledge. Once we know the genes responsible for a biomarker, we can then query environmental nucleic acid databases to understand whether the organisms we think contain the biomarker are present in the environments in which we find it. Thus, by exploring the biological underpinnings of biomarkers we can make them more robust tools in geobiology.

Physiology is another area in which geobiologists are poised to make great strides using tools of molecular biology. Perturbations on very different scales, from altering genes to experimenting with conditions in environmental mesocosms, have the potential to reveal the function of specific biomarker molecules. Ultimately, by understanding the function of a biomarker in an organism we can better interpret the presence of that biomarker in both the modern world and in deep time.

Acknowledgements

We are grateful for support from the NASA Astrobiology and Exobiology Programs, the NSF Biocomplexity, EAR and Chemical Oceanography Programs and the Agouron Institute. Jochen Brocks and Ann Pearson provided reviews which significantly improved the original manuscript.

References

Abelson PH, Hoering TC (1961) Carbon isotopic fractionation in formation of amino acids by photosynthetic organisms. *Proceedings of the National Academy of Sciences of the USA* **47**(5), 623–632.

Adam P, Schneckenburger P, Schaeffer P, Albrecht P (2000) Clues to early diagenetic sulfurization processes from mild chemical cleavage of labile sulfur-rich geomacromolecules. *Geochimica et Cosmochimica Acta* **64**(20), 3485–3503.

Allen MA, Burns BP, Neilan BA, Jahnke LL, Summons RE (2010) Lipid biomarkers in Hamelin Pool microbial mats and stromatolites. *Organic Geochemistry* 41(11), 1207–1218.

Amrani A, Sessions AL, Adkins JF (2009) Compound-specific δ34S analysis of volatile organics by coupled GC/multicollector-ICPMS. *Analytical Chemistry* 81(21), 9027–9034.

Arens NC, Jahren AH (2000) Carbon isotope excursion in atmospheric CO_2 at the Cretaceous-Tertiary boundary: evidence from terrestrial sediments. *PALAIOS* 15(4), 314–322.

Arthur MA, Zachos JC, Jones DS (1987) Primary productivity and the Cretaceous/Tertiary boundary event in the oceans. *Cretaceous Research* 8(1), 43–54.

Bacia K, Schwille P, Kurzchalia T (2005) Sterol structure determines the separation of phases and the curvature of the liquid-ordered phase in model membranes. *Proceedings of the National Academy of Sciences of the USA* 102(9), 3272–3277.

Bao J, Li M (2001) Unprecedented occurrence of novel C26–C28 21-norcholestanes and related triaromatic series in evaporitic lacustrine sediments. *Organic Geochemistry* 32(8), 1031–1036.

Bergquist PR, Karuso P, Cambie RC, Smith DJ (1991) Sterol composition and classification of the Porifera. *Biochemical Systematics and Ecology* 19(1), 17–24.

Berry AM, Harriott OT, Moreau RA, Osman SF, Benson DR, Jones AD (1993) Hopanoid lipids compose the Frankia vesicle envelope, presumptive barrier of oxygen diffusion to nitrogenase. *Proceedings of the National Academy of Sciences of the USA* 90(13), 6091–6094.

Biddle J, Lipp JS, Lever MA, *et al.* (2006) Heterotrophic Archaea dominate sedimentary subsurface ecosystems off Peru. *Proceedings of the National Academy of Sciences of the USA* 103, 3846–3851.

Birgel D, Himmler T, Freiwald A, Peckmann J (2008) A new constraint on the antiquity of anaerobic oxidation of methane: Late Pennsylvanian seep limestones from southern Namibia. *Geology* 36(7), 543–546.

Bloch K (1979) Speculations on the evolution of sterol structure and function. *CRC Critical Reviews of Biochemistry* 7(1), 1–5.

Bloch K (1987) Summing up. *Annual Review of Biochemistry* 56(1), 1–18.

Bloch K (1992) Sterol molecule: structure, biosynthesis, and function. *Steroids* 57(8), 378–83.

Blumenberg M, Seifert R, Reitner J, Pape T, Michaelis W (2004) Membrane lipid patterns typify distinct anaerobic methanotrophic consortia. *Proceedings of the National Academy of Sciences of the USA* 101(30), 11111–11116.

Blumenberg M, Krüger M, Talbot HM, *et al.* (2006) Biosynthesis of hopanoids by sulfate-reducing bacteria (genus *Desulfovibrio*). *Environmental Microbiology* 8, 1220–1227.

Blumenberg M, Seifert R, Michaelis W (2007) Aerobic methanotrophy in the oxic-anoxic transition zone of the Black Sea water column. *Organic Geochemistry* 38(1), 84–91.

Boreham CJ, Fookes CJR, Popp BN, Hayes JM (1990) Origin of petroporphyrins. 2. Evidence from stable carbon isotopes. *Energy & Fuels* 4(6), 658–661.

Bradley A, Hayes J, Summons R (2009) Extraordinary ^{13}C enrichment of diether lipids at the Lost City Hydrothermal Field indicates a carbon-limited ecosystem. *Geochimica et Cosmochimica Acta* 73(1), 102–118.

Brassell SC, Eglinton G, Maxwell JR (1983) The geochemistry of terpenoids and steroids. *Biochemical Society Transactions* 11, 575–586.

Brassell SC, Eglinton G, Marlowe IT, Pflaumann U, Sarnthein M (1986) Molecular stratigraphy: a new tool for climatic assessment. *Nature* 320(6058), 129.

Bravo JM, Perzl M, Hartner T, Kannenberg EL, Rohmer M (2001) Novel methylated triterpenoids of the gammacerane series from the nitrogen-fixing bacterium *Bradyrhizobium japonicum* USDA 110. *European Journal of Biochemistry* 268(5), 1323–1331.

Brazelton WJ, Schrenk MO, Kelley DS, Baross JA (2006) Methane- and sulfur-metabolizing microbial communities dominate the lost city hydrothermal field ecosystem. *Applied and Environmental Microbiology* 72(9), 6257–6270.

Brocks JJ, Pearson A (2005) Building the biomarker tree of life. *Reviews in Mineralogy and Geochemistry* 59(1), 233–258.

Brocks JJ, Summons RE (2003) Sedimentary hydrocarbons, biomarkers for early life. In: *Treatise in Geochemistry*, Vol. Ch. 8.03 (eds Holland HD, Turekian K), pp. 65–115.

Brocks JJ, Logan GA, Buick R, Summons RE (1999) Archean molecular fossils and the early rise of eukaryotes. *Science* 285(5430), 1033–1036.

Brocks JJ, Buick R, Logan GA, Summons RE (2003a) Composition and syngeneity of molecular fossils from the 2.78 to 2.45 billion-year-old Mount Bruce Supergroup, Pilbara Craton, Western Australia. *Geochimica et Cosmochimica Acta* 67(22), 4289–4319.

Brocks JJ, Buick R, Summons RE, Logan GA (2003b) A reconstruction of Archean biological diversity based on molecular fossils from the 2.78 to 2.45 billion-year-old Mount Bruce Supergroup, Hamersley Basin, Western Australia. *Geochimica et Cosmochimica Acta* 67(22), 4321.

Brocks JJ, Love GD, Summons RE, Knoll AH, Logan GA, Bowden SA (2005) Biomarker evidence for green and purple sulphur bacteria in a stratified Palaeoproterozoic sea. *Nature* 437(7060), 866.

Brocks JJ, Grosjean E, Logan GA (2008) Assessing biomarker syngeneity using branched alkanes with quaternary carbon (BAQCs) and other plastic contaminants. *Geochimica et Cosmochimica Acta* 72(3), 871–888.

Bryant DA, Costas AMG, Maresca JA, *et al.* (2007) Candidatus *Chloracidobacterium thermophilum*: an aerobic phototrophic Acidobacterium. *Science* 317(5837), 523–526.

Canfield DE (1989) Sulfate reduction and oxic respiration in marine sediments: implications for organic carbon preservation in euxinic environments. *Deep Sea Research Part A. Oceanographic Research Papers* 36(1), 121–138.

Canfield DE, Poulton SW, Narbonne GM (2007) Late-Neoproterozoic deep-ocean oxygenation and the rise of animal life. *Science* 315(5808), 92–95.

Cao C, Wang W, Jin YG (2002) Carbon isotope excursions across the Permian-Triassic boundary in the Meishan section, Zhejiang Province, China. *Chinese Science Bulletin* 47, 1125–1129.

Cao C, Love GD, Hays LE, Wang W, Shen S, Summons RE (2009) Biogeochemical evidence for a Euxinic Ocean and ecological disturbance presaging the end-Permian mass extinction event. *Earth and Planetary Science Letters* 288, 188–201.

Chikaraishi Y, Matsumoto K, Ogawa NO, Suga H, Kitazato H, Ohkouchi N (2005) Hydrogen, carbon and nitrogen isotopic fractionations during chlorophyll biosynthesis in C3 higher plants. *Phytochemistry* **66**(8), 911–920.

Chikaraishi Y, Kashiyama Y, Ogawa NO, *et al.* (2008) A compound-specific isotope method for measuring the stable nitrogen isotopic composition of tetrapyrroles. *Organic Geochemistry* **39**(5), 510–520.

Collister JW, Summons RE, Lichtfouse E, Hayes JM. (1992) An isotopic biogeochemical study of the Green River oil shale. *Organic Geochemistry* **19**(1–3), 265.

Comita P. B. and Gagosian R. B. (1983) Membrane lipid from deep-sea hydrothermal vent methanogen: a new macrocyclic glycerol diether. *Science* **222**(4630), 1329–1331.

D'Hondt S, Donaghay P, Zachos JC, Luttenberg D, Lindinger M (1998) Organic carbon fluxes and ecological recovery from the Cretaceous-Tertiary mass extinction. *Science* **282**(5387), 276–279.

de la Torre JR, Walker CB, Ingalls AE, Könneke M, Stahl DA (2008) Cultivation of a thermophilic ammonia oxidizing archaeon synthesizing crenarchaeol. *Environmental Microbiology* **10**, 810–818.

De Niro MJ, Epstein S (1977) Mechanism of carbon isotope fractionation associated with lipid synthesis. *Science* **197**, 261–263.

DeLong EF (1992) Archaea in coastal marine environments. *Proceedings of the National Academy of Sciences of the USA* **89**(12), 5685–5689.

DeLong EF (2006) Archaeal mysteries of the deep revealed. *Proceedings of the National Academy of Sciences of the USA* **103**(17), 6417–6418.

DeLong EF, King LL, Massana R, *et al.* (1998) Dibiphytanyl ether lipids in nonthermophilic Crenarchaeotes. *Applied and Environmental Microbiology* **64**(3), 1133–1138.

Dickens GR (2003) Rethinking the global carbon cycle with a large, dynamic and microbially mediated gas hydrate capacitor. *Earth and Planetary Science Letters* **213**(3–4), 169–183.

Doughty DM, Hunter RC, Summons RE, Newman DK (2009) 2-Methylhopanoids are maximally produced in akinetes of *Nostoc punctiforme*: geobiological implications. *Geobiology* **7**(5), 524–532.

Doughty DM, Coleman ML, Hunter RC, Sessions AL, Summons RE, Newman D K (2011) Hopanoid lipids impact cell division by *Rhodopseudomonas palustris* at elevated growth temperatures. *Proceedings of the National Academy of Sciences of the USA* **108** (45), E1045–E1051.

Droser ML, Jensen S, Gehling JG (2002) Trace fossils and substrates of the terminal Proterozoic-Cambrian transition: Implications for the record of early bilaterians and sediment mixing. *Proceedings of the National Academy of Sciences of the USA* **99**(20), 12572–12576.

Dutkiewicz A, Rasmussen B, Buick R (1998) Oil preserved in fluid inclusions in Archaean sandstones. *Nature* **395**(6705), 885–888.

Dutkiewicz A, Volk H, George SC, Ridley J, Buick R (2006) Biomarkers from Huronian oil-bearing fluid inclusions: An uncontaminated record of life before the Great Oxidation Event. *Geology* **34**(6), 437–440.

Eglinton G (1970) *Chemical Fossils*. WH Freeman, New York.

Eglinton G, Hamilton RJ (1967) Leaf epicuticular waxes. *Science* **156**(3780), 1322–1335.

Eigenbrode JL (2007) Fossil lipids for life-detection: a case study from the early Earth record. *Space Science Reviews* **135**(1–4), 161–185.

Eigenbrode J, Summons RE, Freeman KH (2008) Methylhopanes in Archean sediments. *Earth and Planetary Science Letters* **273**, 323–331.

Eisenreich W, Schwarz M, Cartayrade A, Arigoni D, Zenk MH, Bacher A (1998) The deoxyxylulose phosphate pathway of terpenoid biosynthesis in plants and microorganisms. *Chemistry & Biology* **5**(9), R221–R233.

Farrimond P, Talbot HM, Watson DF, Schulz LK, Wilhelms A (2004) Methylhopanoids: molecular indicators of ancient bacteria and a petroleum correlation tool. *Geochimica et Cosmochimica Acta* **68**(19), 3873–3882.

Fike DA, Grotzinger JP, Pratt LM, Summons RE (2006) Oxidation of the Ediacaran Ocean. *Nature* **444**(7120), 744–747.

Fischer WW, Pearson A (2007) Hypotheses for the origin and early evolution of triterpenoid cyclases. *Geobiology* **5**(1), 19–34.

Fischer WW, Summons RE, Pearson A (2005) Targeted genomic detection of biosynthetic pathways: anaerobic production of hopanoid biomarkers by a common sedimentary microbe. *Geobiology* **3**(1), 33–40.

Flesch G, Rohmer M (1988) Prokaryotic hopanoids: the biosynthesis of the bacteriohopane skeleton. Formation of isoprenic units from two distinct acetate pools and a novel type of carbon/carbon linkage between a triterpene and D-ribose. *European Journal of Biochemistry* **175**(2), 405–411.

Frigaard N, Martinez A, Mincer TJ, DeLong EF (2006) Proteorhodopsin lateral gene transfer between marine planktonic Bacteria and Archaea. *Nature* **439**, 847–850.

Fuhrman JA, McCallum K, Davis AA (1992) Novel major archeabacterial group from marine plankton. *Nature* **356**, 148–149.

Gaines SM, Eglinton G, Rullkötter J (2009) *Echoes of Life: What Fossil Molecules Reveal about Earth History*. Oxford University Press, Oxford.

Gattinger A, Schloter M, Munch JC (2002) Phospholipid etherlipid and phospholipid fatty acid fingerprints in selected euryarchaeotal monocultures for taxonomic profiling. *FEMS Microbiology Letters* **213**(1), 133–139.

George SC, Krieger FW, Eadington PJ, *et al.* (1997) Geochemical comparison of oil-bearing fluid inclusions and produced oil from the Toro sandstone, Papua New Guinea. *Organic Geochemistry* **26**(3–4), 155.

George SC, Volk H, Dutkiewicz A, Ridley J, Buick R (2008) Preservation of hydrocarbons and biomarkers in oil trapped inside fluid inclusions for >2 billion years. *Geochimica et Cosmochimica Acta* **72**(3), 844–870.

Gliozzi A, Rolandi R, De Rosa M, Gambacorta A (1983) Monolayer black membranes from bipolar lipids of archaebacteria and their temperature-induced structural changes. *Journal of Membrane Biology* **75**(1), 45–56.

Gold T (1992) The deep, hot biosphere. *Proceedings of the National Academy of Sciences of the USA* **89**(13), 6045–6049.

Graham JE, Lecomte JTJ, Bryant DA (2008) Synechoxanthin, an aromatic C40 xanthophyll that is a major carotenoid in the cyanobacterium *Synechococcus* sp. PCC 7002. *Journal of Natural Products* **71**(9), 1647–1650.

Grantham PJ, Wakefield LL (1988) Variations in the sterane carbon number distributions of marine source rock derived crude oils through geological time. *Organic Geochemistry* **12**, 61–73.

Grice K, Gibbison R, Atkinson JE, Schwark L, Eckardt CB, Maxwell JR (1996) Maleimides (1H-pyrrole-2,5-diones) as molecular indicators of anoxygenic photosynthesis in ancient water columns. *Geochimica et Cosmochimica Acta* **60**(20), 3913–3924.

Grice K, Cao C, Love GD, *et al.* (2005) Photic zone euxinia during the Permian-Triassic superanoxic event. *Science* **307**(5710), 706–709.

Grice K, Fenton S, Bottcher ME, Twitchett RJ, Summons RE, Grosjean E (2006) The biogeochemical cycling of sulfur, carbon and nitrogen across the Permian-Triassic (P-Tr) Hovea-3 borehole (Western Australia) and Schuchert Dal Section (Eastern Greenland). *Geochimica et Cosmochimica Acta* **70**(18, Supplement 1), A218.

Grosjean E, Logan GA (2007) Incorporation of organic contaminants into geochemical samples and an assessment of potential sources: examples from Geoscience Australia marine survey S282. *Organic Geochemistry* **38**, 853–869.

Grosjean E, Love GD, Stalvies C, Fike DA, Summons RE (2009) Origin of petroleum in the Neoproterozoic-Cambrian South Oman Salt Basin. *Organic Geochemistry* **40**(1), 87–110.

Guckert JB, Antworth CP, Nichols PD, White DC (1985) Phospholipid, ester-linked fatty acid profiles as reproducible assays for changes in prokaryotic community structure of estuarine sediments. *FEMS Microbiology Letters* **31**(3), 147–158.

Han J, Calvin M (1969) Hydrocarbon distribution of algae and bacteria and microbial activoty in sediments. *Proceedings of the National Academy of Sciences of the USA* **64**(2), 436–443.

Hayes JM (1993) Factors controlling 13C contents of sedimentary organic compounds: Principles and evidence. *Marine Geology* **113**(1–2), 111–125.

Hayes JM (2001) Fractionation of carbon and hydrogen isotopes in biosynthetic processes. *Reviews in Mineralogy & Geochemistry* **43**, 225–277.

Hayes JM (2004) Isotopic order, biogeochemical processes, and earth history: Goldschmidt lecture, Davos, Switzerland, August 2002. *Geochimica et Cosmochimica Acta* **68**(8), 1691–1700.

Hays LE, Love GD, Foster CB, Grice K, Summons RE (2006) Lipid biomarker records across the Permian–Triassic boundary from Kap Stosch, Greenland. *Eos Transactions AGU, 87*(52), *Fall Meeting Suppl.*, *Abstract PP41B–1203*.

Hays LE, Beatty T, Henderson CM, Love GD, Summons RE (2007) Evidence for photic zone euxinia through the end-Permian mass extinction in the Panthalassic Ocean (Peace River Basin, Western Canada). *Palaeoworld* **16**(1–3), 39–50.

Hebting Y, Schaeffer P, Behrens A, *et al.* (2006) Biomarker evidence for a major preservation pathway of sedimentary organic carbon. *Science* **312**(5780), 1627–1631.

Hedges JI, Hare PE (1987) Amino acid adsorption by clay minerals in distilled water. *Geochimica et Cosmochimica Acta* **51**, 255–259.

Hedges JI, Keil RG (1995) Sedimentary organic matter preservation: an assessment and speculative synthesis. *Marine Chemistry* **49**(2–3), 81–115.

Herndl GJ, Reinthaler T, Teira E, *et al.* (2005) Contribution of Archaea to total prokaryotic production in the deep Atlantic Ocean. *Applied and Environmental Microbiology* **71**, 1715–1726.

Heuer V, Elvert M, Tille S, *et al.* (2006) Online δ^{13}C analysis of volatile fatty acids in sediment/porewater systems by liquid chromatography–isotope ratio mass spectrometry. *Limnology and Oceanography: Methods* **4**, 346–357.

Hinrichs K-U, Hayes JM, Sylva SP, Brewer PG, DeLong EF (1999) Methane-consuming archaebacteria in marine sediments. *Nature* **398**, 802–805.

Hinrichs K-U, Summons RE, Orphan V, Sylva SP, Hayes JM (2000) Molecular and isotopic analysis of anaerobic methane-oxidizing communities in marine sediments. *Organic Geochemistry* **31**, 1685–1701.

Hinrichs K-U, Eglinton G, Engel MH, Summons RE (2001) Exploiting the multivariate isotopic nature of organic compounds. *Geochemistry Geophysics Geosystems* **1**, Paper number 2001GC000142.

Holba AG, Dzou LIP, Masterson WD, *et al.* (1998a) Application of 24-norcholestanes for constraining source age of petroleum. *Organic Geochemistry* **29**(5–7), 1269–1283.

Holba AG, Tegelaar EW, Huizinga BJ, *et al.* (1998b) 24-norcholestanes as age-sensitive molecular fossils. *Geology* **26**(9), 783–786.

Holser WT (1977) Catastrophic chemical events in the history of the ocean. *Nature* **267**(5610), 403–408.

Holser WT, Schonlaub H-P, Attrep M, *et al.* (1989) A unique geochemical record at the Permian/Triassic boundary. *Nature* **337**(6202), 39–44.

Hopmans EC, Schouten S, Pancost RD, *et al.* (2000) Analysis of intact tetraether lipids in archaeal cell material and sediments by high performance liquid chromatography/atmospheric pressure chemical ionization mass spectrometry. *Rapid Communications in Mass Spectrometry* **14**(7), 585–589.

House CH, Schopf JW, Stetter KO (2003) Carbon isotopic fractionation by Archaeans and other thermophilic prokaryotes. *Organic Geochemistry* **34**, 345–356.

Hsu KJ, Oberhansli H, Gao JY, Shu S, Haihong C, Krahenbuhl U (1985) Strangelove ocean before the Cambrian explosion. *Nature* **316**, 809–811.

Ingalls AE, Shah SR, Hansman RL, *et al.* (2006) Quantifying archaeal community autotrophy in the mesopelagic ocean using natural radiocarbon. *Proceedings of the National Academy of Sciences of the USA* **103**, 6442–6447.

Jackson MJ, Powell TG, Summons RE, Sweet IP (1986) Hydrocarbon shows and petroleum source rocks in sediments as old as 1.7 × 10⁹ years. *Nature* **322**(6081), 727–729.

Jaeschke A, Lewan MD, Hopmans EC, Schouten S, Damsté JSS (2008) Thermal stability of ladderane lipids as determined by hydrous pyrolysis. *Organic Geochemistry* **39**(12), 1735–1741.

Jaeschke A, Rooks C, Trimmer M, Nicholls JC, Hopmans EC, Schouten S, Sinninghe Damsté JS (2009) 'Comparison of ladderane phospholipid and core lipids as indicators for anaerobic ammonium oxidation (anammox) in marine sediments. *Geochimica et Cosmochimica Acta* **73**, 2077–2088.

Jahn U, Summons R, Sturt H, Grosjean E, Huber H (2004) Composition of the lipids of *Nanoarchaeum equitans* and their origin from its host *Ignicoccus* sp. strain KIN4/I. *Archives of Microbiology* **182**(5), 404.

Jahnke LL, Eder W, Huber R, *et al.* (2001) Signature lipids and stable carbon isotope analyses of Octopus Spring hyperthermophilic communities compared with those of Aquificales representatives. *Applied Environmental Microbiology* **67**(11), 5179–5189.

Jahnke LL, Embaye T, Hope J, *et al.* (2004) Lipid biomarker and carbon isotopic signatures for stromatolite-forming, microbial mat communities and *Phormidium* cultures from Yellowstone National Park. *Geobiology* **2**(1), 31–47.

Jetten MSM, Sliekers O, Kuypers M, *et al.* (2003) Anaerobic ammonium oxidation by marine and freshwater planctomycete-like bacteria. *Applied Microbiology and Biotechnology* **63**(2), 107–114.

Karner MB, Delong EF, Karl DM (2001) Archaeal dominance in the mesopelagic zone of the Pacific Ocean. *Nature* **409**, 507–510.

Kates M (1978) The phytanyl ether-linked polar lipids and isoprenoid neutral lipids of extremely halophilic bacteria. *Progress in Chemistry: Fats and Other Lipids* **15**(4), 301–342.

Kelley DS, Karson JA, Blackman DK, *et al.* (2001) An off-axis hydrothermal vent field near the Mid-Atlantic Ridge at 30 degrees N. *Nature* **412**(6843), 145–149.

Kelley DS, Karson JA, Fruh-Green GL, *et al.* (2005) A serpentinite-hosted ecosystem: the Lost City hydrothermal field. *Science* **307**(5714), 1428–1434.

Kennett JP, Stott LD (1991) Abrupt deep-sea warming, palaeoceanographic changes and benthic extinctions at the end of the Palaeocene. *Nature* **353**, 225–229.

Klemme HD, Ulmishek GF (1991) Effective petroleum source rocks of the world: stratigraphic distribution and controlling depositional factors. *AAPG Bulletin* **75**, 1809–1851.

Knoll AH, Beukes NJ (2009) Introduction: Initial investigations of a Neoarchean shelf margin-basin transition (Transvaal Supergroup, South Africa). *Precambrian Research* **169**(1–4), 1–14.

Knoll AH, Carroll SB (1999) Early animal evolution: emerging views from comparative biology and geology. *Science* **284**(5423), 2129–2137.

Knoll AH, Summons RE, Waldbauer JR, Zumberge J (2007) The geological succession of primary producers in the oceans. In: *The Evolution of Primary Producers in the Sea* (eds Falkwoski P, Knoll AH). Elsevier, Amsterdam, pp. 133–163.

Kodner RB, Summons RE, Pearson A, King N, Knoll AH (2008) Sterols in a unicellular relative of the metazoans. *Proceedings of the National Academy of Sciences of the USA* **105**(29), 9897–9902.

Koga Y, Morii H (2005) Recent advances in structural research on ether lipids from Archaea including comparative and physiological aspects. *Bioscience, Biotechnology, and Biochemistry* **69**(11), 2019–2034.

Kohnen MEL, Damste JSS, ten Haven HL, de Leeuw JW (1989) Early incorporation of polysulphides in sedimentary organic matter. *Nature* **341**(6243), 640.

Konneke M, Bernhard AE, de la Torre JR, Walker CB, Waterbury JB, Stahl DA (2005) Isolation of an autotrophic ammonia-oxidizing marine archaeon. *Nature* **437**, 543–546.

Kump L (1991) Interpreting carbon-isotope excursions: Strangeloveoceans. *Geology* **19**, 299–302.

Kump LR, Pavlov A, Arthur MA (2005) Massive release of hydrogen sulfide to the surface ocean and atmosphere during intervals of oceanic anoxia. *Geology* **33**, 397–400.

Kuypers MMM, Blokker P, Erbacher J, *et al.* (2001) Massive expansion of marine Archaea during a mid-Cretaceous oceanic anoxic event. *Science* **293**(5527), 92–95.

Kuypers MMM, van Breugel Y, Schouten S, Erba E, Damsté JSS (2004) N_2-fixing cyanobacteria supplied nutrient N for Cretaceous oceanic anoxic events. *Geology* **32**(10), 853–856.

Lewan MD, Winters JC, McDonald JH (1979) Generation of oil-like pyrolyzates from organic-rich shales. *Science* **203**(4383), 897–899.

Lewan MD, Spiro B, Illich H, *et al.* (1985) Evaluation of petroleum generation by hydrous pyrolysis experimentation [and discussion]. *Philosophical Transactions of the Royal Society of London. Series A, Mathematical and Physical Sciences* **315**(1531), 123.

Lichtenthaler HK, Schwender J, Disch A, Rohmer M (1997) Biosynthesis of isoprenoids in higher plant chloroplasts proceeds via a mevalonate-independent pathway. *FEBS Letters* **400**(3), 271–274.

Lipp J, Morono Y, Inagaki F, Hinrichs K (2008) Significant contribution of Archaea to extant biomass in marine subsurface sediments. *Nature* **454**(7207), 991–994.

Love GD, Snape CE, Carr AD, Houghton RC (1995) Release of covalently-bound alkane biomarkers in high yields from kerogen via catalytic hydropyrolysis. *Organic Geochemistry* **23**(10), 981–986.

Love GD, Grosjean E, Stalvies C, *et al.* (2009) Fossil steroids record the appearance of Demospongiae during the Cryogenian period. *Nature* **457**(7230), 718–721.

Macalady JL, Vestling MM, Baumler D, Boekelheide N, Kaspar CW, Banfield JF (2004) Tetraether-linked membrane monolayers in *Ferroplasma* spp: a key to survival in acid. *Extremophiles* **8**(5), 411–419.

Mackenzie AS, Brassell SC, Eglinton G, Maxwell JR (1982) Chemical fossils: the geological fate of steroids. *Science* **217**(4559), 491–504.

Magaritz M, Bar R, Baud A, Holser WT (1988) The carbon-isotope shift at the Permian/Triassic boundary in the southern Alps is gradual. *Nature* **331**, 337–339.

Magaritz M, Krishnamurthy RV, Holser WT (1992) Parallel trends in organic and inorganic carbon isotopes across the Permian/Triassic boundary. *American Journal of Science* **292**(10), 727–739.

Maresca J, Graham J, Bryant D (2008) The biochemical basis for structural diversity in the carotenoids of chlorophototrophic bacteria. *Photosynthesis Research* **97**(2), 121–140.

Marshall CR (2006) Explaining the Cambrian explosion of animals. *Annual Review of Earth and Planetary Sciences* **34**(1), 355–384.

Martin W, Russell MJ (2007) On the origin of biochemistry at an alkaline hydrothermal vent. *Philosophical Transactions of the Royal Society B: Biological Sciences* **362**(1486), 1887–1926.

Martin W, Baross J, Kelley D, Russell MJ (2008) Hydrothermal vents and the origin of life. *Nature Reviews Microbiology* **6**(11), 805.

Massana R, Murray AE, Preston CM, DeLong EF (1997) Vertical distribution and phylogenetic characterization of marine planktonic Archaea in the Santa Barbara Channel. *Applied and Environmental Microbiology* **63**, 50–56.

McCaffrey MA, Michael Moldowan J, Lipton PA, *et al.* (1994) Paleoenvironmental implications of novel C30 steranes in Precambrian to Cenozoic Age petroleum and bitumen. *Geochimica et Cosmochimica Acta* **58**(1), 529.

McCollom TM, Seewald JS (2006) Carbon isotope composition of organic compounds produced by abiotic synthesis under hydrothermal conditions. *Earth and Planetary Science Letters* **243**(1–2), 74–84.

McCollom TM, Seewald JS (2007) Abiotic synthesis of organic compounds in deep-sea hydrothermal environments. *Chemical Reviews (Washington, DC, United States)* **107**(2), 382–401.

McDonald JG, Thompson B, McCrum EC, Russell DW (2007) Extraction and analysis of sterols in biological matrices by high-performance liquid chromatography electrospray ionization mass spectrometry. *Methods in Enzymology* **432**, 143–168.

Metzger P, Largeau C (2005) *Botryococcus braunii*: a rich source for hydrocarbons and related ether lipids. *Applied Microbiology and Biotechnology* **66**(5), 486–496.

Meyer KM, Kump LR, Ridgwell A (2008) Biogeochemical controls on photic-zone euxinia during the end-Permian mass extinction. *Geology* **36**(9), 747–750.

Moldowan JM (1984) C30-steranes, novel markers for marine petroleums and sedimentary rocks. *Geochimica et Cosmochimica Acta* **48**(12), 2767–2768.

Moldowan JM, Talyzina NM (1998) Biogeochemical evidence for dinoflagellate ancestors in the early Cambrian. *Science* **281**(5380), 1168–1170.

Moldowan JM, Lee CY, Watt DS, Jeganathan A, Slougui N-E, Gallegos EJ (1991) Analysis and occurrence of C26-steranes in petroleum and source rocks. *Geochimica et Cosmochimica Acta* **55**(4), 1065–1081.

Moldowan JM, Dahl J, Huizinga BJ, *et al.* (1994) The molecular fossil record of oleanane and its relation to angiosperms. *Science* **265**(5173), 768–771.

Monson KD, Hayes JM (1980) Biosynthetic control of the natural abundance of carbon 13 at specific positions within fatty acids in *Escherichia coli*. Evidence regarding the coupling of fatty acid and phospholipid synthesis. *Journal of Biological Chemistry* **255**(23), 11435–11441.

Monson KD, Hayes JM (1982) Carbon isotopic fractionation in the biosynthesis of bacterial fatty acids. Ozonolysis of unsaturated fatty acids as a means of determining the intramolecular distribution of carbon isotopes. *Geochimica et Cosmochimica Acta* **46**(2), 139–149.

Morii H, Eguchi T, Nishihara M, Kakinuma K, König H, Koga Y (1998) A novel ether core lipid with H-shaped C80-isoprenoid hydrocarbon chain from the hyperthermophilic methanogen *Methanothermus fervidus*. *Biochimica et Biophysica Acta (BBA) – Lipids and Lipid Metabolism* **1390**(3), 339–345.

Murray AP, Edwards D, Hope JM, *et al.* (1998) Carbon isotope biogeochemistry of plant resins and derived hydrocarbons. *Organic Geochemistry* **29**(5–7), 1199–1214.

Narbonne GM, Gehling JG (2003) Life after snowball: the oldest complex Ediacaran fossils. *Geology* **31**(1), 27–30.

Nes W (1974) Role of sterols in membranes. *Lipids* **9**(8), 596–612.

Neunlist S, Bisseret P, Rohmer M (1988) The hopanoids of the purple non-sulfur bacteria *Rhodopseudomonas palustris* and *Rhodopseudomonas acidophila* and the absolute configuration of bacteriohopanetetrol. *European Journal of Biochemistry* **171**(1–2), 245–252.

Nishihara M, Koga Y (1995) Two new phospholipids, hydroxyarchaetidylglycerol and hydroxyarchaetidylethanolamine, from the Archaea *Methanosarcina barkeri*. *Biochimica et Biophysica Acta (BBA) – Lipids and Lipid Metabolism* **1254**(2), 155–160.

Ohkouchi N, Kahiyama Y, Chikaraishi Y, Ogawa NO, Tada R, Kitazato H (2006) Nitrogen isotopic composition of chlorophylls and porphyrins in geological samples as tools for reconstructing paleoenvironment. *Geochimica et Cosmochimica Acta* **70**(18, Supplement 1), A452.

Ohkouchi N, Nakajima Y, Ogawa NO, *et al.* (2008) Carbon isotopic composition of the tetrapyrrole nucleus in chloropigments from a saline meromictic lake: a mechanistic view for interpreting the isotopic signature of alkyl porphyrins in geological samples. *Organic Geochemistry* **39**(5), 521–531.

Ourisson G, Albrecht P (1992) Geohopanoids: the most abundant natural products on Earth? *Accounts of Chemical Research* **25**, 298–402.

Ourisson G, Albrecht P, Rohmer M (1979) The hopanoids. palaeochemistry and biochemistry of a group of natural products. *Pure and Applied Chemistry* **51**(4), 709–729.

Ourisson G, Rohmer M, Poralla K (1987) Prokaryotic hopanoids and other polyterpenoid sterol surrogates. *Annual Review of Microbiology* **41**, 301–333.

Ouverney CC, Fuhrman JA (2000) Marine planktonic archaea take up amino acids. *Applied and Environmental Microbiology* **66**, 1429–143x.

Pancost RD, Sinninghe Damsté JS, de Lint S, van der Maarel MJEC, Gottschal JC, Party MSS (2000) Biomarker evidence for widespread anaerobic methane oxidation in Mediterranean sediments by a consortium of methanogenic Archaea and Bacteria. *Applied and Environmental Microbiology* **66**(3), 1126–1132.

Patience RL, Rowland SJ, Maxwell JR (1978) The effect of maturation on the configuration of pristane in sediments and petroleum. *Geochimica et Cosmochimica Acta* **42**(12), 1871–1875.

Payne JL, Lehrmann DJ, Wei J, Orchard MJ, Schrag DP, Knoll AH (2004) Large perturbations of the carbon cycle during recovery from the end-Permian extinction. *Science* **305**(5683), 506–509.

Pearson A, McNichol AP, Benitez-Nelson BC, Hayes JM, Eglinton TI (2001) Origins of lipid biomarkers in Santa Monica Basin surface sediment: a case study using compound-specific [Delta]14C analysis. *Geochimica et Cosmochimica Acta* **65**(18), 3123–3137.

Pearson A, Huang Z, Ingalls AE, *et al.* (2004) Nonmarine crenarchaeol in Nevada hot springs. *Applied and Environmental Microbiology* **70**(9), 5229–5237.

Pearson A, Page SRF, Jorgenson TL, Fischer WW, Higgins MB (2007) Novel hopanoid cyclases from the environment. *Environmental Microbiology* **9**(9), 2175–2188.

Peretó J, López-García P, Moreira D (2004) Ancestral lipid biosynthesis and early membrane evolution. *Trends in Biochemical Sciences* **29**(9), 469–477.

Peters KE, Walters CC, Moldowan JM (2005) *The Biomarker Guide*, 2nd edn. Cambridge University Press, Cambridge.

Peterson KJ, Butterfield NJ (2005) Origin of the Eumetazoa: testing ecological predictions of molecular clocks against the Proterozoic fossil record. *Proceedings of the National Academy of Sciences of the USA* 102(27), 9547–9552.

Peterson KJ, Summons RE, Donoghue PCJ (2007) Molecular palaeobiology. *Palaeontology* 50(4), 775–809.

Prahl FG, Wakeham SG (1987) Calibration of unsaturation patterns in long-chain ketone compositions for palaeotemperature assessment. *Nature* 330(6146), 367–369.

Proskurowski G, Lilley MD, Seewald JS, *et al.* (2008) Abiogenic hydrocarbon production at Lost City hydrothermal field. *Science* 319(5863), 604–607.

Rampen SW, Schouten S, Abbas B, *et al.* (2007) On the origin of 24-norcholestanes and their use as age-diagnostic biomarkers. *Geology* 35(5), 419–422.

Rashby SE, Sessions AL, Summons RE, Newman DK (2007) Biosynthesis of 2-methylbacteriohopanepolyols by an anoxygenic phototroph. *Proceedings of the National Academy of Sciences of the USA* 104(38), 15099–15104.

Rasmussen B, Fletcher IR, Brocks JJ, Kilburn MR (2008) Reassessing the first appearance of eukaryotes and cyanobacteria. *Nature* 455, 1101–1104.

Rattray J, van de Vossenberg J, Hopmans E, *et al.* (2008) Ladderane lipid distribution in four genera of anammox bacteria. *Archives of Microbiology* 190(1), 51.

Raymond J, Segrè D (2006) The effect of oxygen on biochemical networks and the evolution of complex life. *Science* 311(5768), 1764–1767.

Riccardi AL, Arthur MA, Kump LR (2006) Sulfur isotopic evidence for chemocline upward excursions during the end-Permian mass extinction. *Geochimica et Cosmochimica Acta* 70, 5740–5752.

Rohmer M (2003) Mevalonate-independent methylerythritol phosphate pathway for isoprenoid biosynthesis. Elucidation and distribution. *Pure and Applied Chemistry* 75(2–3), 375–387.

Rohmer M, Bouvier P, Ourisson G (1979) Molecular evolution of biomembranes – structual equivalents and phylogenetic precursors of sterols. *Proceedings of the National Academy of Sciences of the USA* 76(2), 847–851.

Rohmer M, Bouvier-Nave P, Ourisson G (1984) Distribution of hopanoid triterpanes in prokaryotes. *Journal of General Microbiology* 130, 1137–1150.

Rohmer M, Knani M, Simonin P, Sutter B, Sahm H (1993) Isoprenoid biosynthesis in bacteria: a novel pathway for the early steps leading to isopentenyl diphosphate. *Biochemical Journal* 295, 517–524.

Rontani J-F, Beker B, Volkman JK (2004) Long-chain alkenones and related compounds in the benthic haptophyte *Chrysotila lamellosa* Anand HAP 17. *Phytochemistry* 65, 117–126.

Rosell-Melé A, Eglinton G, Pflaumann U, Sarnthein M (1995) Atlantic core-top calibration of the U37K index as a sea-surface palaeotemperature indicator. *Geochimica et Cosmochimica Acta* 59(15), 3099–3107.

Rowland SJ (1990). Production of acyclic isoprenoid hydrocarbons by laboratory maturation of methanogenic bacteria. *Organic Geochemistry* 15, 9–16.

Rullkötter J, Aizenshtat Z, Spiro B (1984) Biological markers in bitumens and pyrolyzates of Upper Cretaceous bituminous chalks from the Ghareb Formation (Israel). *Geochimica et Cosmochimica Acta* 48(1), 151–157.

Rutters H, Sass H, Cypionka H, Rullkotter J (2002) Phospholipid analysis as a tool to study complex microbial communities in marine sediments. *Journal of Microbiological Methods* 48(2–3), 149.

Schimmelmann A, Lewan MD, Wintsch RP (1999) D/H isotope ratios of kerogen, bitumen, oil, and water in hydrous pyrolysis of source rocks containing kerogen types I, II, IIS, and III. *Geochimica et Cosmochimica Acta* 63(22), 3751–3766.

Schoell M (1988) Multiple origins of methane in the Earth. *Chemical Geology* 71(1–3), 1–10.

Schouten S, Van Der Maarel MJEC, Huber R, Damsté JSS (1997) 2,6,10,15,19-Pentamethylicosenes in *Methanolobus bombayensis*, a marine methanogenic archaeon, and in *Methanosarcina mazei*. *Organic Geochemistry* 26(5–6), 409–414.

Schouten S, Rijpstra WIC, Kok M, *et al.* (2001) Molecular organic tracers of biogeochemical processes in a saline meromictic lake (Ace Lake). *Geochimica et Cosmochimica Acta* 65(10), 1629.

Schouten S, Hopmans EC, Schefuß E, Sinninghe Damsté JS (2002) Distributional variations in marine crenarchaeotal membrane lipids: a new organic proxy for reconstructing ancient sea water temperatures? *Earth and Planetary Science Letters* 204, 265–274.

Schouten S, Baas M, Hopmans EC, Sinninghe Damsté JS (2008a) An unusual isoprenoid tetraether lipid in marine and lacustrine sediments. *Organic Geochemistry* 39(8), 1033–1038.

Schouten S, Hopmans EC, Baas M, *et al.* (2008b) Intact membrane lipids of 'candidatus *Nitrosopumilus maritimus*,' a cultivated representative of the cosmopolitan mesophilic Group I Crenarchaeota. *Applied and Environmental Microbiology* 74(8), 2433–2440.

Schouten S, Özdirekcan S, van der Meer MTJ, *et al.* (2008c) Evidence for substantial intramolecular heterogeneity in the stable carbon isotopic composition of phytol in photoautotrophic organisms. *Organic Geochemistry* 39(1), 135–146.

Schwark L, Empt P (2006) Sterane biomarkers as indicators of palaeozoic algal evolution and extinction events. *Palaeogeography, Palaeoclimatology, Palaeoecology* 240(1–2), 225–236.

Sephton MA, Looy CV, Veefkind RJ, Brinkhuis H, De Leeuw JW, Visscher H (2002) Synchronous record of $\delta^{13}C$ shifts in the oceans and atmosphere at the end of the Permian. *Geological Society of America Special Paper* 356, 455–462.

Sephton MA, Looy CV, Brinkhuis H, Wignall PB, de Leeuw JW, Visscher H (2005) Catastrophic soil erosion during the end-Permian biotic crisis. *Geology* 33(12), 941–944.

Sessions AL, Burgoyne TW, Schimmelmann A, Hayes JM (1999) Fractionation of hydrogen isotopes in lipid biosynthesis. *Organic Geochemistry* 30(9), 1193–1200.

Sessions AL, Sylva SP, Summons RE, Hayes JM (2004) Isotopic exchange of carbon-bound hydrogen over geologic timescales. *Geochimica et Cosmochimica Acta* 68(7), 1545–1559.

Sessions AL, Doughty DM, Welander PV, Summons RE, Newman DK (2009) The continuing puzzle of the Great Oxidation Event. *Current Biology* 19(14), R567–R574.

Shah SR, Mollenhauer G, Ohkouchi N, Eglinton TI, Pearson A (2008) Origins of archaeal tetraether lipids in sediments: insights from radiocarbon analysis. *Geochimica et Cosmochimica Acta* 72(18), 4577–4594.

Sherman L S, Walbauer JR, Summons RE (2007) Methods for biomarker analyses of high maturity Precambrian rocks. *Organic Geochemistry* 38, 1987–2000.

Sherwood Lollar B, Lacrampe-Couloume G, Slater GF, *et al.* (2006) Unravelling abiogenic and biogenic sources of methane in the Earth's deep subsurface. *Chemical Geology* 226(3–4), 328–339.

Sikes EL, Volkman JK (1993) Calibration of alkenone unsaturation ratios (Uk'37) for paleotemperature estimation in cold polar waters. *Geochimica et Cosmochimica Acta* 57(8), 1883–1889.

Sikes EL, Volkman JK, Robertson LG, Pichon J-J (1997) Alkenones and alkenes in surface waters and sediments of the Southern Ocean: Implications for paleotemperature estimation in polar regions. *Geochimica et Cosmochimica Acta* 61(7), 1495–1505.

Silva CJ, Wunsche L, Djerassi C (1991) Biosynthetic studies of marine lipid 35. The demonstration of de novo sterol biosynthesis in sponges using radiolabeled isoprenoid precursors. *Comparative Biochemistry and Physiology Part B: Biochemistry and Molecular Biology* 99(4), 763.

Simonson BM, Sumner DY, Beukes NJ, Johnson S, Gutzmer J (2009) Correlating multiple Neoarchean-Paleoproterozoic impact spherule layers between South Africa and Western Australia. *Precambrian Research* 169(1–4), 100–111.

Sinninghe Damsté JS, Muyser G, Abbas B, *et al.* (2004a) The rise of the rhizosolenid diatoms. *Science* 304(5670), 584–587.

Sinninghe Damsté JS, Rijpstra WIC, Schouten S, Fuerst JA, Jetten MSM, Strous M (2004b) The occurrence of hopanoids in planctomycetes: implications for the sedimentary biomarker record. *Organic Geochemistry* 35(5), 561–566.

Sinninghe Damsté JS, Rijpstra WIC, Geenevasen JAJ, Strous M, Jetten MSM (2005) Structural identification of ladderane and other membrane lipids of planctomycetes capable of anaerobic ammonium oxidation (anammox). *FEBS Journal* 272(16), 4270–4283.

Sinninghe Damsté JS, Rijpstra W, Hopmans E, Schouten S, Balk M, Stams A (2007) Structural characterization of diabolic acid-based tetraester, tetraether and mixed ether/ester, membrane-spanning lipids of bacteria from the order Thermotogales. *Archives of Microbiology* 188(6), 629.

Smittenberg RH, Pancost RD, Hopmans EC, Paetzel M, Sinninghe Damsté JS (2004) A 400-year record of environmental change in an euxinic fjord as revealed by the sedimentary biomarker record. *Palaeogeography, Palaeoclimatology, Palaeoecology* 202(3–4), 331–351.

Sofer Z (1984) Stable carbon isotope compositions of crude oils: application to source depositional environments and petroleum alteration. *AAPG Bulletin* 68, 31–49.

Sprott GD, Ekiel I, Dicaire C (1990) Novel, acid-labile, hydroxydiether lipid cores in methanogenic bacteria. *Journal of Biological Chemistry* 265(23), 13735–13740.

Stadnitskaia A, Muyser G, Abbas B, *et al.* (2005) Biomarker and 16S rDNA evidence for anaerobic oxidation of methane and related carbonate precipitation in deep-sea mud volcanoes of the Sorokin Trough, Black Sea. *Marine Geology* 217(1–2), 67–96.

Sternberg LO, Deniro MJ, Johnson HB (1984) Isotope ratios of cellulose from plants having different photosynthetic pathways. *Plant Physiology* 74(3), 557–561.

Sturt HF, Summons RE, Smith K, Elvert M, Hinrichs K-U (2004) Intact polar membrane lipids in prokaryotes and sediments deciphered by high-performance liquid chromatography/electrospray ionization multistage mass spectrometry – new biomarkers for biogeochemistry and microbial ecology. *Rapid Communications in Mass Spectrometry* 18, 617–628.

Summons RE, Volkman JK, Boreham CJ (1987) Dinosterane and other steroidal hydrocarbons of dinoflagellate origin in sediments and petroleum. *Geochimica et Cosmochimica Acta* 51(11), 3075.

Summons RE, Thomas J, Maxwell JR, Boreham CJ (1992) Secular and environmental constraints on the occurrence of dinosterane in sediments. *Geochimica et Cosmochimica Acta* 56(6), 2437.

Summons RE, Jahnke LL, Hope JM, Logan GA (1999) 2-Methylhopanoids as biomarkers for cyanobacterial oxygenic photosynthesis. *Nature* 400(6744), 554–557.

Summons RE, Bradley AS, Jahnke LL, Waldbauer JR (2006) Steroids, triterpenoids and molecular oxygen. *Philosophical Transactions of the Royal Society B–Biological Sciences* 361(1470), 951–968.

Talbot HM, Squier AH, Keely BJ, Farrimond P (2003) Atmospheric pressure chemical ionisation reversed-phase liquid chromatography/ion trap mass spectrometry of intact bacteriohopanepolyols. *Rapid Communications in Mass Spectrometry* 17(7), 728–737.

Talbot HM, Rohmer M, Farrimond P (2007) Rapid structural elucidation of composite bacterial hopanoids by atmospheric pressure chemical ionisation liquid chromatography/ion trap mass spectrometry. *Rapid Communications in Mass Spectrometry* 21(6), 880–892.

Talbot HM, Summons RE, Jahnke LL, Cockell CS, Rohmer M, Farrimond P (2008) Cyanobacterial bacteriohopanepolyol signatures from cultures and natural environmental settings. *Organic Geochemistry* 39(2), 232–263.

Teira E, van Aken H, Veth C, Herndl GJ (2006) Archaeal uptake of enantiomeric amino acids in the meso- and bathypelagic waters of the North Atlantic. *Limnology and Oceanography* 51, 60–69.

Thiel V, Peckmann J, Seifert R, Wehrung P, Reitner J, Michaelis W (1999) Highly isotopically depleted isoprenoids: molecular markers for ancient methane venting. *Geochimica et Cosmochimica Acta* 63(23/24), 2959–3966.

Tissot BP, Welte DH (1978) *Petroleum Formation and Occurrence: A New Approach to Oil and Gas Exploration.* Springer-Verlag, Berlin.

Tornabene TG, Langworthy TA, Holzer G, Oro J (1979) Squalenes, phytanes and other isoprenoids as major neutral lipids of methanogenic and thermoacidophilic 'Archaebacteria'. *Journal of Molecular Evolution* 13(1), 73–83.

Treibs A (1936) Chlorophyll- und Häminderivate in organischen Mineralstoffen. *Angewandte Chemie* 49(38), 682–686.

van Kaam-Peters HME, Köster J, van der Gaast SJ, Dekker M, de Leeuw JW, Sinninghe Damsté JS (1998) The effect of clay minerals on diasterane/sterane ratios. *Geochimica et Cosmochimica Acta* 62(17), 2923.

Van Mooy BAS, Rocap G, Fredricks HF, Evans CT, Devol AH (2006) Sulfolipids dramatically decrease phosphorus demand by picocyanobacteria in oligotrophic marine environments. *Proceedings of the National Academy of Sciences of the USA* **103**(23), 8607–8612.

Vilcheze C, Llopiz P, Neunlist S, Poralla K, Rohmer M (1994) Prokaryotic triterpenoids: new hopanoids from the nitrogen-fixing bacteria *Azotobacter vinelandii, Beijerinckia indica* and *Beijerinckia mobilis. Microbiology* **140**(10), 2749–2753.

Volkman JK (2005) Sterols and other triterpenoids: source specificity and evolution of biosynthetic pathways. *Organic Geochemistry* **36**(2), 139–159.

Volkman JK, Maxwell JR (1984) Acyclic isoprenoids as biological markers. In: *Biological Markers in the Sedimentary Record* (ed. Johns RB). Elsevier, Amsterdam, pp. 1–42.

Volkman JK, Eglinton G, Corner EDS, Sargent JR (1980) Novel unsaturated straight-chain C37–C39 methyl and ethyl ketones in marine sediments and a coccolithophore *Emiliania huxleyi. Physics and Chemistry of the Earth* **12**, 219–227.

Volkman JK, Barrerr SM, Blackburn SI, Sikes EL (1995) Alkenones in *Gephyrocapsa oceanica*: implications for studies of paleoclimate. *Geochimica et Cosmochimica Acta* **59**(3), 513–520.

Wagner F, Rottem S, Held H-D, Uhlig S, Zähringer U (2000) Ether lipids in the cell membrane of *Mycoplasma fermentans. European Journal of Biochemistry* **267**(20), 6276–6286.

Wakeham SG, Amann R, Freeman KH, *et al.* (2007) Microbial ecology of the stratified water column of the Black Sea as revealed by a comprehensive biomarker study. *Organic Geochemistry* **38**, 2070–2097.

Waldbauer JR, Sherman LS, Sumner DY, Summons RE (2009) Late Archean molecular fossils from the Transvaal Supergroup record the antiquity of microbial diversity and aerobiosis. *Precambrian Research* **169**, 28–47.

Wang C (2007) Anomalous hopane distributions at the Permian-Triassic boundary, Meishan, China – evidence for the end-Permian marine ecosystem collapse. *Organic Geochemistry* **38**(1), 52–66.

Wang C, Visscher H (2007) Abundance anomalies of aromatic biomarkers in the Permian-Triassic boundary section at Meishan, China – evidence of end-Permian terrestrial ecosystem collapse. *Palaeogeography, Palaeoclimatology, Palaeoecology* **252**(1–2), 291–303.

Welander PV, Hunter RC, Zhang L, Sessions AL, Summons RE, Newman DK (2009) Hopanoids play a role in membrane integrity and pH homeostasis in *Rhodopseudomonas palustris* TIE-1. *Journal of Bacteriology* **191**(19), 6145–6156.

Welander PV, Coleman M, Sessions AL, Summons RE, Newman DK (2010) Identification of a methylase required for 2-methylhopanoid production and implications for the interpretation of sedimentary hopanes. *Proceedings of the National Academy of Sciences of the USA* **107**, 8537–8542.

Wuchter CS, Schouten S, Boschker HTS, Sinninghe Damsté JS (2003) Bicarbonate uptake by marine Crenarchaeota. *FEMS Microbiology Letters* **219**, 203–207.

Xiao S, Zhang Y, Knoll AH (1998) Three-dimensional preservation of algae and animal embryos in a Neoproterozoic phosphorite. *Nature* **391**(6667), 553–558.

Xie S, Pancost RD, Yin H, Wang H, Evershed RP (2005) Two episodes of microbial change coupled with Permo/Triassic faunal mass extinction. *Nature* **434**, 494–497.

Yin L, Zhu M, Knoll AH, Yuan X, Zhang J, Hu J (2007) Doushantuo embryos preserved inside diapause egg cysts. *Nature* **446**(7136), 661–663.

Zachos JC, Arthur MA, Dean WE (1989) Geochemical evidence for suppression of pelagic marine productivity at the Cretaceous/Tertiary boundary. *Nature* **337**(6202), 61–64.

Zachos JC, Rohl U, Schellenberg SA, *et al.* (2005) Rapid acidification of the ocean during the Paleocene-Eocene thermal maximum. *Science* **308**(5728), 1611–1615.

Zhang X, Gillespie AL, Sessions AL (2009) Large D/H variations in bacterial lipids reflect central metabolic pathways. *Proceedings of the National Academy of Sciences of the USA* **106**(31), 12580–12586.

Zundel M, Rohmer M (1985a) Hopanoids of the methylotrophic bacteria *Methylococcus capsulatus* and *Methylomonas* sp. as possible precursors of C$_{29}$ and C$_{30}$ hopanoid chemical fossils. *FEMS Microbiology Letters* **28**, 61–64.

Zundel M, Rohmer M (1985b) Prokaryotic triterpenoids. *European Journal of Biochemistry* **150**(1), 23–27.

Zundel M, Rohmer M (1985c) Prokaryotic triterpenoids. 3. The biosynthesis of 2 beta-methylhopanoids and 3 beta-methylhopanoids of *Methylobacterium organophilum* and *Acetobacter pasteurianus* ssp. *pasteurianus. European Journal of Biochemistry* **150**(1), 35–39.

16

THE FOSSIL RECORD OF MICROBIAL LIFE

Andrew H. Knoll

Department of Organismic and Evolutionary Biology, Harvard University, Cambridge MA 02138, USA

16.1 Introduction

For two centuries, paleontologists have studied the fossil record left by plants, animals and microscopic eukaryotes that secrete mineralized skeletons. Reports of both smaller and much older fossils occasionally surfaced in the past, but it was not until 1954 that Stanley Tyler and Elso Barghoorn published the first widely accepted report of microbial fossils nearly four times the age of the oldest trilobites. Since that time, geobiologists have come to appreciate the potential for bacteria and non-skeletal protists to preserve in sedimentary rocks, and we understand the sedimentary and geochemical circumstances that favour microbial fossilization. Moreover, it is now clear that *most* – about 85% – of Earth's paleobiological record is a history of microbial life. Animals and plants are evolutionary latecomers.

16.2 The nature of Earth's early microbial record

In principle, the preservation and interpretation of Earth's early microbial fossil record can be approached much like the better known record of marine animals. For example, 50-million-year-old sedimentary rocks exposed in the Gulf of Mexico region of North America contain abundant fossils of bivalved mollusks. Only mineralized skeletons are preserved – no trace persists of DNA, proteins, or other organic materials, save perhaps for some limited organic matter in the shell matrix. In consequence, we cannot consult a paleogenome to draw conclusions about function or evolutionary relationships. Fortunately, important inferences can be made on the basis of the features that are preserved. Morphology provides diagnostic information on phylogenetic placement – many of the fossils belong to the bivalve family Veneridae. From this, we can draw robust conclusions about anatomy and physiology because many such features are shared by all venerids, or more broadly among all bivalves. Given preserved shells, then, we can reconstruct soft part anatomy in some detail, based on closely related living clams with comparable shell structure. Thus, we can know how the clam fed, how it converted its food into biomass and energy, its relative rates of basal and exercise metabolism, and its physiology and biochemistry of shell precipitation.

The fossils of life before the animals are, like those of the clams mentioned above, a partial record of their makers. Cell walls commonly preserve more readily than cytoplasm, and in groups like the cyanobacteria, extracellular envelopes made of secreted polysaccharides can preserve still more readily. From these we obtain a record of morphology and, if a population can be investigated, life cycles preserved by individuals arrested in various stages of cell division or colony formation. As in conventional fossils, morphology provides key information for phylogenetic interpretation, but in bacteria and unicellular eukaryotic organisms, simple morphology may not be diagnostic for a particular taxonomic group. Phylogenetic interpretation is limited to fossils whose evolutionary relationships can be inferred from preserved morphologies. Diagnostic morphology, in turn, permits physiological inference about ancient microfossils.

Fundamentals of Geobiology, First Edition. Edited by Andrew H. Knoll, Donald E. Canfield and Kurt O. Konhauser.
© 2012 Blackwell Publishing Ltd. Published 2012 by Blackwell Publishing Ltd.

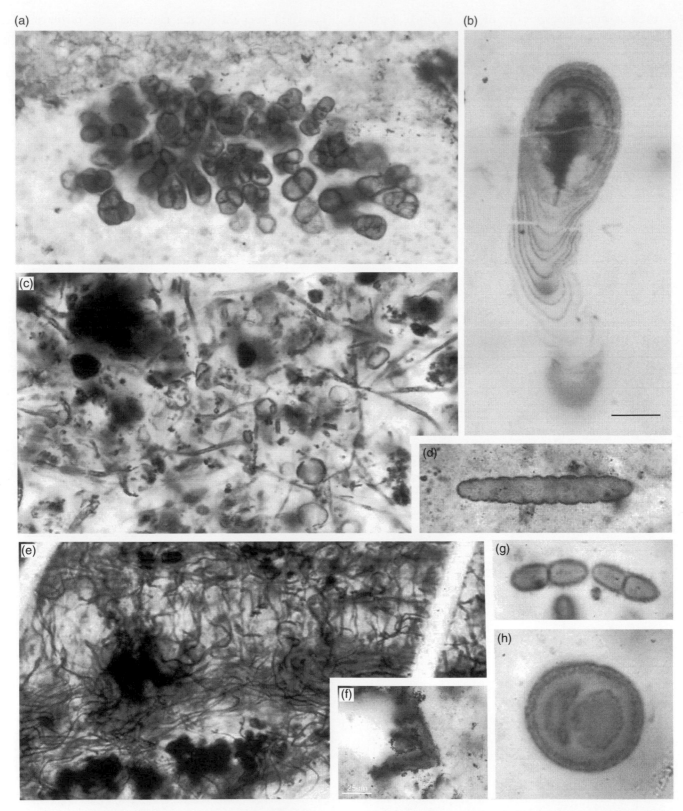

Figure 16.1 Cyanobacterial and other bacterial microfossils in Proterozoic rocks. (a) *Eohyella dichotoma*, an endolithic cyanobacterium preserved in silicified ooids of the 750–800 Ma Upper Eleonore Bay Group, Greenland (see text for discussion). (b) *Polybessurus bipartitus*, a cyanobacterial microfossil preserved in tidal flat deposits of the 750–800 Ma Draken Formation, Spitsbergen. (c) Microfossils in siliceous stromatolites of the ca. 1900 Ma Gunflint Formation, Canada (see text for discussion). (d) *Filiconstrictosus* ex gr. *majusculus*, the cast and mold of a filamentous cyanobacterium,

Bacterial and eukaryotic microfossils can be preserved as organic compressions in shale – cell walls or extracellular envelopes flattened to two dimensions by the weight of accumulating sediments. Microfossils can also be preserved in three-dimensional detail by the precipitation of minerals in sediments before they become compacted.

Permineralized (preserved by precipitated minerals) microfossils occur most commonly in early diagenetic chert (SiO_2), although iron minerals and carbonate precipitates can also preserve microorganisms, and preservation in phosphate minerals can be exquisite. Most commonly, paleontologists study ancient microbial fossils by means of optical microscopy, aided by scanning electron microscopy and, to a lesser extent, transmission electron microscopy. In recent years, new imaging and analytical tools have been applied to the early fossil record; some of these are outlined below.

Permineralized fossils are usually studied in petrographic thin section, in part because the fossils tend to disintegrate when their mineral matrix is dissolved away, but also because in thin sections spatial relationships among fossils are preserved in three dimensions, facilitating the recognition of populations and preserving a record of microbial behavior. Paleontologists sometimes work with thin sections of shale cut parallel to bedding, but microfossils preserved in shales commonly occur at low density, and so are more readily studied by dissolving the rock matrix and plating organic remains onto microscope slides. This enables more fossils to be examined, but forfeits information on spatial orientation.

16.3 Paleobiological inferences from microfossil morphology

16.3.1 A Proterozoic cyanobacterium

In general, morphology provides limited clues to energy metabolism in prokaryotic microorganisms and their fossils. Sizes are generally small, shapes are mostly simple, and associated physiologies are diverse. Fortunately, however, the prokaryotic group of greatest interest to

historical geobiology, the cyanobacteria, displays an unusual degree of morphological variation, including some forms not shared with other Bacteria or Archaea. We can never know with confidence the phylogenetic or physiological properties of 2 μm spheroids preserved in chert, but cyanobacteria with complex morphologies and preserved patterns of behavior provide important tools for interpreting the preserved record of early ecosystems.

Eohyella dichotoma is a population of simple multicellular microfossils preserved within silicified ooids in a 750–800 million carbonate platform from central East Greenland (Fig. 16.1a; Green *et al.*, 1988). As in the case of the Eocene clams, only some features of *E. dichotoma* biology are preserved, in this case extracellular envelopes and cell walls that, fortunately, preserve overall developmental plan and morphology with high fidelity. Patterns of cell division are preserved clearly, as is a conspicuous pattern of behavior.

The fossils consist of coccoidal cells arrayed in filament-like lines that radiate inward into ooid grains from their surfaces. Each so-called pseudofilament formed from equidimensional to elongated cells 8–21 μm in maximum dimension; apparent branching reflects binary cell division in a plane perpendicular to the axis of the pseudofilament, or the slippage and reorientation of cells.

Living cyanobacteria of the genus *Hyella* show closely comparable patterns of morphology and behavior; modern *Hyella* species are endolithic – that is, they bore into carbonate substrates and live within them (Le Campion-Alsumard, 1989). Ooids forming today on the Bahama Banks and elsewhere commonly contain large populations of *Hyella* and other endolithic microorganisms, and the orientation and growth pattern of *E. dichotoma* unambiguously record its systematic affinities and endolithic habit (Green *et al.*, 1988). *Eohyella dichotoma*, then, was a cyanobacterium, placed phylogenetically within the order Pleurocapsales. As a cyanobacterium, *E. dichotoma* photosynthesized using coupled Photosystems I and II. It also secreted weak acids that dissolved carbonate, enabling the thallus to grow beneath the protective surface of its substrate; depth of penetration reflects a behavioral response to

Figure 16.1 Continued. preserved in coastal deposits of the ca. 1400–1500 Ma Kotuikan Formation, Siberia. (e) Alternating vertically and horizontally oriented sheaths of mat-building, filamentous cyanobacteria, Upper Eleonore Bay Group; the orientations indicate that these filaments were mat-building microorganisms. (g) *Eosynechococcus medius*, Draken Formation, rod-like microfossils that are possibly, but not demonstrably, cyanobacterial. (f) Ghost of a carbonate rhomb outlined by organic matter, preserved in the 3430 Ma Strelley Pool Chert, northwestern Australia (courtesy of A. Allwood). (h) *Gloeodiniopsis lamellosa*, a pair of cyanobacterial cells preserved as collapsed extracellular envelopes, both within an encompassing envelope, ca. 800 Ma Bitter Springs Formation, Australia. Bar in B = 75 μm for A, = 15 μm for B, C = 25 μm for D, = 125 μm for E, = 7.5 μm for F and H, and = 40 μm for G.

the gradient of decreasing light availability below grain surfaces.

16.3.2 Diverse Proterozoic cyanobacteria

Schopf and Walter (1982) dubbed the Proterozoic Eon 'the Age of Cyanobacteria,' reflecting the early discovery of numerous microfossils with cyanobacteria-like morphologies in Proterozoic cherts. Years later, it remains the case that cyanobacteria-like fossils occur abundantly in Proterozoic sedimentary successions (Fig. 16.1a, b, d, e, g, h; see Golubic and Seong-Joo, 1999, and Knoll, 2007, for detailed reviews). Confidence in interpreting such fossils as cyanobacteria varies, the principal problem being the one mentioned earlier – simple morphologies occur among phylogenetically distinct and physiologically disparate microorganisms. Nonetheless, many of the Proterozoic microfossils that closely resemble modern cyanobacteria also come from peritidal environments washed by well-oxygenated waters. That is, many of the best preserved Proterozoic microfossils formed in coastal environments that even today support cyanobacterial communities. Given the probable paucity of alternative electron donors, most probably are the remains of ancient cyanobacteria.

Traditional classifications of cyanobacteria divide the phylum into unicellular taxa (Chroococcales), simple cells or pseudofilaments that produce specialized reproductive cells called baeocytes (Pleurocapsales), simple undifferentiated filaments (Oscillatoriales), simple filaments with differentiated cells specialized for nitrogen fixation (heterocysts) and resting stages (akinetes) (Nostocales), and more complex filamentous forms with differentiated cells (Stigonematales). In molecular phylogenies, all cyanobacteria that differentiate multiple cell types form a monophyletic group, with a stigonematalean clade nested within a paraphyletic Nostocales (Tomitani *et al.*, 2006). The Pleurocapsales also turn out to be monophyletic, but chroococcalean and simple oscillatorialean forms are polyphyletic.

Most cyanobacterial morphotypes, including nostocalean filaments and akinetes, occur in rocks as old as 1700 Ma (Tomitani *et al.*, 2006), demonstrating that much of the large scale diversity of cyanobacteria was established early in Earth's history. The oldest unambiguous cyanobacterial fossils are colony-forming unicellular forms in the ca. 2000 Ma Belcher Supergroup, Canada (Hofmann, 1976; Golubic and Hofmann, 1976). Ca. 2500 Ma microfossils from silicified peritidal carbonates of the Transvaal Supergroup, South Africa could be cyanobacteria, but their preservation is not sufficiently good to establish this with confidence (Lanier 1988; Altermann and Schopf, 1995). The body fossil record of cyanobacteria, thus, extends backward

in time nearly to the initial rise in atmospheric oxygen recorded geochemically. Whether cyanobacteria evolved substantially before this environmental transformation remains an important and unsolved issue in paleobiology (see below).

16.3.3 Preservation of other early prokaryotes

Not all prokaryotic microfossils are presumptive cyanobacteria. The ca. 1900 Ma Gunflint Formation, for example, contains abundant microfossils commonly interpreted as iron metabolizing bacteria (Barghoorn and Tyler, 1965; Cloud, 1965; Knoll, 2003). This interpretation originated in morphological comparisons between ancient and modern organisms, but it receives fresh support from careful stratigraphic analysis of fossiliferous cherts in the ca. 1800–1900 Ma Duck Creek Formation, Australia. In the Duck Creek Formation, Gunflint-type microfossil assemblages record life on relatively deep seafloors, close to the environmental boundary between oxygenated surface waters and subsurface water masses rich in reduced iron (Wilson *et al.*, 2010). This suggests that the ferruginous deep waters that persisted until ca. 1800 Ma (Poulton *et al.*, 2004; Wilson *et al.*, 2010) supported physiologically distinct seafloor communities without close counterparts in the modern ocean. More comprehensive insights into the phylogenetic and metabolic diversity of Precambrian Bacteria and Archaea come from isotopic and molecular biomarker geochemistry (see Chapters 14 and 15, and below).

16.3.4 Microscopic eukaryotes in a Proterozoic basin

Deep within the Grand Canyon, Arizona, large dolomite nodules in basinal black shales preserve a remarkable assemblage of vase-shaped microfossils (Fig. 16.2f; Porter and Knoll; 2000; Porter *et al.*, 2003). An ash bed just above the fossiliferous nodules has been dated by U-Pb zircon analysis at 742 ± 6 Ma, sharply constraining their age. The fossils are ca. 100 µm long tests, rounded at one end and exhibiting a distinctly collared opening at the other. No prokaryote makes such structures, so these remains are clearly eukaryotic. Within the eukaryotes, several phylogenetically disparate lineages have evolved vase-like test morphologies. Traditionally, many of these were placed in the testate amoebae, now known to be divided between two distinct clades: filose testate amoebae nest within the Rhizaria, whereas lobose testate amoebae belong to the Amoebozoa, sister to the opisthokonts (fungi, animals, and related unicells). The Grand Canyon rocks contain at least 11 distinct forms. Some preserve tests known from both lobose and filose amoebae today – the simple vases also find convergent comparisons among basal foraminiferan

Figure 16.2 Eukaryotic microfossils in Proterozoic rocks. (a) *Bangiomorpha pubsescens*, a filamentous red alga from the ca. 1200 Ma Hunting Formation, arctic Canada (courtesy of N. Butterfield). (b) *Tappania plana*, a microfossil whose branches indicate that it was a eukaryotic cell with an internal cytoskeleton, 1400–1500 Ma Roper Group, Australia. (c) *Leiosphaeridia* sp., also from the Roper Group – a large complex wall of uncertain phylogenetic affinities. (d) *Alicesphaeridium* sp., wall of a possible metazoan resting cyst, Ediacaran (580–560 Ma) Vychegda Formation, Russia (courtesy of V.N. Sergeev and N.G. Vorob'eva). (e) TEM of wall ultrastructure, *Leiosphaeridia jacutica*, Roper Group, showing preserved ultrastructure that is characteristic of eukaryotes but unknown from morphologically comparable bacteria. (f) *Melanocyrillium hexodiadema*, interpreted as the test of a lobose testate amoeban, 740–750 Ma Kwagunt Formation, Grand Canyon, USA (see text for discussion). Bar in A = 25 μm for A and C, = 50 μm for B, = 100 μm for D, = 1 μm for E, and = 15 μm for F.

taxa. The assemblage, however, also includes several distinctive test forms that compare closely to morphologically diagnostic lobosan species (Porter *et al.*, 2003). Thus, the exceptionally preserved Grand Canyon assemblage, in concert with a more widespread record of testate molds in shallow marine carbonates and cherts, documents a mid-Neoproterozoic radiation of aerobic eukaryotic heterotrophs with preservable tests. These fossils, which probably included both bacteriovores and predators on other protists, place firm minimum dates on the beginnings of at least one major eukaryotic clade.

16.3.5   More Proterozoic eukaryotes

Like the fossils of prokaryotic microorganisms, Proterozoic protists can be related to extant eukaryotic clades with varying degrees of confidence and precision. At one extreme are those small spheroids that cannot be allied to any domain with confidence – these are disturbingly common in Proterozoic rocks! Moving along the interpretational continuum, we find many populations whose morphologies and/or preserved ultrastructure reliably tag them as eukaryotes, but do not permit assignment to smaller clades within the domain. For example, Mesoproterozoic shales deposited in coastal marine environments commonly contain large (>100 μm) organic-walled microfossils that exhibit processes or other forms of surface ornamentation that indicate formation under the control of a sophisticated cytoskeletal-endomembrane system – the very hallmark of eukaryotic biology (Fig. 16.2b; Javaux *et al.*, 2001). Several of these fossils additionally preserve complex, heterogeneously layered wall ultrastructures, observed by transmission electron microscopy (TEM), that further differentiate their makers from known prokaryotes (Fig. 16.2e; Javaux *et al.*, 2004). Late Paleoproterozoic and Mesoproterozoic rocks even preserve macrofossils unlikely to be produced by Bacteria or Archaea (Walter *et al.*, 1990). Collectively, these record modest eukaryotic diversity in the oceans beginning no later than about 1800 Ma; most were probably aerobes, but whether they were photosynthetic or heterotrophic remains unknown.

None of the earliest known eukaryotic body fossils can be related readily to extant phyla or kingdoms. Teyssédre (2006) speculated that large spheroidal microfossils in these ancient rocks (Fig. 16.2c) are the preserved walls of resting cells, or phycomata, made by prasinophyte green algae. To date, however, no Mesoproterozoic or older microfossils have been shown to preserve the distinctive wall ultrastructure common to extant and unambiguous fossil prasinophytes (Arouri *et al.*, 2000), and some candidate phycomata actually preserve ultrastructural features unknown in extant green algae (Javaux *et al.*, 2004). The fossils could record

stem eukaryotes or early, morphologically undiagnostic members of crown group Eukarya (Knoll *et al.*, 2006).

The oldest eukaryotic fossils that can be assigned to an extant phylum with confidence are simple multicellular microfossils, exceptionally preserved in growth position by rapid burial in (now silicified) peritidal carbonate mud (Fig. 16.2a; Butterfield, 2000). Found in the Hunting Formation, Somerset Island, Arctic Canada, these fossils occur abundantly in a succession thought to be ca. 1200 Ma. The fossils are filamentous, with a cellularly differentiated holdfast and distinctive wedge-shaped reproductive cells. In multicellular organization, inferred developmental pattern, life cycle, and taphonomic detail, these fossils closely resemble modern *Bangia* and related red algae (Butterfield, 2000). By inference, then, the phylogenetic divergence of the red and green algae must have taken place by the time that these fossils formed (Knoll, 2003), but fossils that can confidently be identified as green algae are well preserved only in rocks ca. 800 Ma and younger (Butterfield *et al.*, 1994).

16.4   Inferences from microfossil chemistry and ultrastructure (new technologies)

As noted above, optical microscopy has historically been the technique of choice for research in Precambrian micropaleontology. The strengths of optical microscopy are many, including micron scale resolution, comparability to a large catalog of images available for living microorganisms, and, in thin sections at least, the ability to place fossils in petrological context, resolving the orientation and three-dimensional distributions of preserved populations. A limitation is the inability to resolve features much smaller than a micron, including both fine-scale details of larger fossils and the entire structures of smaller prokaryotes. Another limitation is chemical. Organic matter is readily recognized by optical microscopy, and the colour of carbonaceous microfossils provides insights into burial history. Nonetheless, optical images do not permit quantitative analysis of the elemental, isotopic, or molecular composition of fossils, data that would help to resolve the systematic and physiological uncertainties discussed in the previous section.

Over the past decade, catalysed by instrument development for astrobiological exploration, new analytical techniques have been brought to bear on the recognition and interpretation of ancient microfossils. At heart are three questions:

1   Can novel analyses be used to determine the biogenicity of carbonaceous microstructures in ancient rocks?
2   Can novel analyses help us to understand better the processes that preserve microorganisms and the degree to which biological information is lost or retained?

Table 16.1 Some new approaches to the analysis of Precambrian microfossils

Technique	Application	Reference
Confocal laser microscopy	3-D imagery	Schopf *et al.* (2006, 2010b), Lepot *et al.* (2008)
Atomic force microscopy	Submicron structural resolution	Kempe *et al.* 2002
Focused ion beam SEM	Directed sectioning for fine scale imaging	Kempe *et al.* (2005), Lepot *et al.* (2008)
Transmission electron microscopy	Fine structure for taphonomy, systematics	Schopf (1970), Oehler (1977), Talyzina (2000), Javaux *et al.* (2004), Willman and Moczydłowska (2007)
Laser Raman spectroscopy/imagery	Fine-scale mineralogy; submicron mapping and crystallinity of reduced C	Arouri *et al.* (1999), Schopf *et al.* (2002, 2005, 2010a,b) Marshall *et al.* (2005) van Zuilen *et al.* (2007), Cohen *et al.* (2011)
SIMS/nanoSIMS ion microprobe and electron microprobe	Elemental and isotopic composition and mapping	House *et al.* (2000), Ueno *et al.* (2001), Kaufman and Xiao (2003), Boyce *et al.* (2001), Oehler *et al.*(2006), van Zuilen *et al.* (2007), Wacey *et al.* (2008)
X-ray microscopy (SXM)/X-ray absorption near edge spectroscopy (XANES)	Imaging, elemental and molecular mapping; microtomography	Boyce *et al.* (2002), Foriel *et al.* (2004), Hagadorn *et al.* (2006), Lemelle *et al.* (2008), Lepot *et al.* (2008)
Laser pyrolysis GC-MS	Molecular composition	Arouri *et al.* (1999, 2000)
FTIR spectroscopy	Molecular composition	Marshall *et al.* (2005), Igisu *et al.* (2006)

3 Can novel techniques aid in the phylogenetic and/or physiological interpretation of microfossils?

Table 16.1 lists a number of novel techniques that have been used to investigate ancient microfossils. Several of them increase the spatial resolution of images, offering improved understanding of preservational processes and evolutionary relationships. Confocal laser microscopy, atomic force microscopy, laser Raman imaging and synchrotron X-ray tomography all allow rapid and non-destructive three-dimensional imaging of fossils at high spatial resolution.

Transmission electron microscopy was applied to Precambrian microfossils early in the development of the field, but pioneering research on silicified microfossils largely served to underscore the phylogenetic uncertainties and complex preservational histories of these remains (Schopf, 1970; Oehler, 1977). More recent focus on organic-walled microfossils compressed in shale shows that biologically informative details of wall ultrastructure can be preserved in Proterozoic fossils (e.g. Talyzina, 2000; Arouri *et al.*, 2000; Javaux *et al.*, 2004; Willman and Moczydłowska, 2007), contributing to phylogenetic and, therefore, physiological interpretation.

Microanalytical techniques that map elements on a submicron scale can also be used in high resolution imaging (e.g. Schopf *et al.*, 2002, 2005; Lemelle *et al.*, 2008), enabling researchers to create three-dimensional images that can be rotated around any axis, facilitating morphological analysis (e.g. Schopf *et al.*, 2010a). The greatest potential of microanalysis, however, may lie in elucidating the chemical composition of preserved

fossils. For example, Cohen *et al.* (2011) used laser Raman imaging and spectroscopy to demonstrate the phosphatic composition of protistan scales preserved in mid-Neoproterozoic cherts from northwestern Canada. This discovery, in turn, prompted Cohen *et al.* to dissolve carbonates adjacent to the chert nodules, freeing well-preserved scale fossils by the thousands that could then be imaged using SEM.

Organic geochemical research on sedimentary organic matter has yielded important insights into the biological diversity and metabolic pathways in ancient microbial ecosystems (see Chapter 3), and the ability to constrain the molecular composition of individual microfossils could help greatly to resolve the phylogenetic relationships of fossils that lack diagnostic morphology. Such studies are in their infancy (e.g. Arouri *et al.*, 1999, 2000; Marshall *et al.*, 2005) but will grow in importance as the precision of analytical techniques improves. Equally, our ability to interpret molecular signatures will improve as we gain a better understanding of the phylogenetic distribution of cell wall and envelope constituents and the degree to which molecular signatures are altered during decay and diagenesis.

An ideal research program addresses four questions, in sequence:

1 How does the chemical composition of morphologically preservable structures vary among living microorganisms? Molecular signatures illuminate phylogenetic relationships only to the degree that they are unique to specific clades.

2 To what extent do potentially illuminating signatures become incorporated into sediments?

3 Are chemical signatures preserved over geologic time scales?
4 Can the identification of preserved signatures elucidate the biology of previously problematic fossils in Proterozoic or Paleozoic rocks?

As an example, consider a group of eukaryotic microorganisms mentioned earlier. Some prasinophyte green algae produce a large resting cell called the phycoma that can become fossilized. Phycoma walls have a distinctive ultrastructure, and microanalysis might reveal, as well, a distinctive chemical signature (Table 16.1). If comparative studies convince us that microchemical signatures reliably distinguish phycomata from preservable wall structures made by other living organisms, we can conduct experiments to establish whether post-mortem alteration preserves or obfuscates these signatures. If a signature is preserved in the short term, we can analyse, say, Jurassic phycomata identified on the basis of wall morphology and ultrastructure to evaluate chemical preservation over geologic time scales. The ultimate pay-off comes if and when we can analyse the ultrastructure and microchemistry of previously problematic microfossils in Proterozoic rocks and recognize a preserved chemical signature of prasinophyte algae.

16.4.1 Evaluating the biogenicity of Early Archean microstructures

In the opening decade of the twenty-first century, the biogencity of purported 3500 Ma microfossils has been a hot-button issue. Debate has focused, in particular, on carbonaceous microstructures in cherts from the Warrawoona Group, Australia, originally reported by Schopf and Packer (1987) and Schopf (1993). Brasier *et al.* (2002) reinterpreted these structures as abiological structures formed diagenetically in cherts that formed in hydrothermal plumbing systems. Debate continues (Schopf *et al.*, 2002; Brasier *et al.*, 2005; Schopf, 2006), shaped in part by microanalyses of the disputed structures.

Laser Raman imaging and spectroscopy show that the Warrawoona microstructures are carbonaceous and that the carbon is present largely as altered organic materials, not graphite (Schopf *et al.*, 2002). This demonstrates that the structures are not graphite inclusions formed at high temperature, but does not by itself indicate a biological origin (Pasteris and Wopena, 2003). Ueno *et al.* (2001) have also measured the C-isotopic composition of carbonaceous microstructures in Warrawoona rocks. Their reported $\delta^{13}C$ values of −42 to −32‰ are also consistent with a biological origin, but once again, do not demonstrate that the microstructures are biogenic.

To understand why these analyses fall short of their goal, we must make a sharp distinction between the bio-

genicity of organic matter in debated microstructures and the origin of the structures themselves. A single observation illustrates the problem. Figure 16.1f shows the ghost of a diagenetic carbonate crystal preserved in Warrawoona chert. Organic matter coats the rhomb, either plowed outward by the growing crystal or adsorbed onto its surface (e.g. van Zuilen *et al.*, 2007). Very likely, both carbon-isotopic and microRaman analysis of this organic matter would be consistent with the analyses reported above, but no one would argue that this rhombohedral microstructure is a fossil.

The Warrawoona rhomb highlights the key issue. Microchemical analysis might convince us that the organic matter imaged in Warrawoona microstructures originated biologically, but it cannot demonstrate that the organic material originated in the wall or envelope of a microorganism. In the end, biogenicity must be demonstrated or refuted by the time-honored methods of optical microscopy, aided by newer forms of imaging. Morphology and petrographic context remain critical components of interpretation.

16.4.2 Insights into preservational history

Among other things, novel imaging techniques demonstrate, in many microfossils, carbon is only patchily preserved on a submicron scale (e.g. Schopf, 1970; Oehler, 1977; Kempe *et al.*, 2002; Schopf *et al.*, 2006). Other taphonomic insights come from elemental mapping by ion and electron microprobe. For example, element maps of organic walled fossils show a close spatial correspondence of S and C in fossil walls (Boyce *et al.*, 2001; Lemelle *et al.*, 2008), supporting the hypothesis that organically bound sulfur promotes the preservation of organic matter on geologic timescales (Sinninghe Damsté and de Leeuw, 1990). Continuing progress will link the tools of microanalysis to experimental research (e.g. Bartley, 1996), enabling us to relate better than we can at present the biological and physical processes that preserve microbial remains with microtextural and microchemical patterns observable in fossils. In the future, it should be possible to quantify the preservational state of fossil materials.

16.4.3 Insights into microfossil physiology and phylogeny

As noted above, TEM images are increasingly being brought to bear in phylogenetic interpretations of organic walled microfossils, especially those preserved in shale. Studies completed to date indicate that wall ultrastructure can be preserved in compressed Proterozoic microfossils, permitting tests of phylogenetic hypotheses made on the basis of overall morphology. TEM images confirm a prasinophyte algal origin for

Figure 16.3 Stromatolites, thrombolites, precipitated structures and microbially induced sedimentary structures in Proterozoic and Paleozoic rocks. (a) Columnar stromatolites in a patch reef, 750–800 Ma Draken Formation, Spitsbergen. (b) Microbially induced wrinkle structures coating ripple marks, Ediacaran (549–544 Ma) Nama Group, Namibia; circular structure reflects attachment of an Ediacaran microorganism. (c) Columnar precipitate structures in Paleoproterozoic (ca. 1900 Ma) Rocknest Formation, Canada; bar = 3 cm (courtesy of J. Grotzinger). (d) Domal thrombolite, Lower Ordovician Boat Harbour Formation, Newfoundland, Canada. (e) Detail of (d) showing mesoclots that make up the fabric of thrombolite (each unit in scale = 1 mm).

some the ubiquitous spheroidal microfossils known as leiosphaerids (e.g. Talzina and Moczydłowska, 2000; Arouri *et al.*, 2000), but rule it out for other, older Proterozoic populations (Javaux *et al.*, 2004). TEM has also proven useful in supporting the hypothesis (Yin *et al.*, 2007) that at least some of the large, process-bearing microfossils preserved globally in mid-Ediacaran rocks (Fig. 16.2d; Grey, 2005) record resting stages in the life cycles of early animals (Cohen *et al.*, 2009). Further, synchrotron X-ray tomography has supplied phylogenetically informative, three-dimensional reconstructions of Cambrian (Donoghue *et al.*, 2006) and Ediacaran (Hagadorn *et al.*, 2006) embryos.

In parallel with TEM investigations, Marshall *et al.* (2005) used micro-Fourier transform infrared (FTIR) spectroscopy to characterize the chemical structure of Mesoproterozoic microfossils from the Roper Group, Australia. FTIR spectroscopy produces spectra that record the abundance and variety of chemical bonds in fossil kerogen. To date, such spectra have been used to demonstrate chemical differences between populations in the same sample and to draw consistency arguments about biological origin; their broader promise will be fulfilled when fossil spectra can be evaluated using the four step process outlined above.

Carbon isotopes in ancient organic matter have also provided a rich source of geobiological data (see Chapter 14), and improvements in ion microprobes now permit physiologically informative C-isotopic composition to be ascertained for individual Precambrian microfossils, an impressive analytical achievement (House *et al.*, 2000; Ueno *et al.*, 2001; Kaufman and Xiao, 2003). As different biochemical mechanisms for carbon fixation fractionate C-isotopes to varying degrees, isotopic measurements have the potential (if diagenetic effects are understood) to constrain the metabolic capacities of ancient organisms. The few analyses made to date of unambiguous microfossils in Precambrian rocks (House *et al.*, 2000; Kaufman and Xiao, 2003) are consistent with carbon fixation using the enzyme Rubsico. The fossils could, thus, be photosynthetic microorganisms that used the Rubisco pathway (algae, cyanobacteria, photosynthetic proteobacteria), chemosynthetic prokaryotes that fixed carbon using Rubisco, or heterotrophs that consumed organic matter produced by Rubisco-utilizing autotrophs. The demonstration of consistent isotopic differences among different populations in the same assemblage will permit more extensive inferences about metabolism.

In a different physiological application, Kaufman and Xiao (2003) used ion microprobe analysis of individual microfossils to infer carbon dioxide levels in the Mesoproterozoic atmosphere. Careful $\delta^{13}C$ measurements were plugged into a biogeochemical algorithm based on living algae to estimate that 1400 Ma air contained >10 to 200 times present day levels of CO_2, the order or magnitude range in their estimate reflecting uncertainties about diagenetic history and paleotemperature. Additional uncertainties arise from two further assumptions: (1) that the fossils are the vegetative walls of algae with a cytoplasmic volume equivalent to the internal volume of the fossil; and (2) that preserved walls have a C-isotopic ratio equal to that of the total organic matter in the original cells. Continuing phylogenetic and taphonomic studies will provide tests of these assumptions; whatever their results, studies like that of Kaufman and Xiao (2003) show that novel imaging and analytical techniques will play key roles in future micropaleontological research.

16.5 Inferences from microbialites

The reefs that fringe tropical shorelines in the modern world accrete through the metabolic activities of skeleton-forming corals and algae. Neither has a pre-Ediacaran fossil record, yet reefs have built upward from the seafloor for billions of years. The biological architects of Precambrian reefs were microorganisms that trapped, bound, and precipitated sediments. Thus, in addition to microfossils and chemical records, microorganisms in Precambrian oceans have left a third, altogether more apparent signature in macroscopic structures preserved in sedimentary rocks. Where precipitating minerals permitted laminated, three dimensional structures to accrete, **stromatolites** formed (Walter, 1976; Grotzinger and Knoll, 1999; Dupraz *et al.*, 2009). But even where accretion was not favoured by local chemistry, distinctive features called **microbially induced sedimentary structures,** or MISS, imparted a microbial signature to sand beds (Noffke 2010; Schieber *et al.*, 2004).

16.5.1 Stromatolites

Figure 16.3a shows a candelabrum-like stromatolite, part of a Neoproterozoic patch reef preserved in Spitsbergen. The unit structures are convex-upward, mm-scale carbonate (now dolomite) layers, or laminae. One upon another, these laminae accreted through time to form columns, separated from one another by migrating sediment grains and occasionally branching to introduce increased morphological complexity. Petrological observation shows that the laminae consist of more or less featureless dolomite crystals – no microfossils are preserved, although rare sand grains demonstrate that sediment was trapped and bound into accumulating structures. Despite the absence of microfossils, however, sedimentary geologists agree that these stromatolites record the interplay between mat-forming microbial communities and carbonate sediments.

This conclusion provides a classic example of inference in historical geobiology – patterns in ancient sedimentary rocks are interpreted in terms of the (unobserved) processes that formed them. In this case, the inference is based on studies of modern environments where stromatolites still form, places like the Bahama Banks (e.g. Andres and Reid, 2006) and Shark Bay, Australia (Reid *et al.*, 2003, and references therein). In such environments, geobiologists can observe stromatolite accretion in progress, providing the critical link between pattern and process that guides interpretation of ancient examples.

Classically, stromatolite accretion occurs through the trapping and binding of (mostly) carbonate sediments by mat-forming microbial communities. Fine-grained sediments form a thin layer on top of mats, through which mat populations grow upward to reestablish a coherent new mat at the sediment–water interface, thereby binding sediment particles into a lamina. Modern mat builders are mostly cyanobacterial and are overwhelmingly photosynthetic, but there is no guarantee that this was the case in early oceans. Microbial activities, mostly heterotrophic metabolism within the mat, also facilitate the precipitation of carbonate cements from pore waters. Lamina shape and thickness depend on the nature of the local substrate, the growth potential of mat building microorganisms, the capacity of mat heterotrophs to drive cementation, and the physical environment (Seong-Joo *et al.*, 2000). The diverse morphologies of stromatolites in ancient carbonate rocks reflect the varying interactions between physical and biological processes. Stromatolites have sometimes been classified using quasi-Linnean binomial names. While these provide a useful shorthand for discussion, they can provide a misleading impression that stromatolites were ancient organisms. They were not: stromatolites record the influence of microbial communities on accumulating sediments and so are more similar in concept to trace fossils (tracks, trails, burrows) than to clam shells or trilobite skeletons. A number of models have been put forward to explain how specific stromatolite morphologies accrete (e.g. Grotzinger and Rothman, 1996; Batchelor *et al.*, 2005); these make predictions that can be tested via sedimentological and petrologic investigations of ancient examples.

Because physical processes do not easily mimic the macroscopic and microtextural features of stromatolites built by trapping and binding, such structures provide reliable indicators of ancient life. As was the case for cyanobacteria-like microfossils, stromatolites that formed in well-oxygenated environments along Proterozoic shorelines very likely record mat communities in which cyanobacteria played a major role. In modern mats, anoxygenic photosynthetic bacteria commonly contribute to primary production at depths within the mat that receive sunlight but no oxygen, and this was probably the case in Proterozoic mats, as well. The cementation of trapped and bound sediments, associated today with the metabolic activities of heterotrophs, especially sulfate-reducing bacteria beneath the photic layer, probably also drove stromatolite cementation and, hence, accretion on Proterozoic seafloors. Physiological inference is less certain for trapped and bound Archean stromatolites that formed before oxygen began to accumulate in the atmosphere and surface ocean.

If all stromatolites formed by trapping and binding, the interpretation of ancient structures as biogenic would be straightforward. It is not, primarily because some stromatolites accrete by precipitation alone, and these can be difficult to differentiate from precipitated structures that accreted without the influence of microbial mats (Fig. 16.3c). Early Archean stromatolites are predominantly precipitated structures (Allwood *et al.*, 2006, 2009), compounding the challenge of detecting biological signatures in the oldest known sedimentary successions. In contrast, most Neoproterozoic stromatolites accreted by trapping and binding.

16.5.2 A word about thrombolites

In many Cambrian and Ordovician carbonate successions, distinctive microbialites occur in subtidal facies that lacked sandy or muddy sediments. Called **thrombolites**, these structures can be domal or columnar, branched or unbranched, but they lack the laminated fabric that defines stromatolites (Kennard and James, 1986). In contrast, thrombolites built upward by the accretion of cm-scale patches (called mesoclots) of precipitated carbonate minerals (Fig. 16.3d, e). Modern microbialites provide clues to the formation of their ancient counterparts. Where continual inundation by sand or unusual chemistry inhibit the growth of seaweeds, cyanobacteria thrive and the laminated fabrics of stromatolites develop. Where seaweeds can grow, however, algae cover microbialite surfaces, resulting in clotted, poorly laminated fabrics like those in ancient thrombolites (Feldmann and McKenzie, 1998; Andres and Reid, 2006). The distributions of thrombolites in Cambrian and Ordovician carbonate successions are consistent with this recent pattern, suggesting similar controls on Paleozoic microbialite development. Thus, the appearance near the Proterozoic–Cambrian boundary of microbialites with thrombolitic fabrics may record the spread of small seaweeds across carbonate platform environments (Grotzinger *et al.*, 2000).

16.5.3 Microbially induced sedimentary structures

The sedimentary signatures of microbial communities in siliciclastic rocks are more subtle than those in carbonate successions. We know, however, from studies of coastal sand flats along the present day North Sea, Tunisia, and elsewhere that microorganisms can leave a discernible record in the surface features and internal textures of sand beds (Gerdes *et al.*, 2000; Noffke *et al.*, 2003). At the microscopic scale, coarse sand grains 'floating' in finer-grained layers document microbial trapping and binding. At a scale visible to the naked eye, mechanical interactions between microbial communities and physical transport processes can result in biologically mediated patterns of sedimentation and erosion. For example, mechanical erosion initiated at point sources within microbially stabilized surfaces leads to ragged-edged erosional pockets in sand beds. Multidirectional ripple marks on the surface of a sand bed also attest to locally patchy patterns of erosion and deposition where mats bind accumulating sands. Moreover, desiccating mats that cover sand beds can impart a distinctive 'wrinkle' structure to surfaces (Fig. 16.3b). Even shales can reveal a microbial influence, largely through petrographic details and erosional features related to the cohesion imparted by mats (see Schieber *et al.*, 2004, and Noffke, 2010, for comprehensive reviews of microbial influences on siliciclastic rocks).

MISS are, unsurprisingly, most apparent in times and places where seaweeds and animals had little influence on sediment surfaces. Microbially induced structures occur abundantly in sandstones deposited near the Proterozoic–Cambrian boundary (e.g. Hagadorn and Bottjer, 1997), and they occur throughout the Precambrian record, as well. They do not, however, occur ubiquitously in older rocks. There are rules of occurrence, worked out by Noffke *et al.* (2002; see also Porada *et al.*, 2008) in a study of MISS in Ediacaran sandstones from Namibia. In their words: 'The facies distribution of observed structures reflects the superposition of a taphonomic window of mat preservation on the ecological window of mat development. Mat colonization is favoured by clean, fine-grained, translucent quartz sands deposited at sites where hydrodynamic flow is sufficient to sweep mud from mat surfaces but insufficient to erode biostabilized sand layers. Mat preservation is facilitated by subsequent sedimentary events that bury the microbial structures without causing erosional destruction.'

Noffke *et al.* (2008) have shown that distinctive microbial indicators are preserved in Archean siliciclastic rocks, describing a rich inventory of MISS in 2900 Ma sandstones from the Pongola Supergroup, South Africa.

16.6 A brief history, with questions

16.6.1 *Early Archean life*

Stable isotopes provide what may be the most compelling evidence that life existed 3500 Ma or earlier.

Abundant organic matter in Lower Archean shales consistently yields $\delta^{13}C$ values of −25 to −35‰, pretty much what would be expected for carbon cycles driven by Rubisco-based autotrophy (see Schopf, 2006, for review). Total sulfur isotopic measurements of early diagenetic pyrite in Lower Archean rocks similarly indicate a microbial sulfur cycle that included dissimilatory sulfate or sulfur reduction (Philippot *et al.*, 2007; Johnston, 2011). Paleoenvironmental inferences suggest that early biogeochemical cycles operated in the absence of free oxygen and in the presence of ferrous iron, constraining the range of energy metabolisms available in early marine ecosystems.

Other lines of evidence corroborate the presence of life in early Archean oceans, but place few additional constraints on its properties. Microbially influenced sedimentary structures called roll-ups occur in nearly 3500 Ma cherts from South Africa (Tice and Lowe, 2004), and coeval carbonates from Australia preserve stromatolites, at least some of which have textural features suggestive of biological influence (Fig. 16.4b; Allwood *et al.*, 2009). Biologically informative molecular fossils have not been reported from Lower Archean rocks, but the careful application of multiple spectroscopic techniques identified a series of alkanes and polycyclic aromatic hydrocarbons covalently bound into kerogen in 3500 Ma cherts from the Warrawoona Group, Australia (Marshall *et al.*, 2007). The molecular signatures are at least consistent with thermal alteration of biological materials.

Microfossils have been reported from early Archean cherts for decades (see Schopf, 2006, for review), but, as noted above, their interpretations remain controversial and, to date, these structures have provided little insight into phylogeny or physiology beyond that gleaned from other sources. Recently, more compelling records of 3200–3000 Ma microfossils have been reported (e.g. Schopf *et al.*, 2010b), of which the remarkable preservation of compressed microfossils in 3200 Ma shales from South Africa is notable (Javaux *et al.*, 2010). The physiological and phylogenetic interpretation of these fossils may be uncertain, but it is hard to dispute the claim that these are bona fide microfossils preserved in rocks more than 3 billion years old. Tiny trace fossils of endolithic bacteria have been also reported from Archean volcanic rocks (Furnes *et al.*, 2007); what types of microbes constructed these trails and what, if any, nutrients they extracted from their substrates remain uncertain.

In short, we know that life existed at least 3500 Ma, but have only limited insights into the nature of early Archean ecosystems. The key question, then, is clear: can continuing discovery and application of microanalytical tools resolve the biogenicity and, further, elucidate the physiological properties or phylogenetic relationships of Early Archean microfossils? Time will tell.

(a) (b)

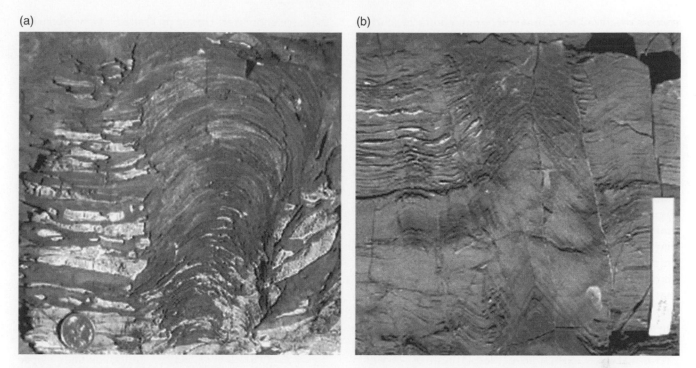

Figure 16.4 (a) Columnar stromatolite that accreted by trapping and binding, Late Archean, ca. 2720 Ma, Fortescue Group, Australia. (b) Conoidal stromatolite, ca. 3430 Ma Strelley Pool Chert, Warrawoona Group, Australia (scale is 15 cm long).

16.6.2 Late Archean life

As we move forward in time, both the quantity and quality of sedimentary rocks improve; late Archean sedimentary rocks deposited 2800–2500 Ma unmistakably contain robust biological signatures. As noted above, microfossils of unambiguous biological origin but uncertain phylogenetic relationships occur in peritidal and deeper water cherts of ca. 2600–2500 Ma rocks from South Africa (Klein *et al.*, 1987; Lanier, 1988; Altermann and Schopf, 1995). There are a few other reports of late Archean microfossils, but their interpretation remains uncertain (see Knoll, 1996, for discussion of individual reports).

Microbially accreted stromatolites and MISS both occur in the 2900–3000 Ma Pongola Supergroup, South Africa (Beukes and Lowe, 1989; Noffke *et al.*, 2008). More generally, stromatolites are common in late Archean carbonate successions, and these include trapped and bound domes and columns of unquestioned microbial mat origin (Fig. 16.4a; see Schopf, 2006, for a catalog of occurrences). Carbon isotopes indicate a microbial carbon cycle driven by organisms that fixed carbon via Rubisco, but also included both methanogenic archaea and methanotrophic bacteria (Hayes, 1994; Hayes and Waldbauer, 2006). S-isotopes similarly record a microbial sulfur cycle that included bacterial sulfate reduction (Canfield, 2004).

The chief question for late Archean paleobiology is whether primary producers included cyanobacteria.

All published records of stromatolites, stable isotopes and molecular biomarkers are consistent with the presence of cyanobacteria in the late Archean photic zone. None, however, requires such an interpretation, and geochemical indicators of redox conditions tell us that oxygen was, at best, a transient and trace constituent of the atmosphere and surface oceans (Kopp *et al.*, 2005). Total sulfur isotopes and Mo abundances in latest Archean shales suggest 'a whiff of oxygen' 2500–2600 Ma (Anbar *et al.*, 2007; Kaufman *et al.*, 2007; Wille *et al.*, 2007), and such paleoenvironmental probes may provide our best means of establishing the antiquity of cyanobacteria. At the moment, however, the question of when oxygenic photosynthesis – the metabolism that reshaped the world – originated remains unanswered.

16.6.3 Life in Proterozoic oceans

By 2450–2320 Ma, when oxygen first began to accumulate permanently in the atmosphere (Bekker *et al.*, 2004), cyanobacteria were clearly a physiological presence in the oceans. Both microfossils and biomarker molecules, however, indicate that other bacterial autotrophs continued to play important roles in marine ecosystems. As noted above, the current best reading of the microfossil assemblages found in the Gunflint and other late Paleoproterozoic (ca. 1900–1800 Ma) iron formations is that these communities thrived in subsurface water masses where energy could still be gained via the oxidation of iron. A limited amount of molecular biomarker evidence further suggests that even when iron was swept clean from subsurface oceans, primary producers other than cyanobacteria remained important. Notably, organic geochemical results for ca. 1640 Ma basinal black shales from northern Australia indicate the presence of purple and green photosynthetic bacteria in a sulfidic oxygen minimum zone beneath a shallow redoxcline (Brocks *et al.*, 2005).

Microfossils indicate that eukaryotes participated in Paleoproterozoic ecosystems, but molecular biomarkers suggest that algae played at best a limited role in primary production. What role did these early eukaryotes play in contemporaneous ecosystems? Here, new techniques may provide decisive insights. The question is: can microchemical and ultrastructural research on the eukaryotic microfossils in Paleoproterozoic and Mesoproterozoic rocks sharpen constraints on their phylogenetic and physiological interpretation (e.g. Marshall *et al.*, 2005)?

If eukaryotes existed in earlier Proterozoic oceans, why did they not dominate primary production on continental shelves and platforms? And why did the diversity of preservable protists remain low for so long? Physiology provides the rudiments of answers. Mounting geochemical evidence supports the hypothesis (Canfield, 1998) that beginning ca. 1800–1900 Ma, the oxygen minimum zone beneath oxygenated surface waters of the oceans exhibited a pronounced statistical tendency to become anoxic, and at times sulfidic. This must have had direct physiological consequences for eukaryotes insofar as sulfide that mixed into surface waters would be toxic to most eukaryotes (Martin *et al.*, 2003). In contrast, many cyanobacteria and other photosynthetic bacteria could not only tolerate sulfide but use it as an electron source for photosynthesis (see Johnston *et al.*, 2009).

Nitrogen availability would also favour prokaryotic organisms, at least among primary producers. It has been proposed that fixed nitrogen would have been at a premium in the photic zone of mid-Proterozoic oceans (Anbar and Knoll, 2002; Fennel *et al.*, 2004); nitrate would have been scarce, and a good proportion of the ammonia returned to the surface from deep waters would be consumed by nitrification and annamox metabolism. Moreover, ammonia that did return to the base of the photic zone would commonly have been sequestered by anoxygenic photosynthetic bacteria (Johnston *et al.*, 2009). Under such conditions, broad dominance of bacteria in primary production and the generally slow rates of eukaryotic diversification make good physiological and ecological sense. There remains a great need for research of all kinds on the historical geobiology of Proterozoic oceans – much of what we think we know is still based on limited observations. For microfossils, the question is the same as that posed earlier in this section: can emerging techniques provide new insights into the hypothesized dominance of prokaryotes in primary production and into the timing and environmental context of early divergence among eukaryotic organisms?

16.6.4   *The dawn of the modern*

Microfossil assemblages show signs of change beginning about 800 Ma (Knoll *et al.*, 2006; Cohen *et al.*, 2011). Vase-shaped testate amoebae, introduced earlier as records of heterotrophic protists, appear globally in shallow marine successions (Porter and Knoll, 2000; Porter *et al.*, 2003); other protists began to leave a record of mineralized scales (Allison and Hilgert, 1986; Cohen *et al.*, 2011); and unusual protists with complex patterns of branching and fusion appeared (Butterfield, 2004, 2005). Biomarkers in bitumens show a terminal Proterozoic rise of green algae to prominence in primary production (Knoll *et al.*, 2007b), and following collapse of the global Marinoan ice sheets at 635 Ma, fossils of developmentally complex eukaryotes, both animals (Narbonne, 2005) and algae (Xiao *et al.*, 2002, 2004) become widespread. Not surprisingly, these evolutionary events appear to accompany the oceans' transition toward a more modern redox state (Fike *et al.*, 2006; Canfield *et al.*, 2007, 2008; McFadden *et al.*, 2008; Scott *et al.*, 2008; Shen *et al.*, 2008; Dahl *et al.*, 2010).

The physiological bridge between late Proterozoic evolutionary and environmental changes is in many ways the converse of that argued for mid-Proterozoic oceans and life. The demise of persistently anoxic (and sulfidic) oxygen minimum zones would remove a barrier to eukaryotic diversification, whereas increasing oxygen tensions would remove another, specifically for organisms with thick tissues and high demands of exercise metabolism. Increasing nitrate availability would both favour primary production by algae and provide a positive feedback for environmental transition (Johnston *et al.*, 2009).

By the end of the Proterozoic Eon, the modern biological world was taking shape, catalysed by and perhaps fomenting environmental transformation. Nonetheless, to paraphrase Shakespeare, the course of environmental evolution never did run smooth. We live in a world in which deep oceans are persistently well oxygenated, and this has been the case for most of the last 65 million years. During parts of the Paleozoic and Mesozoic eras, however, anoxia transiently re-established itself in deep ocean waters. During the most profound of these events – at the Proterozoic–Cambrian (Wille *et al.*, 2008) and Permian–Triassic (Knoll *et al.*, 2007a) boundaries – ancient inhibitors of eukaryotic biology returned, eliminating much of standing diversity and reorienting the vector of evolution. Indeed, the relationship between Paleozoic oceanic redox conditions and evolution provides a fertile but under-investigated arena for research in historical geobiology. Microfossils have much to tell us in geobiological studies of Proterozoic and Phanerozoic oceans, and, once again, TEM and microchemistry will let us know to what extent organic-walled microfossils complement or duplicate the skeletal and trace fossil records of animal evolution. Can the same tools with the potential to elucidate phylogenetic and physiological relationships of mid-Proterozoic fossils be used to chart their later Proterozoic and Paleozoic diversification and rise to ecological prominence?

16.7 Conclusions

Fossils of skeletons, wood and leaves provide some of the best evidence for biological evolution. Microbial fossils and trace fossils extend this record deep into our planet's past, demonstrating that Earth has been a biological planet since its infancy. Physiology provides the conceptual link needed to interpret the fossil record in the context of Earth's dynamic environmental history. Physiological inference can be especially challenging for the simple microfossils abundant in Precambrian sedimentary rocks, but success will allow us to understand how the microbial diversity and biogeochemical cycles that underpin present day ecosystems came to be.

Acknowledgements

I thank Abby Allwood, Nick Butterfield, Emmanuelle Javaux, Vladimir Sergeev, Nataliya Vorob'eva, John Grotzinger, and Susannah Porter for some of the images used in this chapter and two anonymous reviewers for helpful criticisms of the manuscript. Research leading to this paper was supported in part by the NASA Astrobiology Institute and NSF Grant EAR-0420592.

References

Allison CW, Hilgert JW (1986) Scale microfossils from the early Cambrian of northwest Canada. *Journal of Paleontology* **60**, 973–1015.

Allwood AC, Walter MR, Kamber BS, Marshall CP, Burch IW (2006) Stromatolite reef from the Early Archaean era of Australia. *Nature* **441**, 714–718.

Allwood AC, Grotzinger JP, Knoll AH, *et al.* (2009) Controls on development and diversity of Early Archean stromatolites. *Proceedings of the National Academy of Sciences, USA* **106**, 9548–9555.

Altermann W, Schopf JW (1995) Microfossils from the Neoarchean Campbell Group, Griqualand West Sequence of the Transvaal Supergroup, and their paleoenvironmental and evolutionary implications. *Precambrian Research* **75**, 65–90.

Anbar A, Knoll AH (2002) Proterozoic ocean chemistry and evolution: a bioinorganic bridge? *Science* **297**, 1137–1142.

Anbar AD, Duan Y, Lyons TW, *et al.* (2007) A whiff of oxygen before the Great Oxidation Event? *Science* **317**, 1903–1906.

Andres MS, Reid RP (2006) Growth morphologies of modem marine stromatolites: A case study from Highborne Cay, Bahamas. *Sedimentary Geology* **185**, 319–328.

Arouri K, Greenwood PF, Walter MR (1999) A possible chlorophycean affinity of some Neoproterozoic acritarchs. *Organic Geochemistry* **30**, 1323–1337.

Arouri K, Greenwood PF, Walter MR (2000) Biological affinities of Neoproterozoic acritarchs from Australia: microscopic and chemical characterization. *Organic Geochemistry* **31**, 75–89.

Batchelor MT, Burne RV, Henry BI, Slatyer T (2005) Statistical physics and stromatolite growth: new perspectives on an ancient dilemma. *Physica A-Statistical Mechanics and its Applications* **350**, 6–11.

Barghoorn ES, Tyler SA (1965) Microorganisms from the Gunflint Chert. *Science* **147**, 563–577.

Bartley JK (1996) Actualistic taphonomy of Cyanobacteria: implications for the Precambrian fossil record. *Palaios* **11**, 571–586.

Bekker A, Holland HD, Wang PL, Rumble D, Stein HJ, Hannah JL, Coetzee LL, Beukes NJ (2004) Dating the rise of atmospheric oxygen. *Nature* **427**, 117–120.

Beukes NJ, Lowe DR (1989) Environmental control on diverse stromatolite morphologies in the 3000 Myr Pongola Supergroup, South Africa. *Sedimentology* **36**, 383–397.

Boyce CK, Hazen RM, Knoll AH (2001) Nondestructive, in situ cellular-scale mapping of elemental abundances including organic carbon in permineralized fossils. *Proceedings of the National Academy of Sciences, USA* **98**, 5970–5974.

Brasier MD, Green OR, Jephcoat AP, *et al.* (2002) Questioning the evidence for Earth's oldest fossils. *Nature* **416**, 76–91.

Brasier MD, Green OR, Lindsay JF, McLoughlin N, Steele A, Stoakes C (2005) Critical testing of Earth's oldest putative fossil assemblage from the ~3.5 Ga Apex chert, Chinaman Creek, Western Australia. *Precambrian Research* **140**, 55–102.

Brocks JJ, Love GD, Summons RE, Knoll AH, Logan GA, Bowden S (2005) Biomarker evidence for green and purple sulfur bacteria in an intensely stratified Paleoproterozoic ocean. *Nature* **437**, 866–870.

Butterfield NJ (2000) *Bangiomorpha pubescens* n. gen. n. sp.: implications for the evolution of sex, multicellularity and the Mesoproterozoic/Neoproterozoic radiation of eukaryotes. *Paleobiology* **26**, 386–404.

Butterfield NJ (2004) A vaucheriacean alga from the middle Neoproterozoic of Spitsbergen: implications for the evolution of Proterozoic eukaryotes and the Cambrian explosion. *Paleobiology* **30**, 231–252.

Butterfield NJ (2005) Probable Proterozoic fungi. *Paleobiology* **31**, 165–182.

Butterfield NJ, Knoll AH, Swett K (1994) Paleobiology of the Upper Proterozoic Svanbergfjellet Formation, Spitsbergen. *Fossils and Strata* **34**, 1–84.

Canfield DE (1998) A new model for Proterozoic ocean chemistry. *Nature* **396**, 450–453.

Canfield DE (2004) The evolution of the Earth surface sulfur reservoir. *American Journal of Science* **304**, 839–861.

Canfield DE, Poulton SW, Narbonne GM (2007) Late-Neoproterozoic deep-ocean oxygenation and the rise of animal life. *Science* **315**, 92–95.

Canfield DE, Poulton SW, Knoll AH, *et al.* (2008) Ferruginous conditions dominated later Neoproterozoic deep water chemistry. *Science* **321**, 949–952.

Cohen PA, Kodner R, Knoll AH (2009) Large spinose acritarchs in Ediacaran rocks as animal resting cysts. *Proceedings of the National Academy of Sciences, USA* **106**, 6519–6524.

Cohen, PA, Schopf JW, Butterfield NJ, Kudryavtsev AB, Macdonald FA (2011) Phosphate biomineralization in mid-Neoproterozoic protists. *Geology* **39**, 539–542.

Cloud PE (1965) Significance of the Gunflint (Precambrian) microflora. *Science* **148**, 27–35.

Dahl TW, Hammarlund E, Gill BC, *et al.* (2010) Devonian rise in atmospheric oxygen correlated to the radiations of terrestrial plants and large predatory fish. *Proceedings of the National Academy of Sciences, USA* **107**, 17853–18232.

Donoghue PCJ, Bengtson S, Dong XP (2006) Synchrotron X-ray tomographic microscopy of fossil embryos. *Nature* **442**, 680–683.

Dupraz C, Reid RP, Braissant O, Decho AW, Norman RS, Visscher PT (2009) Processes of carbonate precipitation in modern microbial mats. *Earth-Science Reviews* **96**, 141–162.

Feldmann M, McKenzie JA (1998) Stromatolite–thrombolite associations in a modern environment, Lee Stocking Island, Bahamas. *Palaios* **13**, 201–212.

Fennel K, Follows M, Falkowski PG (2005) The co-evolution of the nitrogen, carbon and oxygen cycles in the Proterozoic ocean. *American Journal of Science* **305**, 526–545.

Fike DA, Grotzinger JP, Pratt LM, Summons RE (2006) Oxidation of the Ediacaran ocean. *Nature* **444**, 744–747.

Furnes H, Banerjee NR, Muehlenbachs K, Kontinen A (2007) Preservation of biosignatures in metaglassy volcanic rocks from the Jormua ophiolite complex, Finland. *Precambrian Research* **136**, 125–137.

Gerdes G, Noffke N, Klenke T, Krumbein WE (2000) Microbial signatures in peritidal sediments – a catalogue. *Sedimentology* **47**, 279–308.

Golubic S, Hofmann HJ (1976) Comparison of modern and mid-Precambrian Entophysalidaceae (Cyanophyta) in stromatolitic algal mats: cell division and degradation. *Journal of Paleontology* **50**, 1074–1082.

Golubic S, Seong-Joo L (1999) Early cyanobacterial fossil record: preservation, palaeoenvironments and identification. *European Journal of Phycology* **34**, 339–348.

Green J, Knoll AH, Swett K (1988) Microfossils in oolites and pisolites from the Upper Proterozoic Eleonore Bay Group, central East Greenland. *Journal of Paleontology* **62**, 835–852.

Grey K (2005) Ediacaran palynology of Australia. *Association of Australasian Palaeontologists Memoir* **31**, 1–439.

Grotzinger JP, Knoll AH (1999) Proterozoic stromatolites: evolutionary mileposts or environmental dipsticks? *Annual Review of Earth and Planetary Science* **27**, 313–358.

Grotzinger JP, Rothman DH (1996) An abiotic model for stromatolite morphogenesis. *Nature* **383**, 423–425.

Grotzinger JP, Watters WA, Knoll AH (2000) Calcified metazoans in thrombolite–stromatolite reefs of the terminal Proterozoic Nama Group, Namibia. *Paleobiology* **26**, 334–359.

Hagadorn JW, Bottjer DJ (1997) Wrinkle structures: microbially mediated sedimentary structures in siliciclastic settings at the Proterozoic–Phanerozoic transition. *Geology* **25**, 1047–1050.

Hagadorn JW, Xiao S, Donoghue PCJ, *et al.* (2006) Cellular and subcellular structure of Neoproterozoic animal embryos. *Science* **314**, 291–294.

Hayes JM (1994) Global methanotrophy at the Archean–Proterozoic transition. In: *Early Life on Earth*. Nobel Symposium 84. (ed Bengtson S). Columbia University Press, New York, pp. 220–236.

Hayes JR, Waldbauer JR (2006). The carbon cycle and associated redox processes through time. *Philosophical Transactions of the Royal Society, B – Biological Sciences* **361**, 931–950.

Hofmann HJ (1976) Precambrian microflora, Belcher Islands, Canada – significance and systematics. *Journal of Paleontology* **50**, 1040–1073.

House CH, Schopf JW, McKeegan KD, Coath CD, Harrison TM, Stetter KO (2000) Carbon isotopic composition of individual Precambrian microfossils. *Geology* **28**, 707–710.

Igisu M, Nakashima S, Ueno Y, Awramik SM, Maruyama S (2006) In situ infrared microspectroscopy of ~850 million-year-old prokaryotic fossils. *Applied Spectroscopy* **60**, 1111–1120.

Javaux EJ, Knoll AH, Walter MR (2001) Morphological and ecological complexity in early eukaryotic ecosystems. *Nature* **412**, 66–69.

Javaux E, Knoll AH, Walter MR (2004) TEM evidence for eukaryotic diversity in mid-Proterozoic oceans. *Geobiology* **2**, 121–132.

Javaux EJ, Marshall CP, Bekker A (2010) Organic-walled microfossils in 3.2-billion-year-old shallow-marine siliciclastic deposits. *Nature* **463**, 934–938.

Johnston DT (2011) Multiple sulfur isotopes and the evolution of Earth's surface sulfur cycle. *Earth-Science Reviews* **106**, 161–183.

Johnston, DT, Wolf-Simon F, Pearson A, Knoll AH (2009) Anoxygenic photosynthesis modulated Proterozoic oxygen and sustained Earth's middle age. *Proceedings of the National Academy of Sciences, USA* **106**, 16925–16929.

Kaufman AJ, Xiao SH (2003) High CO_2 levels in the Proterozoic atmosphere estimated from analyses of individual microfossils. *Nature* **425**, 279–282.

Kaufman AJ, Johnston DT, Farquhar J, *et al.* (2007) Late Archean biospheric oxygenation and atmospheric evolution. *Science* **317**, 1900–1903.

Kempe A, Schopf JW, Altermann W, Kudryavtsev AB, Heckl WM (2002) Atomic force microscopy of Precambrian microscopic fossils. *Proceedings of the National Academy of Sciences, USA* **99**, 9117–9120.

Kempe A, Wirth R, Altermann W, Stark RW, Schopf JW, Heckl WA (2005) Focussed ion beam preparation and in situ nanoscopic study of Precambrian acritarchs. *Precambrian Research* **140**, 36–54.

Kennard JM, James NP (1986) Thrombolites and stromatolites; two distinct types of microbial structures. *Palaios* **1**, 492–503.

Klein C, Beukes NJ, Schopf JW (1987) Filamentous microfossils in the early Proterozoic Transvaal Supergroup: their morphology, significance, and paleoenvironmental setting. *Precambrian Research* **36**, 81–94.

Knoll AH (1996) Archean and Proterozoic paleontology. In *Palynology: Principles and Applications*, Volume 1 (ed Jansonius J, McGregor DC). American Association of Stratigraphic Palynologists Foundation, Tulsa OK, pp. 51–80.

Knoll AH (2003) *Life on a Young Planet*. Princeton University Press, Princeton NJ.

Knoll AH (2007) Cyanobacteria and Earth history. In *The Cyanobacteria: Molecular Biology, Genomics and Evolution* (eds Herrero A, Flores E). Horizon Scientific Press, Heatherset UK, pp. 1–19.

Knoll AH, Javaux EJ, Hewitt D, Cohen P (2006) Eukaryotic organisms in Proterozoic oceans. *Philosophical Transactions of the Royal Society, B– Biological Sciences* **361**, 1023–1038.

Knoll AH, Bambach RK, Payne J, Pruss S, Fischer W (2007a) A paleophysiological perspective on the end-Permian mass extinction and its aftermath. *Earth and Planetary Science Letters* **256**, 295–313.

Knoll AH, Summons RE, Waldbauer J, Zumberge J (2007b) The geological succession of primary producers in the oceans. In *The Evolution of Primary Producers in the Sea* (eds Falkowski P, Knoll AH). Elsevier, Burlington MA, pp. 133–163.

Kopp RE, Kirschvink JL, Hilburn IA, Nash CZ (2005) The Paleoproterozoic snowball Earth: A climate disaster triggered by the evolution of oxygenic photosynthesis. *Proceedings of the National Academy of Sciences, USA* **102**, 11131–11136.

Lanier WP (1988) Approximate growth rates of Early Proterozoic microstromatolites as deduced by biomass productivity. *Palaios* **1**, 525–542.

Le Campion-Alsumard, T (1989) Endolithic marine cyanobacteria. *Bulletin de la Société Botanique de France-Actualites Botaniques* **136**, 99–112.

Lemelle L, Labrot P, Salome M, Simionovici A, Viso M, Westall F (2008) In situ imaging of organic sulfur in 700–800 My-old Neoproterozoic microfossils using X-ray spectromicroscopy at the SK-edge. *Organic Geochemistry* **39**, 188–202.

Lepot K, Benzerara K, Brown GE, Philippot P (2008) Microbially influenced formation of 2,724-million-year-old stromatolites. *Nature Geoscience* **1**, 118 – 121.

Marshall CP, Javaux EJ, Knoll AH, Walter MR (2005) Combined micro-Fourier transform infrared (FTIR) spectroscopy and Micro-Raman spectroscopy of Proterozoic acritarchs: a new approach to palaeobiology. *Precambrian Research* **138**, 208–224.

Marshall CP, Love GD, Snape CE, *et al.* (2007) Structural characterization of kerogen in 3.4 Ga Archaean cherts from the Pilbara Craton, Western Australia. *Precambrian Research* **155**, 1–23.

Martin W, Rotte C, Hoffmeister M, *et al.* (2003) Early cell evolution, eukaryotes, anoxia, sulfide, oxygen, fungi first (?), and a tree of genomes revisited. *IUBMB* **55**, 193–204.

McFadden KA, Huang J, Chu XL (2008) Pulsed oxidation and biological evolution in the Ediacaran Doushantuo Formation. *Proceedings of the National Academy of Sciences, USA* **105**, 3197–3202.

Narbonne GM (2005) The Ediacara biota: Neoproterozoic origin of animals and their ecosystems. *Annual Review of Earth and Planetary Science* **33**, 421–442.

Noffke N (2010) *Geobiology: Microbial Mats in Sandy Deposits from the Archean Era to Today*. Springer-Verlag, Berlin.

Noffke N, Gerdes G, Klenke T (2003) Benthic cyanobacteria and their influence on the sedimentary dynamics of peritidal depositional systems (siliciclastic, evaporitic salty, and evaporitic carbonatic). *Earth Science Reviews* **62**, 163–176.

Noffke N, Knoll AH, Grotzinger JP (2002) Sedimentary controls on the formation and preservation of microbial mats in siliciclastic deposits: a case study from the Upper Neoproterozoic Nama Group, Namibia. *Palaios* **17**, 533–544.

Noffke N, Beukes N, Bower D, Hazen RM, Swift DJP (2008) An actualistic perspective into Archean worlds – cyanobacterially induced sedimentary structures in the siliciclastic Nhlazatse Section, 2.9 Ga Pongola Supergroup, South Africa. *Geobiology* **6**, 5–20.

Oehler DZ (1977) Pyrenoid-like structures in Late Precambrian algae from the Bitter Springs Formation. *Journal of Paleontology* **51**, 885–901.

Oehler DZ, Robert F, Mostefaoui S, Meibom A, Selo M, McKay D (2006) Chemical mapping of Proterozoic organic matter at submicron spatial resolution. *Astrobiology* **6**, 838–850.

Pasteris JD, Wopenka B (2003) Necessary, but not sufficient: Raman identification of disordered carbon as a signature of ancient life. *Astrobiology* **3**, 727–738.

Philippot P, Van Zuilen M, Lepot K, Thomazo C, Farquhar J, Van Kranendonk MJ (2007) Early Archaean microorganisms preferred elemental sulfur, not sulfate. *Science* **317**, 1534–1537.

Porada H, Ghergut, J, Bouougri (2008) Kinneyia-type wrinkle structures – critical review and model of formation. *Palaios* **23**, 65–77.

Poulton SW, Fralick PW, Canfield DE (2004) The transition to a sulphidic ocean similar to 1.84 billion years ago. *Nature* **431**, 173–177.

Porter SM, Knoll AH (2000) Testate amoebae in the Neoproterozoic Era: evidence from vase-shaped microfossils in the Chuar Group, Grand Canyon. *Paleobiology* **26**, 360–385.

Porter SM, Meisterfeld R, Knoll AH (2003) Vase-shaped microfossils from the Neoproterozoic Chuar Group, Grand Canyon: a classification guided by modern testate amoebae. *Journal of Paleontology* **77**, 205–225.

Reid RP, James NP, Macintyre IG, Dupraz CP, Burne RV (2003) Shark Bay stromatolites: Microfabrics and reinterpretation of origins. *Facies* **49**, 299–324.

Schieber J, Bose P, Eriksson PG, *et al.* (2004) Atlas of Microbial Mat Features Preserved within the Siliciclastic Rock Record. Elsvier, Amsterdam.

Schopf JW (1970) Electron microscopy of organically preserved Precambrian microorganisms. *Journal of Paleontology* **44**, 1–6.

Schopf JW (1993) Microfossils of the Early Archean Apex Chert – New evidence of the antiquity of life. *Science* **260**, 640–646.

Schopf JW (2006) Fossil evidence of Archaean life. *Philosophical Transactions of the Royal Society, B-Biological Sciences* **361**, 869–885.

Schopf JW, Kudryavtsev AB (2005) Three-dimensional Raman imagery of Precambrian microscopic organisms. *Geobiology* **3**, 1–12.

Schopf JW, Packer BM (1987) Early Archean (3.3-billion to 3.5-billion-year-old) microfossils from Warrawoona Group. Australia *Science* **237**, 70–74.

Schopf JW, Walter MR (1982) Origin and early evolution of cyanobacteria: the geological evidence. In The Biology of Cyanobacteria (eds Carr NG, Whitton BA). University of California Press, Berkeley, pp. 543–564.

Schopf JW, Kudryavtsev AB, Agresti DG, Wdowiak TJ, Czaja AD (2002) Laser-Raman imagery of Earth's earliest fossils. *Nature* **416**, 73–76.

Schopf JW, Kudryavtsev AB, Agresti DG, Czaja AD, Wdowiak TJ (2005) Raman imagery: A new approach to assess the geochemical maturity and biogenicity of permineralized Precambrian fossils. *Astrobiology* **5**, 333–371.

Schopf JW, Tripathi AB, Kudryavtsev AB (2006) Three-dimensional confocal optical imagery of Precambrian microscopic organisms. *Astrobiology* **6**, 1–16.

Schopf JW, Kudryavtsev AB, Sergeev VN (2010a) Confocal laser scanning microscopy and Raman imagery of the late Neoproterozoic Chichkan Microbiota of South Kazakhstan. *Journal of Paleontology* **84**, 402–416.

Schopf JW, Kudryavtsev AB, Sugitani K, Walter MR (2010b) Precambrian microbe-like pseudofossils: A promising solution to the problem. *Precambrian Research* **179**, 191–205.

Scott C, Lyons TW, Bekker A, *et al.* (2008) Tracing the stepwise oxygenation of the Proterozoic ocean. *Nature* **452**, 456–459.

Seong-Joo L, Browne KM, Golubic S (2000) On stromatolite lamination. In Microbial Sediments (ed. Riding R, Awramik SM). Springer-Verlag, Heidelberg, pp. 16–24.

Shen Y, Zhang T, Hoffman PF (2008) On the coevolution of Ediacaran oceans and animals. *Proceedings of the National Academy of Sciences, USA* **105**, 7376–7381.

Sinninghe Damsté JS, de Leeuw, JW (1990) Analysis, structure and geochemical significance of organically-bound sulphur in the geosphere: state of the art and future research. *Organic Geochemistry* **16**, 1077–1101.

Talyzina N (2000) Ultrastructure and morphology of *Chuaria circularis* (Walcott, 1899) Vidal and Ford (1985) from the Neoproterozoic Visingső Group, Sweden. *Precambrian Research* **102**, 123–134.

Talyzina N, Moczydłowska M (2000) Morphological and ultrastructural studies of some acritarchs from the Lower Cambrian Lukati Formation, Estonia. *Review of Palaeobotany and Palynology* **112**, 1–21.

Teyssédre B (2006) Are the green algae (phylum Viridiplantae) two billion years old? Carnets de Geologie, Article 2006/03 (CG2006_A03).

Tice MM, Lowe DR (2004) Photosynthetic microbial mats in the 3.5 Ga old ocean. *Nature* **431**, 549–550.

Tomitani A, Knoll AH, Cavanaugh CM, Ohno T (2006) The evolutionary diversification of cyanobacteria: molecular phylogenetic and paleontological perspectives. *Proceedings of the National Academy of Sciences, USA* **103**, 5442–5447.

Tyler SA, Barghoorn ES (1954) Occurrence of structurally preserved plants in pre-Cambrian rocks of the Canadian Shield. *Science* **119**, 606–608.

Ueno Y, Isozaki Y, Yurimoto H, Maruyama S (2001) Carbon isotopic signatures of individual Archean microfossils (?) from Western Australia. *International Geology Review* **43**, 196–212.

van Zuilen MA, Chaussidon M, Rollion-Bard C (2007) Carbonaceous cherts of the Barberton Greenstone Belt, South Africa: Isotopic, chemical and structural characteristics of individual microstructures. *Geochimica et Cosmochimica Acta* **71**, 655–669.

Wacey D, Kilburn MR, McLoughlin N, Parnell J, Stoakes CA, Grovenor CRM, Brasier MD (2008) Use of NanoSIMS in the search for early life on Earth: ambient inclusion trails in a c. 3400 Ma sandstone. *Journal of the Geological Society* **165**, 43–53.

Walter MR (1976) Stromatolites. Elsevier, Amsterdam, 802 pp.

Walter MR, Du R, Horodyski RJ (1990) Coiled carbonaceous megafossils from the Middle Proterozoic of Jixian (Tianjin) and Montana. *American Journal of Sciience* **290A**, 133–148.

Wille M, Kramers JD, Nagler TF, *et al.* (2007) Evidence for a gradual rise of oxygen between 2.6 and 2.5 Ga from Mo isotopes and Re-PGE signatures in shales. *Geochimica et Cosmochimica Acta* **71**, 2417–2435.

Wille M, Nagler TF, Lehmann B, Schröder S, Kramers JD (2008) Hydrogen sulphide release to surface waters at the Precambrian/Cambrian boundary. *Nature* **453**, 767–769.

Willman S, Moczydłowska M (2007) Wall ultrastructure of an Ediacaran acritarch from the Officer Basin, Australia. *Lethaia* **40**, 111–123.

Wilson JP, Fischer WW, Johnston DT, *et al.* (2010) Geobiology of the Paleoproterozoic Duck Creek Dolomite, Western Australia. *Precambrian Research,* **179**, 135–149.

Xiao S, Knoll AH, Yuan X, Pueschel C (2004) Phosphatized multicellular algae in the Neoproterozoic Doushantuo Formation, China, and the early evolution of florideophyte red algae. *American Journal of Botany* **91**, 214–227.

Xiao S, Yuan X, Steiner M, Knoll AH (2002) Carbonaceous macrofossils in a terminal Proterozoic shale: a systematic reassessment of the Miaohe biota, South China. *Journal of Paleontology* **76**, 347–376.

Yin L, Zhu M, Knoll AH, Yuan X, Zhang J, Hu J (2007) Doushantuo embryos preserved within diapause egg cysts. *Nature* **446**, 661–663.

17

GEOCHEMICAL ORIGINS OF LIFE

Robert M. Hazen

Geophysical Laboratory, Carnegie Institution of Washington,
5251 Broad Branch Road NW, Washington, DC 20015, USA

17.1 Introduction

Life's origins were a sequence of geochemical events, the consequence of interactions among atmosphere, oceans, and rocks in an energetic prebiotic environment. While the details of that ancient transformation from geochemistry to biochemistry remain elusive, the general outlines of life's origins are gradually emerging through a combination of laboratory experiments, observations of living cells, and theoretical analysis.

The experimental investigation of the origins of life commenced in earnest more than a half-century ago with the pioneering work of Miller (Miller, 1953, 1955; Wills and Bada, 2000; Lazcano, 2010), who synthesized many of life's molecular building blocks under what were then thought to be plausible prebiotic conditions. Despite an initial euphoric sense that the origin mystery would be solved quickly, scientists soon realized that the natural transition from rock, water and gas to living cells, though completely consistent with the laws of chemistry and physics, would not quickly be deduced by the scientific method.

The central challenge of origins research lies in replicating in a laboratory setting the extraordinary increase in complexity that is required to evolve from isolated small molecules to a living cell. This chapter reviews some of the efforts by origins-of-life researchers to induce such increases in complexity. A unifying theme of these studies, and hence a useful organizing framework for this review, is the principle of emergence – the natural process by which complexity arises. We will find repeatedly that geochemical complexities – notably thermal and compositional gradients, fluid fluxes,

cycles, and interfaces – are essential attributes of early Earth's geochemical environment that promote the emergence of chemical complexification (Hazen, 2009; Hazen and Sverjensky, 2010).

17.2 Emergence as a unifying concept in origins research

Life's origins can be modeled as a sequence of so-called 'emergent' events, each of which added new structure and chemical complexity to the prebiotic Earth. Observations of numerous everyday phenomena reveal that new patterns commonly emerge when energy flows through a collection of many interacting particles (Prigogine, 1984; Nicolis and Prigogine, 1989; Holland, 1995, 1998; Morowitz, 2002; Hazen, 2005, 2009; Hazen and Sverjensky, 2010).

17.2.1 What is emergence?

In the words of John Holland, an influential leader in the study of emergent systems, 'It is unlikely that a topic as complicated as emergence will submit meekly to a concise definition, and I have no such definition' (Holland, 1998). Nevertheless, emergent systems display three distinctive characteristics: (1) they arise from the interactions of many particles or 'agents;' (2) energy flows through those systems of particles; and (3) they display new patterns or behaviours that are not manifest by the individual agents. Sand grains, for example, interact under the influence of wind or waves to form dune and ripple structures (Bagnold, 1941, 1988). Similarly, ants interact to form colonies, neurons interact to produce consciousness, and people interact to form societies (Solé and Goodwin, 2000; Camazine *et al.*, 2001).

The history of the universe can be viewed as a progression of emergent events, from hydrogen to stars, from stars to the periodic table of chemical elements, from chemical elements to planets and life, and from life to consciousness (Chaisson, 2001; Morowitz, 2002). The inexorable stepwise transition from simplicity to emergent complexity is an intrinsic characteristic of the cosmos. In such a pregnant universe, one need not resort to divine intervention or to intelligent design for life's origin. Thus, in spite of the lack of a precise definition, the recognition and description of these varied emergent systems provides an important foundation for origins of life research, for life is the quintessential emergent phenomenon. The first cell emerged as vast collections of lifeless molecules interacted under the influence of such energy sources as solar radiation, lightning, cosmic rays, and Earth's inner heat. Understanding the underlying principles governing such emergent systems thus provides insights to our experimental and theoretical efforts to understand life's origins.

17.2.2 Geochemical complexities and emergence

One key to understanding life's origin is to recognize the critical role of complex geochemical environments that are in disequilibrium. Many familiar natural systems lie close to chemical and physical equilibrium and thus do not display emergent behaviour. For example, water gradually cooled to below the freezing point equilibrates to become a clear chunk of ice, whereas water gradually heated above the boiling point similarly equilibrates by converting to steam. Dramatically different (emergent) behaviour occurs far from equilibrium. Water subjected to the strong temperature gradient of a boiling pot displays complex turbulent convection (Prigogine, 1984; Nicolis and Progogine, 1989). Water flowing downhill in the gravitational gradient of a river valley interacts with sediments to produce the emergent landforms of braided streams, meandering rivers, sandbars, and deltas. Emergent systems seem to share a common origin; they arise away from equilibrium when energy flows through a collection of many interacting agents. Such systems tend spontaneously to become more ordered and display new, often surprising behaviours. The whole is more than the simple sum of the parts.

Framing the origins-of-life problem in terms of emergence is more than simply providing a new label to parts of the origin story we don't understand. Emergent systems share key characteristics that can inform origin models and experiments. For example, all emergent systems, including collections of sand, molecules, cells, or stars, require a minimum number of interacting agents

before new patterns arise (Camazine *et al.*, 2001; Hazen, 2005, 2009). Thus origins experiments require minimum concentrations of simple organic molecules such as amino acids, sugars, and lipids, which provided the agents of life's emergence. These systems also require a source of energy within specific limits – too little energy and no patterning occurs, but too much energy and patterns are destroyed (Chaisson, 2001).

Observations of natural and experimental emergent systems reveal that such geochemical complexities as cycles (including wet-dry, hot-cold, and day-night, as well as intermittent lightning strikes and asteroid impacts), chemical and thermal gradients and associated fluid fluxes (notably at or near deep-ocean hydrothermal vents), and solid-fluid or fluid-fluid interfaces enhance emergent complexity (Hansen *et al.*, 2001; Kessler and Werner, 2003; Hazen, 2009; Hazen and Sverjensky, 2010). An understanding of these and other general characteristics of emergent systems thus informs experimental design and theoretical analysis of life's origins. Indeed, origins-of-life researchers have long incorporated such complexities in their experiments. The transformational experiments of Miller and Urey (Miller, 1953, 1955; Miller and Urey 1959) employed thermal gradients and fluid fluxes in association with spark discharges to achieve organic synthesis. As we shall see in the following pages, recent theoretical models (Wächtershäuser, 1988a, 1992; de Duve, 1995a; Russell and Hall, 1997, 2002; Hunding *et al.*, 2006; Root-Bernstein, 2009; Benner *et al.*, 2010) and experiments (Lahav *et al.*, 1978; Ferris *et al.*, 1996; Huber and Wächtershäuser, 1997; McCollum and Simoneit, 1999; Klussmann *et al.*, 2006, 2007; Mansy *et al.*, 2008; Budin *et al.*, 2009; Whitfield, 2009; Chen and Walde, 2010) have incorporated aspects of molecular complexity that may emerge from multi-component geochemical systems with cycles, gradients, fluxes, and interfaces.

17.2.3 Life's origins as a sequence of emergent steps

The overarching problem with studying life's origins is that even the simplest known life form is vastly more complex than any non-living components that might have contributed to it. How does such astonishing, intricate complexity arise from lifeless raw materials? What now appears to us as a yawning divide between non-life and life reflects the fact that the chemical evolution of life must have occurred as a stepwise sequence of successively more complex stages of emergence. When modern cells emerged, they quickly consumed virtually all traces of these earlier stages of chemical evolution.

The challenge, therefore, is to establish a progressive hierarchy of emergent steps that leads from a pre-biotic ocean enriched in organic molecules, to functional clusters of molecules perhaps self-assembled or arrayed on a mineral surface, to self-replicating molecular systems that copied themselves from resources in their immediate environment, to encapsulation and eventually cellular life. The nature and sequence of these steps may have varied in different environments and we may never deduce the exact sequence that occurred on the early Earth. Yet many researchers suspect that the inexorable direction of the chemical path is similar on any habitable planet or moon (De Duve, 1995a; Morowitz, 2002).

Such a stepwise scenario informs attempts to define life, because the exact point at which such a system of gradually increasing complexity becomes 'alive' is intrinsically arbitrary. The evolutionary path to cellular life must have featured a rich variety of intermediate, complex, self-replicating emergent chemical systems. Each of those steps represents a distinctive, fundamentally important stage in life's molecular synthesis and organization. Each step requires independent experimental study, and perhaps a distinctive name in a taxonomic scheme richer than 'living' versus 'non-living' (Cleland and Chyba, 2002; Hazen, 2005; Cleland and Copley, 2005).

Ultimately, the key to defining the progressive stages between non-life and life lies in experimental studies of relevant chemical systems under plausible geochemical environments. The concept of emergence simplifies this experimental endeavor by reducing an immensely complex historical process to a more comprehensible succession of measurable steps. Each emergent step provides a tempting focus for laboratory experimentation and theoretical modeling.

This view of life's origins, as a stepwise transition from geochemistry to biochemistry, is of special relevance to the search for life elsewhere in the universe. Before an effective search can be undertaken, it is necessary to know what to look for. It is plausible, for example, that Mars, Europa, and other bodies progressed only part way along the chemical path to cellular life. If each emergent step in life's origins produces distinctive and measurable isotopic, molecular, and/or structural signatures in its environment, and if such markers can be identified, then these chemical features represent important observational targets for planned space missions. It is possible, for example, that characteristic isotopic, molecular, and structural 'fossils' survive only if they have not been eaten by more advanced cells. We may find that those distinctive molecular structures can serve as extraterrestrial 'abiomarkers' – clear evidence that molecular organization and evolution never progressed beyond a certain precellular stage. As scientists search for life elsewhere in the universe, they may be able to characterize extraterrestrial environments according to their degree of emergence

along this multi-step path (Hazen *et al.*, 2002; Hazen, 2005; Cleland, 2007).

17.3 The emergence of biomolecules

The first vital step in life's emergence on Earth must have been the synthesis and accumulation of abundant carbon-based biomolecules. In the beginning, life's raw materials consisted of water, rock, and simple gases – predominantly carbon dioxide and nitrogen, but with local concentrations of hydrogen, methane, ammonia and other species. Decades of experiments have revealed that diverse suites of organic molecules can emerge from plausible geochemical environments of the early Earth.

The experimental pursuit of geochemical organic synthesis, arguably the best understood aspect of life's origin, began a half-century ago with the pioneering studies of University of Chicago graduate student Stanley Miller and his distinguished mentor Harold Urey (Miller, 1953, 1955; Miller and Urey, 1959). Together they established the potential role of organic synthesis that occurred in Earth's primitive atmosphere and ocean as they were subjected to bolts of lightning and the Sun's intense radiation. This experimental investigation of life's origin is a surprisingly recent pursuit. Two centuries ago most scientists accepted the intuitively reasonable idea that life is generated spontaneously all around us, all the time. The question of life's ancient origins was not asked, at least not in the modern experimental sense (Lahav, 1999; Fry, 2000; Strick, 2000; Dick and Strick, 2004).

By the early 20th century many scientists would have agreed that life's origins, wherever and however it occurs, depends on three key resources. First, all known life forms on Earth require liquid water. All living cells, even those that survive in the driest desert ecosystem, are formed largely of water. In spite of creative attempts to postulate alien biochemistries (e.g. Ward and Benner, 2007; Wolf-Simon *et al.*, 2009), surely the first cells on Earth arose in a watery environment. Life also needs a ready source of energy. The radiant energy of the Sun provides the most obvious supply for life today, but bolts of lightning, impacts of asteroids, Earth's inner heat and the chemical energy of minerals have also been invoked as life-triggering energy sources. And, third, life depends on a variety of chemical elements. All living organisms consume atoms of carbon, oxygen, hydrogen, and nitrogen, with a bit of sulfur and phosphorus and other elements as well. These elements combine in graceful geometries to form essential biomolecules.

In spite of the intrinsic importance of the topic, it was not until the 1920s that such scientific speculation took a more formal guise. Most notable among the modern school of origin theorists was the Russian biochemist Alexander Oparin (Oparin, 1924). In 1924, while still in his 20s, Oparin elaborated on the idea that life arose from a body of water that gradually became enriched in organic molecules – what was to become known as the 'primordial soup' (Haldane, 1929). Somehow, he posited, these molecules clustered together and became self-organized into a chemical system that could duplicate itself.

Many of Oparin's postulates were echoed in 1929 by the independent ideas of British biochemist and geneticist J.B.S. Haldane, whose brief, perceptive article focused on the production of large carbon-based molecules under the influence of the Sun's ultraviolet radiation (Haldane, 1929). Given such a productive chemical environment, Haldane envisioned the first living objects as self-replicating, specialized molecules.

17.3.1   The Miller–Urey Experiment

Oparin and Haldane offered original and intriguing ideas that were subject to experimental testing, but Oparin and his contemporaries didn't try to replicate experimentally the prebiotic formation of biomolecules. Not until the years after World War II were the landmark experiments of Miller and Urey devised (Miller and Urey, 1959; Wills and Bada, 2000). They mimicked aspects of Earth's early surface by sealing water and simple gases into a tabletop glass apparatus and subjecting the contents to electric sparks, while gently boiling the water and circulating the contents (Fig. 17.1). When the experiment began the water was pure and clear, but within days the solution turned yellowish and a black residue had begun to accumulate near the electrodes. Reactions of water and simple gases had produced organic molecules, most notably a suite of amino acids (the building blocks of proteins), in abundance.

The Miller–Urey Experiment transformed the science of life's chemical origins. For the first time an experimental protocol duplicated a plausible life-forming process. It should be noted that the success of the Miller–Urey process depended in no small measure on the clever incorporation of the kinds of environmental complexities that characterized early Earth: fluid fluxes, thermal gradients, cyclic electric sparks, and interfaces among the solid, liquid and gas components of the experiment.

Given such an exciting finding, other groups jumped at the chance to duplicate the amino acid feat. More than a dozen amino acids were synthesized from scratch, along with other key biomolecules: membrane-forming lipids, energy-rich sugars and other carbohydrates, and metabolic acids. Other experiments employed ultraviolet radiation or alpha particles to reveal similar synthetic pathways (Garrison *et al.*, 1951; Abelson, 1966). So facile were these reactions that some researchers assumed that the problem of prebiotic chemical evolution had been solved.

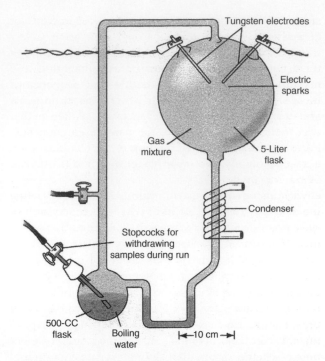

Figure 17.1 The Miller–Urey Experiment, which attempted to simulate an early Earth environment with water, methane, hydrogen and ammonia, incorporated temperature gradients, fluid fluxes, interfaces, and periodic electric spark discharges (Miller 1953; Miller and Urey 1959; Wills and Bada 2000).

Enthusiasm grew as other scientists discovered promising new chemical pathways. In 1960 chemist John Oró demonstrated that a hot, concentrated hydrogen cyanide solution produces adenine, a crucial biomolecule that plays a role in both genetic material and in metabolism (Oró, 1960, 1961). Other chemists, following the lead of studies conducted a century earlier (Butlerow, 1861), performed similar experiments starting with relatively concentrated solutions of formaldehyde, a molecule thought to be common in some prebiotic environments (Shapiro, 1988; Weber, 2007). Their simple experiments produced a rich variety of sugar molecules, including the critical compound ribose. Gradually, researchers filled in gaps in the prebiotic inventory of life's molecules.

As exciting and important as these results may have been, seemingly intractable problems remain. Within a decade of Miller's triumph serious doubts began to arise about the true composition of Earth's earliest atmosphere. Miller exploited a highly reactive, chemically reducing atmosphere of methane (CH_4), ammonia (NH_3), hydrogen (H_2), and water (H_2O) – an atmosphere distinctly lacking in oxygen. But by the 1960s, new geochemical calculations and data from ancient rocks pointed to a much less reducing early atmosphere composed primarily of nitrogen (N_2) and carbon dioxide (CO_2), though such a neutral atmosphere also yields significant molecules of biological interest in a Miller apparatus (Cleaves *et al.*, 2008).

For decades Miller and his supporters have countered with a pointed argument: The molecules of life match those synthesized in the original Miller experiment with great fidelity. Miller's advisor, Harold Urey, is said to have often quipped, 'If God did not do it this way, then he missed a good bet' (Wills and Bada, 2000). Most geochemists discount the possibility of more than a trace of atmospheric methane or ammonia at the time of life's emergence, except perhaps in local volcanic environments (Kasting and Siefert, 2002; Kasting, 2003), though Tian and colleagues recently (and controversially) proposed that hydrogen may have comprised as much as 30% of the Hadean atmosphere (Tian *et al.*, 2005; Chyba, 2005).

Added to this atmospheric concern is the fact that the molecular building blocks of life created by Miller represent only the first step on the long road to life. Living cells require that such small molecules be carefully selected and then linked together into vastly more complex structures – lipid molecules organized into cell membranes, amino acids joined to form protein enzymes, and other so-called 'macromolecules.' Even under the most optimistic estimates, the prebiotic ocean was an extremely dilute solution of countless thousand different kinds of organic molecules, most of which play no known role in life. By what processes were just the right molecules selected and organized?

The Miller–Urey scenario suffers from another nagging problem. Macromolecules tend to fragment rather than form when subjected to lightning, cosmic radiation, or the Sun's ultraviolet light (Chyba and Sagan, 1992). These so-called 'ionizing' forms of energy are useful for making highly reactive molecular fragments that combine into modest-sized molecules like amino acids. Combining many amino acids into an orderly chain-like enzyme, however, is best accomplished in a less destructive energy domain. Emergent complexity relies on a flow of energy, to be sure, but not too much energy. Could life have emerged in the harsh glare of daylight, or was there perhaps a different, more benign origins environment?

17.3.2 Deep origins

All living cells require a continuous source of energy. Until recently, scientists claimed that the metabolic pathways of virtually all life forms rely directly or indirectly on photosynthesis. This view of life changed in February 1977, with the discovery of an unexpected deep-ocean ecosystem (Corliss *et al.*, 1979). These thriving communities, cut off from the Sun, subsist at least in part on geothermal energy supplied by Earth's inner heat. Microbes serve as primary energy producers in these deep zones; they play the same ecological role as plants at Earth's sunlit surface. These single-celled vent organisms exploit the fact that the cold oxygen-infused ocean water, the hot volcanic water, and the sulfur-rich mineral surfaces over which these mixing fluids flow are not in chemical equilibrium. The unexpected discovery of this exotic ecosystem quickly led to speculation that a hydrothermal vent might have been the site of life's origins (Corliss *et al.*, 1981; Baross and Hoffman, 1985).

New support for the idea gradually consolidated, as hydrothermal ecosystems were found to be abundant along ocean ridges of both the Atlantic and Pacific. Researchers realized that, at a time when Earth's surface was blasted by comets and asteroids, deep ocean ecosystems would have provided a much more benign location than the surface for life's origins and evolution (Sleep *et al.*, 1989; Holm, 1992; Gold, 1999). New discoveries of abundant primitive microbial life in the deep continental crust further underscored the viability of deep, hot environments (Parkes *et al.*, 1993; Stevens and McKinley, 1995; Frederickson and Onstott, 1996). By the 1990s, the deep-origin hypothesis had become widely accepted as a viable, if unsubstantiated, alternative to the Miller surface scenario.

Following the revolutionary hydrothermal origins proposal, numerous scientists began the search for life in deep, warm, wet environments. Everywhere they look, it seems – in deeply buried sediments, in oil wells, even in porous volcanic rocks more than a mile down – microbes abound. Microbes survive under miles of Antarctic ice and deep in dry desert sand (Pedersen, 1993; Madigan and Marrs, 1997; Gerday and Glansdorff, 2007). These organisms seem to thrive on mineral surfaces, where water–rock interactions provide the chemical energy for life.

Another incentive exists for looking closely at the possibility of deep origins. If life is constrained to form in a sun-drenched pond or ocean surface then Earth, and perhaps ancient Mars or Venus, are the only possible places where life could have begun in our Solar System. If, however, cells can emerge from deeply buried wet zones, then life may be much more widespread than previously imagined. The possibility of deep origins raises the stakes in our exploration of other planets and moons.

Recent experiments bolster aspects of the deep hydrothermal origins hypothesis. The most fundamental biological reaction is the incorporation of carbon atoms (starting with the gas carbon dioxide) into organic molecules. Many common minerals, including most minerals of iron, nickel, cobalt or copper, promote carbon addition (Blöchl *et al.*, 1992; Heinen and Lauwers, 1996; Huber and Wächtershäuser, 1997, 1998; Cody *et al.*, 2000, 2001, 2004; Ji *et al.*, 2008). Furthermore, under some conditions minerals are found to increase the thermal stability of organic molecules (Hoang *et al.*, 2003; Scorei and Cimpoiaşu, 2006), though under other conditions minerals may accelerate destruction of organic molecules (Strasak and Sersen, 1991; Rimola *et al.*, 2009). One conclusion seems certain. Mineral-rich hydrothermal systems contributed to the early Earth's varied inventory of bio-building blocks.

It now appears that anywhere energy and simple carbon-rich molecules are found together, from the deep ocean to deep space, a suite of interesting organic molecules is sure to emerge (Chyba and Sagan, 1992; Ehrenfreund and Charnley, 2000; Martins *et al.*, 2007; Pasek and Lauretta, 2008; Ehrenfreund and Cami, 2010; Pizzarello and Shock, 2010). In spite of the polarizing advocacy for one favoured environment or another (Bada *et al.*, 1995; Wills and Bada, 2000; Bada and Lazcano, 2002), experiments point to the likelihood that there was no single dominant source. By 4 billion years ago Earth's oceans must have become a complex, albeit dilute, soup of life's building blocks. Though not alive, this chemical system was poised to undergo a sequence of increasingly complex emergent stages of molecular organization and evolution.

17.4 The emergence of macromolecules

Prebiotic processes produced a bewildering diversity of molecules. Some of those organic molecules were poised to serve as the essential starting materials of life, but most of that molecular jumble played no role whatsoever in the dawn of life. The emergence of concentrated suites of just the right mix remains a central puzzle in origins of life research.

Life's simplest molecular building blocks – amino acids, sugars, lipids, and more – emerged inexorably through facile, inevitable chemical reactions in numerous prebiotic environments. A half-century of synthesis research has elaborated on Miller's breakthrough experiments. Potential biomolecules from many sources must have littered the ancient Earth.

What happened next? Individual biomolecules are not remotely life-like. Life requires the assembly of just the right combination of small molecules into much larger collections – 'macromolecules' with specific functions. Making macromolecules is complicated by the fact that for every potentially useful small molecule in the prebiotic soup, dozens of other molecular species had no obvious role in biology. Life is remarkably selective in its building blocks; the vast majority of carbon-based molecules synthesized in prebiotic processes have no biological use whatsoever.

17.4.1 Life's idiosyncrasies

Consider sugar molecules, for example (Nelson and Cox, 2004; Weber and Pizarello, 2006). All living cells rely on two kinds of 5-carbon sugar molecules, ribose and deoxyribose (the 'R' and 'D' in RNA and DNA, respectively). Several plausible prebiotic synthesis pathways yield a small amount of these essential sugars, but for every ribose molecule produced many other 5-carbon sugar species also appear – xylose, arabinose, and lyxose, for example. Adding to this chemical complexity is a bewildering array of 3-, 4-, 6- and 7-carbon sugars, in chain, branch and ring structures. What is more, many sugar molecules, including ribose and deoxyribose, come in mirror-related pairs. These left- and right-handed varieties possess the same chemical formula and many of the same physical properties, but they differ in shape like left and right hands. All known prebiotic synthesis pathways yield equal amounts of left- and right-handed sugars, but cells employ only the right-handed sugar varieties. Consequently, many origins researchers have shifted their focus to the processes by which just the right molecules might have been selected, concentrated and organized into the essential structures of life.

Four key types of molecules – sugars, amino acids, bases and lipids – exemplify life's chemical parsimony. Perhaps the most compelling aspect of Stanley Miller's experiment and the discoveries of subsequent prebiotic chemical research was that they synthesized representatives of all four groups – the building blocks of biological carbohydrates, proteins, nucleic acids and lipid membranes. The greatest challenge in understanding the chemical emergence of life lies in finding mechanisms by which just the right combination of these smaller molecules was selected, concentrated and organized into the larger macromolecular structures vital to life.

The oceans are of little help because they are so vast – a volume of almost 1.4 billion cubic kilometres. No matter how much organic matter was made, the oceans formed a hopelessly dilute soup. In such a random, weak solution, it would have been difficult for just the right combination of molecules to bump into one another and make anything useful in the chemical path to life. By what process were the right molecules selected?

17.4.2 Molecular selection

Lipid molecules, the building blocks of cell membranes, accomplish the selection trick in a striking way. One end of these long, slender molecules is hydrophilic (water loving), whereas the rest of the molecule is hydrophobic (water hating). Consequently, when placed in water, life's lipids spontaneously self-organize into tiny cell-like spheres, called vesicles (Fig. 17.2). This selection process is rapid and spontaneous (Luisi and Varela, 1989; Deamer, 1997; Segré *et al.*, 2001; Hanczyc *et al.*, 2007; Hazen and Deamer, 2007; Chen and Walde, 2010).

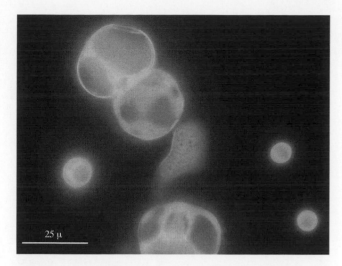

Figure 17.2 Lipid molecules can self-organize in aqueous solutions to form cell-like enclosures called vesicles (from Hazen and Deamer 2007).

Table 17.1 Selected rock-forming minerals and their proposed roles in life's origins

Mineral	Formula	Proposed roles and comments
Pyrrhotite	$Fe_{1-x}S$	Reductant in prebiotic reactions
Pyrite	FeS_2	Adsorb and concentrate organics
Millerite	NiS	Reductant in prebiotic reactions
Calcite	$CaCO_3$	Molecular selection, concentration; chiral molecular selection
Feldspar	(K,Na,Ca) Al-silicates	Adsorb, organize and protect organics
Clays	Various Mg-Al-layer silicates	Adsorb and organize organics, promote polymerization
Zeolites	Hydrous Al-silicates (mesoporous)	Molecular selection and polymerization
Rutile	TiO_2	Nitrogen reduction; adsorb and concentrate organics
Uraninite	UO_2	Prebiotic energy source for molecular synthesis

Lipid molecules typically require concentrations >1 mM before vesicle formation can occur (Deamer, 1997). However, geochemical complexities can greatly facilitate the process in a number of surprising ways. Budin *et al.* (2009) discovered that lipid solutions can be concentrated more than 100-fold when placed in a narrow capillary with a temperature gradient, as might be found in the cracks and fissures of a hydrothermal system. Lipids aggregate and self-organize in the coolest portions of such systems. The addition of clay minerals, which would have been common fine-grained minerals at the time of life's origins, also triggers lipid self-organization at much reduced concentrations, though the exact mechanisms of this effect remain uncertain (Hanczyc *et al.*, 2007). Yet another factor that can enhance the formation and stability of vesicles is use of a mixture of different lipids (Mansy *et al.*, 2008). The latter finding supports the idea that molecular complementarity may have played a significant role in several aspects of life's origin (Hunding *et al.*, 2006; Root-Bernstein, 2009).

In spite of these geochemical tricks, most molecules do not self organize. Consequently, many scientists have focused on surfaces as an alternative solution. Chemical complexity can arise at surfaces, where different molecules can congregate and interact. The surface of the ocean where air meets water is one promising interface, where a primordial oil slick might have concentrated organic molecules (Lasaga *et al.*, 1971; Rajamani *et al.*, 2008). Evaporating tidal pools, where rock and water meet and cycles of evaporation concentrate stranded chemicals, provide another appealing scenario for origin-of-life chemistry (Lahav *et al.*, 1978). Deep within the crust and in hydrothermal volcanic zones, mineral surfaces may have had a similar role in concentrating and organizing molecules on their periodic crystalline surfaces (Hazen, 2006, 2009; Table 17.1).

Solid rocks provide especially attractive surfaces for concentrating and assembling molecules. Experiments reveal that amino acids concentrate and polymerize on clay particles to form small, clump-like molecules called 'protenoids,' while layered minerals also have the ability to adsorb and assemble the building blocks of RNA or soak up small organic molecules in the rather large spaces between layers (Ferris, 1993, 1999, 2005; Orgel, 1998; Smith, 1998; Pitsch *et al.*, 1995; Benetoli *et al.*, 2007; Lambert, 2008). Once confined and concentrated, these small molecules tend to undergo reactions to form larger molecular species that are not otherwise likely to emerge from the soup.

In the case of experiments on mineral–molecule interactions in an aqueous environment, most experiments to date have focused on a single well-characterized mineral with one solute in water with at most a single electrolyte at room conditions (e.g. Jonsson *et al.*, 2009; 2010). Such experiments are essential to obtain baseline information on the magnitude and geometry of adsorption for various mineral–molecule pairs. Nevertheless, these studies do not replicate prebiotic complexities, including the multiple solutes of seawater, numerous competing organic species in the prebiotic soup, and numerous competing

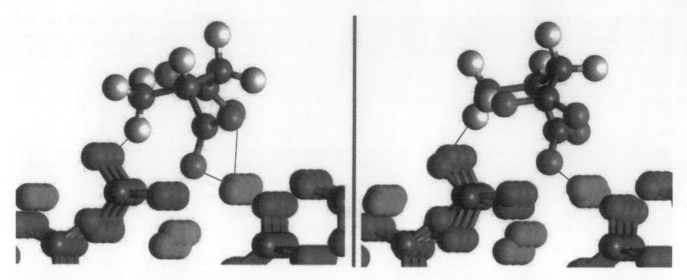

Figure 17.3 The most stable configurations for L- and D-aspartate on the calcite (21–34) surface (left and right, respectively). The D enantiomer, which requires significantly less calcite surface relaxation and aspartate distortion, is favoured by 8 kcal mol^{-1} – the largest known enantiospecific effect (after Hazen 2006).

mineral phases and surfaces, all present over a range of temperature, pressure, pH, and solute concentrations.

In the most extreme version of the mineral origins approach, Scottish researcher Graham Cairns-Smith has speculated that fine-grained crystals of clay, themselves, might have been the first life on Earth (Cairns-Smith, 1968, 1982, 1985a, b). The crux of the argument rests on a simple analogy. Cairns-Smith likens the origins of life to the construction of a stone archway, with its carefully fitted blocks and crucial central keystone that locks the whole structure in place. But the arch cannot be built by just piling one stone atop another. He proposes that a simple support structure like a scaffolding facilitates the construction and then is removed. 'I think this must have been the way our amazingly 'arched' biochemistry was built in the first place. The parts that now lean together surely used to lean on something else – something low tech' (Cairns-Smith 1985a). That something, he argues, was a clay mineral.

17.4.3 Right and left

One of the most intriguing and confounding examples of prebiotic molecular selection was the incorporation of handedness. Many of the most important biomolecules, amino acids and sugars included, come in 'chiral,' or mirror image, pairs. These left- and right-handed molecules have virtually the same energies and physical properties, and all known prebiotic synthesis pathways produce chiral molecules in essentially 50:50 mixtures. Thus, no obvious inherent reason exists why left or right should be preferred, yet living cells display the most exquisite selectivity, choosing right-handed sugars over left, and left-handed amino acids over right

(Bonner, 1991, 1995; Hazen and Sholl, 2003; Blackmond, 2010).

Some analyses of chiral amino acids in carbonaceous meteorites point to the possibility that Earth was seeded by amino acids that already possessed a left-handed bias (Cronin and Pizzarello, 1983; Engel and Macko, 1997; Pizzarello and Cronin, 2000; Pizzarello *et al.*, 2008). According to one scenario, left-handed molecules could have been concentrated if circularly polarized synchrotron light from a rapidly rotating neutron star selectively photolysed right-handed amino acids in the solar nebula (Bailey *et al.*, 1998; Clark, 1999; Podlech, 1999; Bargueño and Pérez de Tudela, 2007; Gusev *et al.*, 2007). However, it is also difficult to eliminate entirely the possibility of a left-handed overprint imposed in the laboratory during the difficult extractions and analyses of meteorite organics.

Alternatively, many origins-of-life researchers argue that the chirality of life occurred as a chance event – the result of an asymmetric local physical environment on Earth (Blackmond, 2010). Such local chiral environments abounded on the prebiotic Earth, both as chiral molecules, themselves, and in the form of asymmetric mineral surfaces (Hazen, 2004, 2006). Experiments show that left- and right-handed mineral surfaces provide one possible solution for separating a 50:50 mixture of L and D molecules (Hazen *et al.*, 2001; Hazen and Sholl 2003; Castro-Puyana *et al.*, 2008). Minerals often display chiral crystal faces, which might have provided templates for the assembly of life's molecules (Fig. 17.3). Whether local or global in scale, once a slightly chiral enrichment was established then any of a number of chiral amplification mechanisms could have produced local environments of near chiral purity (Pizzarello and

Weber, 2004; Weber and Pizzarello, 2006; Klussmann *et al.*, 2006, 2007; Blackmond, 2007, 2010).

In spite of these insights, many aspects of prebiotic molecular selection and organization remain uncertain and thus represent topics of intense ongoing research. Even so, the emergence of macromolecular structures is but one increment in the stepwise progression from geochemistry to biochemistry. Life requires that macromolecules be incorporated into a self-replicating system.

17.5 The emergence of self-replicating systems

Four billion years ago the seeds of life had been firmly planted. The Archean Earth boasted substantial repositories of serviceable organic molecules. These molecules must have become locally concentrated, where they assembled into vesicles and polymers of biological interest. Yet accumulations of organic molecules, no matter how highly selected and intricately organized, are not alive unless they also possess the ability to reproduce. Devising a laboratory experiment to study chemical self-replication has proven to be vastly more difficult than the prebiotic synthesis of biomolecules or the selective concentration and organization of those molecules into membranes and polymers. In a reproducing chemical system, one small group of molecules must multiply again and again at the expense of other molecules, which serve as food.

17.5.1 *Metabolism versus genetic mechanisms*

A fundamental debate on the origins of life relates to the timing of two essential biological processes, metabolism versus replication of genetic material (Orgel, 1986; Fox, 1988; Morowitz, 1992; De Duve 1995a, b; Dyson, 1999; Wills and Bada, 2000; Russell and Hall, 2002; Hazen, 2005; Deamer and Weber, 2010). Metabolism is the ability to manufacture biomolecular structures from a source of energy (such as sunlight) and matter scavenged from the surroundings (usually in the form of small molecules). An organism cannot survive and grow without an adequate supply of energy and matter. Genetic mechanisms, by contrast, involve the transfer of biological information from one generation to the next – a blueprint for life *via* the mechanisms of DNA and RNA. An organism cannot reproduce without a reliable means to pass on this genetic information.

The problem for understanding origins is that metabolism and genetic mechanisms constitute two separate, chemically distinct systems in cells. Nevertheless, metabolism and genetic mechanisms are inextricably linked in modern life. DNA holds genetic instructions to make hundreds of molecules essential to metabolism, while metabolism provides both the energy and the basic building blocks to make DNA and other genetic materials. Like the dilemma of which came first, the chicken or the egg, it is difficult to imagine back to a time when metabolism and genetic mechanisms were not intertwined. Consequently, origins-of-life researchers engage in an intense ongoing debate about whether these two aspects of life arose simultaneously or independently and, if the latter, which one came first.

Most experts seem to agree that the simultaneous emergence of metabolism and genetic mechanisms is unlikely. The chemical processes are just too different, and they rely on completely different sets of molecules. It's much easier to imagine life arising one small step at a time, but what is the sequence of emergent steps?

Those who favour genetics first argue on the basis of life's remarkable complexity; they point to the astonishing intricacy of even the simplest living cell. Without a genetic mechanism there would be no way to insure the faithful reproduction of all that complexity. Metabolism without a genetic mechanism may be viewed as nothing more than a collection of overactive chemicals.

Other scientists are persuaded by the principle that life emerged through stages of increasing complexity. Metabolic chemistry, at its core, is vastly simpler than genetic mechanisms because it requires relatively few small molecules that work in concert to duplicate themselves. In this view the core metabolic cycle – the citric acid cycle that lies at the heart of every modern cell's metabolic processes – survives as a biochemical fossil from life's beginning. This comparatively simple chemical cycle is an engine that can bootstrap all of life's biochemistry, including the key genetic molecules.

Origins-of-life scientists are not shy about voicing their opinions on the metabolism- vs. genetics-first problem, which will probably remain a central controversy in the field for some time. Meanwhile, as this debate fuels animated discussions, several groups of creative researchers are attempting to shed light on the issue by devising self-replicating chemical systems – metabolism in a test tube.

17.5.2 *Self-replicating molecular systems*

The simplest imaginable self-replicating system consists of one type of molecule that makes exact copies of itself (Wilson, 1998). Under just the right chemical environment, such an isolated molecule will become two copies, then four, then eight molecules and so on in a geometrical expansion. Such an 'autocatalytic' molecule must act as a template that attracts and assembles smaller building blocks from an appropriate chemical broth. Single self-replicating molecules are intrinsically complex in structure, but organic chemists have managed to devise several varieties of these curious beasts, including small

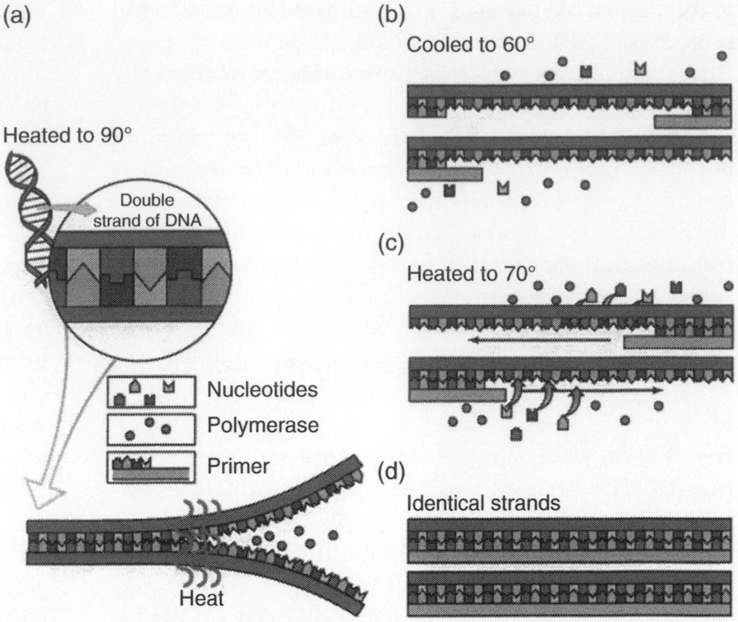

Figure 17.4 The polymerase chain reaction (PCR) copies a sequence of DNA. (a) A strand of DNA is mixed in solution with DNA nucleotides (precursors), a primer that targets a specific piece of DNA, and an enzyme (polymerase) that helps to assemble DNA. The mix is heated to about 90 °C to separate DNA strands. (b) When cooled to about 60 °C, primers attach to the DNA strands. (c) At 70 °C, nucleotides begin to attach to the DNA strands. (d) At the end you have two copies of the desired DNA (from Trefil and Hazen 2009).

peptides (made of amino acids) and short strands of DNA (Von Kiedrowski, 1986; Sievers and Von Kiedrowski, 1994; Rebek, 1994; Lee *et al.*, 1996; Yao *et al.*, 1997).

Nevertheless, these self-replicating molecules don't meet anyone's minimum requirements for life on at least two counts. These systems require a steady input of smaller highly specialized molecules – synthetic chemicals that must be supplied by the researchers. Under no plausible natural environment could these idiosyncratic 'food' molecules arise independently in nature. Furthermore – and this is a key point in distinguishing life from non-life – these particular self-replicating molecules can't change and evolve, any more than a photocopy can evolve from an original.

More relevant to biological metabolism are systems of two or more molecules that form a self-replicating cycle or network. In the simplest system, two molecules (call them AA and BB) form from smaller feedstock molecules A and B. If AA catalyses the formation of BB, and BB in turn catalyses the formation of AA, then the system will sustain itself as long as researchers maintain a reliable supply of food molecules A and B. The well-known polymerase chain reaction (PCR) of crime scene investigation fame is an important example of such a self-replicating molecular system (Mullis *et al.*, 1994; Trefil and Hazen, 2009; Fig. 17.4). A strand of DNA is added to a solution of nucleotides, a primer that targets a specific DNA sequence, and the enzyme polymerase that helps to assemble DNA. Each cycle of heating (to separate DNA strands) and cooling (to trigger assembly of double-stranded DNA) doubles the number of copies of DNA. Twenty cycles can thus produce more than a

million copies of a single piece of DNA. Theorists elaborate on such a model with networks of many molecules, each of which promotes the production of another species in the system (Eigen and Schuster, 1979; Kauffman, 1993; Goldstein, 2006).

What separates living systems from simple self-replicating collections of molecules? The answer in part is complexity; living systems require numerous interacting molecules. In addition, a living metabolic cycle must incorporate a certain degree of sloppiness. Only through such copying 'mistakes' can the system experiment with new, more efficient reaction pathways and thus evolve – an essential attribute of living systems.

A dramatic gap exists between plausible theory and actual experiment. Metabolism is a special kind of cyclical chemical process with two requisite inputs. Living cells undergo chemical reactions, not unlike burning in which two chemicals (oxygen and fuel) react and release energy. However, the trick in metabolism, unlike an open fire, is to capture part of that released energy to make new useful molecules that reinforce the cycle. So metabolism requires a sequence of chemical reactions that work in concert.

17.5.3 The iron–sulfur world

The theoretical and experimental pursuit of the metabolism-first viewpoint is exemplified by Günter Wächtershäuser's iron–sulfur world hypothesis (Wächtershäuser, 1988a, b, 1990a, b, 1992). This origin-of-life scenario incorporates a strikingly original autotrophic, metabolism-first model. According to Wächtershäuser, all of life's essential biomolecules are manufactured in place, as needed from the smallest of building blocks: CO, H_2O, NH_3, H_2S, and so forth. Chemical synthesis is accomplished step-wise, a few atoms at a time.

The contrast between a heterotrophic versus an autotrophic origin is profound, and represents a fundamental point of disagreement among origins-of-life researchers. On the one hand, supporters of heterotrophic origins point out that it is much easier for an organism to use the diverse molecular products that are available in the prebiotic soup, rather than make them all from scratch. Why go to the trouble of synthesizing lots of amino acids if they are already available in the environment? In this sense, heterotrophic cells can be chemically and structurally simpler than autotrophic cells because they do not require all of the biochemical machinery to manufacture amino acids, carbohydrates, lipids and so forth. One can thus argue that simpler heterotrophy should come first.

Autotrophic advocates are equally insistent that true simplicity lies in building molecules a few atoms at a time with just a few simple types of chemical reactions (e.g. Morowitz *et al.*, 2000; Cody, 2004). Supporters of such a mechanism, furthermore, argue that it possesses a significant philosophical advantage: autotrophism, if based on a few robust and universal reaction pathways, is deterministic. Thus, rather than depending on the idiosyncrasies of a local environment for biomolecular components, autotrophic organisms make them from scratch the same way every time, on any viable planet or moon, in a predictable chemical path. Such a philosophy leads to a startling conclusion. For advocates of autotrophy, the prebiotic soup is irrelevant to the origins of life. With autotrophy, biochemistry is hard wired into the universe. The self-made cell emerges from geochemistry as inevitably as basalt or granite.

As with any metabolic strategy, the iron-sulfur world model requires a source of energy, a source of molecules, and a self-replicating cycle. Wächtershäuser's model relies on the abundant chemical energy of minerals that find themselves out of equilibrium with their environment. He begins by suggesting that iron monosulfide (FeS), a common mineral deposited in abundance at the mouths of many deep-sea hydrothermal vents, is unstable with respect to the surrounding seawater. As a consequence, iron monosulfide combines with the volcanic gas hydrogen monosulfide (H_2S) to produce the mineral pyrite (FeS_2) plus hydrogen gas (H_2) and energy.

$$FeS + H_2S \rightarrow FeS_2 + H_2 + Energy$$

Given that energetic boost, hydrogen reacts immediately with carbon dioxide (CO_2) to synthesize organic molecules such as formic acid (HCOOH).

$$Energy + H_2 + CO_2 \rightarrow HCOOH$$

Wächtershäuser envisions cascades of these reactions coupled to build up essential organic molecules from CO_2 and other simple gases.

Recent experiments lend support to the iron-sulfur world hypothesis, which makes the unambiguous prediction that minerals promote a variety of organic reactions. Experiments by Wächtershäuser and colleagues in Germany used iron, nickel and cobalt sulfides to synthesize acetate, an essential metabolic compound with two carbon atoms that plays a central role in countless biochemical processes (Huber and Wächtershäuser 1997). They expanded on this success by adding amino acids to their experiments and making peptides – yet another essential step to life (Huber and Wächtershäuser, 1998; Huber *et al.*, 2003). These and other experiments lead to a firm conclusion: common sulfide minerals have the ability to promote a variety of interesting organic synthesis reactions.

A centrepiece of the iron-sulfur world hypothesis is Günter Wächtershäuser's conviction that life began with a simple, self-replicating cycle of compounds similar to the one that lies at the heart of every cell's metabolism – the 'reductive citric acid cycle.' But how did the first metabolic cycle operate without enzymes? Some of the requisite steps, such as combining carbon dioxide and pyruvate to make oxaloacetate, are energetically unlikely; if one step fails then the whole cycle fails. One of the clever proposals in Günter Wächtershäuser's model is that sulfide minerals promote primordial metabolic reactions. In fact, many modern metabolic enzymes have at their core a small cluster of iron or nickel and sulfur atoms – clusters that look like tiny bits of sulfide minerals. Perhaps ancient minerals played the role of enzymes.

Wächtershäuser's iron–sulfur world was manifest as flat life, which predated the emergence of cells. The first self-replicating entity was a thin layer of chemical reactants on a sulfide mineral surface. The film grew laterally, spreading from mineral grain to mineral grain as an invisibly thin organic film. Bits of these layers could break off and reattach to other rocks to make clone-like colonies. Given time, different minerals might induce variations in the film, fostering new 'species' of flat life.

17.5.4   The RNA world

In spite of Wächtershäuser's elaborate iron–sulfur world scenario and several competing variants of that model (notably Russell and Hall, 1997, 2002), most origin experts dismiss a purely metabolic life form in favour of a genetics-first scenario. Even the simplest known cell must pass volumes of information from one generation to the next, and the only known way to store and copy that much information is with a genetic molecule like DNA or RNA. California chemist Leslie Orgel stated that the central dilemma in understanding a genetic origin of life is the identification of a stable, self-replicating genetic molecule – a polymer that simultaneously carries the information to make copies of itself *and* catalyses that replication (Orgel, 1986). Accordingly, he catalogs four broad approaches to the problem of jumpstarting such a genetic organism.

One possibility is a self-replicating peptide – a short sequence of amino acids that emerged first and then 'invented' DNA. That is an appealing idea because amino acids are thought to have been widely available in the soup. The problem is that, while cells have learned how to form ordered chain-like polymers, amino acids on their own link together in irregular clusters of no obvious biological utility. Alternatively, the simultaneous evolution of proteins and DNA seems even less likely because it requires the emergence of two improbable molecules. Cairns-Smith's clay world scenario provides an intriguing third option, though one totally unsupported by experimental evidence.

Consequently, the favoured genetics origin model is based on a molecule like RNA – a polymer that acted both as both carrier of information and catalyst that promoted self-replication (Woese, 1967; Crick, 1968; Orgel, 1968; Cheng and Unrau, 2010). This idea received a boost in the early 1980s with the discovery of ribozymes – genetic material that also acts as a catalyst (Kruger *et al.*, 1982; Guerrier-Takeda *et al.*, 1983; Zaug and Cech, 1986). Modern life relies on two complexly intertwined molecules: DNA to carry information and proteins to perform chemical functions. This interdependence leads to the chicken-and-egg dilemma: Proteins make and maintain DNA, but DNA carries the instructions to make proteins. Which came first? RNA, it turns out, has the potential to do both jobs.

The 'RNA world' theory that quickly emerged champions the central role of genetic material in the dual tasks of catalyst *and* information transfer (Joyce, 1989, 1991; Joyce and Orgel, 1993). Over the years 'RNA world' has come to mean different things to different people, but four precepts are common to all versions: (1) once upon a time RNA carried genetic information, (2) RNA replication followed the same rules as modern DNA, by matching pairs of bases such as C-G and A-T (or A-U in the case of RNA), (3) RNA can act as a catalyst, and (4) proteins were not involved in the process. The first life form in this scenario was simply a self-replicating strand of RNA, perhaps enclosed in a protective lipid membrane. According to most versions of this hypothesis, metabolism emerged later as a means to make the RNA replication process more efficient.

The RNA world model is not without its difficulties. Foremost among these problems is the exceptional difficulty in the prebiotic synthesis of RNA (Joyce, 1991; Joyce and Orgel, 1993; Wills and Bada, 2000; Orgel, 2003; Sutherland, 2010). Many of the presumed protometabolic molecules are easily synthesized in experiments that mimic prebiotic environments. RNA nucleotides, by contrast, have never been synthesized from scratch, though creative efforts continue (Sutherland, 2007; Anastasi *et al.*, 2008), and recent efforts have succeeded in producing some RNA nucleotides from small precursor molecules (Powner *et al.*, 2009).

Even if a prebiotic synthetic pathway to nucleotides could be found, a plausible geochemical mechanism to link those nucleotides into an RNA strand has not been demonstrated. It is not obvious how useful catalytic RNA sequences would have formed spontaneously in any prebiotic environment. Perhaps, some scientists speculate, a simpler nucleic acid preceded RNA (Nielsen, 1993; Eschenmoser, 1999, 2004; Hazen, 2005).

Whatever the scenario – metabolism first or genetics first, on the ocean's surface or at a deep ocean vent – the origins of life required more than the replication of chemicals. For a chemical system to be alive, it must display evolution by the process of natural selection.

17.6 The emergence of natural selection

Molecular selection, the process by which a few key molecules earned key roles in life's origins, proceeded on many fronts. Some molecules were inherently unstable or highly reactive and so they quickly disappeared from the scene. Other molecules easily dissolved in the oceans and so were effectively removed from contention. Still other molecular species may have sequestered themselves by bonding strongly to surfaces of chemically unhelpful minerals or clumped together into tarry masses of little use to emerging biology.

In every geochemical environment, each kind of organic molecule had its dependable sources and its inevitable sinks. For a time, perhaps for hundreds of millions of years, a kind of molecular equilibrium was maintained as the new supply of each species was balanced by its loss. Such equilibrium features nonstop competition among molecules, to be sure, but the system does not evolve.

The first self-replicating molecules changed that equilibrium. Even a relatively unstable collection of molecules can be present in significant concentration if it learns how to make copies of itself. The first successful metabolic cycle of molecules, for example, would have proven vastly superior to its individual chemical neighbors at accumulating atoms and harnessing energy. But success breeds competition. Inevitable slight variations in the self-replicating cycle, triggered by the introduction of new molecular species or by differences in environment, initiated an era of increased competition.

More efficient cycles flourished at the expense of the less efficient. Evolution by natural selection had begun on Earth.

The simplest self-replicating molecular systems are not alive. Given all the time in the world such systems won't evolve. A more interesting possibility is an environment in which two or more self-replicating suites of molecules compete. These dueling molecular networks vie for resources, mimicking life's unceasing struggle for survival. A stable system without competition, by contrast, has no selective pressure to evolve.

Two common processes – variation and selection – provide the powerful mechanism by which self-replicating systems evolve. For a system to evolve it must first display a range of variations. Natural systems display random variations through molecular mutations, which are undirected changes in the chemical makeup of key biomolecules. Most variations harm the organism and are doomed to failure. Once in a while, however, a random mutation leads to an improved trait – a more efficient metabolism, better camouflage, swifter locomotion, or greater tolerance for extreme environmental conditions. Such beneficial variations are more likely to survive in the competitive natural world – such variations fuel the process of natural selection.

Competition drives the emergence of natural selection. Such behaviour appears to be inevitable in any self-replicating chemical system in which resources are limited and some molecules have the ability to mutate. Over time, more efficient networks of autocatalytic molecules will increase in concentration at the expense of less efficient networks. In such a competitive milieu, the emergence of increasing molecular complexity is inevitable; new chemical pathways overlay the old. So it is that life has continued to evolve over the past four billion years of Earth history.

17.6.1 *Molecular evolution in the lab*

Laboratory studies of molecular evolution have led to methods for the artificial evolution of molecules that accomplish specific chemical tasks. Harvard biologist Jack Szostak, for example, has explored this approach in attempts to design an evolving synthetic laboratory life form (Szostak *et al.*, 2001). Szostak focuses on devising RNA strands that perform one key task such as copying other RNA molecules (Joyce, 1992).

Szostak and coworkers (Doudna and Szostak, 1989; Green and Szostak, 1992; Bartel and Szostak, 1993; Szostak and Ellington, 1993) tackle RNA evolution by first generating a collection of trillions of random RNA sequences, each about 120 nucleotides (RNA 'letters') long. The vast majority of these chains do nothing of interest, but a few are able to perform a specific task, for example grabbing onto another molecule that has been attached to glass beads in a beaker (Hazen *et al.*, 2007). Flushing out the beaker with water removes all of the unattached RNA strands and thus selects and concentrates RNA sequences bonded to the target. These select strands are then copied with a controlled rate of mutational errors to produce a new batch of trillions of RNA sequences, some of which are even better suited to attach to the target molecules. Repeating this process improves the speed and accuracy with which the RNA latches onto the targets. Note that this and other processes of molecular evolution (e.g. Seelig and Szostak, 2007), relying as they do on chemical interfaces, fluid fluxes, and selective cycles, mimic processes that must have occurred on the prebiotic Earth.

An enticing, yet elusive, target of molecular evolution research is self-replicating RNA – a molecule that would carry genetic information, catalyse its own reproduction, and mutate and evolve, as well. But no plausible geochemical environment could have fed such an unbound molecule, nor would it have survived long under most natural chemical conditions. Several synthetic biology research groups ultimately hope to encapsulate such a replicative RNA strand in a lipid vesicle to synthesize a self-replicating cell-like entity (Szostak *et al.*, 2001; Monnard *et al.*, 2008).

The synthesis of an 'RNA organism' would provide a degree of credibility to the RNA world hypothesis, that a strand of RNA (or some chemically more stable precursor genetic molecule) formed the basis of the first evolving, self-replicating chemical system. But such a synthetic life form will not have emerged spontaneously from chemical reactions among the simple molecular building blocks of the prebiotic Earth, nor would such an object be capable of reproduction without a steady supply of laboratory chemicals. The exact sequence of steps that led to life's origins thus remains unknown.

17.7 Three scenarios for the origins of life

What do we know so far? Abundant organic molecules must have been synthesized and accumulated in a host of complex and productive prebiotic geochemical environments. Subsequently, biomolecular systems including lipid membranes and genetic polymers might have formed through self-organization, as well as by selection on mineral surfaces. As molecular complexity increased, it seems plausible that simple metabolic cycles of self-replicating molecules emerged, as did self-replicating genetic polymers.

The greatest gap in understanding life's origins lies in the transition from a more-or-less static geochemical world with an abundance of interesting organic molecules, to an evolving biochemical world. How that transition occurred may be reduced to a choice among three possible scenarios.

1. *Life began with metabolism and genetic molecules were incorporated later*: Following the hypothesis of Wächtershäuser, Morowitz, Russell and others, life may have begun autotrophically. In this scenario, life's first building blocks were the simplest of molecules, while minerals provided chemical energy. A self-replicating chemical cycle akin to the reverse citric acid cycle became established on a mineral surface. All subsequent chemical complexities, including genetic mechanisms and encapsulation into a cell-like structure, arose through natural selection as variants of the cycle competed for resources and the system became more efficient. In this scenario, life first emerged as an evolving chemical coating on rocks.

2. *Life began with self-replicating genetic molecules and metabolism was incorporated later*: In the case of the RNA world hypothesis, life began heterotrophically. Organic molecules in the prebiotic soup, perhaps aided by clays or some other mineral template, self-organized into information-rich polymers. Eventually, one of these polymers, perhaps surrounded by a lipid membrane, acquired the ability to self-replicate. All subsequent chemical complexities, including metabolic cycles, arose through natural selection as variants of the genetic polymer became more efficient at self-replication. In this scenario, life emerged as an evolving polymer with a functional genetic sequence.

3. *Life began as a cooperative chemical phenomenon between metabolism and genetics*: A third possible scenario rests on the possibility that neither primitive metabolic cycles (which lack the means of faithful self-replication) nor primitive genetic molecules (which are not very stable and lack a reliable source of chemical energy) could have progressed far by themselves. If, however, a crudely self-replicating genetic molecule became attached to a crudely functioning surface-bound metabolic coating, then a kind of cooperative chemistry might have kicked in. The genetic molecule might have used chemical energy produced by metabolites to make copies of itself, while protecting itself by binding to the surface. Any subsequent variations of the genetic molecule that fortuitously offered protection for itself or for the metabolites, or improved the chemical efficiency of the system, would have been preserved preferentially. Gradually, both the genetic and metabolic components would have become more efficient and more interdependent.

Scientists do not yet know the answer, but they are poised to find out. Whatever the process, ultimately competition began to drive the emergence of ever more elaborate chemical cycles by the process of natural selection. Inexorably, life emerged, never to relinquish its foothold on Earth.

Acknowledgements

This chapter is adapted in part from: Hazen RM (2005) *Genesis: The Scientific Quest for Life's Origins*. National Academy of Sciences, Joseph Henry Press, Washington DC.

This work was supported in part by grants from the NASA Astrobiology Institute, the NASA Exobiology and Evolutionary Biology Program, the National Science Foundation and the Carnegie Institution of Washington.

References

Abelson PH (1966) Chemical events on the primitive Earth. *Proceedings of the National Academy of Sciences USA*, **55**, 1365–1372.

Anastasi C, Buchet FE, Crowe MA, Helliwell M, Raftery J, Sutherland JD (2008) The search for a potentially prebiotic synthesis of nucleotides via arabinose-3-phosphate and its cyanamide derivative. *Chemistry-A European Journal*, **14**, 2375–2388.

Bada JL, Lazcano A (2002) Some like it hot, but not the first biomolecules. *Science*, **296**, 1982–1983.

Bada JL, Miller SL, Zhao M (1995) The stability of amino acids at submarine hydrothermal vent temperatures. *Origins of Life and Evolution of the Biosphere*, **25**, 111–118.

Bagnold RA (1941) *The Physics of Blown Sand and Desert Dunes*. London: Chapman & Hall.

Bagnold RA (1988) *The Physics of Sediment Transport by Wind and Water*. New York: American Society of Civil Engineers.

Bailey J, Chrysostomou A, Hough JH, *et al.* (1998) Circular polarization in star-formation regions: Implications for biomolecular homochirality. *Science*, **281**, 672–674.

Bargueño P, Pérez de Tudela R (2007) The role of supernova neutrinos on molecular homochirality. *Origins of Life and Evolution of the Biosphere*, **37**, 253–257.

Baross JA, Hoffman SE (1985) Submarine hydrothermal vents and associated gradient environments as sites for the origin and evolution of life. *Origins of Life and Evolution of the Biosphere*, **15**, 327–345.

Bartel DP, Szostak JW (1993) Isolation of new ribozymes from a large pool of random sequences. *Science*, **261**, 1411–1418.

Beinert H, Holm RH, Münck E (1997) Iron-sulfur clusters: Nature's modular, multipurpose structures. *Science*, **277**, 653–659.

Benetoli LOB, de Souza CMD, da Silva KL, *et al.* (2007) Amino acid interaction with and adsorption on clays: FT-IR and Mössbauer spectroscopy and x-ray diffraction investigations. *Origins of Life and Evolution of the Biosphere*, **37**, 479–493.

Benner SA, Kim H-J, Kim M-J, Ricardo A (2010) Planetary organic chemistry and origins of biomolecules. In DW Deamer and JW Szostak (eds), *Origins of Cellular Life*. Cold Springs Harbor Perspectives in Biology. Cold Spring Harbor: Cold Spring Harbor Laboratory Press, pp. 67–88.

Blackmond DG (2007) Chiral 'amnesia' as a driving force for solid-phase homochirality. *Chemistry*, **13**, 3290–3295.

Blackmond DG (2010) The origin of biochemical homochirality. In DW Deamer and JW Szostak (eds), *Origins of Cellular Life*. Cold Springs Harbor Perspectives in Biology. Cold

Spring Harbor: Cold Spring Harbor Laboratory Press, pp.123–140.

Blöchl E, Keller M, Wächtershäuser G, Stetter KO (1992) Reactions depending on iron sulfide and linking geochemistry with biochemistry. *Proceedings of the National Academy of Sciences USA*, **89**, 8117–8120.

Bonner WA (1991) The origin and amplification of biomolecular chirality. *Origins of Life and Evolution of the Biosphere*, **21**, 59–111.

Bonner WA (1995) Chirality and life. *Origins of Life and Evolution of the Biosphere*, **25**, 175–190.

Budin I, Bruckner R, Szostak J (2009) Formation of protocell-like vesicles in a thermal diffusion column. *Journal of the American Chemical Society*, **131**, 9628–9629.

Butlerow A (1861) Formation of monosaccharides from formaldehyde. [in French] *Comptes Rendu Hebdomadaires des Scéances de l'Académie des Sciences (Paris)*, 53, 145–167.

Cairns-Smith AG (1968) The origin of life and the nature of the primitive gene. *Journal of Theoretical Biology*, **10**, 53–88.

Cairns-Smith AG (1982) *Genetic Takeover and the Mineral Origins of Life*. Cambridge: Cambridge University Press.

Cairns-Smith AG (1985) The first organisms. *Scientific American*, **252(6)**, 90–100.

Cairns-Smith AG (1985) *Seven Clues to the Origin of Life*. Cambridge: Cambridge University Press.

Camazine S, Deneubourg J-L, Franks NR, Sneyd J, Theraulaz G, Bonabeau E (2001) *Self-Organization in Biological Systems*. Princeton, NJ: Princeton University Press.

Castro-Puyana M, Salgado A, Hazen RM, Crego AL, Marina ML (2008) Investigation of the enantioselective adsorption of 3-carboxy adipic acid on minerals by capillary electrophoresis. *Electrophoresis*, **29**, 1548–1555.

Chaisson EJ (2001) *Cosmic Evolution: The Rise of Complexity in Nature*. Cambridge, MA: Harvard University Press.

Chen IA, Walde P (2010) From self-assembled vesicles to protocells. In DW Deamer and JW Szostak (eds), *Origins of Cellular Life*. Cold Springs Harbor Perspectives in Biology. Cold Spring Harbor: Cold Spring Harbor Laboratory Press, pp. 179–192.

Cheng LKL, Unrau PJ (2010) Closing the circle: replicating RNA with RNA. In DW Deamer and JW Szostak (eds), *Origins of Cellular Life*. Cold Springs Harbor Perspectives in Biology. Cold Spring Harbor: Cold Spring Harbor Laboratory Press, pp. 229–244.

Chyba CF (2005) Rethinking Earth's early atmosphere. *Science*, **308**, 962–963.

Chyba CF, Sagan C (1992) Endogenous production, exogenous delivery, and impact-shock synthesis of organic molecules: an inventory for the origins of life. *Nature*, **355**, 125–132.

Clark S (1999) Polarized starlight and the handedness of life. *American Scientist*, **87**, 336–343.

Cleaves HJ, Chalmers JH, Lazcano A, Miller SL, Bada JL (2008) A reassessment of prebiotic organic synthesis in neutral planetary atmospheres. *Origins of Life and Evolution of the Biosphere*, **38**, 105–115.

Cleland CE (2007) Epistemological issues in the study of microbial life: alternative terran biospheres? *Studies in the History and Philosophy of the Biological and Biomedical Sciences*, **38**, 847–861.

Cleland CE, Chyba C (2002) Defining life. *Origins of Life and Evolution of the Biosphere*, **32**, 387–393.

Cleland CE, Copley SD (2005) The possibility of alternative microbial life on Earth. *International Journal of Astrobiology*, **4**, 165–173.

Cody GD (2004) Transition metal sulfides and the origins of metabolism. *Annual Review of Earth and Planetary Sciences*, **32**, 569–599.

Cody GD, Boctor NZ, Filley TR, *et al*. (2000) Primordial carbonylated iron-sulfur compounds and the synthesis of pyruvate. *Science*, **289**, 1337–1340.

Cody GD, Boctor NZ, Hazen RM, Brandes JA, Morowitz HJ, Yoder HS Jr. (2001) Geochemical roots of autotrophic carbon fixation: Hydrothermal experiments in the system citric acid-H_2O-(±FeS)-(±NiS). *Geochimica et Cosmochimica Acta*, **65**, 3557–3576.

Cody GD, Boctor NZ, Brandes JA, Filley TR, Hazen RM, Yoder HS Jr (2004) Assaying the catalytic potential of transition metal sulfides for prebiotic carbon fixation. *Geochimica et Cosmochimica Acta*, **68**, 2185–2196.

Corliss JB, Dymond J, Gordon LI, *et al*. (1979) Submarine thermal springs on the Galapagos rift. *Science*, **203**, 1073–1083.

Corliss JB, Baross JA, Hoffman SE (1981) An hypothesis concerning the relationship between submarine hot springs and the origin of life on earth. In *Proceedings of the 26th International Geological Congress, Geology of the Oceans Symposium, Paris 1980* (X. Le Pichon, J. Debyser and F. Vine, eds). *Oceanologica Acta* **4**(supplement), 59–69.

Crick FHC (1968) The origin of the genetic code. *Journal of Molecular Biology*, **38**, 367–379.

Cronin JR, Pizzarello S (1983) Amino acids in meteorites. *Advances in Space Research*, **3**, 5–18.

Deamer DW (1997) The first living systems: A bioenergetic perspective. *Microbiology and Molecular Biology Review*, **61**, 239–261.

Deamer D, Weber AL (2010) Bioenergetics and life's origins. In DW Deamer and JW Szostak (eds), *Origins of Cellular Life*. Cold Springs Harbor Perspectives in Biology. Cold Spring Harbor: Cold Spring Harbor Laboratory Press, pp. 141–156.

De Duve C (1995a) *Vital Dust: Life as a Cosmic Imperative*. New York: Basic Books.

De Duve C (1995b) The beginnings of life on earth. *American Scientist* **83**, 428–437.

Dick SJ, Strick JE (2004) *The Living Universe: NASA and the Development of Astrobiology*. New Brunswick, NJ: Rutgers University Press.

Doudna JA, Szostak JW (1989) RNA catalysed synthesis of complementary strand RNA. *Nature*, **339**, 519–522.

Dyson F (1999) *Origins of Life*. Cambridge: Cambridge University Press.

Ehrenfreund P, Cami J (2010) Cosmic carbon chemistry: from the interstellar medium to the early Earth. In DW Deamer and JW Szostak (eds), *Origins of Cellular Life*. Cold Springs Harbor Perspectives in Biology. Cold Spring Harbor: Cold Spring Harbor Laboratory Press, pp. 21–34.

Ehrenfreund P, Charnley SB (2000) Organic molecules in the interstellar medium, comets, and meteorites. *Annual Review of Astronomy and Astrophysics*, **38**, 427–483.

Eigen M, Schuster P (1979) *The Hypercycle: A Principle of Natural Self-Organization*. Berlin: Springer-Verlag.

Engel MH, Macko SA (1997) Isotopic evidence for extraterrestrial non-racemic amino acids in the Murchison meteorite. *Nature*, **296**, 837–840.

Eschenmoser A (1999) Chemical etiology of nucleic acid structure. *Science*, **284**, 2118–2124.

Eschenmoser A (2004) The TNA-family of nucleic acid systems: Properties and prospects. *Origins of Life and Evolution of the Biosphere*, **34**, 277–306.

Ferris JP (1993) Catalysis and prebiotic synthesis. *Origins of Life and Evolution of the Biosphere*, **23**, 307–315.

Ferris JP (1999) Prebiotic synthesis on minerals: bridging the prebiotic and RNA worlds. *Biology Bulletin*, **196**, 311–314.

Ferris JP (2005) Mineral catalysis and prebiotic synthesis: Montmorillonite-catalysed formation of RNA. *Elements*, **1**, 145–149.

Ferris JP, Hill AR Jr, Liu R, Orgel LE (1996) Synthesis of long prebiotic oligomers on mineral surfaces. *Nature*, **381**, 59–61.

Fox SW (1988) *The Emergence of Life: Darwinian Evolution from the Inside*. New York: Basic Books.

Frederickson JK, Onstott TC (1996) Microbes deep inside the Earth. *Scientific American*, **275**(4), 68–73.

Fry I (2000) *The Emergence of Life on Earth: A Historical and Scientific Overview*. New Brunswick, NJ: Rutgers University Press.

Garrison WM, Morrison DC, Hamilton JG, Benson AA, Calvin M (1951) Reduction of carbon dioxide in aqueous solutions by ionizing radiation. *Science*, **114**, 416–418.

Gerday C, Glansdorff N (2007) *Physiology and Biochemistry of Extremophiles*. Washington: American Society for Microbiology.

Gold T (1999) *The Deep Hot Biosphere*. New York: Copernicus.

Goldstein RA (2006) Emergent robustness in competition between autocatalytic chemical networks. *Origins of Life and Evolution of the Biosphere*, **36**, 381–389.

Green R, Szostak JW (1992) Selection of a ribozyme that functions as a superior template in a self-copying reaction. *Science*, **258**, 1910–1915.

Guerrier-Takada C, Gardiner K, Marsh T, Pace N, Altman S (1983) The RNA moiety of ribonuclease P is the catalytic subunit of the enzyme. *Cell*, **35**, 849–857.

Gusev GA, Saito T, Tsarev VA, Uryson AV (2007) A relativistic neutron fireball from a supernova explosion as a possible source of chiral influence. *Origins of Life and Evolution of the Biosphere*, **37**, 259–266.

Haldane JBS (1929) The origin of life. *The Rationalist Annual*, **1929**, 3–10.

Hansen JL, van Hecke M, Haaning A, *et al.* (2001) Instabilities in sand ripples. *Nature*, **410**, 324.

Hanczyc MM, Mansay SS, Szostak JW (2007) Mineral surface directed membrane assembly. *Origins of Life and Evolution of the Biosphere*, **37**, 67–82.

Hazen RM (2004) Chiral crystal faces of common rock-forming minerals. In: *Progress in Biological Chirality* (eds Palyi G, Zucchi C, Caglioti L). Oxford: Elsevier, pp. 137–151.

Hazen RM (2005) *Genesis: The Scientific Quest for Life's Origin*. Washington, DC: Joseph Henry Press.

Hazen RM (2006) Mineral surfaces and the prebiotic selection and organization of biomolecules. *American Mineralogist*, **91**, 1715–1729.

Hazen RM (2009) The emergence of patterning in life's origin and evolution. *International Journal of Developmental Biology*, doi: 10.1387/ijdb.092936rh

Hazen RM, Deamer DW (2007) Hydrothermal reactions of pyruvic acid: Synthesis, selection, and self-assembly of amphiphilic molecules. *Origins of Life and Evolution of the Biosphere*, **37**, 143–152.

Hazen RM, Sholl DS (2003) Chiral selection on inorganic crystalline surfaces. *Nature Materials*, **2**, 367–374.

Hazen RM, Sverjensky DA (2010) Mineral surfaces, geochemical complexities, and the origins of life. In: *Origins of Cellular Life* (eds Deamer DW, Szostak JW). Cold Springs Harbor Perspectives in Biology. Cold Spring Harbor NY: Cold Spring Harbor Laboratory Press, pp. 157–177.

Hazen RM, Filley T, Goodfriend GA (2001) Selective adsorption of L- and D-amino acids on calcite: implications for biochemical homochirality. *Proceedings of the National Academy of Sciences USA*, **98**, 5487–5490.

Hazen RM, Steele A, Toporski J, Cody GD, Fogel M, Huntress WT Jr. (2002) Biosignatures and abiosignatures. *Astrobiology*, **2**, 512–513.

Hazen RM, Griffin P, Carothers JM, Szostak JW (2007) Functional information and the emergence of biocomplexity. *Proceedings of the National Academy of Sciences USA*, **104**, 8574–8581.

Heinen W, Lauwers AM (1996) Organic sulfur compounds resulting from interaction of iron sulfide, hydrogen sulfide and carbon dioxide in an aerobic aqueous environment. *Origins of Life and Evolution of the Biosphere*, **26**, 131–150.

Hoang QQ, Sicheri F, Howard AJ, Yang DSC (2003) Bone recognition mechanism of porcine osteocalcin from crystal structure. *Nature*, **425**, 977–980.

Holland JH (1995) *Hidden Order*. Reading, MA: Helix Books.

Holland JH (1998) *Emergence: From Chaos to Order*. Reading, MA: Helix Books.

Holm NG (1992) Why are hydrothermal systems proposed as plausible environments for the origin of life? *Origins of Life and Evolution of the Biosphere*, **22**, 5–14.

Huber C, Wächtershäuser G (1997) Activated acetic acid by carbon fixation on (Fe,Ni)S under primordial conditions. *Science*, **276**, 245–247.

Huber C, Wächtershäuser G (1998) Peptides by activation of amino acids with CO on (Ni,Fe)S surfaces: implications for the origin of life. *Science*, **281**, 670–672.

Huber C, Eisenreich W, Hecht S, Wächtershäuser G (2003) A possible primordial peptide cycle. *Science*, **301**, 938–940.

Hunding A, Kepes F, Lancet D, *et al.* (2006) Compositional complementarity and prebiotic ecology in the origin of life. *BioEssays*, **28**, 399–412.

Ji F, Zhou H, Yang Q (2008) The abiotic formation of hydrocarbons from dissolved CO_2 under hydrothermal conditions with cobalt-bearing magnetite. *Origins of Life and Evolution of the Biosphere*, **38**, 117–125.

Jonsson C, Jonsson CL, Sverjensky D, Cleaves HJ, Hazen RM (2009) Attachment of L-glutamate to rutile (TiO_2): A potentiometric, adsorption and surface complexation study. *Langmuir*, **25**, 12127–12135.

Jonsson C, Jonsson CL, Estrada C, Sverjensky D, Cleaves HJ, Hazen RM (2010) Adsorption of L-aspartate to rutile (TiO_2): experimental and theoretical surface complexation studies. *Geochemica et Cosmochemica Acta*, **74**, 2356–2367.

Joyce GF (1989) RNA evolution and the origins of life. *Nature*, **338**, 217–224.

Joyce GF (1991) The rise and fall of the RNA world. *New Biology*, **3**, 399–407.

Joyce GF (1992) Directed molecular evolution. *Scientific American*, **267(6)**, 90–97.

Joyce GF, Orgel LE (1993) Prospects for understanding the RNA world. In: *The RNA World* (eds Gesteland RF, Atkins JF). Cold Spring Harbor, NY: Cold Spring Harbor Laboratory Press, pp. 1–25.

Kasting JF (2003) Methane-rich proterozoic atmosphere? *Geology*, **31**, 87–90.

Kasting JF, Siefert JL (2002) Life and the evolution of Earth's atmosphere. *Science*, **296**, 1006–1068.

Kauffman SA (1993) *The Origins of Order: Self-Organization and Selection in Evolution*. New York: Oxford University Press.

Kessler MA, Werner BT (2003) Self-organization of sorted patterned ground. *Science*, **299**, 380–383.

Klussmann M, Iwamura H, Mathew SP, *et al.* (2006) Thermodynamic control of asymmetric amplification in amino acid catalysis. *Nature*, **441**, 621–623.

Klussmann M, Izumi T, White AJ, Armstrong A, Blackmond DG (2007) Emergence of solution-phase homochirality via engineering of amino acids. *Journal of the American Chemical Society*, **129**, 7657–7660.

Kruger K, Grabowski PJ, Zaug AJ, Sands J, Gottschling DE, Cech TR (1982) Self-splicing RNA: Autoexcision and autocyclization of the ribosomal RNA intervening sequence of Tetrahymena. *Cell*, **31**, 147–157.

Lahav N (1999) *Biogenesis*. Oxford: Oxford University Press.

Lahav N, White D, Chang S (1978) Peptide formation in the prebiotic era: Thermal condensation of glycine in fluctuating clay environments. *Science*, **201**, 67–69.

Lambert J-F (2008) Adsorption and polymerization of amino acids on mineral surfaces: a review. *Origin of Life and Evolution of the Biosphere*, **38**, 211–242.

Lasaga AC, Holland HD, Dwyer MJ (1971) Primordial oil slick. *Science*, **174**, 53–55.

Lazcano A (2010) Historical development of origins research. In: *Origins of Cellular Life* (eds Deamer DW, Szostak JW). Cold Springs Harbor Perspectives in Biology. Cold Spring Harbor, NY: Cold Spring Harbor Laboratory Press, pp. 5–20.

Lazcano A, Miller SL (1999) On the origin of metabolic pathways. *Journal of Molecular Evolution*, **49**, 424–431.

Lee DH, Granja JR, Martinez JA, Severin K, Ghadiri MR (1996) A self-replicating peptide. *Nature*, **382**, 525–528.

Luisi PL, Varela FJ (1989) Self-replicating micelles: a chemical version of a minimal autopoietic system. *Origins of Life and Evolution of the Biosphere*, **19**, 633–643.

Madigan MT, Marrs BL (1997) Extremophiles. *Scientific American*, **276(4)**, 82–87.

Mansy SS, Schrum JP, Krishnamurthy M, Tobe S, Treco DA, Szostak JW (2008) Template-directed synthesis of a genetic polymer in a model protocell. *Nature*, **454**, 122–126.

Martins Z, Alexander CMO'D, Orzechowska GE, Fogel ML, Ehrenfreund P (2007) Indigenous amino acids in primitive CR meteorites. *Meteoritics and Planetary Science*, **42**, 2125–2136.

McCollom TM, Simoneit BR (1999) Abiotic formation of hydrocarbons and oxygenated compounds during thermal decomposition of iron oxalate. *Origin of Life and Evolution of the Biosphere*, **29**, 167–186.

Miller SL (1953) Production of amino acids under possible primitive earth conditions. *Science*, **17**, 528–529.

Miller SL (1955) Production of some organic compounds under possible primitive earth conditions. *Journal of the American Chemical Society*, **77**, 2351–2361.

Miller SL, Urey HC (1959) Organic compound synthesis on the primitive earth. *Science*, **130**, 245–251.

Monnard PA, DeClue MS, Ziock H-J (2008) Organic nanocompartments as biomimetic reactors and protocells. *Current Nanoscience*, **4**, 71–87.

Morowitz HJ (1992) *The Beginnings of Cellular Life: Metabolism Recapitulates Biogenesis*. New Haven: Yale University Press.

Morowitz HJ (2002) *The Emergence of Everything*. New York: Oxford University Press.

Morowitz HJ, Kostelnik JD, Yang J, Cody GD (2000) The origins of intermediary metabolism. *Proceedings of the National Academy of Sciences USA*, **97**, 7704–7708.

Mullis KB, Ferré F, Gibbs RA (1994) *The Polymerase Chain Reaction* Boston: Birkhäuser Press.

Nelson DL, Cox MM (2004) *Lehninger's Principles of Biochemistry*, 4th edition. New York: Worth Publishers.

Nicolis G, Prigogine I (1989) *Exploring Complexity: An Introduction*. New York: W.H. Freeman and Company.

Nielsen PE (1993) Peptide nucleic acid (PNA): a model structure for the primordial genetic material? *Origins of Life and Evolution of the Biosphere*, **23**, 323–327.

Oparin AI (1924) *Proiskhozhdenie Zhizny* (in Russian). Moscow: Rabochii. (An English translation appears in: Bernal JD (1967) *The Origin of Life*. London: Weidenfeld & Nicholson.

Orgel LE (1968) Evolution of the genetic apparatus. *Journal of Molecular Biology*, **38**, 381–393.

Orgel LE (1986) RNA catalysis and the origin of life. *Journal of Theoretical Biology*, **123**, 127–149.

Orgel LE (1998) Polymerization on the rocks: theoretical introduction. *Origins of Life and Evolution of the Biosphere*, **28**, 227–234.

Orgel LE (2003) Some consequences of the RNA world hypothesis. *Origins of Life and Evolution of the Biosphere*, **33**, 211–218.

Oró J (1960) Synthesis of adenine from ammonium cyanide. *Biochemical and Biophysical Communications*, **2**, 407–412.

Oró J (1961) Mechanism of synthesis of adenine from hydrogen cyanide under possible primitive earth conditions. *Nature*, **191**, 1193–1194.

Parkes RJ, Craig BA, Bale SJ, *et al.* (1993) Deep bacterial biosphere in Pacific Ocean sediments. *Nature*, **371**, 410–413.

Pasek M, Lauretta D (2008) Extraterrestrial flux of potentially prebiotic C, N, and P to the early Earth. *Origins of Life and Evolution of the Biosphere*, **38**, 5–21.

Pedersen K (1993) The deep subterranean biosphere. *Earth Science Reviews*, **34**, 243–260.

Pitsch S, Eschenmoser A, Gedulin B, Hui S, Arrhenius G (1995) Mineral induced formation of sugar phosphates. *Origins of Life and Evolution of the Biosphere*, **25**, 297–334.

Pizzarello S, Cronin JR (2000) Non-racemic amino acids in the Murray and Murchison meteorites. *Geochimica et Cosmochimica Acta*, **64**, 329–338.

Pizzarello S, Shock E (2010) The organic composition of carbonaceous meteorites: the evolutionary story ahead of biochemistry. In: *Origins of Cellular Life* (eds Deamer DW, Szostak JW). Cold Springs Harbor Perspectives in Biology. Cold Spring Harbor: Cold Spring Harbor Laboratory Press, pp. 89–108.

Pizzarello S, Weber A (2004) Prebiotic amino acids as asymmetric catalysts. *Science*, **303**, 1151.

Pizzarello S, Huang Y, Alexandre MR (2008) Molecular asymmetry in extraterrestrial chemistry: Insights from a pristine meteorite. *Proceedings of the National Academy of Sciences USA*, **105**, 3700–3704.

Podlech J (1999) New insight into the source of biomolecular homochirality: An extraterrestrial origin for molecules of life. *Angewandt Chemie International Edition English*, **38**, 477–478.

Powner MW, Gerland B, Sutherland JD (2009) Synthesis of activated pyrimidine ribonucleotides in prebiotically plausible conditions. *Nature*, **459**, 239–242.

Prigogine I (1984) *Order out of Chaos: Man's New Dialogue with Nature*. Toronto: Bantam Books.

Rajamani S, Vlassov A, Benner S, Coombs A, Olasagasti F, Deamer D (2008) Lipid-assisted synthesis of RNA-like polymers from mononucleotides. *Origins of Life and Evolution of the Biosphere*, **38**, 57–74.

Rebek J Jr (1994) Synthetic self-replicating molecules. *Scientific American*, **271**(1), 48–55.

Rimola A, Ugliengo P, Sodupe M (2009) Formation versus hydrolysis of the peptide bond from a quantum-mechanical viewpoint: the role of mineral surfaces and implications for the origin of life. *International Journal of Molecular Sciences*, **10**, 746–760.

Root-Bernstein R (2009) An 'ecosystems first' theory of the origin of life based on molecular complementarity. In VA Basiuk (ed.) *Astrobiology: Emergence, Search and Detection*. Los Angeles: American Scientific Publishers, pp. 1–30.

Russell MJ, Hall AJ (1997) The emergence of life from iron monosulphide bubbles at a submarine hydrothermal redox and pH front. *Journal of the Geological Society of London*, **154**, 377–402.

Russell MJ, Hall AJ (2002) From geochemistry to biochemistry: chemiosmotic coupling and transition element clusters in the onset of life and photosynthesis. *The Geochemical News*, **#113**, 6–12.

Scorei R, Cimpoiaşu VM (2006) Boron enhances the thermostability of carbohydrates. *Origins of Life and Evolution of the Biosphere*, **36**, 1–11.

Seelig B, Szostak JW (2007) Selection and evolution of enzymes from a partially randomized non-catalytic scaffold. *Nature*, **448**, 828–831.

Segré S, Deamer DW, Lancet D (2001) The lipid world. *Origins of Life and Evolution of the Biosphere*, **31**, 119–145.

Shapiro R (1988) Prebiotic ribose synthesis: a critical analysis. *Origins of Life and Evolution of the Biosphere*, **18**, 71–85.

Sievers D, von Kiedrowski G (1994) A self-replication of complementary nucleotide-based oligomers. *Nature*, **369**, 221–224.

Sleep NH, Zahnle K, Kasting JF, Morowitz HL (1989) Annihilation of ecosystems by large asteroid impacts on the early Earth. *Nature*, **342**, 139–142.

Smith JV (1998) Biochemical evolution. I. Polymerization on internal, organophilic silica surfaces of dealuminated zeolites and feldspars. *Proceedings of the National Academy of Sciences USA*, **95**, 3370–3375.

Solé R, Goodwin B (2000) *Signs of Life: How Complexity Pervades Biology*. New York: Basic Books.

Stevens TO, McKinley JP (1995) Lithoautotrophic microbial ecosystems in deep basalt aquifers. *Science*, **270**, 450–454.

Strasak M, Sersen F (1999) An unusual reaction of adenine and adenosine on montmorillonite: a new way of prebiotic synthesis of some purine nucleotides? *Naturwissenschaften*, **78**, 121–122.

Strick JE (2000) *Sparks of Life: Darwinism and the Victorian Debate over Spontaneous Generation*. Cambridge, MA: Harvard University Press.

Sutherland JD (2007) Looking beyond the RNA structural neighborhood for potentially primordial genetic systems. *Angew Chemie International Edition English*, **46**, 2354–2356.

Sutherland JD (2010) Ribonucleotides. In DW Deamer and JW Szostak (eds) *Origins of Cellular Life*. Cold Springs Harbor Perspectives in Biology. Cold Spring Harbor NY: Cold Spring Harbor Laboratory Press, pp. 109–122.

Szostak JW, Ellington AD (1993) In vitro selection of functional RNA sequences. In RF Gesteland and JF Atkins (eds) *The RNA World*. Cold Spring Harbor, NY: Cold Spring Harbor Laboratory Press, pp. 511–533.

Szostak JW, Bartel DP, Luisi PL (2001) Synthesizing life. *Nature*, **409**, 387–390.

Tian F, Toon OB, Pavlov AA, De Sterck H (2005) A hydrogen-rich early Earth atmosphere. *Science*, **308**, 1014–1016.

Trefil J, Hazen RM (2009) *The Sciences: An Integrated Approach*, 6th edition. John Wiley & Sons, Inc.: Hoboken, NJ.

Von Kiedrowski G (1986) A self-replicating hexadeoxynucleotide. *Angewandt Chemie International Edition English*, **25**, 932–935.

Wächtershäuser G (1988a) Before enzymes and templates: theory of surface metabolism. *Microbiology Review*, **52**, 452–484.

Wächtershäuser G (1988b) Pyrite formation, the first energy source for life: a hypothesis. *Systematic Applied Microbiology*, **10**, 207–210.

Wächtershäuser G (1990a) The case for the chemoautotrophic origin of life in an iron-sulfur world. *Origins of Life and Evolution of the Biosphere*, **20**, 173–176.

Wächtershäuser G (1990b) Evolution of the first metabolic cycles. *Proceedings of the National Academy of Sciences USA*, **87**, 200–204.

Wächtershäuser G (1992) Groundworks for an evolutionary biochemistry: the iron-sulfur world. *Progress in Biophysics and Molecular Biology*, **58**, 85–201.

Ward PD, Benner SA (2007) Alien biochemistries. In WT Sullivan and JT Baross (eds) *Planets and Life*. New York: Cambridge, pp. 537–544.

Weber AL (2007) The sugar model: Autocatalytic activity of the triose-ammonia reaction. *Origins of Life and Evolution of the Biosphere*, **37**, 105–111.

Weber AL, Pizzarello S (2006) *Proceedings of the National Academy of Sciences USA*, **103**, 12713.

Whitfield J (2009) Nascence man. *Nature*, **459**, 316–319.

Wills C, Bada JL (2000) *The Spark of Life: Darwin and the Primeval Soup*. Cambridge, MA: Perseus.

Wilson EK (1998) Go forth and multiply. *Chemical and Engineering News*, **76** (December 7, 1998), 40–44.

Woese CR (1967) *The Genetic Code*. New York: Harper and Row.

Wolf-Simon F, Davies PCW, Anbar AD (2009) Did nature also choose arsenic? *International Journal of Astrobiology*, doi:10.1017/S1473550408004394.

Yao S, Ghosh I, Zutshi R, Chmielewski J (1997) A pH-modulated, self-replicating peptide. *Journal of the American Chemical Society*, **119**, 10559–10560.

Zaug AJ, Cech TR (1986) The intervening sequence RNA of Tetrahymenia is an enzyme. *Science*, **231**, 470–475.

18

MINERALOGICAL CO-EVOLUTION OF THE GEOSPHERE AND BIOSPHERE

Robert M. Hazen[1] and Dominic Papineau[2]

[1]Geophysical Laboratory, Carnegie Institution of Washington,
5251 Broad Branch Road NW, Washington, DC 20015, USA
[2]Department of Earth and Environmental Sciences, Boston College,
140 Commonwealth Avenue, Chestnut Hill, MA 02467, USA

18.1 Introduction

A central theme of the emerging science of geobiology is the close relationship between geological and biological processes at or near Earth's surface. Feedbacks between geology and biology are especially evident in the >3.5 billion year history of the chemical evolution of Earth's oceans and atmosphere (Chapters 2–7). Less well recognized, but equally profound, are biologically mediated changes in the solid Earth – the co-evolution of Earth's mineralogy and life.

The subject of mineral evolution (Hazen et al., 2008, 2009; Hazen, 2010; Hazen and Ferry, 2010) considers changes over geological time in several aspects of the mineral realm. Most dramatic, and central to the theme of increased heterogeneity of Earth's surface environment, is mineralogical diversity. Thus, over a span of 4.56 billion years, the number of different minerals has increased from about a dozen to more than 4500 known types (with about 50 newly described species added yearly; e.g. http://rruff.info/ima). Mineral evolution also encompasses such changes as the relative abundances of near-surface minerals, the range of mineral compositional variations (including solid solutions and minor or trace elements), and the sizes and shapes of mineral grains. We find that all of these aspects of Earth's near-surface mineralogy have been profoundly affected by biological influences.

The near-surface mineralogy of any terrestrial planet or moon evolves as a result of varied physical, chemical and biological processes. Accordingly, Hazen et al. (2008) recognized three eras of Earth's mineral evolution. The first era commenced with gravitational collapse of the pre-Solar nebula and ignition of the Sun – events that spanned a relatively brief 10-million-year period during which dust and gas formed planetesimals. Prestellar molecular clouds contain dust particles that incorporate approximately a dozen minerals, all of which form in the high-temperature envelopes of exploding stars. These few phases – the graphite and diamond forms of carbon, silicon carbide, silicon and titanium nitrides, and oxides and silicates of Al, Mg, Ca, and Ti – constituted a starting point for Earth's mineral evolution. Flash heating events of the young Sun in the protoplanetary disc produced perhaps 60 different mineral species, which are the primary refractory constituents of chondritic meteorites. Subsequent alteration of these chondrites in planetessimals resulted in many new meteorite types and a consequent increase in mineral diversity, to as many as 250 different species, all of which are still found today in unweathered meteorite samples.

The second era of Earth's mineral evolution, following the epic moon-forming impact event, was a time when elements and molecules in the crust and upper mantle were subjected to repeated, strictly physical and chemical reworking by a variety of igneous and metamorphic processes. Igneous fractional crystallization, crystal settling, partial melting, and associated fluid-rock interactions resulted in the separation and concentration of elements and led to a marked diversification of the terrestrial mineral realm. In addition, plate tectonic reworking of the upper mantle and crust led to a host of new minerals associated with massive hydrothermal ore deposits and surface exposure of high-pressure metamorphic terrains.

Fundamentals of Geobiology, First Edition. Edited by Andrew H. Knoll, Donald E. Canfield and Kurt O. Konhauser.
© 2012 Blackwell Publishing Ltd. Published 2012 by Blackwell Publishing Ltd.

The third era of mineral evolution, perhaps unique to Earth in our Solar System, is associated with biological activity and the coevolution of the geosphere and biosphere. Life caused gradual changes in the compositions of oceans and atmosphere, most dramatically through the rise in atmospheric oxygen fostered by photosynthetic microorganisms. Microbes, with their varied metabolic redox strategies, are also able to concentrate metals and create and sustain chemical gradients in a variety of geochemical environments, thus leading to precipitation of minerals at scales from microenvironments to regional terrains. The Phanerozoic innovation of bioskeletons of carbonate, phosphate or silica resulted in new mechanisms of mineralization that continue to the modern era. This chapter thus reviews Earth's mineral evolution, with an emphasis on the role of life in our planets evolving geosphere.

18.2   Prebiotic mineral evolution I – evidence from meteorites

Hazen *et al.* (2008) recognized 10 stages of Earth's mineral evolution, based on a number of significant irreversible mineralogical events, each triggered by new physical, chemical and/or biological processes (Table 18.1; Fig. 18.1). The first five of these evolutionary stages, which appear to be deterministic aspects of the evolution of any Earth-like planet, diversified Earth's mineralogy by producing an increasingly wide range of localized compositions and environmental conditions (e.g. P, T, a_{H2O}, a_{CO2}, a_{O2}, etc.) from which minerals form.

The mineral evolution of terrestrial planets begins in so-called 'dense molecular clouds,' which are breeding grounds of the protoplanetary disks where stars and planets form. Material from these objects, composed of gas (primarily H and He) with widely dispersed dust grains, has been recovered in the form of individual presolar interstellar grains in the matrix of primitive meteorites. These dust grains incorporate nanometre- to micron-sized grains of a handful of refractory minerals, including carbides, nitrides, oxides and silicates (Brearley and Jones, 1998; Nittler, 2003; Messenger *et al.*, 2003, 2006; Stroud *et al.*, 2004; Mostefaoui and Hoppe, 2004; Vollmer *et al.*, 2007).

The first stage of mineral evolution involved flash heating and accretion of fine-grained nebular material into primitive planetesimals, evidence of which is preserved in the least-altered (type 3.0) chondritic meteorites. These meteorites host abundant chondrules, which are small spherical objects (typically ~1 mm diameter) that represent molten droplets formed in space (Rubin, 2000; Desch and Connolly, 2002). Chondritic meteorites are complex aggregations of chondrules, along with calcium-aluminium-rich inclusions and other small (~1 mm diameter) refractory objects that formed in the

Table 18.1 Ten stages of mineral evolution of terrestrial planets, with possible timing on Earth, examples of minerals, and estimates of the cumulative number of different mineral species (after Hazen *et al.*, 2008; Hazen and Ferry, 2010). Note that the timings of some of these stages overlap, and several stages continue to the present

Stage	Age (Ga)	~ Cumulative number of species
0. 'Ur-minerals' in presolar grains	>4.56 Ga	12
1. Primary chondrite minerals	>4.56 Ga	60
2. Achondrite and planetesimal alteration	>4.56 to 4.55 Ga	250
3. Igneous rock evolution	4.55 to 4.0 Ga	350–500
4. Granite and pegmatite formation	4.0 to 3.5 Ga	1000
5. Plate tectonics	>>3.0 Ga	1500
6. Anoxic biological world	3.9 to 2.5 Ga	1500
7. Great Oxidation Event	2.5 to 1.9 Ga	>4000
8. Intermediate ocean	1.9 to 1.0 Ga	>4000
9. Snowball Earth events	1.0 to 0.542 Ga	>4000
10. Phanerozoic era of biomineralization	0.542 Ga to present	4500+

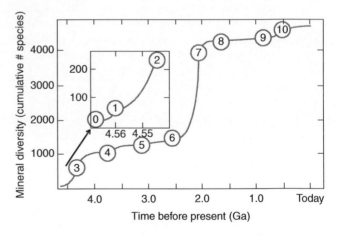

Figure 18.1 Estimated cumulative number of different mineral species versus time, with key events in Earth history. Numbers correspond to stages, as outlined in Table 18.1.

high-temperature protostellar environment by evaporation, condensation and melting of nebular materials. The least altered chondrites are characterized by extreme mineralogical parsimony, with only about 60 different mineral species, many of which are present only as micro- or nano-scale grains. (e.g. Brearley and Jones, 1998; Ebel, 2006; Messenger *et al.*, 2006; MacPherson, 2007).

The most important primary chondrule minerals are olivine (Mg_2SiO_4) and pyroxene ($MgSiO_3$), with kama-cite and taenite (both Fe-Ni metal) and troilite (FeS) as the major iron-bearing phases, and chromite ($FeCr_2O_4$) and pentlandite [$(Fe,Ni)_9S_8$] as important but volumetrically minor accessory minerals. In addition, at least 30 accessory minerals occur in the fine-grained silicate matrix, and as opaque minerals, commonly as micro- or nanoscale phases. The primary minerals of chondritic meteorites represent the earliest stage in the mineral evolution of terrestrial planets. It should be noted that, in spite of the lack of mineral diversity, chondrites contain all of the chemical complexity of terrestrial planets, including all of the 83 stable geochemical elements. Most of these elements are present in solid solutions, but are too widely dispersed to form their own discrete mineral phases.

As gravitation caused chondrite meteorites to clump together into larger planetesimals, aqueous and thermal alteration – the second stage of mineral evolution – led to new suites of minerals (Brearley and Jones, 1998; Krot *et al.*, 2006; Brearley, 2006; MacPherson, 2007). Aqueous alteration products of silicates include the first clay minerals and other layer silicates, hydroxides, sulfates and carbonates (Brearley and Jones, 1998). Thermal alteration within planetessimals produced many more new minerals, such as the first significant occurrences of albite feldspar ($NaAlSi_3O_8$), feldspathoids (alkali aluminosilicates), and a variety of micas, amphiboles and pyroxenes. All told, chondritic meteorites display approximately 150 different mineral species (Rubin, 1997a, 1997b; Brearley and Jones, 1998; Brearley, 2006; MacPherson, 2007).

Mineralogical diversity further increased through igneous processes in planetesimals greater than ~200 km in diameter. Achondrite meteorites reveal different extents of partial melting and differentiation under the influence of variable composition and impact events (McCoy *et al.*, 2006; McCoy, 2010). Planetesimals that underwent extensive melting experienced separation of stony (crust and mantle) and metallic (core) components, as well as fractionation of silicate magmas (Shukolyukov and Lugmair, 2002; McCoy *et al.*, 2006; Wadhwa *et al.*, 2006).

Some stony achondrites, such as the common eucrites, represent planetesimal crust. These meteorites are similar to terrestrial basalts and feature the first significant appearance of a number of important rock-forming minerals, including the quartz form of SiO_2, potassium feldspar ($KAlSi_3O_8$), titanite ($CaTiSiO_5$) and zircon ($ZrSiO_4$). Iron–nickel meteorites, by contrast, represent the core material of differentiated planetesimals and are dominated by Fe-Ni alloys (kamacite, taenite), with significant metal sulfides (troilite, daubreelite), carbides (cohenite and haxonite), and graphite.

A total of ~250 mineral species are known from meteorites (e.g. Mason, 1967; Rubin, 1997a, b; Brearley and Jones, 1998; Gaffey *et al.*, 2002; MacPherson, 2007; McCoy, 2010). Even though many of these phases are unstable at near-surface Earth conditions, it is safe to assume that all varieties of meteorites have fallen throughout Earth's 4.5-billion-year history. As a result, all of these diverse meteorite minerals have been present continuously in Earth's near-surface environment. Thus, relatively few meteorite minerals, predominately magnesium silicates, iron sulfide and Fe-Ni metal, provided raw materials for Earth and other terrestrial planets.

18.3 Prebiotic mineral evolution II – crust and mantle reworking

By 4.5 billion years ago, Earth and the other terrestrial planets had formed and differentiated into core, mantle and crust. The impact of a Mars-sized body at about that time disrupted and largely melted Earth's outer layers and resulted in the formation of the Moon (Tonks and Melosh, 1993; Ruzicka *et al.*, 1999; Touboul *et al.*, 2007). Subsequently, as Earth's surface solidified, near-surface mineralogical diversity increased through the reworking of crust and mantle by igneous processes.

In spite of the relative paucity and extensive alteration of the early rock record, many of Earth's earliest mineral-forming processes can be inferred from the existing rock record and knowledge of petrologic, geochemical and geodynamic principles (Papineau, 2010a). Thus, the third stage of mineral evolution, which involved processing of igneous rocks, likely dominated Earth's near-surface environment for more than half a billion years following planetary formation. Mafic and ultramafic lithologies, notably basalts composed primarily of plagioclase and pyroxene and cumulate rocks formed by crystal settling, would have dominated the earliest crustal igneous rocks. This wet mafic igneous veneer would have been repeatedly cycled by fractional melting, fractional crystallization, and magma immiscibility, leading to a diversity of igneous lithologies – gabbro, diorite, granodiorite, and granite – in part following Bowen's classic reaction series (Bowen, 1928).

These igneous rocks may represent the end point of mineral evolution for relatively small (<5000 km diameter), volatile-poor terrestrial bodies, including the Moon (Jolliff *et al.*, 2006) and Mercury (Clark, 2007). Accordingly, Hazen *et al.* (2008) estimate that the mineralogical diversity of these bodies is limited to no more than about 350 distinct phases.

By contrast, igneous activity on the volatile-rich Earth, Venus and Mars quickly led to a greater diversity as a consequence of volcanic outgassing and fluid-rock interactions associated with the formation of the atmosphere and hydrosphere (e.g. Holland, 1984; Kump *et al.*, 2001;

Catling and Claire, 2005). Minerals formed in this very early stage include hydrous silicates and hydroxides, notably the first extensive production of clay minerals via hydrothermal alteration of basalt. As Earth's climate cooled, the mineral ice (crystalline H_2O) would have appeared for the first time at Earth's surface along with the first evaporite deposits, including halide and sulfate minerals. It is likely that Venus and Mars progressed to this third stage of mineral evolution, with perhaps 500 different mineral species (Hazen *et al.*, 2008). However, these planets may not have experienced three additional key processes – granite production, plate tectonics, and anoxic biological metabolism – that greatly expanded Earth's mineralogical diversity.

Earth's fourth stage of mineral evolution saw the extensive production of granite-like rocks (granitoids) and the consequent development of the volatile-rich lithosphere (Smithies and Champion, 2000; Sandiford and McLaren, 2002; Smithies *et al.*, 2003). Granitoids arise from the partial melting of wet basalts or clay-rich sediments, and thus approximate the minimum melting composition in the SiO_2-Al_2O_3-Na_2O-K_2O system (e.g. White and Chappell, 1983; Hess, 1989; Kemp *et al.*, 2007; Eiler, 2007). Repeated melting of rocks at the base of the crust can thus generate large volumes of granitoid rocks in the overlying continental crust (Parman 2007; Pearson *et al.* 2007).

Granitoids resulted in the formation of many distinctive minerals, including quartz (SiO_2), K-feldspar ($KAlSi_3O_8$), albite feldspar ($NaAlSi_3O_8$), and a variety of micas, amphiboles and alkali pyroxenes. The common granite accessory phases, such as titanite ($CaTiSiO_5$), zircon ($ZrSiO_4$), monazite [$(Ce,La,Y,Th)PO_4$], apatite [$Ca_5(PO_4)_3(OH,F)$] and fluorite (CaF_2), would also have been produced in significantly greater abundance than previously. Of special interest was the enrichment of two-dozen rare elements that concentrate in residual aqueous fluids and lead to the formation of pegmatite minerals (Foord, 1982; Moore, 1982; Černy 1982; Ewing and Chakoumakos, 1982; London, 2008; Grew and Hazen, 2009, 2010a, b; Grew *et al.*, 2011). Development of complex pegmatites marked the first occurrences of distinctive minerals of lithium, cesium, boron, beryllium, and Nb-Ta, as well as a host of Zr-Hf, Ga, Sn, rare-earth elements and U minerals. As many as 500 minerals are unique to those pegmatitic environments.

While granite formation likely represents the mineralogical end point for some terrestrial planets, plate tectonics – the fifth stage of mineral evolution on Earth – led to significant additional mechanisms of mineral diversification (Parnell, 2004). Experts disagree as to the time when plate tectonics began, but most agree that some form of episodic subduction was active significantly before 3.0 Ga, and possibly before 4.0 Ga (Harrison *et al.*, 2005; Smithies *et al.*, 2005; Witze, 2006; Silver *et al.*, 2006;

Silver and Behm, 2008). But from their onset at least as early as 3.9 Ga (Shirey *et al.*, 2008), plate tectonics had significant mineralogical consequences, most significantly owing to extensive hydrothermal processing of the upper mantle and crust. For example, magmatic and volcanogenic processes at subduction zones and at ridges produced massive sulfide deposition and associated precious metal concentrations (Sangster, 1972; Hutchinson, 1973). These deposits host dozens of sulfide minerals, as well as more than 100 associated minor phases including selenides, tellurides, arsenides, antimonides, and numerous sulfosalts (Dana, 1958). Another mineralogical consequence of plate tectonics was the uplift and exposure of regional metamorphic terrains with many characteristic high-pressure phases.

Taken together, these varied mineral-forming mechanisms – evolution of igneous rocks, fluid–rock interactions, metamorphism, and uplift – can result in at least 1500 different mineral species. However, many of the 4500 known minerals at or near Earth's surface today may have required an additional evolutionary innovation – the origin of cellular life.

18.4   The anoxic Archean biosphere

Microbial life had become well established by the Archean Eon (3.8 to 2.5 Ga), though microorganisms appear to have had minimal influence on Earth's mineral diversity during this period (stage 6 of Hazen *et al.*, 2008). Most notably, while modern life persists and thrives in oxygen-rich environments, geochemical and mineralogical evidence indicate that Earth's atmosphere was essentially devoid of free oxygen during the Archean and early Paleoproterozoic (Holland, 1984; Farquhar *et al.*, 2001; 2007; Sverjensky and Lee, 2010). Rounded detrital grains of pyrite (FeS_2) and uraninite (UO_2) have been found in Archean clastic sedimentary rocks in the South African Witwatersrand Group and in the early Paleoproterozoic Canadian Blind River (Rasmussen and Buick, 1999; England *et al.*, 2001, 2002; Kumar and Srinivasan 2002). These minerals rapidly degrade in the presence of free oxygen, so their occurrence in subaerial clastic sediments points to erosion and transport in an oxygen-deficient environment (Frimmel 2005). Multiple sulfur isotope ratios in sedimentary rocks also indicate that the atmosphere was mostly anoxic from the beginning of the sedimentary rock record at about 3.85 Ga until the early Paleoproterozoic (Farquhar *et al.*, 2000, 2007; Ono *et al.*, 2003; Bekker *et al.*, 2004; Papineau *et al.*, 2007).

The most convincing microbially-mediated Archean rocks are the stromatolites (Fig. 18.2), which are precipitated by microorganisms that date from at least 3.49 Ga (Walter *et al.*, 1980; Lowe, 1980; Buick *et al.*, 1981; Byerly *et al.*, 1986; Walter, 1994; Grotzinger and Knoll, 1999;

Figure 18.2 (a) The oldest known stromatolitic dolomite from the Trendall locality of the ~3.35 Ga Strelley Pool Formation, Western Australia. (b) Three-dimensional view of stomatolitic chert-barite from the 3.49 Ga Dresser Formation, Western Australia (photos by D. Papineau).

Allwood *et al.*, 2006; Van Kranendonk, 2006, 2007; see Chapter 16). Stromatolites are organo-sedimentary layered rock structures predominantly accreted by sediment trapping and binding and by biomineralization due to the growth and metabolic activity of microorganisms. Stromatolites are composed of common rock-forming minerals, including carbonates, chert, phosphates and/or iron minerals, and mineral diversity in stromatolites became richer in the Paleoproterozoic (Fig. 18.3). Modern stromatolites are known to form by lithification of microbial mats by photosynthetic cyanobacteria (Reid *et al.*, 2000), though cyanobacteria constitute only a small proportion of microorganisms in stromatolitic communities (Papineau *et al.*, 2005a). Although chemically precipitated stromatolites can be difficult to interpret unambiguously (Allwood *et al.*, 2009), these distinctive ancient fossil structures provide the most unambiguous case for localized Paleoarchean biomineralization (Fig. 18.2). In spite of these occurrences, extensive sedimentary carbonates are uncommon in the geological record until ~3.0 Ga, when the first significant dolostone [CaMg(CO₃)₂] units appear. The ~2.95 Ga Steep Rock Group of northwestern Ontario incorporates a thick carbonate platform with giant stromatolites (up to 10 m) with aragonite ($CaCO_3$) fans, calcite (also $CaCO_3$), and rare gypsum ($CaSO_4 \cdot 2H_2O$) moulds (Jolliffe, 1955; Wilks and Nisbet, 1988; Tomlinson *et al.*, 2003). However, extensive development of limestone platforms did not occur until ~2.7 Ga (Sumner, 1997). Dolomitic stromatolites, which occur in western Australia's 2.72 Ga Tumbiana Formation, contain the oldest primary aragonite in the form of nanocrystals closely associated with organic globules (Lepot *et al.*, 2008).

In addition to stromatolites, a few other distinctive rock types may point to Archean biological influences. Banded iron formations (BIFs), which are among the earliest kinds of sedimentary rocks, are found in formations older that 3.8 Ga (Papineau, 2010a). Both environmental and compositional factors influence the composition of BIFs, which have been classified into three general families according to their mineralogy (Fig. 18.4; Klein, 2005): (1) an oxide facies [with jasper (a form of finely crystalline SiO_2), hematite (Fe_2O_3) and/or magnetite (Fe_3O_4)], a sulfide facies [often organic-rich and containing pyrite and marcasite (both FeS_2), pyrrhotite ($Fe_{1-x}S$) and chalcopyrite ($CuFeS_2$)], and a carbonate facies {including siderite ($FeCO_3$), ankerite [$FeCa(CO_3)_2$], Fe-dolomite [$Ca(Mg,Fe)CO_3$] and calcite ($CaCO_3$)}. While the oxide facies is most abundant in the Archean, sulfide and carbonate facies also occur. The secular trend in BIF occurrences throughout the Archean and their extensive development at the Archean–Proterozoic boundary has been related to geochemical changes in the environments of deposition and to the development of large igneous provinces (LIPs) (Isley and Abbott, 1999; Klein, 2005; Bekker *et al.*, 2010).

Uncertainty remains about the origin of these chemically precipitated sedimentary rocks. Most minerals in BIFs can precipitate in the absence of O_2, and none has so far been identified as an unambiguous biosignature. But the hypothesis that microorganisms were responsible for their formation (e.g. LaBerge, 1973; Widdel *et al.*, 1993) continues to have support, which points to an important role by oxygenic photosynthesizers and/or Fe^{2+} oxidizers (Anbar and Holland, 1992; Konhauser *et al.*, 2002, 2007; Kappler *et al.*, 2005). However, the case for microbial mediation in the origins of BIFs is not

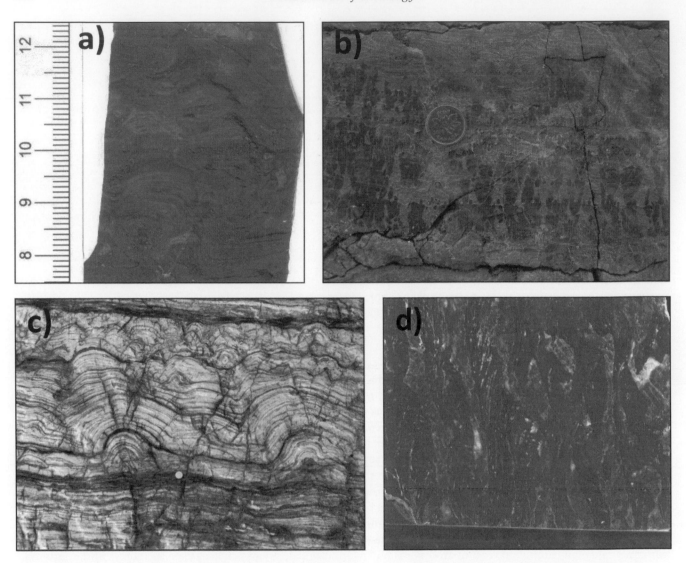

Figure 18.3 (a) Slab of a stromatolitic jasper banded iron formation from the ~1.88 Ga Biwabik Formation, Minnesota, USA, (b) stromatolitic chert-dolomite from the ~1.9 Ga McLeary Formation, Belcher Islands, Canada, (c) stromatolitic dolomite from the 2.06–2.09 Ga Rantamaa Formation, Finland, (d) stromatolitic phosphorite (carbonate-fluorapatite and dolomite) from the 2.0 Ga Jhamarkotra Formation, Rajasthan, India (photos by D. Papineau).

unambiguous, since significant abiotic components of BIFs appear to originate from hydrothermal sources (Jacobsen and Pimentel-Klose, 1988; Bau and Möller, 1993; Klein, 2005; Bekker *et al.*, 2010). These observations are not in conflict with a biological involvement for BIF precipitation and may give clues on the depositional environment and microbial community composition where they formed.

The Paleoarchean sulfur cycle was undoubtedly influenced by biology, as well. Evidence from sulfur isotopes indicates that microbial sulfur metabolisms were active as early as 3.5 Ga in Western Australia (Shen *et al.*, 2001; 2009; Phillippot *et al.*, 2007; Wacey *et al.*, 2010). Accordingly, Huston and Logan (2004) suggested that biological processes played a role in sulfate deposition associated with some banded iron formations. Sulfate minerals, notably the least soluble sulfate barite ($BaSO_4$), occur as massive bands, finely wrinkled laminated beds, and veins in some Archean volcano-sedimentary successions. Sulfate is the ocean's second most abundant anion today, but its concentration has varied over time. Kah *et al.* (2004) estimate that sulfate levels in the Proterozoic ocean were only 5 to 15% of modern levels, the primary source coming from oxidative weathering of continental sulfides (Habicht *et al.*, 2002). However sulfate and oxygen levels in some restricted Mesoproterozoic basins were higher than oceanic levels, which favoured the microbially mediated oxidation of sulfides, themselves derived from microbial sulfate reduction, and thus points to the existence of complex microbial ecosystems in environments with distinct redox states (Parnell *et al.*, 2010).

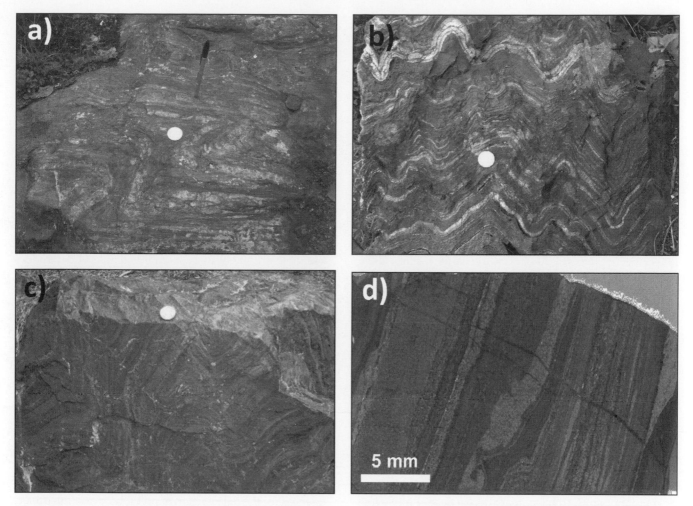

Figure 18.4 (a) Chert-magnetite-siderite banded iron formation from the Neoarchean Hunter Mine Group, Canada, (b) chert-magnetite banded iron formation from the Neoarchean Bababudan Supracrustal Belt, India, (c) jasper-chert banded iron formation from the Paleoproterozoic Negaunee Formation, Michigan, United States, (d) sulfide facies banded iron formation from the ~2.0 Ga Pathavaara Formation, Finland. All coins are about 2.5 cm in diameter (photos by D. Papineau).

Increased abundance of sulfate in seawater thus relates to periods of widespread oxidative weathering of the crust, which lead to favourable conditions for sulfate reducing microorganisms, such as in the Paleo- and Neoproterozoic.

Several authors have proposed that, even in the absence of free oxygen, microbial activity could have triggered significant mineralogical consequences. Rosing *et al.* (2006) made the intriguing observation that plate tectonics and biological activity in the early Archean could have combined to increase dramatically the rate of granite production, and hence the stabilization of Earth's earliest continents prior to 3.0 Ga. This idea follows from the fact that granite forms as a consequence of subduction and dehydration of altered basalt and sediments. Since microbial activity increases rates of silicate weathering by an order of magnitude over abiotic processes (Bennet *et al.*, 1996; Paris *et al.*, 1996;

Barker *et al.*, 1998; Tazaki, 2005), surface microbial communities on an anoxic world may have indirectly affected the rate of sediment accumulation, and thus continent formation. Note, however, that Edmond and colleagues (e.g. Edmond and Huh 1997) have argued that abiological chemical weathering has represented the dominant mode of continental weathering in Earth history.

In spite of the likely influences of cellular life on the nature and distribution of Archean carbonates, sulfates, and even granites, life on the anoxic early Earth played a relatively minor role in modifying Earth's surface mineralogy. Hazen *et al.* (2008) estimated that by 2.5 Ga Earth's near-surface environment hosted approximately 1500 different mineral species, most of which could form in the near-surface environment of any volatile-rich anoxic terrestrial planet that had experienced Earth-like cycles of granite formation and plate tectonics. However, large-scale biologically mediated changes in ocean and atmospheric chemistry more than 2.2 Ga changed this situation dramatically.

18.5 The Great Oxidation Event

For the past 2.45 billion years minerals at or near Earth's surface co-evolved with life at an accelerated pace. Indeed, Hazen *et al.* (2008) conclude that most of Earth's mineral diversity today may be an indirect consequence of the biosphere. The rise of atmospheric oxygen from about 2.45 to 2.0 billion years ago was perhaps the most significant mineralogical event in Earth's history and has been the subject of intense recent research (Canfield *et al.*, 2000; Kump *et al.*, 2001; Kasting, 2001; Kasting and Siefert, 2002; Towe, 2002; Holland, 2002; Bekker *et al.*, 2004; Barley *et al.*, 2005; Catling and Claire, 2005; Papineau *et al.*, 2005b, 2007; Kump and Barley, 2007; Sverjensky and Lee, 2010). While the series of events that led to the Great Oxidation Event is still incompletely understood, elevated tectonic activity and rifting in the early Paleoproterozoic may have triggered a series of major glaciations and thus perturbed the greenhouse-controlled climate. Increased tectonic activity coupled with rapid climate change might have accelerated erosion, which in turn may have provided a critical source of seawater nutrients that stimulated primary productivity (Papineau *et al.*, 2007, 2009, 2010b). Another major control on climate is the potent greenhouse gas methane, which may have decreased in response to increasing atmospheric oxygen or due to a decrease in the activity of methanogenic Archaea near the Archean–Proterozoic boundary (Konhauser *et al.*, 2009; Chapter 7).

This dramatic episode of atmospheric oxygenation was principally a consequence of photosynthesis by cyanobacteria (Papineau, 2010b). Sedimentological studies of Paleoproterozoic sediments reveal an unprecedented increase in taxonomic diversity and expansion of stromatolites (Melezhik *et al.*, 1997). Oxygenic photosynthetic cyanobacteria were most likely to have been active long before 2.45 Ga, but other factors served to delay onset of atmospheric oxidation. For example, methanogenesis, both microbial and abiotic (from serpentinization), may have impeded Archean atmospheric oxidation. Atmospheric methane concentrations may have attained levels as much as 600 times present day concentrations, which may have suppressed the rise of free oxygen (Sherwood-Lollar *et al.*, 1993, 2006; Catling *et al.*, 2001; Kasting and Siefert, 2002; Scott *et al.*, 2004; Ueno *et al.*, 2006).

The Great Oxidation Event had a profound effect on Earth's near-surface mineralogy, as measured by an extreme change in near-surface oxygen fugacity. Thermochemical data point to important mineralogical consequences of this change. At the low oxygen activities before 2.2 billion years, numerous minerals that are common today would have been unstable (Garrels and

Christ, 1965; Bowers *et al.*, 1984; Hazen *et al.*, 2009; Sverjensky and Lee, 2010). Indeed, more than half of all known minerals are oxidized and hydrated weathering products of other minerals (*http://rruff.info*). For example, 256 of 321 known copper oxide minerals are hydrated species, most of which likely resulted from weathering in an oxygen-rich environment. Thermochemical calculations (Garrels and Christ, 1965, figures 7–27*a* and *b*; Bowers *et al.*, 1984) reveal that such distinctive copper minerals as cuprite (Cu_2O), tenorite (CuO), malachite [$Cu_2(CO_3)(OH)_2$], and perhaps azurite [$Cu_3(CO_3)_2(OH)_2$] are unstable at oxygen fugacities less than 10^{-40}, and thus may not have occurred to any significant extent prior to biological oxygenesis. Hundreds of other minerals for which no thermochemical data are available are likely to be even less stable under reducing surface conditions. These unstable minerals include as many as 202 of 220 different uranium minerals (Hazen *et al.* 2009), 319 of 451 minerals with Mn and O, 47 of 56 minerals with Ni and O, and 582 of 790 minerals with Fe and O are hydrated and oxidized. While many of these rare species are known from only a handful of specimens, it is evident that the biologically mediated occurrence of perhaps >2500 new mineral phases after 2.2 Ga would have marked a significant rise in Earth's mineralogical diversity. It thus appears likely that the majority of mineral species on Earth are biologically mediated – the consequence of microbially-driven atmospheric oxidation.

The Great Oxidation Event also had important consequences for the distribution of previously existing minerals at or near Earth's surface. For example, after the major Paleoproterozoic glaciations between about 2.2 and 1.9 Ga, carbonates preserve a remarkable diversity of stromatolite types, and various kinds of shallow-marine environments became host to mineralogically diverse stromatolites (Fig. 18.3). Atmospheric oxygenation also led to new large-scale episodes of hematitic BIFs deposition as well as massive manganese oxide deposits in marine environments, with numerous distinctive iron and manganese minerals (Leclerc and Weber 1980; James and Trendall, 1982; Dasgupta *et al.*, 1992; Tsikos and Moore, 1997; Barley *et al.*, 1997; Roy, 2006).

The rise of oxygen also had profound effects on the chemistry of coastal seawater. Oxidative weathering of sulfides in the continental crust led to increased delivery of sulfate via rivers to the ocean. Consequently, sulfate minerals such as anhydrite and gypsum became more abundant in shallow marine environments (Cameron, 1983; El Tabakh *et al.*, 1999; Bekker *et al.*, 2006). The combination of high weathering rates and climate change was also probably responsible for the formation of shallow marine phosphatic sediments, as revealed by the oldest significant sedimentary phosphorite deposits (Banerjee, 1971; Chauhan, 1979; Bekker *et al.*, 2003; Papineau, 2010b).

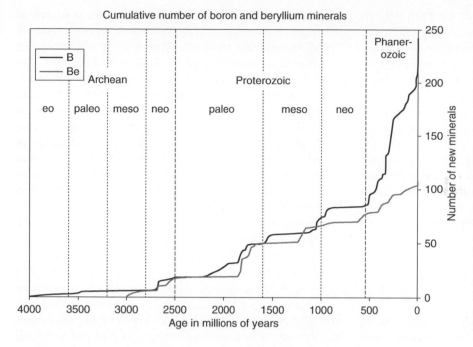

Cumulative number of boron and beryllium minerals

Figure 18.5 Plot of the reported oldest occurrences of 106 Be minerals and 261 B minerals based on literature search (Grew and Hazen, 2009, 2010a, 2010b). The plot is cumulative because each reported new appearance is added to the number of minerals having been reported prior the age of the appearance. The plot is not meant to indicate the totality of minerals forming in the Earth's near surface at any given time, including the present; i.e. some minerals formed once or over a limited time interval, and have not formed since. Note the approximate doubling of Be and B mineral species during the time interval from 2.0 to 1.7 Ga. (courtesy of Edward Grew).

Thus, the Paleoproterozoic rise in atmospheric oxygen, coupled with oxygenation of near-surface groundwater, likely represents the single most important event in the diversification of Earth's mineralogy.

18.6 A billion years of stasis

Hazen *et al.* (2008) identify the billion-year interval commencing at about 1.9 billion years ago, a period of relative stasis, as Earth's eighth stage of mineral evolution. A relatively abrupt cessation of banded iron formations at ca. 1.85 Ga may point to a significant increase in deep-ocean oxidation, which caused the removal of the source of upwelling Fe^{2+} that fueled BIF deposition (Cloud, 1972; Holland, 1984). Alternatively, the end of BIF deposition could have been due to an increased supply of H_2S from microbial sulfate reduction, which overcame the hydrothermal flux of iron to the deep ocean and resulted in sulfidic bottom waters (Canfield, 1998; Poulton *et al.*, 2004). However, the spatial distribution of euxinic basins at that time varied considerably (Poulton *et al.*, 2010) and the global extent of such environments with sulfidic bottom waters remains to be determined. In either case, the change in ocean chemistry to an intermediate oxidation state – the 'intermediate ocean' (Anbar and Knoll, 2002) – appears to have resulted as a consequence of prior surface oxidation and enhanced weathering.

Another contributing factor to deep ocean anoxia was probably the raining out of organic carbon, also the consequence of biological activity. Deposition of organic-rich shales in marine environments reached a peak in development around 2.1 billion years ago (Condie *et al.*, 2001). The deep ocean thus would have become starved of Fe and other metals through precipitation of insoluble metal sulfides, rather than by oxidation. As Anbar and Knoll (2002) pointed out, this scenario has important implications for life in sulfidic deep-ocean conditions, which would have scavenged biologically important metals such as Fe and Mo (Scott *et al.*, 2008), as well as phosphorus (Berner, 1972; Wheat *et al.*, 1996). Consequently, by restricting biological productivity, sulfide mineralization may have deterred the radiation of life for a billion years.

The billion-year period following the Great Oxidation Event may have been a time of minimal mineralogical innovation. Most physical, chemical and biological mineral forming processes had evolved, and atmospheric and ocean chemistries did not change sufficiently to trigger significant new modes of mineralization. On the other hand, Grew and Hazen (2009, 2010a) note significant increases in the diversity of preserved beryllium and boron minerals between 2.0 and 1.7 Ga – pulses of mineralization that may reflect bias in the preservation of rare minerals, but might also point to the significant time required to select and concentrate these rare elements in mineral-forming fluids (Fig. 18.5).

18.7 The snowball Earth

Numerous lines of geological and geochemical evidence reveal that Earth experienced another sequence of global interconnected fluctuations in climate and atmospheric composition between about 1.0 and 0.542 Ga. This ninth stage of mineral evolution (Hazen et al., 2008) encompasses at least two major global-scale glaciations, or 'snowball Earth' events, which occurred between about 0.75 and 0.58 Ga (Hoffman et al., 1998; Kennedy et al., 1998; Jacobsen, 2001). Geological evidence includes extensive near-sea level, low-latitude glacial deposits (Young, 1995; Hoffman and Prave, 1996; Halverson, 2005; Evans, 2006) and the occurrence of cap carbonates stratigraphically overlying glacial deposits (Fairchild, 1993; Kennedy, 1996; Halverson et al., 2005). These geological observations are amplified by carbon and sulfur isotope excursions between and after glaciations (Knoll et al. 1986; Hoffman et al., 1998; Gorjan et al., 2000; Hurtgen et al., 2005; Fike et al., 2006; Halverson and Hurtgen, 2007; Fike and Grotzinger, 2008; Halverson et al., 2010), the distribution of iron minerals (Young, 1976; Canfield et al., 2007, 2008), and anomalies in concentrations of the element iridium (Bodiselitsch et al., 2005).

According to the snowball Earth scenario, glacial cycles occurred during a period when continents were clustered near the equator (Kirschvink, 1992; Hoffman and Schrag, 2000). Decreased atmospheric CO_2 levels (or other greenhouse gases) are likely to have triggered these severe glaciations, which would then have been amplified by a runaway 'albedo feedback' caused by snow and ice (Ridgwell et al., 2003; Donnadieu et al., 2004). Ultimately, Earth became covered in ice or slush. During these global glaciations events, life was more restricted than previously in aerial extent and total biomass, but phototrophic organisms and complex microbial communities persisted in some marine environments (Olcott et al., 2005).

These snowball or slushball episodes are estimated to have lasted on the order of 10 million years, based on volcanic CO_2 fluxes at convergent margins (Caldeira, 1991) and the accumulated iridium from the steady rain of meteoritic dust in glacial deposits (Bodiselitsch et al., 2005). During those glacial intervals, the hydrological cycle slowed significantly owing to reduced evaporation. Meanwhile, atmospheric concentrations of volcanic CO_2 increased to as high as 0.12 atmospheres, because the ice-covered continents prevented the normal sequestration of carbon dioxide by crustal weathering (Caldeira and Kasting, 1992; Pierrehumbert 2004). Atmospheric levels of CO_2 and the associated oceanic pH during the older interglacial periods were high, although they were not as elevated as those of the younger deglaciation, perhaps indicating increasingly colder snowball Earth events (Bao et al., 2008, 2009; Kasemann et al., 2010). Ultimately, this increased atmospheric CO_2 led to relatively rapid greenhouse warming, melting of the reflective ice cover, and a dramatic episode of carbonate mineral deposition from seawater that had become highly supersaturated in CO_2.

An important consequence of Neoproterozoic events in the Earth system was an increase in atmospheric oxygen (Fike et al., 2006; Canfield et al., 2007). Increased nutrient availability in postglacial periods of the Neoproterozoic may have provided nutrients for algal blooms (Nagy et al., 2009; Elie et al., 2007; Papineau, 2010b) to produce significant quantities of oxygen in the atmosphere. Higher delivery of nutrients to seawater by Neoproterozoic rivers could have been the result of intense greenhouse climate during interglacial periods between snowball Earth events, and/or possibly combined with rifting of Rodinia (Papineau, 2010b). This increase of atmospheric oxygen was essential to the subsequent rise of large oxygen-breathing animals (e.g. Runnegar, 1991; Canfield et al., 2007).

The snowball Earth events may have had significant effects on surface mineralogy. One novel proposal, the so-called 'clay mineral factory' hypothesis of Kennedy and coworkers (2006), builds on the well established ability of microbes to enhance clay mineral production, for example by the bio-weathering of feldspar and mica (Schwartzman and Volk, 1989; Bennet et al., 1996; Paris et al., 1996; Barker et al., 1998; Ueshima and Tazaki, 1998; Ueshima et al., 2000; Tazaki, 2005). Accordingly, Kennedy et al. (2006) document a significant increase in clay mineral deposition after 1.0 Ga that may have been the result of increased microbial activity in soils (however, see Tosca et al., 2009). Such an increase in clay production may have had important atmospheric consequences, because clay mineral surfaces are efficient in adsorbing and sequestering organic carbon (Hedges and Kiel, 1995; Mayer et al., 2004). Because the oxidation of organic carbon is an important sink for atmospheric oxygen, the clay-mediated removal of organic carbon could have contributed to the rise in atmospheric oxygen.

The geochemical cycle of phosphorus points to other feedback mechanisms among minerals, oceans, and microbes. Phosphorus supplied by rivers to the oceans can episodically increase bioproductivity of photosynthetic organisms and biomineralization. Thus, large sedimentary phosphorite deposits are observed in intervals immediately following snowball Earth events, when a relatively stagnant ocean was followed by periods of vigorous ocean circulation (Cook and Shergold, 1984; Donnelly et al., 1990). Major Neoproterozoic and Cambrian post-glacial phosphorite deposits occur in North and South America, Africa, Europe, Asia and Australia (Cook and Shergold, 1984; Notholt and Sheldon, 1986; Howard, 1986; Yueyan, 1986; Dardenne et al., 1986; Flicoteaux and Trompette, 1998; Misi et al., 2006).

18.8 The rise of skeletal mineralization

Since the rise of animals with mineral skeletons in the early Cambrian Period (0.542 billion years ago), biology has dominated near-surface mineralogy. This tenth stage in Earth's mineral evolution (Hazen *et al.*, 2008; Dove, 2010) saw the abrupt rise of all major skeletal minerals (the carbonates calcite, aragonite, and magnesian calcite; the calcium phosphate apatite; and the opal form of silica), and few new types of structural biominerals have appeared since then (Runnegar, 1987; Knoll, 2003; Dove *et al.*, 2003). Calcium carbonate minerals, which represent the most extensive and diverse group of biominerals, played many roles. The planktonic calcifiers, for example, for the first time provided a steady source of $CaCO_3$ to deep ocean sediments and thus reduced the episodic formation of shallow-water sedimentary carbonates. This buffering of ocean carbonate-ion concentration, in turn, moderated glacial events and reduced the likelihood of snowball Earth events during the past half-billion years (Ridgwell *et al.*, 2003).

An important mineralogical innovation was the variable biomineralization of calcium carbonate by corals, mollusks and other invertebrates (Knoll, 2003). Calcite is the predominant biocarbonate from the Cambrian through early Carboniferous periods, but a shift to the aragonite form of $CaCO_3$ is observed in the late Paleozoic Era through the mid-Jurassic Period. Calcite again became the dominant carbonate biomineral after the mid-Jurassic (Stanley and Hardie, 1998), though both calcite and aragonite are common skeletal minerals in modern invertebrates. Variations in the skeletal mineralogy of individual organisms are controlled by macromolecules (glycoproteins) in modern mollusks (Falini *et al.*, 1996; Cohen and McConnaughey, 2003), whereas changes over time in the relative abundance of calcite versus aragonite appear to correlate with variations in ocean chemistry: Aragonite and Mg-calcite are favoured over calcite when Mg/Ca mole ratio is greater than 2 (Stanley and Hardie, 1998).

More than 20 different phosphate minerals, notably hydroxylapatite [$Ca_5(PO_4)_3(OH)$] and whitlockite [$Ca_9H(Mg,Fe)(PO_4)_7$], occur as biominerals in both vertebrates, notably as the principal minerals in teeth and bones, and invertebrates, for example, in the shells of inarticulate brachiopods (Knoll, 2003). Microbial precipitation of phosphates played the dominant role in the formation of what are now commercial phosphorite deposits (Zhao *et al.*, 1985; Burnett and Riggs, 1990; Cook and Shergold, 1990).

Clay mineralization provides another important example of the increasing importance of biological activity through time. Production of clay minerals in soils, especially since the advent of land plants with root systems in the Siluro-Devonian periods, is estimated to have been at least an order of magnitude greater than that of previous eras (Schwartzman and Volk, 1991; Bennet *et al.*, 1996; Paris *et al.*, 1996; Barker *et al.*, 1998; Tazaki, 2005). This increase, which may be seen as an

extension of the 'clay mineral factory' hypothesis of Kennedy *et al.* (2006), represents a distinctive characteristic of post-Ordovician mineralogy. Indeed, evidence suggests that atmospheric oxygen concentrations continued to rise perhaps up to 30% during the Phanerozoic, perhaps partly due to increased rates of organic carbon sequestration on clay mineral surfaces (Berner *et al.*, 2000) or to periods of increased primary biological productivity.

Microbes are likely to have played many significant roles in modifying Earth's near-surface mineralogy during the past half-billion years. Indeed, the geochemical cycles of most elements are now known to influence Earth's near-surface microbiology (Banfield and Nealson, 1997; Banfield *et al.*, 1998; Frankel and Bazylinski, 2003). For example, Christner *et al.* (2008) discovered that airborne microbes may play a role in the nucleation and growth of ice particles in clouds and in subsequent snowfall. Microbial colonies are known to form biofilms and colloids with large negatively charged reactive surface areas that may adsorb metals and precipitate a variety of minerals. Thus, microbes that catalytically oxidize pyrite (and produce acidic waters as a byproduct) may increase dissolution and precipitation reaction rates by six orders of magnitude (Singer and Stumm, 1970; Nordstrom and Southam, 1997). Even the precipitation of gold is strongly affected by biological activity; Reith *et al.* (2006) observe that microbes can both solubilize gold and trigger the precipitation of secondary gold grains from aqueous $AuCl_4^-$ in shallow crustal environments. Thus, some gold deposits may have been enriched by microbial activity.

18.9 Summary

Three principal mechanisms drive the evolution of Earth's near-surface mineralogy. First are physical and chemical processes, including planetary differentiation, outgassing, fractional crystallization, partial melting, crystal settling and leaching by aqueous fluids, all of which separate, select, and concentrate elements. These processes will inevitably lead to diversification of an initially homogeneous and complex element distribution into a broad spectrum of bulk compositions on any terrestrial planet or moon. New suites of minerals invariably will emerge from this separation and concentration of elements into new compositional regimes.

The second mechanism of mineral evolution is the subjection of varying bulk compositions to an increased range of intensive variables, including temperature, pressure, and the activities of volatiles such as H_2O, CO_2, and O_2. Any terrestrial planet will experience a range of environments, such as ice caps, dry lakes, high-pressure metamorphic terrains, deep-ocean hydrothermal systems, and impact sites, each of which promotes the genesis of distinctive suites of minerals.

The third and most dramatic driving force behind mineral evolution is the influence of living organisms. Life creates and sustains local- to global-scale compositional gradients that promote reaction pathways that lead to new minerals. In particular, thousands of minerals occur as a consequence of microbially induced atmospheric oxidation, which is the single most important cause for mineralogical diversification at or near Earth's surface.

These general principles for the emergence of mineralogical complexity on Earth will apply equally to any differentiated asteroid, moon, or terrestrial planet. Mineral evolution will always occur in a logical progression as a result of local, regional, and global processes. The stage to which a body advances in mineralogical diversity will directly reflect the extent to which cyclic processes, including igneous differentiation, granitoid formation, plate tectonics, atmospheric and oceanic reworking (including weathering), and biological influences, have affected the body's history. Each planet or moon can thus be characterized according to its stage of mineralogical evolution. Furthermore, the stage of mineral evolution of a distant planet or moon, if it can be accurately determined by remote sensing or surface sampling, may eventually point to the existence other living worlds.

Acknowledgements

We thank the Editors for the opportunity to contribute to this volume. We gratefully acknowledge valuable discussions and constructive reviews by Russell Hemley, James Kasting, Scott McLennen, and Dimitri Sverjensky. This work was supported in part by grants from the NASA Astrobiology Institute, the NASA Exobiology and Evolutionary Biology Program, the National Science Foundation, and the Carnegie Institution of Washington.

References

Allwood AC, Walter MR, Kamber BS, Marshall CP, Burch IW (2006) Stromatolite reef from the Early Archaean era of Australia. *Nature*, **441**, 714–718.

Allwood AC, Grotzinger JP, Knoll AH, *et al.* (2009) Controls on development and diversity of Early Archean stromatolites. *Proceedings of the National Academy of Sciences of the USA*, **106**, 9548–9555.

Anbar AD, Holland HD (1992) The photochemistry of manganese and the origin of banded iron formations. *Geochimica et Cosmochimica Acta*, **56**, 2595–2603.

Anbar AD, Knoll AH (2002) Proterozoic ocean chemistry and evolution: A bioinorganic bridge? *Science*, **297**, 1137–1142.

Banerjee DM (1971) Precambrian stromatolitic phosphorites of Udaipur, Rajasthan, India. *GSA Bulletin*, **82**, 2319–2329.

Banfield JF, Nealson KH (eds) (1997) Geomicrobiology: Interactions between microbes and minerals. *Reviews in Mineralogy and Geochemistry*, **35**, 448 p.

Banfield JF, Welch SA, Edwards KJ (1998) Microbes as geochemical agents. *The Geochemical Society News*, **96**, 11–17.

Bao H, Lyons JR, Zhou C (2008) Triple oxygen isotope evidence for elevated CO_2 levels after a Neoproterozoic glaciation. *Nature*, **453**, 504–506.

Bao H, Fairchild IJ, Wynn, PM, Spotl C (2009) Stretching the envelope of past surface environments: Neoproterozoic glacial lakes from Svalbard. *Science*, **323**, 119–122.

Barker WW, Welch SA, Banfield JF (1998) Experimental observations of the effects of bacteria on aluminosilicate weathering. *American Mineralogist*, **83**, 1551–1563.

Barley ME, Pickard AL, Sylvester PJ (1997) Emplacement of a large igneous province as a possible cause of banded iron formation 2.45 billion years ago. *Nature*, **385**, 55–58.

Barley ME, Bekker A, Krapez B (2005) Late Archean to early Paleoproterozoic global tectonics, environmental change and the rise of atmospheric oxygen. *Earth and Planetary Science Letters*, **238**, 156–171.

Bau M, Möller P (1993) Rare earth element systematics of the chemically precipitated component in Early Precambrian iron-formations and the evolution of the terrestrial atmosphere-hydrosphere–lithosphere system. *Geochimica et Cosmochimica Acta*, **57**, 2239–2249.

Bekker A, Karhu JA, Eriksson KA, Kaufman AJ (2003) Chemostratigraphy of Paleoproterozoic carbonate successions of the Wyoming Craton: Tectonic forcing of biogeochemical change? *Precambrian Research*, **120**, 279–325.

Bekker A, Holland HD, Wang P-L, *et al.* (2004) Dating the rise of atmospheric oxygen. *Nature*, **427**, 117–120.

Bekker A, Karhu JA, Kaufman AJ (2006) Carbon isotope record for the onset of the Lomagundi carbon isotope excursion in the Great Lakes area, North America. *Earth and Planetary Science Letters*, **148**, 145–180.

Bekker A, Slack JF, Planavsky N, *et al.* (2010) Iron formations: The sedimentary product of a complex interplay among mantle, tectonic, oceanic, and biospheric processes. *Economic Geology*, **105**, 467–508.

Bennet PC, Hiebert FK, Choi WJ (1996) Microbial colonization and weathering of silicates in petroleum-contaminated groundwater. *Chemical Geology*, **132**, 45–53.

Berner RA (1972) Phosphate mineral removal from sea water by adsorption on volcanogenic ferric oxides. *Earth and Planetary Science Letters*, **18**, 77–86.

Berner RA, Petsch SA, Lake JA, *et al.* (2000) Isotope fractionation and atmospheric oxygen: implications for Phanerozoic O_2 evolution. *Science*, **287**, 1630–1633.

Bodiselitsch B, Koeberl C, Master S, Reimold WU (2005) Estimating duration and intensity of Neoproterozoic snowball glaciations from Ir anomalies. *Science*, **308**, 239–242.

Bowen NL (1928) The Evolution of the Igneous Rocks. Princeton University Press, Princeton, New Jersey.

Bowers TS, Jackson KJ, Helgeson HC (1984) Equilibrium Activity Diagrams for Coexisting Minerals and Aqueous Solutions at Pressures and Temperatures to 5kb and 600°C. Springer, New York.

Brearley AJ (2006) The action of water. In *Meteorites and the Early Solar System II* (eds Lauretta DS, McSween HY Jr). University of Arizona Press, Tucson, pp. 587–624.

Brearley AJ, Jones RH (1998) Chondritic meteorites. In *Planetary Materials* (ed Papike JJ). *Reviews in Mineralogy and Geochemistry*, **36**, 3.1–3.398.

Buick R, Dunlop JSR, Groves DI (1981) Stromatolite recognition in ancient rocks: an appraisal of irregularly laminated structures in an early Archean chert-barite unit from North-Pole, Western-Australia. *Alcheringa*, **5**, 161–181.

Burnett WC, Riggs SR (eds) (1990) *Phosphate Deposits of the World: Vol. 3, Genesis of Neogene to Recent Phosphorites.* Cambridge University Press, New York.

Byerly GR, Lowe DR, Walsh MM (1986) Stromatolites from the 3,300–3,500-Myr Swaziland Supergroup, Barberton Mountain Land, South Africa. *Nature*, **319**, 489–491.

Caldeira K (1991) Continental-pelagic carbonate partitioning and the global carbonate-silicate cycle. *Geology*, **19**, 204–206.

Caldeira K, Kasting JF (1992) Susceptibility of the early Earth to irreversible glaciation caused by carbon dioxide clouds. *Nature*, **359**, 226–228.

Cameron EM (1983) Evidence from early Proterozoic anhydrite for sulfur isotopic partitioning in the Precambrian oceans. *Nature*, **304**, 54–56.

Canfield DE (1998) A new model for Proterozoic ocean chemistry. *Nature*, **396**, 450–453.

Canfield DE, Habicht KS, Thamdrup B (2000) The Archean sulfur cycle and the early history of atmospheric oxygen. *Science*, **288**, 658–661.

Canfield DE, Poulton SW, Narbonne GM (2007) Late-Neoproterozoic deep-ocean oxygenation and the rise of animal life. *Science*, **315**, 92–95.

Canfield DE, Poulton SW, Knoll AH, *et al.* (2008) Ferruginous condition dominated later Neoproterozoic deep-water chemistry. *Science*, **321**, 949–952.

Catling DC, Claire MW (2005) How Earth's atmosphere evolved to an oxic state: A status report. *Earth and Planetary Science Letters*, **237**, 1–20.

Catling D, Zahnle K, McKay C (2001) Biogenic methane, hydrogen escape, and the irreversible oxidation of early Earth. *Science*, **293**, 839–843.

Černy P (1982) Mineralogy of rubidium and cesium. *MAC Short Course Handbook*, **8**, 145–162.

Chauhan DS (1979) Phosphate-bearing stromatolites of the Precambrian Aravalli phosphorite deposits of the Udaipur region, their environmental significance and genesis of phosphorite. *Precambrian Research*, **8**, 95–126.

Christner BC, Morris CE, Foreman CM, Cai R, Sands DC (2008) Ubiquity of biological ice nucleators in snowfall. *Science*, **319**, 1214.

Claire MW, Catling DC, Zahnle KJ (2006) Biogeochemical modelling of the rise in atmospheric oxygen. *Geobiology*, **4**, 239–269.

Clark PE (2007) *Dynamic Planet: Mercury in the Context of its Environment.* Springer, New York.

Cloud P (1972) A working model of the primitive Earth. *American Journal of Science*, **272**, 537–548.

Cohen AL, McConnaughey TA (2003) Geochemical perspectives on coral mineralization. In *Biomineralization* (eds. Dove PM, DeYoreo JJ, Weiner S). *Reviews in Mineralogy and Geochemistry*, **54**, 151–187.

Condie KC, DesMarais DJ, Abbott D (2001) Precambrian superplumes and supercontinents: a record in black shales, carbon isotopes, and paleoclimates? *Precambrian Research*, **106**, 239–260.

Cook PJ, Shergold JH (1984) Phosphorus, phosphorites and skeletal evolution at the Precambrian-Cambrian boundary. *Nature*, **308**, 231–236.

Cook PJ, Shergold JH (1990) *Phosphate Deposits of the World. Volume 1, Proterozoic and Cambrian Phosphorites.* Cambridge University Press, New York.

Dana ES (1958) *A Textbook of Mineralogy*, 4th edition. John Wiley & Sons, New York.

Dardenne MA, Trompette R, Magalhaes LF, Soares LA (1986) Proterozoic and Cambrian phosphorites – regional review: Brazil. In *Phosphate Deposits of the World*, Volume 1 (eds. Cook PJ, Shergold JH). Cambridge University Press, New York, pp. 116–131.

Dasgupta S, Roy S, Fukuoka M (1992) Depositional models for manganese oxide and carbonate deposits of the Precambrian Sausar Group, India. *Economic Geology*, **87**, 1412–1418.

Desch SJ, Connolly HC Jr (2002) A model of the thermal processing of particles in solar nebula shocks: Application to the cooling rates of chondrules. *Meteoritics & Planetary Science*, **37**, 183–207.

Donnadieu Y, Goddéris Y, Ramstein G, Nédélec A, Meert J (2004) A 'snowball Earth' climate triggered by continental break-up through changes in runoff. *Science*, **428**, 303–306.

Donnelly TH, Shergold JH, Southgate PN, Barnes CJ (1990) Events leading to global phosphogenesis around the Proterozoic/Cambrian boundary. *Geological Society Special Publication (Phosphorite Research and Development)*, **52**, 273–287.

Dove PM (2010) The rise of skeletal biomineralization. *Elements*, **6** (1), 37–42.

Dove PM, DeYoreo JJ, Weiner S (eds.) (2003) *Biomineralization. Reviews in Mineralogy and Geochemistry* **54**. Mineralogical Society of America, Chantilly, Virginia.

Ebel DS (2006) Condensation of rocky material in astrophysical environments. In *Meteorites and the Early Solar System II* (eds Lauretta DS, McSween HY Jr). University of Arizona Press, Tucson, pp. 253–277.

Edmond JM, Huh Y (1997) Chemical weathering yields in hot and cold climates. In *Tectonic Uplift and Climate Change*. (eds Ruddiman WF, Prell W). Plenum, New York, pp. 329–351.

Eiler JM (2007) On the origin of granites. *Science*, **315**, 951–952.

El Tabakh M, Grey K, Pirajno F, Schreiber BC (1999) Pseudomorphs after evaporitic minerals interbedded with 2.2 Ga stromatolites of the Yerrida basin, Western Australia: origin and significance. *Geology*, **27**, 871–874.

Elie M, Noueira ACR, Nédélec A, Trindade RIF, Kenig F (2007) A red algal bloom in the aftermath of the Marinoan snowball Earth. *Terra Nova*, **19**, 303–308.

England GL, Rasmussen B, Krapez B, Groves DL (2001) The origin of uraninite, bitumen nodules, and carbon seams in Witwatersrand gold-uranium-pyrite ore deposits, based on a Permo-Triassic analogue. *Economic Geology*, **96**, 1907–1920.

England GL, Rasmussen B, Krapez B, Groves DL (2002) Paleoenvironmental significance of rounded pyrite in siliciclastic sequences of the Late Archean Witwatersrand Basin: Oxygen-deficient atmosphere or hydrothermal alteration? *Sedimentology*, **49**, 1133–1136.

Evans DAD (2006) Proterozoic low orbital obliquity and axial-dipolar geomagnetic field from evaporite palaeolatitudes. *Nature*, **44**, 51–55.

Ewing RC, Chakoumakos BC (1982) Lanthanide, X, Th, U, Zr and Hf minerals, selected structure descriptions. *MAC Short Course Handbook*, **8**, 239–266.

Fairchild IJ (1993) Balmy shores and icy wastes: the paradox of carbonates associated with glacial deposits in Neoproterozoic times. In *Sedimentology Review 1* (ed Wright VP). Blackwell, Oxford, pp. 1–16.

Falini G, Albeck S, Weiner S, Addadi L (1996) Control of aragonite or calcite polymorphism by mollusk shell macromolecules. *Science*, **271**, 67–69.

Farquhar J, Bao H, Thiemens MH (2000) Atmospheric influence of Earth's earliest sulfur cycle. *Science*, **289**, 756–758.

Farquhar J, Savarino I, Airieau S, Thiemens MH (2001) Observations of wavelength-sensitive, mass-independent sulfur isotope effects during SO_2 photolysis: Implications for the early atmosphere. *Journal of Geophysical Research*, **106**, 1–11.

Farquhar J, Peters M, Johnston DT, *et al.* (2007) Isotopic evidence for mesoarchean anoxia and changing atmospheric sulphur chemistry, *Nature*, **449**, 706–709.

Fike DA, Grotzinger JP (2008) A paired sulfate–pyrite $\delta34S$ approach to understanding the evolution of the Ediacaran–Cambrian sulfur cycle. *Geochimica et Cosmochimica Acta*, **72**, 2636–2648.

Fike DA, Grotzinger JP, Pratt LM, Summons RE (2006) Oxidation of the Ediacaran ocean. *Nature*, **444**, 744–747.

Flicoteaux R, Trompette R (1998) Cratonic and foreland Early Cambrian phosphorites of West Africa: palaeoceanographical and climatical contexts. *Palaeogeography, Palaeoclimatology, Palaeoecology*, **139**, 107–120.

Foord, EB (1982) Minerals of tin, titanium, niobium and tantalum in granitic pegmatites. *MAC Short Course Handbook*, **8**, 187–238.

Frankel RB, Bazylinski DA (2003) Biologically induced mineralization by bacteria. In *Biomineralization* (eds Dove PM, DeYoreo JJ, Weiner S). *Reviews in Mineralogy and Geochemistry*, **54**, 95–114.

Frimmel HE (2005) Archaean atmospheric evolution: evidence from the Witwatersrand gold fields, South Africa. *Earth-Science Reviews*, **70**, 1–46.

Gaffey MJ, Cloutis EA, Kelley MS, Reed KL (2002) Mineralogy of asteroids. In *Asteroids III* (eds. Bottke WF Jr, Cellino A, Paolicchi P, Binzel RP). University of Arizona Press, Tucson, pp. 183–204.

Garrels RM, Christ CL (1965) *Solutions, Minerals and Equilibria*. Freeman, Cooper & Company, San Francisco.

Gorjan P, Veevers JJ, Walter MR (2000) Neoproterozoic sulfur-isotope variation in Australia and global implications. *Precambrian Research*, **100**, 151–179.

Grew ES, Hazen RM (2009) Evolution of the minerals of beryllium, a quintessential crustal element. *Geological Society of America Abstracts with Programs*, **41**, 161608.

Grew E, Hazen RM (2010a) Evolution of the minerals of beryllium, and comparison with boron mineral evolution. *Geological Society of America Abstracts with Programs*, **42**, #176345.

Grew ES, Hazen RM (2010b) Evolution of boron minerals: Has early species diversity been lost from the geological record? *Geological Society of America Abstracts with Programs*, **42**, 92.

Grew ES, Bada JL, Hazen RM (2011) Borate minerals and the origin of the RNA world, Origins *of Life and Evolution of the Biosphere*, **41**, 301–316.

Grotzinger JP, Knoll AH (1999) Stromatolites in Precambrian carbonates: Evolutionary mileposts or environmental dipsticks. *Annual Review of Earth and Planetary Sciences*, **27**, 313–358.

Habicht KS, Gade M, Thandrup B, Berg P, Canfield DE (2002) Calibration of sulfate levels in the Archean ocean. *Science*, **298**, 2372–2374.

Halverson GP (2005) A Neoproterozoic chronology. In *Neoproterozoic Geobiology and Paleobiology* (eds. Xiao S, Kaufman AJ). *Topics in Geobiology* **27**. Kluwer, New York, pp. 231–271.

Halverson GP, Hurtgen MT (2007) Ediacaran growth of the marine sulfate reservoir. *Earth and Planetary Science Letters*, **263**, 32–44.

Halverson GP, Hoffman PF, Schrag DP, Maloof AC, Rice AHN (2005) Toward a Neoproterozoic composite carbon-isotope record. *Geological Society of America Bulletin*, **117**, 1–27.

Halverson GP, Wade BP, Hurtgen MT, Barovich KM (2010) Neoproterozoic chemostratigraphy. *Precambrian Research*, **182**, 337–350.

Harrison TM, Blichert-Toft J, Müller W, Albarede F, Holden P, Mojzsis SJ (2005) Heterogeneous Hadean hafnium: evidence of continental crust at 4.4 to 4.5 Ga. *Science*, **310**, 1947–1950.

Hazen RM (2010) The evolution of minerals. *Scientific American*, **303**(3), 58–65.

Hazen RM, Ferry JM (2010) Mineral evolution: Mineralogiy in the fourth dimension. *Elements*, **6**(1), 9–12.

Hazen RM, Papineau D, Bleeker W, *et al.* (2008) Mineral evolution. *American Mineralogist*, **93**, 1693–1720.

Hazen RM, Ewing RJ, Sverjensky DA (2009) Evolution of uranium and thorium minerals. *American Mineralogist*, **94**, 1293–1311.

Hedges JI, Keil RG (1995) Sedimentary organic matter preservation: An assessment and speculative synthesis. *Marine Chemistry*, **49**, 81–139.

Hess PC (1989) *Origins of Igneous Rocks*. Harvard University Press, Cambridge, Massachusetts.

Hoffman PF, Prave AR (1996) A preliminary note on a revised subdivision and regional correlation of the Otavi Group based on glaciogenic diamictites and associated cap dolostones. *Communications of the Geological Survey of Namibia*, **11**, 77–82.

Hoffman PF, Schrag DP (2000) Snowball Earth. *Scientific American*, **January 2000**, 68–75.

Hoffman PF, Kaufman AJ, Halverson GP, Schrag DP (1998) A Neoproterozoic snowball Earth. *Science*, **281**, 1342–1346.

Holland HD (1984) *The Chemical Evolution of the Atmosphere and Ocean*. Princeton Series in Geochemistry. Princeton University Press, Princeton, New Jersey.

Holland HD (2002) Volcanic gases, black smokers, and the great oxidation event. *Geochimica et Cosmochimica Acta*, **66**, 3811–3826.

Howard PF (1986) Proterozoic and Cambrian phosphorites – regional review: Australia. In *Phosphate Deposits of the World*, Volume 1 (eds Cook PJ, Shergold JH) Cambridge University Press, New York, pp. 20–41.

Hurtgen MT, Arthur MA, Halverson GP (2005) Neoproterozoic sulfur isotopes, the evolution of microbial sulfur species, and the burial efficiency of sulfide as sedimentary pyrite. *Geology*, **33**, 41–44.

Huston DL, Logan BW (2004) Barite, BIFs and bugs: evidence for the evolution of the Earth's early hydrosphere. *Earth and Planetary Science Letters*, **220**, 41–55.

Hutchinson RW (1973) Volcanogenic sulfide deposits and their metallogenic significance. *Economic Geology*, **68**, 1223–1246.

Isley AE, Abbott DH (1999) Plume-related mafic volcanism and the deposition of banded iron formation. *Journal of Geophysical Research*, **104**, 15461–15477.

Jacobsen S (2001) Gas hydrates and deglaciations. *Nature*, **412**, 691–693.

Jacobsen SB, Pimentel-Klose MR (1988) A Nd isotopic study of the Hamersley and Michipicoten banded iron formations: the source of REE and Fe in Archean oceans. *Earth and Planetary Science Letters*, **87**, 29–44.

James HL, Trendall AF (1982) Banded iron-formation: distribution in time and paleoenvironmental significance. In *Mineral Deposits and the Evolution of the Biosphere* (eds Holland HD, Schidlowski M). Springer-Verlag, New York, pp. 199–217.

Jolliff BL, Wieczorek MA, Shearer CK, Neal CR (eds) (2006) *New Views of the Moon. Reviews in Mineralogy & Geochemistry*, **60**.

Jolliffe AW (1955) Geology and iron ores of Steep Rock Lake. *Economic Geology*, **50**, 373–398.

Kah LC, Lyons TW, Frank TD (2004) Low marine sulphate and protracted oxygenation of the Proterozoic biosphere. *Nature*, **431**, 834–837.

Kappler A, Pasquero C, Konhauser KO, Newman DK (2005) Deposition of banded iron formations by photoautotrophic Fe(II)-oxidizing bacteria. *Geology*, **33**, 865–868.

Kasemann SA, Prave AR, Fallick AE, Hawkesworth CJ, Hoffmann KH (2010) Neoproterozoic ice ages boron isotope, and ocean acidification: implications for a snowball Earth. *Geology*, **38**, 775–778.

Kasting JF (2001) The rise of atmospheric oxygen. *Science*, **293**, 819–820.

Kasting JF, Siefert JL (2002) Life and the evolution of Earth's atmosphere. *Science*, **296**, 1066–1068.

Kemp AIS, Hawkesworth CJ, Foster GL, *et al.* (2007) Magmatic and crustal differentiation history of granitic rocks from Hf-O isotopes in zircon. *Science*, **315**, 980–983.

Kennedy MJ, Runnegar B, Prave AR, Hoffmann KH, Arthur MA (1998) Two or four Neoproterozoic glaciations? *Geology*, **26**, 1059–1063.

Kennedy MJ, Droser M, Mayer LM, Pevear D, Mrofka D (2006) Late Precambrian oxygenation; inception of the clay mineral factory. *Science*, **311**, 1446–1449.

Kirschvink JL (1992) A paleogeographic model for Vendian and Cambrian time. In *The Precambrian Biosphere* (eds Schopf JW, Klein C). Cambridge University Press, New York, pp. 51–52.

Klein C (2005) Some Precambrian banded iron-formations (BIFs) from around the world: Their age, geologic setting, mineralogy, metamorphism, geochemistry, and origin. *American Mineralogist*, **90**, 1473–1499.

Knoll AH (2003) Biomineralization and evolutionary history. In *Biomineralization* (eds Dove PM, DeYoreo JJ, Weiner S). *Reviews in Mineralogy and Geochemistry*, **54**, 329–356.

Knoll AH, Hayes JM, Kaufman AJ, Swett K, Lambert IB (1986) Secular variations in carbon isotope ratios from Upper Proterozoic successions of Svalbard and East Greenland. *Nature*, **321**, 832–838.

Konhauser KO, Hamade T, Raiswell R, *et al.* (2002) Could bacteria have formed the Precambrian banded iron-formations? *Geology*, **30**, 1079–1082.

Konhauser KO, Amskold L, Lalonde SV, Posth NR, Kappler A, Anbar A (2007) Decoupling Photochemical Fe(II) Oxidation from Shallow-Water BIF Deposition. *Earth and Planetary Science Letters*, **258**, 87–100.

Konhauser KO, Pecoits E, Lalonde SV, *et al.* (2009) Oceanic nickel depletion and a methanogen famine before the Great Oxidation Event. *Nature*, **458**, 750–753.

Krot AN, Hutcheon ID, Brearley AJ, Pravdivtseva OV, Petaev MI, Hohenberg CM (2006) Timescales and settings for alteration of chondritic meteorites. In *Meteorites and the Early Solar System II* (eds Lauretta DS, McSween HY Jr). University of Arizona Press, Tucson, pp. 525–553.

Kumar PS, Srinivasan R (2002) Fertility of late Archaean basement granite in the vicinity of U-mineralized Neoproterozoic Bhima basin, peninsular India. *Current Science*, **82**, 571–576.

Kump LR, Barley ME (2007) Increased subaerial volcanism and the rise of atmospheric oxygen 2.5 billion years ago. *Nature*, **448**, 1033–1036.

Kump LR, Kasting JF, Barley ME (2001) Rise of atmospheric oxygen and the 'upside down' Archean mantle. *Geochemistry, Geophysics, Geosystems*, **2**, Paper #2000GC000114.

LaBerge GL (1973) Possible biological origin of Precambrian iron-formations. *Economic Geology*, **68**, 1098–1109.

Leclerc J, Weber F (1980) Geology and genesis of the Moanda manganese deposits. In *Geology and Geochemistry of Manganese*, Volume 2 (eds Varentsov IM, Grasselly G) E. Schweizerbart'sche Verlagsbuchhandlung, Stuttgart, pp. 89–109.

Lepot K, Benzerara K, Brown GE, Philippot P (2008) Microbially influenced formation of 2,724-million-year-old stromatolites. *Nature Geoscience*, **1**, 1–4.

London D (2008) *Pegmatites*. Mineralogical Association of Canada, Quebec, *Special Publication*, **10**, 347 p.

Lowe DR (1980) Stromatolites 3,400-Myr old from the Archean of Western Australia. *Nature*, **284**, 441–443.

MacPherson GJ (2007) Calcium-aluminum-rich inclusions in chondritic meteorites. In *Treatise on Geochemistry*, Volume 1 (eds Holland HD, Turekian KK). Elsevier, San Diego, pp. 201–246.

Mason B (1967) Extraterrestrial mineralogy. *American Mineralogist*, **52**, 307–325.

Mayer LM, Schtik LL, Hardy KR, Wagai R, McCarthy J (2004) Organic matter in small mesopores in sediments and soils. *Geochimica et Cosmochimica Acta*, **68**, 3863–3872.

McCoy TJ (2010) Mineralogical evolution of meteorites. *Elements*, **6**(1), 19–24.

McCoy TJ, Mittlefehldt DW, Wilson L (2006) Asteroid differentiation. In *Meteorites and the Early Solar System II* (eds Lauretta DS, McSween HY Jr). University of Arizona Press, Tucson, pp. 733–746.

Melezhik VA, Fallick AE, Makarikhin VV, Lyubtsov VV (1997) Links between Palaeoproterozoic palaeogeography and rise and decline of stromatolites: Fennoscandian Shield. *Precambrian Research*, **82**, 311–348.

Messenger S, Keller LP, Stadermann FJ, Walker RM, Zinner E (2003) Samples of stars beyond the solar system: Silicate grains in interplanetary dust. *Science*, **300**, 105–108.

Messenger S, Sandford S, Brownlee D (2006) The population of starting materials available for solar system construction. In *Meteorites and the Early Solar System II* (eds Lauretta DS, McSween HY Jr). University of Arizona Press, Tucson, pp. 187–207.

Misi A, Kaufman AJ, Veizer J, *et al.* (2006) Chemostratigraphic correlation of Neoproterozoic succession in South America. *Chemical Geology*, **237**, 143–167.

Moore PB (1982) Pegmatite minerals of P(V) and B(III). *MAC Short Course Handbook*, **8**, 217–292.

Mostefouai S, Hoppe P (2004) Discovery of abundant in situ silicate and spinel grains from red giant stars in a primitive meteorite. *Astrophysical Journal*, **613**, L149–L152.

Nagy RM, Porter SM, Dehler CM, Shen Y (2009) Biotic turnover driven by eutrophication before the Sturtian low-latitude glaciation. *Nature Geoscience*, **2**, 415–418.

Nittler LR (2003) Presolar stardust in meteorites: Recent advances and scientific frontiers. *Earth and Planetary Science Letters*, **209**, 259–273.

Nordstrom DK, Southam G (1997) Geomicrobiology of sulfide mineral oxidation. In *Geomicrobiology: Interactions between Microbes and Minerals* (eds Banfield JF, Nealson KH). *Reviews in Mineralogy*, **35**, 361–390.

Notholt AJG, Sheldon R. (1986) Proterozoic and Cambrian phosphorites – regional review: world resources. In *Phosphate Deposits of the World, Volume 1* (eds Cook PJ, Shergold JH). Cambridge University Press, New York, pp. 9–19.

Olcott AN, Sessions AL, Corsetti FA, Kaufman AJ, Flavio de Olivera T (2005) Biomarker evidence for photosynthesis during Neoproterozoic glaciation. *Science*, **310**, 471–474.

Ono S, Eigenbrode JL, Pavlov AA, *et al.* (2003) New insights into Archean sulfur cycle from mass-independent sulfur isotope records from the Hamersley Basin, Australia. *Earth and Planetary Science Letters*, **213**, 15–30.

Papineau D, Walker JJ, Mojzsis SJ, Pace NR (2005a) Composition and structure of microbial communities from stromatolites of Hamelin Pool in Shark Bay, Western Australia. *Applied and Environmental Microbiology*, **71**, 4822–4832.

Papineau D, Mojzsis SJ, Coath CD, Karhu JA, McKeegan KD (2005b) Multiple sulfur isotopes of sulfides from sediments in the aftermath of Paleoproterozoic glaciations. *Geochimica et Cosmochimica Acta*, **69**, 5033–5060.

Papineau D, Mojzsis SJ, Schmitt AK (2007) Multiple sulfur isotopes from Paleoproterozoic Huronian interglacial sediments and the rise of atmospheric oxygen. *Earth and Planetary Science Letters*, **255**, 188–212.

Papineau D, Purohit A, Goldberg T, *et al.* (2009) High productivity and nitrogen cycling after the Paleoproterozoic phosphogenic event in the Aravalli Supergroup, India. *Precambrian Research*, **171**, 37–56.

Papineau D (2010a) Mineral environments on the earliest Earth. *Elements*, **6**(1), 9–12.

Papineau D (2010b) Global biogeochemical changes at both ends of the Proterozoic: Insights from phosphorites. *Astrobiology*, **10**, 165–181.

Paris F, Bottom B, Lapeyrie F (1996) In vitro weathering of phlogopite by ectomycorrhizal fungi. *Plant and Soil*, **179**, 141–150.

Parman SW (2007) helium isotopic evidence for episodic mantle melting and crustal growth. *Nature*, **446**, 900–903.

Parnell J (2004) Plate tectonics, surface mineralogy, and the early evolution of life. *International Journal of Astrobiology*, **3**, 131–137.

Parnell J, Boyce AJ, Mark D, Bowden S, Spinks S (2010) Early oxygenation of the terrestrial environments during the Mesoproterozoic. *Nature*, **468**, 290–293.

Pearson DG, Parman SW, Nowell GM (2007) A link between large mantle melting events and continent growth seen in osmium isotopes. *Nature*, **449**, 202–205.

Phillippot P, Van Zuilen M, Lepot K, Thomazo C, Farquhar J, van Kranendonk MJ (2007) Early Archean microorganisms preferred elemental sulfur, not sulfate. *Science*, **317**, 1534–1537.

Pierrehumbert RT (2004) High levels of atmospheric carbon dioxide necessary for the termination of global glaciation. *Nature*, **429**, 646–648.

Poulton SW, Frallick PW, Canfield DE (2004) The transition to a sulphidic ocean ~1.84 billion years ago. *Nature*, **431**, 173–177.

Poulton SW, Frallick, PW, Canfield, DE (2010) Spatial variability in oceanic redox structure 1.8 billion years ago. *Nature Geoscience*, **3**, 486–490.

Rasmussen B, Buick R (1999) Redox state of the Archean atmosphere: Evidence from detrital heavy minerals in ca.3250–2750 Ma sandstones from the Pilbara Craton, Australia. *Geology*, **27**, 115–118.

Reid RP, Visscher PT, Decho AW, *et al.* (2000) The role of microbes in accretion, lamination and early lithification of modern marine stromatolites. *Nature*, **406**, 989–992.

Reith F, Rogers SL, McPhail DC, Webb D (2006) Biomineralization of gold: Biofilms on bacterioform gold. *Science*, **313**, 233–236.

Ridgwell AJ, Kennedy MJ, Caldeira K (2003) Carbonate deposition, climate stability, and Neoproterozoic ice ages. *Science*, **302**, 859–862.

Rosing MT, Bird DK, Sleep NH, Glassley W, Albarede F (2006) The rise of continents – An essay on the geologic consequences of photosynthesis. *Palaeo*, **232**, 99–113.

Roy S (2006) Sedimentary manganese metallogenesis in response to the evolution of the Earth system. *Earth Science Reviews*, **77**, 273–305.

Rubin AE (1997a) Mineralogy of meteorite groups. *Meteoritics & Planetary Science*, **32**, 231–247.

Rubin AE (1997b) Mineralogy of meteorite groups: An update. *Meteoritics & Planetary Science*, **32**, 733–734.

Rubin AE (2000) Petrologic, geochemical and experimental constraints on models of chondrules formation. *Earth Science Reviews*, **50**, 3–27.

Runnegar B (1987) The evolution of mineral skeletons. In *Origin, Evolution, and Modern Aspects of Biomineralization in Plants and Animals* (ed Crick RE). Plenum, New York, pp. 75–94.

Runnegar B (1991) Precambrian oxygen levels estimated from the biochemistry and physiology of early eukaryotes. *Global and Planetary Change*, **97**, 97–111.

Ruzicka A, Snyder GA, Taylor LA (1999) Giant impact hypothesis for the origin of the Moon: A critical review of some geochemical evidence. In *Planetary Petrology and Geochemistry* (eds Snyder GA, Neal CR, Ernst WG). Geological Society of America, Boulder, Colorado, pp. 121–134.

Sandiford M, McLaren S (2002) Tectonic feedback and the ordering of heat producing elements within the continental lithosphere. *Earth and Planetary Science Letters*, **204**, 133–150.

Sangster DF (1972) Precambrian volcanogenic massive sulfide deposits in Canada: A review. *Geological Survey of Canada Paper*, **72–82**, pp. 1–43.

Schwartzman DW, Volk T (1989) Biotic enhancement of weathering and the habitability of Earth. *Nature*, **340**, 457–460.

Schwartzman DW, Volk T (1991) Biotic enhancement of weathering and surface temperatures on Earth since the origin of life. *Palaeogeography Palaeoclimate Palaeoecology*, **90**, 357–371.

Scott C, Lyons TW, Bekker A, Shen Y, Poulton SW, Anbar AD (2008) Tracing the stepwise oxygenation of the Proterozoic ocean. *Nature*, **452**, 456–459.

Scott H, Hemley RJ, Mao HK, *et al.* (2004) Generation of methane in the Earth's mantle: in situ high P-T measurements of carbonate reduction. *Proceedings of the National Academy of Sciences, USA*, **101**, 14023–14026.

Shen Y, Buick R, Canfield DE (2001) Isotopic evidence for microbial sulphate reduction in the early Archaean era. *Nature*, **410**, 77–81.

Shen Y. Farquhar, J., Masterson A, Kaufman AJ, Buick R (2009) Evaluating the role of microbial sulfate reduction in the early Archean using quadrupole isotope systematics. *Earth and Planetary Science Letters*, **279**, 383–391.

Sherwood-Lollar B, Frape SK, Weise SM, Fritz P, Macko SA, Welhan JA (1993) Abiogenic methanogenesis in crystalline rocks. *Geochimica et Cosmochimica Acta*, **57**, 5087–5097.

Sherwood-Lollar B, McCollom TM, Ueno Y, *et al.* (2006) Biosignatures and abiotic constraints on early life – comment and reply. *Nature*, **440**, E18–E19.

Shirey SB, Kamber BS, Whitehouse MJ, Mueller PA, Basu AR (2008) A review of the isotopic and trace element evidence for mantle and crustal processes in the Hadean and Archean: Implications for the onset of plate tectonic subduction. In *When Did Plate Tectonics Start on Earth?* (eds Condie KC, Pease V). Geological Society of America, Boulder, Colorado, *Special Paper*, **440**, pp. 1–29.

Shukolyukov A, Lugmair GW (2002) Chronology of asteroid accretion and differentiation. In *Asteroids III* (eds Bottke WF Jr, Cellino A, Paolicchi P, Binzel RP). University of Arizona Press, Tucson, pp. 687–695.

Silver PG, Behm MD (2008) Intermittent plate tectonics? *Science*, **319**, 85–88.

Silver PG, Behn MD, Kelley K, Schmitz M, Savage B (2006) Understanding cratonic flood basalts. *Earth and Planetary Science Letters*, **245**, 190–201.

Singer PC, Stumm W (1970) Acid mine drainage: the rate determining step. *Science*, **167**, 1121–1123.

Smithies RH, Champion DC (2000) The Archean high-Mg diorite suite: Links to tonalite-trondhjemite-granodiorite magmatism and implications for early Archean crustal growth. *Journal of Petrology*, **41**, 1653–1671.

Smithies RH, Champion DC, Cassidy KF (2003). Formation of Earth's early. Archaean continental crust. *Precambrian Research*, **127**, 89–101.

Smithies RH, Champion DC, Van Kranendonk MJ, Howard HM, Hickman AH (2005) Modern-style subduction processes in the Mesoarchaean: geochemical evidence from the 3.12 Ga Whundo intraoceanic arc. *Earth and Planetary Science Letters*, **231**, 221–237.

Stanley SM, Hardie LA (1998) Secular oscillations in the carbonate mineralogy of reef-building and sediment-producing organisms driven by tectonically forced shifts in seawater chemistry. *Palaeogeography Palaeoclimatology Palaeoecology*, **144**, 3–19.

Stroud RM, Nittler LR, Alexander CMO'D (2004) Polymorphism in presolar Al_2O_3 grains from asymptotic giant branch stars. *Science*, **305**, 1455–1457.

Sumner DW (1997) Carbonate precipitation and oxygen stratification in late Archean seawater as deduced from facies and stratigraphy of the Gamohaan and Frisco Formations, Transvaal Supergroup, South Africa. *American Journal of Science*, **297**, 455–487.

Sverjensky DA, Lee N (2010) The Great Oxidation Event and mineral diversification. *Elements*, **6**(1), 31–36.

Tazaki K (2005) Microbial formation of a halloysite-like mineral. *Clays and Clay Minerals*, **55**, 224–233.

Tomlinson KY, Davis DW, Stone D, Hart T (2003) U-Pb age and Nd isotopic evidence for Archean terrane development and crustal recycling in the south-central Wabigoon Subprovince, Canada. *Contributions to Mineralogy and Petrology*, **144**, 684–702.

Tonks WB, Melosh HJ (1993) Magma ocean formation due to giant impacts. *Journal of Geophysical Research*, **98**, 5319–5333.

Tosca NJ, Johnston DT, Mushegian A, Rothman DH, Knoll AH (2009) Clay mineralogy and organic carbon burial in Proterozoic basins. *Geochimica et Cosmocimica Acta*, **74**, 1579–1592.

Touboul M, Kleine T, Bourdon B, Plame H, Wieler R (2007) Late formation and prolonged differentiation of the Moon inferred from W isotopes in lunar metals. *Nature*, **450**, 1206–1209.

Towe KM (2002) The problematic rise of Archean oxygen. *Science*, **295**, 798–799.

Tsikos H, Moore JM (1997) Petrography and geochemistry of the Paleoproterozoic Hotazel iron formation, Kalahari manganese field, South Africa: implications for Precambrian manganese metallogenesis. *Economic Geology*, **92**, 87–97.

Ueno Y, Yamada K, Yoshida N, Maruyama S, Isozaki Y (2006) Evidence from fluid inclusions for microbial methanogenesis in the early Archaean era. *Nature*, **440**, 516–519.

Ueshima M, Tazaki K (1998) Bacterial bio-weathering of K-feldspar and biotite in granite. *Clay Science Japan*, **38**, 68–92.

Ueshima M, Mogi K, Tazaki K (2000) Microbes associated with bentonite. *Clay Science Japan*, **39**, 171–183.

Van Kranendonk MJ (2006) Volcanic degassing, hydrothermal circulation and the flourishing of early life on Earth: new evidence from the Warrawoona Group, Pilbara Craton, Western Australia. *Earth Science Reviews*, **74**, 197–240.

Van Kranendonk MJ (2007) A review of the evidence for putative Paleoarchean life in the Pilbara Craton. In *Earth's Oldest Rocks* (eds Van Kranendonk MJ, Smithies RH, Bennet V). *Developments in Precambrian Geology* **15**, Elsevier, Amsterdam, pp. 855–896.

Vollmer C, Hoppe P, Brenker FE, Holzapfel C (2007) Stellar MgSiO$_3$ perovskite: A shock-transformed silicate found in a meteorite. *Astrophysical Journal*, **666**, L49–L52.

Wacey D, McLoughlin N, Whitehouse, MJ, Kilburn M (2010) Two coexisting sulfur metabolisms in ca. 3400 Ma sandstone. *Geology*, **38**, 1115–1118.

Wadhwa M, Srinivasan G, Carlson RW (2006) Timescales of planetesimal differentiation in the early solar system. In *Meteorites and the Early Solar System II* (eds Lauretta DS, McSween HY Jr). University of Arizona Press, Tucson, pp. 715–731.

Walter MR (1994) The earliest life on Earth: clues to finding life on Mars. In *Early Life on Earth* (ed Bengtson S). *Nobel Symposium*, **84**, 270–286. Columbia University Press, New York.

Walter MR, Buick R, Dunlop JSR (1980) Stromatolites 3,400–3,500 Myr old from the North-Pole area, Western-Australia. *Nature*, **284**, 443–445.

Wheat CG, Feely RA, Mottl MJ (1996) Phosphate removal by oceanic hydrothermal processes: An update of the phosphorus budget in the oceans. *Geochimica et Cosmochimica Acta*, **60**, 3593–3608.

White AJR, Chappell BW (1983) Granitoid types and their distribution in the Lachlan Fold Belt, southeastern Australia. *Geological Society of America Memoir*, **159**, 21–34.

Widdel F, Schnell S, Heising S, Ehrenreich A, Assmus B, Schink B (1993) Ferrous iron oxidation by anoxygenic phototrophic bacteria. *Nature*, **362**, 834–836.

Wilks ME, Nisbet EG (1988) Stratigraphy of the Steep Rock Group, northwest Ontario: a major Archaean unconformity and Archaean stromatolites. *Canadian Journal of Earth Sciences*, **25**, 370–391.

Witze A (2006) The start of the world as we know it. *Nature*, **442**, 128–131.

Young GM (1976) Iron-formation and glaciogenic rocks of the Rapitan Group, Northwest Territories. *Canada Precambrian Research*, **3**, 137–158.

Young GM (1995) Are Neoproterozoic glacial deposits preserved on the margins of Laurentia related to the fragmentation of two supercontinents? *Geology*, **23**, 153–156.

Yueyan L (1986) Proterozoic and Cambrian phosphorites – regional review: China. In *Phosphate Deposits of the World*, Volume 1 (eds Cook PJ, Shergold JH). Cambridge University Press, New York, pp. 42–62.

Zhao Z, Xing Y, Ma G, Chen Y (1985) *Biostratigraphy of the Yangtze Gorge Area*. Geological Publishing House, Beijing, China.

19

GEOBIOLOGY OF THE ARCHEAN EON

Roger Buick

Department of Earth & Space Sciences and Astrobiology Program,
University of Washington, Seattle WA 98195-1310, USA

19.1 Introduction

The Archean represents the time in Earth history from the oldest rocks (about 4 billion years ago) until 2.5 billion years ago. The geobiology of these ancient times is an inherently difficult to study. There are very few well-preserved rocks of such great age in which to search for evidence of geobiology (Fig. 19.1). Moreover, because life first proliferated during the Archean, primitive geobiological signals might have been very different from those familiar on the modern Earth. Also, the long and complex metamorphic and deformational histories of all Archean geobiological relics have distorted the quality and quantity of the information preserved. Finally, the Archean environment was probably much different, potentially influencing the rates and styles of geobiological processes such that their true nature is now unclear.

Despite these issues, it has become evident that enough records remain to provide at least a qualitative picture of how Archean life influenced its geological setting and vice versa. Indeed, after reviewing the major biogeochemical cycles and microbe-mineral interactions operating on the early Earth, I argue here that sufficient information is now available to claim that by the end of the Archean, most of the major microbial geobiological processes were already established.

19.2 Carbon cycle

Despite the great time difference, it seems that for most of the Archean, the carbon cycle was not greatly dissimilar to that operating now. It clearly consisted of both an inorganic and organic component and, as was first noted by Manfred Schidlowski and colleagues (1979), carbon buried in sediments was partitioned between these two components at a roughly similar ratio (~80% inorganic, ~20% organic) to the present (Hayes and Waldbauer, 2006), indicated by the mean $\delta^{13}C$ of ~0‰ for Archean marine sedimentary carbonate and around −30‰ for marine organic matter (Fig. 19.2). For this to occur, the proportions of dissolved CO_2 converted to organic matter must have been significant, implying active biological carbon fixation. Although there are many potential pathways for carbon fixation, the tendency from the mid-Archean onward for organic carbon isotopes to cluster within a standard deviation of the pre-1.0 Ga mean of −28‰ (Fig. 19.2b) indicates that carbon fixation by the reductive citric acid (Calvin–Benson–Bassham) cycle was probably dominant, used today by oxygenic photosynthesizers but also by a wide range of other microbes. The other common fixation pathways, used by some anoxygenic photosynthesizers and chemolithoautotrophs, generally fractionate organic carbon isotopes to a much greater (>−35‰: reductive acetyl-CoA pathway) or much lesser (<−15‰: reductive citric acid and 3-hydroxypropionate cycles) extent. Thus, it can be concluded that by the mid-Archean, the gross partitioning of carbon into the various components of the cycle was nearly modern in aspect and that the main process transforming inorganic to organic carbon was probably fixation via the Calvin cycle using the enzyme Rubisco.

The late Archean carbon isotopic record also shows the widespread occurrence from 2.8 Ga onwards of extremely light $\delta^{13}C_{org}$ values, down to −65‰ and usually

Fundamentals of Geobiology, First Edition. Edited by Andrew H. Knoll, Donald E. Canfield and Kurt O. Konhauser.
© 2012 Blackwell Publishing Ltd. Published 2012 by Blackwell Publishing Ltd.

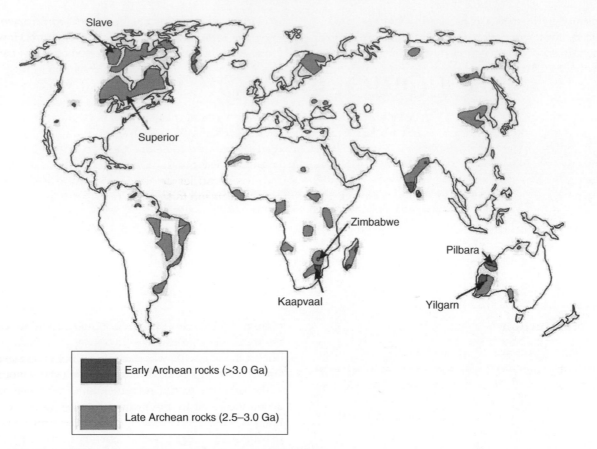

Figure 19.1 Global distribution of Archean rocks showing major cratons.

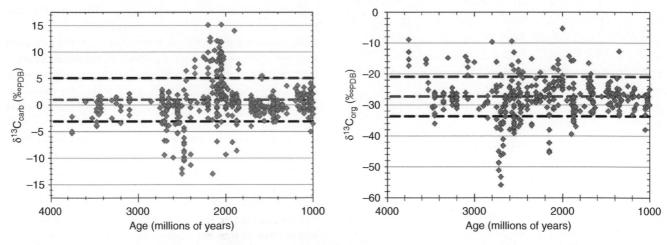

Figure 19.2 Carbon isotope data from marine sedimentary rocks through early Precambrian time: $\delta^{13}C$ of carbonates (left) and organics (right).

below $-40‰$ (Fig. 19.2b). This has been interpreted as indicating a major role for methanogenesis in organic recycling and subsequent fixation of the highly fractionated methane into the marine biomass via methanotrophy (Hayes, 1994; Eigenbrode & Freeman, 2006). Support for this hypothesis comes from studies of hydrocarbon biomarkers, which show an anomalous abundance of 3β-methylhopanes in the isotopically depleted kerogenous shales (Brocks *et al.*, 2003a; Eigenbrode *et al.*, 2008). These molecules are strongly associated with aerobic methanotrophs (Brocks *et al.*, 2003b). Despite recent arguments for contamination (Rasmussen *et al.*, 2008), no plausible source apart from the host rocks and other stratigraphically adjacent, isotopically light, late Archean sediments could have produced these unusual (for the pre-Phanerozoic) hydrocarbons. The suggestion by Hinrichs (2002) that archaeal anaerobic methanotrophy coupled to bacterial sulfate reduction could have been responsible for incorporating the light methanogenic isotopic signature into late Archean kerogens may have been of limited ecological significance due to the generally low sulfate concentrations at this time in Earth history.

Enhanced methane metabolism during the late Archean may have resulted from a limited availability of oxidized substrates for respiration, leaving a greater proportion of recycled organic matter available for degradation during methanogenesis. Though a methanogenic signature occurs in some marine sediments deposited in moderate to considerable water depths through the last 300 Ma of the Archean, it is notable that the most extreme isotopic depletions occur in non-marine sediments. As oxidants for respiratory consumption, especially sulfate, would have been particularly scarce in non-marine environments, this supports the hypothesis that methanogenesis was a significant recycling pathway, allowing reincorporation of the methanogenic carbon into the biomass via aerobic methanotrophy if oxygenic photosynthesis were occurring locally.

The general absence of this signature after the Archean can be explained by the early Paleoproterozoic oxygenation of the atmosphere, which generated more oxidized substrates for respiration, limiting the organic matter available for methanogenic recycling. Why it is not readily apparent in the early Archean record is more surprising, especially when carbon isotopic data from fluid inclusions in ~3.45 Ga old chert veins suggest that methanogenesis had already evolved (Ueno *et al.*, 2004). Perhaps methanotrophy was a late-evolving metabolism, electron acceptors such as oxygen or sulfate for methanotrophy could have been scarce early on, or maybe a comparative dearth of detailed isotopic data at high stratigraphic resolution might be to blame.

A feature widely noted in the Archean sedimentary record is the scarcity of carbonates when compared with younger times. This is especially odd in light of the isotopic evidence for ~4:1 partitioning between carbonate and organic deposition throughout the Archean, particularly when the generally high organic content of Archean shales is also considered. This should translate to even higher volumes of sedimentary carbonate if the average amount of organic carbon buried in mudrocks was greater. There are several possible explanations, none satisfactory by themselves but in combination perhaps capable of accounting for this anomaly:

1 Depositional bias: Archean shallow-marine rocks are rare, and these settings typically contain most carbonates in younger successions. It is widely believed that there was little continental crust until the late Archean,

in which case shallow shelf settings would have been geographically restricted and temporally rare. However, extensive shallow-marine carbonates occur as far back as ~3.32 Ga in the Strelley Pool Formation of the Pilbara Craton (Allwood *et al.*, 2006) which was deposited on pre-existing continental crust (Buick *et al.*, 1995a; Green *et al.*, 2000), and broad spreads in mantle-extraction and crustal-crystallization age spectra for detrital zircons (e.g. Pietranik *et al.*, 2008; Harrison, 2009) show that continental crust must have been reasonably abundant throughout earlier times, so by itself this is clearly not a sufficient reason for the dearth of carbonates.

2 Preservational bias: In volcanic-dominated Archean greenstone belts, carbonates have generally been hydrothermally silicified (Lowe, 1983; Buick and Barnes, 1984), so a mask of silicification may be hiding them elsewhere. However, although cherts are comparatively abundant in Archean terrains, even assuming that all of them had carbonate protoliths would not contribute enough carbonate to provide mass-balance for carbon isotopes.

3 Subduction removal: It is possible that the locus of carbonate deposition shifted after the Archean, with most Archean carbonate precipitating in the pelagic photic zone but accumulating on deep-water oceanic crust, in contrast to post-Archean carbonates (and organic-rich shales) which were largely deposited in shallower shelf and slope settings. Its ultimate fate would thus have been subduction, with very little being incorporated into the continental record via obduction in ophiolites, as Archean ophiolites are apparently limited to only one convincing example (Furnes *et al.*, 2008). In support of this model is the recent discovery of fine-grained deep-marine carbonates from the >3.5 Ga Coonterunah Group in the Pilbara Craton of northwest Australia (Buick, 2007) with relatively heavy carbon isotopes indicating a shallow source. This indicates the potential for a deep-sea subductable carbonate reservoir. The Archean flux of pelagic carbonate to the deep-sea floor may have been enhanced due to the relative absence of oxidized substrates for respiratory consumption of organic matter and remineralization to CO_2 during sinking. If so, then dissolved CO_2 concentrations could have been lower in the Archean deep oceans, leading to a greater carbonate compensation depth. Thus, the pelagic rain of carbonate may have largely reached subductable depths without dissolution, unlike the modern ocean.

4 Basalt carbonatization: If atmospheric pCO_2 were higher in the Archean, then more would have dissolved in seawater, potentially allowing submarine basalts to undergo extensive carbonatization during low-temperature hydrothermal alteration (Zahnle and Sleep, 2002). There is some analytical evidence that this was indeed a significant process (Nakamura and Kato, 2004), and if Archean sea-floor spreading rates were faster and

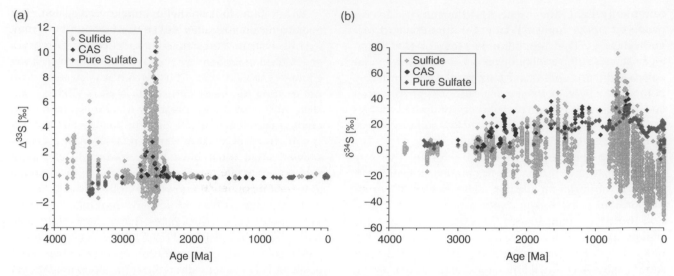

Figure 19.3 Sulfur isotope data from marine sedimentary rocks through time: (a) $\Delta^{33}S$ from sulfides, sulfate minerals and carbonate-associated sulfate (CAS); (b) $\delta^{34}SCDT$ from sulfides, sulfate minerals and carbonate-associated sulfate (CAS).

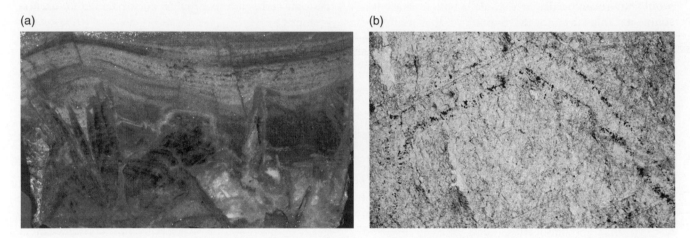

Figure 19.4 Barite from the ~3.48 Ga Dresser Formation, Warrawoona Group, Pilbara Craton, Australia: (a) Barite crystals draped by sediment showing original gypsum morphology; (b) Thin-section of a barite crystal showing microscopic pyrite grains (black) with similar $\Delta^{33}S$ and $\delta^{34}S$ fractionated by up to –25‰ with respect to surrounding sulfate.

ocean area greater than now as many have speculated, then this sink may account for much of the carbonate burial implied by the isotopic ratios.

19.3 Sulfur cycle

The principal difference between the Archean and modern sulfur cycles is in the relative importance of atmospheric sulfur cycling. At present, this is not particularly significant and leaves little isotopic trace in the geological record; then, atmospheric photochemistry of volcanic sulfur gases produced pronounced mass-independent fractionations of the rare isotopes ^{33}S and ^{36}S with respect to the commoner ^{32}S and ^{34}S (see Chapters 5 and 14). These fractionations differ from the usual mass-dependent fractionations occurring at the Earth's surface in that the isotope effect is not proportional to the mass difference between the isotopes and are thus

expressed as Δ^{33} or $\Delta^{36}S$ which record the deviation in per mill (‰) of a sample from the mass-dependent fractionation trend. Since ~2.3 Ga this deviation has been consistently negligible, attributable to the continued existence of sufficient atmospheric oxygen to produce an ozone shield that prevents the deep penetration of high-energy UV radiation and thus the photolytic reactions of volcanic SO_2. Before, however, photolysis produced opposite mass-independent fractionation of S_8 and SO_4^{2-}, forming sulfides showing $\Delta^{33}S$ of up to +12‰ and sulfates with $\Delta^{33}S$ down to –2‰, which are transmitted through subsequent mass-dependent fractionations on the Earth's surface (Fig. 19.3).

From $\delta^{34}S$ fractionations alone, it has been proposed that dissimilatory sulfate reduction evolved as early as ~3.49 Ga (Shen *et al.*, 2001). Sedimentary barite from North Pole in the Pilbara Craton of Australia, which initially precipitated as evaporative gypsum indicated by conserved crystal interfacial angles, contains microcrystalline pyrites defining crystal growth faces (Fig. 19.4). These have $\delta^{34}S$ fractionations as light as $-22\permil$, whereas the host barite has consistent $\delta^{34}S$ values of $+4\permil$ to $+5\permil$. This was interpreted as indicating microbial reduction of the soluble sulfate in gypsum to sulfide in pyrite, as no evidence of alteration by oxidized hydrothermal fluids is evident in these rocks (the other potential source of such large isotopic fractionations). Because microbial sulfate reduction usually causes fractionations of $-10\permil$ to $-45\permil$ when sulfate is abundantly available, this would imply that the absence of such large fractionations in most Archean sedimentary rocks is due to the scarcity of dissolved sulfate in the primordial oceans (Habicht *et al.*, 2002). However, in the unusual North Pole environment where evaporating brine pools locally concentrated sulfate, microbes were able to express the large $\delta^{34}S$ fractionations typical of this metabolism.

Recent re-analysis of barite and pyrite from the same locality by *in-situ* methods (NanoSIMS) has led to the suggestion that these sulfur isotopic fractionations were induced by sulfur disproportionation rather than sulfate reduction (Philippot *et al.*, 2007). In this metabolism, intermediate redox states of sulfur (elemental sulfur S^0, sulfite SO_3^- and thiosulfate $H_2SO_3^-$) are disproportionated into sulfide and sulfate, also causing a small but significant $\delta^{34}S$ fractionation in the reduced species of $-5\permil$ to $-11\permil$. The rationale for this interpretation depend on using the mass-independent fractionation $\Delta^{33}S$ as a metabolic tracer. The North Pole barite shows a consistent $\Delta^{33}S$ of around $-1\permil$, so if the pyrite with fractionated $\delta^{34}S$ values were derived by microbial reduction of this material, it too should show similar $\Delta^{33}S$ values. But Philippot and colleagues (2007) observed a population of microcrystalline pyrites with differing $\Delta^{33}S$ of up to $+6\permil$. They interpreted this to mean that the ultimate sulfur source for these pyrites was elemental sulfur with positive $\Delta^{33}S$ derived from the atmosphere, which was then repeatedly disproportionated and reoxidized to produce large $\delta^{34}S$ fractionations of up to $-22\permil$. However, two subsequent studies of the same rocks (Ueno *et al.*, 2008; Shen *et al.*, 2009) failed to find positive $\Delta^{33}S$ associated with negative $\delta^{34}S$ in barite-hosted pyrite. Instead, such pyrites had very similar but slighter higher $\Delta^{33}S$ values than the range seen in their host barite and together these formed a mass-dependently fractionated trend on a $\delta^{34}S/\Delta^{33}S$ diagram. These relationships resemble those imparted by microbial sulfate reduction but not those produced by hydrothermal or disproportionation processes. Thus, these new data are most consistent with microbial sulfate reduction imparting a mass-dependent fractionation upon sulfate sulfur with an initial mass-independent fractionation and if microbial sulfur disproportionation did indeed occur, it was insignificant compared with microbial sulfate reduction.

In addition to photolytic sulfate production, the anoxygenic photosynthesis of sulfide (see chapter 5) may have also been a significant sulfate source in the Archean. Based on molecular phylogenies, it is thought that the evolution of anoxygenic photosynthesis preceded oxygenic photosynthesis (Xiong *et al.*, 2000; Raymond *et al.*, 2002), and with abundant Archean volcanic and hydrothermal H_2S sources, this metabolism could have been a significant mode of primary production (Canfield *et al.*, 2006). Indeed, it has been hypothesized that the early Archean ecosystem locally resembled a sulfuretum (Nisbet and Fowler, 1999) where autotrophic and heterotrophic microbes cycled sulfur between reduced and oxidized states. As many of the enzymes in the phototrophic sulfide oxygenation pathway are the same as those employed during dissimilatory sulfate reduction, only operating in reverse (Meyer and Kuever, 2007), and given the existence of microbial sulfate reduction at North Pole at ~3.49 Ga (see above), this hypothesis is certainly plausible (see also Chapter 5). However, it would need confirmation from diagnostic biomarker assemblages (see Brocks *et al.*, 2005) as quadruple sulfur isotope systematics have not yet revealed a large and diagnostic signature for sulfide photosynthesis (Zerkle *et al.*, 2009).

19.4 Iron cycle

Along with carbon and sulfur, iron is the third major element whose burial controls the overall redox state of Earth's surface environments. It is also a key nutrient for life, contributing to haem-proteins and iron–sulfur proteins such as ferredoxins and cytochromes. As the latter are important during photosynthesis, iron deficiency can inhibit biological productivity. Iron is mainly supplied to the oceans by weathering and erosion of exposed landmasses and by hydrothermal input into the deep sea. It is removed in an oxidized state as iron oxyhydroxides that transform to the oxides magnetite (Fe_3O_4) or hematite (Fe_2O_3) during diagenesis, or in a reduced state as iron monosulfides (that eventually convert to pyrite, FeS_2) or iron carbonates ($FeCO_3$).

At present, iron entering the oceans via fluvial runoff and aeolian dust is in its most oxidized and insoluble form as Fe^{3+}, whereas hydrothermal fluids supply reduced and soluble Fe^{2+}. In the Archean, iron was delivered dominantly as Fe^{2+}, as weathering under an anoxic atmosphere would have limited oxidation to Fe^{3+} on land. Indeed, Archean soils on iron-rich rocks show extreme iron leaching indicating removal of soluble Fe^{2+} by groundwater (Rye and Holland, 1998). Furthermore, unlike today, hydrothermal Fe^{2+} would not have been oxidized to Fe^{3+} in the anoxic Archean deep seas. Thus, levels of dissolved Fe^{2+} should have been high in Archean ocean waters.

How was this dissolved iron removed? It is likely that some hydrothermal Fe^{2+} was removed at or near vents

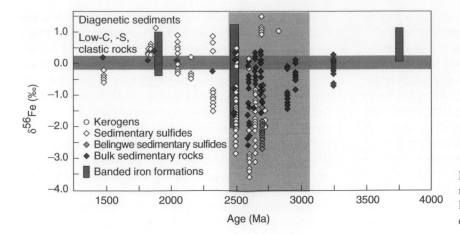

Figure 19.5 Iron isotope data from marine sedimentary rocks through early Precambrian time. Diagram from Johnson *et al.*, 2008b.

by abiotic pyrite precipitation, because if oceanic crust was basaltic, as the oldest plausible example of an ophiolite (Furnes *et al.*, 2007) implies, sulfide would also be a component of the hydrothermal fluid. However, low Archean levels of seawater sulfate would have limited hydrothermal sulfide to that derived from magmatic sources (Kump and Seyfried, 2005), in which case Fe output would have exceeded that of sulfide. The same low sulfate levels would have also diminished more distal pyrite precipitation by microbial sulfate reduction, supported by the low pyrite abundance in mid-Archean kerogenous shales (Buick, 2008a). So for much of the Archean, at least until ~2.7 Ga when pyrite in kerogenous shales became abundant, the principal sink of dissolved marine iron was as oxides, carbonates or silicates in banded iron formations (BIF).

BIF is a characteristic, though volumetrically minor, sedimentary rock from the earliest Archean through to the mid-Paleoproterozoic. It is marked by cyclic, planar to undulose banding at many scales (from microns to decameters) of siliceous chert alternating with iron oxides (hematite or magnetite) and lesser carbonates (siderite or ankerite) or silicates (varying according to metamorphic grade). Its origin has been endlessly debated, but almost all models agree that the iron was precipitated from seawater. The classical explanation (Cloud, 1973) invoked anoxic deep oceans as a reservoir of dissolved iron with cyanobacterial oxygenic photosynthesis in the shallow oceans causing iron oxidation at the interface between deep and shallow water masses. The cyclical alternations were ascribed to varying cyanobacterial activity and/or iron upwelling against a background of constant silica precipitation. Thus, BIF formation was, in this model, a fundamentally geobiological phenomenon. However, there are several plausible alternatives. Abiotic photolysis of dissolved Fe^{2+} by ultraviolet light to $Fe(OH)^+$ and ultimately to iron oxides after diagenetic dehydration has been argued (Cairns-Smith, 1978; Braterman *et al.*, 1983; Francois, 1986) to

have been sufficiently effective to account for the large volumes of iron oxides deposited in early Precambrian BIF. But recent experiments and thermodynamic modeling (Konhauser *et al.*, 2007a) have shown that photochemistry has an insignificant effect on iron precipitation from seawater of presumed Archean composition. Thus, biotic intervention is required in order to precipitate the ferric oxides in BIF.

Were cyanobacteria responsible for Fe^{2+} oxidation, as envisaged in Cloud's (1973) classic model? Not necessarily, as various bacteria can oxidize Fe^{2+} to Fe^{3+} during anoxygenic photosynthesis (Widdel *et al.*, 1993) in a process known as photoferrotrophy. This metabolism can precipitate Fe^{3+} at rates sufficient to account for the largest late Archean and early Paleoproterozoic BIFs using the available micro-nutrients (Konhauser *et al.*, 2002; Kappler *et al.*, 2005). Moreover, photoferrotrophs thrive in modern environments closely resembling postulated Archean oceanic conditions (Canfield *et al.*, 2006; Crowe *et al.*, 2008).

So which was it, oxygenic photosynthesis or photoferrotrophy, that was responsible for oxidizing the iron in Archean BIFs? Unfortunately, neither metabolism has a diagnostic biosignature that can be preserved in BIFs, as photoferrotrophy produces an iron isotope fractionation similar to that of abiotic iron oxidation (Croal *et al.*, 2004). However, it should be noted that >C_{31} 2α-methylhopanes, a class of hydrocarbon biomarkers almost exclusively associated with oxygenic photosynthetic cyanobacteria (Summons *et al.*, 1999, but see also Rashby *et al.*, 2007; Welander *et al.*, 2010), occur abundantly in kerogenous shale interbeds in the ~2.6 Ga Marra Mamba Iron Formation (Brocks *et al.*, 1999), perhaps the largest Archean BIF.

Iron isotopes (Fig. 19.5) reveal that that once oxidized, some BIF Fe-oxyhydroxides were recycled back to dissolved Fe^{2+} by dissimilatory iron reduction, a metabolism performed by a wide range of Bacteria and Archaea that imparts a more marked negative fractionation to the resulting Fe^{2+} than other biotic and abiotic iron redox reactions (Johnson *et al.*, 2008a). This isotopic signature of $\delta^{56}Fe < -1‰$ is only evident in late Archean BIFs and pyritic shales, disappearing after atmospheric oxygenation ~2.3 Ga and absent before ~3.0 Ga (Johnson *et al.*, 2008b). Dissimilatory iron reduction coupled to dissimilatory sulfate reduction may also be recorded by covarying $\delta^{56}Fe$ and $\delta^{34}S$ values in 2.7 Ga pyrite from Belingwe, Zimbabwe (Archer and Vance, 2006). It has been estimated from iron cycle modeling that perhaps 70% of the Fe^{3+} initially precipitated was recycled by dissimilatory iron reduction during diagenesis (Konhauser *et al.*, 2005).

19.5 Oxygen cycle

The previous three cycles ultimately control the oxygen cycle. To allow oxygen to accumulate in the atmosphere, two preconditions are required:

1 that a source of oxygen is available, and
2 that reduced redox species able to back-react with oxygen are removed from surface environments by burial (organic carbon, sulfide, ferrous iron) or escape to space (hydrogen gas).

The only significant oxygen source is oxygenic photosynthesis, performed by cyanobacteria or their descendants the endosymbiotic chloroplasts within eukaryote cells. Though UV photolysis of water (Rosenqvist and Chassefière, 1995) and photochemical peroxide formation above glacial ice (Liang *et al.*, 2006) could have provided some prebiotic oxygen, the amounts would have been minor and a vigorous biogeochemical oxygen cycle could not have begun until cyanobacterial oxygenic photosynthesis evolved. There are three views about the timing of this evolutionary event:

1 It happened just before the 2.4–2.3 Ga 'Great Oxidation Event' (after which the atmosphere remained moderately to highly oxygenated), because with ubiquitous reactants and high energy yields the ecological advantage it conferred upon cyanobacteria would have caused their populations to explode immediately (Kopp *et al.*, 2005);

2 It occurred hundreds of millions of years before permanent atmospheric oxygenation because it took eons to oxidize the large reservoirs and continued production of reduced volcanic gases, hydrothermal fluids and volcanic minerals (Catling and Claire, 2005); and

3 It took place very early in Earth's history, before the start of the geological record, such that the atmosphere was always highly oxygenated (Ohmoto, 1997).

For model 3, there is now much evidence that the Archean environment was not significantly oxygenated (also see Chapters 5 and 7), most notably as anomalous mass-independent fractionation of the rare sulfur isotopes ^{33}S and ^{36}S is widespread in Archean sedimentary rocks. Such mass-independent fractionation is induced by UV photolysis of volcanic SO_2 gas (Farquhar *et al.*,

2000) in an atmosphere with $<10^{-5}$ of the present level of O_2 (Pavlov and Kasting, 2002). Though it has recently been argued that thermochemical reduction of sulfate minerals by kerogen can also induce such fractionations (Watanabe *et al.*, 2009), it has yet to be shown that this occurs in natural geological environments, and that the characteristic Archean pattern of small negative $\Delta^{33}S$ fractionations to sulfates and large positive $\Delta^{33}S$ to sulfides is produced. Moreover, the thermochemical products scatter widely on $\Delta^{33}S/\delta^{34}S$ and $\Delta^{33}S/\Delta^{36}S$ plots, unlike most temporally sorted Archean sulfide data (Farquhar *et al.*, 2007). Lastly, though laboratory experiments show minimal anomalous fractionation of $\Delta^{33}S$ accompanying microbial sulfate reduction (Johnston *et al.*, 2005, 2007), Archean sulfides with isotopic signatures indicating production by this process and not by thermochemical sulfate reduction show significant mass-independent $\Delta^{33}S$ fractionation (Ueno *et al.*, 2008; Shen *et al.*, 2009), strongly implying an oxygen-poor Archean atmosphere.

For the other end-member model 1, any signs of environmental oxygen prior to the 2.4–2.3 Ga 'Great Oxidation Event' would invalidate it. And indeed, there is a wide range of evidence suggesting that cyanobacteria, oxygenic photosynthesis and transient environmental oxygenation all appeared well before this time. First, hydrocarbon biomarker molecules in kerogenous shales as old as 2.72 Ga contain abundant $>C_{31}$ 2α-methylhopanes and diverse steranes (Brocks *et al.*, 2003a), the former almost exclusively produced by cyanobacteria metabolizing by oxygenic photosynthesis (Summons *et al.*, 1999) and the latter almost exclusively produced by eukaryotes requiring abundant oxygen for their synthesis (Summons *et al.*, 2006). Though there has been a recent controversy over whether these biomarkers are indigenous to their host rocks or younger contaminants (Rasmussen *et al.*, 2008), a similar suite of biomarker molecules also occurs in contamination-proof fluid inclusions in ~2.45 Ga rocks predating the 'Great Oxidation Event' and likely sourced from immediately overlying shales (Dutkiewicz *et al.*, 2006; George *et al.*, 2007). Thus, organic geochemistry points towards the existence of oxygen-producing and -consuming organisms prior to general atmospheric oxygenation.

Second, there is growing evidence that oxygen appeared transiently and locally in the ocean well before the atmosphere became highly oxygenated. In the 2.5 Ga Mt McRae Shale, a temporary spike in redox-sensitive metal (Mo, Re) abundance (Anbar *et al.*, 2007) coinciding with a shift to negatively fractionated $\delta^{34}S$ values (Kaufman *et al.*, 2007) and a large positive excursion in $\delta^{15}N_{org}$ values (Garvin *et al.*, 2009) are all consistent with a transient oxygen 'whiff' to levels between 10^{-6} and 10^{-5} present atmospheric level. This implies substantial oxygen production well before the Great Oxidation Event and, if positive $\delta^{15}N_{org}$ values as old as 2.67 Ga also

(a)

(b)

Figure 19.6 Pyritic kerogenous shale from the 2.5 Ga Mt McRae Shale (a) contrasted with non-sulfidic kerogenous shale from the 3.2 Ga Gorge Creek Group (b).

represent the effects of nitrification and denitrification (Godfrey and Falkowski, 2009), long before.

Third, stromatolites from 2.72 Ga lakes have palimpsest fabrics indicating construction by photo-autotrophic microbes which, judging from the lack of associated iron- and sulfur-rich sediments, were apparently not metabolizing by ferrotrophic or sulfuro-trophic anoxygenic photosynthesis (Buick, 1992). As the paucity of serpentinized olivine in the basaltic substrate suggests limited abiogenic hydrogen fluxes into these lakes from below, autotrophic methanogenesis and hydrogenotrophic anoxygenic photosynthesis were also unlikely modes of primary production (Buick, 2008a). However, if H_2 were a major component of the Archean atmosphere, as claimed by Tian *et al.* (2005), then anoxygenic photosynthesis converting this gas to water may have fixed carbon abundantly (Canfield *et al.*, 2006), leaving no diagnostic geological trace. But this claim for high pH_2 has been disputed (Catling, 2005; Catling and Claire, 2005) on the grounds that hydrogen escape to space would have precluded significant atmospheric build-up. Thus, oxygenic photosynthesis is left as the most reasonable explanation for the productivity powering this prolific Archean ecosystem.

Older rocks from the ~3.2 Ga Gorge Creek Group in the Pilbara Craton of northwest Australia provide some further evidence. These black kerogenous shales (up to 12% organic carbon) are widespread (over hundreds of kilometres), thick (over hundreds of metres), and generally lack sedimentary and diagenetic iron or sulfur minerals (Fig. 19.6; Buick, 2008a). With no obvious source of reducing power from Fe^{2+} or H_2S (these would have left

abundant iron and sulfur minerals), anoxygenic photosynthesis using these compounds was probably not the principal autotrophic pathway. As mentioned above, hydrogentrophic photosynthesis was unlikely to have been a major contributor to biomass. So it seems that oxygenic photosynthesis was the only viable process for producing the large quantities of kerogen in these shales (Buick, 2008a).

Lastly, in ~3.8 Ga graphitic metasediments from Greenland, highly radiogenic Pb isotopes indicate that the rocks originally had high U abundances relative to Th showing that during deposition U was environmentally mobile whereas Th was not (Rosing and Frei, 2004). If so, this could mean that the water column through which the sediment settled was, at least in part, relatively oxidized, because U is soluble in oxygenated water but immobile under reducing conditions, in contrast to Th which is immobile under either redox regime (Rosing and Frei, 2004). However, similar isotopic patterns have not been observed in other less-metamorphosed Archean sedimentary rocks deposited after other indicators of oxygenic photosynthesis appear in the geological record, so it is not yet clear just how robust this interpretation is.

Taken in total, it seems very likely that oxygen oases temporarily appeared in shallow water bodies well before the atmosphere became permanently oxygenated and thus that model 1 (post-Archean evolution of oxygenic photosynthesis) is probably also incorrect. Thus, the intermediate model 2 in which oxygen fluxes were consumed by reducing sinks for hundreds of millions of years before the Great Oxidation Event fits the evidence best. It therefore appears that the Earth had a biogeochemical oxygen cycle extending back for some time into the Archean, that oxygenic photosynthesis evolved well before atmospheric oxygen accumulated, and that by the Archean/Proterozoic boundary the system was transiently trying to flip from a stable anoxic atmosphere to a permanently oxic state (Goldblatt *et al.*, 2006; Claire *et al.*, 2006).

19.6 Nitrogen cycle

The biogeochemical nitrogen cycle is presumably very ancient, as N is an essential nutrient for all organisms, constituting a key component of the amino acids in proteins and the purine and pyrimidine bases in nucleic acids. Its evolution can be monitored through fractionations of its stable isotopes ^{14}N and ^{15}N. As most organisms cannot satisfy their nitrogen needs by fixation (the conversion of N_2 gas to organic nitrogen), most cellular N must be assimilated from the environment, in the form of either dissolved ammonium (NH_4^+) or nitrate (NO_3^-). Thus, sedimentary organic matter generally has a $\delta^{15}N$ value reflecting that of environmental dissolved N, which in turn reflects the extent of the main process in the nitrogen cycle that fractionates N isotopes: denitrification. Because microbial dissimilatory denitrification (the reduction of nitrate or nitrite to nitrogen gas to yield energy) shows a strong preference for the light isotope, the residual marine nitrate pool is left enriched in ^{15}N. Upon assimilatory uptake of this heavy nitrate, marine organic matter is similarly enriched in the heavy isotope, such that most modern marine organic matter ranges in $\delta^{15}N$ from +5‰ to +10‰, varying somewhat according to latitude (Wada, 1980). Other steps in the N-cycle exert much smaller isotopic fractionations, such that in ancient geological samples they would be obscured by the effects of sedimentary homogenization, diagenesis and metamorphism. However anammox, a recently discovered alternative denitrification pathway wherein certain planctomycete bacteria react nitrite with ammonium to produce hydrazine and then N_2, imparts a similar isotopic fractionation to denitrification and thus has the potential to leave some sort of geological record.

Nitrogen fixation by abiotic processes is presumably as ancient as the Earth's atmosphere. It involves lightning shock-heating atmospheric N_2 and CO_2 to produce NO and eventually nitrite and nitrate (Mancinelli and McKay, 1988) that rains out into the ocean. However, this source could only sustain a biosphere a small fraction of the size of the modern biota and so was probably only significant in the very earliest stages of the evolution of life. By the mid-Archean, when there is evidence for voluminous and widespread biological productivity in the form of thick beds of highly kerogenous black shales extending regionally across sedimentary basins (see above), biological nitrogen fixation must have overtaken this abiotic pathway as the principal source of

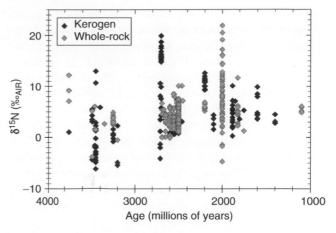

Figure 19.7 Nitrogen isotope data from marine sedimentary and low-grade meta-sedimentary rocks through early Precambrian time, showing whole-rock values and those from separated kerogen.

bio-available N (Falkowski, 1997; Glass *et al.*, 2009). This process imparts a small positive fractionation to $\delta^{15}N$ of about +1‰, which is near the mean value of sedimentary organic matter from the earlier Archean (Fig. 19.7). Beaumont and Robert (1989) interpreted this to reflect a nitrogen cycle dominated by biological nitrogen fixation with limited (if any) nitrification and denitrification. Papineau and colleagues (2005) noted that some Archean $\delta^{15}N$ values from ammonium in micas were much more positive than this, but concluded that as these came from metamorphosed rocks, they were likely to result from metamorphic resetting. As a result, Falkowski and Godfrey (2008) considered that the minimum $\delta^{15}N_{org}$ value for any particular time more faithfully records the pre-metamorphic marine nitrogen isotopic signature. For the earlier Archean, these are all +1 ±1‰ or less (Fig. 19.7), further indicating a fixation-dominated nitrogen cycle with limited nitrification/denitrification until the atmosphere became permanently oxygenated in the early Paleoproterozoic (Fennel *et al.*, 2005). Some early Archean cherts from hydrothermal dykes have lighter $\delta^{15}N_{org}$, down to −2‰, which have been interpreted by Pinti *et al.* (2001) as recording nitrogen assimilation of hydrothermal ammonium by chemosynthetic microbes.

By the end of the Archean, denitrification is clearly expressed in the nitrogen isotopic record. In the 2.5 Ga Mt McRae Shale of the Pilbara Craton in Australia, a stratigraphically restricted positive excursion in $\delta^{15}N_{org}$, from +1‰ up to +8‰ and then back to +2‰ over ~30 m in homogeneous kerogenous black shale, coincides with a spike in Mo abundance and a switch from positive to negative $\delta^{34}S_{sulfide}$. As the trace metal and sulfur isotopic data has been interpreted as recording a transient 'whiff' of free oxygen in the atmosphere and/or upper ocean (Anbar *et al.*, 2007; Kaufman *et al.*, 2007), it seems reasonable to regard the nitrogen isotopic excursion as a corollary. As oxygen levels rose, nitrification commenced.

This provided the nitrate to drive denitrification, fractionating nitrogen isotopes and making the residual marine nitrate pool and thus sedimentary organic nitrogen heavier (Garvin *et al.*, 2009). It is unclear why $\delta^{15}N_{org}$ returned to light values thereafter, but perhaps the onset of euxinic (anoxic sulfidic) conditions as marine sulfate levels rose and sulfate reduction increased (Reinhard *et al.*, 2009) then inhibited nitrification and denitrification, returning the cycle to a state dominated by nitrogen fixation. If so, then this implies that by the end of the Archean, the nitrogen cycle had both its primitive anaerobic and more modern aerobic components. Godfrey and Falkowski (2009) found nitrogen isotopic signs of this anaerobic to aerobic transition as early as 2.67 Ga.

19.7 Phosphorus cycle

Phosphorus is an essential macronutrient for life, but with only one stable isotope, its biogeochemical cycle is difficult to monitor through time. Its source in the marine environment is ultimately from the weathering of subaerially exposed rocks, and as all igneous rocks contain about the same levels of phosphorus, its fluvial supply to the oceans should not have changed much over Earth's history unless weathering and erosion rates or land area have differed. It is removed from the ocean by hydrothermal alteration of oceanic crust and by sedimentary deposition. As these parameters may have changed through time, it is worthwhile examining whether phosphorus bioavailability may have also varied.

First, have P sources varied? Several studies have argued that Archean weathering rates were higher than today's (Wronkiewicz and Condie, 1987; Fedo *et al.*, 1996; Sugitani *et al.*, 1996; Lowe and Tice, 2004), based on extreme depletions of labile elements in weathering profiles or their eroded products. The cause is usually attributed to higher atmospheric pCO_2 and/or higher temperatures. But in such circumstances, P supply to Archean oceans would have been enhanced only if erosion rates and the land area exposed to faster weathering were similar to today. However, low strontium isotope ratios in Archean carbonates (Veizer, 1989) imply that continental erosion was relatively less significant than at present. Though some models of continental evolution imply roughly constant land areas through time (e.g. Armstrong, 1991), most suggest that continental crustal volume, and hence exposed land area, was less before the late Archean. If so, P supply to the oceans may have been lower than today.

So, have P sinks varied? Bjerrum and Canfield (2002) argued that Archean biological productivity was limited by phosphate adsorption onto iron oxides that were ultimately deposited as banded iron formation. As BIFs are voluminous in Archean basins, this burial sink could have lead to oceans with P concentrations of only 10–25% modern levels. However, Konhauser *et al.* (2007) countered this model by demonstrating that dissolved silica, which was evidently at high concentrations in Archean seawater judging from the abundance of precipitated cherts, out-competes phosphate for sorption sites on iron oxyhydroxides. Thus, despite its abundance, BIF may not have been a critical sink that changed the dynamics of the phosphorus cycle. But another sedimentary rock type may have been. Shallow-marine sandstones, including those of Archean age, contain minute (~1 μm) crystals of aluminophosphate minerals such as florencite (REEAlPO), crandallite (CaAlPO), gorceixite (BaAlPO) and goyazite (SrAlPO) that form during diagenesis and are extremely recalcitrant to subsequent dissolution. This constitutes a major but rarely considered portion of the modern marine burial flux of phosphate, perhaps as much as 50% (Rasmussen, 1996). Thus, any variation in either the AlPO content or abundance of shallow-marine sandstones in the Archean might have a significant effect on the ocean phosphorus budget. Though only a few Archean and Paleoproterozoic sandstones have been analysed for their AlPO phosphorus content (Rasmussen *et al.*, 1998), they seem similar to younger rocks. However, shallow-marine sandstones may have been scarcer in a tectonically immature Archean world with less extensive continental shelves than today. Moreover, burial efficiency of phosphorus is widely thought to be lower in sediment deposited beneath anoxic waters (Ingall and Jahnke, 1997), as would have prevailed in most areas during most of the Archean. If so, bioavailable phosphorus may not have been limited by sink size but more by source area and weathering/erosion rates. However, recent studies (Goldhammer *et al.*, 2010) suggest that biogenic calcium phosphate burial is enhanced under anoxic conditions, so the converse may have applied. Clearly, phosphorus cycling in the Archean remains an issue of considerable importance but great uncertainty.

19.8 Bioaccretion of sediment

19.8.1 Stromatolites

The most obvious evidence of Archean geobiological activity is in the form of stromatolites, laminated sedimentary structures often composed of carbonate and often with convex-upwards flexures that accrete as a result of microbial growth, movement or metabolism (Chapter 16). Thus, they are trace fossils produced by interactions between microbes and clastic or chemical sediments. Their shapes vary, but predominantly convex-upward lamina flexures often form domes or columns, with the converse producing conical forms. If accreted by photosynthetic microbes, laminae thicken over flexure crests because the constructing microbes grow more successfully on topographic highs in brighter light. The microbes accrete sediment by three processes: trapping, binding and precipitation.

Many assemblages of stromatolites are now known from rocks older then 2.5 Ga (Hofmann, 2000). Most of

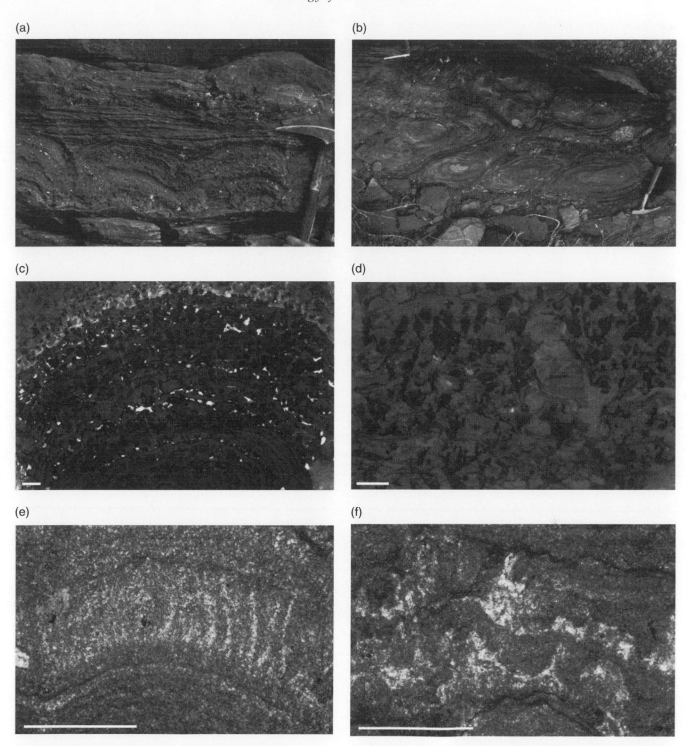

Figure 19.8 Stromatolites from the ~2.72 Ga Tumbiana Formation, Fortescue Group, Pilbara Craton, Australia: (a) Pseudocolumnar domes in cross-section showing wrinkly lamination; (b) Same structures in plan view; (c) Same structures in a close-up view of a cut slab, showing fenestral fabric of subspherical voids filled by diagenetic carbonate (scale = 10 mm); (d) Digitate microstructure in turbinate columns (scale = 10 mm); (e) Pale palimpsests of vertically-oriented filaments (scale = 100 μm); (f) Pale palimpsests of filament tufts (scale = 100 μm).

those of late Archean age (<3.0 Ga) show strong evidence of biological activity during their formation. For instance, the ~2.72 Ga Tumbiana Formation from Australia has a diverse assemblage of stromatolites (at least 10 distinct morphologies) in both tuffaceous and carbonate sedi-ments that were apparently deposited in evaporative, fault-bounded lakes (Buick, 1992; Awramik and Buchheim, 2009). As well as the large-scale structural features typical of biogenic accretion, they also have internal fabrics such as fenestrae (precipitate-filled gas bubbles produced by

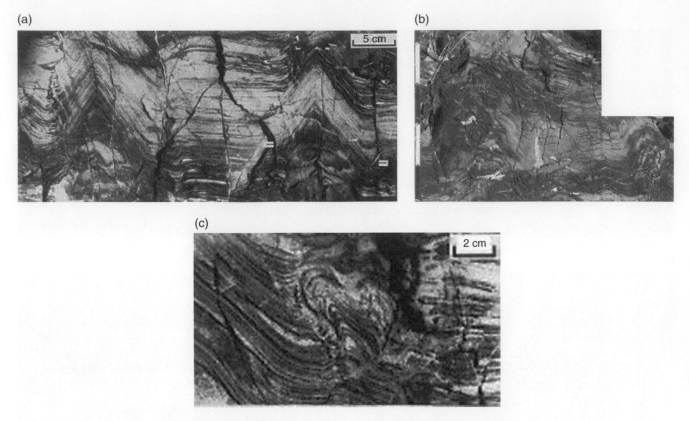

Figure 19.9 Stromatolites from the ~3.35 Ga Strelley Pool Formation, Kelly Group, Pilbara Craton, Australia; (a) Conical pseudocolumnar stromatolites; (b) Rounded conical pseudocolumnar stromatolites (scale in decimeters); (c) Branching pseudocolumnar stromatolite. Images from Hofmann *et al.*, 1999.

metabolic activity; Fig. 19.8c), cuspate lamination (tufted layers formed by gliding filaments congregating on topographic high-points; Fig. 19.8d,f), and filament palisades (carpet-like palimpsests of vertically-oriented filaments thickening on substrate highs; Fig. 19.8e). There can be little doubt that these structures were accreted by microbes that responded in some way to sunlight.

The origin of older Archean stromatolite-like structures is more controversial. Indeed, Lowe (1994) proposed that all such structures >3.2 Ga are abiogenic in origin, formed as evaporative precipitates, hydrothermal deposits, or deformational features. This analysis, however, neglected perhaps the most complex structures from the ~3.49 Ga Warrawoona Group at North Pole, Australia. These show a range of large-scale features also seen in younger, clearly biogenic stromatolites, such as pseudocolumnar structure (stacked, domically flexed layers that are laterally continuous), semicolumnar structure (stacked domical flexures that are only continuous in one direction), and complex wrinkly lamination that thickens over flexure crests (Buick *et al.*, 1981). Moreover, they defy categorization as evaporative, hydrothermal, or deformational structures. They are not simple chemical precipitates, because they show lateral variation in lamina thickness, merely moderate inherit-

ance of lamina irregularities, and several orders of flexures that are not self-similar. Moreover, the presence of lenses filled by desiccated wrinkly-laminated intraclasts in troughs between domes indicates that the structures are not deformational but had primary relief above the sea-floor. Lastly, the existence of kerogenous microlaminae in well-preserved outcrops implies microbial inhabitants. As a result of the many apparently biotic features displayed by these structures, Lowe's (1994) conclusion that all early Archean stromatolites are abiogenic is perhaps unjustified (Buick *et al.*, 1995b).

Supporting this reasoning, Hofmann *et al.* (1999) and Allwood *et al.* (2006) have added another ~3.32 Ga stromatolite assemblage from ~5 km higher in the North Pole stratigraphy to the list of potentially biogenic structures from the early Archean. These Strelley Pool Formation stromatolites have more complex large-scale structures than any others known from such ancient rocks, consisting of linked conical pseudocolumns sometimes with domical pseudocolumns branching from them (Fig. 19.9c), similar to some assuredly biogenic Proterozoic stromatolite forms. There is a diversity of morphology within this general pattern, with Allwood *et al.* (2006) recognizing seven distinct morphotypes found in discrete environments. Furthermore, some

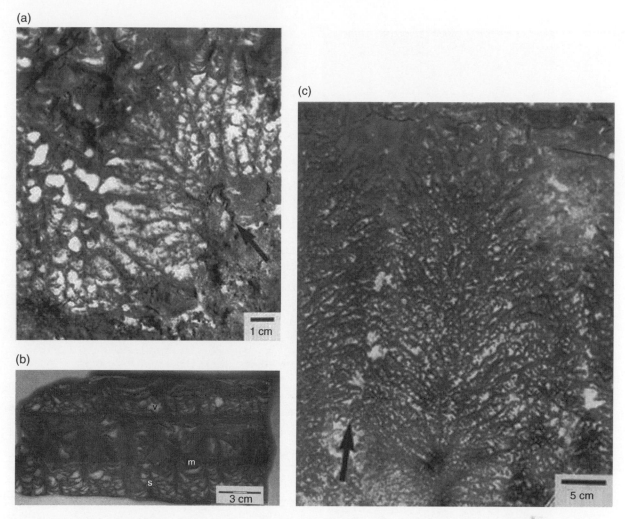

Figure 19.10 Irregular microbialites from the ~2.52 Ga Gamohaan Formation, Transvaal Supergroup, Kaapvaal Craton, South Africa; (a) Plumose microbialites separated by cement-filled troughs (arrow); (b) Cuspate microbialite of laminated mat (m) and supports (s) with voids (v) filled by bladed calcite; (c) Plumose microbialite with compacted support (arrow). Images from Sumner, 1997a.

conical structures are partially mantled by consistently oriented slump bulges, implying that the initial sediment was soft but coherent as a result of microbial binding or mucilaginous adhesion. Thus it seems that at least some early Archean stromatolites do indeed represent the activities of primordial microbial communities and that microbes were responsible for accreting sedimentary minerals very early in Earth's history.

19.8.2 Other carbonate structures

Other Archean carbonate structures have also been attributed to a microbial origin. Perhaps the most striking are roll-up structures which are fine-grained laminae with curled, circular or even spiral geometries (Simonson and Carney, 1999). These occur in deep-water clastic carbonate successions, apparently resulting from gentle scouring or syn-depositional deformation of microbially bound sediments, and are known from

several late Archean formations. Their existence implies that biofilms could form at depths below the photic zone and thus they provide evidence of Archean geobiology in deep-water environments. Somewhat surprisingly, oncolites, the subspherical concentrically-laminated equivalents of stromatolites accreted by microbes on mobile substrates, are hardly known from the Archean (Henderson, 1975; Lowe, 1983). This is strange given the abundance of stromatolites in the late Archean, with many clearly forming in settings influenced by waves and currents where oncolites often develop. Also somewhat similar to stromatolites are cuspate and plumose microbialites (Sumner, 1997a; Schröder *et al.*, 2009) that have vertical support structures and draping concave-upward laminae forming a framework around infrequent lenticular (cuspate) to abundant irregular (plumose) cement-filled voids, producing an overall tented (cuspate) to branching (plumose) structure (Fig. 19.10). These are thought to be produced by flexible gelatinous

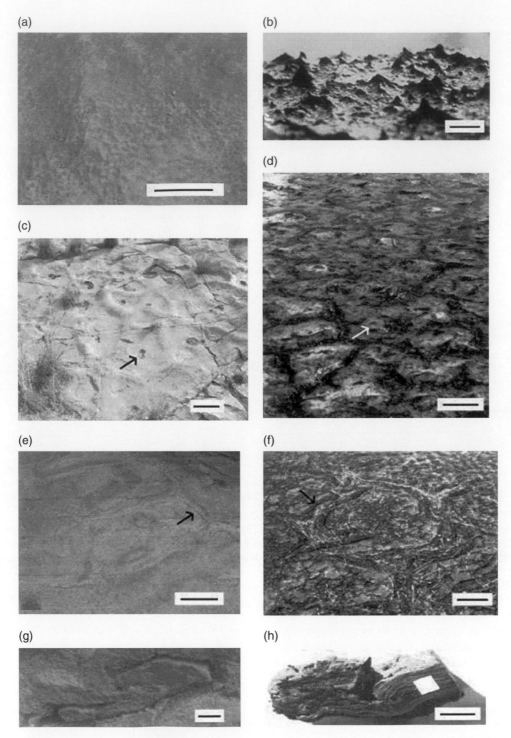

Figure 19.11 Microbially induced sedimentary structures (MISS) from the ~2.9 Ga Pongola Supergroup, Kaapvaal Craton, South Africa (left) and modern equivalents from tidal flats of Tunisia (right); (a) Pongola: bedding plane with tufted structures up to 3 mm high (scale = 10 cm); (b) Tunisia: tufts of vertically orientated cyanobacterial filaments on a modern microbial mat (scale = 1.5 cm); (c) Pongola: polygons with central holes (arrow) a few centimeters in diameter and up to 3 cm deep (scale = 25 cm); (d) Tunisia: polygons of desiccated modern microbial mat with central holes (arrow) from collapsed gas domes (scale = 20 cm); (e) Pongola: oscillation cracks (arrow) in lithified microbial mat (scale = 10 cm); (f) Tunisia: oscillation cracks (arrow) in modern microbial mat formed by expansion and shrinkage of gas domes (scale = 10 cm); (g) Pongola: overfolded clast in cross-section (scale = 5 cm); (h) Tunisia: overfolded clast of modern microbial mat deposited after a storm (scale = 5 cm). Images from Noffke *et al.*, 2008.

biofilms deformed by and draping over gas bubbles, though the origin of the support structures is unclear (Sumner, 1997a). Another unusual structure attributed to microbial precipitation occurs in late Archean platform carbonates adjacent to stromatolitic bioherms. Straight or curved dolomitic tubes up to 2 cm across and 15 cm long with multiple internal and external encrusting microlaminae are variably oriented with respect to bedding and have been interpreted as fluid-escape structures that were coated by carbonate-precipitating microbial biofilms (Murphy and Sumner, 2008).

19.8.3 Microbially induced sedimentary structures

Microbially induced sedimentary structures, or MISS, occur in siliciclastic sediments where thin microbial mats or biofilms bind the sediment surface preserving such structures as gas domes, wrinkles, multidirectional ripple marks, erosional pockets, mat-curls, oriented grains etc. (Noffke *et al.*, 2001). They are similar to some of the aforementioned structures formed in clastic carbonates (roll-ups) but differ from stromatolites in being surface features rather than three-dimensional growth structures, as they are constructed by microbial communities less than 3 cm thick (Bose and Chafetz, 2009). This is apparently because they tend to form in very shallow water or intermittently exposed settings where sediment accumulation is limited. Textural preservation indicates that they were accreted by filamentous microbes wrapping around sediment grains, and the frequent extreme inflexion of sedimentary laminae implies that contractile extracellular mucilage was responsible for sediment binding. Their record in the Archean is quite good (Fig. 19.11), extending back to the ~3.2 Ga Moodies Group of South Africa (Noffke *et al.*, 2006a), and occurrences as morphologically diverse as modern equivalents have been found in the ~2.9 Ga Witwatersrand and Pongola Supergroups of South Africa (Noffke *et al.*, 2006b, 2008). Indeed, the latter have been specifically equated with modern bacterial, or even cyanobacterial, mat structures, but in the absence of well-preserved microfossil or hydrocarbon biomarker evidence, such taxonomic certitude seems unfounded as filamentous microbes secreting extracellular mucilaginous sheaths are not restricted to these groups.

19.9 Bioalteration

19.9.1 Basalt microboring

Geobiological interactions between microbes and rocks can be destructive. In many modern settings, microbes bore into igneous minerals, volcanic glass and carbonates, leaving empty cavities, often tubular, or altered zones where more oxidized, more hydrated or more

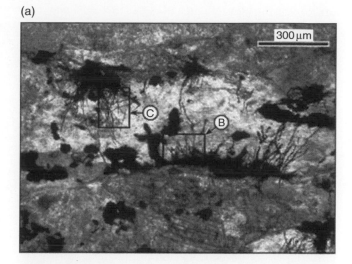

(a)

300 µm

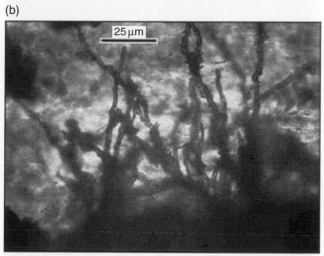

(b)

25 µm

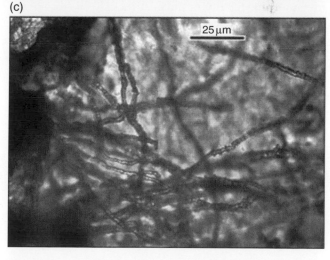

(c)

25 µm

Figure 19.12 Tubular microstructures in interpillow hyaloclastites from the ~3.35 Ga Euro Basalt, Kelly Group, Pilbara Craton: (a) Black patches of titanite mark the healed boundary between originally glassy fragments from which tubular structures ramify; (b and c) Enlargements of the boxed areas B and C. Images from Furnes *et al.*, 2007.

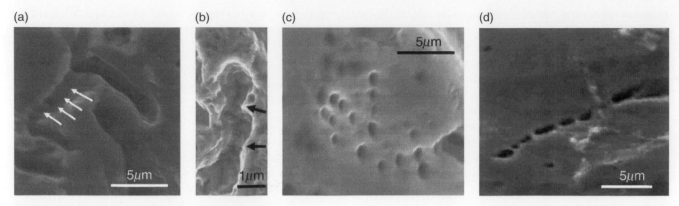

Figure 19.13 Pits and channels on detrital pyrite grain surfaces from the ~3.35 Ga Strelley Pool Formation, Kelly Group, Pilbara Craton, thought to be formed by microbial bioalteration: (a, b) Secondary electron SEM images of channels on pyrite grains showing evenly spaced indentations (arrows) suggesting attachment by chains of cells; (c) Secondary electron SEM image of a cluster of ~1 μm etch pits; (d) Secondary electron SEM image of a chain of etch pits, developing into a channel-like structure. Images from Wacey *et al.*, 2010.

crystalline minerals replace the original phase. This has been termed 'bioalteration' and can be discriminated from abiotic features by preferential orientation of tubular cavities towards or away from particular minerals in a way that might result from microbial mining for metallic micro-nutrients (Walton, 2008; Buick 2008b). In the Archean, similar features have been discovered in metamorphosed volcanic glass surrounding basalt pillows (Fig. 19.12) at several locations ~3.3 Ga in age (Furnes *et al.*, 2004; Staudigel *et al.*, 2006). To ensure that these bioalteration microstructures are indeed of Archean age and not recent, formed during Cenozoic exhumation and exposure, the minerals filling the cavities or generated during alteration need to be congruent with the metamorphic history of the host rock. In the case of the Archean basalt glasses, this appears to be the case. To show that they are biogenic and not the result of enhanced abiotic alteration along fractures, either some evidence of behaviour such as preferential orientation or some evidence of internal organic matter is required. Both apply to the examples in Archean pillow basalts. Thus, it appears that already in the Archean, microbes had evolved the ability to dissolve and alter endolithic mineral substrates. For the Archean examples, it is unclear whether this was to exploit metallic micronutrients, to avoid hostile environmental conditions or to secure a stable substrate in turbulent water.

19.9.2 Sediment microboring

Though many examples of endolithic microbial boring into carbonates are known from the Proterozoic (e.g. Knoll *et al.*, 1986), none have been discovered in Archean rocks. This is somewhat surprising, because Archean shallow-water carbonates with suitable ooids, peloids and intraclasts for endolithic boring are known to exist (e.g. Buick, 1992; Sumner, 1997b). Possibly the higher

degree of metamorphic recrystallization in these Archean rocks has lead to the obliteration of traces of any such bioalteration activity.

However, an apparently unique example of Archean bioalteration has been recognized in chert clasts within ~3.35 Ga shallow-marine sandstone (Wacey *et al.*, 2006). These tubular micro-structures penetrate grains that were clearly silicified prior to sedimentation, eliminating metallic micronutrient mining as a plausible motive, in contrast to basaltic bioalteration. Moreover, the biochemicals used in chert dissolution must have been substantially different to those used in carbonate bioalteration. Structurally, however, the chert microtubules are very similar to those seen in basalt and carbonate. The age of the microstructures is not absolutely clear, as the tubules contain hydrous minerals that were probably precipitated after peak metamorphism, but their biogenic origin is not in doubt because they contain traces of organic carbon. In the same rocks, detrital pyrite grains show surface pits and channels (Fig. 19.13) similar to microbial etchings in more modern rocks. These have been attributed to bioalteration by syngenetic iron-oxidizing microbes (Wacey *et al.*, 2010).

Bioalteration studies of Archean sediments are still in their infancy, and it is likely that other examples from different substrates and settings will be discovered. These may provide a useful window upon the potential for a voluminous subsurface microbiota very early in Earth's history when surface conditions would have frequently been inimical to habitation due to the vagaries of meteoritic impacts, high radiation doses and fluctuating environmental conditions.

19.10 Conclusions

From the foregoing, it seems likely that most major microbially mediated geobiological interactions were established by the end of the Archean, with signs that several important processes, including microbial sulfate reduction, stromatolitic sediment accretion, basalt bioalteration and perhaps even oxygenic photosynthesis, existed well before then. Shortly after the end of the Archean, the Great Oxidation Event allowed the aerobic parts of biogeochemical cycles to flourish, but it was not until the Meso- to Neoproterozoic rise to ecological dominance of the eukaryotes that the Archean pattern of microbial geobiology became largely subordinated.

References

Allwood AC, Walter MR, Kamber BS, Marshall CP, Burch IW (2006) Stromatolite reef from the early Archaean era of Australia. *Nature*, **441**, 714–718.

Anbar AD, Duan Y, Lyons TW, *et al.* (2007) A whiff of oxygen before the Great Oxidation Event? *Science*, **317**, 1903–1906.

Archer C, Vance D (2006) Coupled Fe and S isotope evidence for Archean microbial Fe(III) and sulfate reduction. *Geology*, **34**, 153–156.

Armstrong RL (1991) The persistent myth of crustal growth. *Australian Journal of Earth Sciences*, **38**, 613–630.

Awramik SM, Buchheim HP (2009) A giant, late Archean lake system: the Meentheena Member (Tumbiana Formation, Fortescue Group), Western Australia. *Precambrian Research*, **174**, 215–240.

Beaumont V, Robert F (1999) Nitrogen isotope ratios of kerogens in Precambrian cherts: a record of the evolution of atmospheric chemistry? *Precambrian Research*, **96**, 63–82.

Bjerrum CJ, Canfield DE (2002) Ocean productivity before about 1.9 Gyr ago limited by phosphorus adsorption onto iron oxides. *Nature*, **417**, 159–162.

Bose S, Chafetz HS (2009) Topographic control on distribution of modern microbially induced sedimentary structures (MISS): a case study from Texas coast. *Sedimentary Geology*, **213**, 136–149.

Braterman PS, Cairns-Smith AG, Sloper RW (1983) Photo-oxidation of hydrated Fe^{2+} – significance for banded iron formations. *Nature*, **303**, 163–164.

Brocks JJ, Logan GA, Buick R, Summons RE (1999) Archean molecular fossils and the early rise of eukaryotes. *Science*, **285**, 1033–1036.

Brocks JJ, Buick R, Logan GA, Summons RE (2003a) Composition and syngeneity of molecular fossils from the 2.78 to 2.45 billion-year-old Mount Bruce Supergroup, Pilbara Craton, Western Australia. *Geochimica et Cosmochimica Acta*, **67**, 4289–4319.

Brocks JJ, Buick R, Summons RE, Logan GA (2003b) A reconstruction of Archean biological diversity based on molecular fossils from the 2.78 to 2.45 billion-year-old Mount Bruce Supergroup, Hamersley Basin, Western Australia. *Geochimica et Cosmochimica Acta*, **67**, 4321–4335.

Brocks JJ, Love GD, Summons RE, Knoll AH, Logan GA, Bowden SA (2005) Biomarker evidence for green and purple sulphur bacteria in a stratified Palaeoproterozoic ocean. *Nature*, **437**, 866–870.

Buick R (1992) The antiquity of oxygenic photosynthesis: evidence from stromatolites in sulphate-deficient Archaean lakes. *Science*, **255**, 74–77.

Buick R (2007) The earliest records of life. In Sullivan WT III, Baross JA (eds) 'Planets and Life: The Emerging Science of Astrobiology', Cambridge University Press, Cambridge, pp. 237–264.

Buick R (2008a) When did oxygenic photosynthesis evolve? *Philosophical Transactions of the Royal Society B: Biology*, **363**, 2731–2743.

Buick R (2008b) Journal club: an astrobiologist considers the implications of microbes' mining abilities. *Nature*, **455**, 569.

Buick R, Barnes KR (1984) Cherts in the Warrawoona Group: early Archean silicified sediments deposited in shallow-water environments. Publications of the Geology Department and Extension Service, University of Western Australia, **9**, 37–53.

Buick R, Dunlop JSR, Groves DI (1981) Stromatolite recognition in ancient rocks: an appraisal of irregularly laminated structures in an early Archaean chert-barite unit from North Pole, Western Australia. *Alcheringa*, **5**, 161–181.

Buick R, Thornett JR, McNaughton NJ, Smith JB, Barley ME, Savage M (1995a) Record of emergent continental crust ~3.5 billion years ago in the Pilbara Craton of Australia. *Nature*, **375**, 574–577.

Buick R, Groves DI, Dunlop JSR (1995b) Abiological origin of described stromatolites older than 3.2 Ga: comment. *Geology*, **23**, 191.

Cairns-Smith AG (1978) Precambrian solution photochemistry, inverse segregation, and banded iron formations. *Nature*, **76**, 807–808.

Canfield DE, Rosing MT, Bjerrum C (2006) Early anaerobic metabolisms. **361**, 1819–1836.

Catling DC (2005) Comment on 'A hydrogen-rich early Earth atmosphere'. *Science*, **311**, 38a.

Catling DC, Claire MW (2005) How the Earth's atmosphere evolved to an oxic state: a status report. *Earth and Planetary Science Letters*, **237**, 1–20.

Claire MW, Catling DC, Zahnle KJ (2006) Biogeochemical modelling of the rise in atmospheric oxygen. *Geobiology*, **4**, 239–269.

Cloud P (1973) Paleoecological significance of the banded iron formation. *Economic Geology*, **68**, 1135–1143.

Croal LR, Johnson CM, Beard BL, Newman DK (2004) Iron isotope fractionation by Fe(II)-oxidizing photoautrophic bacteria. *Geochimica et Cosmochimica Acta*, **68**, 1227–1242.

Crowe SA, Jones CA, Katsev S, *et al.* (2008) Photoferrotrophs thrive in an Archean ocean analogue. *Proceedings of the National Academy of Sciences, USA*, **105**, 15938–15943.

Dutkiewicz A, Volk H, George SC, Ridley J, Buick R (2006) Biomarkers from Huronian oil inclusions: an uncontaminated record of life before the Great Oxidation Event. *Geology*, **34**, 437–440.

Eigenbrode JL, Freeman KH (2006) Late Archean rise of aerobic microbial ecosystems. *Proceedings of the National Academy of Sciences, USA*, **103**, 15759–15764.

Eigenbrode JL, Freeman KH, Summons RE (2008) Methylhopane biomarker hydrocarbons in Hamersley Province sediments provide evidence for Neoarchean aerobiosis. *Earth and Planetary Science Letters*, **273**, 323–331.

Falkowski PG (1997) Evolution of the nitrogen cycle and its influence on the biological sequestration of CO_2 in the ocean. *Nature*, **387**, 272–275.

Falkowski PG, Godfrey LV (2008) Electrons, life and the evolution of Earth's oxygen cycle. *Philosophical Transactions of the Royal Society B: Biology*, **363**, 2705–2716.

Farquhar J, Bao H, Thiemens M (2000) Atmospheric influence of Earth's earliest sulfur cycle. *Science*, **289**, 756–758.

Farquhar J, Peters M, Johnston DT, *et al.* (2007) Isotopic evidence for Mesoarchaean anoxia and changing atmospheric sulfur chemistry. *Nature*, **449**, 706–709.

Fedo CM, Eriksson KA, Krogstad EJ (1996) Geochemistry of shales from the Archean (~3.0 Ga) Buhwa Greenstone Belt, Zimbabwe: implications for provenance and source-area weathering. *Geochimica et Cosmochimica Acta*, **60**, 1751–1763.

Fennel K, Follows M, Falkowski PG (2005) The co-evolution of the nitrogen, carbon and oxygen cycles in the Proterozoic ocean. *American Journal of Science*, **305**, 526–545.

Francois LM (1986) Extensive deposition of banded iron formations was possible without photosynthesis. *Nature*, **320**, 352–354.

Furnes H, Banerjee NR, Muehlenbachs K, Staudigel H, de Wit M (2004) Early life recorded in Archean pillow lavas. *Science*, **304**, 578–581.

Furnes H, Banerjee NR, Staudigel H, *et al.* (2007) Comparing petrographic signatures of bioalteration in recent to Mesoarchean pillow lavas: tracing subsurface life in oceanic igneous rocks. *Precambrian Research*, **158**, 156–176.

Furnes H, de Wit M, Staudigel H, Rosing M, Muehlenbachs K (2008) A vestige of Earth's oldest ophiolite. *Science*, **315**, 1704–1707.

Garvin J, Buick R, Anbar AD, Arnold GL, Kaufman AJ (2009) Isotopic evidence for an aerobic nitrogen cycle in the late Archean. *Science*, **323**, 1045–1048.

George SC, Volk H, Dutkiewicz A, Ridley J, Buick R (2007) Preservation of hydrocarbons and biomarkers in oil trapped inside fluid inclusions for >2 billion years. *Geochimica et Cosmochimica Acta*, **72**, 842–870.

Glass JB, Wolfe-Simon F, Anbar AD (2009) Coevolution of metal availability and nitrogen assimilation in cyanobacteria and algae. *Geobiology*, **7**, 100–123.

Godfrey LV, Falkowski PG (2009) The cycling and redox state of nitrogen in the Archaean ocean. *Nature Geoscience*, **2**, 725–729.

Goldblatt C, Lenton TM, Watson AJ (2006) Bistability of atmospheric oxygen and the Great Oxidation. *Nature*, **443**, 683–686.

Goldhammer T, Brüchert V, Ferdelman TG, Zabel M (2010) Microbial sequestration of phosphorus in anoxic upwelling sediments. *Nature Geoscience*, **3**, 557–561.

Green MG, Sylvester PJ, Buick R (2000) Growth and recycling of early Archean continental crust: geochemical evidence from the Coonterunah and Warrawoona Groups, Pilbara Craton, Australia. *Tectonophysics*, **322**, 69–88.

Habicht KS, Gade M, Thamdrup B, Berg P, Canfield DE (2002) Calibration of sulfate levels in the Archean ocean. *Science*, **298**, 2372–2374.

Harrison TM (2009) The Hadean crust: evidence from >4 Ga zircons. *Annual Review of Earth and Planetary Sciences*, **37**, 479–505.

Hayes JM (1994) Global methanotrophy at the Archean-Proterozoic transition. In Bengtson S (ed.) 'Early life on Earth.' Columbia University Press, New York, pp. 220–236.

Hayes JM, Waldbauer JR (2006) The carbon cycle and associated redox processes through time. *Philosophical Transactions of the Royal Society B: Biology*, **361**, 931–950.

Henderson JB (1975) Archean stromatolites in the northern Slave Province, Northwest Territories, Canada. *Canadian Journal of Earth Sciences*, **12**, 1619–1630.

Hinrichs K-U (2002) Microbial fixation of methane carbon at 2.7 Ga: was an anaerobic mechanism possible? Geochemistry Geophysics Geosystems, 3, 1042, doi:10.1029/2001GC000286.

Hofmann HJ (2000) Archean stromatolites as microbial archives. In Riding R, Awramik SM (eds) 'Microbial Sediments.' Springer, Heidelberg, pp. 315–328.

Hofmann HJ, Grey K, Hickman AH, Thorpe RI (1999) Origin of 3.45 Ga coniform stromatolites in Warrawoona Group, Western Australia. *Geological Society of America Bulletin*, **111**, 1256–1262.

Ingall E, Jahnke R (1997) Influence of water-column anoxia on the elemental fractionation of carbon and phosphorus during sediment diagenesis. *Marine Geology*, **139**, 219–229.

Johnson CM, Beard BL, Klein C, Beukes NJ, Roden EE (2008a) Iron isotopes constrain biologic and abiologic processes in banded iron formation genesis. *Geochimica et Cosmochimica Acta*, **72**, 151–169.

Johnson CM, Beard BL, Roden EE (2008b) The iron isotope fingerprints of redox and biogeochemical cycling in modern and ancient Earth. *Annual Review of Earth and Planetary Sciences*, **36**, 457–493.

Johnston DT, Farquhar J, Wing BA, Kaufman AJ, Canfield DE, Habicht KS (2005) Multiple sulfur isotope fractionations in microbial systems: a case study with sulfate reducers and sulfur disproportionators. *American Journal of Science*, **305**, 645–660.

Johnston DT, Farquhar J, Canfield DE (2007) Sulfur isotope insights into microbial sulfate reduction: when microbes meet models. *Geochimica et Cosmochimica Acta*, **71**, 3929–3947.

Kappler A, Pasquero C, Konhauser KO, Newman DK (2005) Deposition of banded iron formations by anoxygenic phototrophic Fe(II)-oxidizing bacteria. *Geology*, **33**, 865–868.

Kaufman AJ, Johnston DT, Farquhar J, *et al.* (2007) Late Archean biospheric oxygenation and atmospheric evolution. *Science*, **317**, 1900–1903.

Knoll AH. Golubic S, Green J, Swett K (1986) Organically preserved microbial endoliths from the late Proterozoic of East Greenland. *Nature*, **321**, 856–857.

Konhauser KO, Hamade T, Raiswell R, *et al.* (2002) Could bacteria have formed the Precambrian banded iron formations? *Geology*, **30**, 1079–1082.

Konhauser KO, Newman DK, Kappler A (2005) The potential significance of microbial Fe(III) reduction during deposition of Precambrian banded iron formations. *Geobiology*, **3**, 167–177.

Konhauser KO, Amskold L, Lalonde SV, Posth NR, Kappler A, Anbar A (2007a) Decoupling photochemical Fe(II) oxidation from shallow-water BIF deposition. *Earth and Planetary Science Letters*, **258**, 87–100.

Konhauser KO, Lalonde SV, Amskold L, Holland HD (2007b) Was there really an Archean phosphate crisis? *Science*, **315**, 1234.

Kopp RE, Kirschvink JL, Hilburn IA, Nash CZ (2005) The Paleoproterozoic snowball earth: a climate disaster triggered by the evolution of oxygenic photosynthesis. *Proceedings of the National Academy of Sciences, USA*, **102**, 11131–11136.

Kump LR, Seyfried WE Jr (2005) Hydrothermal Fe fluxes during the Precambrian: effect of low oceanic sulfate concentration and low hydrostatic pressure on the composition of black smokers. *Earth and Planetary Science Letters*, **235**, 654–662.

Liang M-C, Hartman H, Kopp RE, Kirschvink JL, Yung YL (2006) Production of hydrogen peroxide in the atmosphere of a Snowball Earth and the origin of oxygenic photosynthesis. *Proceedings of National Academy of Sciences, USA*, **103**, 18896–18899.

Lowe DR (1983) Restricted shallow-water sedimentation of early Archean stromatolitic and evaporitic strata of the Strelley Pool Chert, Pilbara Block, Western Australia. *Precambrian Research*, **19**, 239–283.

Lowe DR (1994) Abiological origin of described stromatolites older than 3.2 Ga. *Geology*, **22**, 387–390.

Lowe DR, Tice MM (2004) Geologic evidence for Archean atmospheric and climatic evolution: fluctuating levels of CO_2, CH_4 and O_2 with an overriding tectonic control. *Geology*, **32**, 493–496.

Mancinelli RL, McKay CP (1988) The evolution of nitrogen cycling. *Origins of Life and Evolution of the Biosphere*, **18**, 311–325.

Meyer B, Kuever J (2007) Phylogeny of the alpha and beta subunits of the dissimilatory adenosine-5'-phosphosulfate (APS) reductase from sulfate reducing prokaryotes – origin and evolution of the dissimilatory sulfate reduction pathway. *Microbiology*, **153**, 2026–2044.

Murphy MA, Sumner DY (2008) Tube structures of probable microbial origin in the Neoarchean Carawine Dolomite, Hamersley Basin, Western Australia. *Geobiology*, **6**, 83–93.

Nakamura K, Kato Y (2004) Carbonatization of oceanic crust by the seafloor hydrothermal activity and its significance as a CO_2 sink in the early Archean. *Geochimica et Cosmochimica Acta*, **68**, 4595–4618.

Nisbet EG, Fowler CMR (1999) Archaean metabolic evolution of microbial mats. *Proceedings of the Royal Society B: Biology*, **266**, 2375–2382.

Noffke N, Gerdes G, Klenke T, Krumbein WE (2001) Microbially induced sedimentary structures: a new category within the classification of primary sedimentary structures. *Journal of Sedimentary Research*, **71**, 649–656.

Noffke N, Eriksson KA, Hazen RM, Simpson EL (2006a) A new window into early Archean life: microbial mats in Earth's oldest siliciclastic tidal deposits (3.2 Ga Moodies Group, South Africa). *Geology*, **34**, 253–256.

Noffke N, Beukes N, Gutzmer J, Hazen R (2006b) Spatial and temporal distribution of microbially induced sedimentary structures: a case study from siliciclastic storm deposits of the 2.9 Ga Witwatersrand Supergroup, South Africa. *Precambrian Research*, **146**, 35–44.

Noffke N, Beukes NJ, Bower D, Hazen RM, Swift DJP (2008) An actualistic perspective into Archean worlds – (cyano-) bacterially induced sedimentary structures in the siliciclastic Nhlazatse section, 2.9 Ga Pongola Supergroup, South Africa. *Geobiology*, **6**, 5–20.

Ohmoto H (1997) When did the Earth's atmosphere become oxic? Geochemical News, **93**, 12–13 and 26–27.

Papineau D, Mojzsis SJ, Karhu JA, Marty B (2005) Nitrogen isotopic composition of ammoniated phyllosilicates: case studies from Precambrian metamorphosed sedimentary rocks. *Chemical Geology*, **216**, 37–58.

Pavlov AA, Kasting JF (2002) Mass-independent fractionation of sulfur isotopes in Archean sediments: strong evidence for an anoxic Archean atmosphere. *Astrobiology*, **2**, 27–41.

Philippot P, van Zuilen M, Lepot K, Thomazo C, Farquhar J, van Kranendonk MJ (2007) Early Archaean microorganisms preferred elemental sulfur, not sulfate. *Science* **317**, 1534–1537.

Pietranik AB, Hawkesworth CJ, Storey CD, *et al.* (2008) Episodic mafic crust formation from 4.5 to 2.8 Ga: new evidence from detrital zircons, Slave Craton, Canada. *Geology*, **36**, 875–878.

Pinti DL, Hashizume K, Matsuda J-I (2001) Nitrogen and argon signatures in 3.8 to 2.8 Ga metasediments: clues on the chemical state of the Archean ocean and the deep biosphere. *Geochimica et Cosmochimica Acta*, **65**, 2301–2315.

Rashby SE, Sessions AL, Summons RE, Newman DK (2007) Biosynthesis of 2-methylbacteriohopanepolyols by an anoxygenic phototroph. *Proceedings of National Academy of Sciences, USA*, **104**, 15099–15104.

Rasmussen B (1996) Early-diagenetic REE-aluminophosphate minerals (florencite, gorceixite, crandallite and xenotime) in marine sandstones: a major sink for oceanic phosphorus. *American Journal of Science*, **296**, 601–632.

Rasmussen B, Buick R, Taylor WR (1998) Removal of oceanic REE by authigenic precipitation of phosphatic minerals. *Earth and Planetary Science Letters*, **164**, 135–149.

Rasmussen B, Fletcher IR, Brocks JJ, Kilburn MR (2008) Reassessing the first appearance of eukaryotes and cyanobacteria. *Nature*, **455**, 1101–1104.

Raymond J, Zhaxybayeva O, Gogarten JP, Gerdes SY, Blankenship RE (2002) Whole-genome analysis of photosynthetic prokaryotes. *Science*, **298**, 1616–1620.

Reinhard CT, Raiswell R, Scott C, Anbar AD, Lyons TW (2009) A late Archean sulfidic sea stimultated by early oxidative weathering of the continents. *Science*, **326**, 713–716.

Rosenqvist J, Chassefière E (1995) Inorganic chemistry of O_2 in a dense prebiotic atmosphere. *Planetary and Space Science*, **43**, 3–10.

Rosing MT, Frei R (2004) U-rich Archean seafloor sediments from Greenland: indications of >3700 Ma oxygenic photosynthesis. *Earth and Planetary Science Letters*, **217**, 237–244.

Rye R, Holland HD (1998) Paleosols and the evolution of atmospheric oxygen: a critical review. *American Journal of Science*, **298**, 621–672.

Schidlowski M, Appel PWU, Eichmann R, Junge CE (1979) Carbon isotope geochemistry of the 3.7×10^9 yr-old Isua sediments, West Greenland: implications for the Archaean carbon and oxygen cycles. *Geochimica et Cosmochimica Acta*, **43**, 189–199.

Schröder S, Beukes NJ, Sumner DY (2009) Microbialite-sediment interactions on the slope of the Campbellrand carbonate platform (Neoarchean, South Africa). *Precambrian Research*, **169**, 68–79.

Shen Y, Buick R, Canfield DE (2001) Isotopic evidence for microbial sulphate reduction in the early Archaean era. *Nature*, **410**, 77–81.

Shen Y, Farquhar J, Masterson A, Kaufman AJ, Buick R (2009) Evaluating the role of microbial sulfate reduction in the early

Archean using quadruple isotope systematics. *Earth and Planetary Science Letters*, **279**, 383–391.

Simonson B, Carney KE (1999) Roll-up structures: evidence of in situ microbial mats in late Archean deep shelf environments. *Palaios*, **14**, 13–24.

Staudigel H, Furnes H, Banerjee NR, Dilek Y, Muehlenbachs K (2006) Microbes and volcanoes: a tale of oceans, ophiolites, and greenstone belts. GSA Today, October **2006**, 1–10.

Sugitani K, Horiuchi Y, Adachi M, Sugisaki, R (1996) Anomalously low Al_2O_3/TiO_2 values for Archean cherts from the Pilbara Block, Western Australia: possible evidence for extensive chemical weathering on the early Earth. *Precambrian Research*, **80**, 49–76.

Summons RE, Jahnke LL, Hope JM, Logan GA (1999) 2-Methylhopanoids as biomarkers for cyanobacterial oxygenic photosynthesis. *Nature*, **400**, 554–557.

Summons RE, Bradley AS, Jahnke LL, Waldbauer JR (2006) Steroids, triterpenoids and molecular oxygen. *Philosophical Transactions of the Royal Society B: Biology*, **361**, 951–968.

Sumner DY (1997a) Late Archean calcite-microbe interactions: two morphologically distinct microbial communities that affected calcite nucleation differently. *Palaios*, **12**, 302–318.

Sumner DY (1997b) Carbonate precipitation and oxygen stratification in late Archean seawater as deduced from facies and stratigraphy of the Gamohaan and Frisco Formations, Transvaal Supergroup, South Africa. *American Journal of Science*, **297**, 455–487.

Tian F, Toon OB, Pavlov AA, de Sterck H (2005) A hydrogen-rich early Earth atmosphere. *Science*, **308**, 1014–1017.

Ueno Y, Yamada K, Yoshida N, Maruyama S, Isozaki Y (2004) Evidence from fluid inclusions for microbial methanogenesis in the early Archaean era. *Nature*, **440**, 516–519.

Ueno Y, Ono S, Rumble D, Maruyama S (2008) Quadruple sulfur isotope analysis of ca. 3.5 Ga Dresser Formation: new evidence for microbial sulfate reduction in the early Archean. *Geochimica et Cosmochimica Acta*, **72**, 5675–5691.

Veizer J (1989) Strontium isotopes in seawater through time. *Annual Review of Earth and Planetary Sciences*, **171**, 141–167.

Wacey D, McLoughlin N, Green OR, Parnell J, Stoakes CA, Brasier MD (2006) The ~3.4 billion-year-old Strelley Pool Sandstone: a new window into early life on Earth. *International Journal of Astrobiology*, **5**, 333–342.

Wacey D, Saunders M, Brasier MD, Kilburn MR (2010) Earliest microbially mediated pyrite oxidation in ~3.4 billion-year-old sediments. *Earth and Planetary Science Letters*, **301**, 393–402.

Wada E (1980) Nitrogen isotope fractionation and its significance in biogeochemical processes occurring in marine environments. In Goldberg ED *et al.* (eds) 'Isotope Marine Chemistry', Uchida Rokahuko, Tokyo, pp. 375–398.

Walton AW (2008) Microtubules in basalt glass from Hawaii Scientific Drilling Project #2 phase 1 core and Hilina slope, Hawaii: evidence of the occurrence and behavior of endolithic microorganisms. *Geobiology*, **6**, 351–364.

Watanabe Y, Farquhar, J, Ohmoto H (2009) Anomalous fractionations of sulfur isotopes during thermochemical sulfate reduction. *Science*, **324**, 370–373.

Welander PV, Coleman ML, Sessions AL, Summons RE, Newman DK (2010) Identification of a methylase required for 2-methylhopanoid production and implications for the interpretation of sedimentary hopanes. *Proceedings of the National Academy of Sciences, USA*, **107**, 8537-

Widdel F, Schnell S, Heising S, Ehrenreich A, Assmus B, Schink B (1993) Ferrous iron oxidation by anoxygenic phototrophic bacteria. *Nature*, **362**, 834–836.

Wronkiewicz DJ, Condie KC (1987) Geochemistry of shales from the Archean Witwatersrand Supergroup, South Africa: source-area weathering and provenance. *Geochimica et Cosmochimica Acta*, **51**, 2401–2416.

Xiong J, Fischer WM, Inoue K, Nakahara M, Bauer CE (2000) Molecular evidence for the early evolution of photosynthesis. *Science*, **289**, 1724–1730.

Zahnle K, Sleep NJ (2002) Carbon dioxide cycling through the mantle and implications for the climate of ancient Earth. Geological Society of London, Special Publication, **199**, 231–257.

Zerkle AL, Farquhar J, Johnston DT, Cox RP, Canfield DE (2009) Fractionation of multiple sulfur isotopes during phototrophic oxidation of sulfide and elemental sulfur by a green sulfur bacterium. *Geochimica et Cosmochimica Acta*, **73**, 291–306.

20

GEOBIOLOGY OF THE PROTEROZOIC EON

Timothy W. Lyons[1], Christopher T. Reinhard[1], Gordon D. Love[1] and Shuhai Xiao[2]

[1]Department of Earth Sciences, University of California, Riverside, California 92521 USA
[2]Department of Geosciences, Virginia Polytechnic Institute and State University, Blacksburg, Virginia 24061 USA

20.1 Introduction

The Proterozoic Eon, spanning from 2.5 to 0.54 billion years ago, is Earth's great middle age – bridging the beginnings of life with the biotic world we see today. Amidst all this change, a long interval in the centre of the Proterozoic is noted for unusual geochemical and tectonic stability. But no matter how you slice it, the Proterozoic was a time of transitions – marking the establishment of diverse eukaryotic communities and the origins of animals, both linked through cause-and-effect relationships to the rise of oxygen in the atmosphere and ocean.

Our story begins with a big step in biospheric evolution, the initial oxygenation of the atmosphere about 2.4 billion years ago. We follow with a discussion of the early stages of that oxic world, marked by uncertain conditions in the deep ocean, a record-breaking carbon isotope excursion, and the first of our great ice ages. Then comes a time of relative geochemical stability, with a deep ocean that likely remained poor in oxygen and rich in hydrogen sulfide and, perhaps dominantly, iron. Oxygen in the atmosphere increased but waited almost 2 billion years to rise to levels more familiar to those living in the Phanerozoic. In the face of this assumed stability, it was a time no less important for the evolution of life, including our first record of eukaryotic fossils that can be linked unambiguously to extant eukaryotic clades. We end with a discussion of tectonic upheaval, global ice cover, a second major step in oxygenation of the atmosphere and deep ocean, and the emergence of animals under these more oxic conditions.

Basic remaining questions centre on the timing and patterns of eukaryotic diversification; the chemistry and specifically the redox state of the deep ocean long after appreciable oxygen first accumulated in the atmosphere; and the oxygen history of the atmosphere and its relationship with the appearance of animals, the great Proterozoic ice ages, and coevolving life in the ocean. But despite these lingering questions, the last decade of research has brought us much closer to understanding the Proterozoic environment and its life.

20.2 The Great Oxidation Event

The Proterozoic began 2.5 billion years ago as a vestige of the Archean and its vanishingly low oxygen in the atmosphere and ocean. Evidence points to slight and possibly transient increases in oxygen at or before the Archean–Proterozoic boundary (Anbar et al., 2007; Kaufman et al., 2007; Wille et al., 2007; Frei et al., 2009; Garvin et al., 2009; Godfrey & Falkowski, 2009; Reinhard et al., 2009; Kendall et al., 2010; Thomazo et al., 2011), but appreciable and persistent oxygen came later, accumulating for the first time within the first 100 to 200 million years of the Proterozoic. Although atmospheric oxygen content must have wavered up and down from this point forward, this milestone represents the first major step in biospheric oxygenation – the irreversible appearance of oxygen in the atmosphere now known as the 'Great Oxidation Event' (GOE; Holland, 2002).

Evidence for the GOE is wide ranging, including the disappearance of fluvial sediments with detrital minerals (pyrite, siderite, and uraninite) stable only under an anoxic atmosphere, fundamental shifts in the redox states inferred for ancient soils, sulfur isotope evidence from pyrite for increasing seawater sulfate

Fundamentals of Geobiology, First Edition. Edited by Andrew H. Knoll, Donald E. Canfield and Kurt O. Konhauser.

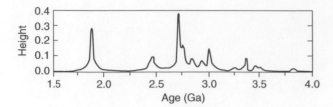

Figure 20.1 Temporal distribution of iron formations. 'Height' is an approximation of the relative abundance of IFs, taking into account the frequency of occurrence and uncertainties in the age estimates. Ages are reported in billions of years before present (Ga). After Isley and Abbott (1999).

concentrations, and the post-GOE appearance of red beds (Cameron, 1982; Holland, 1984; Rye and Holland, 1998; Rasmussen and Buick, 1999). We also see a dramatic decrease in the deposition of iron formations (IFs; often banded, i.e. BIFs), whose earlier abundances mirror the anoxic, iron-dominated Archean ocean and, importantly, time-varying but episodically copious inputs of hydrothermal iron along mid-ocean ridges (Fig. 20.1; Isley and Abbott, 1999; Kump and Seyfried, 2005; Bekker *et al.*, 2010). This and other proxy evidence for early anoxia and the GOE are reviewed in Canfield (2005) and Holland (2006). Among all the evidence for atmospheric oxygenation, however, none is more compelling than the temporal distribution of mass-independent fractionation (MIF) of sulfur isotopes, including the relatively rare isotopes ^{33}S and ^{36}S.

Also known as non-mass-dependent or NMD fractionation, this diagnostic deviation from purely mass-dependent behaviour abounds in the Archean and earliest Paleoproterozoic as a product of atmospheric, photochemical reactions involving ultraviolet (UV) radiation and SO_2 released by volcanoes (Farquhar *et al.*, 2000). (The Paleoproterozoic Eon extended from 2.5 to 1.6 billion years ago [Ga]). Despite a wide range of potential controls on the NMD signal, very low oxygen levels estimated at less than 2 ppmv are the common thread (i.e. $<10^{-5}$ or 0.001% of the present atmospheric level or PAL; Pavlov and Kasting, 2002). A rise in atmospheric oxygen to concentrations above at least 2 ppmv is constrained between ~2.45 and 2.32 Ga by the permanent loss of NMD behaviour (Fig. 20.2; Farquhar and Wing, 2003; Bekker *et al.*, 2004; Papineau *et al.*, 2005). This loss has refined our positioning of the GOE, defined originally by diverse, independent proxies for paleoredox.

Less clear than when the atmosphere became oxygenated are the details of why. We know that significant oxygen inputs to the atmosphere reflect biological activity – specifically oxygenic photosynthesis with incomplete respiratory consumption of the resulting biomass. Appreciable and persistent O_2 increases thus reflect large-scale burial of organic matter typically recorded in positive shifts in the $\delta^{13}C$ of marine carbonate rocks. In

this light, two paradoxes emerge in our story. One is that there does not appear to be a distinct shift in $\delta^{13}C$ at the GOE (Fig. 20.3) (Goldblatt *et al.*, 2006). The other inconsistency is the suggestion at ~2.7–2.5 Ga of oxygenic photosynthesis by cyanobacteria and/or eukaryotes in the form of co-occurring 2-methylhopane and 24-alkylated sterane biomarkers, respectively, some 100 to 300 million years before the GOE (Brocks *et al.*, 1999, 2003; Summons *et al.*, 1999; Dutkiewicz *et al.*, 2006; Eigenbrode *et al.*, 2008; George *et al.*, 2008; Waldbauer *et al.*, 2009). The precursor sterols for the sterane fossil molecules should have required molecular oxygen for their synthesis (Summons *et al.*, 2006) despite speculations to the contrary (Raymond and Blankenship, 2004; Kopp *et al.*, 2005), suggesting that some aquatic environments in the Archean contained sufficiently high dissolved oxygen to permit sterol synthesis – although likely at still very low levels and in the surface ocean only (e.g. Waldbauer *et al.*, 2011). Others have challenged these biomarker data as representing later contamination, suggesting that the first biological production of oxygen was later, coincident with the GOE (Kopp *et al.*, 2005; Rasmussen *et al.*, 2008). If one accepts the earlier photosynthetic origins of oxygen, and many researchers do, with arguments bolstered by wide ranging geochemical data (e.g. Anbar *et al.*, 2007; Kaufman *et al.*, 2007; Wille *et al.*, 2007; Frei *et al.*, 2009; Garvin *et al.*, 2009; Reinhard *et al.*, 2009; Duan *et al.*, 2010; Kendall *et al.*, 2010; Scott *et al.*, 2011), the goal then becomes to explain the delay in atmospheric oxygenation. Here, we must rely on reactions that buffer the atmosphere against rising oxygen, such as an interplay between O_2 and hydrothermal Fe^{2+} and/or reduced gases that include volcanogenic H_2 and biogenic CH_4 (e.g. Catling *et al.*, 2001; Claire *et al.*, 2006; Kump and Barley, 2007; Konhauser *et al.*, 2009; Gaillard *et al.*, 2011). The GOE, from this angle, reflects a fundamental loss in this buffering capacity – in combination, perhaps, with increasing oxygen production. We can agree to disagree about the specifics, but it is clear that atmospheric oxygen began its rise at the GOE and never returned to its earlier low, NMD-supporting levels.

20.3 The early Proterozoic: Era geobiology in the wake of the GOE

20.3.1 *Greenhouse gases, evolving redox, and the first great glaciations*

The Precambrian atmosphere must have had greenhouse gas properties that enabled it to offset the effects of lower solar luminosity and sustain liquid water (Sagan and Mullen, 1972). Early models for this 'faint young Sun' paradox invoked attenuated carbonate-silicate weathering on a relatively cool Earth and a corresponding increase in atmospheric pCO_2 (Walker *et al.*, 1981; Kasting, 1987). However, empirical

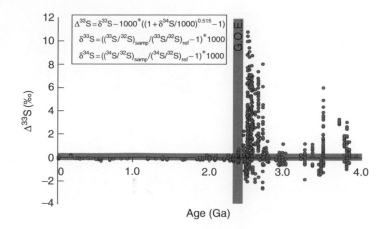

Figure 20.2 Compilation of Δ^{33}S data from many sources (available on request). Clearly delineated is the disappearance of non-mass-dependent (NMD) fractionation at the Great Oxidation Event (GOE), which fingerprints the first persistent and appreciable accumulation of oxygen in the atmosphere.

$$\Delta^{33}S = \delta^{33}S - 1000^{*}((1 + \delta^{34}S/1000)^{0.515} - 1)$$
$$\delta^{33}S = ((^{33}S/^{32}S)_{samp}/(^{33}S/^{32}S)_{ref} - 1)^{*}1000$$
$$\delta^{34}S = ((^{34}S/^{32}S)_{samp}/(^{34}S/^{32}S)_{ref} - 1)^{*}1000$$

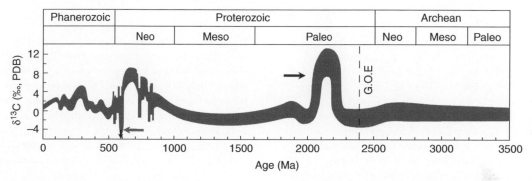

Figure 20.3 δ^{13}C of marine carbonates through time as a proxy for the isotopic composition of dissolved inorganic carbon in the ocean. After Karhu (1999). Recent work on the late Neoproterozoic suggests an extended period of strongly negative δ^{13}C values (even lower than those shown) – comprising the 'Shuram–Wonoka' anomaly (left arrow).

Also apparent is the long-lived positive Paleoproterozoic anomaly of the Lomagundi Event (right arrow) and the extended period of mid-Proterozoic δ^{13}C stability. The approximate position of the Great Oxidation Event (GOE) is indicated. Ages are reported in millions of years before present (Ma).

constraints from paleosols and weathering rinds on riverine gravels suggest that CO_2 alone would not have been able to compensate for lower solar luminosity during the Precambrian (Rye *et al.*, 1995; Hessler *et al.*, 2004; Sheldon, 2006), although such reconstructions are somewhat challenging (Sheldon and Tabor, 2009).

The solution to the problem of insufficient CO_2 is now generally expressed in terms of increased atmospheric CH_4 concentrations during the Precambrian, making microbial methanogenesis the linchpin that connects global climate, biological activity, and changes in Earth surface redox (Pavlov *et al.*, 2000, 2001; Catling *et al.*, 2007). There is both molecular phylogenetic (e.g. Woese, 1977; Barns *et al.*, 1996; Battistuzzi *et al.*, 2004) and stable carbon isotope (Hayes, 1994; Ueno *et al.*, 2006) evidence to suggest that methanogenesis is an ancient metabolic pathway, perhaps present by the early Archean. Limited respiratory oxidants in the ocean would have promoted the ecological significance of methanogenesis during organic matter remineralization, and low oceanic concentrations of both oxygen

and sulfate would have inhibited global rates of biological methane oxidation (Canfield *et al.*, 2000; Habicht *et al.*, 2002). (Sulfate is a key substrate for the anaerobic oxidation of methane.)

A large body of evidence indicates that atmospheric oxygen concentrations increased significantly over a geologically short interval of time during the Paleoproterozoic (Fig. 20.4; discussed above), and it has long been known that there is a close association between this rise in O_2 and a series of extensive glaciogenic deposits (Roscoe, 1973). The oldest of three glacial diamictites in the Huronian Supergroup of southern Canada is underlain by the Matinenda Formation (~2.45 Ga), which contains detrital pyrite (FeS_2) and uraninite (UO_2) indicating weathering and transport beneath an oxygen-deficient atmosphere (Young *et al.*, 2001). The youngest Huronian diamictite, the Gowganda Formation (~2.22 Ga), is overlain by a series of redbeds indicating weathering under oxidizing conditions (Roscoe *et al.*, 1973; Young *et al.*, 2001). Sulfur isotope analyses suggest that the disappearance of NMD sulfur isotope fractionation from authigenic

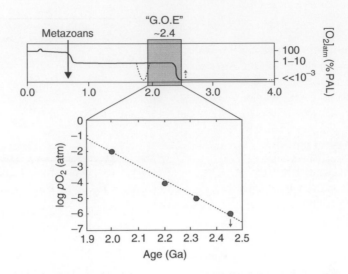

Figure 20.4 Estimates of atmospheric O_2 content (pO_2) through time (Ga) reported as % of the present atmospheric level (PAL). The arrow at 2.5 Ga indicates the pre-GOE 'whiff' of oxygen described in the text. The suggestion of a decrease at ~2.0 to 1.8 Ga reflects the recurrence of IFs and the associated Cr isotope arguments of Frei *et al.* (2009). Modified from Canfield (2005), Kump (2008), and Lyons and Reinhard (2009b). Proterozoic pO_2 values are not well constrained and may have been substantially lower than those shown.

The inset is a summary of literature constraints, perhaps better described as best guesses given the inherent uncertainties, on pO_2 as a function of age – at and immediately following the GOE. The estimates are as follows: $pO_2 = 10^{-6}$ atm at 2.45 Ga (the arrow indicates that pre-GOE pO_2 may have been lower than indicated; Pavlov and Kasting, 2002; Bekker *et al.*, 2004), $pO_2 = 10^{-5}$ atm at 2.32 Ga (Bekker *et al.*, 2004), $pO_2 = 10^{-4}$ atm at 2.2 Ga (Rye and Holland, 1998), and $pO_2 = 10^{-2}$ atm at 2.0 Ga (Rye and Holland, 1998).

marine sulfides occurs some time before 2.32 Ga in correlative units from the Transvaal basin in South Africa – that is, between the second and third of the Huronian glacial deposits (Bekker *et al.*, 2004). Importantly, this series of glaciations culminates with the deposition of the Makganyene diamictite in the Transvall Supergroup, which has been interpreted to reflect extensive low-latitude glaciation (Evans *et al.*, 1997).

Although the temporal association between these Paleoproterozoic glaciations and the rise in atmospheric oxygen has long been known, a refined appreciation for the role of CH_4 cycling during this period provides a specific mechanism for this linkage. It is straightforward, then, to surmise that an increase in atmospheric O_2 would have decreased the atmospheric residence time of CH_4, thereby causing Earth's first extensive glaciations. Indeed, most models linking the Paleoproterozoic glacial events to atmospheric composition presume that a decline in atmospheric CH_4 was caused by the rise in O_2 (Pavlov *et al.*, 2000; Kasting, 2005; Kopp *et al.*, 2005; Bekker and Kaufman, 2007). Recently, however, Konhauser *et al.* (2009) gave us a different way to think about early CH_4 cycling: atmospheric O_2 increased at the GOE as a result of reduced CH_4 production with declining seawater concentrations of Ni, an enzymatic staple for methanogenesis. Here the suggestion is that the reduction in atmospheric methane was a cause rather than an effect of rising oxygen (e.g. Zahnle *et al.*, 2006). One way or the other, it seems that

a decline in atmospheric CH_4 was mechanistically associated with the inception of Earth's first great ice ages, linking global climate with the geochemical effects of oxygenic photosynthesis and biological methane cycling, and it has been suggested that CH_4 still had an important role to play in global climate for much of the Proterozoic (Pavlov *et al.*, 2003).

20.3.2 *The Lomagundi Event and implications for oxygen in the atmosphere and sulfate in the ocean*

Following initial reports of isotopically anomalous carbonates from the Lomagundi Group in Zimbabwe (Schidlowski *et al.*, 1975, 1976), a number of workers have shown that shallow-water sedimentary carbonates from Paleoproterozoic basins worldwide exhibit a pronounced enrichment in ^{13}C between ~2.22 and 2.06 Ga (Melezhik and Fallick, 1996; Buick *et al.*, 1998; Bekker *et al.*, 2006; Melezhik *et al.*, 2007). This isotope excursion, the Lomagundi Event, is characterized by values for $\delta^{13}C_{PDB}$ of up to +28‰, and commonly greater than 6–8‰, for ~140 million years (Fig. 20.3; Melezhik *et al.*, 2007).

The carbon isotope composition of sedimentary carbonates in a steady-state carbon cycle, to first order, is a function of the isotopic composition of inputs to the ocean ($\delta^{13}C_{in}$), the fraction of carbon buried in sediments in organic form (f_{org}), and the isotopic offset between carbonate minerals and organic carbon (ε_{TOC}), simplified as:

$$\delta^{13}C_{carb} = \delta^{13}C_{in} + f_{org}\,\varepsilon_{TOC}. \qquad (20.1)$$

The conventional interpretation of positive carbon isotope excursions is that they primarily reflect increases in the relative fraction of carbon exiting the surface Earth as organic matter (i.e. f_{org}; Kump and Arthur, 1999). It was therefore suggested initially (Karhu and Holland, 1996) that the Lomagundi carbon isotope excursion reflects substantial burial of organic matter in marine environments. The amount of oxygen released by such an organic matter burial event would equal 12–22 times the present atmospheric O_2 inventory, thus providing a direct link between the Lomagundi Event and the Paleoproterozoic rise in atmospheric oxygen (Karhu and Holland, 1996). The problem, however, is that refined constraints on the timing of the GOE (Bekker *et al.*, 2004) indicate that it came well before the Lomagundi Event. The lack of direct evidence within the patchy sedimentary record for copious burial of organic matter adds to the confusion.

Is there evidence for a big rise in atmospheric oxygen during the Lomagundi Event? Perhaps. In an oxic world, sulfate in the ocean derives primarily from the oxidative weathering of sulfide minerals on the continents beneath an atmosphere containing free O_2. It is not clear how, if at all, sulfate delivery and atmospheric pO_2 correlate beyond low threshold levels of oxygenation – for example, the pO_2 threshold that triggers pyrite oxidation is potentially very low (Canfield *et al.*, 2000; Anbar *et al.*, 2007; Reinhard *et al.*, 2009). Nevertheless, suggestions of appreciable gypsum deposition during the Lomagundi interval imply a substantial transient increase in seawater sulfate concentrations compared to times before and after the event (Melezhik *et al.*, 2005; Schröder *et al.*, 2008). This gypsum and the implicitly high sulfate concentrations could be a hint of the purported boost in O_2 during the Lomagundi and correspondingly high sulfate delivery (and/or muted sulfate-consuming pyrite burial under more pervasively oxic conditions) (Figs 20.4 and 20.5; Canfield, 2005).

Hayes and Waldbauer (2006) assumed, instead, that the Lomagundi Event does not correlate to anomalous burial of sedimentary organic matter, implicitly challenging the notion of a big rise in oxygen during the event. Their model is based on the observation that some forms of microbial methanogenesis produce ^{13}C-enriched CO_2. During the Archean, methanogenesis would have been a more ecologically significant process, with methanogens likely inhabiting the water column and shallow sediments rather than being relegated to deeper sedimentary environments as they are today in the presence of abundant O_2 and SO_4^{2-}. This spatial relationship, they argued, would allow the ^{13}C-enriched CO_2 to exchange with the vast oceanic pool of dissolved inorganic carbon (DIC), effectively buffering the isotopic composition of sedimentary carbonates away from

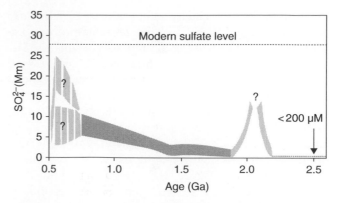

Figure 20.5 Highly schematic summary of seawater sulfate concentrations for the Proterozoic. Apparently conflicting estimates for sulfate in the Neoproterozoic range from high to very low, and the model emerging for the Paleoproterozoic predicts temporal swings between high and low values. Modified from Lyons and Gill (2010).

the ^{13}C enrichment. Following the GOE, however, increasing O_2 and SO_4^{2-} availability in the ocean caused the retreat of methanogenic organisms into deeper sediments, resulting in more direct isotopic exchange between methanogenic CO_2 and carbonate minerals thanks to diagenetic reactions during burial, increasing $\delta^{13}C$ values. As the oxidant budget of the ocean increased further, methanogenesis became inhibited by other respiratory metabolisms to the extent that its isotopic influence of carbonate minerals was small – corresponding to the falling limb of the Lomagundi excursion (Fig. 20.3). Recently, Bekker *et al.* (2008) suggested something different. Their observation of a roughly constant ε_{TOC} of ~30‰ in shallow marine environments during the Lomagundi implies that the isotopic excursion was also recorded in organic carbon, and this, they surmised, occurred because of a perturbation to the entire oceanic DIC pool, rather than local, diagenetic controls. In any case, the Lomagundi Event records a profound perturbation of the global carbon cycle linked to secular changes in the mode and magnitude of organic matter remineralization in the aftermath of the GOE.

20.4 The mid-Proterozoic: a last gasp of iron formations, deep ocean anoxia, the 'boring' billion, and a mid-life crisis

20.4.1 Introduction

By about 2.0 Ga the ocean once again favoured deposition of voluminous IF after an interval from ~2.4 to 2.0 marked by a distinct paucity (Fig. 20.1). One explanation is that oxygen levels decreased in the atmosphere to the point of reducing sulfate delivery to the ocean (Figs 20.4 and 20.5; Canfield, 2005; Frei *et al.*, 2009). Recall that the interval corresponding to the Lomgundi Event may have been characterized by comparatively high levels of seawater sulfate. If the deep ocean remained oxygen-free (Holland, 2006), bacterial reduction of sulfate to H_2S and its inevitable reaction with Fe would have lowered iron solubility. In fact, there are data indicating at least some presence of anoxic and sulfidic (euxinic) conditions in the ocean during this IF gap (Scott *et al.*, 2008). If seawater sulfate decreased at roughly 2.0 Ga, the deep, anoxic ocean may have returned to IF-favouring ferruginous conditions. (A ferruginous water column is defined as being anoxic and containing appreciably more dissolved iron than hydrogen sulfide.) Possibilities of euxinic and even oxic deep waters aside, our best guess is that the lack of IF between 2.4 to 2.0 Ga and the subsequent return at ~2.0 Ga (Fig. 20.1) must partly, if not dominantly, be a product of varying inputs of hydrothermal iron (Isley and Abbott, 1999; Bekker *et al.*, 2010).

By 2.0 Ga we see evidence for a period noted for its long-lived geochemical stability, with little variation in the $\delta^{13}C$ of the ocean (Fig. 20.3; Brasier and Lindsay, 1998) and, by inference, the oxygen content of the ocean and atmosphere (Canfield, 2005). This is the so-called 'boring billion' – positioned by some estimates between about 2.0 and 0.8 Ga (Holland, 2006), while other studies have highlighted increasing isotopic variability as early as 1.3 Ga (Frank *et al.*, 2003; Bartley and Kah, 2004).

To some researchers, the isotopic stability reflects tectonic inactivity between two periods of supercontinent formation (Brasier and Lindsay, 1998). Mountain building, in contrast, favours nutrient delivery to the ocean, weathering of ancient organic matter sequestered in black shales, and burial of modern organic matter along continental margins receiving large amounts of river-borne detritus. When occurring on grand scales, each has an effect on the carbon isotope composition of the ocean – rather than 'boring' $\delta^{13}C$ stability. Other researchers have viewed tectonic quiescence as an oversimplification given the ample tectonism suggested for this interval (Bartley *et al.*, 2001; Campbell and Allen, 2008). Regardless, this interval may have witnessed low productivity and diminished organic carbon burial, perhaps facilitated by limited supplies of nutrients and micronutrients (Buick *et al.*, 1995; Anbar and Knoll, 2002; Scott *et al.*, 2008; also Bjerrum and Canfield, 2002; cf., Konhauser *et al.*, 2007). For example, efficient recycling in the water column as a function of lower organic settling rates (Holland, 2006) would have limited organic carbon burial, although poor oxidant availability (e.g. O_2 and SO_4^{2-}), in contrast, would have inhibited recycling. Still others have argued for a large pool of dissolved inorganic carbon (DIC or ΣCO_2; Grotzinger and Kasting, 1993) that may have remained isotopically insensitive to burial of ^{12}C-enriched organic matter by virtue of its large buffering capacity (Bartley and Kah, 2004; also Kah and

Bartley, 2011). The details can be debated, but it is likely that all these factors played a role in stabilizing the $\delta^{13}C$ of the ocean and perhaps the pO_2 of the atmosphere for as long as a billion or more years.

20.4.2 Oxygen in the ocean

There is little doubt that oxygen (O_2) first accumulated in the atmosphere about 2.4 billion years ago, many proxies point to that, but how the deep ocean responded to atmospheric oxygenation is less clear. Some researchers link the disappearance of IFs at about 1.8 Ga to oxygenation of the deep ocean – hundreds of millions of years after the GOE (Holland, 2006) – while others favour deep anoxia well beyond that point and therefore IF distributions dictated by different controls (Canfield, 1998) (Fig. 20.6). Now we explore the conditions of the deep ocean beginning at ~1.8 Ga and extending into and beyond the Mesoproterozoic (1.6 to 1.0 Ga).

Using a simple three-box ocean model of the type developed by Sarmiento and Toggweiler (1984) and Sarmiento *et al.* (1988), Canfield (1998) first posited that deep-water anoxia was possible well beyond 1.8 Ga. This approach assumes that concentrations of dissolved phosphate and its cycling were similar to the modern ocean and that atmospheric oxygen contents were much lower, maybe even a few orders of magnitude lower, than those of the Phanerozoic. Although we can debate the specifics, reasonable assumptions lead to model predictions for deep ocean anoxia during the mid-Proterozoic. One important implication of an anoxic deep ocean is that its waters could have supported bacterial sulfate reduction (BSR) and thus euxinia (free H_2S in the bottom waters) if sulfate delivery to the ocean and organic carbon flux were sufficiently high. The disappearance of IFs could then reflect iron's insolubility in the presence of ubiquitous H_2S rather than oxygen in what has come to be known as the 'Canfield ocean.' Another possibility, not mutually exclusive, is that hydrothermal inputs of Fe decreased.

Loss of NMD fractionations is not the only sulfur isotope signature of the GOE. Pyrite samples younger than ~2.4 Ga also show a substantial broadening of their $\delta^{34}S$ properties away from the roughly 0‰ value of the mantle, including substantially ^{34}S-depleted samples (Fig. 20.7). This transition points to an increasing sulfate content, since large fractionations during BSR demand sulfate concentrations approaching 1% or more of present-day levels (Canfield *et al.*, 2010), and such values are not suggested for the ocean prior to the GOE (Habicht *et al.*, 2002). Greater sulfate delivery from weathering continents is expected with increasing atmospheric oxygenation and the corresponding instability of pyrite and other reduced-S-bearing minerals. The fabric of this early rise in sulfate is hard to constrain,

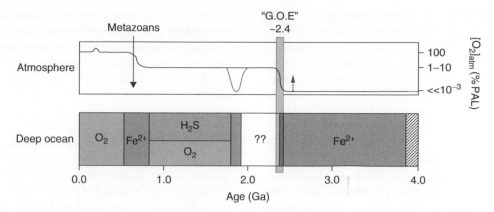

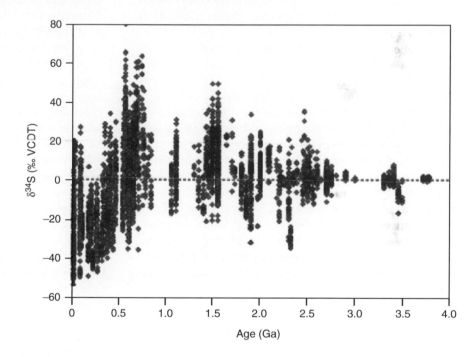

Figure 20.6 Classic views of evolving deep-ocean chemistry arguing that either oxic (Holland, 2006) or euxinic conditions (the so-called 'Canfield ocean,' Canfield, 1998) prevailed over much of the Proterozoic. Note the return to ferruginous conditions assumed for the latter part of the Neoproterozoic (Canfield *et al.*, 2008). These conceptual models are displayed relative to the backdrop of best estimates for the evolving O₂ content of the atmosphere (described in Fig. 20.4). Recent work, discussed at length in the text, is pointing to a more textured deep-ocean redox across time and space, including pervasive if not dominant ferruginous conditions in the deep mid-Proterozoic ocean. Absolute atmospheric oxygen levels are only approximations included to highlight the relative trends through time.

Figure 20.7 Summary of sedimentary pyrite sulfur isotope data. The data are presented using the standard $\delta^{34}S$ notation, and VCDT refers to the Vienna Canyon Diablo troilite standard. Modified from Canfield (2005). See the text and Lyons and Gill (2010) for additional details. Data are from many sources (available on request).

including our earlier suggestions of a kick to anomalously high concentrations during the Lomagundi interval (Fig. 20.5). Following a possible post-Lomagundi decline, sulfate delivery may have ramped up again in earnest at 1.8 Ga, marking a jump in (or return to) oxygen levels and corresponding sulfide production that dealt a deathblow to IFs (Canfield, 1998; Figs 20.4 and 20.5; Frei *et al.*, 2009). More precisely, though, varying hydrothermal delivery of Fe may have been the real driver of IF distributions (e.g. Bekker *et al.*, 2010) with only secondary ties to sulfate/sulfide relationships in a deep ocean that was largely ferruginous rather than euxinic. The balance between hydrogen sulfide, oxygen, and iron availability

in the deep ocean during this interval is an active area of research (Lyons *et al.*, 2009a, b; Lyons and Gill, 2010; Planavsky *et al.*, 2011; Poulton and Canfield, 2011).

What does the geologic record tell us? Poulton *et al.* (2004) provided an important piece of the puzzle by quantifying the species of reactive iron in the ~1.8-billion-year-old Gunflint IF and overlying shales of the Rove Formation from the Superior Province, Canada. The data pointed to a transition from (1) sediments enriched in highly reactive iron relative to total iron contents (high Fe_{HR}/Fe_T) but poor in pyrite to (2) sediments also showing Fe_{HR} enrichment but with near-complete pyritization of the reactive iron pool. The authors argued that their data captured a global transition to euxinia coincident with the cessation of IF deposition.

More recently, Poulton *et al.* (2010) reassessed the same sequence in North America along an inferred paleo-water-depth gradient and argued for contemporaneous ferruginous and euxinic conditions, with the latter occurring as a mid-depth wedge of sulfide extending basinward from the margin (analogous to Reinhard *et al.*, 2009, and Li *et al.*, 2010). The extent to which these hypothesized Gunflint-Rove paleoenvironmental conditions reflect the open ocean has been questioned (Pufahl *et al.*, 2010). Shen *et al.* (2002, 2003) also used iron paleoredox proxies in their evaluation of ~1.7–1.5-billion-year-old sediments from the McArthur Basin, northern Australia. The authors found the signature high Fe_{HR}/Fe_T ratios and high degrees of pyritization of euxinic deposition and specifically documented a basin-to-shelf transition from euxinic to oxic conditions in the 1.4–1.5-billion-year-old Roper Group. The basin's intracratonic setting again demands that we pay careful attention to the extent to which inferred chemical conditions reflect the open ocean.

Brocks *et al.* (2005) reported organic biomarker evidence for abundant green and purple sulfur bacteria inputs to 1.64-billion-year-old rocks of the Barney Creek Formation, also from the McArthur Basin. These data were assumed to reflect euxinia in the photic zone of the water column, because the green and purple sulfur bacteria that oxidize H_2S in the absence of oxygen are phototrophic. Again, continuity between these conditions and the open ocean is not a given, and other interpretations of these biomarker data are possible (discussed below). More generally, the 1.8 Ga loss of IF is roughly coincident with the large-scale emergence of sedimentary exhalative (Sedex) Pb-Zn-sulfide mineralization. This event could mark increasing sulfate and therefore H_2S availability in the ocean and a likely wealth of the oxygen-poor bottom waters required to preserve massive amounts of pyrite and other sulfide minerals accumulating on or just below the seafloor (Lyons *et al.*, 2006). At the same time, the prevailing redox conditions in the deep ocean seem to have buffered sulfate concentrations to still relatively low levels, albeit higher than the trace levels typical of the Archean.

Despite potentially higher sulfate concentrations in the early part of the Proterozoic and suggestions of strong riverine delivery (Fig. 20.5), the frequently heavy (^{34}S-enriched) pyrite seen in mid-Proterozoic sediments and sediment-hosted ore deposits is best explained by sulfate concentrations in the ocean that were only a small fraction of the 28 mM present today (Figs 20.5 and 20.7; Canfield *et al.*, 1998; Lyons *et al.*, 2000; Luepke and Lyons, 2001; Shen *et al.*, 2002; Lyons *et al.*, 2006; Farquhar *et al.*, 2010). Recent work constrains sulfate levels at less than 1 mM to explain these heavy pyrites (Canfield *et al.*,

2010), although larger fractionations between parent sulfate and product hydrogen sulfide, as preserved in isotopically lighter pyrite, are also observed during this interval. Also diagnostic of the mid-Proterozoic are rapid rates of S isotope variability inferred for seawater sulfate (Kah *et al.*, 2004; Gellatly and Lyons, 2005; see Hurtgen *et al.*, 2005, for the later Proterozoic). Although exceptions are common (e.g. Gill *et al.*, 2007, 2011), the Phanerozoic is characterized instead by much broader temporal variability, spread over intervals of 10^7 to 10^8 million years (Kampschulte and Strauss, 2004). The rapid S isotope variation of the earlier ocean is a symptom of low sulfate content and thus a short residence time and correspondingly high sensitivity to the sulfate input/output processes and the isotopic compositions of each flux.

Oxidative weathering of sulfides exposed on the continents and riverine transport of the resulting sulfate is key to the euxinia story (Canfield, 1998). If appreciable sulfate was introduced to the still-anoxic bottom waters, H_2S production might have overwhelmed the dissolved Fe of the earlier ocean. However, burial of the resulting pyrite beneath the deep seafloor and its eventual subduction would have maintained relatively low sulfate levels in the ocean (Canfield, 2004). Overall, low sulfate in the mid-Proterozoic ocean is consistent with widespread euxinic conditions in the deep ocean. But a question remains: how widespread?

Molybdenum isotope ($\delta^{98/95}$Mo) data from ~1.7–1.5-billion-year-old euxinic shales are appreciably lighter than modern seawater and the euxinic muds from the Black Sea that record the seawater value. These mid-Proterozoic data, again from the McArthur Basin, suggest a greater proportion of Mo uptake under euxinic (rather than oxic) conditions (Arnold *et al.*, 2004). However, the Mo isotope data do not demand a global Black Sea (see also Kendall *et al.*, 2009) – far from it. Furthermore, the role of suboxia remains poorly constrained, including still-large portions of the seafloor suggested by Arnold *et al.* (2004) to be oxic that could also be viewed, in a molybdenum isotope sense, as suboxic (Anbar *et al.*, 2005). (We are using suboxia to imply very low concentrations of dissolved oxygen in bottom waters, with H_2S confined to the pore waters.)

The initial model of Arnold *et al.* (2004) assumed a $\delta^{98/95}$Mo value for the riverine flux to the ocean of 0‰ (Arnold *et al.*, 2004). Recent work, however, suggests an input closer to 0.7‰ (Archer and Vance, 2008), based on modern river analysis, which is similar to the $\delta^{98/95}$Mo recorded in the shales of the McArthur Basin. If the same value applies to Proterozoic rivers, a vastly greater portion of mid-Proterozoic Mo uptake could be modeled as euxinic compared to the estimate of Arnold *et al.* (2004), but there may be a compromise: a dominantly anoxic but ferruginous deep ocean with still widespread

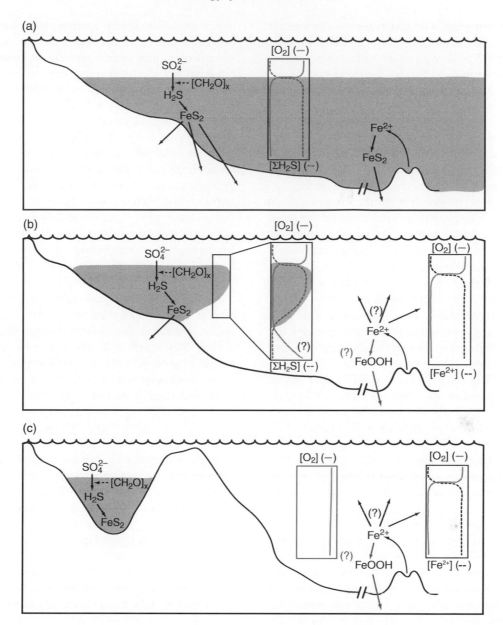

Figure 20.8 Conceptual models for the redox structure of Proterozoic ocean ranging from (a) pervasively and persistent 'whole-ocean' deep euxinia to (b) recent arguments for more localized and perhaps transient euxinia, possibly as mid-water 'wedges' (similar to modern oxygen minimum zones) that rimmed the global ocean as controlled, for example, by upwelling regions of high biological productivity. In the latter case, the sulfidic waters may have given way to dominantly ferruginous deeper waters throughout the Proterozoic. Alternatively, euxinia may have been deep but limited to restricted, marginal marine basins analogous to the modern Black Sea (c). In each case, surface waters were likely to have been well oxygenated. We can predict that all three possibilities existed at different times and likely at the same time in different places of the ocean. Modified from Lyons *et al.* (2009b). Additional details are available in the text and Lyons *et al.* (2009a, b).

sulfidic conditions limited mostly to the ocean margins and restricted basins with limited connection to the open ocean (Fig. 20.8). Such conditions could minimize fractionation of Mo isotopes without demanding global euxinia (Lyons *et al.*, 2009b). Importantly, however, it is certainly possible if not likely that the mean Mo isotope composition of rivers has changed through time

(e.g. Neubert *et al.*, 2011), and we also know little about possible Mo fractionations in ferruginous settings.

The redox state of the deep mid-Proterozoic ocean undoubtedly remains a topic of discussion (e.g. Lyons *et al.*, 2009a, b; Lyons and Gill, 2010). For example, the Mo mass balance of Scott *et al.* (2008) – that is, specifically muted Mo enrichments in euxinic shales – argues for a mid-to-deep ocean that was more commonly euxinic than today's – but nowhere near globally euxinic. Approximately 1.7-billion-year-old rocks from Arizona suggest suboxic (non-H_2S) conditions in deep seawater during the mid Proterozoic (Slack *et al.*, 2007). The final resolution to this debate likely lies with a model for the Proterozoic ocean wherein free H_2S was confined to mid-water-column depths, analogous to modern oxygen minimum zones but exacerbated by lower oxygen levels in the atmosphere, with suboxic, oxic, and perhaps dominantly anoxic-ferruginous water masses at depth.

Such a stratified model has gained popularity at either end of the Proterozoic (extending into the latest Archean), with mid-water-column sulfidic layers or wedges sandwiched between overlying oxic waters and deeper anoxic and ferruginous conditions (Fig. 20.8) (Reinhard *et al.*, 2009; Li *et al.*, 2010; Kendall *et al.*, 2010; Poulton *et al.*, 2010; also consistent with Wilson *et al.*, 2010). This style of redox layering could well hold true for much of the Proterozoic, as suggested by our newest data, which point to a dominantly ferruginous mid-Proterozoic ocean (Planavsky *et al.*, 2011). The localization of sulfidic conditions would reflect either enhanced BSR supported by elevated levels of organic matter along productive margins, as in a sulfate minimum zone (Reinhard *et al.*, 2009; see also Logan *et al.*, 1995; Johnston *et al.*, 2010), and/or by greater sulfate availability on the ocean margin tied to riverine delivery from the continents under particularly low overall marine sulfate conditions (Li *et al.*, 2010) or upwelling (Poulton *et al.*, 2010).

Reinhard *et al.* (2009), in their study of a 2.5 Ga black shale, argued for an augmented riverine flux of sulfate to the ocean as the result of an early 'whiff' of atmospheric oxygen, which raised sulfate concentrations throughout the ocean and allowed for rates of microbial sulfate reduction to scale more directly with organic matter flux in productive regions of the ocean. In general, the sulfidic waters, if they extended into the photic zone, could have supported appreciable anoxygenic primary production, as a positive feedback sustaining anoxia at depth (Johnston *et al.*, 2009; discussed in Lyons and Reinhard, 2009a). Proof of such a model awaits further evidence for widespread, very shallow euxinia during this time period. Again, deep anoxic but not sulfidic waters would satisfy the recent Mo isotope constraints of Archer and Vance (2008), but without demanding a deep ocean that was dominantly euxinic, and such conditions are equally consistent with the model predictions of Canfield (1998). Additional details are reviewed in Lyons *et al.* (2009a, b) and Lyons and Gill (2010).

Despite remaining questions, the data point persuasively toward a more reducing deep ocean as first argued by Canfield (1998), with widespread euxinia perhaps dominantly or only on the ocean margins and within restricted marginal basins. Even something much less than whole-ocean euxinia, however, would have favoured enhanced sequestration of Fe, Mo, S, and diverse other biologically relevant elements and thus the potential for depleted ocean reservoirs. For example, Anbar and Knoll (2002) suggested that patterns of eukaryotic ecology and evolution for the mid-Proterozoic (discussed below; Javaux *et al.* 2001; Knoll *et al.*, 2006) could mirror the redox state of the deep ocean. Low oxygen availability favours biologically mediated loss of fixed nitrogen through microbial denitrification processes, and euxinia might have been sufficiently frequent to deplete the inventories of metals insoluble in the presence of H_2S but essential for diverse enzymatic pathways. And these conditions could have set the tone for the evolution of mid-Proterozoic life (Saito *et al.*, 2003; Dupont *et al.*, 2006, 2010; Zerkle *et al.*, 2006; Buick, 2007; Anbar, 2008; Morel, 2008; David and Alm, 2011).

The timing of eukaryotic emergence remains controversial, and, although early ancestral eukaryotes most likely possessed mitochondria (Williams *et al.*, 2002), it does not follow that the earliest eukaryotes were necessarily aerobic. Anaerobic forms of mitochondria exist (hydrogenosomes and some mitosomes), and anaerobic organisms are distributed across all of the eukaryotic phylogenetic tree (as reviewed in Theissen *et al.*, 2003; Embley and Martin, 2006; Mentel and Martin, 2008). As a cautionary note, ancient eukaryotes thought to have been anaerobic might have lived within the surface, oxic waters of an otherwise largely anoxic ocean. Regardless of whether the deep oceans were ferruginous or sulfidic, mid-Proterozoic surface oceans were likely oxic and could support planktonic microalgae given sufficient nutrient availability, as along ocean margins.

The extent to which the Proterozoic ocean was truly limited in its supplies of bioessential elements remains an area ripe with research opportunities. The greatest promise seems to lie with recent development of methods that provide direct information about elemental concentrations rather than inferences hanging on assumed ocean redox conditions. Specifically, focused whole-rock analyses of IFs and shales can yield data that scale quantitatively with the ambient concentrations of the key elements dissolved in ancient seawater (Konhauser *et al.*, 2007, 2009; Scott *et al.*, 2008; Planavsky *et al.*, 2010).

In considering all these details we must not forget that controls on primary production may have limited the extents of anoxia and specifically euxinia through negative, nutrient-dependent feedbacks, that is, loss of nitrate through denitrification and limited N-fixation under metal-limited conditions (Canfield, 2006; Scott *et al.*, 2008). Given the relatively high organic requirement to sustain euxinia via BSR, the deep, central, largely oligotrophic ocean may in general behave as a binary system between oxic and at least weakly ferruginous conditions – varying, among other factors, as a function of O_2 availability in the atmosphere and hydrothermal inputs of Fe. Nevertheless, we imagine that H_2S was still widespread in the Proterozoic, if only on the productive margins, and even a few percent additional euxinic ocean compared to today would have gone a long way in pulling-down metal supplies in seawater (e.g. Scott *et al.*, 2008).

20.5 The history of Proterozoic life: biomarker records

20.5.1 An overview: hydrocarbon patterns from Proterozoic rocks and oils

Molecular geochemical analyses of rocks and oils that have experienced a reasonably mild thermal history generally reveal a wide diversity of hydrocarbon compounds. High yields of bitumen extracts and biomarker hydrocarbons can be recovered from many Proterozoic rocks (e.g. Summons *et al.*, 1988a,b; Summons and Walter, 1990; Pratt *et al.*, 1991; Logan *et al.*, 1995, 1997; Peng *et al.*, 1998; Höld *et al.*, 1999; Li *et al.*, 2003; Brocks *et al.*, 2005; Olcott *et al.*, 2005; McKirdy *et al.*, 2006; Grosjean *et al.*, 2009; Love *et al.*, 2009). Some characteristic molecular features can broadly distinguish many Proterozoic alkane profiles from those obtained from Phanerozoic-age rocks and oils, such as (1) a higher ratio of monomethyl branched alkanes relative to the dominant *n*-alkanes and (2) often a more prominent unresolved complex mixture of diverse branched and cyclic alkanes (Summons and Walter, 1990). These compositional differences are often attributed to a higher input of bacteria over eukaryotes in the Proterozoic, which is supported by generally higher ratios of hopanes/steranes in Paleo- and Mesoproterozoic rocks (e.g. Pratt *et al.*, 1991; Dutkiewicz *et al.*, 2003; Li *et al.*, 2003; Brocks *et al.*, 2005).

Stable carbon isotope relationships between bitumens (soluble organics), kerogens (insoluble sedimentary organic matter), and alkane components of the bitumen are also fundamentally different for Proterozoic and Phanerozoic marine rocks (Fig. 20.9). These isotope patterns and their temporal switching were first recognized by Logan *et al.* (1995, 1997) in their comparisons between the Late Neoproterozoic and Early Cambrian rocks of the Centralian Superbasin, Australia. These authors proposed that the observed switch marked a major reorganization of geochemical cycles instigated by the divergence and increasing ecological prominence of grazing pelagic zooplankton, which could make fecal pellets, a repackaging that may have increased the rate and efficiency of transport of biological organic matter to the seafloor and could have resulted in enhanced burial of reduced carbon in sediments. However, the first convincing fossil evidence for macrozooplankton is not recorded until well into the Cambrian around ~520 Ma (Vannier and Chen, 2000; Peterson *et al.*, 2005), and the role of

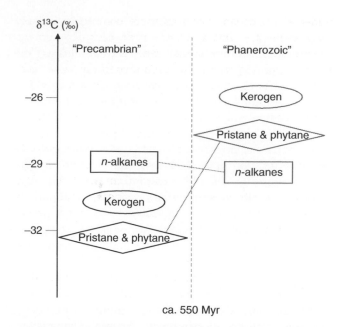

Figure 20.9 A fundamental switch in stable carbon isotopic patterns ($\delta^{13}C$) is found for hydrocarbon constituents of extractable rock bitumen (free *n*-alkanes and the acyclic isoprenoids, pristine and phytane) versus kerogen (from Logan *et al.*, 1995) for ancient sedimentary rocks. The switch in isotopic ordering appears to occur around 550 million years ago (Kelly *et al.*, 2008). Absolute $\delta^{13}C$ (‰) values vary from sample to sample.

zooplankton fecal pellets in overall carbon export is now generally considered to be somewhat minor (e.g. Turner, 2002). More recently, a similar carbon isotope transition was observed in well-dated shallow marine sedimentary rocks from South Oman Salt Basin, with the shift occurring around 550 Ma (Kelly *et al.*, 2008). Importantly, this timing corresponds to the onset of the last of the three stages of oxygenation inferred for the Ediacaran ocean based on carbon and sulfur chemostratigraphic records from the same region (Fike *et al.*, 2006).

Much interest lies in obtaining robust sterane records for Proterozoic rocks as a way of characterizing evolving eukaryotic community structure. Overwhelming evidence from lipid surveys of extant microorganisms combined with genomic data (Volkman, 2003, 2005; Brocks and Pearson, 2005; Summons *et al.*, 2006; Kodner *et al.*, 2008) and temporal patterns of ancient steranes from the Proterozoic (e.g. Summons and Walter, 1990; McCaffrey *et al.*, 1994; Brocks *et al.*, 2005; Olcott *et al.*, 2005; McKirdy *et al.*, 2006; Grosjean *et al.*, 2009; Love *et al.*, 2009) and Phanerozoic (e.g. Grantham and Wakefield, 1988; Summons *et al.*, 1992; Schwark and Empt, 2006) suggests that 24-alkylated steroids are a robust marker for eukaryotes. In the Proterozoic rock record, microalgae are most likely a major source of these steroids. Furthermore, sterol biosynthesis is an oxygen-intensive process, and multiple lines of evidence indicate that an ancestral anaerobic pathway (Kopp *et al.*, 2005) is highly unlikely (Summons *et al.*, 2006).

The most commonly biosynthesized steroids are the C_{27}, C_{28}, and C_{29} sterols, which are preserved over geologic time as C_{27}, C_{28}, and C_{29} steranes, respectively. The most abundant regular sterane constituents of any particular Proterozoic rock or oil are either C_{27} or C_{29} compounds. In the marine realm, the proportion of C_{27} steranes (cholestanes) over total C_{27-29} abundances likely records variation in the balance of red over green algal contributions to microbial communities, as red algal clades generally biosynthesize C_{27} sterols preferentially (Volkman, 2003; Kodner *et al.*, 2008). Prior to the appearance of terrestrial plants in the Paleozoic, a C_{29} (stigmastane) dominance generally reflects high green algal inputs (Volkman, 2003).

More distinctive steranes with a Proterozoic fossil record include C_{30} compounds with unusual side-chain alkylation patterns. Examples of commonly found C_{30} steranes include dinosteranes. In the Phanerozoic, dinosteranes are particularly abundant in marine rocks and oils of Triassic age and younger (Summons *et al.*, 1992), tracking a later radiation of dinoflagellates, but these same structures are also found in Proterozoic strata as old as ~1.5 to 1.2 Ga from the Beidajian Formation of the Ruyang Group in north China (Meng *et al.*, 2005) and from ~1.4-billion-year-old shales of Roper Group from the McArthur Basin, northern Australia (Moldowan *et al.*, 1996, 2001). Dinosteranes, and thus evidence for dinoflagellates, have been reported as free hydrocarbons from the 1.1-bilion-year-old Nonesuch Formation, USA (Pratt *et al.*, 1991), and are commonly found in Neoproterozoic rock extracts, such as from the Officer (Zang and McKirdy, 1993) and Amadeus basins (Summons, 1992) in Australia. Other distinctive C_{30} sterane structures found in the Proterozoic rock record include 24-*n*-propylcholestanes, which are reliable markers for marine pelagophyte algae and appear to be ubiquitous in restricted or open marine depositional settings (Moldowan *et al.*, 1990).

20.5.2   A case study: a mixed signal of ancient anoxia

Dolomitic, carbonaceous and pyritic siltstones and shales from thermally well-preserved strata from drillcore of the ~1.6 Ga Barney Creek Formation in the McArthur Group of the McArthur Basin in northern Australia have yielded the most intriguing assemblage of biomarkers reported for the Proterozoic (Brocks *et al.*, 2005). Solvent-extractable hydrocarbon biomarker data from Barney Creek strata seem consistent with a euxinic mid-Proterozoic ocean, specifically a highly stratified and sulfidic water column with abundant green and purple sulfur bacterial communities thriving in the photic zone – but with

only trace eukaryotic sterane signals. But this may not be the end of the story.

Many researchers favour a dominantly deeper-water, subwave-base facies model for the Barney Creek Formation with variable amounts of carbonate mud delivered from the shallow shelf by lateral transport as small-volume debris flows (Bull, 1998), particularly for the northern depocenter of the Glyde Sub-basin (Davidson and Dashlooty, 1993), which has been the focus of biomarker work. We agree and note that the association of green and purple sulfur bacterial markers occurs dominantly in transported shallow-water carbonate facies, rather than as planktonic communities thriving below the oxycline in distal marine settings (as reported by Brocks *et al.*, 2005). We should not be surprised. The idea of H_2S sustained in the photic zone in the open ocean at the very shallow depths required by planktonic purple sulfur bacteria (12– 20 m) is hard to reconcile with normal surface-ocean mixing processes beneath an oxygenated atmosphere. In this light, along with other biomarker evidence for restricted saline, likely shallow initial conditions, the aromatic carotenoids in the carbonate-rich Barney Creek Formation are likely markers for green and purple sulfur bacteria reworked from layered benthic microbial mats. These mats must have grown on the surface of sediments in shallow, likely oxic, saline carbonate-rich settings that were episodically eroded and transported by turpid currents into deeper waters. The deep, basinal waters, based on organic and inorganic proxy archives in the shales, were anoxic/ferruginous and locally euxinic, including suggestions of green but not purple sulfur bacteria in the water column in the Glyde Sub-basin and thus H_2S extending only into the deeper portions of the photic zone.

It is also possible that our view of global mid-Proterozoic ocean chemistry and microbial community structure is biased by results from the McArthur Basin, perhaps reflecting conditions in a large but restricted basin with uncertain and varying geochemical continuity with the broader ocean. Compared to the younger Roper Group, restricted exchange is a relatively easier case to make for the Barney Creek Formation. These strata typically yield extremely low, perhaps anomalous sterane abundances; such aquatic environments would have been especially hostile to eukaryotes, even for phytoplankton living in well-mixed oxic surface waters.

20.5.3   Other organic records of mid-Proterozoic marine microbial communities and ocean chemistry

In contrast to the McArthur Basin, a full range and high abundances of C27-C29 sterane biomarkers has been reported in solvent extracts (Peng et al., 1998; Li et al., 2003) for the 1.7 Ga Chuanlinggou Formation and younger strata of the Yanshan Basin in North China (1.7–0.85 Ga), which suggests the presence of diverse and abundant marine eukaryotic communities throughout a large interval of the mid-Proterozoic. Hopane/sterane ratios are higher (2.2–7.1; Li *et al*, 2003) in these rocks compared to the Phanerozoic marine average of 0.5–2.0, suggesting high bacterial primary productivity, especially in the more saline and restricted depositional environments. Probable eukaryotic fossil remains have been reported from the same region – consistent with this tantalizing biomarker evidence for abundant marine eukaryotic communities in mid-Proterozoic seas in the Yanshan Basin. Importantly though, none of the free sterane compounds reported have yet been robustly confirmed as syngenetic with the host rocks from analysis of the corresponding kerogen-bound biomarker pool. There exists the possibility then that the free steranes are the result of younger contamination.

20.6 The history of Proterozoic life: mid-Proterozoic fossil record

Frequent euxinia in the mid-Proterozoic, even if concentrated on the ocean margins, may have shaped the evolutionary pattern of eukaryotes through the modulated availability of certain bioessential trace metals (Anbar and Knoll, 2002; Anbar, 2008). As discussed above, iron and molybdenum, for example, are important micronutrients particularly for eukaryote metabolism, but they might not have been readily available in seawater if euxinia was far more common than today. Could the deficiency in bioessential trace metals have affected eukaryote evolution? The fossil record provides some clues, but the overall picture is still murky. Eukaryotes most likely diverged before or at the beginning of the 'boring billion', but large scale diversification of modern-style eukaryotes apparently did not get started until much later, after 'the billion' – suggesting evolutionary stasis of eukaryotes that may be linked to widespread sulfidic conditions in the water column and the overarching redox state of the ocean-atmosphere system more generally. Recall, eukaryotic biomarkers from 2.7 Ga shales in the Pilbara Craton of Australia (Brocks *et al*, 1999) provide a minimum, albeit controversial, age for the initial divergence of eukaryotes; the authenticity of these biomarkers has been questioned (Kirschvink and Kopp, 2008; Rasmussen *et al.*, 2008). There are reports of possible eukaryote (*Grypania*-like) fossils from the ~1.9 Ga Nagaunee Formation in northern Michigan (Han and Runnegar, 1992), although subsequently challenged, and probable eukaryote fossils are known from the 1.8–1.6 Ga Changzhougou, Chuanlinggou, and Tuanshanzi formations in the Yanshan Basin, North China (Zhang, 1986, 1997; Yan and Liu, 1993; Zhu *et al.*, 1995; Lamb *et al.*, 2009; Peng *et al.*, 2009).

Unambiguous eukaryote fossils are known from rocks deposited during the 'boring billion'. For example, micro- and ultrastructures uniquely characteristic of living eukaryotes occur in microfossils from the 1.5 Ga Roper Group in northern Australia and the >1.0 Ga Ruyang Group in North China (Xiao *et al.*, 1997; Javaux *et al.*, 2001, 2004). However, the taxonomic diversity of mid-Proterozoic eukaryote fossils is relatively low, and only a few have been resolved phylogenetically into modern eukaryote sub-groups (Fig. 20.10), including a ~1.2 Ga red alga (Butterfield, 2000) and a >1.0 Ga xanthophyte chromoalveolate (Hermann, 1990; Rainbird *et al.*, 1998). Eukaryotes evolved a moderate diversity toward the end of the mid-Proterozoic, between 0.85 and 0.75 Ga (Knoll, 1994; Vidal and Moczydłowska-Vidal, 1997; Knoll *et al.*, 2006), and included green algae (Butterfield *et al.*, 1994), xanthophytes (Butterfield, 2004), rhizarian amoebae (Porter and Knoll, 2000), amoebozoans (Porter and Knoll, 2000), and possibly fungi (Butterfield, 2005). This moderate fossil diversity (Fig. 20.10), however, was dwarfed after the great glaciations receded about 635 million years ago.

A few case studies have shown that mid-Proterozoic eukaryote diversity was concentrated in marginal marine settings (Butterfield and Chandler, 1992; Javaux *et al.*, 2001). This distribution pattern is similar to that of modern marine ecosystems; modern eukaryotic phytoplankton are more competitive than cyanobacterial phytoplankton in coastal regions where nutrients are more available due to riverine input and oceanic upwelling. Could this same distribution pattern in the mid-Proterozoic have been driven by similar controls on nutrient availability in an anoxic and widely euxinic ocean?

Compilation of the Proterozoic eukaryote fossil record reveals that mid-Proterozoic eukaryotes displayed not only relatively low diversity but also relatively slow turnover rates (Knoll, 1994). We can now ask whether this evolutionary stasis was related to pervasive anoxia and euxinia. But the answers to these questions require more experimental data to better understand the physiological and ecological sensitivities of eukaryotes to the predicted ocean conditions, and we need more well-dated paleobiological data generated within a rigorous paleoenvironmental context from multiple basins, including a better grasp of the oceanic extents of euxinia. Molecular biomarker studies have inferred low eukaryotic inputs to rocks of the 1.4–1.6 Ga McArthur Basin (such as the Roper Group and Barney Creek formations) deposited under oxygen-poor marine conditions. Specifically, the data point to low abundances of steranes, particularly in comparison to the bacterially derived hopane biomarker series (Summons *et al.*, 1988a; Brocks *et al.*, 2005). This relationship is in contrast to the more abundant and diverse sterane record reported for sedimentary rocks from the 1.7–0.85 Ga Yanshan Basin in North China. We are left wondering whether the frequent paleontological and

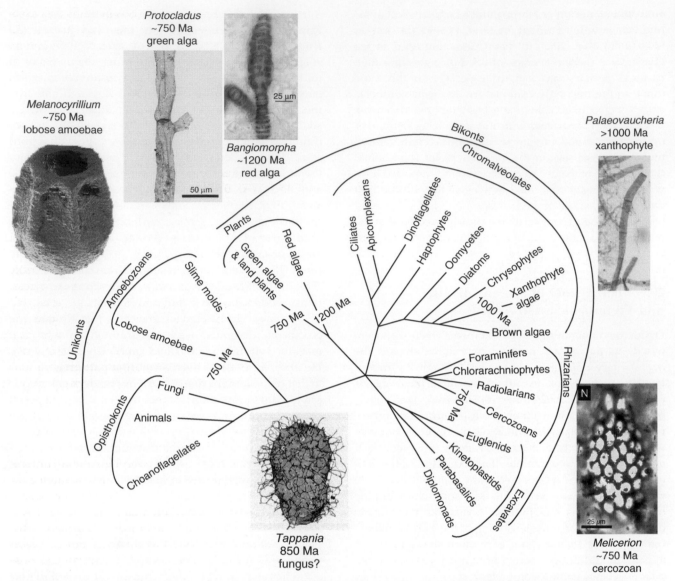

Figure 20.10 Eukaryote phylogeny based on molecular and cytological data. The following lineages are photosynthetic: green algae and land plants, red algae, dinoflagellates, haptophytes, diatoms, chrysophytes, xanthophyte algae, brown algae, chlorarachniophytes, and euglenids (some ciliates and apicomplexans may have been photosynthetic but subsequently lost their photosynthetic capabilities). Six mid-Proterozoic fossils are used to date the divergence time of major eukaryote lineages. Modified from Baldauf (2000) and Porter (2006). Fossil images courtesy of N.J. Butterfield, A.H. Knoll, and S.M. Porter.

geochemical emphasis on rocks from the McArthur Basin, which is known for at least episodes of local restriction, may be biasing our perspective. The result could be underestimated global eukaryotic diversity and perhaps overestimated extents of euxinia in the mid-Proterozoic ocean. As an intriguing backdrop to all this work, putative multicellular organisms of uncertain affinity, including *Grypania*, have been described in very old rocks, with records recently extended back to 2.1 Ga (El Albani *et al.*, 2010) – 1.5 billion years before the rapid expansion of multicellularity in the form of animals.

20.7 The late Proterozoic: a supercontinent, oxygen, ice, and the emergence of animals

20.7.1 Overview

By the beginning of the Neoproterozoic, 1.0 billion years ago, the supercontinent Rodinia was assembled, and the sediments shed from the resulting highlands yielded efficient burial of organic matter in shallow marine settings along the continental margins. With organic burial, which is more assumed than quantified or even confidently fingerprinted, oxygen would have risen in the atmosphere. In chorus, the eroding mountains fed phosphorous and bioessential trace metals to the ocean, spawning accelerated primary production and carbon burial (Campbell and Allen, 2008; Campbell and Squire, 2010). However, other factors must have favoured the burial of organic matter that drove rising oxygen, and each must be viewed in a complex framework of positive and negative feedbacks set into motion by single or multiple external drivers. For example, weathering mountains yield nutrients and sediments that promote carbon burial, but at the same time organic matter weathered in the same uplifted rocks is an important oxygen sink (Berner, 2003).

Other factors have been suggested to explain Neoproterozoic oxygenation, including enhanced organic burial bolstered by organic-preserving interactions with an increasing flux of clay minerals (Kennedy *et al.*, 2006; compare Tosca *et al.*, 2010). Delivery of essential phosphorus to the ocean may have jumped dramatically in the wake of global-scale glaciation (Planavsky *et al.*, 2010), favouring organic production and burial and concomitant O_2 release to the atmosphere (see also Papineau, 2010). Important oxygen sinks in the ocean and atmosphere may have diminished, such as dissolved organic carbon in the ocean (Rothman *et al.*, 2003) and methane in the ocean and atmosphere (Pavlov *et al.*, 2003). Once thresholds were achieved, feedbacks may have sealed the deal. Regardless of mechanisms, we can be quite certain that oxygen increased in the late Neoproterozoic in the biosphere's second great step in oxygenation (Des Marais *et al.*, 1992). Nevertheless, the fabric of this transition; the interplay of physical, chemical, and biological controls (Meyer and Kump, 2008; Lyons *et al.*, 2009a, b; Lyons and Gill, 2010); and relationships with earlier and later redox in the atmosphere and ocean were likely complex (Fike *et al.*, 2006) and remain unresolved.

The first widespread ecological expansion and increasing morphological/ecological complexity of early animals may be linked to an external environmental trigger, such as increases in atmosphere/dissolved oxygen over a critical threshold value and/or fundamental changes in ocean redox structure. At least low levels of dissolved oxygen were required by multicellular benthic macroorganisms living on the seafloor, with shallow continental shelves being colonized initially (Sperling *et al.*, 2007). The biosynthesis of collagen – an important structural protein that occurs in animals but not protists, requires molecular oxygen (Towe, 1970). Physiological consideration also predicts that the evolution of macroscopic animals with thick tissues, ultimately large size, and metabolically active lifestyles critically depends on the availability of molecular oxygen (Knoll and Carroll, 1999). The genetic toolkit required for animal development was also assembled earlier in the Proterozoic (Knoll and Carroll, 1999; King, 2004).

The origin of the gut and nervous systems was associated with the divergence of the eumetazoans (all animals except sponges) shortly after the Marinoan glaciation in the Early Ediacaran, as predicted from molecular clocks (Peterson and Butterfield, 2005; Sperling *et al.*, 2007). Yet, the fossil record of planktic predation (Butterfield, 1994, 1997; Vannier and Chen, 2000) suggests that planktonic bilaterian predators producing fecal pellets of sufficient size to allow improved transport and burial of organic matter on the seafloor would not likely have appeared until after the Tommatian (~530 Ma), well into the Cambrian (Peterson *et al.*, 2005).

It is impossible to ignore other drivers of prokaryotic and eukaryotic diversification and extinction coincident with, and likely coupled to, biospheric oxygenation, such as the great ice ages of the late Neoproterozoic 'snowball Earth' and related environmental stress. Neoproterozoic climate was extreme, with at least two extended deep freezes that may have enveloped the continents and vast portions of the oceans in ice. The coupled biogeochemical cycles among oxygen, methane, and carbon dioxide in the atmosphere and ocean are certain to have dominated the beginnings and ends of the 'snowball' episodes.

20.7.2 The snowball Earth

The climate extremes of the Neoproterozoic are among the major stories to emerge in early Earth research over the last decade, and the many details have been documented and debated in a voluminous literature covering the subject. Much of the story begins with Kirschvink (1992) who, based on a global distribution of glaciomarine diamictites and associated iron deposits, hypothesized a snowball Earth with extensive glacial deposition at or near the paleoequator. This work was followed by detailed documentation of the chemostratigraphy (Kaufman *et al.*, 1997) and the important development of a broadly encompassing model designed to include the diverse geochemical, stratigraphic, sedimentologic, paleoclimatic, and paleobiologic chapters of a complex tale (Hoffman *et al.*, 1998). Beyond the unusual conditions of low elevation, low latitude glaciers on the continents, researchers further conjecture that the ocean may have been frozen to appreciable depth even at the equator, although the lateral continuity of this sea ice is a source of particular disagreement. Many models now imagine something closer to a 'slushball,' with large ice-free areas at equatorial latitudes (Hyde *et al.*, 2000; Poulsen *et al.*, 2001; Poulsen, 2003; Peltier *et al.*, 2007) or relatively thin tropical sea ice (McKay, 2000; Pollard and Kasting, 2005), but the stability of such conditions is an important realm of future research (e.g. Lewis *et al.*, 2007).

Extensive data now point to two globally distributed glaciations: the Sturtian (~0.716–0.712 Ga; Macdonald *et al.*, 2010, and references therein) followed by the younger Marinoan (~0.635 Ga for the well-dated glacial termination; Condon *et al.*, 2005), although the global synchroneity of many so-called Sturtian diamictites is debated. A still younger glaciation, the Gaskiers (~0.580 Ga), has a less clear global distribution but is certain to have been large-scale. It is notable that the glacial deposits are typically overlain by 'cap carbonates,' which have received much attention. Cap carbonates overlying the Marinoan glacial deposits, for example, are usually 5–20 m thick (Hoffman *et al.*, 2007) and consist of dolostone with evidence for strong ^{13}C depletions (Kaufman *et al.*, 1997; Hoffman *et al.*, 1998, 2007; Kennedy *et al.*, 1998; Hoffman and Schrag, 2002; Jiang *et al.*, 2003; Zhou and Xiao, 2007). Competing and complementary models exist for the origins of the caps, and nested within these arguments are fundamental assumptions about the extremity of the 'snowball' conditions. If, for example, the Neoproterozoic earth was covered by ice, the hydrological cycle and continental weathering may have been suppressed for millions of years. In the absence of weathering reactions, accumulation of volcanogenic and metamorphic CO_2 in the atmosphere could have led to the greenhouse conditions that ultimately triggered deglaciation. It follows that carbon dioxide transferred from the atmosphere to the ocean via silicate and carbonate weathering stimulated cap carbonate precipitation during the deglaciations (Hoffman *et al*, 1998; Higgins and Schrag, 2003).

The upwelling model of Grotzinger and Knoll (1995) for the cap carbonate demands physical stratification of the ocean with a strong surface-to-deep carbon isotope gradient (also Hurtgen *et al.*, 2006). Under such conditions, postglacial upwelling or flooding attendant to rising sea level could have delivered alkalinity-rich deep water with a light carbon isotope composition to continental shelves and interior basins, catalyzing cap carbonate precipitation with diagnostic δ^{13}C signatures. As a bridge between the two models, the lower cap could reflect upwelling, with the upper cap deriving from weathering-related CO_2 inputs. Kennedy *et al.* (2001, 2008) argued something very different, suggesting that the cap carbonates and associated isotopic anomalies are a product of methane release and oxidation triggered by postglacial warming and sea-level rise. By inference, the intensities of glacial and post-glacial processes need not have been as extreme as those suggested by the original snowball Earth hypothesis. Jiang *et al.* (2003) and Wang *et al.* (2008) reported a small set of strongly depleted δ^{13}C values consistent with at least some contribution from methane release and oxidation, although the extent to which methane-related processes dominate the cap carbonates seems challenged by the relative paucity of extremely ^{13}C-depleted data despite copious analyses of cap carbonates from many localities worldwide.

It is not our goal to cover the many details and arguments about the snowball Earth hypothesis that have emerged since 1998; these are covered in numerous primary sources and review papers (e.g. Hoffman and Schrag, 2000; Fairchild and Kennedy, 2007; Allen and Etienne, 2008). It is important, however, to comment further on relationships between these great ice ages and the patterns of Neoproterozoic biota, emphasizing the interrelationships between the coevolving environment and life. Certainly, many lineages must have survived the Sturtian and Marinoan glaciations, as required by continuity of evolution. Molecular clock estimates, although complicated by calibration issues, imply that major eukaryotic clades diverged prior to and survived the snowball Earth events (Douzery *et al.*, 2004). The fossil record may be incomplete, but the positive identification of cyanobacteria and several photosynthetic eukaryote clades (e.g. red algae, green algae, and yellow green algae or xanthophytes; reviewed in Porter, 2004) in rocks predating the Sturtian glaciation requires that some members of these photosynthetic groups must have survived the Sturtian and Marinoan glaciations (Corsetti *et al.*, 2003, 2006; Xiao, 2004; Moczydłowska, 2008a, b). There is also paleontological evidence for heterotrophic protists (e.g. testate amoebae) dating from ~0.750 Ga (Porter and Knoll, 2000) and a geochemical record of primitive animals in a 100-million-year archive of demosponge sterane biomarkers from rock bitumens and kerogens from the Huqf Supergroup, South Oman. The biomarker data begin in the late Cryogenian but below the Marinoan cap carbonate (discussed below; Love *et al.*, 2009). Other findings include putative metazoan carbonate reefs as old as 1.083–0.779 Ga reported from the early Neoproterozoic Little Dal Formation in Canada (Neuweiler *et al.*, 2009; compare Planavsky, 2009). (The Cryogenian Period [0.850–0.635 Ga] is currently defined as the interval between 850 million years ago and the beginning of the Ediacaran Period (635 Ma); it includes the Sturtian and Marinoan ice ages.)

The evolutionary continuity of high-ranking clades across Neoproterozoic glaciations may not be an appropriate test of the severity of these glaciations; the above mentioned clades survived all Phanerozoic mass extinctions, which nonetheless caused major diversity changes at lower taxonomic levels. The evolutionary response to Neoproterozoic glaciations is better examined at the levels of ecological abundance, ecological diversity, and species/genus diversity. The fossil record of acritarchs (traditionally interpreted as eukaryotes) does suggest a drop in the species diversity during the Cryogenian glaciation intervals, followed by significant post-glacial rediversification in the Ediacaran Period (Knoll, 1994;

Vidal and Moczydłowska-Vidal, 1997; Knoll *et al.*, 2006; Moczydłowska, 2008a; the Ediacaran Period is defined as the interval following the Marinoan glaciation extending to the base of the Paleozoic: 0.635 to 0.542 Ma). Although these paleobiological patterns should be viewed cautiously because of preservational biases and unresolved taxonomic issues, the big picture does seem to suggest that species diversity was generally low during the Sturtian and Marinoan glaciations and the interglacial they bracket.

20.7.3 Geochemistry of the Neoproterozoic atmosphere and ocean: a puzzling record of oxygenation

By most accounts, the late Neoproterozoic saw the second significant rise in atmospheric oxygenation (Fig. 20.4; Des Marais *et al.*, 1992; Canfield, 2005; Catling and Claire, 2005; cf., Campbell and Allen, 2008). It is hard to pinpoint precisely when and how this transition occurred. For example, we can imagine that multiple substeps were involved, not unlike models emerging for the GOE, and oxygenation of the deep ocean may have continued to lag behind increasing oxygen contents in the atmosphere and surface ocean. Ironically, despite its relatively young age, the second 'step' is less well constrained than the GOE, and we rely on proxies other than NMD fractionation of sulfur isotopes that speak indirectly and at best semi-quantitatively to atmospheric compositions and specifically the rise in oxygen (e.g. Frei *et al.*, 2009). These arguments often depend on risky generalizations about redox conditions in the global ocean, and this conversation is not always consistent. What seems clear, however, is that this increase in biospheric oxygenation coincided with the rise of animals and set the stage for the radiation events that defined the Phanerozoic. However, we are often at risk of circularity when we default to the rise of animals as our most convincing evidence for the rise in oxygen, when at the same time we explain the appearance of animals by rising oxygen.

Unlike the preceding mid-Proterozoic, the Neoproterozoic is characterized by famously wide variation in the $\delta^{13}C$ composition of the ocean, including a ubiquity of positive values consistent with an overall increase in organic carbon burial, as well as long-lived negative excursions, shorter-term isotopic volatility that characterizes the glacial episodes, and unusually large fluctuations within and outside the glacial intervals (Fig. 20.3). An ocean with waning DIC content, based on hypothesized decreases in carbonate saturation states in the ocean with increasing DIC uptake/burial by planktonic productivity, would have been more sensitive to the inputs and outputs that drove the variability (Bartley and Kah, 2004). The focus of Rothman *et al.* (2003) was drawn to the uniqueness of these large fluctuations and

specifically the significant negative excursions and unusual apparent decoupling between carbonate and organic $\delta^{13}C$ records (Fike *et al.*, 2006; Swanson-Hysell *et al.*, 2010). In response, they suggested something quite novel: nonsteady-state behaviour in the presence of a large and long-lived pool of dissolved/suspended organic carbon (DOC) in the deep, anoxic ocean – a conceptual model that is sometimes referred to as the 'Rothman Ocean.' In such an ocean, large and rapid negative excursions could reflect remineralization of DOC at depth, a process that would be tied ultimately to protracted oxidation of the deep ocean and availability of sulfate in seawater (Fike *et al*, 2006; Li *et al.*, 2010; Swanson-Hysell *et al.*, 2010). Levels of sulfate, the other primary oxidant in the ocean, correlate positively but likely not linearly with biospheric oxygenation.

Global-scale glaciation is not a prerequisite for exceptional $\delta^{13}C$ variation in the Neoproterozoic (Kaufman and Knoll, 1995; Rothman *et al.*, 2003; Dehler *et al.*, 2005), although such variations, when observed in the postglacial cap carbonates immediately overlying the glacial diamictites, are a classic fingerprint of the snowball Earth (Kaufman *et al.*, 1997). Conceptual models for those variations and specifically the exceptionally negative excursions observed in the cap carbonates range from voluminous releases of methane with attendant oxidation to DIC (Jiang *et al.*, 2003) to the formation of ^{13}C-depleted DIC in an anoxic synglacial ocean through BSR and concomitant oxidation of ^{12}C-enriched organic C (Knoll *et al.*, 1996; Hurtgen *et al.*, 2006).

Anoxic and likely euxinic conditions in the synglacial oceans are widely assumed (e.g. Hurtgen *et al.*, 2006); however, the temporal extent of Neoproterozoic deep-ocean anoxia and its relationship to the oxygen-deficiencies hypothesized for the mid-Proterozoic ocean and the increasing oxygen contents in the Neoproterozoic atmosphere warrant further discussion. Also salient to our discussion is the sulfate content of the ocean and its evolution, since riverine inputs of sulfate are tied qualitatively to oxygen in the atmosphere and oxidative weathering of the continents, although scaling relationships are not well known, and any correlation may abate at post-GOE O_2 levels. The removal term, pyrite burial, is initiated by reduction of sulfate by bacteria that require anoxia, and pyrite burial is particularly efficient under euxinic conditions (e.g. Lyons, 1997).

The large isotope fractionations that occur during BSR may top out at 40 to 45‰, enriching the product hydrogen sulfide in the light isotope, ^{32}S, although recent work suggests that larger fractionations may be possible by BSR alone (e.g. Wortmann *et al.*, 2001; Werne *et al.*, 2003; Brunner and Bernasconi, 2005; Canfield *et al.*, 2010). Canfield and Thamdrup (1994) showed us that additional fractionation can also result from further S cycling via pathways wherein intermediate S species such as elemental S are disproportionated, thus explaining the >45‰ fractionations that are common in natural systems (e.g. Lyons, 1997). Canfield and Teske (1996) went a step further and conjectured that such large fractionations, common in the Phanerozoic, first appeared (or proliferated) in the $\delta^{34}S$ record for pyrite in the Neoproterozoic (Fig. 20.7). Such an increase could capture a greater role by S disproportionators and the increasing presence of S intermediates produced by non-photosynthetic sulfide oxidizing bacteria. These bacteria may have evolved or at least thrived in the oxidizing shallow Neoproterozoic ocean beneath an atmosphere with oxygen estimated at 5 to 18% PAL (Canfield and Teske, 1996). It is likely that disproportionation occurred in the earlier ocean (Johnston *et al.*, 2005), but large fractionations were muted by low seawater sulfate contents, perhaps exacerbated under locally restricted marine conditions (Hurtgen *et al.*, 2005; Lyons *et al.*, 2006; Canfield *et al.*, 2010) and/or lower availability of S intermediates. Hurtgen et al. (2005) argued specifically that fractionations beyond those possible from BSR alone did not appear until 0.58 Ga and that this shift marked increasing seawater sulfate contents rather than an enhanced contribution from disproportionation (also Halverson and Hurtgen, 2007).

Fike *et al.* (2006) and Fike and Grotzinger (2008) attributed increasing fractionations captured in ~0.55 billion-year-old samples from Oman to an increasing role for disproportionation, although increasing sulfate availability cannot be precluded. Comparable $\delta^{34}S$ relationships in the similarly aged Doushantuo Formation (McFadden *et al*, 2008) could be attributed to an analogous combination of increasing sulfate and disproportionation. Sulfate concentrations inferred from fluid inclusions in halite deposited in Oman during the terminal Proterozoic argue for levels in seawater that were two-thirds or more of the modern value (Horita *et al.*, 2002) – consistent with gypsum accumulation in the same sequence and similar-aged halite inclusion data from other parts of the world (Kovalevych *et al.*, 2006) – although these concentration estimates vary significantly with required assumptions about the calcium content of the ocean.

In contrast to arguments for increasing sulfate in the ocean, rocks of the same age from Namibia instead show evidence for super heavy (^{34}S-enriched) pyrite that is almost certainly a product of very low levels of seawater sulfate (Ries *et al.*, 2010). Rapid S isotope variability and heavy pyrite in cap carbonates formed in the immediate wake of Marinoan glaciation also point to low sulfate (Hurtgen *et al.*, 2002; Kah *et al.*, 2004; Shen Y *et al.*, 2008; see also Shen B *et al.*, 2008). Finally, recent arguments by Canfield *et al.* (2008) for a persistently and perhaps pervasively iron-rich, deep anoxic ocean may suggest only trace levels of sulfate for much of the late

Neoproterozoic beginning around 0.70 Ga. Importantly, however, availability of organic matter is also a/the key ingredient in producing euxinia, and, as we suggest above, ferruginous deep ocean waters were likely common if not dominant throughout the Proterozoic.

Already low sulfate levels could be lowered further by the diminished weathering inputs expected under glacial conditions and the exaggerated burial of pyrite beneath an anoxic synglacial ocean. However, a protracted persistence of only trace sulfate concentrations well beyond the glacial episodes requires another explanation. For example, sulfate deficiencies over a long interval of the Neoproterozoic and a return to ferruginous conditions in the deep ocean could indicate an expanded Proterozoic history of pyrite burial beneath a deep sulfidic ocean (Canfield, 1998) and subsequent subduction of the pyritic facies (Fig. 20.6; Canfield, 2004; Canfield *et al.*, 2008; discussed in Lyons, 2008). Subduction would preclude later recycling by removing S from the crustal reservoir – eventually reducing S inputs to the ocean. It will be important to revisit this model in light of the possibility of a ferruginous deep ocean for much of the Proterozoic and any differences in the amount of pyrite buried in deep, subductable seafloor. Finally, the contrasting evidence for both high and low sulfate contents in the Neoproterozoic ocean (Fig. 20.5), while contradictory at first blush, may make sense in light of emerging models for lateral and vertical heterogeneities in the Neoproterozoic ocean. Analogous variability is expressed in the inferred patterns for oxygen availability.

From the same sections from Oman studied for their sulfur isotope properties (Fike *et al.*, 2006; Fike and Grotzinger, 2008), Fike *et al.* (2006) documented an outstanding example of the 'Shuram' carbonate-C isotope excursion, expressed as persistently negative $\delta^{13}C$ values between 0 and −12‰ for as long as 30 million years during the middle–late portion (~0.580 to 0.550 Ga) of the Ediacaran Period, although the duration of the event is not well dated. What is better constrained, however, is a full appreciation of the global extent of this feature, including the Wonoka Formation of Australia and the uppermost Doushantuo Formation of South China (Fike *et al.*, 2006; Le Guerroué *et al.*, 2006; McFadden *et al.*, 2008), with diagnostic decoupling of bulk organic and inorganic $\delta^{13}C$ records and unusually light carbonate values. Swanson-Hysell *et al.* (2010) recently reported analogous isotopic behaviour even deeper in the Neoproterozoic. Fike *et al.* (2006) interpreted the Shuram and its global equivalents as a product of deep ocean oxygenation and specifically progressive oxidation of the ^{12}C-depleted DOC pool posited by Rothman *et al.* (2003) for the Neoproterozoic. Among the vexing aspects of the Shuram are the seemingly decoupled $\delta^{13}C$ records for organic and inorganic C, which couple by the end of the event and may mark final exhaustion of the deep reduced carbon pool by ~0.550 Ga, although further work is required to explore the temporal persistence of the coupling and the specifics of the prior 'decoupling.' Also curious are the oxidant demands (principally oxygen and sulfate) that may have exceeded their availability in the ocean and atmosphere (Bristow and Kennedy, 2008). The Shuram and coeval anomalies and their implications for ocean-atmosphere redox remain red hot topics in Ediacaran research (Grotzinger *et al.*, 2011), including evolving views on primary oceanographic/atmospheric drivers, such as the role of massive methane release (Bjerrum and Canfield, 2011), and recent claims arguing instead for diagenetic and/or sea level controls (Knauth and Kennedy, 2009; Derry, 2010; Swart & Kennedy, 2011).

Consistent with the deep-ocean oxygenation model for the Shuram–Wonoka anomaly, Canfield *et al.* (2007) argued, based on distributions of reactive iron, for at least a locally oxic deep ocean recorded in fine-grained sediments in Newfoundland about 0.580 Ga. Similarly, the distributions of molybdenum in euxinic black shales described by Scott *et al.* (2008) point to dramatic reductions in the global extent of oxygen-poor to euxinic deep seafloor by ~0.550 Ga. Inconsistent, however, are arguments based on reactive iron in rocks spanning much of the late Neoproterozoic from about 0.70 Ga forward that suggest persistently anoxic and likely ferruginous conditions over much of this window at widely spaced locations (Canfield *et al.*, 2008; Li *et al.*, 2010). Greater hydrothermal iron inputs may also have been a factor – but at levels insufficient to produce persistent, appreciable IF deposition across this interval. (There was, however, a return to IF deposition specifically associated with Neoproterozoic glacial conditions.)

What should be clear from the above discussion of the sulfur and oxygen properties of the Neoproterozoic ocean is that no simple argument for a pervasively and persistently sulfate- and oxygen-rich or -poor deep ocean is possible. Instead, we are left with only one conclusion: ocean chemistry was heterogeneous, which is anything but surprising. Heterogeneity in the ocean is an expected product of low concentrations of dissolved iron, sulfate, and oxygen, and thus their short residence times, as well as local restriction (Fig. 20.8), which is easy to imagine given global tectonic conditions that spawned widespread rifting related, for example, to the breakup of Rodinia.

Equally important is the possibility of strong geochemical stratification in the late Proterozoic water column, a condition again favoured by generally low seawater concentrations of iron, sulfate, and oxygen. We can imagine, for example, oxygen minimum zones analogous to those present today, or perhaps mid-water-column layers of euxinia (sulfate minimum zones; Logan *et al.*, 1995), underlain by deep waters that were anoxic (ferruginous) – or even oxygenated, by analogy to the modern ocean. Also, we cannot ignore the impact a snowball Earth would have on the chemistry of the ocean during and in the wake of possible global-scale ice cover and varying continental delivery of sulfate, nutrients, and sediment load. Not surprisingly, recent studies are suggesting mid-water sulfidic zones underlain by ferruginous waters, as controlled by sulfate delivery (Li *et al.*, 2010; Poulton *et al.*, 2010; see also Pufahl *et al.*, 2010), enhanced rates of BSR associated within organic-rich zones on ocean margins (analogous to Johnston *et al.*, 2010), or some combination of both (analogous to Reinhard *et al.*, 2009) (Fig. 20.8). We are left speculating that these stratified conditions may have prevailed throughout much of the Proterozoic (and even the Archean; e.g. Reinhard *et al.*, 2009) under low, near-threshold concentrations of iron, sulfate, and oxygen. At the same time, preferred tectonic preservation of marginal, often restricted settings and the local conditions they record must be biasing our interpretation of the global Proterozoic ocean (see also reviews in Lyons *et al.*, 2009a, b; Lyons and Gill, 2010).

20.7.4 The emergence of animals

One of the most significant geobiological events in the last 200 million years of the Proterozoic Eon is the rise of animals. Molecular phylogeny of living animals makes predictions about the sequence of animal divergences: sponge-grade animals are among the earliest diverging metazoans, followed by cnidarian-grade diploblastic animals, and finally triploblastic bilaterian animals. The fossil record agrees broadly with the molecular predictions. C_{30} sterane (24-isopropylcholestane) fossils from the Huqf Supergroup in South Oman suggest demosponges in certain shallow marine environments from the Late Cryogenian–Early Cambrian (Love *et al.*, 2009). Carbonate textures in early Neoproterozoic fossil reefs (1.083–0.779 Ga) from the Little Dal Group, Canada, have been interpreted as evidence for calcification of ancient sponge extracellular collagenous matrix and hence the oldest existing fossil evidence for animals (Neuweiler *et al.*, 2009), but Planavsky (2009) argued that such microstructures could also have derived from lithification of microbialite reefs with no metazoan colonies. Recent molecular clock analyses that take into account heterogeneities in molecular substitution rates among lineages, such as between invertebrates and vertebrates, predict a Cryogenian origin of Metazoa (Aris-Broscou and Yang, 2003; Peterson *et al.*, 2004, 2008). Divergence estimates of 0.664–0.654 Ga for the last common ancestor of all animals (Urmetazoa, which appear to have been sponge-grade animals) based on certain molecular clocks (Peterson and Butterfield, 2005; Sperling *et al.*, 2007) correspond well with the first appearance of

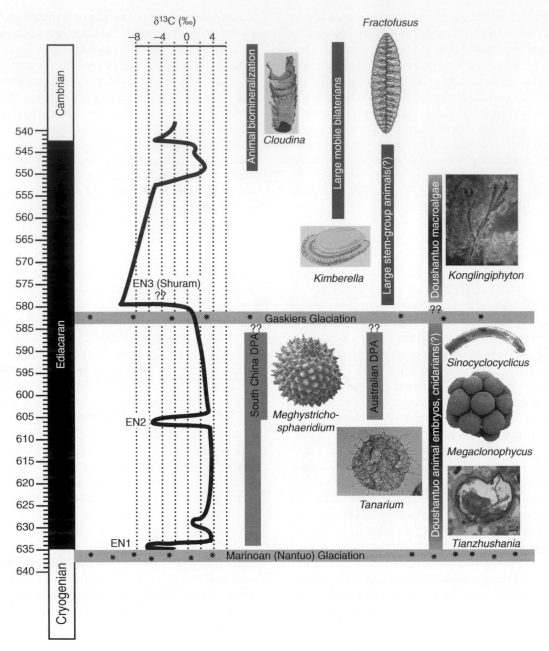

Figure 20.11 A conjectural model of Ediacaran geobiological events. Uncertainties about geochronological constraints and phylogenetic interpretations are indicated by question marks. Modified from Xiao (2008).

demosponge steranes in the sedimentary record in South Oman between the Sturtian and Marinoan glaciations. Possible microscopic stem-group cnidarians may have evolved in the middle Ediacaran Period (Xiao *et al.*, 2000; Chen *et al.*, 2002; Liu *et al.*, 2008).

The early-middle Ediacaran Period is also characterized by an unprecedented diversity of so-called Doushantuo-Pertatataka acritarchs (Fig. 20.11) – large (200–1000 μm in diameter) spiny organic-walled microfossils. Some of these have been interpreted as resting eggs of early metazoans (Yin *et al.*, 2007; Cohen *et al.*,

2009). These putative resting eggs, together with cellularly preserved animal embryos (Xiao *et al.*, 1998; Hagadorn *et al.*, 2006), provide information about the physiology and developmental biology of early animals in the early-middle Ediacaran Period. By the late Ediacaran Period (0.575–0.541 Ga), a variety of macroscopic animals are represented in the classical soft-bodied Ediacara Biota (Narbonne, 2005; Xiao and Laflamme, 2009). The Ediacara biota most likely represents a cross-section of the phylogenetically diverse biosphere in the late Ediacaran Period, including stem-group animals (e.g.

rangeomorphs; Fig. 20.11), sponge-grade animals, diploblasts, bilaterians (e.g. *Kimberella*; Fig. 20.11), as well as various algae and protists.

Toward the end of the Ediacaran Period, about 0.550 Ga, biomineralizing animals evolved (Fig. 20.11), probably driven by ecological pressure from early predators (Bengtson and Yue, 1992; Hua *et al.*, 2003). At the same time, the complexity of ecological interactions – although still much simpler than modern marine ecosystems – rose to a new level (Bambach *et al.*, 2007), including predation, burrowing, tiering, suspension feeding, osmotrophic feeding (Laflamme *et al.*, 2009), deposit feeding, grading, and mining.

Ecological feedbacks (e.g. a predation-induced 'arm race') almost certainly played an important role in driving the explosive evolution of animals in the Cambrian Period that followed the Ediacaran (Stanley, 1973; Marshall, 2006). Our focus here, however, is on the environmental backdrop that permitted the rise of animals in the Ediacaran Period. A number of environmental factors, including temperature and salinity (Knauth, 2005), have been proposed as important in setting the stage for animal evolution. But the most often cited environmental factor is the rise of oxygen in the Ediacaran atmosphere and oceans (Knoll, 2003). Although some extant animals can tolerate low-oxygen conditions (Budd, 2008; Danovaro *et al.*, 2010), the active lifestyles (e.g. predation) that permitted the ecological escalation responsible for the explosive radiation of macro-metazoans almost universally require high oxygen availability. Thus, it is natural on theoretical grounds to hypothesize that atmosphere/ocean oxygenation was a permissive factor for animal evolution.

The question is whether the evolutionary patterns of early animals agree with the redox history of Neoproterozoic oceans. Although it is easy to argue for oxygen in the Ediacaran shallow ocean, much if not most of the deep ocean may have remained anoxic through much of the Neoproterozoic (Canfield *et al.*, 2008; Li *et al.*, 2010), although contrary evidence exists for the later Ediacaran (Fike *et al.*, 2006; Canfield *et al.*, 2007; McFadden *et al.*, 2008; Scott *et al.*, 2008; Shen Y *et al.*, 2008). On balance, this apparently conflicting evidence suggests a high degree of spatial heterogeneity and temporal volatility of Ediacaran redox conditions. The redox instability would have been particularly enhanced in continental shelf and platform environments through the upwelling of anoxic deep-ocean waters. It is thus possible that Ediacaran benthic macrometazoans and other oxygen-demanding organisms locally tracked favourable redox conditions in space and time, leading to their patchy distribution and limited value in global biostratigraphic correlation (McFadden *et al.*, 2008; Li *et al.*, 2010). This heterogeneity may help reconcile the much debated biostratigraphic correlations

of Ediacaran acritarchs (some of which may be animal resting eggs) between South China and Australia (Grey, 2005; Zhou *et al.*, 2007) (Fig. 20.11).

Ediacaran oceanic anoxia could also place physiological and ecological constraints on Ediacaran animals. The dominance of animal eggs in the early-middle Ediacaran Period with a resting stage – a physiological strategy used by modern animals to cope with hostile and/or variable environmental conditions – may be a physiological adaptation to frequent anoxia in continental shelf environments influenced by upwelling of anoxic deep waters (Cohen *et al.*, 2009). The late Ediacaran disappearance of this physiological strategy coincides roughly with the diversification of macroscopic animals in the Ediacara biota. Both of these evolutionary patterns – the abandonment of a resting stage and the increase in body size – could have been driven by the increased availability of oxygen after 0.575 Ga (Canfield *et al.*, 2007).

The inferred geobiological links between Ediacaran metazoans and oxygen are intriguing. However, as in the case of causal links between mid-Proterozoic eukaryotes and euxinia, much of the detail remains unresolved, including the global picture of Ediacaran redox and concomitant animal evolution, the precise temporal relationship between environments and evolution, and quantitative constraints on redox conditions using geochemical proxies and experimental physiology data. Also, a number of studies suggest that the impacts of low oxygen conditions on patterns of evolution in the ocean may have been felt even in the early parts of the Paleozoic (e.g. Gill *et al.*, 2011).

20.7.5 Biomarker evidence for early animals

Currently, the earliest biomarker record of animals (metazoans) is from anomalously elevated amounts of 24-isopropylcholestanes, signifying a chemical fossil record of demosponges, in thermally well-preserved late Cryogenian–Early Cambrian strata from South Oman (Love *et al.*, 2009). Sedimentary 24-isopropylcholestanes, the hydrocarbon remains of distinctive C_{30} sterols produced in abundance by certain genera of marine demosponges (McCaffrey *et al.*, 1994; Love *et al.*, 2009), were found in significant quantities in all formations of the Huqf Supergoup of the South Oman Basin (SOSB) and suggest that demosponges first achieved ecological prominence in shallow marine settings in certain basins in the late Cryogenian (Love *et al.*, 2009).

The demosponge steranes first occur in a lime mudstone from the late Cryogenian Ghadir Manquil Formation, which underlies the Marinoan cap carbonate (>0.635 Ga) and, based on its stratigraphic position, was deposited after the end of the Sturtian glaciation dated at ~0.713 Ga in Oman (Bowring *et al.*, 2007). Thus, the sample placement suggests that the sponge biomarker record commenced between 0.713 and 0.635 Ga. These data, the first robust evidence for animals predating the termination of the Marinoan glaciation, recently gained additional credence thanks to 'sponge-grade metazoan' fossils described from pre-Marinoan strata in South Australia (Maloof *et al.*, 2010). Biomarker analysis has yet to yield a convincing case for ancient demosponges predating the Sturtian glaciation, as no significant quantities of 24-isopropylcholestanes have been detected above typical background levels in rock bitumens from older rocks. Indeed, an obvious temporal pulse of 24-isopropylcholestanes has been detected in many Ediacaran to Early Ordovician sedimentary rocks and petroleum compared with older and younger samples (McCaffrey *et al.*, 1994), suggesting that demosponges must have made a significant contribution to biomass in this interval (Sperling *et al.*, 2007; Peterson *et al.*, 2008). Benthic sponges and other basal animals may have been confined to the shallowest, ventilated shelf environments for much of the late Cryogenian and Ediacaran.

20.8 Summary

The Proterozoic Eon, spanning two billion years (or more than 40%) of Earth's history, is best known for the flurries of activity at its beginning and end. The beginning is marked by the first accumulation of oxygen in the atmosphere beyond trace levels – the Great Oxidation Event (GOE) about 2.4–2.3 billion years ago. In a cause or effect relationship with this rise (both models are tenable), methane stability in the atmosphere crashed and, with this decline, its vital role as a greenhouse gas. Extensive low-latitude glaciation was the inevitable consequence. Not long after, we see the longest-lived positive carbon isotope excursion in Earth history – the Lomagundi event from roughly 2.2 to 2.0 billion years ago. If triggered by persistent, widespread burial of organic carbon, as many suggest, we would predict a large jump in O_2 content in the ocean and atmosphere and perhaps a corresponding rise in sulfate availability in the ocean. Dramatically increasing sulfur isotope fractionations after the GOE indicate generally higher levels of seawater sulfate fed by oxidative weathering on the continents, and there are indications, including abundant sulfate evaporites, that favour predictions for particularly high concentrations during the Lomagundi.

The distribution of iron formations is also telling us something important about the first few hundred million years following the GOE. Their disappearance at or near the GOE may point to oxic or euxinic (anoxic and sulfidic) conditions in the deep ocean. Waning hydrothermal delivery of iron to the ocean is also key, with a decline in IF-forming levels of dissolved Fe in the ocean instead leading and thus contributing to the rise in atmospheric oxygen. Conditions in the global deep ocean between 2.4 and 2.0 Ga are not well known, but oxic, euxinic, and ferruginous (anoxic, Fe-dominated) conditions are all possible; indeed, it is likely that the deep ocean varied among these conditions over time and space. What we know with certainty is that IFs returned in earnest around 2.0 Ga, reflecting increasing hydrothermal delivery to the ocean, decreasing oxygen in the ocean and atmosphere, and perhaps lower levels of sulfate and therefore infrequent euxinia. Our money is on all three.

About the same time, the Earth began a long, more than 1-billion-year period of apparent stability – a remarkably unremarkable time of low variation in the carbon isotope composition in the ocean and, by inference, oxygen content of the atmosphere. There is a substantive debate about when the first eukaryotic organisms appeared but little disagreement about the low levels of taxonomic and ecological diversity during this so-called 'boring billion.' Discussions abound over the mechanisms behind this stasis, but a favoured model hypothesizes increasing sulfate delivery to an anoxic deep ocean, leading to widespread euxinia via bacterial sulfate reduction.

With expansive euxinia came corresponding reduction in nutrient availability tied to the nitrogen cycle, including decreases in enzymatically critical trace metals. Globally euxinic conditions, however, are not required to weaken nutrient availability, nor are they necessarily predicted by diverse paleoredox proxy data; the deep waters may have been ferruginous under conditions of low sulfate and/or low organic carbon availability, with euxinia at mid-water depths along the margins. (There is ample evidence for generally low sulfate levels during this interval, favoured in part by efficient euxinic pyrite burial.) In fact, a new model ties shallow euxinia to pervasive anoxygenic photosynthesis, which may have helped, as a feedback, to throttle oxygen in the atmosphere at still low levels and sustain anoxia in the deep ocean. The loss of IFs at 1.8 Ga nonetheless points to expansion of these euxinic conditions on still large scales and very likely to a decline in hydrothermal inputs. We should never forget that much of our understanding of ocean chemistry comes from marginal marine settings whose preservation is favoured over the easily subducted deep seafloor. These areas are vulnerable to restricted conditions, which might have exacerbated oxygen deficiency, for example, that was global. The degree to which environments recorded in these settings capture the global ocean must be questioned.

By the later part of the Proterozoic, carbon isotope data from the ocean once again show wide variability, including what may well be the Earth's longest lived, highly negative marine excursion. To many, this 'Shuram–Wonoka' event is tied to increasing atmospheric oxygenation through oxidation of dissolved organic carbon stored in the deep ocean – eventually lost completely to pervasive deep oxygenation. To nearly everyone, the preponderance of light inorganic C during this interval remains perhaps the most vexing mystery in studies of Precambrian geobiology, whether controlled by primary chemical oceanography or later diagenesis. Despite this and other suggestions of deep-ocean oxygenation, there is also evidence for persistent deep anoxia within this time window, and these waters may have been ferruginous at great depth and euxinic at mid-depths along the ocean margins. With these varying views of the ocean comes equally diverse evidence for sulfate levels in the late Proterozoic ocean, ranging from very low, perhaps Archean-like, to near-modern concentrations. Some of this disagreement speaks to the work that remains. But much of the varied texture of the record and it implications is likely a reflection of true spatial and temporal heterogeneity in an ocean operating near still low, threshold concentrations of dissolved iron, sulfate, and oxygen that readily allowed spatiotemporal switches among the redox end members.

Helping to sustain and reset these thresholds were snowball Earth events, a hallmark of the later Proterozoic. By many estimates, ice covered much of the ocean and continents during these events, even at low elevations and low latitudes. Finally, animals rose in diversity and abundance following the last of the two great glacial episodes 0.635 Ga but show geochemical fingerprints even earlier. An almost unquestioned but in truth poorly known trigger to the rise of animals was a big rise in oxygen in the oceans and atmosphere as a milestone of the terminal Proterozoic.

Acknowledgements

We thank the NASA Exobiology Program and the US National Science Foundation for financial support. The authors are grateful to J. Abelson and the Agouron Institute for their support and to our many colleagues for their shared insights, including A. Anbar, A. Bekker, B. Berner, R. Buick, D. Canfield, X. Chu, J. Farquhar, D. Fike, B. Gilhooly, B. Gill, J. Grotzinger, D. Holland, M. Hurtgen, D. Johnston, L. Kah, J. Kaufman, A. Kelly, A. Knoll, L. Kump, C. Li, J. Owens, N. Planavsky, S. Poulton, R. Raiswell, A. Scott, C. Scott, S. Severmann, R. Summons. We thank D. Johnston for his comments on this manuscript. D. Canfield, A. Knoll, and K. Konhauser are appreciated for their many editorial efforts in putting this book together.

References

Allen PA, Etienne JL (2008) Sedimentary challenge to Snowball Earth. *Nature Geoscience* **1**, 817–825.

Anbar AD (2008) Elements and evolution. *Science* **322**, 1481–1483.

Anbar AD, Knoll AH (2002) Proterozoic ocean chemistry and evolution: A bioinorganic bridge? *Science* **297**, 1137–1142.

Anbar AD, Arnold GL, Lyons TW, Barling J (2005) Response to comment on 'Molybdenum isotope evidence for widespread anoxia in mid-Proterozoic oceans. *Science* **309**, 1017d.

Anbar AD, Duan Y, Lyons TW, *et al.* (2007) A whiff of oxygen before the Great Oxidation Event? *Science* **317**, 1903–1906.

Archer C, Vance D (2008) The isotopic signature of the global riverine molybdenum flux and anoxia in the ancient oceans. *Nature Geoscience* **1**, 597–600.

Aris-Broscou S, Yang Z (2003) Bayesian models of episodic evolution support a late Precambrian diversification of the Metazoa. *Molecular Biology and Evolution* **20**, 1947–1957.

Arnold GL, Anbar AD, Barling J, Lyons TW (2004) Molybdenum isotope evidence for widespread anoxia in mid-Proterozoic oceans. *Science* **304**, 87–90.

Baldauf SL (2000) The deep roots of eukaryotes. *Science* **300**, 1703–1706.

Bambach RK, Bush AM, Erwin DH (2007) Autecology and the filling of ecospace: Key metazoan radiations. *Palaeontology* **50**, 1–22.

Barns SM, Delwiche CF, Palmer JD, Pace NR (1996) Perspectives on archaeal diversity, thermophyly and monophyly from environmental rRNA sequences. *Proceedings of the National Academy of Sciences, USA* **93**, 9188–9193.

Bartley JK, Semikhatov MA, Kaufman AJ, Knoll AH, Pope MC, Jacobsen SB (2001) Global events across the Mesoproterozoic-Neoproterozoic boundary: C and Sr isotopic evidence from Siberia. *Precambrian Research* **111**, 165–202.

Bartley JK, Kah LC (2004) Marine carbon reservoir, C_{org}-C_{carb} coupling, and the evolution of the Proterozoic carbon cycle. *Geology* **32**, 129–132.

Battistuzzi FU, Feijao A, Hedges SB (2004) A genomic timescale of prokaryote evolution: insights into the origin of methanogenesis, phototrophy, and the colonization of land. *BMC Evolutionary Biology* **4**, 44–57.

Bekker A, Kaufman AJ (2007) Oxidative forcing of global climate change: A biogeochemical record across the oldest Paleoproterozoic ice age in North America. *Earth and Planetary Science Letters* **258**, 486–499.

Bekker A, Holland HD, Wang PL, *et al.* (2004) Dating the rise of atmospheric oxygen. *Nature* **427**, 117–120.

Bekker A, Karhu JA, Kaufman AJ (2006) Carbon isotope record for the onset of the Lomagundi carbon isotope excursion in the Great Lakes area, North America. *Precambrian Research* **148**, 145–180.

Bekker A, Holmden C, Beukes NJ, Kenig F, Eglington B, Patterson WP (2008) Fractionation between inorganic and organic carbon during the Lomagundi (2.22–2.1 Ga) carbon isotope excursion. *Earth and Planetary Science Letters* **271**, 278–291.

Bekker A, Slack JF, Planavsky N, *et al.* (2010) Iron formation: The sedimentary product of a complex interplay among mantle, tectonic, oceanic, and biospheric processes. *Economic Geology* **105**, p. 467–508.

Bengtson S, Yue Z (1992) Predatorial borings in late Precambrian mineralized exoskeletons: *Science* **257**, 367–369.

Berner RA (2003) The long-term carbon cycle, fossil fuels and atmospheric composition. *Nature* **426**, 323–326.

Bjerrum CJ, Canfield, DE (2002) Ocean productivity before about 1.9 Gyr ago limited by phosphorus adsorption onto iron oxides. *Nature* **417**, 159–162.

Bjerrum CJ, Canfield DE (2011) Towards a quantitative understanding of the late Neoproterozoic carbon cycle. *Proceedings of the National Academy of Sciences, USA* doi:10.1073/pnas.1101755108

Bowring SA, Grotzinger JP, Condon DJ, Ramezani J, Newall M, Allen PA (2007) Geochronologic constraints on the chronostratigraphic framework of the Neoproterozoic Huqf Supergroup, Sultanate of Oman. *American Journal of Science* **307**, 1097–1145.

Brasier MD, Lindsay JF (1998) A billion years of environmental stability and the emergence of eukaryotes: New data from northern Australia. *Geology* **26**, 555–558.

Bristow TF, Kennedy MJ (2008) Carbon isotope excursions and the oxidant budget of the Ediacaran atmosphere and ocean. *Geology* **36**, 863–866.

Brocks JJ, Pearson A (2005) Building the biomarker tree of life. *Reviews in Mineralogy & Geochemistry* **59**, 233–258.

Brocks JJ, Logan GA, Buick R, Summons RE (1999) Archean molecular fossils and the early rise of eukaryotes. *Science* **285**, 1033–1036.

Brocks JJ, Buick R, Logan GA, Summons RE (2003) Composition and syngeneity of molecular fossils from the 2.78 to 2.45 billion-year old Mount Bruce Supergroup, Pilbara Craton, Western Australia. *Geochimica et Cosmochimica Acta* **67**, 4289–4319.

Brocks JJ, Love GD, Summons RE, Knoll AH, Logan GA, Bowden SA (2005) Biomarker evidence for green and purple sulphur bacteria in a stratified Palaeoproterozoic sea. *Nature* **437**, 866–870.

Brunner B, Bernasconi SM (2005) A revised isotope fractionation model for dissimilatory sulfate reduction in sulfate reducing bacteria. *Geochimica et Cosmochimica Acta* **69**, 4759–4771.

Budd G (2008) The earliest fossil record of the animals and its significance. *Philosophical Transactions of the Royal Society, B – Biological Sciences* **363**, 1425–1434.

Buick IS, Uken R, Gibson RL, Wallmach T (1998) High-δ^{13}C Paleoproterozoic carbonates from the Transvaal Supergroup, South Africa. *Geology* **26**, 875–878.

Buick R (2007) Did the Proterozoic 'Canfield Ocean' cause a laughing gas greenhouse? *Geobiology* **5**, 97–100.

Buick R, Des Marais DJ, Knoll AH (1995) Stable isotopic compositions of carbonates from the Mesoproterozoic Bangemall Group, northwestern Australia. *Chemical Geology* **123**, 153–171.

Bull SW (1998) Sedimentology of the Paleoproterozoic Barney Creek Formation in DDH BMR McArthur 2, southern McArthur Basin, Northern Territory. *Australian Journal of Earth Sciences* **45**, 21–31.

Butterfield NJ (1994) Burgess Shale-type fossils from a Lower Cambrian shallow-shelf sequence in northwestern Canada. *Nature* **369**, 477–479.

Butterfield NJ (1997) Plankton ecology and the Proterozoic–Phanerozoic transition. *Paleobiology* **23**, 247–262.

Butterfield NJ (2000) *Bangiomorpha pubescens* n. gen., n. sp.: Implications for the evolution of sex, multicellularity, and the Mesoproterozoic/Neoproterozoic radiation of eukaryotes. *Paleobiology* **26**, 386–404.

Butterfield NJ (2004) A vaucheriacean alga from the middle Neoproterozoic of Spitsbergen: Implications for the evolution of Proterozoic eukaryotes and the Cambrian explosion. *Paleobiology* **30**, 231–252.

Butterfield NJ (2005) Probable Proterozoic fungi. *Paleobiology* **31**, 165–182.

Butterfield NJ, Chandler FW (1992) Paleoenvironmental distribution of Proterozoic microfossils, with an example from the Agu Bay Formation, Baffin Island. *Palaeontology* **35**, 943–957.

Butterfield NJ, Knoll AH, Swett K (1994) Paleobiology of the Neoproterozoic Svanbergfjellet Formation, Spitsbergen. *Fossils and Strata* **34**, p. 1–84.

Cameron EM (1982) Sulphate and sulphate reduction in early Precambrian oceans. *Nature* **296**, 145–148.

Campbell IH, Allen CM (2008) Formation of supercontinents linked to increases in atmospheric oxygen. *Nature Geoscience* **1**, 554–558.

Campbell IH, Squire RJ (2010) The mountains that triggered the Late Neoproterozoic increase in oxygen: The second Great Oxidation Event. *Geochimica et Cosmochimica Acta* **74**, 4187–4206.

Canfield DE (1998) A new model for Proterozoic ocean chemistry. *Nature* **396**, 450–453.

Canfield DE (2004) The evolution of the Earth surface sulfur reservoir. *American Journal of Science* **304**, 839–861.

Canfield DE (2005) The early history of atmospheric oxygen: Homage to Robert M. Garrels. *Annual Review of Earth and Planetary Sciences* **33**, 1–36.

Canfield DE (2006) Models of oxic respiration, denitrification and sulfate reduction in zones of coastal upwelling. *Geochimica et Cosmochimica Acta* **70**, 5753–5765.

Canfield DE, Thamdrup B (1994) The production of 34S-depleted sulfide during bacterial disproportionation of elemental sulfur. *Science* **266**, 1973–1975.

Canfield DE, Teske A (1996) Late Proterozoic rise in atmospheric oxygen concentration inferred from phylogenetic and sulphur-isotope studies. *Nature* **382**, 127–132.

Canfield DE, Habicht KS, Thamdrup B (2000) The Archean sulfur cycle and the early history of atmospheric oxygen. *Science* **288**, 658–661.

Canfield DE, Poulton SW, Narbonne GM (2007) Late Neoproterozoic deep-ocean oxygenation and the rise of animal life. *Science* **315**, 92–95.

Canfield DE, Poulton SW, Knoll AH, et al. (2008) Ferruginous conditions dominated later Neoproterozoic deep-water chemistry. *Science* **321**, 949–952.

Canfield DE, Farquhar J, Zerkle AL (2010) High isotope fractionations during sulfate reduction in a low-sulfate euxinic ocean analog. *Geology* **38**, 415–418.

Catling DC, Zahnle KJ, McKay CP (2001) Biogenic methane, hydrogen escape, and the irreversible oxidation of early Earth. *Science* **293**, 839–843.

Catling DC, Claire MW (2005) How Earth's atmosphere evolved to an oxic state: A status report. *Earth and Planetary Science Letters* **237**, 1–20.

Catling DC, Claire MW, Zahnle KJ (2007) Anaerobic methanotrophy and the rise of atmospheric oxygen. *Philosophical Transactions of the Royal Society, A – Mathematical, Physical & Engineering Sciences* **365**, 1867–1888.

Chen J, Oliveri P, Gao F, et al. (2002) Precambrian animal life: probable developmental and adult cnidarian forms from southwest China. *Developmental Biology* **248**, 182–196.

Claire MW, Catling DC, Zahnle KJ (2006) Biogeochemical modelling of the rise in atmospheric oxygen. *Geobiology* **4**, 239–269.

Cohen PA, Knoll AH, Kodner RB (2009) Large spinose microfossils in Ediacaran rocks as resting stages of early animals. *Proceeding of the National Academy of Sciences, USA* **106**, 6519–6524.

Condon D, Zhu M, Bowring S, Wang W, Yang A, Jin Y (2005) U-Pb ages from the Neoproterozoic Doushantuo Formation, China. *Science* **308**, 95–98.

Corsetti FA, Awramik SM, Pierce D (2003) A complex microbiota from snowball Earth times: Microfossils from the Neoproterozoic Kingston Peak Formation, Death Valley, USA. *Proceedings of the National Academy of Sciences, USA* **100**, 4399–4404.

Corsetti FA, Olcott AN, Bakermans C (2006) The biotic response to Neoproterozoic snowball Earth. *Palaeogeography Palaeoclimatology Palaeoecology* **232**, 114–130.

Danovaro R, Dell'Anno A, Pusceddu A, Gambi C, Heiner I, Kristensen RM (2010) The first metazoa living in permanently anoxic conditions. *BMC Biology* **8**, doi:10.1186/1741.

David LA, Alm EJ (2011) Rapid evolutionary innovation during an Archaean genetic expansion. *Nature* **469**, 93–96.

Davidson GJ, Dashlooty SA (1993) The Glyde Sub-basin: A volcaniclastic-bearing pull-apart basin coeval with the McArthur River base-metal deposit, Northern Territory. *Australian Journal of Earth Sciences* **40**, 527–543.

Dehler CM, Elrick M, Bloch JD, Karlstrom KE, Crossey LJ, Des Marais DJ (2005) High-resolution $\delta^{13}C$ stratigraphy of the Chuar Group (ca. 770–742 Ma), Grand Canyon: Implications for mid-Neoproterozoic climate change. *Geological Society of America Bulletin* **117**, 32–45.

Derry LA (2010) A burial diagenesis origin for the Ediacaran Shuram-Wonoka carbon isotope anomaly. *Earth and Planetary Science Letters* **294**, 152–162.

Des Marais DJ, Strauss H, Summons RE, Hayes JM (1992) Carbon isotope evidence for the stepwise oxidation of the Proterozoic environment. *Nature* **359**, 605–609.

Douzery EJP, Snell EA, Bapteste E, Delsuc F, Philippe H (2004) The timing of eukaryotic evolution: Does a relaxed molecular clock reconcile proteins and fossils? *Proceedings of the National Academy of Sciences, USA* **101**, 15386–15391.

Duan Y, Anbar AD, Arnold GL, Lyons TW, Gordon GW, Kendall B (2010) Molybdenum isotope evidence for mild environmental oxygenation before the Great Oxidation Event. *Geochimica et Cosmochimica Acta* **74**, 6655–6668.

Dupont CL, Yang S, Palenik B, Bourne PE (2006) Modern proteomes contain putative imprints of ancient shifts in trace metal geochemistry. *Proceedings of the National Academy of Sciences, USA* **103**, 17822–17827.

Dupont CL, Butcher A, Valas RE, Bourne PE, Caetano-Anollés G (2010) History of biological metal utilization inferred through phylogenomic analysis of protein structures. *Proceedings of the National Academy of Sciences, USA* **107**, 10567–10572.

Dutkiewicz A, Volk H, Ridley J, George SC (2003) Biomarkers, brines and oil in the Mesoproterozoic Roper Superbasin, Australia. *Geology* **31**, 937–957.

Dutkiewicz A, Volk H, George SC, Ridley J, Buick R (2006) Biomarkers from Huronian oil-bearing fluid inclusions: An uncontaminated record of life before the Great Oxidation Event. *Geology* **34**, 437–440.

Eigenbrode JL, Freeman KH, Summons RE (2008) Methylhopane biomarker hydrocarbons in Hamersley Province sediments provide evidence for Neoarchaen aerobiosis. *Earth and Planetary Science Letters* **273**, 323–331.

El Albani A, Bengtson S, Canfield DE, *et al.* (2010) Large colonial organisms with coordinated growth in oxygenated environments 2.1 Gyr ago. *Nature* **466**, 100–104.

Embley TM, Martin W (2006) Eukaryotic evolution, changes and challenges. *Nature* **440**, 623–630.

Evans DA, Beukes NJ, Kirschvinnk JL (1997) Low-latitude glaciation in the Palaeoproterozoic era. *Nature* **386**, 262–266.

Fairchild IJ, Kennedy MJ (2007) Neoproterozoic glaciation in the Earth System. *Journal of the Geological Society* **164**, 895–921.

Farquhar J, Wing BA (2003) Multiple sulfur isotopes and the evolution of the atmosphere. *Earth and Planetary Science Letters* **213**, 1–13.

Farquhar J, Bao H, Thiemens M (2000) Atmospheric influence of Earth's earliest sulfur cycle. *Science* **289**, 756–758.

Farquhar J, Wu N, Canfield DE, Oduro H (2010) Connections between sulfur cycle evolution, sulfur isotopes, sediments, and base metal sulfide deposits. *Economic Geology* **105**, 509–533.

Fike DA, Grotzinger JP (2008) A paired sulfate-pyrite $\delta^{34}S$ approach to understanding the evolution of the Ediacaran-Cambrian sulfur cycle. *Geochimica et Cosmochimica Acta* **72**, 2636–2648.

Fike DA, Grotzinger JP, Pratt LM, Summons RE (2006) Oxidation of the Ediacaran ocean. *Nature* **444**, 744–747.

Frank TD, Kah LC, Lyons TW (2003) Changes in organic matter production and accumulation as a mechanism for isotopic evolution in the Mesoproterozoic ocean. *Geological Magazine* **140**, 1–24.

Frei R, Gaucher C, Poulton SW, Canfield DE (2009) Fluctuations in Precambrian atmospheric oxygenation recorded by chromium isotopes. *Nature* **461**, 250–253.

Gaillard F, Scaillet B, Arndt NT (2011) Atmospheric oxygenation caused by a change in volcanic degassing pressure. *Nature* **478**, 229–232.

Garvin J, Buick R, Anbar AD, Arnold GL, Kaufman AJ (2009) Isotopic evidence for an aerobic nitrogen cycle in the latest Archean. *Science* **323**, 1045–1048.

Gellatly AM, Lyons TW (2005) Trace sulfate in mid-Proterozoic carbonates and the sulfur isotope record of biospheric evolution. *Geochimica et Cosmochimica Acta* **69**, 3813–3829.

George SC, Volk H, Dutkiewicz A, Ridley J, Buick R (2008) Preservation of hydrocarbons and biomarkers in oil trapped inside fluid inclusions for >2 billion years. *Geochimica et Cosmochimica Acta* **72**, 844–870.

Gill BC, Lyons TW, Saltzman MR (2007) Parallel, high-resolution carbon and sulfur isotope records of the evolving Paleozoic marine sulfate reservoir. *Palaeogeography, Palaeoclimatology, Palaeoecology* **256**, 156–173.

Gill BC, Lyons TW, Young SA, Kump LR, Knoll AH, Saltzman MR (2011) Geochemical evidence for widespread euxinia in the Later Cambrian ocean. *Nature* **469**, 80–83.

Godfrey LV, Falkowski PG (2009) The cycling and redox state of nitrogen in the Archean ocean. *Nature Geoscience* **2**, 725–729.

Goldblatt C, Lenton TM, Watson AJ (2006) Bistability of atmospheric oxygen and the Great Oxidation. *Nature* **443**, 683–686.

Grantham PJ, Wakefield LL (1988) Variations in the sterane carbon number pattern distributions of marine source rock derived crude oils through geological times. *Organic Geochemistry* **12**, 61–77.

Grey K (2005) Ediacaran palynology of Australia. *Memoirs of the Association of Australasian Palaeontologists* **31**, 1–439.

Grosjean E, Love GD, Stalyies C, Fike DA, Summons RE (2009) Origin of petroleum in the Neoproterozoic-Cambrian South Oman Salt Basin. *Organic Geochemistry* **40**, 87–110.

Grotzinger JP, Kasting JF (1993) New constraints on Precambrian ocean composition. *Journal of Geology* **101**, 235–243.

Grotzinger JP, Knoll AH (1995) Anomalous carbonate precipitates: Is the Precambrian the key to the Permian? *Palaios* **10**, 578–596.

Grotzinger JP, Fike DA, Fischer WW (2011) Enigmatic origin of the largest-known carbon isotope excursion in Earth's history. *Nature Geoscience* **4**, 285–292.

Habicht KS, Gade M, Thamdrup B, Berg P, Canfield DE (2002) Calibration of sulfate levels in the Archean ocean. *Science* **298**, 2372–2374.

Hagadorn JW, Xiao S, Donoghue PCJ, et al. (2006) Cellular and subcellular structure of Neoproterozoic embryos: *Science* **314**, 291–294.

Halverson GP, Hurtgen MT (2007) Ediacaran growth of the marine sulfate reservoir. *Earth and Planetary Science Letters* **263**, 32–44.

Han TM, Runnegar B (1992) Megascopic eukaryotic algae from the 2.1 billion-year-old Negaunee Iron-Formation, Michigan. *Science* **257**, 232–235.

Hayes JM (1994) Global methanotrophy at the Archean-Proterozoic transition. In: *Early life on Earth* (ed Bengston S). *Nobel Symposium*. Columbia University Press, New York, NY. p. 220–236.

Hayes JM, Waldbauer JR (2006) The carbon cycle and associated redox processes through time. *Philosophical Transactions of the Royal Society, B – Biological Sciences* **361**, 931–950

Hermann TN (1990) Organic World Billion Year Ago: Leningrad, Nauka, 49 p.

Hessler AM, Lowe DR, Jones RL, Bird DK (2004) A lower limit for atmospheric carbon dioxide levels 3.2 billion years ago. *Nature* **428**, 736–738.

Higgins JA, Schrag DP (2003) Aftermath of a snowball Earth. *Geochemistry, Geophysics, Geosystems* **4**, doi:10.1029/2002 GC000403.

Hoffman PF, Kaufman AJ, Halverson GP, Schrag DP (1998) A Neoproterozoic snowball Earth. *Science* **281**, 1342–1346.

Hoffman PF, Schrag DP (2000) Snowball Earth. *Scientific American* **282**, 68–75.

Hoffman PF, Schrag DP (2002) The snowball Earth hypothesis: testing the limits of global change. *Terra Nova* **14**, 129–155.

Hoffman PF, Halverson GP, Domack EW, Husson JM, Higgins JA, Schrag DP (2007) Are basal Ediacaran (635 Ma) post-glacial 'cap dolostones' diacronous? *Earth and Planetary Science Letters* **258**, 114–131.

Höld IM, Schouten S, Jellema J, Sinninghe Damsté JS (1999) Origin of free and bound mid-chain methyl alkanes in oils, bitumens and kerogens of the marine, Infra-cambrian Huqf Formation (Oman). *Organic Geochemistry* **30**, 1411–1428.

Holland HD (1984) *The Chemical Evolution of the Atmosphere and Oceans*. Princeton University Press, Princeton, NJ, 598p.

Holland HD (2002) Volcanic gases, black smokers, and the Great Oxidation Event. *Geochimica et Cosmochimica Acta* **66**, 3811–3826.

Holland HD (2006) The oxygenation of the atmosphere and oceans. *Philosophical Transactions of the Royal Society, B – Biological Sciences* **361**, 903–915.

Horita J, Zimmermann H, Holland HD (2002) Chemical evolution of seawater during the Phanerozoic: Implications from the record of marine evaporites. *Geochimica et Cosmochimica Acta* **66**, 3733–3756.

Hua H, Pratt BR, Zhang L (2003) Borings in *Cloudina* shells: Complex predator–prey dynamics in the terminal Neoproterozoic. *Palaios* **18**, 454–459.

Hurtgen MT, Arthur MA, Suits N, Kaufman AJ (2002) The sulfur isotopic composition of Neoproterozoic seawater sulfate: implications for a snowball Earth? *Earth and Planetary Science Letters* **203**, 413–429.

Hurtgen MT, Arthur MA, Halverson GP (2005) Neoproterozoic sulfur isotopes, the evolution of microbial sulfur species, and the burial efficiency of sulfide as sedimentary pyrite. *Geology* **33**, 41–44.

Hurtgen MT, Halverson GP, Arthur MA, Hoffman PF (2006) Sulfur cycling in the aftermath of a 635-Ma snowball glaciation: Evidence for a syn-glacial sulfidic deep ocean. *Earth and Planetary Science Letters* **245**, 551–570.

Hyde WT, Crowley TJ, Baum SK, Peltier WR (2000) Neoproterozoic 'snowball Earth' simulations with a coupled climate/ice-sheet model. *Nature* **405**, 425–429.

Isley AE, Abbott DH (1999) Plume-related mafic volcanism and the deposition of banded iron formation. *Journal of Geophysical Research* **104**, 15461–15477.

Javaux EJ, Knoll AH, Walter MR (2001) Morphological and ecological complexity in early eukaryotic ecosystems. *Nature* **412**, 66–69.

Javaux EJ, Knoll AH, Walter MR (2004) TEM evidence for eukaryotic diversity in mid-Proterozoic oceans. *Geobiology* **2**, 121–132.

Jiang G, Kennedy MJ, Christie-Blick N (2003) Stable isotopic evidence for methane seeps in Neoproterozoic postglacial cap carbonates. *Nature* **426**, 822–826.

Johnston DT, Wing BA, Farquhar J, *et al.* (2005) Active microbial sulfur disproportionation in the Mesoproterozoic. *Science* **310**, 1477–1479.

Johnston DT, Wolfe-Simon F, Pearson A, Knoll AH (2009) Anoxygenic photosynthesis modulated Proterozoic oxygen and sustained Earth's middle age. *Proceedings of the National Academy of Sciences, USA* **106**, 16925–16929.

Johnston DT, Poulton SW, Dehler C, *et al.* (2010) An emerging picture of Neoproterozoic ocean chemistry: Insights from the Chuar Group, Grand Canyon, USA. *Earth and Planetary Science Letters* **290**, 64–73.

Kah LC, Lyons TW, Frank TD (2004) Low marine sulphate and protracted oxygenation of the Proterozoic biosphere. *Nature* **431**, 834–838.

Kah LC, Bartley JK (2011) Protracted oxygenation of the Proterozoic biosphere. *International Geology Review* **53**, 1424–1442.

Kampschulte A, Strauss H (2004) The sulfur isotopic evolution of Phanerozoic seawater based on the analysis of structurally substituted sulfate in carbonates. *Chemical Geology* **204**, 255–286.

Karhu JA, Holland HD (1996) Carbon isotopes and the rise of atmospheric oxygen. *Geology* **24**, 867–870.

Karhu J (1999) Carbon isotopes. In: *Encyclopedia of Geochemistry* (eds Marshall CP, Fairbridge RW). Kluwer Academic Publishers, Dordrecht, p. 67–73.

Kasting JF (1987) Theoretical constraints on oxygen and carbon dioxide concentrations in the Precambrian atmosphere. *Precambrian Research* **34**, 205–228.

Kasting JF (2005) Methane and climate during the Precambrian era. *Precambrian Research* **137**, 119–129.

Kaufman AJ, Knoll AH (1995) Neoproterozoic variations in the C-isotopic composition of seawater: stratigraphic and biogeochemical implications. *Precambrian Research* **73**, 27–49.

Kaufman AJ, Knoll AH, Narbonne GM (1997) Isotopes, ice ages, and terminal Proterozoic earth history. *Proceedings of the National Academy of Sciences USA* **94**, 6600–6605.

Kaufman AJ, Johnston DT, Farquhar J, *et al.* (2007) Late Archean biospheric oxygenation and atmospheric evolution. *Science* **317**, 1900–1903.

Kelly AE, Rothman DR, Love GD, Fike DA, Zumberge JE, Summons RE (2008) Environmental Implications of Ediacaran C-isotopic Shifts. *Eos Transactions of the AGU Fall Meet. Suppl*, **89**(53), Abstract PP33A–1524.

Kendall B, Creaser RA, Gordon GW, Anbar AD (2009) Re-Os and Mo isotope systematics of black shales from the Middle Proterozoic Velkerri and Wollogorang Formations, McArthur Basin, northern Australia. *Geochimica et Cosmochimica Acta* **73**, 2534–2558.

Kendall B, Reinhard CT, Lyons TW, Kaufman AJ, Poulton SW, Anbar AD (2010) Pervasive oxygenation along late Archaean ocean margins. *Nature Geoscience* **3**, 647–652.

Kennedy MJ, Runnegar B, Prave AR, Hoffmann KH, Arthur MA (1998) Two or four Neoproterozoic glaciations? *Geology* **26**, 1059–1063.

Kennedy MJ, Christie-Blick N, Sohl LE (2001) Are Proterozoic cap carbonates and isotopic excursions a record of gas hydrate destabilization following Earth's coldest intervals? *Geology* **29**, 443–446.

Kennedy MJ, Droser M, Mayer LM, Pevear D, Mrofka D (2006) Late Precambrian oxygenation; Inception of the clay mineral factory. *Science* **311**, 1446–1449.

Kennedy MJ, Mrofka D, von der Borch C (2008) Snowball Earth termination by destabilization of equatorial permafrost methane clathrate. *Nature* **453**, 642–645.

King N (2004) The unicellular ancestry of animal development. *Developmental Cell* **7**, 313–325.

Kirschvink JL (1992) Late Proterozoic low-latitude global glaciations: the snowball earth. In: *The Proterozoic Biosphere: A Multidisciplinary Study* (eds Schopf JW, Klein C). Cambridge University Press, Cambridge, MA, p. 51–52.

Kirschvink JL, Kopp RE (2008) Palaeoproterozoic ice houses and the evolution of oxygen-mediating enzymes: the case for a late origin of photosystem. *Philosophical Transactions of the Royal Society of London, B – Biological Sciences* **363**, 2755–2765.

Knauth LP (2005) Temperature and salinity history of the Precambrian ocean: implications for the course of microbial evolution. *Palaeogeography, Palaeoclimatology, Palaeoecology* **219**, 53–69.

Knauth LP, Kennedy MJ (2009) The late Precambrian greening of the Earth. *Nature* **460**, 728–732.

Knoll AH (1994) Proterozoic and Early Cambrian protists: Evidence for accelerating evolutionary tempo. *Proceedings of the National Academy of Sciences, USA* **91**, 6743–6750.

Knoll AH (2003) *Life on a Young Planet: The First Three Billion Years of Evolution on Earth*. Princeton University Press, Princeton, NJ, 304p.

Knoll AH, Carroll SB (1999) Early animal evolution: emerging views from comparative biology and geology. *Science* **284**, 2129–2137.

Knoll AH, Bambach RK, Canfield DE, Grotzinger JP (1996) Comparative earth history and Late Permian mass extinction. *Science* **273**, 452–457.

Knoll AH, Javaux EJ, Hewitt D, Cohen P (2006) Eukaryotic organisms in Proterozoic oceans. *Philosophical Transactions of the Royal Society of London, B – Biological Sciences* **361**, 1023–1038.

Kodner RB, Pearson A, Summons RE, Knoll AH (2008) Sterols in red and green algae quantification, phylogeny, and relevance for the interpretation of geologic steranes. *Geobiology* **6**, 411–420.

Konhauser KO, Lalonde SV, Amskold L, Holland HD (2007) Was there really an Archean phosphate crisis? *Science* **315**, 1234.

Konhauser KO, Pecoits E, Lalonde SV, *et al.* (2009) Oceanic nickel depletion and a methanogen famine before the Great Oxidation Event. *Nature* **458**, 750–753.

Kopp RE, Kirschvink JL, Hilburn IA, Nash CZ (2005) The Paleoproterozoic snowball Earth: A climate disaster triggered by the evolution of oxygenic photosynthesis. *Proceedings of the National Academy of Sciences, USA* **102**, 11131–11136.

Kovalevych VM, Marshall T, Peryt TM, Petrychenko OY, Zhukova SA (2006) Chemical composition of seawater in Neoproterozoic: Results of fluid inclusion study of halite from Salt Range (Pakistan) and Amadeus Basin (Australia). *Precambrian Research* **144**, 39–51.

Kump LR (2008) The rise of atmospheric oxygen. *Nature* **451**, 277–278.

Kump LR, Arthur MA (1999) Interpreting carbon-isotope excursions: Carbonates and organic matter. *Chemical Geology* **161**, 181–198.

Kump LR, Barley ME (2007) Increased subaerial volcanism and the rise of atmospheric oxygen 2.5 billion years ago. *Nature* **448**, 1033–1036.

Kump LR, Seyfried WE (2005) Hydrothermal Fe fluxes during the Precambrian: Effect of low oceanic sulfate concentrations and low hydrostatic pressure on the composition of black smokers. *Earth and Planetary Science Letters* **235**, 654–662.

Laflamme M, Xiao S, Kowalewski M (2009) Osmotrophy in modular Ediacara organisms. *Proceedings of the National Academy of Sciences, USA* **106**, 14438–14443.

Lamb DM, Awramik SM, Chapman DJ, Zhu S (2009) Evidence for eukaryotic diversification in the ~1800 million-year-old Changzhougou Formation, North China. *Precambrian Research* **173**, 93–104.

Le Guerroué E, Allen PA, Cozzi A, Etienne JL, Fanning M (2006) 50 Myr recovery from the largest negative $\delta^{13}C$ excursion in the Ediacaran ocean. *Terra Nova* **18**, 147–153.

Lewis JP, Weaver AJ, Eby M (2007) Snowball versus slushball Earth: Dynamic versus nondynamic sea ice? *Journal of Geophysical Research* **112**, C11014, doi:10.1029/2006JC004037.

Li C, Peng P, Sheng G, Fu JM, Yuzhong Y (2003) A molecular and isotopic geochemical study of Meso- to Neoproterozoic (1.73–0.85 Ga) sediments from the Jixian section, Yanshan Basin, North China. *Precambrian Research* **125**, 337–356.

Li C, Love GD, Lyons TW, Fike DA, Sessions AL, Chu X (2010) A stratified redox model for the Ediacaran ocean. *Science* **328**, 80–83.

Liu P, Xiao S, Yin C, Zhou C, Gao L, Tang F (2008) Systematic description and phylogenetic affinity of tubular microfossils from the Ediacaran Doushantuo Formation at Weng'an, South China. *Palaeontology* **51**, 339–366.

Logan GA, Hayes JM, Hieshima GB, Summons RE (1995) Terminal Proterozoic reorganization of biogeochemical cycles. *Nature* **376**, 53–56.

Logan GA, Summons RE, Hayes JM (1997) An isotopic biogeochemical study of Neoproterozoic and early Cambrian sediments from the Centralian Superbasin, Australia. *Geochimica et Cosmochimica Acta* **61**, 5391–5409.

Love GD, Grosjean E, Stalvies C, *et al.* (2009) Fossil steroids record the appearance of Demospongiae during the Cryogenian period. *Nature* **457**, 718–721.

Luepke JJ, Lyons TW (2001) Pre-Rodinian (Mesoproterozoic) supercontinental rifting along the western margin of Laurentia: geochemical evidence from the Belt-Purcell Supergroup. *Precambrian Research* **111**, 79–90.

Lyons TW (1997) Sulfur isotopic trends and pathways of iron sulfide formation in upper Holocene sediments of the anoxic Black Sea. *Geochimica et Cosmochimica Acta* **61**, 3367–3382.

Lyons TW (2008) Ironing out ocean chemistry at the dawn of animal life. *Science* **321**, 923–924.

Lyons TW, Gill BC (2010) Ancient sulfur cycling and oxygenation of the early biosphere. *Elements* **6**, 93–99.

Lyons TW, Reinhard CT (2009a) An early productive ocean unfit for aerobics. *Proceedings of the National Academy of Sciences, USA* **106**, 18045–18046.

Lyons TW, Reinhard CT (2009b) Oxygen for heavy-metal fans. *Nature* **461**, 179–181.

Lyons TW, Luepke JJ, Schreiber ME, Zieg GA (2000) Sulfur geochemical constraints on Mesoproterozoic restricted marine deposition: lower Belt Supergroup, northwestern United States. *Geochimica et Cosmochimica Acta* **64**, 427–437.

Lyons TW, Gellatly AM, McGoldrick PJ, Kah LC (2006) Proterozoic sedimentary exhalative (SEDEX) deposits and links to evolving global ocean chemistry. *GSA Memoirs* **198**, 169–184.

Lyons TW, Anbar AD, Severmann S, Scott C, Gill BC (2009a) Tracking euxinia in the ancient ocean: A multiproxy perspective and Proterozoic case study. *Annual Review of Earth and Planetary Sciences* **37**, 507–534.

Lyons TW, Reinhard CT, Scott C (2009b) Redox redux. *Geobiology* **7**, 489–494.

Macdonald FA, Schmitz MD, Crowley JL, *et al.* (2010) Calibrating the Cryogenian. *Science* **327**, 1241–1243.

Maloof AC, Rose CV, Beach R, *et al.* (2010) Possible animal-body fossils in pre-Marinoan limestones from South Australia. *Nature Geoscience* **3**, 653–659.

Marshall CR (2006) Explaining the Cambrian 'explosion' of animals. *Annual Review of Earth and Planetary Sciences* **34**, 355–384.

McCaffrey MA, Moldowan JM, Lipton PA, *et al.* (1994) Paleoenvironmental implications of novel C_{30} steranes in Precambrian to Cenozoic Age petroleum and bitumen. *Geochimica et Cosmochimica Acta* **58**, 529–532.

McFadden KA, Huang J, Chu X, *et al.* (2008) Pulsed oxygenation and biological evolution in the Ediacaran Doushantuo Formation. *Proceedings of the National Academy of Sciences, USA* **105**, 3197–3202.

McKay CP (2000) Thickness of tropical ice and photosynthesis on a snowball Earth. *Geophysical Research Letters* **27**, 2153–2156.

McKirdy DM, Webster LJ, Arouri KR, Grey KG, Gostin VA (2006) Contrasting sterane signatures in Neoproterozoic marine rocks of Australia before and after the Acraman asteroid impact. *Organic Geochemistry* **37**, 189–207.

Melezhik VA, Fallick AE (1996) A widespread positive $\delta^{13}C$ anomaly at around 2.33–2.06 Ga on the Fennoscandian Shield: a paradox? *Terra Nova* **8**, 141–157.

Melezhik VA, Fallick AE, Rychanchik DV, Kuznetsov AB (2005) Palaeoproterozoic evaporates in Fennoscandia: implications for seawater sulphate, $\delta^{13}C$ excursions and the rise of atmospheric oxygen. *Terra Nova* **17**, 141–148.

Melezhik VA, Huhma H, Condon DJ, Fallick AE, Whitehouse MJ (2007) Temporal constraints on the Paleoproterozoic Lomagundi-Jatuli carbon isotopic event. *Geology* **35**, 655–658.

Meng F, Zhou C, Yin L, Chen Z, Yuan X (2005) The oldest known dinoflagellates: Morphological and molecular evidence from Mesoproterozoic rocks at Yongji, Shanxi Province. *Chinese Science Bulletin* **50**, 1230–1234.

Mentel M, Martin W (2008) Energy metabolism among eukaryotic anaerobes in light of Proterozoic ocean chemistry. *Philosophical Transactions of the Royal Society of London, B – Biological Sciences* **363**, 2717–2729.

Meyer KM, Kump LR (2008) Oceanic euxinia in Earth history: Causes and consequences. *Annual Review of Earth and Planetary Sciences* **36**, 251–288.

Moczydłowska M (2008a) The Ediacaran microbiota and the survival of snowball Earth conditions. *Precambrian Research* **167**, 1–15.

Moczydłowska M (2008b) New records of late Ediacaran microbiota from *Poland. Precambrian Research* **167**, 71–92.

Moldowan JM, Fago FJ, Lee CY, *et al.* (1990) Sedimentary 24-*n*-propylcholestanes, molecular fossils diagnostic of marine algae. *Science* **247**, 309–312.

Moldowan JM, Dahl J, Jacobson SR, *et al.* (1996) Chemostratigraphic reconstruction of biofacies: Molecular evidence linking cyst-forming dinoflagellates with pre-Triassic ancestors. *Geology* **24**, 159–162.

Moldowan JM, Jacobson JR, Dahl J, Al-Hajii A, Huizinga BJ, Fago FJ (2001) Molecular fossils demonstrate Precambrian origin of dinoflagellates. In Zhuralev A, Riding R, eds., *Ecology of the Cambrian Radiation*. Columbia University Press, New York, NY, p. 474–493.

Morel FMM (2008) The co-evolution of phytoplankton and trace element cycles in the oceans. *Geobiology* **6**, 318–324.

Narbonne GM (2005) The Ediacara Biota: Neoproterozoic origin of animals and their ecosystems. *Annual Review of Earth and Planetary Sciences* **33**, 421–442.

Neubert N, Heri AR, Voegelin AR, Nägler TF, Schlunegger F, Villa IM (2011) The molybdenum isotopic composition in river water: Constraints from small catchments. *Earth and Planetary Science Letters* **304**, 180-190.

Neuweiler F, Turner EC, Burdidge DJ (2009) Early Neoproterozoic origin of the metazoan clade recorded in carbonate rock texture. *Geology* **37**, 475–478.

Olcott AN, Sessions AL, Corsetti FA, Kaufman AJ, de Oliviera TF (2005) Biomarker evidence for photosynthesis during Neoproterozoic glaciations. *Science* **310**, 471–474.

Papineau D, Mojzsis SJ, Coath CD, Karhu JA, McKeegan KD (2005) Multiple sulfur isotopes from sulfides from sediments in the aftermath of Paleoproterozoic glaciations. *Geochimica et Cosmochimica Acta* **69**, 5033–5060.

Papineau D (2010) Global biogeochemical changes at both ends of the Proterozoic: Insights from phosphorites. *Astrobiology* **10**, 165–181.

Pavlov AA, Kasting JF (2002) Mass-independent fractionation of sulfur isotopes in Archean sediments: Strong evidence for an anoxic Archean atmosphere. *Astrobiology* **2**, 27–41.

Pavlov AA, Kasting JF, Brown LL, Rages KA, Freedman R (2000) Greenhouse warming by CH_4 in the atmosphere of early Earth. *Journal of Geophysical Research* **105**, 11981–11990.

Pavlov AA, Brown LL, Kasting JF (2001) UV shielding of NH_3 and O_2 by organic hazes in the Archean atmosphere. *Journal of Geophysical Research* **106**, 23267–23287.

Pavlov AA, Hurtgen MT, Kasting JF, Arthur, MA (2003) Methane-rich Proterozoic atmosphere? *Geology* **31**, 87–90.

Peltier WR, Liu Y, Crowley JW (2007) Snowball Earth prevention by dissolved organic carbon remineralization. *Nature* **450**, 813–818.

Peng P, Sheng G, Fu J, Yan Y (1998) Biological markers in 1.7 billion year old rock from the Tuanshanzi Formation, Jixian strata section, North China. *Organic Geochemistry* **29**, 1321–1329.

Peng Y, Bao H, Yuan X (2009) New morphological observations for Paleoproterozoic acritarchs from the Chuanlinggou Formation, North China. *Precambrian Research* **168**, 223–232.

Peterson KJ, Lyons KB, Nowak KS, Takacs CM, Wargo MJ, McPeek MA (2004) Estimating metazoan divergence times with a molecular clock. *Proceedings of the National Academy of Sciences, USA* **101**, 6536–6541.

Peterson KJ, Butterfield NJ (2005) Origin of the Eumetazoa: Testing ecological predictions of molecular clocks against the Proterozoic fossil record. *Proceedings of the National Academy of Sciences, USA* **101**, 6536–6541.

Peterson KJ, McPeek MA, Evans DAD (2005) Tempo and mode of early animal evolution: inferences from rocks, Hox, and molecular clocks. *Paleobiology* **31**, Supplement, 36–55.

Peterson KJ, Cotton JA, Gehling JG, Pisani D (2008) The Ediacaran emergence of bilaterians: Congruence between the genetic and the geological fossil records. *Philosophical Transactions of the Royal Society of London, B – Biological Sciences* **363**, 1435–1443.

Planavsky N (2009) Early Neoproterozoic origin of the metazoan clade recorded in carbonate rock texture: COMMENT. *Geology* **37**, E195.

Planavsky NJ, Rouxel OJ, Bekker A, et al. (2010) The evolution of the marine phosphate reservoir. *Nature* **467**, 1088–1090.

Planavsky NJ, McGoldrick P, Scott CT, Li C, Reinhard CT, Kelly AE, Chu X, Bekker A, Love GD, Lyons TW (2011) Widespread iron-rich conditions in the mid-Proterozoic ocean. *Nature* **477**, 448-451.

Pollard D, Kasting JF (2005) Snowball Earth: A thin-ice solution with flowing sea glaciers. *Journal of Geophysical Research* **110**, C07010, doi:10.1029/2004JC002525.

Porter SM, Knoll AH (2000) Testate amoebae in the Neoproterozoic Era: Evidence from vase-shaped microfossils in the Chuar Group, Grand Canyon. *Paleobiology* **26**, 360–385.

Porter SM (2004) The fossil record of early eukaryotic diversification. In: *The Paleontological Society Papers 10: Neoproterozoic-Cambrian Biological Revolutions* (eds Lipps JH, Waggoner B). Paleontological Society, New Haven, CT, p. 35–50.

Porter SM (2006) The Proterozoic fossil record of heterotrophic eukaryotes. In: *Neoproterozoic Geobiology and Paleobiology* (eds Xiao S, Kaufman, AJ). Springer, Dordrecht, p. 1–21.

Poulsen CJ, Pierrehumbert RT, Jacob RL (2001) Impact of ocean dynamics on the simulation of the Neoproterozoic 'snowball Earth'. *Geophysical Research Letters* **28**, 1575–1578.

Poulsen CJ (2003) Absence of a runaway ice-albedo feedback in the Neoproterozoic. *Geology* **31**, 473–476.

Poulton SW, Canfield DE (2011) Ferruginous conditions: A dominant feature of the ocean through Earth's history. *Elements* **7**, 107–112,

Poulton SW, Fralick PW, Canfield DE (2004) The transition to a sulphidic ocean ~1.84 billion years ago. *Nature* **431**, 173–177.

Poulton SW, Fralick PW, Canfield DE (2010) Spatial variability in oceanic redox structure 1.8 billion years ago. *Nature Geoscience* **3**, 486–490.

Pratt LM, Summons RE, Hieshima GB, Hayes JM (1991) Sterane and triterpane biomarkers in the Precambrian Nonesuch Formation, North American Midcontinent Rift. *Geochimica et Cosmochimica Acta* **55**, 911–916.

Pufahl PK, Hiatt EE, Kyser TK (2010) Does the Paleoproterozoic Animikie Basin record the sulfidic ocean transition? *Geology* **38**, 659–662.

Rainbird RH, Stern RA, Khudoley AK, Kropachev AP, Heaman LM, Sukhorukov VI (1998) U-Pb geochronology of Riphean sandstone and gabbro from Southeast Siberia and its bearing on the Laurentia-Siberia connection. *Earth and Planetary Science Letters* **164**, 409–420.

Rasmussen B, Buick R (1999) Redox state of the Archean atmosphere: Evidence from detrital heavy minerals in ca. 3250–2750 Ma sandstones form the Pilbara Craton, Australia. *Geology* **27**, 115–118.

Rasmussen B, Fletcher IR, Brocks JJ, Kilburn MR (2008) Reassessing the first appearance of eukaryotes and cyanobacteria. *Nature* **455**, 1101–1104.

Raymond AC, Blankenship RE, (2004) Biosynthetic pathways, gene replacement and the antiquity of life. *Geobiology* **2**, 199–203.

Reinhard CT, Raiswell R, Scott C, Anbar AD, Lyons TW (2009) A late Archean sulfidic sea stimulated by early oxidative weathering of the continents. *Science* **326**, 713–716.

Ries JB, Fike DA, Pratt LM, Lyons TW, Grotzinger JP (2009) Superheavy pyrite ($\delta^{34}S_{pyr} > \delta^{34}S_{CAS}$) in the terminal Proterozoic

Nama Group, southern Namibia: A consequence of low sea-water sulfate at the dawn of animal life. *Geology* **37**, 743–746.

Roscoe SM (1973) The Huronian Supergroup: A Paleoaphebian succession showing evidence of atmospheric evolution. *Geological Society of Canada Special Paper* **12**, 31–48.

Rothman DH, Hayes JM, Summons RE (2003) Dynamics of the Neoproterozoic carbon cycle. *Proceedings of the National Academy of Sciences, USA* **100**, 8124–8129.

Rye R, Kuo PH, Holland HD (1995) Atmospheric carbon dioxide concentrations before 2.2 billion years ago. *Nature* **378**, 603–605.

Rye R, Holland HD (1998) Paleosols and the evolution of atmospheric oxygen: A critical review. *American Journal of Science* **298**, 621–672.

Sagan C, Mullen G (1972) Earth and Mars: Evolution of atmospheres and surface temperatures. *Science* **177**, 52–56.

Saito MA, Sigman DM, Morel FMM (2003) The bioinorganic chemistry of the ancient ocean: the co-evolution of cyanobacterial metal requirements and biogeochemical cycles at the Archean-Proterozoic boundary? *Inorganica Chimica Acta* **356**, 308–318.

Sarmiento JL, Toggweiler JR (1984) A new model for the role of the oceans in determining atmospheric pCO_2. *Nature* **308**, 621–624.

Sarmiento JL, Herbert TD, Toggweiler JR (1988) Causes of anoxia in the world ocean. *Global Biogeochemical Cycles* **2**, 115–128.

Schidlowski M, Eichmann R, Junge CE (1975) Precambrian sedimentary carbonates: carbon and oxygen isotope geochemistry and implications for the terrestrial oxygen budget. *Precambrian Research* **2**, 1–69.

Schidlowski M, Eichmann R, Junge CE (1976) Carbon isotope geochemistry of the Precambrian Lomagundi carbonate province, Rhodesia. *Geochimica et Cosmochimica Acta* **40**, 449–455.

Schröder S, Bekker A, Beukes NJ, Strauss H, van Niekerk HS (2008) Rise in seawater sulphate concentration associated with the Paleoproterozoic positive carbon isotope excursion: evidence from sulphate evaporates in the ~2.2–2.1 Gyr shallow-marine Lucknow Formation, South Africa. *Terra Nova* **20**, 108–117.

Schwark L, Empt P (2006) Sterane biomarkers as indicators of Palaeozoic algal evolution. *Palaeogeography, Palaeoclimatology, Palaeoecology* **240**, 225–236.

Scott C, Lyons TW, Bekker A, *et al.* (2008) Tracing the stepwise oxygenation of the Proterozoic ocean. *Nature* **452**, 456–459.

Scott CT, Bekker A, Reinhard CT, *et al.* (2011) Late Archean euxinic conditions before the rise of atmospheric oxygen. *Geology* **39**, 119–122.

Sheldon ND (2006) Precambrian paleosols and atmospheric CO_2 levels. *Precambrian Research* **147**, 148–155.

Sheldon ND, Tabor NJ (2009) Quantitative paleoenvironmental and paleoclimatic reconstruction using paleosols. *Earth Science Reviews* **95**, 1–52.

Shen B, Xiao X, Bao H, Kaufman AJ, Zhou C, Wang H (2008) Stratification and mixing of the post-glacial Neoproterozoic ocean: Evidence from carbon and sulfur isotopes in a cap dolostone from northwest China. *Earth and Planetary Science Letters* **265**, 209–228.

Shen Y, Canfield DE, Knoll AH (2002) Middle Proterozoic ocean chemistry: Evidence from the McArthur Basin, northern Australia. *American Journal of Science* **302**, 81–109.

Shen Y, Knoll AH, Walter MR (2003) Evidence for low sulphate and anoxia in a mid-Proterozoic marine basin. *Nature* **423**, 632–635.

Shen Y, Zhang T, Hoffman PF (2008) On the coevolution of Ediacaran oceans and animals. *Proceedings of the National Academy of Sciences, USA* **105**, 7376–7381.

Slack JF, Grenne T, Bekker A, Rouxel OJ, Lindberg PA (2007) Suboxic deep seawater in the late Paleoproterozoic: Evidence from hematitic chert and iron formation related to seafloor-hydrothermal sulfide deposits, central Arizona, USA. *Earth and Planetary Science Letters* **255**, 243–256.

Sperling EA, Peterson KJ, Pisani D (2007) The importance of poriferan paraphyly in Precambrian (and Cambrian) paleobiology. In: *The Rise and Fall of the Ediacaran Biota* (eds Vicker-Rich P, Kamarower P). Geological Society Special Publications, p. 355–368.

Stanley SM (1973) An ecological theory for the sudden origin of multicellular life in the late Precambrian. *Proceedings of the National Academy of Sciences, USA* **70**, 1486–1489.

Summons RE, Powell TG, Boreham CG (1988a) Petroleum geology and geochemistry of the Middle Proterozoic McArthur Basin, Northern Australia. III Composition of extractable hydrocarbons. *Geochimica et Cosmochimica Acta* **51**, 3075–3082.

Summons RE, Brassell SC, Eglinton G, *et al.* (1988b) Distinctive hydrocarbon biomarkers from fossiliferous sediment of the Late Proterozoic Walcott Member, Chuar Group, Grand Canyon, U.S.A. *Geochimica et Cosmochimica Acta* **52**, 2625–2637.

Summons RE (1992) Proterozoic biogeochemistry: abundance and composition of extractable organic matter. In: *The Proterozoic Biosphere: A Multidisciplinary Study* (eds Schopf JW, Klein C). Cambridge University Press, Cambridge, p. 101–115.

Summons RE, Walter MR (1990) Molecular fossils and microfossils of prokaryotes and protists from Proterozoic sediments. *American Journal of Science* **290-A**, 212–244.

Summons RE, Thomas J, Maxwell JR, Boreham CJ (1992) Secular and environmental constraints on the occurrence of dinosterane in sediments. *Geochimica et Cosmochimica Acta* **56**, 2437–2444.

Summons RE, Jahnke LL, Hope JM, Logan GA (1999) 2-Methylhopanoids as biomarkers for cyanobacterial oxygenic photosynthesis. *Nature* **400**, 554–556.

Summons RE, Bradley AS, Jahnke LL, Waldbauer JR (2006) Steroids, triterpenoids and molecular oxygen. *Philosophical Transactions of the Royal Society, B – Biological Sciences* **361**, 951–968.

Swanson-Hysell NL, Rose CV, Calmet CC, Halverson GP, Hurtgen MT, Maloof AC (2010) Cryogenian glaciation and the onset of carbon-isotope decoupling. *Science* **328**, 608–611.

Swart PK, Kennedy MJ (2011) Does the global stratigraphic reproducibility of $\delta^{13}C$ in Neoproterozoic carbonates require a marine origin? A Pliocene-Pleistocene comparison. *Geology* 40, 87–90.

Theissen U, Hoffmeister M, Grieshaber M, Martin W (2003) Single eubacterial origin of eukaryotic sulfide: quinone oxidoreductase, a mitochondrial enzyme conserved from the early evolution of eukaryotes during anoxic and sulfidic times. *Molecular Biology and Evolution* 20, 1564–1574.

Thomazo C, Ader M, Phillipot P (2011) Extreme ^{15}N-enrichments in 2.72-Gyr-old sediments: evidence for a turning point in the nitrogen cycle. *Geobiology* 9, 107-120.

Tosca NJ, Johnston DT, Mushegian A, Rothman DH, Summons RE, Knoll AH (2010) Clay mineralogy, organic carbon burial, and redox evolution in Proterozoic oceans. *Geochimica et Cosmochimica Acta* 74, 1579–1592.

Towe KM (1970) Oxygen-collagen priority and the early metazoan fossil record. *Proceedings of the National Academy of Sciences, USA* 65, 781–788.

Turner JT (2002) Zooplankton fecal pellets, marine snow and sinking phytoplankton blooms. *Aquatic Microbial Ecology* 27, 57–102.

Ueno Y, Yamada K, Yoshida N, Maruyama S, Isozaki Y (2006) Evidence from fluid inclusions for microbial methanogenesis in the early Archaean era. *Nature* 440, 516–519.

Vannier J, Chen JY (2000) The Early Cambrian colonization of pelagic niches exemplified by Isoxys (Arthropoda). *Lethaia* 33, 295–311.

Vidal G, Moczydłowska-Vidal M (1997) Biodiversity, speciation, and extinction trends of Proterozoic and Cambrian phytoplankton. *Paleobiology* 23, 230–246.

Volkman JK (2003) Sterols in microorganisms. *Applied Microbiology and Biotechnology* 60, 495–506.

Volkman JK (2005) Sterols and other triterpenoids: source specificity and evolution of biosynthetic pathways. *Organic Geochemistry* 36, 139–159.

Waldbauer JR, Sherman LS, Sumner DY, Summons RE (2009) Late Archean molecular fossils from the Transvaal Supergroup record the antiquity of microbial diversity. *Precambrian Research* 169, 28–47.

Waldbauer JR, Newman DK, Summons, RE (2011) Microaerobic steroid biosynthesis and the molecular fossil record of Archean life. *Proceedings of the National Academy of Sciences, USA* 108, 13409–13414.

Walker JCG, Hays PB, Kasting JF (1981) A negative feedback mechanism for the long-term stabilization of Earth's surface temperature. *Journal of Geophysical Research* 86, 9776–9782.

Wang J, Jiang G, Xiao S, Li Q, Wei Q (2008) Carbon isotope evidence for widespread methane seeps in the ~635Ma Doushantuo cap carbonate in South China. *Geology* 36, 347–350.

Werne JP, Lyons TW, Hollander DJ, Formolo MJ, Damsté JSS (2003) Reduced sulfur in euxinic sediments of the Cariaco Basin: Sulfur isotope constraints on organic sulfur formation. *Chemical Geology* 195, 159–179.

Wille W, Kramers JD, Nägler TF, *et al.* (2007) Evidence for a gradual rise of oxygen between 2.6 and 2.5Ga from Mo isotopes and Re-PGE signatures in shales. *Geochimica et Cosmochimica Acta* 71, 2417–2435.

Williams BAP, Hirt RP, Lucoq JM, Embley TM (2002) A mitochondrial remnant in the microsporidian *Trachipleistophora hominis. Nature* 418, 865–869.

Wilson JP, Fischer WW, Johnston DT, *et al.* (2010) Geobiology of the late Paleoproterozoic Duck Creek Formation, Western Australia. *Precambrian Research* 179, 135–149.

Woese CR (1977) A comment on methanogenic bacteria and the primitive ecology. *Journal of Molecular Evolution* 9, 369–371.

Wortmann UG, Bernasconi SM, Böttcher ME (2001) Hypersulfidic deep biosphere indicates extreme sulfur isotope fractionation during single-step microbial sulfate reduction. *Geology* 29, 647–650.

Xiao S (2004) Neoproterozoic glaciations and the fossil record, In: *The Extreme Proterozoic: Geology, Geochemistry, and Climate* (eds Jenkins GS, McMenamin M, Sohl LE, McKay CP). Washington DC, American Geophysical Union (AGU), pp. 199–214.

Xiao S (2008) Geobiological events in the Ediacaran Period. In: *From Evolution to Geobiology: Research Questions Driving Paleontology at the Start of a New Century* (eds Kelly PH, Bambach RK). The Paleontological Society, New Haven, CT, p. 85–104.

Xiao S, Laflamme M (2009) On the eve of animal radiation: Phylogeny, ecology and evolution of the Ediacara biota. *Trends in Ecology & Evolution* 24, 31–40.

Xiao S, Knoll AH, Kaufman AJ, Yin L, Zhang Y (1997) Neoproterozoic fossils in Mesoproterozoic rocks? Chemostratigraphic resolution of a biostratigraphic conundrum from the North China Platform. *Precambrian Research* 84, 197–220.

Xiao S, Zhang Y, Knoll AH (1998) Three-dimensional preservation of algae and animal embryos in a Neoproterozoic phosphorite. *Nature* 391, 553–558.

Xiao S, Yuan X, Knoll AH (2000) Eumetazoan fossils in terminal Proterozoic phosphorites? *Proceedings of the National Academy of Sciences, USA* 97, 13684–13689.

Yan Y, Liu Z (1993) Significance of eukaryotic organisms in the microfossil flora of the Changcheng System. *Acta Micropalaeontologica Sinica* 10, 167–180.

Yin L, Zhu M, Knoll AH, Yuan X, Zhang J, Hu J (2007) Doushantuo embryos preserved inside diapause egg cysts. *Nature* 446, 661–663.

Young GM, Long DGF, Fedo CM, Nesbitt HW (2001) Paleoproterozoic Huronian basin: Product of a Wilson cycle punctuated by glaciations and meteorite impact. *Sedimentary Geology* 141–142, 233–254.

Zahnle K, Claire M, Catling D (2006) The loss of mass-independent fractionation in sulfur due to a Palaeoproterozoic collapse of atmospheric methane. *Geobiology* 4, 271–283.

Zang W, McKirdy DM (1993) Microfossils and molecular fossils from the Neoproterozoic Alinya Formation – a possible new source rock in the eastern Officer Basin, In: *Central Australian Basins Workshop* (eds Alexander EM, Gravestock DI), Alice Springs, Abstracts, p.62–63.

Zerkle AL, House CH, Cox RP, Canfield DE (2006) Metal limitation of cyanobacterial N_2 fixation and implications for the Precambrian nitrogen cycle. *Geobiology* 4, 285–297.

Zhang Z (1986) Clastic facies microfossils from the Chuanlinggou Formation (1800Ma) near Jixian, North China. *Journal of Micropalaeontology* 5, 9–16.

Zhang Z (1997) A new Palaeoproterozoic clastic-facies microbiota from the Changzhougou Formation, Changcheng Group, Jixian, north China. *Geological Magazine* **134**, 145–150.

Zhou C, Xiao S (2007) Ediacaran δ¹³C chemostratigraphy of South China. *Chemical Geology* **237**, 89–108.

Zhou C, Xie G, McFadden K, Xiao S, Yuan X (2007) The diversification and extinction of Doushantuo-Pertatataka acritarchs in South China: Causes and biostratigraphic significance. *Geological Journal* **42**, 229–262.

Zhu SX, Sun SF, Huang XG, *et al.* (1995) Megascopic multicellular organisms from the 1700-milion year old Tuanshanzi Formation in the Jixian area, North China. *Science* **270**, 620–622.

21

GEOBIOLOGY OF THE PHANEROZOIC

Steven M. Stanley

Department of Geology and Geophysics, University of Hawaii,
1680 East-West Road, Honolulu, Hawaii 96822, USA

21.1 The beginning of the Phanerozoic Eon

Other chapters of this book trace important themes of geobiology through time. This chapter explores such themes as well, but it provides what amounts to a 'horizontal' rather than 'vertical' treatment, guiding the reader on a trip through time that focuses on groups of significant geobiologic events that occurred during particular intervals of Earth's history.

In reviewing the history of Phanerozoic geobiology, it is appropriate to begin with the phenomenon that gave the Phanerozoic its name: the polyphyletic evolution of skeletons that ushered in the Cambrian Period. After a long interval of 'aragonite seas' in the Proterozoic, a shift to 'calcite seas' came early in Cambrian time, when the molar Mg/Ca ratio of seawater dropped below 2. Possibly the elevation of $[Ca^{2+}]$ that contributed to this shift promoted the calcification of marine animals by increasing the supersaturation of seawater with respect to $CaCO_3$ (Brennan *et al.*, 2004). There is evidence that the transition to calcite seas also led to the origins of skeletons consisting of low-Mg calcite, whereas the earliest Cambrian taxa produced skeletons of aragonitic or high-Mg calcite (Porter, 2007; Zhuravlev and Wood, 2008).

The expansion of animal activity in the oceans had important consequences for marine geochemistry and sedimentology. For example, the polyphyletic production of skeletons in Early Cambrian time inevitably changed the $CaCO_3$ budget of the ocean. One result would have been a reduction of non-skeletal precipitation of $CaCO_3$. Because silica occurs at a low concentration in seawater, the advent of siliceous biomineralization

must also have strongly affected the silica budget in the ocean (Maliva *et al.*, 1989). In the absence of silica-secreting organisms, silica was relatively abundant in the ocean during Precambrian time, and as a consequence, early diagenetic cherts formed abundantly in peritidal marine sediments, possibly through microbial activity. Although demosponges, which produce spicules of silica, invaded offshore habitats early in the Paleozoic, they failed to suppress the precipitation of cherts in peritidal environments. On the other hand, the initial evolutionary radiation of the Radiolaria resulted in enough silica sequestration that cherts no longer formed in peritidal environments after Ordovician time.

Cambrian strata typically exhibit low levels of bioturbation (Droser and Bottjer, 1988). In the absence of heavy browsing by animals, microbial mats carpeted many areas of shallow Proterozoic seafloors. Thus, stromatolites were widespread, as were 'elephant skin' sedimentary surfaces that formed when microbial mats crinkled. The advent of effective grazing in Cambrian time reduced the production of these structures (Garrett, 1970; Hagadorn and Bottjer, 1997). That microbial mats still formed sporadically in the Cambrian is indicated by the common occurrence of thrombolites. These are stromatolite-like forms that lack layering because of disruption by burrowers or borers that did not exist before the Cambrian, or by obstructing seaweeds, which may also have been new on the scene (Aitkin, 1967; Grotzinger *et al.*, 2005). Also present in Cambrian rocks are occasional stromatolites and flat-pebble conglomerates. The latter contain platy carbonate clasts that resulted from storm wave fracturing of well-laminated, often

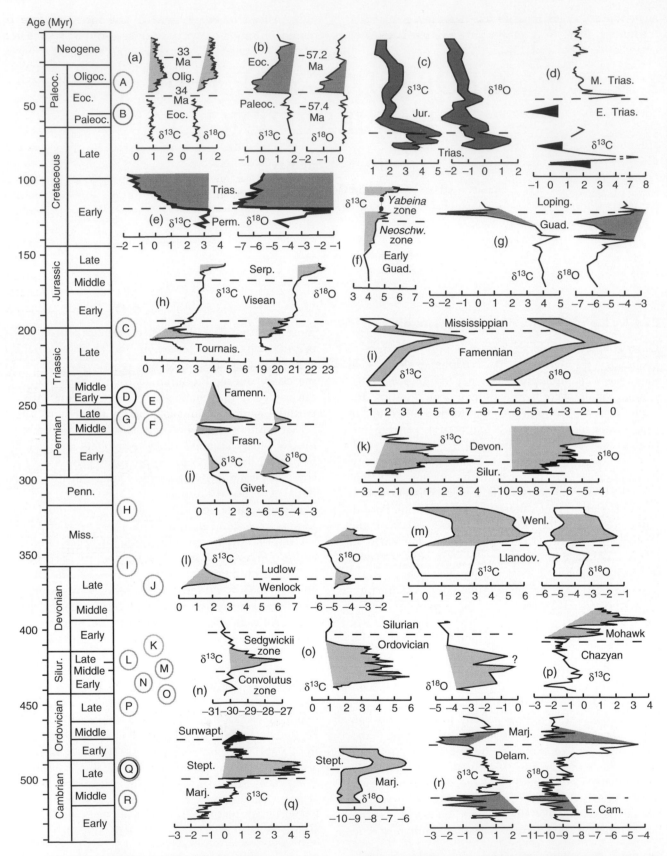

Figure 21.1 Stable isotope excursions that have been documented in shallow marine strata in association with mass extinctions. Eighteen intervals (A–R) contain a total of 26 such δ13C excursions. Corresponding to these, and trending in the

algal-bound, strata that in the absence of extensive burrowing had been partly lithified by submarine cementation (Sepkoski, 1982).

21.2 Cambrian mass extinctions

Several major extinctions occurred during the Cambrian Period. The first came at the end of the Early Cambrian, when redlichiid trilobites died out (Zhu *et al.*, 2004), as did nearly all archaeocyathid (sponge) reef builders (Hill, 1972). The Delamaran/Marjuman stage boundary of the Middle Cambrian also marks a major extinction of trilobites, as do the Marjuman/Steptoean and Steptoan/ Sunwaptan stage boundaries of the Late Cambrian (Palmer, 1998). It has been suggested that the three Middle and Late Cambrian mass extinctions resulted from episodic upward expansion of cold, poorly oxygenated waters that not only caused extinction of taxa in shallow waters but also permitted the migration into these waters of trilobites that had previously occupied deeper habitats (Stitt, 1975; Palmer, 1984; Perfetta *et al.*, 1999). While cooling may have caused these extinctions of shallow-water taxa, it is unlikely that reduced oxygen contributed because waters above wave base are always oxygenated by the atmosphere.

The first three Cambrian mass extinctions illustrate a pattern that characterizes major extinctions for the entire Phanerozoic: they coincide with sharp excursions for carbon and oxygen isotopes (shifts of $\delta^{13}C$ and $\delta^{18}O$) for skeletal carbonates (Fig. 21.1q, r). Numerous *ad hoc* explanations have been offered to explain these various excursions, nearly all quite reasonably focusing on one or more factors that have changed the rate of burial of organic carbon, which is isotopically light. It appears, however, that a unifying explanation can largely account for all of the excursions except a small number associated with global oceanic anoxia (Stanley, 2010). The most important factor is the rate of respiration of bacteria, which increases exponentially with temperature. Because about 90% of carbon burial in the oceans takes place along continental margins (Reimers *et al.*, 1992), these locations are where climatic changes have their greatest impact on bacterial respiration. When global temperatures rise, so do bacterial respiration rates, and therefore a larger proportion of carbon in particulate organic matter is returned to the ocean in the form of

CO_2 instead of being buried. The increased rate of remineralization of isotopically light carbon results in a global decline in $\delta^{13}C$ for seawater. On the other hand, when global temperatures fall, the rate of burial of organic carbon rises and so does $\delta^{13}C$ for seawater. Significantly, $\delta^{18}O$ in calcium carbonate follows the same pattern, because of fractionation by organisms and also, if glaciers expand, because the H_2O containing the light oxygen isotope, ^{16}O, evaporates preferentially and is preferentially locked up in glaciers. Every pair of global carbon and oxygen isotope excursions coinciding with a mass extinction has been either positive or negative, reflecting global climate change (Fig. 21.1). Global climate change must have played a role in nearly all of these mass extinction, the most significant of which will be discussed below.

Three secondary climate-related aspects of the marine ecosystem must also have contributed to the carbon isotope excursions during times of global climate change (Stanley, 2010): (1) growth or melting of clathrates (icy materials along continental margins that contain methane, which is isotopically very light carbon); (2) the positive correlation between temperature and degree of fractionation of carbon isotopes by phytoplankton, although this relationship is weak at temperatures above ~15° C (Freeman and Hayes, 1992); (3) increased phytoplankton productivity during 'icehouse' conditions, when strong latitudinal temperature gradients have strengthened the upwelling of nutrient-rich waters.

The only conspicuous exceptions to the rule described above for mass extinctions and stable isotopes are positive excursions for $\delta^{13}C$ for intervals of global warming such as the those of the Toarcian (Jurassic) and latest Aptian and Cenomanian (Cretaceous), when a global oceanic anoxia developed and huge amounts of isotopically light organic carbon were buried.

21.3 The terminal Ordovician mass extinction

Marine life diversified dramatically as the Ordovician progressed, but then at the end of this period suffered one of the largest mass extinctions of the Phanerozoic. This crisis has been convincingly connected to a brief expansion of continental glaciers in Gondwanaland, reflected by tillites in many regions of Gondwanaland and what is

---

Figure 21.1 Continued.

same direction, are 19 published $\delta^{18}O$ excursions, which are displayed in the plots to the right of those depicting $\delta^{13}C$. Encircled letters on the left indicate temporal positions of excursions. Blue indicates association with global cooling and red, with global warming; black indicates absence of published evidence of associated climate change. Horizontal scales represent magnitudes of $\delta^{13}C$ and $\delta^{18}O$ excursions in ‰. Light $\delta^{13}C$ in N is for organic carbon rather than carbonates, and heavy $\delta^{18}O$ in H is for conodonts rather than bulk or skeletal carbonate. Ordinates represent stratigraphic positions of samples and are neither precisely linear with respect to time nor scaled the same for all graphs (after Stanley, 2010).

now southern Europe (review by Diaz-Martinez and Grahn, 2007) and also by eustatic sealevel lowering, documented in many stratigraphic sections around the world. A strong positive shift for $\delta^{18}O$ in the ocean presumably resulted from both cooling and expansion of glaciers and was paralleled by a shift for $\delta^{13}C$ (Brenchley et al., 1994; Saltzman and Young, 2005) (Fig. 21.1o). The extinction took place in two pulses: warm-adapted taxa died out preferentially in the first pulse, which was the larger of the two, as cold-adapted taxa migrated from deep water and high latitudes to shallow seas positioned at lower latitudes (Berry et al., 1995; Sheehan, 2001). These patterns point to cooling, presumably associated with increased seasonality, as the primary agent of extinction. Cold-adapted taxa died out preferentially in the second pulse, which took place at the very end of Ordovician time as the ice age waned, perhaps 2 my after the first pulse.

21.4   The impact of early land plants

The spread of early land plants during Silurian time altered terrestrial landscapes, but vascular plants did not appear until Late Silurian time, and not until late in the Devonian did land plants first form forests. The evolution of seeds near the end of Devonian time liberated land plants from moist environments and thus added another major step in the transformation of terrestrial landscapes. This ecological expansion had two major consequences for the physical environment. First, plants' root systems stabilized river banks. Whereas braided streams, which produced gravelly, cross-bedded deposits, prevailed on continents before the Devonian, meandering rivers with firm banks first became widespread during the Devonian, producing characteristic point bar cycles with coarse (channel) sediment at the base and fine (floodplain) sediment at the top. Second, the initial global expansion of forests accelerated weathering because the roots of land plants secrete acids and other compounds that break down silicate minerals. Such chemical weathering consumes CO_2, and it appears that accelerated weathering led to climatic cooling and continental glaciation in the Late Devonian through reduction of greenhouse warming (Retallack, 1997).

21.5   Silurian biotic crises

Each of four positive excursions for $\delta^{13}C$ in Silurian marine carbonates, the last at the very end of the period, occurred immediately after a marine biotic crisis (Saltzman, 2001, 2002). These excursions coincided with glacial episodes and positive oxygen isotope excursions (Loydell, 2007) (Fig. 21.1k–n). Thus, the Silurian crises appear to have resulted at least in part from climatic cooling.

21.6   Devonian mass extinctions

Three large mass extinctions struck during the Devonian. The Givetian crisis, which marked the end of the Middle Devonian, has been little studied, but it eliminated many marine taxa, including numerous rugose coral families (House, 2002). The Frasnian crisis of the Late Devonian spanned perhaps 3 million years, nearly eliminating the previously flourishing coral–stromatoporoid reef community (Copper, 2002). The Famennian crisis, which was briefer but more severe, occurred at the end of the Devonian, eliminating not only many invertebrate marine taxa but also the heavily armored marine placoderm fishes and a variety of terrestrial plants. All three Devonian biotic crises struck tropical taxa preferentially and were associated with abrupt sea level declines and positive $\delta^{13}C$ and $\delta^{18}O$ excursions in the ocean (Joachimski and Buggisch, 2002; Buggisch and Joachimski, 2006) (Fig. 21.1 j, k). Glacial deposits in eastern North America confirm the expansion of glaciers in late Famennian time (Brezinski et al., 2008). Unlike the great terminal Ordovician crisis, the Devonian mass extinctions markedly restructured the marine ecosystem, in part by destroying the coral-stromatoporoid reef community (Droser et al., 1997).

21.7   Major changes of the global ecosystem in Carboniferous time

There was renewed continental glaciation during the first (Tournasian) age of the Mississippian (early Carboniferous) (Isaacson et al., 2008), accompanied by positive shifts for $\delta^{13}C$ and $\delta^{18}O$ (Fig. 21.1, h). Then climates warmed. Coal swamps, colonized primarily by lycopod plants and seed ferns, spread broadly over lowland areas of the world early in Pennsylvanian (late Carboniferous) time. Because anaerobic conditions and tannic acid in these swamps excluded decomposing bacteria, reduced organic carbon was buried with little decay. Furthermore, termites had not yet evolved, so that, although subject to attack by fungi, trunks of dead trees often fell into swamp waters largely intact (Labandeira et al., 1997). As a result, a large amount of organic carbon was buried, rather than being returned to the atmosphere as CO_2 via respiration by decomposers. The consequent reduction of greenhouse warming led to the largest glacial episode of the entire Phanerozoic, with massive ice sheet growth in the Southern Hemisphere. Thus, both $\delta^{13}C$ and $\delta^{18}O$ in the ocean increased at the start the Serpukhovian, the last age of the Mississippian (Fig. 21.1h). Also resulting from the glacial expansion were eustatic sea level oscillations, which produced the cyclical deposits on cratons known as cyclothems in North America and coal measures in Europe.

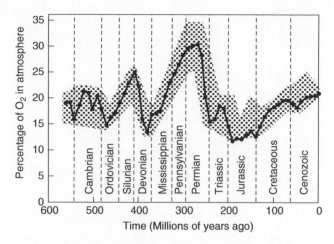

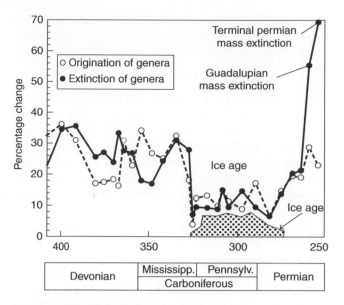

Figure 21.2 Estimated changes in the volume of the oxygen reservoir in the atmosphere during the Phanerozoic. (a) Changes in the relative percentage of ^{13}C in seawater, estimated from the isotopic composition of limestones. (b) Estimated changes in the portion of Earth's atmosphere consisting of free oxygen. Percentages for particular intervals are based on estimates of the concentration of unoxidized carbon and sulfur in sediments, the burial of which causes oxygen to build up in the atmosphere. The broad band depicts uncertainties in calculations. (A after Berner, 1987; B after Berner, 2006.)

Figure 21.3 Reduction of rates of origination and extinction of marine genera at the start of the late Paleozoic ice age to their lowest levels in all of Phanerozoic time. These rates returned to normal levels precisely when the ice age ended, partway through the Permian Period (after Stanley and Powell, 2003).

Burial excludes organic matter from consumption by aerobic consumers and bacteria, leaving behind in the atmosphere oxygen that would otherwise have been consumed in respiration. Therefore, the increased burial of reduced carbon during the Carboniferous resulted in a buildup of atmospheric oxygen. Today, oxygen constitutes 21% of atmospheric oxygen, and it has been estimated that this percentage rose to 35% during Pennsylvanian time (Berner, 2006) (Fig. 21.2). This rise appears to explain the evolution of giant insects, including dragonflies with wingspans of 60 cm, during the Pennsylvanian (insects' assimilation of oxygen is limited by the absorption area of their spiracles) (Graham *et al.*, 1995). High atmospheric oxygen levels probably also increased the incidence of wildfires.

The late Paleozoic ice age not only had a biotic trigger, but also major biotic consequences. Its initiation resulted in a mass extinction near the end of Mississippian time (the seventh largest such crisis of the Phanerozoic Eon), and a new state of the marine ecosystem. For every major marine taxon, rates of origination and extinction dropped at the start of this ice age and remained low until its end (Fig. 21.3). This pattern reflected the preferential loss of narrowly adapted tropical taxa and the survival of taxa with broad thermal tolerances that were resistant to extinction and that, because of their widespread geographic distributions, did not readily produce isolated populations that might emerge as new species (Stanley and Powell, 2003; Powell, 2005).

More generally, the reduction of atmospheric CO_2 that began in Devonian time and continued into the Carboniferous altered the physiology of land plants. Beerling and Berner (2005) concluded that a series of feedbacks occurred. Stomata, the pores through which gases pass to and from leaves, increase in density with a decrease in atmospheric CO_2 because more stomata are needed for CO_2 uptake. An increase in stomatal density results in an increase in water loss. As atmospheric CO_2 declined beginning in the Devonian, stomatal density increased, and this would have increased heat loss from leaves via evapotranspiration of water to the atmosphere. Large leaves, because of their low surface-to-volume ratio, are prone to lethal overheating, and the increased heat loss from leaves as CO_2 declined apparently permitted the increase in maximum leaf size that has been documented for large plants during the Devonian–Mississippian interval.

21.8 Low-elevation glaciation near the equator

A variety of evidence in the American Southwest indicates that glaciers were well-developed at low elevations within about 8° of the equator in Late Pennsylvanian and Early Permian time: a glaciated valley in the ancestral Rocky Mountains, diamictite containing striated clasts, and widespread loessites (Soreghan *et al.*, 2008). The remarkable cooling of climates near the equator at this time has yet to be explained, but it is certainly reminiscent of the so-called snowball Earth intervals of the Proterozoic.

21.9 Drying of climates

The Permian period was marked by a drying of climates on a global scale. This was at least partly a result of the assembly of all continental regions of the world into the supercontinent Pangaea: broad landlocked areas became orographic deserts. This climatic change caused coal swamps to shrink and seed plants, such as conifers, to expand their ecological role. This floral transition actually began at high latitudes late in the Carboniferous and did not reach the tropics until the early Permian (DiMichele *et al.*, 2001). With the burial rate for wood reduced, weathering (oxidation) of buried carbon exposed by erosion eventually elevated the concentration of atmospheric CO_2 to the degree that the ice age ended (although a few small continental glaciers apparently survived beyond the Sakmarian, the second age of the Permian). Possibly, then, the end of the ice age had a plate tectonic trigger.

21.10 A double mass extinction in the Permian

For many years it appeared that the crisis at the end of the Permian, the largest mass extinction of all time, was a protracted event. A number of patterns indicate that there was actually a separate mass extinction at the end of the penultimate (Guadalupian) age of the Permian (Jin *et al.*, 1994; Stanley and Yang, 1994), about 9 million years before the terminal Permian event. For example, all fusulinids that were relatively large or possessed a honeycomb-like wall structure disappear at the end of the Guadalupian Stage. These forms are just as preservable as other fusulinids, so that the observed disappearances clearly represent actual extinction.

Life on the land experienced two Permian transformations that coincided with those in the marine realm. Therapsids (informally termed mammal-like reptiles, although they were not reptiles) experienced two pulses of extinction, and terrestrial floras simultaneously underwent major changes (Retallack *et al.*, 2006). During the terminal Permian event, the *Glossopteris* flora of the Southern Hemisphere died out, and the coal that it had produced in moist environments ceased to form. Coniferous floras also declined dramatically. *Dicroidium*, a plant genus adapted to warm climates, spread poleward. Terrestrial sediments indicate that climates in many areas became drier, probably in part because warmer temperatures elevated evaporation rates.

Marine deposits in Japan indicate that the ocean also became increasingly stratified during the Permian. A block of Central Pacific seafloor that contains the

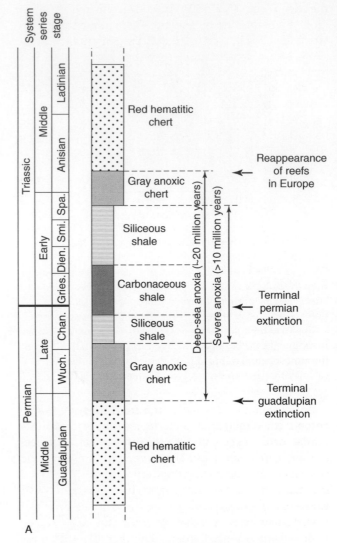

A

Figure 21.4 Obducted rocks in Japan that illustrate the episode of deep-water anoxia that took place in Late Permian time. When anoxia began at the end of Guadalupian (Middle Permian) time, gray chert replaced hematitic (highly oxidized) red chert. An interval of severe anoxia, represented by even darker sediments, began at the time of the terminal Permian extinction. Deposition of hematitic chert resumed in Middle Triassic time, and at this time reefs began to grow again in shallow water (after Isosaki, 1997).

Permo-Triassic boundary was obducted onto the island of Japan during the Jurassic (Isozaki, 1997). The Guadalupian beds of this block consist of cherts formed from radiolarian tests and stained red by ferric oxide (Fig. 21.4). The deep sea throughout most of Guadalupian time was obviously well oxygenated, presumably by cold waters descending at the poles. There appear to have been two phases of mass extinction during the Guadalupian, the first entailed cooling associated with a positive shift for marine $\delta^{13}C$ (Fig. 21.1, f) (Isozaki *et al.*, 2007). At the time of the second Guadalupian extinction, the deep sea sediments turned from red to gray, indicating weaker oxygenation. Then, at the time of the terminal Permian crisis, the these sediments turned black, indicating that the ocean became highly stratified, and respiration by aerobic bacteria soon eliminated free oxygen in the deep sea. This pattern supports the terrestrial evidence that climates twice became warmer on a global scale, with the strongest pulse of warming being associated with the terminal Permian crisis. Polar regions became too warm to ventilate the deep sea with cold downwelling waters. Negative $\delta^{13}C$ and $\delta^{18}O$ excursions for shallow marine carbonates (Fig. 21.1.e and 21.1g) reflect these two steps of global warming.

Currently in favour is the idea that the volcanism that produced the Siberian Traps – the largest continental volcanic outpouring of the Phanerozoic – led to the terminal Permian crisis. Many of the rocks thus produced date to 251 Ma, the precise time of the mass extinction. The lavas erupted through vast coal deposits, and it is thought that large quantities of CO_2 suddenly entered Earth's atmosphere not only from Earth's deep interior but also from the heating and burning of coal, which would also have released methane. It has been suggested that a major volcanic episode in China was similarly the ultimate cause of the Guadalupian crisis. In any event, release of greenhouse gases from coal, perhaps augmented by a submarine release of methane hydrates, may have contributed to the pronounced global shift toward isotopically light carbon that is recorded in both marine and terrestrial sediments at the time of the terminal Permian mass extinction (Berner, 2002).

Three hypothesized kill mechanisms remain viable, at least for some of the terminal Permian losses. Perhaps CO_2 that built up in the stagnant deep sea during the Permian suddenly erupted to the surface, killing marine life even in shallow water (Knoll *et al.*, 2007a). Or Possibly hydrogen sulfide built up in the stagnant deep sea and suddenly erupted (Kump *et al.*, 2005). The simplest idea is that the observed climatic warming – and on the land the attendant increase in aridity – caused the great mass extinction. Perhaps reflecting this agent of extinction was the almost total destruction of low-latitude floras (Rees, 2002), which may have been subjected to lethally high temperatures.

21.11 The absence of recovery in the early Triassic

From beginning to end, the Early Triassic, which encompassed about 6 my, was characterized by highly reduced terrestrial and marine biotic diversity. Aulochthonous deep-sea deposits in Japan show that the deep sea became well oxygenated again precisely at the end of the Early Triassic, indicating the final return of climates

to something resembling their previous state (Isozaki, 1997). Nonetheless, there is evidence that pulses of extinction, rather than a continued inhospitable state, held back biotic recovery. Following the two negative $\delta^{13}C$ and $\delta^{18}O$ excursions associated with the two Permian mass extinctions, three similar excursions occurred early in the Triassic, the last at the end of Early Triassic time (Payne *et al.*, 2004) (Fig. 21.1d). Most marine taxa recover so slowly from crises that their fossil records have as yet failed to reveal pulses of Early Triassic extinction, but the rapidly evolving ammonoids and conodonts clearly experienced severe mass extinctions that were more-or-less coincident with the carbon isotope spikes, followed by rapid recoveries (Fig. 21.5) (Stanley, 2009b). Comprehensive oxygen isotope analyses have not yet been conducted for the Early Triassic, but it is likely that the three mass extinctions were associated with pulses of global warming.

21.12 The terminal Triassic crisis

The Triassic Period ended with one of the largest mass extinctions of the Phanerozoic. The disappearance of nearly all therapsids, which had benefited from an evolutionary head start on the dinosaurs, permitted the latter to rise to dominance on the land (Benton, 1983; Olsen *et al*, 2002). The terrestrial impact of the terminal Triassic event is indicated by a sudden 60% reduction of pollen species accompanied by a 'spore spike,' which probably represented the opportunistic spread of ferns across terrestrial habitats (Fowell *et al.*, 1994).

The timing of Late Triassic extinctions has been controversial, partly because of problematical stratigraphic correlations. Nonetheless, it is evident that marine extinctions occurred over a substantial interval of time and that many marine taxa actually died out during or at the end of the penultimate (Norian) age of the Late Triassic rather than during the final (Rhaetian) stage (review by Tanner *et al.*, 2004).

The Triassic–Jurassic biotic transition on the land is recorded by sediments in Eastern North America that accumulated in rift basins produced in the early stages of the breakup of Pangaea that created the Atlantic Ocean. The Triassic ended very close to 200 Ma. Massive volcanism took place at this time within the Central Atlantic Magmatic Province (CAMP), spanning an interval of perhaps 3 my (Marzoli *et al.*, 1999; Knight *et al.*, 2004; Whiteside *et al.*, 2007) (Fig. 21.6). Furthermore, massive volcanism is indicated by an increase in the osmium-187/osmium-188 ratio in marine mudrocks, accompanied by an increase in the total abundance of osmium and rhenium (Cohen and Coe, 2002). This synchronicity has led to the suggestion that volcanic CO_2 emissions triggered the terminal Triassic mass extinction via greenhouse warming. It remains uncertain whether

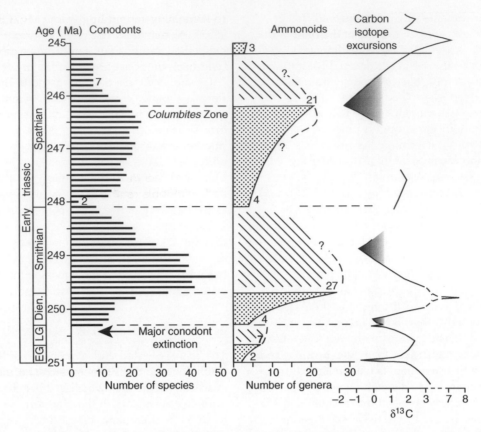

Figure 21.5 Similar patterns of radiation and mass extinction for early Triassic ammonoids and conodonts, with mass extinctions coinciding with negative carbon isotope excursions. Numbers of species and genera are from global compilations (after Stanley, 2009b).

the earlier (Norian) marine extinctions might also have been associated with very early CAMP eruptions or had some independent cause. In any event, stomatal densities provide independent evidence of global warming across the Triassic–Jurassic boundary. They decreased for fossil leaves from Greenland and Sweden, indicating a mean annual temperature increase of 3–4 °C (Fig. 21.7); simultaneously, average leaf width declined, presumably representing an adaptive shift that increased heat loss and thus reduced thermal death of leaves exposed to higher environmental temperatures (McElwain *et al.*, 1999). As would be expected, negative $\delta^{13}C$ and $\delta^{18}O$ excursions in the marine record (Fig. 21.1c) reflect this global warming event. Possibly a sharp drop in atmospheric pO_2 coincident with the rise of CO_2 operated in concert with global warming to cause extinctions of terrestrial vertebrates (Huey and Ward, 2005).

21.13 The rise of atmospheric oxygen since early in Triassic time

Falkowski *et al.* (2005) have documented a general secular increase over the past 205 my for $\delta^{13}C$ in both marine organic carbon and marine carbonates. They

have attributed this trend largely to an increase in the biomass of marine phytoplankton – and, hence, in carbon burial – resulting from the diversification of coccolithophores and diatoms. They have also suggested that an attendant increase in atmospheric pO_2 permitted the late Mesozoic appearance of placental mammals, which require a high level of ambient O_2 to oxygenate embryos.

21.14 The Toarcian anoxic event

A subzone of the Toarcian (the final Jurassic stage), is characterized globally by deep-marine organic-rich black shales; at their base is a negative shift of $\delta^{13}C$, so large (–6‰) that it has been thought necessarily to reflect the release of methane hydrate from continental margins (Hesselbo *et al.*, 2000; Beerling *et al.*, 2002). Although not on the scale of a major biotic crisis, heavy extinction occurred at this time for marine taxa living in basins and on continental shelves at depths greater than perhaps 50 meters (Jenkyns, 1988). Apparently, the oxygen minimum layer rose to this general level, with lethal effects on animal life. The Toarcian anoxic event spanned perhaps only 200 000 years.

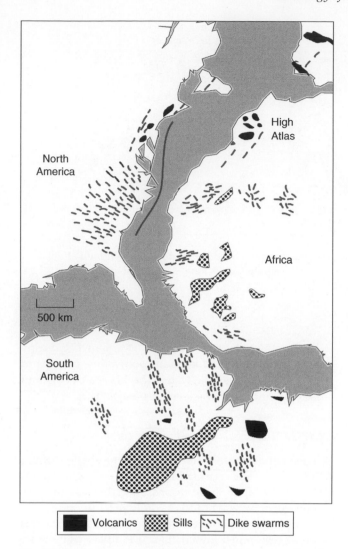

Figure 21.6 The widespread distribution of igneous rocks of the Central Atlantic Magmatic Province, which formed at the end of Triassic time. Continental basalts of this province were even more extensive than shown here because many have been eroded away (after Marzoli *et al.*, 1999).

21.15 Phytoplankton, planktonic foraminifera, and the carbon cycle

The high abundance of C_{29} sterenes in Paleozoic organic matter suggests that green algae played a larger planktonic role in Paleozoic than post-Paleozoic seas (Knoll *et al.*, 2007b). These 'green plastid' forms declined in importance during the Mesozoic, while 'red plastid' phytoplankton (dinoflagellates, coccolithophores, and diatoms) rose to dominance (Falkowski *et al.*, 2004)

The coccolithophores arose in late Triassic time, but only a single species is known to have survived into Jurassic time, and then their diversity rose dramatically until set back by the terminal Cretaceous mass extinction (Bown, 2005). During the Jurassic and Cretaceous, detached coccoliths were a major component of pelagic sediments. Favoured by the low Mg/Ca ratio and high $[Ca^{2+}]$ of seawater, coccolithophores flourished especially in Cretaceous seas, forming the chalk that gave the Cretaceous its name (Stanley *et al.*, 2005).

Planktonic foraminiferans arose in Jurassic time and diversified greatly during the Cretaceous, and they too began to contribute considerable amounts of pelagic carbonate sediment. A consequence of the expansion of calcifying plankton was a huge increase in the conveyor-belting of $CaCO_3$ to subduction zones, where its burial ultimately led to volcanic release of CO_2. This release has significantly supplemented the release of CO_2 by the metamorphism of shallow-water carbonates.

21.16 Diatoms and the silica cycle

Diatoms have become the most successful 'red plastid' phytoplankton group, in part because of their highly efficient system for CO_2 uptake, low quotas for trace metals, and ability to store nutrients in a central vacuole (Knoll *et al.*, 2007b). The diversification and ecological expansion of marine diatoms during the Cretaceous resulted in a reduction of the concentration of silica in the ocean (Maliva *et al.*, 1989).

21.17 Cretaceous climates

There has been much controversy about Cretaceous climates. Cool winter temperatures for the North Slope of Alaska during the Cretaceous are indicated by the presence of dinosaurs, which were endothermic, and the absence of reptiles, which are ectothermic (Clemens and Nelmes, 1993). Nonetheless, terrestrial floras from the North Slope indicate maximum summer temperatures of ~13 °C and winter temperatures no lower than 2–8 °C (Parrish and Spicer, 1988). It is universally agreed that climates were warmest in Cenomanian–Turonian (mid-Cretaceous) time, and floras close to the Arctic Ocean indicate that this polar body of water was at or above the freezing temperature of freshwater not only during the Turonian but also during Coniacean (late Cretaceous) time (Herman and Spicer, 1996).

It now appears that the global latitudinal temperature gradient during much of Cretaceous time was fairly pronounced, and yet the mean global temperature was quite high during mid-Cretaceous (Cenomanian–Turonian) time. Oxygen isotopes of marine fish teeth and pristine (diagenetically unaltered) planktonic foraminiferans indicate, respectively, for the mid-Cretaceous shallow seas temperatures of ~32 °C and 28 °C in the tropics (compared to 24–28 °C today) and ~25 °C and 20 °C at a paleolatitude of 40° (Pucéat *et al.*, 2007). Tetraether lipids of marine Crenarchaeota (prokaryotic plankton), which change their chemical composition with temperature and are resistant to

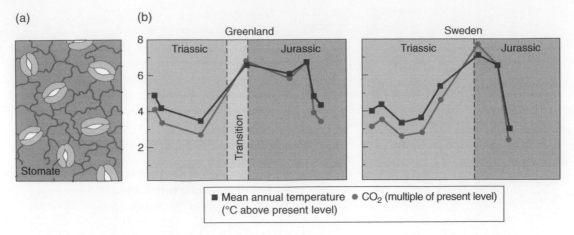

Figure 21.7 Evidence from stomates of increases in atmospheric CO_2 and mean annual temperature on Earth at the end of the Triassic. (a) Illustration of stomatal cells in a leaf. (b) Increases in the proportion of stomata in fossil ginkgo and cycad leaves, indicating a rise in atmospheric CO_2 levels (after McElwain *et al.*, 1999).

diagenesis, indicate temperatures of 32–36°C for the tropical Atlantic during the Cenomanian–Turonian compared to 27–32°C for the preceding, Albian, age (Schouten *et al.*, 2003).

21.17.1 Mid-Cretaceous anoxia

Massive eruptions of submarine lavas in the Pacific Ocean began slightly before 125 Ma (late Barremian time) and continued until ~80 Ma (mid-Campanian time). These eruptions not only reduced the Mg/Ca ratio of seawater, thus favouring the calcification of coccolithophores and other taxa with calcitic, as opposed to aragonitic, skeletons, but they also sent a substantial amount of CO_2 into the atmosphere, accentuating greenhouse warming. The Cenomanian–Turonian episode of extreme global warming (~100–89 Ma) was in the middle of this interval. High rates of seafloor production elevated sea level, and sluggish ocean circulation (an absence of descending cold, oxygenating polar waters) led to expansion of the oxygen minimum zone. Black muds were deposited extensively even in relatively deep waters of epicontinental seas (Larson, 1991) (Fig. 21.8).

The relative abundance in black shales of 2-methylhopanoids, which are membrane lipids found in cyanobacteria and some other bacteria, indicate that during major oceanic anoxic events of the Aptian and Cenomanian, prokaryotes dominated many oceanic phytoplankton assemblages (Kuypers *et al.*, 2004). Low $\delta^{15}N$ values for the organic matter in these black shales apparently reflects a dominance of cyanobacterial nitrogen fixation (air is characterized by light N). In contrast, because the N/P ratio in the ocean was low and upwelling was weak, eukaryotic phytoplankton were unable to flourish.

Rates of extinction were elevated somewhat during the Cenomanian–Turonian transition, a time of upward expansion of dysaerobic waters (Leckie *et al.*, 2002), but losses did not rise to the level of a major crisis.

21.17.2 The puzzle of reef-building corals

A substantial contribution of corals to shallow-water reefs in the tropics during Jurassic and early Cretaceous time is puzzling for two reasons. First, this was an interval of calcite seas, yet today corals produce aragonite. Second, the concentration of atmospheric CO_2 for this interval was much higher than it is today (review by Royer, 2003), and experiments have shown elevated CO_2 to have a negative effect on the calcification of many modern coral species (Marubini and Thake, 1999; Renegar and Riegl, 2005). Nonetheless, calculations show that the high concentration of calcium in late Mesozoic seawater may have compensated for the elevated CO_2, making the saturation state of $CaCO_3$ nearly the same as today (Stanley *et al.*, 2005). Also, experiments have shown that three species of modern corals produce calcium carbonate consisting of about 30% calcite in Cretaceous seawater (Ries *et al.*, 2006); production of such skeletal material in the late Mesozoic would have enhanced coral skeletal growth. It is also possible that late Mesozoic corals differed physiologically from modern corals.

21.17.3 The terminal Cretaceous extraterrestrial event

The biotic crisis that brought the Mesozoic Era to an end was only the fifth most destructive of the Phanerozoic for marine life, but it has always been granted special attention because of having eliminated

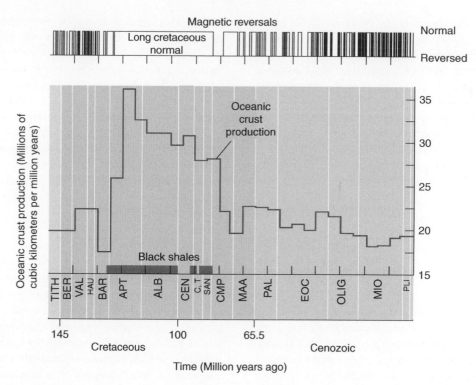

Figure 21.8 Black muds that became black shales accumulated in moderately deep waters in many regions during mid-Cretaceous time, when there was also a high rate of oceanic crust production and an absence of magnetic reversals (after Larson, 1991).

the dinosaurs. It also resulted in the sudden extinction of a large percentage of gymnosperm and angiosperm land plants from regions as far apart as North America (Johnson and Hickey, 1990) and Japan (Saito *et al.*, 1986), and it resulted in the immediate ecological expansion of ferns, as indicated by an abrupt decline of pollen and increase of spores in terrestrial sediments (Tschudy and Tschudy, 1986).

In 1980 the geologist Walter Alvarez, along with his father, Luis (a Nobel Laureate in Physics), and Helen Michel, announced the discovery of an iridium anomaly, a high concentration of the heavy metal iridium, at the level of the terminal Cretaceous crisis. They recognized this as an extraterrestrial signal because iridium is very rare in Earth materials and relatively more abundant in meteorites (Alvarez *et al.*, 1980). Also soon discovered at the level of the extinction were shocked mineral grains, which are products of extraterrestrial impacts on Earth (Bohor *et al.*, 1984); microtectites, which are glassy spheroidal grains produced by the rapid cooling of liquid droplets of materials blasted into the atmosphere by an impact (Montonari *et al.*, 1983), and minute diamonds, which can be produced only at extremely high pressures (Carlisle and Braman, 1991).

The ultimate confirmation of the extraterrestrial event – presumably an asteroid impact – at the end of the Cretaceous stands as a major triumph for geology. This was the discovery that the Chicxulub crater, which

borders Mexico's Yucatan Peninsula and is imaged from geophysical gravity data, formed at exactly the time of the terminal Cretaceous mass extinction: igneous rocks in the crater produced by the heat of the impact date precisely to the Cretaceous–Paleocene transition (Swisher *et al.*, 1992).

A global negative excursion of $\delta^{13}C$ in sediments at the Cretaceous–Paleocene boundary has been taken to indicate a collapse of phytoplankton productivity, and hence a sharp reduction of light carbon burial in the deep sea. Furthermore, carbon isotopic ratios ceased to display the normal gradient from relatively high values for planktonic taxa to relatively low values for deep-sea benthos (D'Hondt *et al.*, 1998) – a gradient reflecting the preferential removal of $\delta^{12}C$ from the photic zone by phytoplankton and transmission of isotopically light carbon to the deep sea: here, too, is evidence of decreased productivity by phytoplankton. Isotopic data from deep-sea foraminifera indicate that recovery of biomass by phytoplankton required about 3 my.

The immediate agent or agents of death in the terminal Cretaceous crisis remain under debate. The area where the asteroid struck contains large volumes of sulfate evaporites and limestones, which should have released large amounts of SO_2 and CO_2 at the time of impact. These compounds, along with production of nitric acid by heating of N_2 and O_2 in the atmosphere, would have adversely affected life by producing strongly acid rain (D'Hondt *et al.*, 1994). In addition, an 'impact winter' may immediately have developed, as particles blasted into the atmosphere screened out the sun's rays (Pope *et al.*, 1994). On the other hand, as these particles descended to Earth, friction in the atmosphere would have generated enormous heat (Melosh *et al.*, 1990). Support for dramatic, sudden warming at the end of the Cretaceous comes from stomatal densities for leaves, which indicate global warming by ~7.5 °C within just 10 000 years (Beerling *et al.*, 2002b).

21.17.5 The ascendancy of mammals and angiosperms: beneficiaries of the terminal Cretaceous crisis

Clearly the dinosaurs' extinction opened the way for the diversification of mammals. Mammals remained relatively small in body size even in late Cretaceous time, about 150 my after their origin. Dinosaurs had the jump on mammals, however, having originated earlier in Triassic time. Although the traditional view has been that dinosaurs suppressed Mesozoic mammals via competition, it is much more likely that the suppression was via predation. Anyone who has seen the movie '*Jurassic Park*,' with its reconstruction of the relatively small but vicious predatory dinosaur *Velociraptor*, will appreciate the victimization that mammals faced throughout Mesozoic time. Supporting the idea the predation held mammals back is the evidence that most Mesozoic mammals had refugial life habits, many being small burrowers or climbers or being active nocturnally.

More recently has it become evident that, following the terminal Cretaceous mass extinction, terrestrial vegetation underwent a change paralleling that of terrestrial quadrupeds. Although angiosperms (flowering plants, including grasses and hardwood trees) experienced considerable taxonomic diversification following their mid-Cretaceous origin, their earliest representatives were largely restricted to unstable habitats along rivers (Doyle and Hickey, 1976). A flora well preserved over a large area in central Wyoming by a sudden eruption of volcanic ash suggests that gymnosperms and spore plants dominated many undisturbed habitats even in latest Cretaceous time (Wing *et al.*, 1993). It was not until the Paleocene that angiosperms first came to dominate most terrestrial landscapes. Thus, the angiosperms, like the mammals, were serendipitous beneficiaries of the meteorite impact that brought the Mesozoic Era to a close.

21.18 The sudden Paleocene–Eocene climatic shift

Isotopic evidence from foraminiferans points to a dramatic change in the thermal structure of the ocean at the very end of the Paleocene Epoch (Kennett and Stott, 1991). Throughout most of Paleocene time, oxygen isotope ratios in foraminiferan skeletons were heavier for deep-sea species than for shallow-water species, indicating colder temperatures in the deep sea. At the very end of Paleocene time, a dramatic shift occurred, indicating that even close to Antarctica, the deep sea suddenly warmed to temperatures close to those of surface waters; cool, dense waters were no longer descending to the deep sea, and deep-sea foraminiferans suffered mass extinction. It appears that at this time Earth experienced a sudden pulse of global warming that lasted less than 3000 years. At the same time $\delta^{13}C$ in soil organic matter and skeletons of foraminiferans at all depths in the ocean experienced a sudden negative shift (Magioncalda *et al.*, 2004; Wing *et al.*, 2005) (Fig. 21.1b). This shift was so abrupt that some workers have attributed it at least in part to the release of isotopically light carbon from methane hydrates along continental shelves (Dickens *et al.*, 1995; Kennett *et al.*, 2003). Methane is a powerful greenhouse gas, and although in about a decade it almost entirely oxidizes to form CO_2, a weaker greenhouse gas, if released over thousands of years at the end of the Paleocene it could have substantially enhanced greenhouse warming caused by elevation of atmospheric pCO_2. Also contributing to the negative $\delta^{13}C$ shift would have been a positive feedback: the increased rate of bacterial respiration along continental margins (Stanley, 2010).

The magnesium content of calcite in planktonic foraminiferans, which increases with temperature, together with oxygen isotopes of this calcite, indicates a sudden temperature increase of 4–5 °C in the tropical Pacific Ocean at the end of the Paleocene (Zachos *et al.*, 2003). Acidification of the ocean also occurred, with the calcite compensation depth (the depth at which solution of calcite begins) shoaling by more than 2 km; isotopic evidence indicates that the thermal structure of the ocean then recovered gradually during less than 50 000 years (Zachos *et al.*, 2005).

Latest Paleocene floras of unique taxonomic composition have been discovered in Wyoming, and analysis of their leaf morphologies has suggested that mean annual temperature in this region increased by about 5 °C in less than 10 000 years (Wing *et al.*, 2005). In this same region, mammalian faunas underwent major changes that entailed the initial arrival from the Old World of artiodactyls and perissodactyls, the two major groups of hoofed herbivores in the modern world (Clyde and Gingerich, 1998).

21.18.1 A warm climate in the Eocene, but why?

Relatively warm climates persisted into the Eocene, well after the terminal Paleocene warming subsided. It has long been recognized that palm trees grew in Wyoming during the Eocene and that alligators were able to exist within the Arctic Circle. What remains to be determined is to what extent greenhouse warming by high levels of atmospheric CO_2 was responsible for the persistence of a remarkably warm global climate after the initial pulse of global warming.

Analyses of stomatal densities on terrestrial ginkgo leaves suggest that atmospheric CO_2 levels were only slightly above present levels during the Eocene (Royer *et al.*, 2001). On the other hand, analyses of alkenones produced by planktonic coccolithophores suggest that atmospheric CO_2 levels were about four times their modern level (Pagani *et al.*, 2005). Alkenones are carbon compounds produced by coccolithophores that are refractory to diagenesis, and coccolithophores fractionate carbon isotopes of CO_2 used in photosynthesis in a manner that varies with the ambient concentration of CO_2, which reflects the atmospheric concentration of CO_2. The results of the alkenone analysis are likely to be valid because they accord with independent evidence of very warm Eocene climates.

The remarkably warm temperatures at high latitudes during the Eocene appear to require a special explanation. Extremely low $\delta^{18}O$ values from *Metasequoia* wood preserved at Axel Heiberg Island, inside the Arctic Circle, apparently reflect transport of moisture northward from the Pacific Coast of Mexico, and progressive fractionation via loss of ^{18}O through precipitation; this northward flow of moist air would have transported much heat (Jahren and Sternberg, 2002).

21.18.2 The origin of the modern climatic regime

Climates cooled on a global scale at the end of the Eocene, as reflected in positive shifts of $\delta^{13}C$ and $\delta^{18}O$ in the ocean (Fig. 21.1a). In many regions climates also became drier, because cooler oceans contributed less water to the atmosphere through evaporation. Terrestrial floras first indicated this climatic shift. There is a linear relationship between mean annual temperature and the percentage of species in angiosperm floras that have smooth-margined (as opposed to jagged-margined or lobed) leaves. Although the slope of the leaf-margin curve may have varied somewhat through time, any substantial change in the percentage of smooth-margined leaves in fossil floras provides a clear indication of a change in mean annual temperature. A major decline in this percentage took place in North America from the Gulf Coast to Alaska at the end of the Eocene (Wolfe, 1971). Seeds from the London Clay of England indicate that slightly earlier in the Eocene a similar transition occurred from a flora resembling that of modern Malaysia to a temperate flora (Collinson *et al.*, 1981).

The Eocene–Oligocene transition ushered in the modern world, in which as climates became drier in many regions, grasslands expanded at the expense of forests. (Trees require a consistent supply of water, whereas grass taxa typically tolerate seasonal drought.)

A mammalian fauna of the Mongolian Plateau records the replacement of forested habitats by open habitats during the Eocene–Oligocene transition (Meng and McKenna, 1998). Many medium-sized hoofed animals, which are most common in forested habitats, disappeared, as did tree-climbing taxa such as primates. At the same time, species of rodents, rabbits, and open-country taxa with teeth adapted for feeding on harsh grasses appeared, along with large herbivores having the stamina to outrun predators in open terrain.

In the marine realm, a second-order mass extinction occurred in Late Eocene time, with a preferential loss of warm-adapted molluscan taxa (Hansen, 1987; Hickman, 2003). Oxygen isotopes of mollusks and fish otoliths (ear bones) from coastal plain deposits of the Mississippi Embayment indicate a decline from tropical temperatures between early Eocene and early Oligocene time, with a winter reduction of ~5°C and a summer reduction of ~3°C (Kobashi *et al.*, 2001, 2004). In other words, the climate change entailed an increase in seasonality. Similarly, from the late Eocene into the Oligocene, radiolarians in the equatorial Pacific experienced numerous extinctions of purely tropical species and an increase in cosmopolitan taxa with relatively broad thermal adaptations (Funakawa *et al.*, 2006). Planktonic foraminiferans underwent stepwise extinction during the same interval (Keller, 1983).

21.19 The cause of the Eocene–Oligocene climatic shift

Traditionally the Eocene–Oligocene climatic shift has been attributed to the formation of the Circumantarctic Current (Kennett *et al.*, 1975). This current traps water that consequently becomes very cold. In the present ocean, the relatively high density of this cold water, enhanced by an elevation of salinity through sea ice formation, causes downward convection, producing the cold bottom layer of the ocean by spreading to the far north in both the Atlantic and Pacific. The Circumantarctic Current formed when Antarctica became isolated over the South Pole as South America and Australia broke away from it. Thus, the Drake Passage and Tasmanian Gateway formed – and with them the modern polar gyre came into being. Upward mixing of cold, deep waters that formed in the vicinity of Antarctica would have cooled climates throughout the globe.

Coincidentally, at ~35 Ma (close to the Eocene–Oligocene transition) the tectonic deepening of the Greenland-Iceland-Faeroes Ridge permitted downward cold-water convection in the North Atlantic (Davies

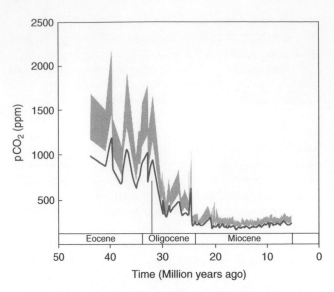

Figure 21.9 Estimates of the concentration of CO_2 in Earth's atmosphere from Eocene through Miocene time, based on the carbon isotopic composition of alkenones in calcareous nanoplankton. The top of the stippled band represents the maximum estimate, the bottom of this band represents an intermediate estimate, and the dashed line represents a minimum estimate. This analysis indicates that the level of atmospheric CO_2 was very high in the Eocene and earliest Oligocene and began to drop precipitously about 32 million years ago (vertical red line), some 2 million years after glaciers expanded in Antarctica and climates changed throughout the world (after Pagani *et al.*, 2005).

et al., 2001). Today, the water descending in the North Atlantic spreads throughout the ocean above dense Antarctic bottom water. A portion of the present Antarctic ice sheet had formed and was producing glacial marine deposits by the start of the Oligocene (Ivany *et al.*, 2006). In the north, ice-rafted debris began to reach the Norwegian-Greenland Sea by 38 Ma, meaning that at least some isolated glaciers had formed by this time on Greenland (Eldrett *et al.*, 2007).

A tectonic evaluation of the opening of the Drake Passage, based on seismology, indicates that the passage began to form during middle Eocene time (Eagles *et al.*, 2006). In addition, a variety of evidence indicates that the Tasmanian Gateway began to form slightly before the end of Eocene time (~Ma) (Stickley *et al.*, 2004). The implication is that the Circumantarctic Current arose during the latter part of the Eocene. Neodymium isotopes in fossil fish teeth support this timing (Scher and Martin, 2006). The $^{143}Ne/^{144}Ne$ ratio has long been higher in the Pacific than in the Atlantic, reflecting circumpacific volcanism, and yet with mixing between the two oceans today there is only a small difference between them in this ratio. Early in the Cenozoic, the difference was much larger, but during the middle Eocene, apparently in response to the formation of an

incipient Circumantarctic Current, this ratio diminished dramatically.

Alternatively, it has been suggested that a decrease in greenhouse warming, via lowering of atmospheric pCO_2, produced the Eocene–Oligocene climatic change. However, alkenones in coccolithophores appear to indicate that although pCO_2 dropped in mid-Cenozoic time, it did not do so until ~32 Ma, some 2 my after the global climatic change occurred (Pagani *et al.*, 2005) (Fig. 21.9). Possibly the timing of this pCO_2 decline will be revised in the future.

21.20   The re-expansion of reefs during Oligocene time

Despite heavy losses in the terminal Cretaceous mass extinction, reef-building scleractinian corals retained substantial taxonomic diversity at the start of the Cenozoic. They nonetheless produced very few reefs of any size during the Paleocene or Eocene. Something prevented scleractinians from flourishing until Oligocene time. Three possibilities are evident:

1 The chemistry of the oceans shifted from calcite to aragonite seas close to the Eocene-Oligocene transition, favouring calcification by scleractinians (Stanley and Hardie, 1998).
2 Atmospheric CO_2 declined markedly during early Oligocene time (Pagani *et al.*, 2005), and this favored the precipitation of calcium carbonate in the ocean.
3 Possibly until Oligocene time Cenozoic corals lacked the symbiotic algae that today promote their calcification. The molecular clock indicates that the symbiotic algae of modern reef-building corals originated during the Eocene (Pochon *et al.*, 2006). The implication is that the terminal Cretaceous mass extinction eliminated more ancient types of symbiotic algae in reef-building corals, and corals were not recolonized by algae until at least Eocene time.

21.21   Drier climates and cascading evolutionary radiations on the land

The fossil record of phytoliths, silica bodies secreted by plants, indicates that the modern taxa of grasses adapted to open habitats diversified in late Oligocene and early Miocene time (Strömberg, 2004). The expansion of open habitats that began at this time produced cascading evolutionary radiations of plant and animal taxa adapted to these habitats and led to the high diversity of these taxa in the present world (Stanley, 1990) (Fig. 21.10). Not only have grasses diversified since this time, but also weeds (the family Compositae), which opportunistically occupy open spaces in grasslands. In addition, the Muridae (Old World rats and mice) and songbirds, both

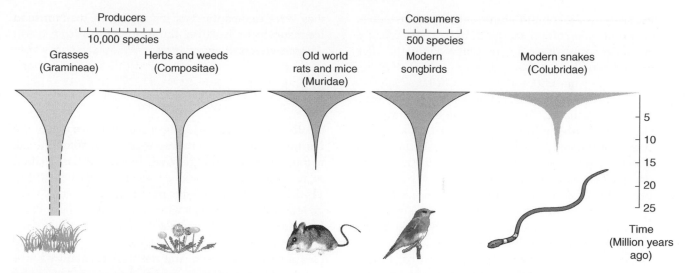

Figure 21.10 Cascading evolutionary radiations of terrestrial taxa during the past 25 million years. Grasses and weeds diversified dramatically, and their prolific production of seeds was partly responsible for a great expansion of rats, mice, and songbirds. Colubrid snakes, which feed on rats and mice and the eggs and chicks of songbirds also experienced a remarkable radiaton (after Stanley, 1990).

of which contain many species that feed on seeds of grasses and weeds, began spectacular evolutionary radiations. Finally, this was the time when the radiation of the snake family Colubridae began. Snakes can slither along branches to consume songbirds' eggs and chicks and can make their way down small rodent holes. The family Colubridae, which contains most species of modern snakes that are not constrictors and includes all venomous forms, arose and began a spectacular evolutionary radiation in the Miocene.

21.21.1 Climate change, extinction, and the spread of C4 grasses

There is evidence of widespread aridification related to cooling at ~7–6 Ma (close to the end of the Miocene). The global volume of glacial ice increased, causing the Messinian sea level fall of at least 30 m (Aharon *et al.*, 1993). At the same time, the oceans cooled at high and middle latitudes in both hemispheres (Poore and Berggren, 1975) and grasslands replaced woodlands in many regions (Webb, 1977; Bernor *et al.*, 1996; Gentry and Heizmann, 1996). The largest extinction event of the past 30 million years for North American mammals occurred at this time, largely in response to the spread of grasslands (Webb, 1984).

Carbon isotope ratios of the teeth of herbivores reflect the isotope ratios of the food that they eat, although fractionation occurs as the food is assimilated. A marked increase in carbon $\delta^{13}C$ occurred on a global scale in mammal teeth preserved in sediments ranging from ~7 to 6 Ma (Cerling *et al.*, 1993) (Fig. 21.11). This change reflected the worldwide spread of C4

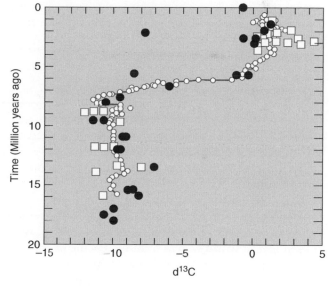

Figure 21.11 Major shifts in carbon isotopes between 7 million and 6 million years ago, indicating the spread of C_4 grasses. The plotted values are carbon isotope ratios from ancient soils and mammal teeth from Pakistan and North America (after Cerling *et al.*, 1993).

grasses, a group that utilizes a different photosynthetic pathway than C3 grasses and fractionates carbon isotopes in such a way that their tissues contain a higher percentage of ^{13}C. Warm, seasonally dry savannah habitats favour C4 grasses. In contrast, C3 grasses require perennial moisture in the tropics or a cool, moist growing season in nontropical regions; Mediterranean climates and northern temperate climates provide the latter conditions.

Because grasses contain abundant phytoliths (silica bodies) that wear down the teeth of grazers, grazing mammals generally have molars that are initially taller than those of mammals that browse on softer leafy vegetation. During the Miocene, hypsodont horse species (ones with tall molars) and then ones classified as 'very hypsodont' increased in numbers, while mesodont species (those with medium-tall molars) declined; after extinction of the last North American mesodont forms at 11–12 Ma, only species adapted for grazing remained (Hulbert, 1993). This net trend developed as a result of the expansion of grasslands during the Miocene (Webb, 1984). Then there was an abrupt shift for American horses toward very hypsodont teeth at 7–6 Ma at the time when C4 grasses proliferated. Because C4 grasses contain on average about five times as many phytoliths per volume of tissue as C3 grasses, it seems evident that very tall teeth were at a premium for horses that fed on C4 grasses (Stanley, 2009a, p. 458). Horses employ inefficient hind-gut digestion and are therefore required to feed for more hours every day than other large herbivores. Presumably, North American horse species that lacked very hypsodont molars experienced shortened lifespans as C4 grasses expanded, and their overall birth rates declined to levels that could not sustain populations.

21.21.2   The initiation of the modern ice age

The modern ice age of the Northern Hemisphere, during which we still live, cannot be attributed to greenhouse cooling, because all indications are that there was no decline in atmospheric pCO_2 during the onset of the ice age, between 3.5 and 3.0 my ago. At this time plate tectonic movements emplaced the Isthmus of Panama between North and South America. The Arctic region today is cold because the Arctic Ocean is isolated, with little inflow of warm waters from the Pacific or Atlantic oceans. The most important factor here is that northward-flowing Atlantic waters today sink north of Iceland, depriving the Arctic Ocean of their warmth. These waters descend not only because they become cold, but also because they are unusually saline as a result of high evaporation rates farther south, in the trade wind belt.

Before the emplacement of the Isthmus of Panama, tropical Atlantic waters would have flowed into the Pacific and compensatory flow would have moved water in the opposite direction. Therefore, the Atlantic would have been less saline than it is today. It follows that Atlantic waters would have sunk farther north than they do today, i.e. within the Arctic Ocean. With the inflow of these warm waters, the Arctic would have been much warmer then than it is today. If these inferences are correct, the formation of the Isthmus of Panama

may have caused the modern ice age of the Northern Hemisphere by isolating the Arctic Ocean from warm Atlantic waters (Stanley, 1995).

21.21.3   Biotic consequences of the modern ice age of the Northern Hemisphere

Cooling of the ocean during the modern ice age of the Northern Hemisphere caused extinctions of marine life in the North Atlantic region, where temperature declines were concentrated (Raffi *et al.*, 1985; Stanley, 1986). Elsewhere, extinctions were minor, partly because of the geographic pattern of cooling and partly because Earth had already moved into a glacial mode at the end of the Eocene, so that many biotas were already adapted to cool mean annual temperatures and pronounced seasonality (Stanley, 1986).

Because surface waters of the ocean cooled and yielded less moisture to the atmosphere late in the Pliocene, climates became drier in many regions. In Africa, the result was a contraction of forests and a diversification of antelopes adapted to open habitats (Vrba, 1985). In portions of South America, rainforests also contracted (Hooghiemstra and van der Hammen, 1998).

Human evolution was strongly impacted by vegetational changes in Africa (Stanley, 1992). *Australopithecus*, which was ancestral to the human genus *Homo*, undoubtedly spent considerable time on the ground, but it also possessed numerous adaptations for climbing trees: upward-directed shoulder sockets and also long arms, long fingers, and long toes with the ability to grasp. For animals having a brain size little above that of a chimpanzee and a capacity for only slow movement on the ground, these adaptations were necessary for tree-climbing to avoid vicious African predators. Woodlands contracted in Africa at about 2.5 Ma, when the northern ice age intensified, and *Australopithecus*, which was dependent on woodlands, died out.

At about the time when *Australopithecus* died out, *Homo*, the modern human genus, came into being. The large brain of *Homo* results from delayed development. Monkeys, apes, and humans in the womb grow brains that are about 10% of an embryo's body size. For monkeys and apes, brain size relative to body size falls back after birth, but for humans the 10% relationship persists for about a year after birth. We humans experience a general delay in development that endows us with most of our adult brain size (at an age of about 1 year, we assume the apes' slower postnatal rate of brain growth). Delayed development also saddles humans with relatively immature, helpless infants. Adult and neonate body and brain sizes indicated by fossils show that early *Homo* possessed our delayed development but *Australopithecus* did not (Stanley, 1992). In contrast to

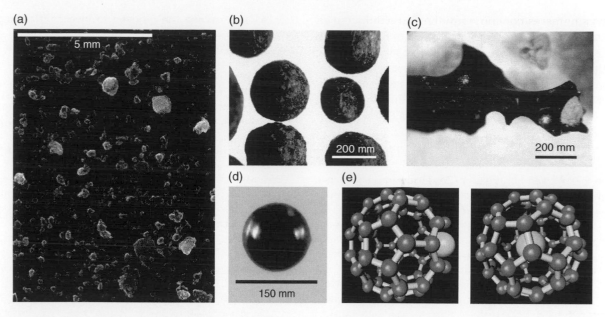

Figure 21.12 Materials found in the YDB layer at many Clovis sites. (a) Soot particles. (b) and (c) Exterior and cross-sectional views of low-density carbon grains. (d) Magnetic microspherule. (e) A model of a fullerene, which is a cage-like molecule formed of carbon atoms; a helium atom is shown entering the fullerene molecule on the left, and one is trapped in the fullerene molecule on the right (after Firestone *et al.*, 2007).

Australopithecus females, early females of our genus could not have climbed trees habitually because they could not have climbed with one arm while carrying a helpless infant in the other. Therefore, *Homo* must have evolved from a population of *Australopithecus* confined to the ground, perhaps after nearly all other populations died out as forests shrank in Africa. We can be thankful that this population happened to survive by evolving into our genus, whose intelligence was undoubtedly essential to its survival amidst many large predatory mammals.

21.21.4 The younger Dryas: evidence for an exraterrestrial impact

About 12 900 years ago, as the Northern Hemisphere was emerging from its most recent glacial maximum, glaciers suddenly re-expanded; they shrank back again, permanently, about 1300 years later. The interval before the glaciers re-expanded is known as the Younger Dryas. Numerous species of large mammals died out in North America at the start of the Younger Dryas, and the Clovis human culture also disappeared. There is now evidence that a comet may have struck North America, producing the Younger Dryas and associated events (Firestone *et al.*, 2007). At numerous sites across North America, there is a black layer (usually less than 3 cm thick) precisely at the level of the mammalian extinctions and termination of the Clovis culture. Containing charcoal, soot, glassy carbon, and carbon spherules, this layer represents widespread burning (Fig. 21.11). At the Murray Hills site in Arizona, the black layer fills mammoth footprints and is draped over a Clovis fireplace and nearly complete mammoth skeleton. The black layer also contains helium, trapped within cagelike fullerene molecules, that has an extraterrestrial isotopic signature (Fig. 21.12e). Also present are magnetite particles. A high concentration of iridium has been reported at some sites, but some researchers (e.g. Paquay *et al.*, 2009) have failed to duplicate the results. These various materials may have come from the core inside the icy exterior of a comet. The compositions of only a few comets have been analysed. They vary considerably in composition and differ from chondritic meteorites, for example in lacking nickel (Lisse *et al.*, 2007). The most compelling evidence for an extraterrestrial impact is supplied by nanodiamonds that are of a type considered to result only from impact events and that occur in vast numbers at many sites where additional evidence is found (Kennett *et al.*, 2009).

A comet has been favoured as the agent of this crisis because shocked minerals and microtectites – signatures of a meteorite impact – are absent. To start numerous fires, a comet would have had to fragment before landing. This would not have been unlikely, because comets are weak bodies, containing about 75% pore space. A large peak in the abundance of ammonia and nitrate in a Greenland ice core dates to 12 900 years ago (dated by counting ice varves), and their protracted decline indicates the occurrence of widespread wildfires for 50 years. How an impact event may have produced the Younger Dryas remains to be more fully explored.

Through the extinction only 12 900 years ago of numerous North American mammals – among them three elephant species, a sabertooth cat, a fast-running bear that stood six feet tall at the shoulder, and a beaver the size of a black bear – we modern humans have inherited a highly impoverished land mammal fauna. This situation should add value to the species that remain.

References

Aharon P, Goldstein SL, Wheeler CW, Jacobson G (1993) Sea-level events in the South Pacific linked with the Messinian salinity crisis. *Geology* **21**, 771–775.

Aitken JD (1967) Classification and environmental significance of cryptalgal limestones and dolomites, with illustrations from the Cambrian and Ordovician of southwestern Alberta. *Journal of Sedimentary Petrology* **37**, 1163–1178.

Alvarez LW, Alvarez W, Asaro F, Michel HV (1980) Extraterrestrial cause for the Cretaceous-Tertiary extinction. *Science, USA* **208**, 1095–1108.

Beerling DJ, Berner RA (2005) Feedbacks and the coevolution of plants and atmospheric CO_2. *Proceedings of the National Academy of Sciences* **102**, 1302–1305.

Beerling DJ, Lomas MR, Groecke DR (2002a) On the nature of methane gas-hydrate dissociation during the Toarcian and Aptian ocean anoxic events. *American Journal of Science* **302**, 28–49.

Beerling DJ, Lomax BH, Royer DL, Upchurch GR, Kump LR (2002b) An atmospheric pCO_2 reconstruction across the Cretaceous-Tertiary boundary. *Proceedings of the National Academy of Sciences, USA* **99**, 7836–7840.

Benton MJ (1983) Dinosaur success in the Triassic; a noncompetitive ecological model. *The Quarterly Review of Biology* **58**, 29–55.

Berner RA (1987) Models for carbon and sulfur cycles and atmospheric oxygen: Application to Paleozoic geologic history. *American Journal of Science* **287**, 177–196.

Berner RA (2002) Examination of hypotheses for the Permo-Triassic boundary extinction by carbon cycle modeling. *Proceedings of the National Academy of Sciences, USA* **99**, 4172–4177.

Berner RA (2006) GEOCARBSULF: A combined model for Phanerozoic atmospheric 02 and C02. *Geochimica et Cosmochimica Acta* **70**, 5653–5664.

Bernor RL, Fahlbusch V, Andrews P, *et al.* (1996) The evolution of western Eurasian Neogene mammal faunas: A chronologic, systematic, biogeographic, and paleoenvironmental synthesis. In: *The Evolution of Western Eurasian Neogene Mammal Faunas* (eds Bernor RL, Fahlbusch V, Mittman H-W). Columbia University Press, New York, pp. 449–459.

Berry WBN, Quinby-Hunt MS, Wilde P (1995) Impact of Late Ordovician glaciation-deglaciation on marine life. In: *Effects of Past Global Change on Life. Studies in Geophysics*. National Academy Press, Washington, DC, pp. 34–46.

Bohor BF, Foord EE, Modreski PJ, Triplehorn DM (1984) Mineralogic evidence for an impact event at the Cretaceous-Tertiary boundary. *Science* **224**, 867–869.

Bond DPR, Wignall PB, Wang W, *et al.* (2010) The mid-Capitanian (Middle Permian) mass extinction and carbon isotope record of South China. *Palaeogeography, Palaeoclimatology, Palaeoecology* **292**: 282–294.

Bown PR (2005) Selective calcareous nannoplankton survivorship at the Cretaceous-Tertiary boundary. *Geology* **33**, 653–656.

Brenchley PJ, Marshall JD, Carden GAF, *et al.* (1994) Bathymetric and isotopic evidence for a short-lived Late Ordovician glaciation in a greenhouse period. *Geology* **22**, 295–298.

Brennan S, Lowenstein TC, Horita J (2004) Seawater chemistry and the advent of biocalcification. *Geology*, **32** 473–476.

Brezinski DK, Cecil CB, Skema VW, Stamm R (2008) Late Devonian glacial deposits from the eastern United States signal an end of the mid-Paleozoic warm period. *Palaeogeography, Palaeoclimatology, Palaeoecology* **268**, 143–151.

Buggisch W, Joachimski MM (2006) Carbon isotope stratigraphy of the Devonian of central and southern Europe. *Palaeogeography, Palaeoclimatology, Palaeoecology* **240**, 68–88.

Carlisle DB, Braman DR (1991) Nanometre-size diamonds in the Cretaceous/Tertiary boundary clay of Alberta. *Nature* **352**, 708–709.

Cerling TE, Wang Y, Quade J (1993) Expansion of C4 ecosystems as an indicator of global ecological change in the late Miocene. *Nature* **361**, 344–345.

Clemens WA, Nelms LG (1993) Paleoecological implications of Alaskan terrestrial vertebrate fauna in latest Cretaceous time at high paleolatitudes. *Geology* **21**, 503–506.

Clyde WC, Gingerich PD (1998) Mammalian community response to the latest Paleocene thermal maximum; an isotaphonomic study in the northern Bighorn Basin, Wyoming. *Geology* **26**, 1011–1014.

Cohen AS, Coe AL (2002) New geochemical evidence for the onset of volcanism in the Central Atlantic magmatic province and environmental change at the Triassic–Jurassic boundary. *Geology* **30**, 267–270.

Collinson ME, Fowler K, Boulter MC (1981) Floristic changes indicate a cooling climate in the Eocene of southern England. *Nature* **291**, 315–317.

Copper P (2002) Silurian and Devonian reefs: 80 million years of global greenhouse between two ice ages. *Society for Sedimentary Geology Special Publication* **72**, 181–238.

Davies R, Cartwright J, Pike J, Line C (2001) Early Oligocene initiation of North Atlantic Deep Water formation. *Nature* **410**, 917–920.

D'Hondt S, Pilson MEQ, Sigurdsson H, Hanson AK, Jr., Carey S (1994) Surface-water acidification and extinction at the Cretaceous–Tertiary boundary. *Geology* **22**, 983–986.

D'Hondt S, Donaghay P, Zachos JC, Luttenberg D, Lindinger M (1998) Organic carbon fluxes and ecological recovery from the Cretaceous–Tertiary mass extinction. *Science* **282**, 276–279.

Diaz-Martinez E, Grahn Y (2007) Early Silurian glaciation along the western margin of Gondwana (Peru, Bolivia and northern Argentina); palaeogeographic and geodynamic setting. *Palaeogeography, Palaeoclimatology, Palaeoecology* **245**, 62–81.

Dickens GR, O'Neil JR, Rea DK, Owen RM (1995) Dissociation of oceanic methane hydrate as a cause of the carbon isotope excursion at the end of the Paleocene. *Paleoceanography* **10**, 965–971.

DiMichele WA, Pfefferkorn HW, Gastaldo RA (2001) Response of Late Carboniferous and Early Permian plant communities to climate change. *Annual Review of Earth and Planetary Sciences* **29**, 461–487.

Doyle JA, Hickey LJ (1976) Pollen and leaves from the mid-Cretaceous Potomac Group and their bearing on early angiosperm evolution. In: *Origin and Early Evolution of the Angiosperms* (ed Beck CB). Columbia University Press, New York, pp. 139–206.

Droser ML, Bottjer DJ (1988) Trends in depth and extent of bioturbation in Cambrian carbonate marine environments, Western United States. *Geology* **16**, 233–236.

Droser ML, Bottjer DJ, Sheehan PM (1997) Evaluating the ecological architecture of major events in the Phanerozoic history of marine invertebrate life. *Geology* **25**, 167–170.

Eagles G, Livermore R, Morris P (2006) Small basins in the Scotia Sea; the Eocene Drake Passage gateway. *Earth and Planetary Science Letters* **242**, 343–353.

Eldrett, JS, Harding IC, Wilson PA, Butler E, Roberts AP (2007) Continental ice in Greenland during the Eocene and Oligocene. *Nature* **446**, 176–179.

Falkowski PG, Katz ME, Knoll AH, *et al.* (2004) The evolution of modern eukaryotic phytoplankton. *Science* **305**, 354–360.

Falkowski PG, Katz ME, Milligan AJ, *et al.* (2005) The rise of oxygen over the past 205 million years and the evolution of large placental mammals. *Science* **309**, 2202–2204.

Firestone RB, West A, Kennett JP, *et al.* (2007) Evidence for an extraterrestrial impact 12,900 years ago that contributed to the megafaunal extinctions and the Younger Dryas cooling. *Proceedings of the National Academy of Sciences* **104**, 16016–16021.

Fowell SJ, Cornet B, Olsen PE (1994) Geologically rapid Late Triassic extinctions; palynological evidence from the Newark Supergroup. *Special Paper – Geological Society of America* **288**, 197–206.

Freeman KH, Hayes JM (1992) Fractionation of carbon isotopes by phytoplankton and estimates of ancient CO_2 levels. *Global Biogeochemical Cycles* **6**,185–198.

Funakawa S, Nishi H, Moore TC, Nigrini CA (2006) Radiolarian faunal turnover and paleoceanographic change around Eocene/Oligocene boundary in the Central Equatorial Pacific, ODP Leg 199, Holes 1218A, 1219A, and 1220A. *Palaeogeography, Palaeoclimatology, Palaeoecology* **230**, 183–203.

Garrett P (1970) Phanerozoic stromatolites; noncompetitive ecologic restriction by grazing and burrowing animals. *Science* **169**, 171–173.

Gentry A, Heizmann EPJ (1996) Miocene ruminants of the central and eastern Paratethys, In: *The Evolution of Western Eurasian Neogene Mammal Faunas* (eds Bernor RL, Fahlbusch V, Mittman H-W). Columbia University Press, New York, pp. 378–395.

Graham JB, Dudley R, Aguilar NM, Gans C (1995) Implications of the late Palaeozoic oxygen pulse for physiology and evolution. *Nature* **375**, 117–120.

Grotzinger J, Adams EW, Schroeder S (2005) Ediacaran–Cambrian paleoecology, sedimentology and stratigraphy of NambiaNamibia. *Geological Magazine* **142**: 499–517.

Hagadorn JW, Bottjer DJ (1997) Wrinkle structures; microbially mediated sedimentary structures common in subtidal silici-clastic settings at the Proterozoic–Phanerozoic transition. *Geology* **25**, 1047–1050.

Hansen TA (1987) Extinction of late Eocene to Oligocene molluscs; relationship to shelf area, temperature changes, and impact events. *Palaios* **2**, 69–75.

Herman AB, Spicer RA (1996) Palaeobotanical evidence for a warm Cretaceous Arctic Ocean. *Nature* **380**, 330–333.

Hesselbo SP, Grocke DR, Jenkyns HC, *et al.* (2000) Massive dissociation of gas hydrate during a Jurassic oceanic anoxic event. *Nature* **406**, 392–395.

Hickman CS (2003) Evidence for abrupt Eocene–Oligocene molluscan faunal changes in the Pacific Northwest. In: *From Greenhouse to Icehouse; the Marine Eocene–Oligocene Transition* (eds Prothero DR, Ivany LC, Nesbitt EA). Columbia University Press, New York, pp. 71–87.

Hill D (1972) *Treatise on Invertebrate Paleontology Part E, Vol. 1 (Revised) Archaeocyatha.* Geological Society of America and University of Kansas: Boulder, Colorado and Lawrence, Kansas), 158 pp.

Hooghiemstra H, van der Hammen T (1998) Neogene and Quaternary development of the neotropical rain forest; the forest refugia hypothesis, and a literature overview. *Earth-Science Reviews* **44**, 147–183.

House MR (2002) Strength, timing, setting and cause of mid-Paleozoic extinctions. *Palaeogeography, Palaeoclimatology, Palaeoecology* **181**, 5–25.

Huey RB, Ward PD (2005) Hypoxia, global warming, and terrestrial Late Permian extinctions. *Science* **308**, 398–401.

Hulbert JRC (1993) Taxonomic evolution in North American Neogene horses (subfamily Equinae): the rise and fall of an adaptive radiation. *Paleobiology* **19**, 216–234.

Isaacson PE, Díaz-Marinez E, Grader GW, Kalvoda J, Babek O, Devuyst FX (2008) Late Devonian-earliest Mississippian glaciation in Gondwanaland and its biogeographic consequences. *Palaeogeography, Palaeoclimatology, Palaeoecology* **268**, 126–142.

Isozaki Y (1997) Anatomy and genesis of a subduction-related orogen: A new view of geotectonic subdivision and evolution of the Japanese Islands. *Science* **276**, 235–238.

Isozaki Y, Kawahata H, Minoshima K (2007) The Capitanian (Kamura) cooling event: the beginning of the Paleozoic-Mesozoic transition. *Palaeoworld* **16**, 16–30.

Ivany LC, van Simaeys S, Domack EW, Samson SD (2006) Evidence for an earliest Oligocene ice sheet on the Antarctic Peninsula. *Geology* **34**, 377–380.

Jahren AH, Sternberg LSL (2002) Eocene meridional weather patterns reflected in the oxygen isotopes of Arctic fossil wood. *GSA Today* **12**, 4–9.

Jenkyns HC (1988) The early Toarcian (Jurassic) anoxic event; stratigraphic, sedimentary and geochemical evidence. *American Journal of Science* **288**, 101–151.

Jin YG, Zhang J, Shang QH (1994) Two phases of the end-Permian mass extinction. *Canadian Society of Petroleum Geologists Memoir* **17**, 813–822.

Joachimski MM, Buggisch W (2002) Conodont apatite $\delta^{18}O$ signatures indicate climatic cooling as a trigger of the Late Devonian mass extinction. *Geology* **30**, 711–714.

Johnson KR, Hickey LJ (1990) Megafloral change across the Cretaceous/Tertiary boundary in the northern Great Plains and Rocky Mountains, USA. *Special Paper - Geological Society of America* **247**, 433–444.

Keller G (1983) Paleoclimatic analyses of middle Eocene through Oligocene planktic foraminiferal faunas. *Palaeogeography, Palaeoclimatology, Palaeoecology* **43**, 73–94.

Kennett, DJ, Kennett JP, West A, *et al.* (2009) Shock-synthesized hexagonal diamonds in Younger Dryas boundary sediments. *Proceedings of the National Academy of Sciences* **106**, 12623–12628.

Kennett JP, Stott LD (1991) Abrupt deep-sea warming, palaeoceanographic changes and benthic extinctions at the end of the Palaeocene. *Nature* **353**, 225–229.

Kennett JP, Cannariato KG, Hendy IL, Behl R (2003) *Methane Hydrates in Quaternary Climate Change; the Clathrate Gun Hypothesis*. American Geophysical Union, Washington, DC.

Kennett JP, Houtz RE, Andrews PB, *et al.* (1975) Cenozoic paleoceanography in the Southwest Pacific Ocean, Antarctic glaciation, and the development of the Circum-Antarctic Current. *Initial Reports of the Deep Sea Drilling Project* **29**, 1155–1169.

Knight KB, Nomade S, Renne PR, Marzoli A, Bertrand H, Youbi N (2004) The Central Atlantic magmatic province at the Triassic-Jurassic boundary; paleomagnetic and $^{40}Ar/^{39}Ar$ evidence from Morocco for brief, episodic volcanism. *Earth and Planetary Science Letters* **228**, 143–160.

Knoll AH, Bambach RK, Payne JL, Pruss S, Fischer WW (2007a) Paleophysiology and end-Permian mass extinction. *Earth and Planetary Science Letters* **256**, 295–313.

Knoll AH, Summons RE, Walbauer JR, Zumberge JE (2007b) The Geological Succession of Primary Producers in the Oceans. In *Evolution of Primary Producers in the Sea* (eds Falkowski PG, Knoll AH). Elsevier Acacemic Press, Amsterdam, pp. 133–163.

Kobashi T, Grossman EL, Yancey TE, Dockery DT, III (2001) Reevaluation of conflicting Eocene tropical temperature estimates; molluskan oxygen isotope evidence for warm low latitudes. *Geology* **29**, 983–986.

Kobashi T, Grossman EL, Dockery DT, III, Ivany LC (2004) Water mass stability reconstructions from greenhouse (Eocene) to icehouse (Oligocene) for the northern Gulf Coast continental shelf (USA). *Paleoceanography* **19**, 16.

Kump LR, Pavlov A, Arthur MA (2005) Massive release of hydrogen sulfide to the surface ocean and atmosphere during intervals of oceanic anoxia. *Geology* **33**, 397–400.

Kuypers MMM, van Breugel Y, Schouten S, Erba E, Sinninghe Damsté JS (2004) N2-fixing cyanobacteria supplied nutrient N for Cretaceous oceanic anoxic events. *Geology* **32**, 853–856.

Labandeira CC, Phillips TL, Norton RA (1997) Oribatid mites and the decomposition of plant tissues in Paleozoic coal-swamp forests. *Palaios* **12**, 319–353.

Larson RL (1991) Latest pulse of Earth; evidence for a Mid-Cretaceous super plume. *Geology* **19**, 547–550.

Leckie RM, Bralower TJ, Cashman R (2002) Oceanic anoxic events and plankton evolution; biotic response to tectonic forcing during the Mid-Cretaceous. *Paleoceanography* **17**, no. 3, 29pp.

Lisse, CM, Kraemer KE, Nuth JA, Li A, Josiak, D (2007) Comparison of the composition of the Tempel 1 ejecta to the dust in Comet C/Hale-Bopp 1995 O1 and YSO HD 100546. *Icarus* **191**, 223–240.

Loydell DK (2007) Early Silurian positive $\delta^{13}C$ excursions and their relationship to glaciations, sea-level changes and extinction events. *Geological Journal* **42**, 531–546.

Magioncalda R, Dupuis C, Smith T, Steurbaut E, Gingerich PD (2004) Paleocene–Eocene carbon isotope excursion in organic carbon and pedogenic carbonate; direct comparison in a continental stratigraphic section. *Geology* **32**, 553–556.

Maliva RG, Knoll AH, Siever R (1989) Secular change in chert distribution: a reflection of evolving biological participation in the silica cycle. *Palaios* **4**, 519–532.

Marubini F, Thake B (1999) Bicarbonate addition promotes coral growth. *Limnology and Oceanography* **44**, 716–720.

Marzoli A, Renne PR, Piccirillo EM, Ernesto M, Bellieni G, De Min A (1999) Extensive 200-million-year-old continental flood basalts of the Central Atlantic Magmatic Province. *Science* **284**, 616–618.

McElwain JC, Beerling DJ, Woodward FI (1999) Fossil plants and global warming at the Triassic-Jurassic boundary. *Science* **285**, 1386–1390.

Melosh HJ, Schneider NM, Zahnle KJ, Latham D (1990) Ignition of global wildfires at the Cretaceous/Tertiary boundary. *Nature* **343**, 251–254.

Meng J, McKenna MC (1998) Faunal turnovers of Paleogene mammals from the Mongolian Plateau. *Nature* **394**, 364–367.

Montanari A, Hay RL, Alvarez W, *et al.* (1983) Spheroids at the Cretaceous–Tertiary boundary are altered impact droplets of basaltic composition. *Geology* **11**, 668–671.

Olsen PE, Kent DV, Sues HD, *et al.* (2002) Ascent of dinosaurs linked to an iridium anomaly at the Triassic–Jurassic boundary. *Science* **296**, 1305–1307.

Pagani M, Zachos JC, Freeman KH, Tipple B, Bohaty S (2005) Marked decline in atmospheric carbon dioxide concentrations during the Paleogene. *Science* **309**, 600–603.

Palmer AR (1984) The biomere problem; evolution of an idea. *Journal of Paleontology* **58**, 599–611.

Palmer AR (1998) A proposed nomenclature for stages and series for the Cambrian of Laurentia. *Canadian Journal of Earth Sciences* **35**, 323–328.

Paquay FS, Goderis S, Ravizza G. *et al.* (2009) Absence of geochemical evidence for an impact event at the Bolling–Allerod/Younger Dryas transition. *Proceedings of the National Academy of Sciences* **106**, 21505–21510.

Parrish JT, Spicer RA (1988) Late Cretaceous terrestrial vegetation; a near-polar temperature curve. *Geology* **16**, 22–25.

Payne JL, Lehrmann DJ, Wei J, Orchard MJ, Schrag DP, Knoll AH (2004) Large perturbations of the carbon cycle during recovery from the end-Permian extinction. *Science* **305**, 506–509.

Perfetta PJ, Shelton KV, Stitt JH (1999) Carbon isotope evidence for deep-water invasion at the marjumiid-pterocephaliid biomere boundary, Black Hills, USA: A common origin for biotic crises on Late Cambrian shelves. *Geology* **27**, 403–406.

Pochon X, Montoya-Burgos JI, Stadelmann B, Pawlowski J (2006) Molecular phylogeny, evolutionary rates, and divergence timing of the symbiotic dinoflagellate genus Symbiodinium. *Molecular Phylogenetics and Evolution* **38**, 20–30.

Poore RZ, Berggren WA (1975) Late Cenozoic planktonic foraminiferal biostratigraphy and paleoclimatology of Hatton-Rockall Basin; DSDP Site 116. *Journal of Foraminiferal Research* **5**, 270–293.

Pope KO, Baines KH, Ocampo AC, Ivanov BA (1994) Impact winter and the Cretaceous/Tertiary extinctions; results of a

Chicxulub asteroid impact model. *Earth and Planetary Science Letters* **128**, 719–725.

Porter SM (2007) Seawater chemistry and early carbonate biomineralization. *Science* **316**, 302.

Powell MG (2005) Climatic basis for sluggish macroevolution during the late Paleozoic ice age. *Geology* **33**, 381–384.

Puceat E, Lecuyer C, Donnadieu Y, *et al.* (2007) Fish tooth $\delta^{18}O$ revising Late Cretaceous meridional upper ocean water temperature gradients. *Geology* **35**, 107–110.

Raffi S, Stanley SM, Marasti R (1985) Biogeographic patterns and Plio-Pleistocene extinction of Bivalvia in the Mediterranean and southern North Sea. *Paleobiology* **11**, 368–388.

Rees, DL (2002) Land-plant diversity and the end-Permian mass extinction. *Geology* **30**: 827–830

Reimers CE, Jahnke RA, McCorkle DC (1992) Carbon fluxes and burial rates over the continental slope and rise off Central California with implications for the global carbon cycle. *Global Biogeochemical Cycles* **6**,199–224.

Renegar DA, Riegl BM (2005) Effect of nutrient enrichment and elevated CO_2 partial pressure on growth rate of Atlantic scleractinian coral *Acropora cervicornis. Marine Ecology Progress Series* **293**, 69–76.

Retallack GJ (1997) Early forest soils and their role in Devonian global change. *Science* **276**, 583–585.

Retallack GJ, Metzger CA, Greaver T, Jahren AH, Smith RMH, Sheldon ND (2006) Middle-Late Permian mass extinction on land. *Geological Society of America Bulletin* **118**, 1398–1411.

Ries JB, Stanley SM, Hardie LA (2006) Scleractinian corals produce calcite, and grow more slowly, in artificial Cretaceous seawater. *Geology* **34**, 525–528.

Royer DL (2003) Estimating latest Cretaceous and Tertiary atmospheric CO_2 from stromatal indices. *Geological Society of America Special Paper* **369**, 79–93.

Royer DL, Berner RA, Beerling DJ (2001) Phanerozoic atmospheric CO_2 change: evaluating geochemical and paleobiological approaches. *Earth-Science Reviews* **54**, 349–392.

Saito T, Yamanoi T, Kaiho K (1986) End-Cretaceous devastation of terrestrial flora in the boreal Far East. *Nature* **323**, 253–255.

Saltzman MR (2001) Silurian $\delta^{13}C$ stratigraphy: A view from North America. *Geology*, **29**, 671–674.

Saltzman MR (2002) Carbon isotope ($\delta^{13}C$) stratigraphy across the Silurian–Devonian transition in North America: evidence for a perturbation of the global carbon cycle. *Palaeogeography, Palaeoclimatology, Palaeoecology* **187**, 83–100.

Saltzman MR, Young SA (2005) Long-lived glaciation in the Late Ordovician? Isotopic and sequence-stratigraphic evidence from western Laurentia. *Geology* **33**, 109–112.

Scher HD, Martin EE (2006) Timing and climatic consequences of the opening of Drake Passage. *Science* **312**, 428–430.

Schouten S, Hopmans EC, Forster A, van Breugel Y, Kuypers MMM, Sinninghe Damsté JS (2003) Extremely high sea-surface temperatures at low latitudes during the Middle Cretaceous as revealed by archaeal membrane lipids. *Geology* **31**, 1069–1072.

Sepkoski JJ (1982) Flat-pebble conglomerates, storm deposits, and the Cambrian bottom fauna. In: *Cyclic and Event Stratification* (eds Einsele G, Seilacher A). Springer-Verlag, Berlin, pp. 371–385.

Sheehan PM (2001) The Late Ordovician mass extinction. *Annual Review of Earth and Planetary Sciences* **29**, 331–364.

Soreghan GS, Soreghan MJ, Poulsen CJ, *et al.* (2008) Anomalous cold in the Pangaean tropics. *Geology* **36**, 659–662.

Stanley SM (1986) Anatomy of a regional mass extinction: Plio-Pleistocene decimation of the Western Atlantic bivalve fauna. *Palaios* **1**, 17–36.

Stanley SM (1990) Adaptive radiation and macroevolution. *Systematics Association Special Volume* **42**, 1–16.

Stanley SM (1992) An ecological theory for the origin of *Homo. Paleobiology* **18**, 237–257.

Stanley SM (1995) New horizons for paleontology, with two examples; the rise and fall of the Cretaceous Supertethys and the cause of the modern ice age. *Journal of Paleontology* **69**, 999–1007.

Stanley SM (2009a) *Earth System History*. W.H.Freeman and Company, New York.

Stanley SM (2009b) Evidence from ammonoids and conodonts for multiple Early Triassic mass extinctions. *Proceedings of the National Academy of Sciences* **106**, 15256–15259.

Stanley SM (2010) Relation of Phanerozoic stable isotope excursions to climate, bacterial metabolism, and major extinctions. *Proceedings of the National Academy of Sciences* **107**, 19185–19189.

Stanley SM, Yang X (1994) A double mass extinction at the end of the Paleozoic Era. *Science* **266**, 1340–1344.

Stanley SM, Hardie LA (1998) Secular oscillations in carbonate mineralogy of reef-building and sediment-producing organisms driven by tectonically forced shifts in seawater chemistry. *Palaeogeography, Palaeoclimatology, Palaeoecology* **144**, 3–19.

Stanley SM, Powell MG (2003) Depressed rates of origination and extinction during the late Paleozoic ice age; a new state for the global marine ecosystem. *Geology* **31**, 877–880.

Stanley SM, Ries JB, Hardie LA (2005) Seawater chemistry, coccolithophore population growth, and the origin of Cretaceous chalk. *Geology* **33**, 593–596.

Stickley CE, Brinkhuis H, Schellenberg SA, *et al.* (2004) Timing and nature of the deepening of the Tasmanian Gateway. *Paleoceanography* **19**, 18.

Stitt JH (1975) Adaptive radiation, trilobite paleoecology, and extinction, ptychaspidid biomere, late Cambrian of Oklahoma. *Fossils and Strata* **4**, 381–390.

Stromberg CAE (2004) Using phytolith assemblages to reconstruct the origin and spread of grass-dominated habitats in the Great Plains of North America during the late Eocene to early Miocene. *Palaeogeography, Palaeoclimatology, Palaeoecology* **207**, 239–275.

Swisher CC, Grajales-Nishimura JM, Montanari A, *et al.* (1992) Coeval 40Ar/39Ar ages of 65.0 million years ago from Chicxulub Crater melt rock and Cretaceous–Tertiary boundary tektites. *Science* **257**, 954–958.

Tanner LH, Lucas SG, Chapman MG (2004) Assessing the record and causes of Late Triassic extinctions. *Earth-Science Reviews* **65**, 103–139.

Tschudy RH, Tschudy BD (1986) Extinction and survival of plant life following the Cretaceous/Tertiary boundary event, Western Interior, North America. *Geology* **14**, 667–670.

Vrba ES (1985) African Bovidae: evolutionary events since the Miocene. *South African Journal of Science* **81**, 263–266.

Webb SD (1977) A history of the savanna vertebrates in the New World; Part I, North America. *Annual Review of Ecology and Systematics* **8**, 355–380.

Webb SD (1984) Ten million years of mammal extinctions in North America. In *Quaternary Extinctions: A Prehistoric Revolution* (eds Martin PS, Klein RG). University of Arizona Press, Tucson, pp. 189–210.

Whiteside JH, Olsen PE, Kent DV, Fowell SJ, Et-Touhami M (2007) Synchrony between the Central Atlantic magmatic province and the Triassic-Jurassic mass-extinction event? *Palaeogeography, Palaeoclimatology, Palaeoecology* **244**, 345–367.

Wing SL, Harrington GJ, Smith FA, Bloch JI, Boyer DM, Freeman KH (2005) Transient floral change and rapid global warming at the Paleocene–Eocene boundary. *Science* **310**, 993–996.

Wing SL, Hickey LJ, Swisher CC (1993) Implications of an exceptional fossil flora for Late Cretaceous vegetation. *Nature* **363**, 342–344.

Wolfe JA (1971) Tertiary climatic fluctuations and methods of analysis of Tertiary floras. *Palaeogeography, Palaeoclimatology, Palaeoecology* **9**, 27–57.

Zachos JC, Roehl U, Schellenberg SA, *et al.* (2005) Rapid acidification of the ocean during the Paleocene–Eocene thermal maximum. *Science* **308**, 1611–1615.

Zachos JC, Wara MW, Bohaty S, *et al.* (2003) A transient rise in tropical sea surface temperature during the Paleocene–Eocene thermal maximum. *Science* **302**, 1551–1554.

Zhu M-Y, Zhang, J-M,Li G-X, Yang, A-H (2004) Evolution of C isotopes in the Cambrian of China: implications for Cambrian subdivision and trilobite mass extinctions. *Geobios* **37**, 287–301.

Zhuravlev AY, Wood R (2008) Eve of biomineralization: Controls on skeletal mineralogy. *Geology* **36**, 923–926.

22
GEOBIOLOGY OF THE ANTHROPOCENE

Daniel P. Schrag

Department of Earth and Planetary Sciences, Harvard University, Cambridge, MA 02138, USA

22.1 Introduction

Homo sapiens first appeared on the Earth somewhere in Africa roughly 200 000 years ago. It happened with little fanfare; few could have imagined that this new species of primate would someday disrupt the Earth system to the point of defining a new geologic epoch around its legacy. Indeed, the first 150 000 years of the natural history of our species, mostly in Africa, were fairly uneventful for reasons still not well understood. But then, with migration out of Africa, things started to change. Having mastered new hunting skills, humans began to perturb their ecosystems, first by overhunting large animals, which also deprived rival predators of adequate food supplies. Then with the development of agriculture in the last 10 000 years, humans began an appropriation of the Earth's surface for food, fuel and fiber that continues to this day. More recently, the industrial revolution, spurred on with cheap, abundant energy from fossil organic carbon, made humans major players in Earth's geochemical cycles, including nitrogen and carbon. The latter now threatens to end the Pleistocene glacial cycles and return the Earth to a state not seen for 35 million years. The future of human interactions with the Earth system remains uncertain, but the impact of human actions already taken will last for more than 100 000 years. Avoiding massive disruptions to geobiological systems in the future is likely to require, ironically, even larger interventions by humans through advanced technology, the final step in a transition to the engineered epoch of Earth history.

22.2 The Anthropocene

The adoption of the term 'Anthropocene' is commonly credited to Paul Crutzen, the Nobel-prize winning chemist, in a speech in 2000, although the recognition of human impact on the Earth and the declaration of a new geological epoch long precedes Crutzen (Zalasiewicz *et al.*, 2011). Already in 1871, Italian geologist Antonio Stoppani used the term 'Anthropozoic' to describe 'new telluric force, which in power and universality may be compared to the greater forces of earth.' Joseph LeConte, in his *Elements of Geology* (1878) uses the term 'Psychozoic' to describe the age of man, characterized by the 'reign of mind.' Even Charles Lyell pondered the enormous impact that humans were having on the Earth, for he recognized that some might see in it a challenge to his arguments about the uniformity of nature's laws. In the original version of his *Principles of Geology* (1830), Lyell discusses the modern origin of humans, and the question of

'whether the recent origin of man lends any support to the same doctrine, or how far the influence of man may be considered as such a deviation from the analogy of the order of things previously established, as to weaken our confidence in the uniformity of the course of nature.'

Lyell states his concern clearly:

'Is not the interference of the human species, it may be asked, such a deviation from the antecedent course of physical events, that the knowledge of such a fact tends to destroy all our confidence in the uniformity of the order of

Fundamentals of Geobiology, First Edition. Edited by Andrew H. Knoll, Donald E. Canfield and Kurt O. Konhauser.
© 2012 Blackwell Publishing Ltd. Published 2012 by Blackwell Publishing Ltd.

nature, both in regard to time past and future? If such an innovation could take place after the earth had been exclusively inhabited for thousands of ages by inferior animals, why should not other changes as extraordinary and unprecedented happen from time to time? If one new cause was permitted to supervene, differing in kind and energy from any before in operation, why may not others have come into action at different epochs? Or what security have we that they may not arise hereafter? If such be the case, how can the experience of one period, even though we are acquainted with all the possible effects of the then existing causes, be a standard to which we can refer all natural phenomena of other periods?'

In the early 19th century, Lyell already recognized the magnitude of human impact on the Earth system.

'When a powerful European colony lands on the shores of Australia, and introduces at once those arts which it has required many centuries to mature; when it imports a multitude of plants and large animals from the opposite extremity of the earth, and begins rapidly to extirpate many of the indigenous species, a mightier revolution is effected in a brief period, than the first entrance of a savage horde, or their continued occupation of the country for many centuries, can possibly be imagined to have produced.'

One can only imagine what Lyell would think were he to see the scale of human activities today.

Lyell's argument was that humans – even with their disruptions to natural ecosystems, even with their morality and their unique ability to interpret nature through an understanding of its natural laws – remain bound by those laws. This is what allowed Lyell to preserve his uniformitarian theory, which he viewed as essential for understanding Earth history. But there are closely related questions looking to the future that Lyell did not address. Is the Anthropocene (to use the modern form) recognizable among other geological epochs? When did it begin and when will it end? And what, among all the many features of the geobiological record of the Anthropocene, will be most recognizable millions of years in the future? In this chapter, I describe some aspects of the geobiology of the Anthropocene in an attempt to address these questions.

22.3   When did the Anthropocene begin?

In a 2000 newsletter of the International Geosphere-Biosphere Program, Crutzen and Stoermer suggested that the transition from the Holocene to the Anthropocene began near the end of the 18th century, coincident with the invention of the steam engine by James Watt (1784)

and the rise in greenhouse gases observed in ice cores (Crutzen and Stoermer, 2000). A different view comes from William Ruddiman, who argued that human interference in the climate system began 7000 years ago with the development of agriculture. Ruddiman pointed to the concentration of atmospheric CO_2, which was 260 ppm approximately 8000 years ago, and suggested that it should have fallen by 20 ppm, synchronous with changes in the Earth's orbit around the sun, as it had during previous interglacials Intervals. Instead, atmospheric CO_2 rose to 280 ppm prior to the industrial revolution, a net difference of 40 ppm that Ruddiman attributed to the release of carbon dioxide from deforestation. It is this reversal of greenhouse gases, Ruddiman claimed, that stabilized the climate of the Holocene and allowed human civilizations to flourish (Ruddiman, 2003, 2007).

Examination of the carbon cycle does not support Ruddiman's hypothesis. Over thousands of years, most of the carbon released from deforestation would dissolve in the ocean, which means that a net change in atmospheric CO_2 of 40 ppm would require roughly 600 billion tonnes of carbon to be released from the land, an amount equivalent to the entire modern terrestrial biosphere. Moreover, such a large release of carbon from biomass would change the isotopic composition of carbon reservoirs, as recorded in shells and ice cores; no such change is observed. Finally, the rise in atmospheric CO_2 has been linear over the last 7000 years, but the expansion of agriculture was not. It seems most likely that early agriculture had a smaller impact on the carbon cycle than Ruddiman claimed.

But that is not to say that early humans had little effect on their environment. It is quite clear that early humans changed their ecosystems by overhunting of large animals long before the invention of agriculture (Alroy, 2001). The extinction of larger mammals and birds in Asia, Europe, Australia, the Americas, and New Zealand and the Pacific Islands immediately followed the spread of human populations across these regions. The timing of the extinctions is diachronous, as predicted by the over-hunting hypothesis; moreover, the disappearance of large animals is a distinctive feature of the extinction across every climatic regime, from the tropics to the temperate zones, and in both hemispheres, refuting a climatic explanation for the extinction as some have proposed. Only in sub-Saharan Africa did megafauna survive the rise of human society, and the reason for their persistence remains a mystery.

In *Guns, Germs and Steel* (Diamond, 1997), Jared Diamond proposed that the co-evolution of African megafauna with humans, as well as with earlier hominids, allowed large mammals to survive – the essential claim is that large animals in Africa learned to be afraid of humans, a behaviour honed by natural selection. This hypothesis predicts good news for future conservation efforts as it suggests that African megafauna have an instinctive key to their own survival – i.e., their fear of humans. An alternative explanation, however, allows less optimism. It seems possible that large mammals in Africa survived not because of co-evolution with humans, but because human occupation of sub-Saharan Africa was never expansive enough to drive these animals to extinction. Tropical diseases, such as malaria and sleeping sickness, are virulent in sub-Saharan Africa, (e.g. Greenwood and Matabingwa, 2002). Perhaps it was the co-evolution of mosquitoes and humans rather than megafauna and humans that prevented human populations in sub-Saharan Africa from reaching a critical level to drive large animals to extinction. If so, it does not bode well for the future of African megafauna, as their demise may be the unintended consequence of modern efforts towards economic development and poverty alleviation.

So, then, what do we identify as the beginning of the Anthropocene? If we define it as the first large impact of humans on the environment, then the megafaunal extinctions provide an excellent candidate. A thorn for stratigraphers, however, is that these extinctions are diachronous across many different regions. A more serious objection is that the extinctions themselves do not foreshadow the extent of human dominance over the Earth system. If the megafaunal extinctions were the major environmental impact over the history of human society, it is not clear that this would rise above the threshold for defining a new geologic epoch; it might only be seen as an ecological bottleneck of some sort. As I will discuss below, human society following the industrial revolution, facilitated by fossil carbon as an energy source, has changed the Earth system in a way that almost challenges Lyell's confidence in the uniformity of nature's laws. One can see this change using many different metrics: economic output, population, energy consumption – all begin to grow exponentially starting in the late 18th century. Of course, earlier events of human history contributed to this accelerated growth, from the classical civilizations of Egypt, Greece and Rome, to the technological achievements of the European renaissance. But if one wants to identify the launching point when humans started down an irreversible path towards a complete transformation of their planet, then Crutzen's choice of 1784 seems appropriate.

22.4 Geobiology and human population

From the time of the Roman Empire until the start of the 18th century, the human population hovered somewhere between 100 million and 1 billion. Around the start of the Anthropocene, it reached 1 billion; by 1930, 2 billion; and, in 1974, human population reached 4 billion. It sits now very close to 7 billion, and most demographic projections predict a peak around 9 billion sometime in the middle of the 21st century. How Earth's ecosystems will fare on a planet with 9 billion human beings is a question we will address here.

Beginning with Thomas Malthus, many have prophesied grimly about how population growth will disrupt human society. The modern version of Malthusian catastrophism is epitomized by *The Population Bomb*, written in 1968 by Paul Ehrlich. In this book, Ehrlich made a series of dire predictions for the near future; for example, he argued that India would not be able to feed 200 million more people by 1980. In general, Ehrlich's predictions have not been accurate; India's population today is nearly 1.2 billion, and it has become the world's largest exporter of rice.

The major reason why Ehrlich's predictions were wrong is the technological innovation in agriculture, commonly referred to as the Green Revolution. Working in Mexico in the 1950s, Norman Borlaug and colleagues developed high-yield varieties of wheat that were resistant to many diseases. When combined with modern agricultural production techniques, these new strains increased wheat yields in Mexico from less than 1 tonne per hectare in 1950 to nearly 5 tonnes per hectare in 2000. India and Pakistan experienced smaller increases in yields, but still enough to make them self-sufficient – something Ehrlich had not envisioned in 1968. Borlaug's approaches were later applied to other crops, including several types of rice. Another component of the Green Revolution involved the industrial production of nitrogen fertilizer, discussed below, and the mining of phosphate that allowed modern industrial agriculture to develop around the world. Overall, the Green Revolution allowed the world's population to grow far beyond what Ehrlich had estimated as the Earth's carrying capacity.

Some take the example of the Green Revolution as proof that there are no environmental constraints on human population, and that Ehrlich's entire approach was wrong. Another perspective is that Ehrlich may have been wrong in his specific predictions, but perhaps only about the time scale; the challenge of exponential population growth in a world with finite resources still exists. In his acceptance speech for the Nobel Peace Prize in 1970, Norman Borlaug warned,

'the green revolution has won a temporary success in man's war against hunger and deprivation; it has given man a breathing space. If fully implemented, the revolution can provide sufficient food for sustenance during the next three decades. But the frightening power of human reproduction must also be curbed; otherwise the success of the green revolution will be ephemeral only' (Borlaug, 1970).

It is true that the growth rate of world population has dropped – much of it due to economic development that leads to a 'demographic transition' in many countries, that is, that women choose to have smaller number of children as their income rises, especially if they are not deprived access to education and employment. But even with a lower growth rate and population stabilization by 2050, it is not clear whether the Earth can support 9 billion people without a large fraction of that population suffering from limited access to food and water. One can ask whether the Green Revolution saved hundreds of millions of people from starvation only to condemn billions of people to a similar fate sometime in the future. Borlaug was keenly aware of the future challenges brought on by a growing population, as well as the increased consumption of meat due to greater affluence in many populous regions of the world. He predicted that we would need to double the world food supply by 2050, and saw this challenge as a major priority for research efforts today. It remains an open question whether the increase in crop yields provided by Borlaug's efforts will continue to grow through genetic modification of plants, especially in the face of human-induced climate change.

In *How Many People Can the Earth Support?* Joel Cohen (1996) examined the full range of constraints on human population including land area, food production, fresh water, and energy. He also explored the history of different ideas of what the carrying capacity of the planet might be. In the end, Cohen concluded that his title asked the wrong question, as the maximum number of people only makes sense if one specifies level of affluence, extent of equality, or the number of people who are allowed to suffer from malnutrition and other impacts of extreme poverty. In Cohen's framing, the statement about the Earth's carrying capacity is a statement about one's values concerning the human condition. Within that framing, one can take a Malthusian perspective that emphasizes exponential population growth in the face of resource limitations or a Borlaugian perspective that emphasizes technology's capacity to remove or at least soften those environmental constraints. We will return to the question of technological innovation below.

The focus of this chapter, involves the impact of human population on the Earth system. We are interested not only in how humans will fare as their population grows, but how 9 billion humans will affect the rest of the geobiological system. The two issues are related, as human societal disruption such as war or famine has its own substantial environmental impacts. However, one can ask how the Earth system will respond to a growing population, and how much of the expected change is actually driven by population itself without answering the moral and ethical questions raised by Cohen. To examine this, it is useful to employ the framework of Paul Ehrlich and John Holdren in their 1971 paper on the *Impact of Population Growth* (Erlich and Holdren, 1971). Ehrlich and Holdren introduced a simple equation: Impact = Population × Affluence × Technology (IPAT) to evaluate the causes of environmental disturbance. At its core, IPAT is a simple identity, but it allows one to identify quickly the factors (population growth, economic growth, or technological change) that are most responsible for creating or solving our environmental problems. A quick analysis of two of the largest drivers of geobiological disturbance, human land-use for agriculture and anthropogenic climate change, reveals the surprising conclusion that in the near term, population growth is not likely to be the major factor driving environmental degradation. For agriculture, the IPAT equation is: land use (area) = population × (GDP/person) × (land use/GDP). Of these terms, population is likely to grow from 7 to 9 billion over the next 40 years, or 29%. Over that same time interval, GDP is expected to grow by 200 to 400%. The technology will also change in the future, driven lower by agricultural innovations that increase crop yields, but driven higher by increases in the amount of meat in the average diet, leading to greater demand for grain as well as land for pastures.

Greenhouse gas emissions tell a similar story: the first two terms are the same, with the final term reflecting the greenhouse gas intensity of our energy systems, as well as how the demand for energy changes with our affluence. One can see from both these examples that economic growth and not population growth is the larger driver of many of our environmental challenges over the next century. The technology term in both cases is uncertain, as it could add to the problem by requiring additional resources, or could reduce human demands for ecosystem services, as in the agronomic discoveries of Borlaug. This is not to say that population growth is not at the root of our current environmental challenges; after all, it was partly the population growth over the past 200 years, growing from 1 billion to 7 billion (world GDP grew by a factor of 40 over this same period), that put the Earth in its current predicament; but the people alive today are more than enough to trigger enormous changes to the Earth system even if population remains constant but consumption continues to rise with prosperity.

22.5 Human appropriation of the Earth

Driven by growing population, by economic development, and by technological innovation fueled with fossil carbon, the scale of human appropriation of the

Earth surface is remarkable. Never before has a single species so closely managed such a vast area of the planet. Roughly 40% of Earth's land surface is used for croplands and pastures (Vitousek *et al.*, 1997). Another 30% remains as forest, capturing much of the terrestrial biodiversity, although roughly 5% of that is heavily managed for human needs including timber and palm oil (Foley *et al.*, 2005). Even where land surfaces are not heavily managed by humans, ecological habitats are occupied by human settlements and industries or subdivided by roads, pipelines and power lines.

The impacts of human land use for agriculture, forestry, roads, cities, and industry has been, thus far, the largest source of damage to biological diversity simply through destruction of habitat. Recent estimates by the Millennium Ecosystem Assessment claim that between 10 and 50% of well-studied higher taxonomic groups (mammals, birds, amphibians, conifers, and cycads) are currently threatened with extinction, based on IUCN–World Conservation Union criteria for threats of extinction (MEA, 2005). They conclude that 12% of bird species, 23% of mammals, 25% of conifers, and 32% of amphibians are threatened with extinction, numbers that may be conservative.

Compared with human appropriation of the terrestrial realm, the marine environment seems vast and untouchable, with many of the ocean's diverse ecosystems barely described. And yet by some measures the devastation of the marine environment is even more extensive than on land. Instead of habitat destruction or pollution as the main cause of the decline of marine ecosystems, the main culprit is overfishing (Pauly *et al.*, 2003). On land, the disappearance of the terrestrial megafauna occurred many thousands of years ago; in the marine realm, it occurred over the last few centuries, so we have historical accounts of what ocean ecosystems were like before the application of modern technology to commercial fishing (Jackson *et al*, 2001). The comparisons are startling. Jackson (2008) compiled an accounting of the percent decline of more than 50 different groups of marine flora and fauna, from corals to sharks to sea turtles, measured at a variety of locations around the world. Many show a loss of more than 90%, in the relevant metric (biomass, percent cover, or catch) and nearly all have sustained losses above 50%. Jackson pointed out that population declines do not result from fishing alone, but also associated disturbances including habitat destruction that comes from trawling the ocean bottom, leaving it flattened like an undersea roadway. The enormous decline in fish populations does not necessarily imply a high extinction rate, and some biologists remain hopeful that biodiversity in the ocean would partially recover if fishing practices were relaxed and other conservation measures implemented (e.g. Lotze

et al, 2006). On the other hand, some argue that, due to a combination of stresses that includes overfishing, pollution with nutrients and toxins, acidification, habitat destruction through trawling, and climate change, a mass extinction in the ocean is unavoidable (Jackson, 2010).

Between land and sea, the future of biological diversity looks ominous. The debate is not about whether extinctions will occur, but how many and how quickly, and whether conservation efforts can be successful. Many studies limit their scope to the next century or so, neglecting to consider the long time associated with the carbon cycle and climate change discussed below (for a longer view, see Myers and Knoll, 2001). For example, the Millennium Assessment admits that their projection 'is likely to be an underestimate as it does not consider reductions due to stresses other than habitat loss, such as climate change and pollution.'

Pollution, of which climate change may be considered a special case, may yet surpass direct habitat destruction (and hunting and fishing) as the largest cause of ecosystem decline. One way to measure the scale of human intervention in the geochemical world is to look at major biogeochemical cycles such as nitrogen, phosphorus and sulfur. (Perturbation to the carbon cycle involves the special case of climate change, which I will discuss later.) For nitrogen, the simplest approach is to consider the entry point into the biogeochemical cycle – the fixation of nitrogen gas from the atmosphere. Through the Haber–Bosch process, humans produce nitrogen fertilizer, mostly in the form of ammonia or urea, at a current rate of 9.5×10^{12} mol per year, almost as large as the 10×10^{12} mol per year fixed by marine organisms (primarily cyanobacteria; Canfield *et al.*, 2010), and slightly larger than the $\sim 8 \times 10^{12}$ mol per year produced via terrestrial nitrogen fixation. Additional human sources include production of nitrogen from fossil fuel combustion as well as cultivation of legumes, primarily soybeans, which have nitrogen-fixing symbionts; each of these produces roughly 2×10^{12} mol per year. Overall, then, human production of fixed nitrogen has almost doubled the overall rate of nitrogen fixation. But this perspective underestimates the magnitude of human intervention. Over-application of fertilizer combined with deposition of nitrogen oxides from fossil fuel combustion (both coal-fired power plants and tailpipes of transportation vehicles) has had much larger impacts on specific regions, particularly aquatic ecosystems such as lakes and estuaries, where runoff can concentrate nitrogen from wide agricultural regions. 'Dead zones' in the open ocean have also been attributed to excess nitrogen discharge. Terrestrial ecosystems can also be affected as the addition of extra nitrogen can greatly upset ecosystem dynamics and inter-species competition.

The story of the phosphorus cycle is similar. Mining of phosphate-bearing rock adds roughly as much phosphorus to the global phosphorus cycle as is released from natural rock weathering (Filippelli, 2002). Like nitrogen, the impacts are highly spatially variable, most concentrated in places where release of phosphorus from fertilization or use of detergents is focused by surface waters. In addition, waste from livestock and poultry farming can concentrate the total phosphorus load from vast agricultural regions, producing devastating impacts on ecosystems downstream, including eutrophication of lakes, rivers and estuaries. Unlike fixed nitrogen, which can be produced from the limitless supply of nitrogen gas in air and depends mostly on the cost of energy required for the Haber–Bosch process, phosphorus production is a mineral resource like copper or iron, and the stability of reserves has been questioned, although phosphate ore is unlikely to be in short supply for many centuries. It is possible that a shift to lower-grade phosphate deposits will raise the price of phosphate, ultimately making agriculture more expensive, but the rate of phosphate extraction is likely to continue to increase for the foreseeable future.

The sulfur cycle has also been perturbed by human activities, primarily from the combustion of coal. In this case, defining the perturbation is more complicated. The vast majority of sulfur released from coal combustion enters the atmosphere as sulfur dioxide, where it is quickly oxidized to sulfate. Deposition occurs through various wet and dry mechanisms, but results in roughly a doubling of dissolved sulfate in rivers relative to pre-anthropogenic loading, mostly from sulfide oxidation during chemical weathering of terrestrial rocks (Bates *et al.*, 1992). From the perspective of the atmosphere, human activities represent an enormous perturbation to a preexisting cycle dominated by volcanic emissions (less than 10% of human emissions) and production of dimethyl sulfide by marine phytoplankton. The release of sulfur has two main impacts on ecosystems. First, oxidation of sulfur dioxide produces acid rain, which can harm terrestrial ecosystems by changing the pH of lakes and soils, possibly affecting the availability of calcium, a critical nutrient for most trees. Second, sulfate aerosols reflect sunlight, offsetting the impacts of greenhouse gases on climate. As discussed below, this unintentional climate intervention is currently masking the impact of human perturbations to the carbon cycle. In the long run, however, the sulfate aerosol effect cannot keep up with sustained release of CO_2 due to the short residence time of sulfur in the troposphere (days to weeks).

Popular awareness of pollution's impact on ecosystems in the United States can be traced to the publication of Rachel Carson's book, *Silent Spring* (Carson, 1962).

Carson focused her attention on the harmful effects of pesticides on the environment, particularly on birds. She began her exposition with 'a fable for tomorrow', a prosperous, rural town in the USA that suddenly lost its birds, its blossoms, its wildlife:

> 'this town does not actually exist, but it might easily have a thousand counterparts in America or elsewhere in the world. I know of no community that has experienced all the misfortunes I describe. Yet every one of these disasters has actually happened somewhere, and many real communities have already suffered a substantial number of them. A grim specter has crept upon us almost unnoticed, and this imagined tragedy may easily become a stark reality we all shall know.'

If one takes Carson's fable in a literal sense, one could argue that her concerns over pesticide use were slightly misplaced, not because pesticides have no substantial effects on wildlife, but because chemical pollution of nature with pesticides is probably not the most harmful way that humans disturb natural ecosystems. It would be interesting to see how Carson would react today were she armed with a deeper understanding of the magnitude and duration of the global threat to ecosystems posed by human-induced climate change.

22.6 The carbon cycle and climate of the Anthropocene

Of all the biogeochemical cycles that humans have affected with industrial activities, carbon dioxide is most significant both for the scale of the disruption and the longevity of its potential impact. The famous Keeling curve, a record of carbon dioxide measured in the atmosphere at Mauna Loa, Hawaii, provides a spectacular demonstration of how the entire atmosphere is affected by human activities, primarily the combustion of fossil fuels with some contribution from deforestation and other land use changes. Viewed in isolation, the Keeling curve understates the scale of human interference. A more appropriate perspective places the Keeling data alongside longer records of atmospheric CO_2 measured in ice cores from Antarctica. Measurements of ancient atmospheric composition extracted from bubbles in the ice have now been extended back 650 000 years before present (Siegenthaler *et al.*, 2005). Over this time interval, CO_2 reached minimum levels of approximately 180 ppm during glacial maxima and peaked below 300 ppm during the interglacials. (Current atmospheric CO_2 concentration as I am writing this chapter is approximately 390 ppm, heading towards a seasonal high of roughly 395 ppm at the beginning of northern spring.) Direct measurement of more ancient CO_2 levels will not be possible unless more ancient ice is identified, but indirectly, through a variety of geochemical measurements, including carbon isotopes of organic molecules (Hendericks and Pagani, 2008), atmospheric CO_2 concentration can be estimated over much longer time scales. These data suggest that atmospheric CO_2 has not been much higher than 300 ppm for the last 34 million years.

What will be the geobiological consequences of higher atmospheric CO_2? Higher CO_2 concentrations do cause direct ecological disruption, particularly in plant communities, as some species are able to take advantage of the higher CO_2 levels by increasing rates of photosynthesis. CO_2 concentrations between 500 and 2000 ppm are unlikely to have strong metabolic effects on most terrestrial animals. In the ocean, however, as CO_2 emissions continue to outpace ocean mixing, a transient lowering of the calcium carbonate saturation state in the surface ocean, commonly called ocean acidification, will put stress on a wide variety of marine organisms, particularly – but not only – those that grow their skeletons out of calcium carbonate. Indeed, coral reefs may be greatly impacted, adding to the multiple stresses on these diverse ecosystems that additionally include overfishing, habitat destruction, runoff of excess nutrients, and warming (e.g. Pandolfi *et al.*, 2005).

But there is no question that the largest impact of human perturbation to the carbon cycle will be in the disruption to the climate system. It remains uncertain exactly how much warming will occur as CO_2 levels rise, mostly due to uncertainty surrounding feedbacks in the climate system that can amplify the direct effects of higher greenhouse gas concentrations, as well as feedbacks in the carbon cycle that can add additional carbon dioxide to the atmosphere. The standard measure of the degree of amplification of radiative forcing is called climate sensitivity, defined as the change in global average temperature for a doubling of atmospheric CO_2. Most general circulation models used to predict future climate change use a climate sensitivity between 1.5 and 4 °C, based on calibration of these models to the observed temperature change over the last century. However, the last century may not be a good predictor of climate sensitivity in the future as CO_2 rises to levels far outside of the calibration period. Several possible feedbacks may only kick in during warmer climates; for example, Kirk-Davidoff *et al.* (2002) proposed that increased stratospheric water vapor due to changes in atmospheric circulation in a warmer climate might lead to enhanced warming at high latitudes in the wintertime due to optically-thick polar stratospheric clouds.

This is where the geologic record of past climate change is especially useful. The last time that we think atmospheric CO_2 was well above 300 ppm, was the Eocene (Hendericks and Pagani, 2008). A range of observations of Eocene climate, including isotopic, chemical, and paleobiological data, reveal a general picture of a very warm world, with globally averaged temperatures elevated by as much as 6 to 10 °C above the present (Zachos *et al.*, 2001). For example, palm trees, plants whose fundamental anatomy makes them intolerant of freezing, grew in continental interiors at mid to high latitudes. From this, one can conclude that winters in these regions were much milder than today (Wing and Greenwood, 1993).

An important difference between the Eocene and climate change over the next few centuries is that the warm climate in the Eocene persisted for millions of years, with higher CO_2 concentrations most likely brought about by higher rates of volcanic outgassing that persisted from the Cretaceous into the Paleocene and Eocene (e.g. Berner *et al.*, 1983). This means that the entire climate system as well as most ecosystems in the Eocene had time to adjust to a warm climate and reach a quasi-equilibrium state, with no ice caps at high latitudes and very warm deep ocean temperatures. In contrast, human perturbations to the atmosphere today are happening so quickly that global ecosystems may have great difficulty adapting to the transient changes.

It is common, when assessing the potential impacts of future climate change, to focus on the climate in 2100 CE, presumably because we assume that people do not care very much about climate change farther out in the future. But if one is concerned with how Earth's ecosystems will be affected on the scale of the paleobiological record of life, then we must look well beyond 2100 CE to appreciate the full impact of the rapid combustion of fossil fuels on the planet.

In a series of papers, David Archer and colleagues elegantly describe the long response-time of the carbon cycle to fossil fuel emissions (e.g. Archer *et al*, 2009). The initial rise in atmospheric CO_2 comes primarily from the fact that humans are burning fossil fuels faster than the uptake by sinks in the ocean and terrestrial biosphere. CO_2 will continue to rise until fossil fuel emissions fall below the natural sinks (Archer and Brovkin, 2008). Consider a hypothetical case in which CO_2 emissions from fossil fuels stop altogether by the end of the 21st century. Atmospheric CO_2 concentration would begin falling as soon as emissions stopped due to continued uptake, primarily by the ocean (assuming that the large stores of carbon in soils in the tundra or in tropical rainforests do not start releasing carbon faster than the natural sinks). As the amount of CO_2 dissolved in the ocean increases, the pH will drop slightly, driving the dissolution of carbonate on the seafloor as chemical equilibration is slowly achieved. After 10 000 years, this chemical exchange, called carbonate compensation, will be essentially complete, but 15 to 25% of the initial CO_2 released from fossil fuel combustion will remain in the atmosphere (Archer and Brovkin, 2008). This amount could be even larger if additional sources, such as release of methane hydrates in the ocean or release of soil carbon from the frozen tundra, were to add substantially to the carbon produced from fossil fuel combustion. Over the next 100 000 to 200 000 years, a slight increase in silicate weathering rates on land, driven by the warmer climate, would eventually convert the remaining CO_2 into calcium carbonate, with some additional uptake into marine organic carbon buried in sediments.

One can think of this residual CO_2 that requires more than 100 000 years for conversion to calcium carbonate as the long tail of human society's impact on the atmosphere. Archer and Brovkin (2008) calculate that if our cumulative CO_2 emissions are 1000 billion tonnes of carbon and released over the next 150 years, then CO_2 will rise to roughly 600 ppm, and will remain near 400 ppm for tens of thousands of years. If cumulative emissions over the next several centuries reach 5000 billion tonnes of carbon, then CO_2 will peak above 1800 ppm, and stay above 1000 ppm for tens of thousands of years. It is important to remember that this tail in the atmospheric CO_2 curve is set solely by the cumulative emissions. It is not sensitive to how quickly the emissions occur over the next millennium. To put it another way, imagine that we were able to reduce global CO_2 emissions from fossil fuel consumption to half of current levels by the middle of the 21st century, but those emissions continued over the following 1000 years as countries slowly used their remaining reserves of coal, natural gas, and petroleum, albeit at a much slower rate. The long-term concentration of CO_2 in the atmosphere in this case is almost the same as if we released that CO_2 all in this century – roughly 1000 ppm – and would remain for tens of thousands of years to more than one hundred thousand years. The only difference between these scenarios comes from the impacts of the transient, century-scale rise in CO_2 on ocean uptake through stratification, and on any potential carbon feedbacks triggered by the extreme CO_2 levels over the next few centuries.

What does this mean for geobiology? Elevated CO_2 at even 400 ppm, much less 1000 ppm, for tens of thousands of years takes the Earth system back to Eocene conditions. It is likely that both polar ice sheets will melt on this timescale; glaciologists argue over specific predictions for how quickly Greenland and Antarctica will lose ice over the next few hundred years (e.g. Vermeer and Rahmstorf, 2009), but over tens of thousands of years, there is no question that most of the ice on Greenland and much of the ice on Antarctica will disappear. The loss of ice will raise sea level by as much as

70 m, with an additional 2 to 10 m coming from the warming of the deep ocean and the thermal expansion of seawater. Disruption to terrestrial and marine ecosystems at virtually all latitudes will be enormous from the sea level rise and submersion of land areas; from changes in the hydrologic cycle, possibly including the migration of the Hadley circulation that causes subsidence and hence aridity in today's subtropical deserts; and from the temperature rise itself, which will disrupt ecosystems in all sorts of direct and indirect ways.

There is one analogy to future climate change in the geologic past that is worth consideration. At the very beginning of the Eocene, 55 million years ago, global temperature warmed by 6 °C in less than 10 000 years (Zachos et al., 2001), coincident with a change in the carbon isotopic composition of seawater that is likely to have been the result of the oxidation of a large amount (i.e. >5000 Gt) of organic carbon (Higgins and Schrag, 2006); a large carbonate dissolution event at that time is the fingerprint of a large and rapid release of CO_2 (Zachos et al, 2005). Compared with Archer's scenarios, the Paleocene–Eocene thermal maximum (PETM) is equivalent to the higher carbon emission scenario; indeed, the carbon and oxygen isotope record in the earliest Eocene shows how the carbon cycle and the warming slowly subsided over the next 200 000 years (Zachos et al, 2001), just as most carbon cycle models predict a long, slow decline in CO_2 concentration following the age of fossil fuel emissions (Archer and Brovkin, 2008).

There are lessons from the PETM that may help us understand how future climate change will affect the geobiological world. By modelling the carbon cycle during this event, Higgins and Schrag (2006) showed that the CO_2 concentration in the atmosphere must have tripled or possibly quadrupled, implying a climate sensitivity of 3 to 4 °C per doubling, on the high end of what most climate models use for predicting the future. Moreover, this may be a low estimate relative to what we may see over the next few centuries because there were no ice sheets or sea ice before the PETM, and the impact of reduced albedo from melting snow and ice on Earth's temperatures over the next few centuries is likely to be significant.

One additional lesson from the PETM regarding the impact of climate change on ecosystems seems quite optimistic, at least on the surface. There is no evidence that either the abrupt warming during the PETM or the direct effects of CO_2 on calcite and aragonite saturation state of the surface ocean drove any large mass extinctions, except for benthic foraminifera (Thomas and Shackleton, 1996) that may have succumbed to acidification, low oxygen levels driven by transient stratification, or perhaps the warming itself. This does not mean that ecosystems were unaffected by the PETM. A Scuba diver observing coastal seas during the event would have witnessed a massive die-off of coral, much like one sees in the Caribbean today (Scheibner and Speijer, 2009). And many land plant species survived PETM warming by migration, a response that is complicated today by cities, croplands, roads and other barriers to migration (Knoll and Fischer, 2011). The persistence of most biodiversity implies that enough refugia existed for species to survive despite the enormous change in environmental conditions.

There are several reasons not to take this as a rosy sign for the future. First, it is possible that additional impacts of human activities, including pollution and land-use changes, in combination with stress from climate change will exceed whatever tolerance most ecosystems have for adapting to rapid changes. For example, as noted above, migration routes for species on land are constrained by roads, cities and farmland – conditions that did not exist during the PETM. Second, an important difference between the PETM and today is that the mean climate state was already quite warm in those times, and had been for tens of millions of years, as discussed above. We are heading toward such warm conditions today from a relatively cold climate. This means that all cold-dwelling plants and animals that currently inhabit polar, sub-polar and even most temperate ecosystems, as well as the many marine ecosystems that live in the colder regions of the oceans, can take no comfort from the resilience of warm-dwelling ecosystems that survived the PETM. Imposing a warmer world on ecosystems from the coldest parts of the Earth may be particularly cruel; even where the human footprint through appropriation of the land and ocean for food, fuel and fiber has been relatively mild and wilderness is abundant, such as Alaska or Siberia, climate change of the scale predicted for the next millennium means that organisms will literally have no place left to go.

22.7   The future of geobiology

In the preceding sections, I have described the enormous scale of human intervention in the Earth system. It seems likely that, as climate change compounds the impacts of human land use, pollution, and overhunting and fishing, the Anthropocene will be seen in the distant geologic future as a time of mass extinction, visible in the fossil record and coincident with evidence for a large warming event and a major marine transgression driven by the temporary deglaciation of the polar continents. One can imagine earth scientists, tens of millions of years in the future, arguing over the connection between the warming and the extinctions, and also over whether the extinction was abrupt or gradual, confused by the earlier and diachronous dates for the extinction of terrestrial megafauna. Whether the global decline in biodiversity over the next many millennia ever comes close to the enormous loss of greater than 90% of species at the end-Permian extinction (Knoll *et al.*, 2007) remains uncertain because we do not know how much carbon will be emitted over the next millennium nor how severe the impacts of climate change will be. In addition, we do not understand the ecological responses to climate changes coupled with all the other stresses discussed above, nor how species extinctions will reduce the resilience of the remaining communities. These are some of the challenges for the future of geobiological research, as we attempt to understand the Earth system and the role of life in sustaining it well enough to inform engineering solutions to anthropogenic impacts. There will be greater and greater demand for such insights as predictive models calibrated to the historical record may be less and less accurate as the world departs from the range of environmental conditions that have persisted for the entirety of the human species. Of course, the outcome of our actions also depends on how humans react to the changes. This may be the true meaning of the Anthropocene, when the geobiological fate of the planet is fundamentally intertwined with the behaviour of human society.

Some have argued that the destruction of nature except for those species or ecosystems that serve some human-centred purpose will ultimately drive the collapse of human society. Some biologists and economists see a focus on 'ecosystem services' as an effective political strategy to encourage conservation and change human behaviour by articulating this possibility, but such a strategy fails to recognize how strong the demand for additional natural resources will be as human population growth and economic development proceeds. The idea that the non-marketed value of natural ecosystems will stand as a barrier to the complete appropriation of nature seems at best naïve. It is possible that humans will fall victim to the environmental destruction they have created, depending on how harmful the impacts of climate change turn out to be, but this probably underestimates the adaptability of the human species. One lesson from the first 200 years of the Anthropocene is that human technology has reduced our dependence on the natural world – or at least changed the terms of its engagement. This is not to deny the possibility that human society could destroy itself. The global nuclear arsenal, if ever used, has the power of more than 6 billion tons of TNT, less than one percent of the power of the Chicxulub impact at the Cretaceous–Tertiary boundary (Bralower *et al.*, 1998), but still large enough to erase most terrestrial ecosystems. But if humans can avoid self-destruction through weapons of mass destruction, their skill at adaptation is likely to allow them to survive environmental degradation – possibly at the cost of many other species.

A more optimistic view for conservation of natural ecosystems, but in some ways a more challenging one, is that humans will not sit back and simply react to climate change and other environmental challenges, but will play an active role in engineering the Earth system to suit their needs. With respect to climate change, this 'geoengineering' has been described as an emergency option if the rate of climate change accelerated over the next few decades, or if consequences looked much worse than anticipated. Recently, such ideas have gained more prominence (Crutzen, 2006), not as a substitute for serious emissions reductions, but in the sober realization that emissions reduction efforts may not be sufficient to avoid dangerous consequences. Adjusting the incoming solar radiation through reflectors in the upper atmosphere (Keith, 2000) appears to come at a very low cost relative to other strategies of climate change mitigation (Schelling, 1996), and may be relatively effective in offsetting the most catastrophic consequences of climate change (Caldeira and Wood, 2008). Archer and Brovkin (2008) argue that, because of the long lifetime of CO_2, sustaining such an engineering system for tens of thousands of years or more is not feasible. This fails to consider that engineering the climate for a few centuries could be combined with a variety of ways of removing CO_2 from the atmosphere, albeit at relatively high cost, so that the problem was completely abated by the end of a millennium or so. Some have expressed consternation at the prospect of engineering the climate for the entire planet, but one can also see it as simply an extension of the wide variety of ways that humans have taken control of the natural world, from artificial fertilizer and pesticides, to genetically modified crops, to large hydroelectric dams that regulate water flow to riverine ecosystems.

Capturing carbon dioxide from the air and pumping it into geological respositories seems fairly straightforward, if we could find a way to do it safely and cheaply. One can see this enterprise simply as an engineering effort aimed at reversing the huge perturbations to the carbon cycle that humans have already imposed on the planet, perhaps by simply speeding up the Earth's way of removing carbon dioxide by silicate weathering (House *et al.*, 2007). Removing carbon dioxide from the atmosphere is fundamentally a slow process; it would take centuries at least to reverse the carbon cycle impacts that humans have already wrought. What is different about solar radiation management is that it is immediate, which brings up a range of questions about ethics and governance. For some reason, such ethical discussions are never raised for changes imposed over longer timescales, such as climate change itself. There is no question that the power to engineer the climate to instantaneously conform to our direction comes with an awesome responsibility, although not fundamentally different than the responsibility that comes with nuclear warheads. How could we engineer the climate in a way that could be failsafe? Which countries would control this effort? Who would decide how much to use, or when? And what would happen if something went wrong, if we discovered some unforeseen consequences that required shutting the effort down once human societies and natural ecosystems depended on it?

Ironically, such engineering efforts may be the best chance for survival for most of the Earth's natural ecosystems – although perhaps they should no longer be called natural if such engineering systems are ever deployed. Those who fight for conservation of nature are faced with a remarkable dilemma. Climate change, when added to all the other human activities that threaten the natural world, has already placed the Earth on the verge of a major extinction, regardless of how effectively we reduce carbon emissions over the next century. Preventing the widespread destruction of natural ecosystems may require an engineering project that transforms the entire Earth into a managed biosphere, like the failed experiment in the Arizona desert. Nearly 50 years ago, Rachel Carson wrote in *Silent Spring,*:

> 'The 'control of nature' is a phrase conceived in arrogance, born of the Neanderthal age of biology and philosophy, when it was supposed that nature exists for the convenience of man. The concepts and practices of applied entomology for the most part date from that Stone Age of science. It is our alarming misfortune that so primitive a science has armed itself with the most modem and terrible weapons, and that in turning them against the insects it has also turned them against the earth.'

What Carson did not realize is that the concepts and practices of the industrial age far beyond 'applied entomology' have brought us to the point of no return. In the Anthropocene, the survival of nature as we know it may depend on the control of nature – a precarious position for the future of society, of biological diversity and of the geobiological circuitry that underpins the Earth system.

Acknowledgements

The ideas in this manuscript grew out of discussions with many individuals over many years. The author thanks Paul Hoffman, Paul Koch, Andy Knoll, Sam Myers, Peter Huybers, John Holdren, Jim McCarthy and Julie Shoemaker for ideas, comments, criticisms and suggestions. Some of these ideas were developed for an Ulam Memorial Lecture in Santa Fe in honor of Murray Gell-Mann's 80th birthday, sponsored by the Santa Fe Institute. Editorial comments from Andy Knoll and Sam Myers greatly improved the manuscript. The author also thanks Henry and Wendy Breck for their support and encouragement.

References

Alroy JA (2001) Multispecies overkill simulation of the end-Pleistocene megafaunal mass extinction. *Science* **292**, 1893–1896.

Archer D, Brovkin V (2008) The millennial atmospheric lifetime of anthropogenic CO_2. *Climatic Change* **90**, 283–297.

Archer, D, Eby, M, Brovkin, V, *et al.* (2009) Atmospheric lifetime of fossil fuel carbon dioxide. *Annual Review of Earth and Planetary Sciences, 37*, 117–134.

Bates, TS, Lamb, BK, Guenther, A, Dignon, J, Stoiber, RE (1992) Sulfur emissions to the atmosphere from natural sources. *Journal of Atmospheric Chemistry*, **14**, 315–337.

Berner, RA, Lasaga, AC, Garrels, RM (1983) The carbonate-silicate geochemical cycle and it effect on atmospheric carbon dioxide over the last 100 million years. *American Journal of Science*, **283**, 641–683.

Bralower, TJ, Paull, CK, Leckie, MR (1998) The Cretaceous–Tertiary boundary cocktail: Chicxulub impact triggers margin collapse and extensive sediment gravity flows. *Geology*, **26** (4), 331–334.

Borlaug, NE. Online. Available from http://nobelprize.org/nobel_prizes/peace/laureates/1970/borlaug-lecture.html (accessed 2 November 2011).

Caldeira, K, Wood, L (2008) Global and Arctic Climate Engineering: Numerical Model Studies. *Philosophical Transactions of the Royal Society A*, **366**, 4039–4056.

Canfield, DE, Glazer, AN, Falkowski, PG (2010) The evolution and future of earth's nitrogen cycle. *Science*, **330**, 192–196.

Carson, Rachel (1962) Silent Spring. Boston: Houghton Mifflin.

Cohen, J (1996) *How Many People Can Support the Earth?* New York: W.W. Norton & Company, Inc.

Crutzen, PJ (2002) Geology of mankind. *Nature*, **415**, 23.

Crutzen, PJ, Stoermer, EF (2000) The Anthropocene. *Global Change Newsletter*, **41**, 17–18.

Crutzen, PJ (2006) Albedo enhancement by stratospheric sulfur injections: a contribution to resolve a policy dilemma? *Climactic Change*, **77** (3–4), 211–220.

Diamond, J (1997) Guns, Germs, and Steel New York: W.W. Norton & Company, Inc.

Ehrlich, PR (1968) The Population Bomb. New York: Ballantine Books.

Ehrlich, PR, Holdren, JP (1971) Impacts of population growth. *Science*, **171**, 1212–1217.

Filippelli, GM (2002) The global phosphorus cycle. *Reviews in Minerology and Geochemistry*, **48** (1), 391–425.

Foley, JA, DeFries, R, Asner, GP, *et al.* (2005) Global consequences of land use. *Science*, **309**, 570–574.

Greenwood, B, Mutabingwa, T (2002) Malaria in 2002. *Nature*, **415** (6872), 670–672.

Henderiks, J, Pagani, M (2008) Coccolithophore cell size and the Paleogene decline in atmospheric CO_2. *Earth and Planetary Science Letters*, **269**, 575–583.

Higgins, JA, Schrag, DP (2006) Beyond methane: towards a theory for Paleocene-Eocene thermal maximum. *Earth and Planetary Science Letters*, **245**, 523–537.

House KZ, House CH, Schrag DP, Aziz MJ (2007) Electrochemical acceleration of chemical weathering as an energetically feasible approach to mitigating anthropogenic climate change. *Environmental Science and Technology*, **41**(24), 8464–8470.

Jackson, JBC (2008) Ecological extinction and evolution in the brave new ocean. *PNAS*, **105** (1), 11458–11456.

Jackson, JBC (2010) The future of the oceans past. *Philosophical Transactions of the Royal Society B*, **365**, 3765–3778.

Jackson, JBC, Kirby, MX, Berger, WH, *et al.* (2001) Historical overfishing and the recent collapse of coastal ecosystems. *Science*, **293**, 629–638.

Keith, DW (2000) Geoengineering the climate: history and prospect. *Annual Review of Energy and the Environment*, **25**, 245–284.

Kirk-Davidoff, DB, Schrag, DP, Anderson, JG (2002) On the feedback of stratospheric clouds on polar climate. *Geophysical Research Letters*, **10**, 1029–1032.

Knoll AH, Fischer WW (2011) Skeletons and ocean chemistry: the long view. In: J.P. Gattuso and L. Hansson, eds., *Ocean Acidification*. Oxford: Oxford University Press.

Knoll, AH, Bambach, RK, Payne, JL, Pruss, S, Fischer, WW (2007) Paleophysiology and End-Permian Mass Extinction. *Earth and Planetary Science Letters*, **256**, 295–313.

Le Conte, J (1878) Elements of Geology: A Text-Book for Colleges and for the General Reader. New York: D. Appleton and Company.

Lotze, HK, Lenihan, HS, Bourque, BJ, *et al.* (2006) Depletion, degradation, and recovery potential of estuaries and coastal seas. *Science*, **312**, 1806–1809.

Lyell, Charles (1830) Principles of Geology, Being an Attempt to Explain the Former Changes of the Earth's Surface, by Reference to Causes Now in Operation. London: John Murray.

Millennium Ecosystem Assessment, (2005) *Ecosystems and Human Well-being: Biodiversity Synthesis*. World Resources Institute, Washington, DC.

Myers N, Knoll AH (2001) The future of evolution. *Proceedings of the National Academy of Sciences, USA* **98**, 5389–5392.

Pandolfi, JM, Jackson, JBC, Baron, N, *et al.* (2005) Are US coral reefs on the slippery slope to slime? *Science*, **307**, 1725–1726.

Pauly, D, Alder, J, Bennett, E, Christensen, V, Tyedmers, P, Watson, R (2003) The future for fisheries. *Science*, **302**, 1359–1361.

Ruddiman, WF (2003) The Anthropogenic greenhouse era began thousands of years ago. *Climatic Change*, **61**, 261–293.

Ruddiman, WF (2007) The early Anthropogenic hypothesis: challenges and responses. *Reviews of Geophysics*, **45**, 1–37.

Scheibner, C, Speijer, RP (2008) Late Paleocene–early Eocene Tethyan carbonate platform evolution – a response to long- and short-term Paleoclimatic change. *Earth-Science Reviews*, **90**, 71–102.

Schelling, TC (1996) The economic diplomacy of geoengineering. *Climatic Change*, **33**, 303–307.

Siegenthaler, U, Stocker, TF, Monnin, E, *et al* (2005) Stable carbon cycle-climate relationship during the late Pleistocene. *Science*, **310**, 1313–1317.

Thomas, E, Shackleton, NJ (1996) The Paleocene–Eocene benthic roraminiferal extinction and stable isotope anomalies. In: *Correlations of the early Paleogene in Northwest Europe*, (eds Knox RW, *et al.*), *Geological Society Special Publications*, **101**, 401–411.

Vermeer, M, Rahmstorf, S (2009) Global sea level linked to global temperature. *PNAS, Early Edition*, 1–6.

Vitousek, PM, Mooney, HA, Lubchenco, J, Melillo, JM (1997) Human domination of Earth's ecosystems. *Science*, **277**, 494–499.

Wing, S, Greenwood, DL (1993) Fossils and fossils climate: the case for equable continental interiors in the Eocene, *Philosophical Transactions of the Royal Society of London*, **341**, 243–252.

Zachos, J, Pagani, M, Sloan, L, Thomas, E, Billups, K (2001) Trends, rhythms, and aberrations in global climate 65Ma to present. *Science*, **292**, 686–693.

Zachos, JC, Röhl, U, Schellenberg, SA, *et al.* (2005) Rapid acidification of the ocean during the Paleocene-Eocene thermal maximum. *Science*, **308**, 1611–1615.

Zalasiewicz AN, Williams M, Haywood A, Ellis M (2011) The Anthropocene: a new epoch of geological time? *Philosophical Transactions of the Royal Society A* **369**, 835–841.

Index

Fundamentals of Geobiology, First Edition. Edited by Andrew H. Knoll, Donald E. Canfield and Kurt O. Konhauser.
© 2012 Blackwell Publishing Ltd. Published 2012 by Blackwell Publishing Ltd.